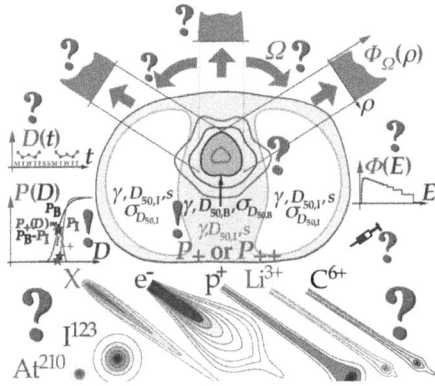

Biologically Optimized Radiation Therapy

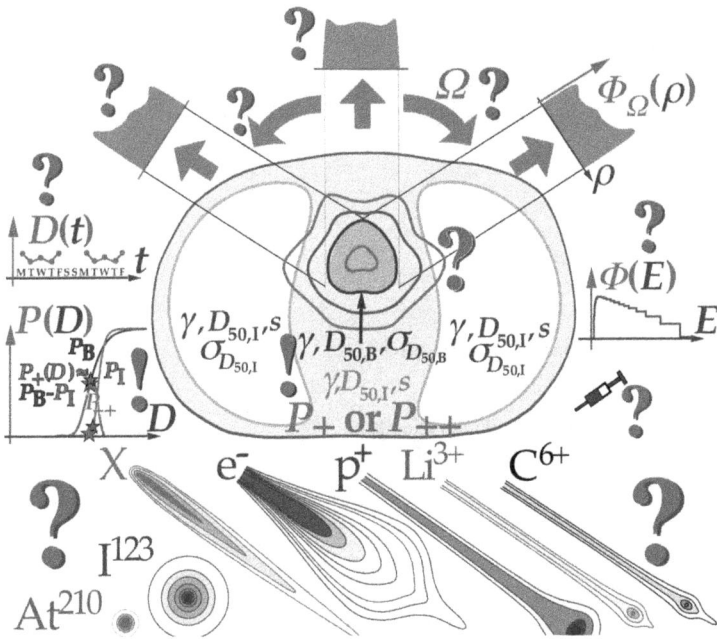

Biologically Optimized Radiation Therapy

Editor

Anders Brahme

Karolinska Institutet, Sweden

 **World Scientific**

NEW JERSEY · LONDON · SINGAPORE · BEIJING · SHANGHAI · HONG KONG · TAIPEI · CHENNAI

Published by

World Scientific Publishing Co. Pte. Ltd.

5 Toh Tuck Link, Singapore 596224

USA office: 27 Warren Street, Suite 401-402, Hackensack, NJ 07601

UK office: 57 Shelton Street, Covent Garden, London WC2H 9HE

Library of Congress Control Number: 2014933427

British Library Cataloguing-in-Publication Data
A catalogue record for this book is available from the British Library.

ISBN 978-981-4277-75-4

Typeset by Stallion Press
Email: enquiries@stallionpress.com

Foreword

As our knowledge about the molecular biology of cancer is rapidly improving we continuously need better tools to improve diagnostic imaging and treatment methods for cancer, one of our major life threatening diseases. Today a Center of Excellence for advanced radiation therapy should be centered on multiple unique developments that will considerably improve our ability to cure cancer patients and maximize their quality of life.

The high sensitivity, high resolution and ultra wide field of view of advanced PET-CT, MRSI and Stereoscopic Phase Contrast tumor imaging methods will be unique for accurate therapy planning and the early detection of tumor spread. Moreover, the resultant initial images of tumor spread could be the base for treatment response monitoring during the early phase of therapy. Primarily through PET-CT imaging it is possible then to visualize the tumor responsiveness and verify that the first delivered treatments had the planed effect and would allow accurate 4 dimensional in vivo predictive assay and biologically based adaptive therapy optimization. To maximize cure and the therapeutic response of the tumor and minimize eventual adverse normal tissue reactions Intensity and radiation Quality modulated light ions and high-energy photons and electrons are the ultimate therapeutic modalities. They can deliver high densities of severe DNA lesions in genetically instable tumor cells and generally only induce a low density of easily repairable lesions in normal tissues.

The past 20 years of significant improvements of treatment results with IMRT and Light Ion therapy not least at NIRS in Chiba Japan makes it now really important to bring these new tools into clinical use. Since the new diagnostic and therapeutic methods both potentially have mm resolution, new proposed Comprehensive Cancer Centers may introduce a quantum leap in our ability to treat and also cure advanced malignant tumors. As important as the new imaging and therapeutic modalities, are the research and development of new effective treatment methods, and to further develop our knowledge in molecular radiation biology of cancer, to stimulate the optimal interaction between fundamental biology, biomedical physics and medical care.

It is very important that such centers will be established as soon as possible to make full use of the fast developments in molecular genomics and proteomics and to make the treatments clinically available to the benefit of our cancer patients. This is one of few areas where a substantial investment in new Imaging and Therapeutic methods is cost effective and rapidly bringing improved treatment results and quality of life and actually reducing the expenditure into our medical care system.

Anders Brahme

Contents

4. Development of High Quality Beams for Uniform and Intensity-Modulated Radiation Therapy   157

Anders Brahme, Roger Svensson, and Bo Nilsson

5. Fundamentals of Physically and Biologically Based Radiation Therapy Optimization

295

Anders Brahme, Johan Löf, and Bengt K. Lind

8. **Physical, Biological, and Clinical Background
    for the Development of Biologically Optimized Light
    Ion Therapy** 499

    Anders Brahme and Hans Svensson

A Brief Introduction to the Development of Radiation Therapy Optimization

1

Anders Brahme

1.1. Introduction

The first radiation treatments of cancer were started within a year after the discovery of X-rays and it did not take long before successful therapeutic results were achieved. Because all treatments were not successful and some adverse reactions were also seen, especially for deep-seated tumors, various ways to improve the efficacy of the treatment were rapidly developed. To treat deeper tumors higher X-ray energies were used, also multiple collimated fixed or dynamic beam portal irradiations as well as various fractionation schedules were tested. As important as the technical development of methods to quantify the treatment in biological (skin erythema) or physical was the development (air ionization) terms. These methods allowed a more accurately quantified prescription of the radiation dose delivered and observation of the clinical response, leading to the first dose response relations used by Holthusen (1936) to describe a successful treatment in terms of the probability of achieving complication-free cure. The conceptual development of Radiation Therapy Optimization particularly when based on radiation biological objectives as well as its mathematical background and clinical potential and benefits will be discussed in more detail in the following chapters and so will some key algorithms and their clinical applicability. Here we will focus on some of the key developments to get a quick overview of the whole field.

1.2. Treatment Units

The treatments started with low-energy X-rays (30–50 kV), which were suitable for shallow tumors such as basal cell cancers which were probably the first tumor documented to have been successfully treated. Then, higher energy X-rays (50–350 kV) was used for deeper therapy and later Coccroft-Walton and

Van der Graaff generators were used to reach the low MV region. In the mid-1930s, neutrons were tested but since their biological effect was far from understood, significant normal tissue damage put too early an end to these trials. To improve the therapeutic efficacy of the high-energy X-rays for deep tumors, different dynamic techniques were rapidly developed such as arc therapy and focused spiral irradiations. At this time just after the war, almost simultaneously, ^{60}Co, betatrons, and the first traveling wave linacs appeared in the clinic during the early 1950s, all making improvements for deep-seated tumors with much milder skin reactions. Interestingly, almost at the same time in the mid-to-late 1950s, protons, helium, and later heavy ions such as neon and argon were tested on tumors also with some initial clinical advantages. From the late 1960s, the first high MV betatrons, standing wave linacs and in the 1970s microtrons (10–50 MeV) were used and high-energy electrons became more commonly used mainly to spare tissues downstream of the tumor (cf. Chapter 4, Figs. 4.2 and 4.8). During this period, high-energy π-mesons generated by very high-energy electrons or protons were tested but with minor clinical improvements compared to their high complexity and cost. Finally in the early 1990s, carbon ions was successfully tested in an extensive dual synchrotron facility in Japan with very significant improvements in the clinical results particularly for tumors that are hard to cure by our more common radiation modalities (cf. Chapter 8, Figs. 8.19–8.24). For the further development, see Secs. 1.9 and 1.10 below.

1.3. Radiation Quality

Classical radiation therapy based on X-rays, ^{60}Co-γ-rays, or high-energy Bremsstrahlung are by definition associated with a relative biological effectiveness (RBE) of close to unity meaning that a given absorbed dose in Gy or J/kg results in essentially the same cell kill. To be exact, this is only true for quasi-relativistic energies $\gtrsim 0.5$ MeV, since then the slowing down spectrum of low-energy δ-ray-electrons per unit energy delivered is closely energy-independent (cf. Chapter 2, Figs. 2.4 and 2.7 and Chapter 8, Fig. 8.6). This is the main reason why the absorbed dose is a sufficient and very useful quantity for describing the biological effect in high-energy electron and photon beams. However, very low-energy photons and electrons have a much higher fluence of low-energy δ-rays per unit dose and may reach an RBE >3 in the low-to-sub-keV region. High-energy protons are also essentially a low RBE radiation, since the secondary electrons that deliver most of the cell kill are of higher energy (RBE $\approx$ 1.1–1.2 and partly due to nuclear interactions). Low-energy protons in the low MeV region will also have a high RBE since their secondary electrons again are in the low and sub-keV energy region.

Unfortunately, this happens only over a few cell diameters for each proton track, so the net effect is very small for high-energy protons since their range straggling, of about 1% of the range or often a few mm, and thus generally totally dilutes the 50 μm high linear energy transfer (LET) portion very effectively. From lithium and above the RBE of about 2 or more is achieved in the Bragg peak and the multiple scatter penumbra is about one-third or less than that for protons making these beams of considerable clinical value (cf. Chapter 8, Figs. 8.2–8.4 and 8.9).

1.4. Radiation Biology

The development of radiation biology has kept good speed to cope with all the new radiation modalities that have been introduced in the clinical arena. Until the 1990s, most cell survival modeling (Zimmer 1961; Kellerer and Rossi 1972, 1978; Scholtz *et al.* 1997) was strongly focused simply on cell kill, as this was an obvious cause for loss of cell viability. However, already in the late 1950s, Elkind and Sutton (1959) clearly showed that something is happening between dose fractions, so after a break in the irradiation many cells seemed to be able to recover and reproduce the full shoulder of the cell survival curve after a rest period of approximately 24 hours. Without the break, the quasi-exponential cell survival curve would continue with almost unchanged slope. Elkind and Sutton correctly interpreted this phenomenon as some of the hit cells that where sublethally injured could repair their damage and be recovered for the next treatment fraction. But it took many years until serious cell survival theory took this fact into account. First, Wolfgang Pohlit and following him Cornelius Tobias and Stan Curtis developed theories where the probability to repair or misrepair potentially lethal radiation damage was explicitly accounted for. The resulting formula were quite complex but had under certain conditions some similarity with the commonly used linear quadratic expressions, but the complexity made them less practical for clinical use. About 15 years later, Lind and coworkers (2003), trying to include low dose hypersensitivity in the formalism, developed a very simple model primarily based on Poisson statistics, namely the Repairable–Conditionally Repairable damage or RCR model. The beauty of this model is its simplicity and its ability to separately specify those cells that survive because they are missed (e^{-aD}) and those that are hit but survive due to the effective cellular repair systems (bDe^{-cD}) that we know in rather great detail today (cf. Chapter 2, Figs. 2.11 and 2.12 and Chapter 8, Figs. 8.7 and 8.8). Interestingly, this last term gives us a way to account for the key cellular repair pathway of Non-Homologous End Joining (NHEJ) and Homologous Recombination (HR), and the expression is based on the assumption that if NHEJ works, then HR may correct eventual misrepair by NHEJ (that is known to be error prone not least at high dose rates).

The total survival is thus, the sum of the above two terms, and for the very simple case, when all single- and double-strand breaks are correctly repaired (this is not far from the truth in normal tissues where close to 99% of the DSBs are repaired), it reduces to $e^{-D/D_0} + D/D_0 e^{-D/D_0} = (1 + D/D_0)e^{-D/D_0}$; hence, for this special degenerate case $a = b = c = 1/D_0$. It is quite clear from this equation, as discussed in more detail in Chapter 8, Section 8.3.6 that this repair capacity explains most of the shoulder of the survival curve as first seen experimentally by Elkind and Sutton in 1959. Interestingly, this simple expression can also be used to describe the LET dependence of effective cross-section of the cell nucleus, the RBE, and the OER, as discussed in further detail in Chapter 8 (cf. Figs. 8.7, 8.8, and 8.5).

1.5. Why Is Radiation Therapy so Curative?

Today, we thus know that around 2 Gy of low LET radiation allow the normal tissue to recover well overnight and that at this common therapeutic dose about 99% of the double-strand breaks in the normal tissues are correctly repaired (cf. Chapter 8, Eqs. (8.5)–(8.7) and Brahme 2011). However, most tumors are genetically instable and practically always have some genes in the growth control and/or DNA damage surveillance pathways mutated (cf. Chapter 8, Fig. 8.25). Therefore, they will not be able to recover their cellular damage as effectively as the normal tissues, particularly when the dose per fractions is the highest in the tumor, for example, using Intensity Modulated Radiation Treatments (IMRT). In fact, the genetic instability characterizing practically all tumors often makes them lack some of the cell cycle blocks essential for high fidelity repair. This is no great surprise since this is probably the principal reason why the tumor was developed in the first place. Most likely it happened after the minor genetic damage that all cells are continuously exposed to, but it was not handled correctly due to possible genetic insufficiencies or predispositions. The tumor cells, therefore, often continue DNA synthesis, and thus instead may accumulate the radiation-induced genomic damage after each treatment fraction. The intact normal tissues, on the other hand, induce well organized and efficient cell cycle blocks that allow high fidelity DNA repair to take place before the S-phase is continued. After each treatment fraction, the tumor cells therefore accumulate more and more damaged DNA in their genome until they lose some essential function and/or cannot divide correctly and reach a state of mitotic catastrophe, *apoptosis or senescence*. Radiation therapy, therefore, makes optimal use of the genomic instability, the Achilles heal of tumor cells, and hit it repeatedly by multiple treatment fractions that the normal tissues largely tolerate quite well. Instead, they are generating steadily increasing amounts of genetic cellular damage to the tumor clonogens. Radiation therapy, thus, makes efficient use of one of the key biological differences between tumors and normal tissues. Through this mechanism, radiation therapy has an important biological advantage over most other therapies, such as surgery, hyperthermia, or coagulation by heat damaging, electrical or chemical reactions on DC or AC electrodes. This is because it has a significantly higher damaging therapeutic effect on the tumor clonogens than on the surrounding normal tissues that repair mild radiation damage very well between dose fractions.

1.6. Treatment Fractionation

In the early days, a single or a few repeated treatments were generally used when minimal tumor effects were seen. During the first half of the 20th century, multiple fraction schedules were developed generally delivering a few Gy five times per week over 5–7 weeks, largely based on trial and error (cf. Thames and Hendry 1987), not knowing about the just described underlying therapeutic mechanism. Furthermore, for some tumors in organs of parallel organization of their functional subunits such as lung, liver, and kidney, multiple quasi–simultaneous beam portals can effectively treat it, in so-called stereotactic irradiations. Due to the parallel organization surrounding normal tissues, as few as 3 fractions can be used quite efficiently on small tumors (≤ 5 cm), since the surrounding normal tissue can take over most of the lost function in the irradiated tumor region due to their parallel organization (cf. Chapter 2, Fig. 2.18 and Brahme 2000). More recently, it has been shown that multiple fixed fractionations may not be the most ideal way to treat. On the first day of the week, almost all the sublethal damage in normal tissues from previous treatments are repaired and a higher dose may preferably be delivered. This is similarly the case on the last day of the week when there is a long repair time for sublethal damage over the weekend. Adjustment of the dose fractions is made to maximize the tumor damage and maximize normal tissue recovery between fractions. This may be a very useful approach to compensate for the common absence of treatments during weekends (Chapter 7, Figs. 7.4 and 7.5 and Brahme 2005).

1.7. Diagnostic Tumor Imaging

To optimize a treatment, the best possible diagnostic information about the location of the tumor and the surrounding organs at risk is essential. It is, therefore, interesting to note that the early development of diagnostic and therapeutic radiology largely went hand-in-hand in trying to maximize the diagnostic information and therapeutic effect by optimal X-ray quality selection for each modality. A quantum leap in diagnostic imaging came with the new contrast agents but even more important was the introduction of computed tomography (CT) in the 1970s and magnetic resonance imaging (MRI) in the late 1980s. The arrival of MRI implied a further improvement in diagnostic information with higher tumor-to-normal tissue contrast in soft tissues and less imaging artifacts near bony structures. The ability to make functional MRI was also very valuable. More recently, the development of MRSI that is spectroscopic imaging, where tumor metabolites like lactate can be detected to visualize tumor spread and dual energy CT to detect atomic number variations. This resulted in significantly improved tumor and normal tissue diagnostics by new mathematical approaches in the imaging field leading to significantly improved target and organ at risk delineation and thereby an improved treatment outcome particularly when biologically optimized radiation therapy planning has been implemented.

It is interesting that the development of the CT principle by Cormack was a result of trying to improve treatment planning with X-ray therapy based on multiple projections of the patient (cf. Cormack 1995). It soon turned out that the mathematics used by Cormack to solve the inverse problem of CT reconstruction to reproduce the full 3D structure of the patient from projections is mathematically related to problems which were much earlier treated by Abel, Radon and Birkhoff and they are all also connected to the inverse problem of radiation therapy planning (cf. Chapter 5, Fig. 5.10 and Brahme 1995). The availability of true 3D data on the location of the normal tissues and often on the tumor location too opened the door for more accurate dose delivery to accurately defined target volumes.

A third quantum leap in diagnostic tumor imaging is taking place today where molecular tumor imaging using Integrated PET-CT units is allowing more accurate tumor diagnostics by positron emission tomography (PET) on a background of the normal tissue anatomy (CT). PET-MRI is a further development in this direction. These units do not only allow accurate tumor diagnostics through different Molecular Tracers but can also determine the degree of hypoxia and vasculature in the tumor and normal tissue anatomy. By repeated PET-CT studies during the first week to 10 days of therapy, it is even possible to derive the radiation resistance of the tumor and detect beam tumor misalignments and even patient set-up and treatment planning errors (cf. Chapter 7, Figs. 7.14–7.19). An equally important development is taking place in the area of accurate docking systems where the patient can be automatically set up in multiple diagnostic and therapeutic units reducing the need for fully integrated therapeutic or diagnostic units (cf. Chapter 8, Figs. 8.38 and 8.39). This reduces the equipment cost and increases the patient throughput, since each unit can be used alone also for patients who do not necessarily need the dual modality.

1.8. Treatment Planning

During the mid-1960s, the first attempts were made to optimize treatments by computer methods applying many beam portals on the same target volume (Hope and Orr 1965; Bahr *et al.* 1967; cf. also the review by Brahme 1995). Unfortunately, these attempts were not too successful since only rectangular wedge filtered beams were used without considering in detail the biological effects on organs at risk. In these studies, linear and quadratic programming were used to try and minimize the deviations from the prescribed tumor dose. As we know today, this is not a too important treatment objective (cf. Chapter 5, Fig. 5.24). In fact, biological treatment optimization is generally based on the displacement of the high tumor dose to regions in the tumor where the normal tissue damage is minimal at the same time as the mean cell survival in the tumor still ensures a desirable level of tumor cure, for example, 0.05 cells for 95% probability for tumor cure (since $e^{-0.05} \approx 0.95$).

In the late 1970s and early 1980s, as numerical computers were increasingly used in treatment planning and more new accurate dose computational methods were developed. These were based on analytical and Monte Carlo calculated point energy deposition kernels which could be integrated to point monodirectional pencil beams and more complex convergent pencil beam energy deposition kernels and thus allowed fast dose calculations by convolution or superposition methods. This resulted in very accurate dose calculations in three dimensions based on CT photon attenuation data and even made it possible to calculate the biologically expected responses in the tumor and surrounding organs at risk. From the mid-1980s, there was an increased interest in using cell-survival based dose–response relations that had been accumulated through the years and by then allowed the estimation of the efficacy of different dose delivery methods. In combination with accurate treatment planning, it became reasonably accurate to calculate tumor cure and normal tissue injury probabilities (cf. Chapter 3, Figs. 4.3 and 4.4) to really design biologically optimized treatment plans (cf. Chapter 5, Figs. 5.7, 5.17, 5.24, and 5.25 and Källman *et al.* 1992).

1.9. IMRT Development

Today the advantages of Intensity Modulated Radiation Therapy (IMRT) are well proven even in many randomized clinical trials (cf e.g. Fang *et al.* (2007), Al-Mamgani *et al.* (2009), Nutting *et al.* (2011) and Staffurth (2010)). Optimized IMRT started in the mid-1970s as a very versatile scanning system was invented which was capable of scanning both high-energy photon and electron beams mainly in order to generate very high quality therapeutic photon and electron beams (cf. Chapter 4, Figs. 4.10, 4.54, and 4.70 and Brahme 1979; Brahme *et al.* 1980). It was well known that high-energy electron beams are difficult to shape and flatten due to their low scattering power and substantial bremsstrahlung production. Similarly high-energy photon beams require very thick flattening filters to make broad uniform photon beams and thereby a large portion of the desired high-energy photons are lost and the beam quality is degraded at the same time as a high-energy electron contamination is generated. Both these two sets of problems are completely solved by the mentioned scanning and purging magnet system and high-quality beams are also obtained at very high energies (cf. Chapter 4, Figs. 4.23, 4.42, and 4.70 and Brahme 1987).

During the early 1980s, the first high-resolution double-focused multileaf collimators also came into clinical use. Since the 1960s, 3–5 cm wide collimator leafs were used for conformation therapy, mainly in Japan. The new multileaf collimators had cm or sub-cm resolution and allowed efficient beam collimation of the irregular beams which were produced for a long time by patient individual beam blocks of low-melting point alloys. The first accelerator to use these new flexible collimation systems was the Racetrack Microtron that produced 2.5–50 MeV scanned electron and photon beams which were collimated by the same double focussed multileaf collimator (cf. Chapter 3, Fig. 4.18 and Chapter 4, Figs. 4.10 and 4.75 and Brahme 1987).

Interestingly, the research and development of this unique new treatment unit opened up two new ways of delivering intensity-modulated beams either by dynamic scanning of narrow electron or photon beams with a fixed multileaf collimator setting or by using a fixed uniform scanning pattern combined by dynamic multileaf collimation. The possibility to deliver intensity modulated beams without fixed material wedge filters or beam compensation filters opened up new ways to improve multi-portal high-dose curative radiation therapy of complex tumors by simultaneously minimizing the amount of normal tissue morbidity (cf Brahme *et al.* 1982). The two ways could also be combined to simultaneously benefit from the high speed and efficiency with scanned beams and the steep dose-distribution gradients possible with double-focused dynamic multileaf collimation, thus opening the way for biologically optimized IMRT.

There are many reasons for using intensity-modulated dose delivery in radiation therapy. First, the gross tumor is generally hypoxic and therefore radiation resistant, whereas the microscopic invasive tumor has fewer cells that are rather well oxygenated and thus should be prescribed a lower dose. Furthermore, some normal tissues are more radiation-sensitive than others requiring the local dose to be kept low. But there are also more physical reasons, for example, if the shape of the body, organs at risk and the tumor are of varying cross-sections, different intensities from different beam directions are generally desirable. Finally, steeper dose gradients are needed between the gross tumor and close to very sensitive organs at risk. The total scenario of radiotherapy optimization therefore, has many complex and interacting reasons for using biologically optimized intensity-modulated beams, for example, to shape dose distributions primarily covering the target volume. However, it is still very difficult to prescribe the optimal treatment just in physical dose distributional terms. To really optimize a treatment, we should aim at a situation where the quality of life is as high as possible for the patient during and primarily after the treatment so that the risk for normal tissue morbidity is simultaneously as low as possible (cf. Chapter 5, Figs. 5.7 and Chapter 6).

The first use of intensity modulation in radiation therapy was probably in connection with perpendicular cross fire with X-rays when wedge filters were used to avoid hotspots at the most superficial edge and under dosage at the deepest edge of the target volume. Often, linear wedges were sufficient with different angles depending on the angle between the incident beams and the inclination of the patient surface. At 180° or parallel opposed beams, no filter was generally needed, since the exponential dose fall off of the two beams compensated each other rather well. Later on, missing tissue compensators were used to get a uniform tumor dose, even if the entrance surface was inclined or non-uniform. For rotation or arc therapy, a special kind of strongly nonlinear wedge filter was required as first derived by algebraic (Lax and Brahme 1982) and analytical (Brahme, Roos and Lax 1982) inverse treatment planning methods (cf. also Chapter 5, Fig. 5.9). This was probably the first time inversion methods were used in external beam radiation therapy. Brachytherapy inversion methods were more natural to use early on, due to the very heterogeneous dose delivery by a single point source, so they were needed in order to find the best possible source locations to get a uniform dose in the target volume (cf. Maynard and Davison 1950).

1.10. Treatment Optimization

The use of an inversion approach to external beam radiation therapy optimization rapidly turned out to be a very powerful way to maximize tumor cure and avoid normal tissue morbidity, as discussed extensively in the present book. In classical radiation therapy, the problem was generally solved by trial and error: how should uniform rectangular or wedge-filtered beams be selected so as to deliver a high quasi-uniform therapeutic dose to the target volume without exceeding the tolerance level of surrounding normal tissues? The classical solutions were to use parallel-opposed or four-field box techniques with rectangular or blocked beams. The solution of this problem using conformation or rotation therapy was to deliver uniform beams from all directions collimated to avoid irradiating normal tissues surrounding the tumor (Takahashi 1965). By inversion methods, a much more general problem could be solved: How should arbitrary incident beams best be shaped and selected to ensure a given sufficiently high dose or dose distribution in the target volume without exceeding the tolerance of the normal tissues? This will generate considerably improved dose distributions, since practically any desired distribution can be generated inside the target volume. However, this cannot generally be done without exceeding the normal tissue tolerance. This means that most often a compromise has to be found between the probability to cure the tumor and the risk of getting severe side effects. As already mentioned, such a compromise is very difficult to formulate accurately in pure dose distributional terms since what we really want is to maximize the number of patients that are cured without severe or even mild morbidity. The real problem of radiation therapy is, therefore, not a true inverse problem but an optimization problem, where the expectation value of the probability to cure each patient without seeing severe or mild side effects should be as high as possible, preferably at the same time as the normal tissue morbidity is negligible or at least as low as possible as shown in Fig. 1.1.

Since we generally do not know exactly the tumor and normal tissue sensitivity to radiation for a given patient, the optimization can in general only be made for a patient of "average" sensitivity. However, when the radiation responsiveness of the patient is not known, it is also possible to optimize the treatment so that it is suitable for a wider range of variability in radiation sensitivity. Such robust treatments can be derived by stochastic optimization where the expectation value of complication-free cure is maximized for a wide range of variability in radiation responsiveness (Kåver et al. 1999). Interestingly, it was shown that such a robust treatment plan could be achieved simply by assuming that the patient had an unusually radiation-resistant tumor (about one-half a standard deviation higher than the mean radioresistance value in the relevant tumor population) and that the normal tissues were more sensitive (also by about one-half a standard deviation in normal tissue sensitivity, cf. Chapter 6).

A further step in the development of optimal radiation therapy would, thus, be to try to determine the responsiveness of the tumor for the patient at hand. For many years people have tried to make a predictive assay based on tumor cell biopsies since the tumor responsiveness is really expected to be associated with the highest uncertainty. Unfortunately, these studies have not been very successful since they are based on growing the tumor biopsies in a medium, which may largely differ from the *in vivo* situation with hypoxia and poor tumor vasculature, etc. Such a method would probably work better in determining the normal tissue responsiveness since it is better vascularized and not prone to hypoxia but since it is generally also less varying from patient to patient, historical response data could be used.

Today, a new noninvasive approach is, therefore, being tested (cf. Chapter 7 and Brahme 2003, 2005, 2009) where the tumor responsiveness in the patient is estimated *in vivo* in the patient during the first part of the treatment. Since the

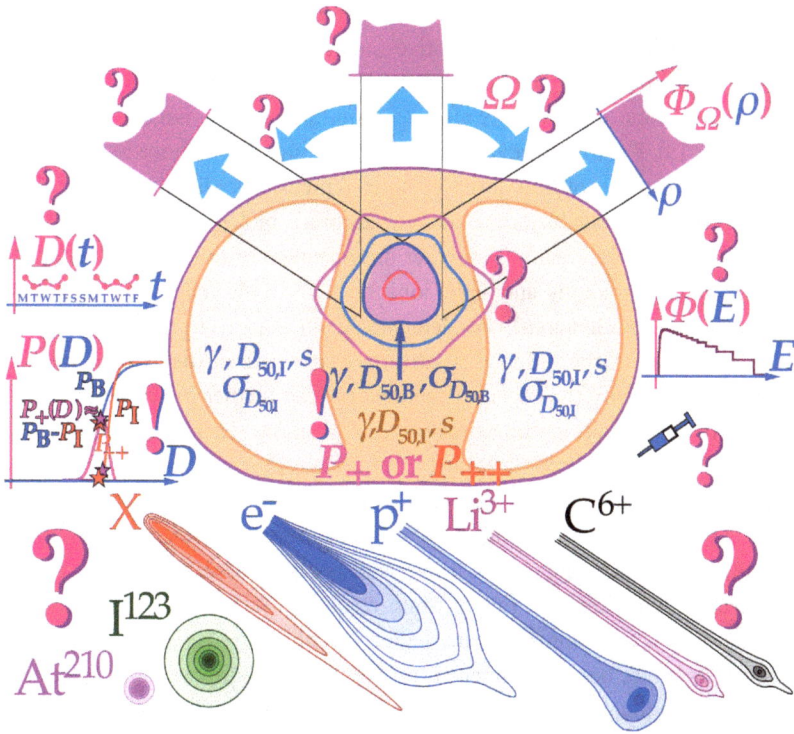

Figure 1.1. Illustration of the fantastic power available with biologically optimized radiation therapy. If we know the approximate sensitivity of the tumor and the normal tissues, it is possible to derive the biologically optimal beam directions (cf. Fig. 5.18) and their intensity modulation (cf. Fig. 6.5(b)) and it is even possible to find the optimal combination of low and high *LET* radiation modalities (cf. Figs. 8.13 and 8.30) and their incident energy spectra (cf. Fig. 6.15) as well as the ideal time dose fractionation (cf. Figs. 7.4 and 7.5) using the P_+ or P_{++} optimization strategies, as discussed in some detail in the present book. If we, in addition, have information about the interaction of the radiation modality of interest with a chemotherapeutic agent of preference, the combined treatment regimen can also be optimized in biological terms.

tumor is then unperturbed and all the tumor clonogens are studied in their real *in vivo* environment, it should be possible to determine more realistic response data, for example, by repeated PET-CT imaging of the tumor uptake. In this way it is not necessary to do fine needle aspiration biopsy to do *in vitro* predictive assay and not necessary to do test irradiation of the cells to determine their intrinsic radiation sensitivity. This so-called biologically optimized *in vivo* predictive assay-based radiation therapy (BIOART approach) starts off using historical dose–response data and after about 1 week of therapy re-optimizes the treatment based on the observed early *in vivo* response of the tumor. Interestingly, this opens up the door for truly biologically optimized adaptive radiation therapy where the treatment is adjusted depending on the actual dose delivery and the observed response of the patient's tumor and, possibly also, surrounding normal tissue (cf. Chapter 7, Figs. 7.14–7.19). Furthermore, the PET-CT imaging of the tumor responsiveness is potentially capable of full three or four dimensional diagnostics, if suitable tumor tracers are available, thus making accurate biologically based treatment optimization a practical reality. A further improvement of this approach is possible with high-energy photons and light ions since then the mean dose delivery to the patient can simultaneously be determined during or after the treatment by PET-CT imaging of the induced positron activity in the patient. This increased ability to *in vivo* verify the average dose delivery can further improve the adaptive approach to biologically optimized treatment planning (cf. Chapter 8, Figs. 8.16 and 8.38).

Interwoven with these broad steps of conceptual development of radiation therapy optimization are a multitude of improvements, which strengthen the whole therapy chain. For example, there is a special value in finding simple more robust treatments where the full sub-mm resolution in intensity modulation can be replaced by a few uniform beam segments, which are easy, fast, and reliable to deliver. Accurate beam alignment is also increasingly important with intensity-modulated dose delivery, and all methods to improve this will potentially improve treatment result. The most forgotten step in the therapy optimization chain is still the continued collection of accurate dose–response data and dose–response relations for biologically based treatment optimization. Similar to the standard road of development of scientific theories, the improvement of clinical dose response models go hand in hand with clinical applications and trial and error testing and modification of the existing models. In this context, radiation therapy optimization works as a magnifying glass allowing maximum utilization of the potential clinical therapeutic windows suggested by the models. When, for example, too much cord-related injury was seen in CHART, this indicated that spinal cord due to its slow cell turnover did not tolerate too large doses per fraction. A very efficient model for collection of improved clinical dose response data is, therefore, to start with the best available historical data and continuously update them, for example, by a Bayesian approach so that the information from patients being treated can be fed back to the model to continuously revise the historical dose–response data set (cf. Chapter 3 and Brahme 2000; Maehle-Schmidt *et al.* 1985). This should work well at least for acute radiation responses, whereas late effects unfortunately need much longer follow-up periods. In the future, clinical trials should always be linked with such dose–response studies to further improve the collection of biological response data to continuously improve the treatment optimization methods. The ultimate development of therapy optimization is the implementation of biologically optimized multimodality adaptive light ion therapy, as discussed extensively in Chapter 8, Figs. 8.12, 8.13, 8.26, 8.27, 8.30, and 8.37.

1.11. Future Possibilities

The widespread use of microarray techniques for studying the genomic characteristics of the tumor and the normal tissues of the patient will in the future most likely help tremendously in improving the selection of treatment approach and appropriate dose–response parameters. Since all cancers are at least linked to mutations in either the Growth Control or the DNA Damage Surveillance pathways (cf. Chapter 8, Fig. 8.25), the identification of these mutations may very well help to identify sensitive or resistant patients (see Chapter 8, Fig. 8.29(b) and Andreassen *et al.* 2003) both with regard to the tumor and the normal tissues. Through such techniques we can, thus, identify very interesting and important ways to further develop and optimize not only the biologically based radiation therapy approaches but also cancer therapy in general. Even if we may develop new molecular treatment methods in the future, we will always need éffective noninvasive methods to debulk the tumor so that the new pharmaca can really reach the tumor stem cells and hopefully further reduce the dose to normal tissues and consequently the morbidity of the treatment. It is particularly desirable to develop new effective molecular treatment methods for the last week of therapy when only a handful of randomly distributed tumor clonogens are left and broad radiation fields may not be the most optimal way to treat spares randomly distributed cells distributed in a background of sensitized normal tissues. Interestingly, the treatment methods developed in the present volume will be useful also for such combined modality treatments, provided the associated dose–response data become available as soon as possible, for example, through clinical trials (Adreassen *et al.* 2003).

Bibliography

Andreassen CN, Alsner J, Overgaard M, Overgaard J (2003) Prediction of normal tissue radiosensitivity from polymorphisms in candidate genes. *Radiother Oncol* 69:127–135.

Bahr GK, Kereiakes JG, Horwitz H, Finney R, Galvin J, Goode K (1968) The method of linear programming applied to radiation treatment planning. *Radiology* 91:686–693.

Brahme A (1979) Scanning system for charged and neutral particles. *Sw Pat* 7904360-0.

Brahme A, Kraepelin T and Svensson H (1980) Electron and photon beams from a 50 MeV Racetrack Microtron. *Acta Rad Oncol* 19:305–319.

Brahme A, Roos JE, Lax I (1982) Solution of an integral equation encountered in rotation therapy. *Phys Med Biol* 27:1221–1229.

Brahme A (1987) Design principle and clinical possibilities with a new generation of radiation therapy equipment. *Acta Oncol* 26:403–412.

Brahme A (1995a) Similarities and differences in radiation therapy optimization and tomographic reconstruction. *Int J Imag Syst Technol* 6:6–13.

Brahme A (1995b) Treatment optimization using physical and biological objective functions. In: Smith A (ed.), *Radiation Therapy Physics*, 209–246. Berlin: Springer.

Brahme A (2000) Development of radiation therapy optimization. *Acta Oncol* 39:579–595.

Brahme A (2003) Biologically optimized 3-dimensional *in vivo* predictive assay based radiation therapy using positron emission tomography-computerized tomography imaging. *Acta Oncol* 42:123–136.

Brahme A (2005) Fractionation & Biologically Optimized IMRT using *in vivo* predictive Assay based Radiation Therapy (BIOART). *Proc Fifth Int Symp on the Lymphatic System*, Limburg, p. 35.

Brahme A (2009) Potential developments of light ion therapy: The ultimate conformal treatment modality. *Radiol Sci* 52(2):8–31 (http://www.nirs.go.jp/info/report/rs-sci/pdf/200902.pdf).

Brahme A, Kraepelin T, Svensson H (1980) Electron and photon beams from a 50 MeV racetrack microtron. *Acta Rad Oncol* 19:305.

Brahme A, Roos J-E, Lax I (1982) Solution of an integral equation encountered in rotation therapy. *Phys Med Biol* 27:1221–1229.

Cormack AM (1995) Some early radiotherapy optimization work. *Int J Imag Syst Technol* 6:2–4.

Curtis SB (1986) Lethal and potential lethal lesions induced by irradiation: A unified repair model. *Radiat Res* 106:252–270.

Elkind MM, Sutton H (1959) X-ray damage and recovery of mammalian cells in culture. *Nature* 184:1293–1295.

Fang F, Tsai W, Chen H, Hsu H, Ching-Yeh Hsiung, Chih-Yen Chien, Sheung-Fat Ko (2007) Intensity-modulated or Conformal Radiotherapy Improves the Quality of Life of Patients With Nasopharyngeal Carcinoma: Comparisons of Four Radiotherapy Techniques *CANCER* 109: 313–321.

Holthusen H (1936) Erfahrunger uber die Vertraglichkeitsgrenze fur Röntgenstrahlen und deren Nutzanwendung zur Verhutung von Schade. *Strahlenther* 57:254.

Hope CS, Orr HS (1965) Computer optimization of 4MeV treatment planning. *Phys Med Biol* 10:365–373.

Källman P, Lind BK, Brahme A (1992) An algorithm for maximizing the probability of complication free tumour control in radiation therapy. *Phys Med Biol* 37:871–890.

Kåver G, Lind BK, Löf J, Liander A, Brahme A (1999) Stochastic optimization of intensity modulated radiotherapy to account for uncertainties in patient sensitivity. *Phys Med Biol* 44:2955–2969.

Kellerer AM, Rossi H (1972) The theory of dual radiation action. *Curr Topics in Radiat Res* 8:85.

Kellerer AM, Rossi H (1978) A generalized formulation of dual radiation action. *Radiat Res* 75:471.

Lax I, Brahme A (1982) Rotation therapy using a novel high-gradient filter. *Radiology* 145:473–478.

Lind BK, Persson LM, Edgren MR, Hedlöf I, Brahme A (2003) Repairable-conditionally repairable damage model based on dual Poisson processes. *Radiat Res* 160:366–375.

Maehle-Schmidt M, Palmgren J, Lind B, *et al.* (1999) A Bayesian sequential model for updating radiobiological parameters in radiation therapy. In: *Second European Conf on Highly Structured Stochastic Systems*, 180–181, Pavia, Book of Abstracts.

Al-Mamgani I, Heemsbergen WD, Peeters STH, Lebesque JV (2009) Role of Intensity-modulated Radiotherapy in Reducing Toxicity in Dose Escalation for Localized Prostate Cancer. *Int J Rad Oncol Biol Phys*, 73: 685–691.

Mayneord WV, Davison B (1950) Some applications of nuclear physics to medicine. *Br J Radiol* 150 (Suppl. 2):197–198.

Nutting CM, Morden JP, Harrington KJ, Guerrero Urbano T, Bhide SA, Clark C, Miles EA, Miah AB, Newbold K, Tanay M, Adab F, Jefferies SJ, Scrase C, Yap BK, A'Hern RP, Sydenham MA, Emson M, Hall E (2011) Parotid-sparing intensity modulated versus conventional radiotherapy in head and neck cancer (PARSPORT): a phase 3 multicentre randomised controlled trial. *Lancet Oncol* 12: 127–136.

Schulz M (1974) The supervoltage story. *Am J Roentgenol* 124:541–559.

Scholz M, Kellerer AM, Kraft-Weyrater W, Kraft G (1997) Computation of cell survival in heavy ion beams for therapy. *Radiat Environ Biophys* 36:59–66.

Staffurth J (2010) A Review of the Clinical Evidence for Intensity-modulated Radiotherapy *Clinical Oncology* 22: 643–657.

Takahashi S (1965) Conformation radiotherapy, rotation techniques as applied to radiography and radiotherapy. *Acta Radiol* Suppl. 242.

Thames HD, Hendry JH (eds) (1987) *Fractionation in Radiotherapy*. London: Taylor & Francis, 1987.

Tobias CA (1985) The repair–misrepair model in radiobiology: comparison to other models. *Radiat Res* 104 (Suppl.):S77–S95.

Zimmer KG (1961) *Studies on Quantitative Radiation Biology*. Edinburgh and London: Oliver and Boyd.

Fundamentals of Clinical Radiation Biology

2

Anders Brahme, Panayiotis Mavroidis, and Bengt K. Lind

2.1. Introduction

The field of radiation biology, by necessity, covers a very broad, interdisciplinary, range of sciences from the interaction of different radiation modalities with matter, via the molecular biology of radiation damage to different types of subcellular structures. From thereon radiation biology also covers the cellular radiation response, and radiation effects on functional subunits (FSU) of different organs and organ systems and the whole organism. In principle, radiation biology covers not only radiation effects on different essential enzymes and other molecules but also the influence of nutrients and protective and sensitizing agents of importance for the life of all different organisms in our surroundings. By necessity the present brief overview will therefore have to be limited to some of the most important radiation effects on humans at the cellular and organ level. To a large extent, the coverage is focused on subjects of importance for the understanding of radiation therapy and to some degree to radiation protection. Since all higher organisms on earth are made up of cells, we will start by analyzing the effect of radiation at the most basic cellular and subcellular level and continue with the response of partially or wholly irradiated organs. Finally, the effect of different radiation modalities depending on their microscopic energy deposition will also be covered.

2.2. Growth Control, Cell Cycle Regulation, and Damage Surveillance and Repair

Almost all cells in an organism are under normal conditions influenced by various forms of growth factors, which make cells divide at more or less regular intervals to replace dying cells and to renew and develop tissues and organs. This process is characterized by the cell cycle, which describes how the cell moves from the S-phase, where a new set of the genomic material with all the molecules of heredity

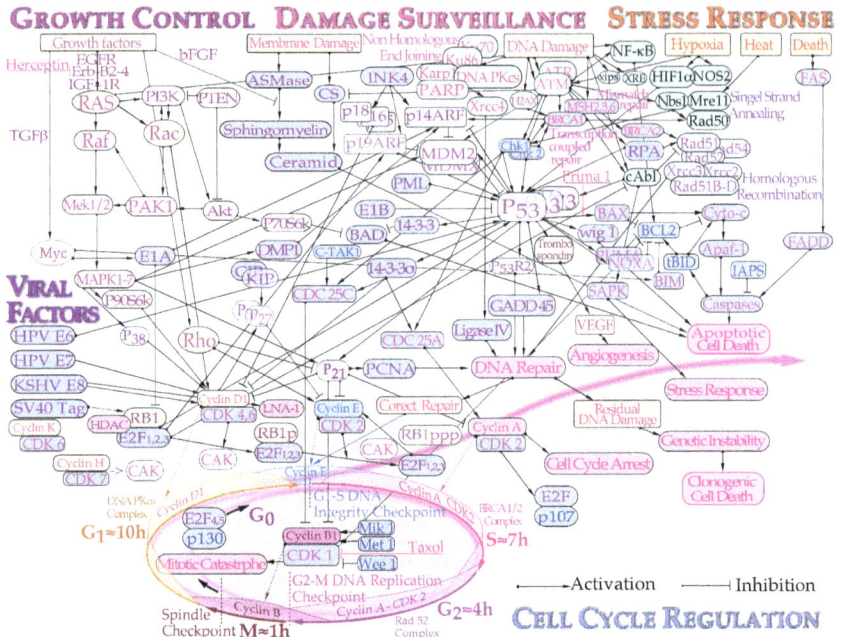

Figure 2.1. The cell cycle controls the growth and renewal of a tissue and it is regulated by the phosphorylation of the retinoblastoma protein (RB) through the influence of cyclins and CDK. Different growth factors stimulate this process, whereas various types of cellular damage and nutritional factors have the ability to halt cell cycle progression through the TP53 and P21 pathways. If the level of damage is too high, the apoptotic pathway may be induced leading to cell annihilation.

is synthesized, so that later on in the mitotic M-phase, there are two copies of the genome. One copy each is needed by the daughter cells that leave the cell division in the M-phase of the cell cycle. The two principal cell phases S and M are separated by two cell cycle gaps, G_1 and G_2, during which the cell prepares for synthesis and cell division, respectively, as illustrated in the lower part in Fig. 2.1. In addition to the G_1 and G_2 gaps, there is a third gap, G_0, where cells may be resting if they are not immediately needed for some function in the organ or tissue where they are born.

Different tissues have different proportions of their cells actively circulating through the cell cycle. The epithelial cells in the intestines have, for example, a fairly rapid turnover, whereas brain cells normally are dividing at a very low rate. This speed of normal cell divisions has important consequences for the radiation sensitivity of the cells as illustrated in Fig. 2.1. Cells that are rapidly proliferating through the cell cycle are driven by various growth factors via the RAS gene to the left in the figure. However, if the cell is now exposed to some external damaging agent, such as chemical, physical, or radiation effects, the integrity primarily of the heredity material in the form of double-stranded DNA (deoxyribonucleic acid, the spiral-like molecule condensed to chromosomes in the cell nucleus) may become severely affected. Then, it is no longer advisable for the cell to start the S-phase to generate a new genome or to split the already doubled genome in the M-phase. This is because severe molecular changes produced by, for example, radiation damage will not ensure a high fidelity of repair, replication, or division of the nuclear material.

For this purpose, a complex surveillance system exists in the cell as illustrated in Fig. 2.1. Some of the main actors are the ATM, TP53 and RB1 gene products, which together with different cyclins (A–G) control the progress of the cell cycle. Under normal circumstances, different growth factors such as RAS stimulate the phosphorylation of the RB1 protein in the RB1–E2F–HDAC complex so that the histone deacetylase (HDAC) is lost. After further phosphorylation by the cyclin-E–cyclin-dependent kinase 2 (CDK2) complex the E2F-transcription factor is released to activate genes required for the start of S-phase. When induced by radiation damage to DNA, the TP53 protein turns on one of its downstream genes, the CDK inhibitor P21, which in turn inhibits the cyclin–CDK complexes and thereby blocks further phosphorylation of RB1. In this way, cells are blocked at the G_1-S and G_2-M checkpoints of the cell cycle to allow repair of induced DNA damage, for example, through the GADD-45 pathway.

If the degree of damage is more severe so that the cellular surveillance system judges it cannot repair all the inflicted damage, TP53 may activate its *apoptotic* pathway through which the whole cell is eliminated by its build-in suicide machinery (upper right corner of Fig. 2.1). This more drastic response is sometimes needed to avoid the replication of cells or division of their DNA, before it is fully repaired, in order to minimize the risk of conserving inflicted DNA damage. Obviously, it is not desirable that all cells follow the apoptotic pathway to avoid total destruction of a tissue. Therefore, an anti-apoptotic control is also present in the cell, for example, through the BCL-2 survival gene. Clearly, this results in cell survival but at the risk that some DNA damage is not repaired. For cells in organs or grown in culture, only a small fraction will have access to the apoptotic pathway.

From the above mechanisms, it is understandable that cells in organs that depend on actively cycling and dividing cells may be more responsive to irradiation since they have a stronger active drive through the cell cycle and less time to repair the inflicted damage. This is also what is seen in radiation accidents and in the clinical use of radiation. The reactions of the intestine and blood-forming organs are often the most severe in accidents and in addition organs such as the lungs and the kidneys are very sensitive during radiation therapy.

2.3. Molecular Biology of Radiation Sensitivity in Tumors and Normal Tissues

Besides the cell cycle control, there are a large number of other genetic factors that can influence the radiation sensitivity of tumors and normal tissues. Obviously, for the tumor the state of proto-oncogenes is most important as they generally encode for proteins that are responsible for the signal transduction cascade of growth factors that normally stimulate cell division or differentiation. When such genes get mutated or erroneously expressed, they can promote tumor development and they are therefore called *oncogenes*. Functionally, there are four main groups of proto-oncogenes depending on how they influence normal cell processes:

- Autocrine growth factors (hst, int, sis)
- Growth factor receptors (erb, fms, sea)
- Signal transduction factors (ras, mos, src)
- Nuclear transcription factors (myc, fos, jun)

In each group some examples of oncogenes belonging to that group are given in parenthesis. There are several mechanisms that can influence the genome to activate oncogenes such as: structural alterations (mutations, deletions), amplification or loss of control mechanism by insertional mutagenesis, transduction, or translocation. Most of these mechanisms are started by external factors such as chemical agents or radiation, others by viruses (viral oncogenes) which can act by insertion or transformation of genes.

Most known proto-oncogenes (about 100 are known today) have a positive stimulating function on cell proliferation and therefore have a dominant influence relative to the other possibly normal allele. There is also a second important group of the so-called *tumor suppressor genes* which instead is recessive in relation to a normal allele, since it promotes neoplasia by loss of function. The classical example here is the RB gene, which is responsible both for the hereditary and non-hereditary sporadic form of the associated eye disease. In the hereditary form, one mutation is somatic and the other is germinally transmitted from an affected parent. However, in the sporadic form, both mutations are somatic. This explains the linear increase with time of the hereditary form, whereas the sporadic form has a much slower parabolic onset with time. More recently, the RB gene has been shown to be affected also in breast and lung carcinoma and osteosarcoma and it has a very important function in the regulation of cell cycle progression as discussed above and clearly shown in Fig. 2.1.

One of the most commonly mutated or lost genes in human cancers, TP53, also belongs to the tumor suppressor group and so do many recently found tumor-specific genes such as BRCA 1, 2, and 3 (breast cancer susceptibility genes), DCC and MCC (deleted or mutated in colon carcinoma, respectively), APC (adenomatous polyposis coli), and WT (Wilms' tumor).

The very interesting TP53 gene product is a DNA-binding transcription factor, which can induce apoptosis or cell cycle arrest in the G_1 phase of the cell cycle as discussed above. It, therefore, seems to influence the decision of the cell whether to rest and repair induced DNA damage or to give in and eliminate itself by apoptosis due to a too severe level of damage. Owing to its central role in handling DNA damage, the TP53 gene is found mutated in many types of cancers such as glioblastoma, astrocytoma, colorectal, breast, brain, and lung carcinomas.

It is interesting to note that if the tumor suppressor gene TP53 was mutated in a cell, the risk is immediately increased that this cell line may develop neoplasia since such cells will have a greater difficulty to handle DNA damage to its genome. Fortunately, for the same reason, radiation therapy may be the ideal way for treating a tumor developed by this process, since the tumor cells will generally still be associated with a poor ability to handle the DNA damage inflicted in a controlled way by the therapeutic beams. This mechanism may explain why radiation therapy has recently been shown to be very useful on node negative breast cancer patients who have certain mutations in their TP53 gene. The situation is not always as simple as this, because TP53 mutations may also decrease the ability of the cells to induce apoptosis. This may show up in an increased survival of the cells after irradiation even though they are damaged and not repaired with a high degree of fidelity.

Beside the oncogenes and tumor suppressor genes that are largely responsible for tumor induction, a large number of other genes may also be affected to promote tumor development and alter the radiation sensitivity of the patient. At least, four gene families can be identified:

(1) Cell cycle control genes (RB1, TP53, cyclin A-G, cdc-2, cdk 2-7, E2F 1-5, GADD-45, cf. Fig. 2.1);
(2) DNA repair or "mutator" genes (XPA-G, ERCC 1-5, XRCC 1-7, ATM, DNA-PK, Ku 70, 86, RAD 1–57, MSH 2, 3, 6, PMS 1, 2, MLH 1, 3, Mut, Hex, RecA, LexA, UvrA);
(3) DNA processing and topology genes (Topo I, IIα, IIβ); and
(4) Detoxification and stress response genes (GS, MRP, HSP).

It is clear that if some of these genes have an impaired function, the processing of normal and damaged DNA may be affected. This may, in turn, promote tumor development and alter the radiation sensitivity of the cells.

As discussed above, the cell cycle control is fundamental for the ability of the cells to halt DNA syntheses and to handle inflicted DNA damage before they continue cycling. For this purpose, the cyclins must function together with RB1, TP53, and the DNA-dependent protein kinase (DNA-PK), and ATM gene products to handle damaged DNA. The large group of DNA repair genes then has to take over trying to eliminate strand breaks and restitute damage sites. Base and nucleotide damage is handled by the excision repair gene products (XP A-G and ERCC), whereas more complex radiation damage is handled by the numerous RAD and XRCC gene products. For mismatch, repair sets of Mut and Hex genes are employed by the cell. During the repair process several DNA-processing genes are active such as the topoisomerases that are active in unwinding the DNA both from the nucleosomes and to separate the strands in order to transcribe DNA or to give access to the enzymes of the repair system to compacted areas of the DNA. Other genes that can influence the radiation sensitivity are the detoxification and stress response genes that may increase the glutathione level to improve radical scavenging or the level of heat shock proteins. It is likely that the status of many of the above genes with regard to polymorphism, amplification, and transcription factors and mutations may be combined as useful genetic predictors for radiation sensitivity both for tumors and probably even more important for normal tissues.

A number of genetic alternations may significantly alter the responsiveness of tumor cells and increase their genomic instability. Based on microarray techniques, these genetic defects can be located and the ideal ion species be used for a biologically optimized treatment schedule (cf. Figs. 2.2, 5.2 and 8.25 and Brahme and Lind 2010).

2.4. Radiation Quality and Radiation Effects

2.4.1. *Physical and Chemical Effects of Radiation Quality*

During the physical interaction stage, charged particle tracks directly ionize the subcellular nuclear targets or water molecules that surround or are bound directly to the DNA. Ionized or excited water molecules dissociate within 10^{-14} s into radical species, while electrons from ionizations are rapidly thermalized and become trapped by surrounding water molecules. During the following chemical stage, radicals diffuse from the sites of energy absorption at a timescale of

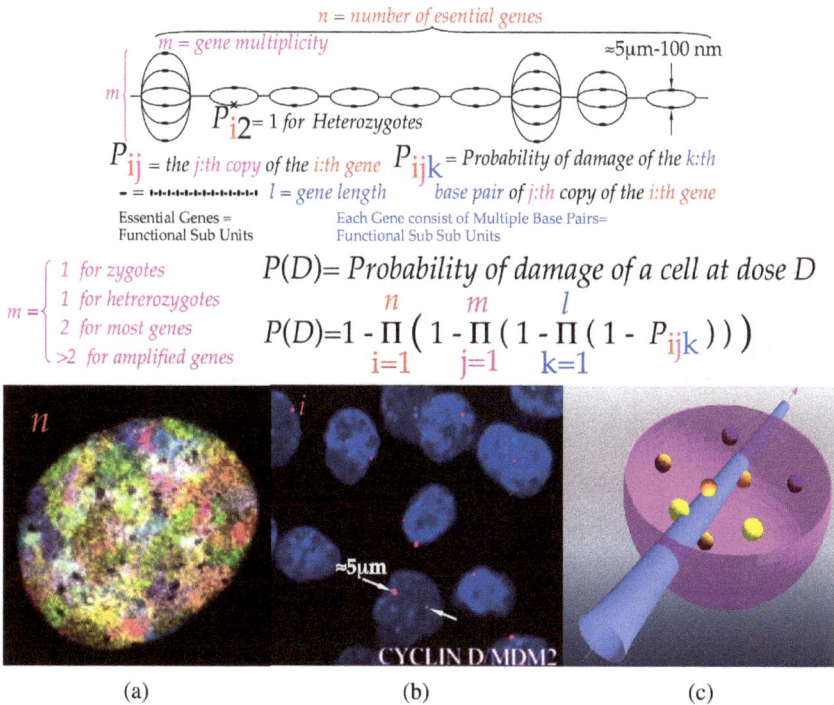

$n = $ number of esential genes

$m = $ gene multiplicity $\approx 5\mu m$-100 nm

m

$P^*_{i2} = 1$ for Heterozygotes

P_{ij} = the j:th copy of the i:th gene P_{ijk} = Probability of damage of the k:th base pair of j:th copy of the i:th gene

$l = $ gene length

Essential Genes = Functional Sub Units   Each Gene consist of Multiple Base Pairs= Functional Sub Subunits

$$m = \begin{cases} 1 & \text{for zygotes} \\ 1 & \text{for hetrerozygotes} \\ 2 & \text{for most genes} \\ >2 & \text{for amplified genes} \end{cases}$$

$P(D)=$ Probability of damage of a cell at dose D

$$P(D)=1 - \prod_{i=1}^{n} \left(1 - \prod_{j=1}^{m} \left(1 - \prod_{k=1}^{l} \left(1 - P_{ijk} \right) \right) \right)$$

n

$\approx 5\mu m$

CYCLIN D MDM2

(a)   (b)   (c)

Figure 2.2.   Description of the functional organization of key genes essential for cell survival. Since these are essential for survival, they are functionally organized in series, even if they may be located at different chromosomes. Some of these genes may be heterozygotes or imprinted with a single active copy, but most are normal and are available in two copies, functionally organized in parallel, whereas others may be amplified up to m-fold multiplicity. In this way, single SNPs and CNPs can be taken into account when estimating the probability for cell kill (cf. Brahme and Lind 2010). Fluorescence *in situ* hybridization photographs of **(a)** chromosomes inside the cell nucleus and **(b)** genes (MDM2 blue and Cyclin D pink) inside nucleus (courtesy T Heiden and A Zetterberg, respectively). The specific cell kill mechanism available with light ions allows correlated inactivation of both alleles of essential genes for tumor cell survival (spheres of same color). Low linear energy transfer (*LET*) or sparsely ionizing radiation has a considerably reduced probability for this type of lethal damage, particularly for hypoxic tumor cells.

about 10^{-8} s and follow three possible reaction pathways. These are: interactions with biologically important molecules causing indirect damage, recombination with each other resulting in H_2, H_2O, and H_2O_2 molecules, or become trapped by various radical scavengers. All the above reactions occur at a timescale $<10^{-6}$ s from the initial physical interaction and before the distribution of radicals reaches a final homogeneous state as a result of diffusion processes.

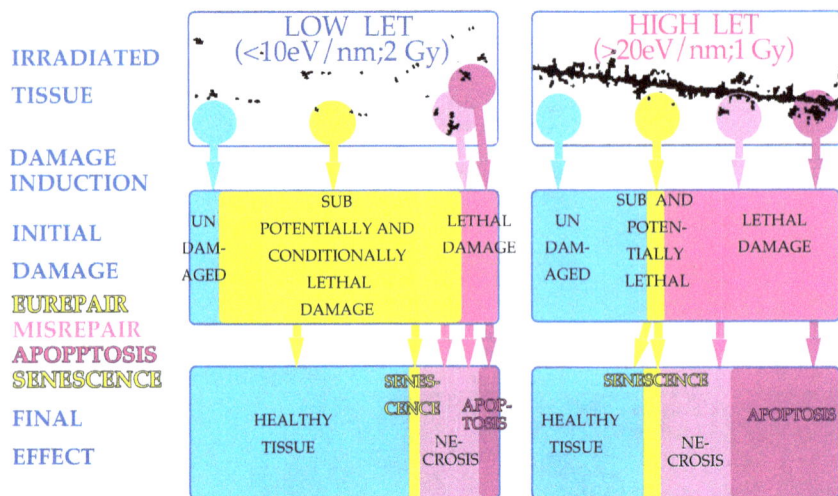

Figure 2.3. Schematic representation of the influence of radiation quality on the type, relative frequency, and effectiveness of the radiation-induced damage and the associated capacity of the repair system. A large part of the initial damage is sublethal and repairable at low *LET* whereas at high *LET* the initial lethal damage is quite large and the sublethal repairable damage much smaller.

Besides the direct hits by secondary electrons from the beam (cf. Fig. 2.3) the indirect action in low-*LET* beams is induced by free radicals, which live some 5 to 10 ns and travel distances of the order of 10 nm. The indirect action, therefore, requires a close proximity between formed radicals and DNA. Owing to this proximity property, it is likely that DNA damage is induced by a combined direct and indirect action close to the interface between the chromatin fiber and the surrounding proteins and water inside the cell nucleus. For low-*LET* radiation beams, about 60% of the total lethal damage is induced by OH· and e_{aq}^- mainly acting on DNA sugar and bases. In high-*LET* densely ionizing beams, the indirect action accounts for <20% of the total lethal damage induction (Fig. 2.3).

The reason for this difference can be found in the dense ionizations in the central secondary electron core of high-*LET* tracks (cf. Fig. 2.4). The lateral dimensions of the core are comparable with the diffusion and recombination distance of the radicals. Consequently, in high-*LET* beams, radical recombination within the dense ion core of the particle track reduces the biological effect by indirect radiation action and a larger part of the biological effect is caused by direct action. In Fig. 2.4, a close-up view of the damage produced by low-energy electron and carbon ions on supercoiled DNA wound two times around nucleosomes in the cell nucleus.

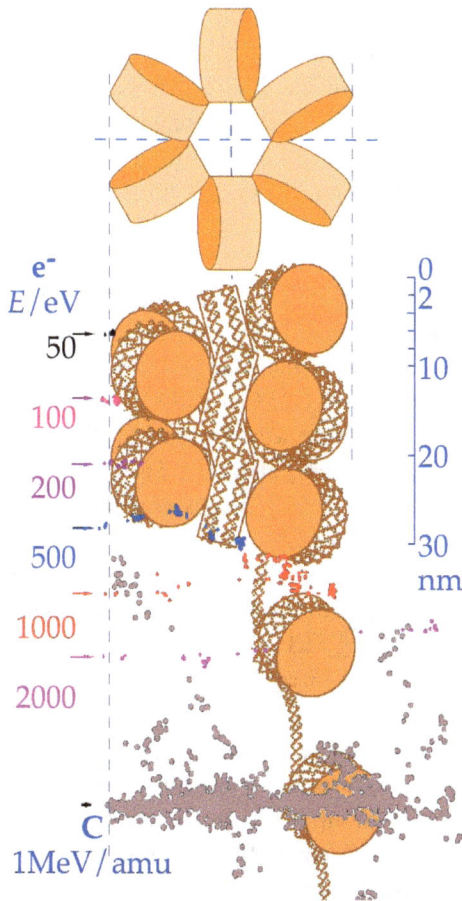

Figure 2.4.   Energy depositions by particle tracks in the chromatin fiber showing that the dominant critical radiation damage to DNA is produced by low-energy electrons and δ rays alone and not least when generated by a primary heavy-charged particle (cf. Fig. 2.5).

Most toxic are the 700 eV electrons, which deposit a dose in the neighborhood of the track of around 10^6 Gy.

The primary biological damage is susceptible to modification particularly by reactions between reducing or oxidizing molecules and organic radicals. For low-*LET* radiations, the indirect mode of action is considerably affected by the presence or absence of oxygen. This is due to the preferential reaction of radicals with oxygen molecules rather than hydrogen donors from thiol-containing protectors and the consequent fixation of the biological damage. Thus, under hypoxic conditions, the induced lethal damage by OH˙ radicals is reduced to about 20% under well-oxygenated conditions. This is seen as a reduced radiation sensitivity and reduced slope of the dose–effect relation particularly for tumors with high concentrations of hypoxic cells. In contrast, the radiation effects of high-*LET* particles are only weakly dependent on the presence or absence of oxygen. Particularly at *LET*s ≥ 100 eV/nm, the oxygen effect disappears almost completely due to the reduced influence of oxygen-related radical species on the biological effect.

Ultra-high dose rates of ionizing radiation delivered within times shorter than the life time of the radicals, that is, of the order of a few nanoseconds, result in a modified biological effectiveness as compared to conventional dose rates. Particularly for normal radiation therapy doses (2–5 Gy) delivered in ≤ 30 ns pulses, the radiation sensitivity is reduced due to simultaneous presence of all radicals and a consequent enhanced recombination. However, at higher doses (10ths of Gy) of low-*LET* radiation, an increased sensitivity may be observed due to increasing probability of closely spaced severe damaged sites similar to the situation along the core of high-*LET* particles.

Critical local damage due to the direct radiation action is most efficiently induced by electrons with energy <1 keV in multi-nanometer size targets. For low-*LET* radiations, almost half the energy deposition is due to electrons with energies below a few kiloelectron-volt (Fig. 2.5).

As is shown in Fig. 2.4 these low-energy electrons have a high probability of inducing severe biological damage to multiple coiled DNA. Furthermore, the similarities <1 keV in the slowing down spectrum per unit dose explain the rather constant relative biological efficiency (RBE) for most low-*LET* beams (upper left panel, Fig. 2.5). For high-*LET* particles, a substantial part of the energy imparted is due to such δ rays from the track core with energies below a few kiloelectron-volt. The energy imparted by these electrons may be of the order of $\geq 80\%$ of the total energy deposition by the track. The similarities in the observed RBE values for low-energy characteristic X-ray photons and α particles also indicate that δ rays are the main cause of biological damage induced by high-*LET* particles. The radial extension of the track depends primarily on the particle energy per nucleon. By increasing the particle energy, more energy is transferred to δ rays in the penumbra region, which then rapidly increases in radii compared to the extension of the dense core. Thus, heavy ions with the same *LET* may have different biological effects due to differences in the radial distribution of the energy deposition. In Fig. 2.6, the microscopic dose deposition as a function of the energy deposition density is shown. It is seen that around beryllium, the highest portion of the dose is delivered at intermediate energy deposition densities.

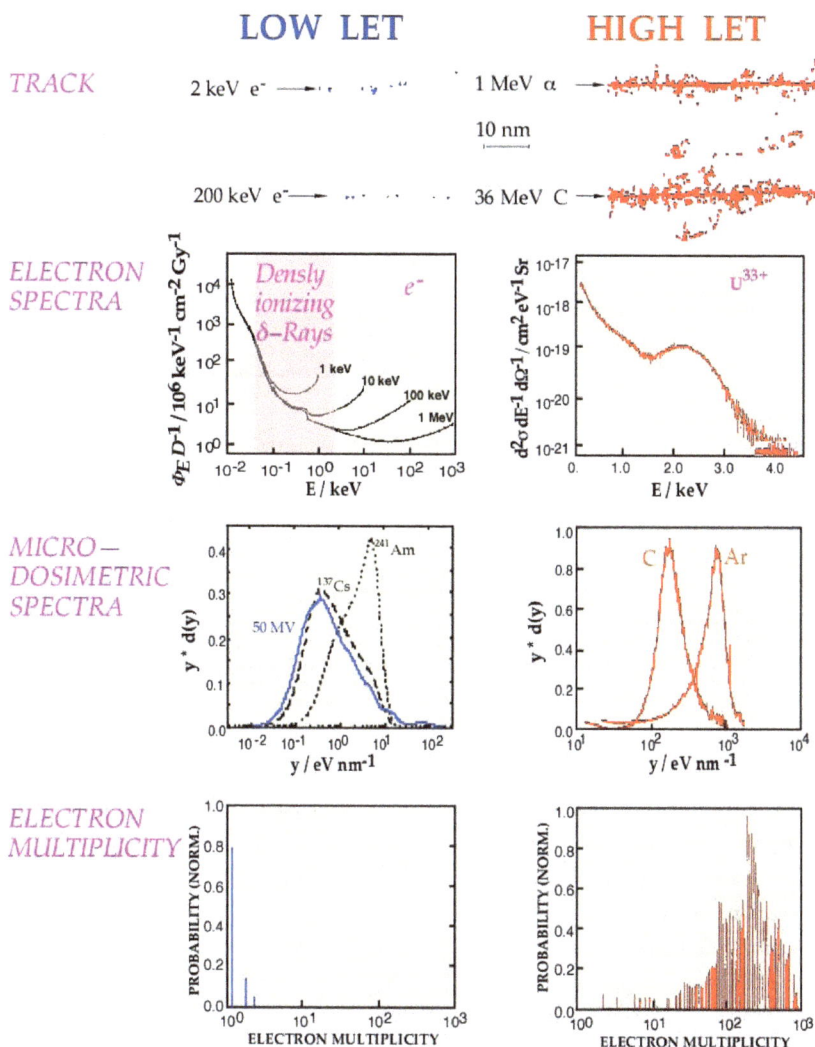

Figure 2.5. Physical description of the energy deposition spectra in low- and high-*LET* beams. In order from the top track segments slowing down spectra, δ ray electron spectra, microdosimetric distributions, and the associated electron multiplicity distributions are illustrated for both low- and high-*LET* beams.

The consequent differences in the microdosimetric distributions to micrometer size objects for low- and high-*LET* particles are shown in the middle panel of Fig. 2.5. These distributions are mainly formed by a combination of *LET* and chord length distributions in the target volume and the additional influence of the energy

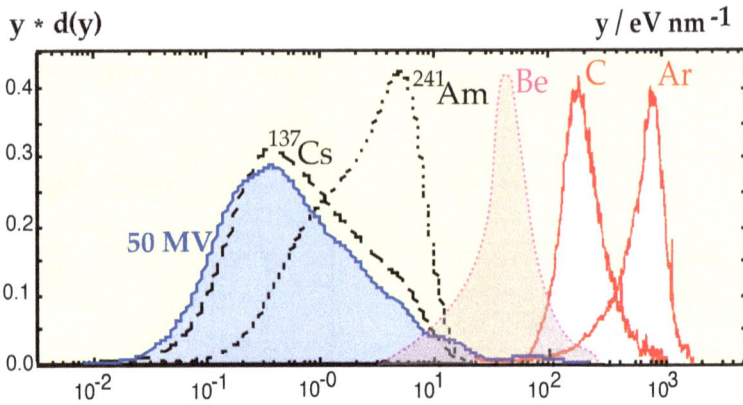

Figure 2.6.   Illustration of microscopic energy deposition spectra of low- and high-energy photons americium on beryllium neutrons and beryllium, carbon, and argon ions. It is the *LET* range around 30–50 eV corresponding to the beryllium peak, that is, of highest interest for hypoxic and apoptotic cell kill.

loss straggling across individual tracks. Low-*LET* radiation results in rather broad distributions with a high variation in the mean energy depositions in the target up to about 30 eV/nm. The mean energy depositions by high-*LET* particles are about 10 to 100 times higher than for low *LET*. The microdosimetric distribution of a high-*LET* particle can be described by the probability distribution of the combined action of multiple low-energy δ rays events from a single track. The microdosimetric distribution of a ^{20}Ne ion beam has been reconstructed in terms of the electron multiplicity distribution of low-energy electrons (Fig. 2.5, lower right panel) using an inversion algorithm.It is thus seen that in micrometer-sized volumes, the energy depositions by high-*LET* ions include a large number of electron events rather than the dominating single event energy depositions commonly seen in low-*LET* beams (Fig. 2.5, lower left panel).

In high-*LET* beams, it is therefore rather difficult to achieve homogeneous irradiation conditions of all the clonogenic tumor cells due to the high-energy deposition concentrated along the track of the incident particles. To ensure a homogeneous microscopic dose delivery on the scale of the cell nucleus heavy ions therefore require substantially higher doses in radiotherapy than those indicated by their low dose RBE which is most relevant for radiation protection purposes.

2.4.2. *Radiobiological Effects of Radiation Quality*

The damage to nuclear DNA by direct and indirect action of ionizing radiation may be expressed in terms of double-strand breaks (DSBs), single-strand breaks

(SSBs), sugar or base damage, DNA–DNA crosslinks, DNA–protein crosslinks, and complex combinations of the same such as multiple damaged sites (MDS) produced by δ rays (cf. Fig. 2.4). Particularly at low *LET*, a significant part of the induced damage is sublethal or potentially lethal and depends on the post-irradiation conditions. Sublethal damage may be repaired by enzymatic and other mechanisms such as mismatch and excision repair if time is allowed before the evaluation of the biological end point. Lack of repair can be attributed to a number of factors such as damage complexity, rejoining fidelity, repair inhibition, and time available for repair. Other factors that will influence the repair capacity are the capability of the cell to block cell cycle progression until all potentially lethal lesions have been detected and repaired (cf. Fig. 2.1), but also the degree of condensation of the DNA and the stage in the cell cycle may be important. For a particular biological system and end point, damage repair is also affected by the *LET* of the radiation beam, the oxygenation of the cells, and the dose rate (cf. Fig. 2.3). In Fig. 2.7, a more functional description of the biological effect of different radiation modalities is, therefore, obtained from the dependence of the biological effect on the energy of the secondary electrons as shown. Below 30 eV, the electrons do not do much harm because they cannot ionize. However, from 100 eV to about 1 keV, the electrons are associated with very dense ionization clusters with an RBE of around 3. As the electron energy is increased

RELATIVE BIOLOGICAL EFFICIENCY

VARIATION WITH δ–ELECTRON ENERGY

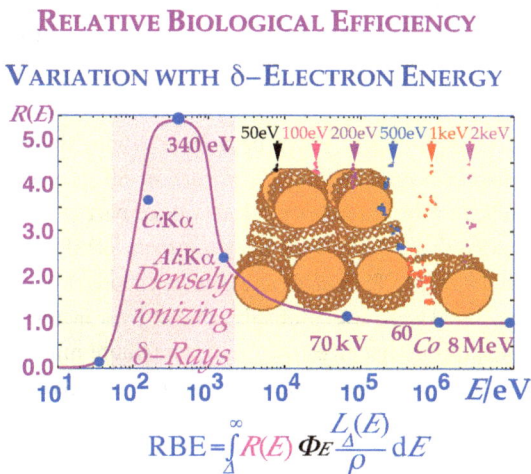

$$\mathrm{RBE} = \int_{\Delta}^{\infty} R(E)\, \Phi_E \frac{L(E)}{\rho}\, dE$$

Figure 2.7. Illustration of the importance of the low-energy δ electrons or electron track ends for the RBE of different radiation modalities from low- and high-linear energy (y). The dots on the electron energy response function $R(E)$ in the upper panel are from different experimental RBE values for photons and electrons. The colored electron paths from 50 eV to 2 keV in the upper panel are shown on top of 30 nm DNA fiber.

above this range, a decreasing portion of the energy deposition is in the form of such low-energy densely ionizing δ electrons.

For mammalian cells, the quasi-exponential repair function has a dominating fast component ($t_{1/2} \approx 10$ min) for low-LET radiation, thus indicating that a large part of the induced sublethal damage is repaired within a few hours. Most often also a slower type of repair is present requiring several hours to repair half of the severe lesions. Probably the more severe lesions, due to the densely ionizing and damaging track ends, existing in low-LET beams may also require use of the complementary allele and therefore requiring more time for completion. From the synergistic effects observed in mammalian cells irradiated with combined high- and low-LET beams, it has been shown that some sublethal damage exists even after irradiation with high-LET ions. However, in high-LET beams, the amount of repairable damage is substantially reduced, as seen by the decreased shoulder and the almost pure exponential shape of the survival curve at LET values of 150–200 eV/nm and disappears completely at LETs $\geq 10^3$ eV/nm. Furthermore, the steepness of the time-dependent repair function is substantially reduced and the time needed for repair is two to five times longer than for low-LET beams. This indicates that the biological damage induced by high-LET beams is more complex than that induced by low-LET radiation (cf. Fig. 2.5).

The damage inflicted by electron track ends on the periphery of a nucleosome may result in dual spatially correlated DSBs reparated by about 80 base pairs. Obviously, such a type of damage is hard to handle for the cell during the repair process when the histones are removed to allow full access for the repair enzymes. The fact that most of these strand breaks are further blunt reduces the probability to have a correct recombination and repair. In addition, there are also several possibilities for misrepair such as loss of one turn or 80 bp of DNA or inversion of the DNA sequence in the same turn. Since a large part of the nuclear DNA is organized in this way, this process may be a common lethal event leading to cell inactivation. As seen in Fig. 2.8, damage sites causing a longer or shorter separation between the strand breaks are less probable, almost by a factor 3, but integrated over all possible separation distances they may contribute more, even though the probability for lethal rearrangements or losses is likely to decrease as the distance increases.

The enormous capability of cells to repair simpler forms of DNA damage results in only about 1% unrejoined DNA breaks per cell and Gray of low-LET radiation. Lethality as a biological end point is likely to be associated with multiple closely spaced DSBs, which may be very complex to repair with a high degree of fidelity due to increased risk for misrepair and loss of DNA. Thus, only about 4% of the unrejoined DNA breaks mentioned above, that is, 0.04% of the initial

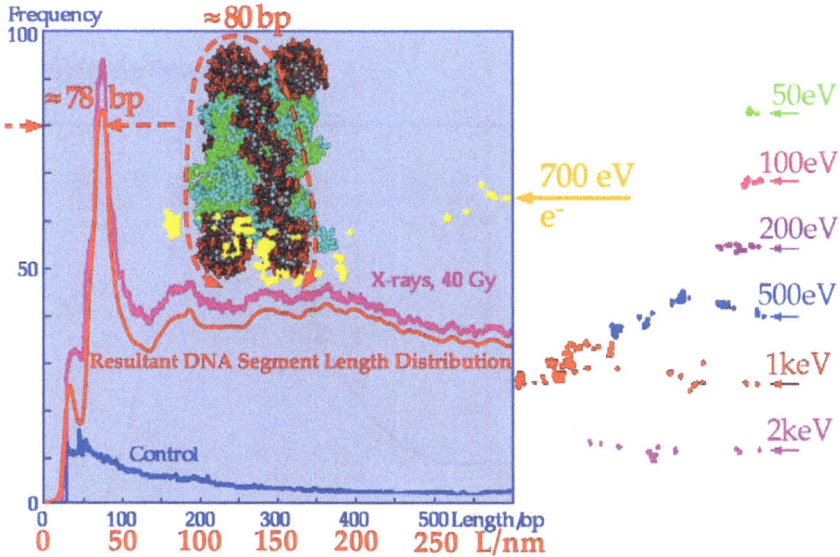

Figure 2.8. Illustration of the effect of conventional X-rays on DNA damage on a cellular system where normal human fibroblasts were irradiated to a total dose of 40 Gy.

DNA damage, will lead to lethality. The formation of critical lesions requires close proximity between severe energy depositions or dense clusters of sublethal or radical damage. However, the differences in RBE obtained by different high-*LET* particles with the same *LET* indicate that closely induced DSBs may be the most severe form of lethal damage even more so than highly complex damage.

For light ions, the ionization density increases about fivefold at the end of its range. Photons, electrons, and protons produce rather low uniform ionization density at all depths and are often called low-*LET* radiation. High-*LET* radiation increases the RBE particularly at the end of the range at the so-called Bragg peak. The RBE is around 2 to 5 for light ions at the Bragg peak with ionization density of around 150 eV/nm and it is generally located in the tumor. This means that a given effect requires that the dose is higher by this factor in comparison to low-*LET* radiation photons, electrons, or protons. In Fig. 2.9, variation of the RBE as a function of the ionization density of light ion beams is shown. The highest biological effectiveness is seen for ionization densities between 25 and 200 eV/nm.

With light ions, the most favorable situation for resistant tumors occurs as the RBE is high at the tumor depth and as low as possible in normal tissues. For tumors, where an intact normal tissue stroma inside the tumor is important for survival, the lightest ions with low to slightly elevated *LET* are most advantageous. An elevated

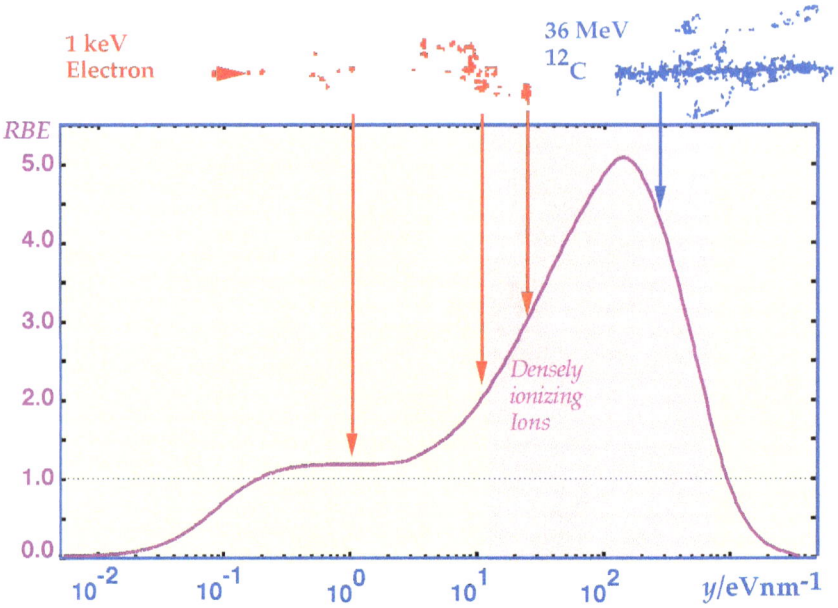

Figure 2.9. Illustration of the biological response functions for high- and low-survival levels (solid and broken curves, respectively) calculated using single event microdosimetric distributions and experimental RBE values for a number of low- and high-*LET* radiation qualities.

LET is also less advantageous when there is considerable life expectancy because of an increased risk of secondary malignancies. However, the high-*LET* peak should then be used only in the gross tumor to high dose levels, so even if a secondary tumor is induced it will be sterilized by the high local dose level.

The biological effect of a radiation beam may be macroscopically described by its RBE and oxygen enhancement ratio (OER) as defined in Fig. 2.10. For a particular biological end point and radiation quality, the RBE is affected by the fractionation scheme and the applied dose rate. The biological effectiveness is furthermore strongly dependent on the stage in the cell cycle (cf. Fig. 2.1). In low-*LET* beams, it is well documented that cells are particularly sensitive during the M- and G_2-phase and resistant during the late S-phase. With increasing *LET*, the variations in the age response during the cell cycle decrease and disappear completely for heavy ions in the *LET* range $>300\,\text{eV/nm}$ (as illustrated in the third panel of Fig. 2.10). The RBE–*LET* function also reaches a maximum at *LET* values between 100 and 300 eV/nm. To be more accurate, the RBE curves show discrete maxima, which are characteristic for each particular ion type and biological end

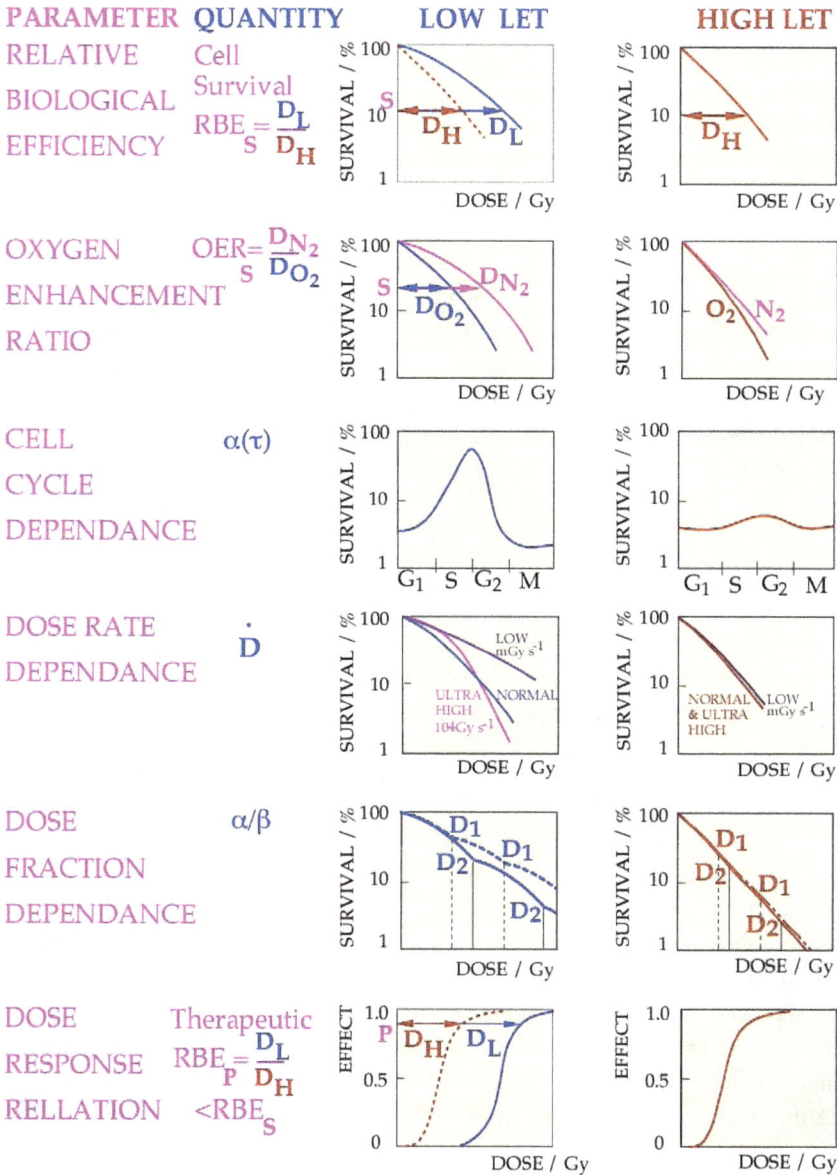

Figure 2.10. The main biological and dosimetric parameters affecting the cell survival and dose–response relation at low- and high-*LET* are shown. Illustration of the difference in cell survival curve shape between low- and high-*LET* beams. In all cases, the larger shoulder of low-*LET* and almost straight exponential fall-off for high-*LET* beams is seen explaining the different biological effects as a function of dose, dose rate, RBE, OER, and time dose fractionation, etc.

point. Above these peaks, the RBE decreases rapidly as seen in Fig. 2.9. This is due to the decreasing interaction cross-section as a result of a shrinking core radius and also due to the associated increase in the recombination probability of radicals and ions due to the extreme proximity and simultaneous presence of the species produced by closely spaced δ electrons.

Until now most data given pertain to low ionization density (LET) electrons, and photons, whereas high-LET radiations such as neutrons and heavy ions have higher RBE and higher microscopic dose heterogeneity and thus shallower dose–response reactions (cf. Fig. 2.10 below). The continued development of the field organ radiation sensitivity is very important both on a local and global basis because the spectrum of individual variations may vary from continent to continent and even between different countries and ethnic groups. Figure 2.10 summarizes the different clinical factors that are affected by the difference in ionization density. Owing to the almost straight cell survival curve with medium-to-high LET and the low dose to organs at risk, dose fractionation is much more flexible and 5–15 fractions are often sufficient (For more details see Chapter 8).

2.5. From Cell Survival Curves to Dose–Response Relations for Organized Tissues

2.5.1. *Cell Survival Models*

The fundamental feature of most cell survival models is that the clonogenic cell death is mainly an exponential process, that is, the number of cell lost is proportional to the dose and the initial number of cells before irradiation. This results in a linear cell survival curve when plotted on a log-linear scale. Over the years, different cell survival models have been presented. However, experimentally observed cell survival curves generally deviate substantially from the basic exponential model due to their significant shoulder at low doses ($D < 5$ Gy). Previous models describing the shoulder were based on target theory, suggesting that cells have certain sensitive areas (targets) and that the number of targets and hits on them are important for the shoulder shape. Single-target single-hit inactivation is then one way to describe cell death, which leads to a purely exponential survival curve:

$$S(D) = e^{-D/D_0}. \tag{2.1}$$

A more complex model is the single-hit multi-target theory, which implies one hit on n targets:

$$S_{\mathrm{SHMT}}(D) = 1 - (1 - \exp(-D/D_0))^n$$

This equation is derived in the approximation of single event cell kill without consideration of repair mechanisms. However, since Elkind's pioneering work (Elkind and Sutton 1960), we know that the shoulder is mainly a repair phenomenon largely unrelated to theoretical target theory assumptions. A better fitting of experimental cell survival data over a somewhat wider range of doses is obtained if the simple exponential cell survival $S(D) = e^{-D/D_0}$ in Eq. (2.1) is replaced by the linear-quadratic expression. With the linear-quadratic model (LQ model), the cell survival is then expressed by

$$S(D) = e^{-(\alpha D + \beta D^2)} = e^{-\alpha D\left(1+\frac{D}{\alpha/\beta}\right)}. \tag{2.2}$$

In the first approximation $\alpha \approx 1/D_0$ in this equation and this parameter is characterizing the slope of the survival curve at low doses. From Eq. (2.2) above, it is also clear that α is proportional to the inactivation cross-section according to $\alpha = \sigma_0/(L_\Delta/\rho)$. In this model, the shape of the shoulder of the cell survival curve is described by the α/β ratio. An α/β value of 3 Gy is common for late responding tissues which means that at 3 Gy, the linear cell kill is equal to the quadratic repairable kill of the D^2 term. Tumors and acutely responding normal tissues have an α/β of about 10 Gy and large doses per fraction are better tolerated. Slowly proliferating normal tissues such as prostate and brain may even have lower α/β ratios then 3 Gy. However, today many people regard the quadratic dose term as a repair term, because it is responsible for the shouldered survival curve shape which was clearly shown by Mortimer Elkind to be due to sublethal damage repair. Unfortunately, this is not logical since a cell line with significant repair capacity then needs a large β term, but this would mean that the total cell survival would decrease because the b term is generally part of a negative exponential as seen in Eq. (2.2). Therefore, the linear α term instead needs to be reduced to get the survival level correct at the same time. This means that the whole idea of dose or fluence being proportional to cell kill at low doses resulting in exponential cell kill at high doses is lost (c.f. Equation 2.1). A more realistic cell survival model therefore needs to take the cellular repair processes more explicitly and accurately into account (cf Eq.2.5 below and Brahme 2011). Of course Eq. (2.2) can still be used but the biological interpretation of the linear and quadratic terms can be strongly misleading!

Since the 1980s, the LQ model has dominated the field. The popularity of this model probably stems from the fact that it fits experimental data rather well at clinical dose levels and it is easy to handle even though it is linked to the same shortcomings as the target theories. Today, it is obvious that the LQ model has disadvantages at both low- and high-dose levels. The LQ model dose not account for the low-dose hypersensitivity and results in a constantly bending curve at high

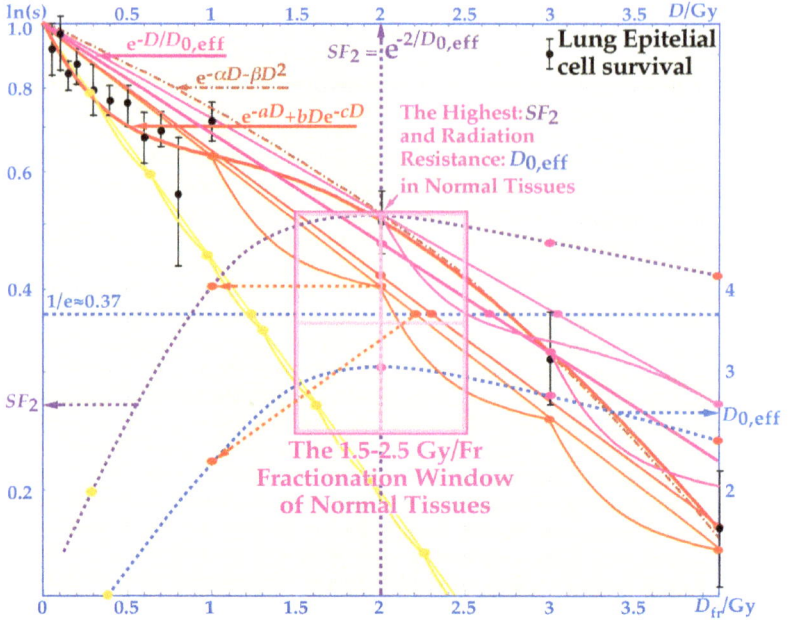

Figure 2.11.　Illustration of the fractionation window caused by low-dose hypersensitivity. Both at low and high doses per fraction,, the normal tissue damage is increased over that at the standard 2 Gy/fraction level used in classical radiotherapy causing fractionation window at around 2 Gy/fraction where the normal tissue damage is minimal.

doses, that is, the radiation becomes more effective per unit dose at higher doses (Fig. 2.11, cf also Ch 8 Eq. 1-38). The repair–misrepair (RMR) model (Tobias 1985) and the lethal–potentially lethal damage model (LPL model) (Curtis 1986) are "repair models." The basis of these models is that ionizing radiation is considered to produce two types of lesions in the cell: repairable and non-repairable. The non-repairable lesions are described by the linear component. Repairable lesions are divided into two groups: lesions that are repaired or misrepaired. The LPL model gives similar results as the LQ model over the first two decades of the survival curve and this model has, therefore, the same problem at low doses, that is, it dose not describe low-dose hypersensitivity. The two population (TP) model is based on the LQ model but the cells are separated into two populations, one sensitive population with α_S and the fraction p of the total number of cells and one radiosensitive cell population with fraction $(1 - p)$, α_R and β_R:

$$S_{TP}(D) = p e^{-\alpha_S D} + (1 - p)e^{-(\alpha_R D + \beta_R D^2)}. \tag{2.3}$$

Another extension of the LQ model has been proposed, which is able to resolve low-dose hypersensitivity as well. The model is called the inducible repair (IR) model and has a modified linear component (Joiner and Johns 1988):

$$S_{IR}(D) = \exp\left\{-\alpha_R\left[1 + \left(\frac{\alpha_S}{\alpha_R} - 1\right)e^{-D/D_0}\right]D - \beta D^2\right\} \qquad (2.4)$$

The TP and IR models are reduced to the LQ model at high doses, that is, they also result in a constantly bending cell survival curve at high doses.

2.5.2. Relationship between the RCR Cell Survival Model and the Classical Models

Low-dose hypersensitivity, that is, a higher cell kill per unit dose at doses up to 0.5 Gy than at higher doses, is an accepted phenomenon today. Several authors have, during the last decade, shown this phenomenon *in vitro* for low- and high-*LET* radiations (Joiner and Johns 1988; Lambin *et al.* 1993; Marples and Joiner 2000; Tsoulou *et al.* 2001). The low-dose hypersensitivity has also been shown *in vivo* in rectum, skin, kidney, and lung of mice (Gasinska *et al.* 1993; Hamilton *et al.* 1996). Turesson *et al.* (2001) have shown similar effects in human skin. The end point used by Turesson *et al.* (2001) was the basal cell density in the epidermis, which means that the result might also be influenced by the cellular repopulation. Contradictory results have been shown on mouse dermal and rat spinal cord. Several explanations of the low-dose phenomenon have been proposed over the years. Variations in sensitivity to radiation between the cell cycle phases were initially proposed as an explanation. Cell cycle delay and apoptosis are also two explanations that have been rejected. Increased repair capacity or repair fidelity is today the most popular explanation. Repair-deficient hamster cell lines have not shown low-dose hypersensitivity as pronounced as their parental cell lines. Pretreatment with hydrogen peroxide or X-rays (free radicals and DNA SSBs) has induced radiation protection, that is, reduced low-dose hypersensitivity. The dominating hypothesis today is that DNA, DSB and the non-homologous end joining (NHEJ) repair pathway are responsible for the low-dose hypersensitivity.

It is well-know today that all classical cell survival models based on target theories have problems in describing the shoulder of the surviving curve since its shape is largely dependent on repair processes rather than the number of targets or hits that are assumed to cause a lethal event. We will, therefore, first quantify the classical exponential cell survival parameters namely the exponential radiation resistance expressed as D_0, the quasi-threshold or wasted dose D_q and the extrapolation number n as well as the LQ parameters α and β using the bi-exponential *abc* form of the RCR model. It has been known for a very long time that the cell kill is purely exponential at very large doses and not quadratic as the LQ relation indicates. More recently, it has been shown that the linear-quadratic model is not so well suited to describe the response at low doses either even though it works quite well in the standard clinical 2 Gy/fraction range. At low doses, the cell kill seems to be much steeper than previously realized. The increased sensitivity at low doses as shown in Fig. 2.11 is more pronounced when the shoulder of the cell survival curve is large such as for late responding tissues with a low α/β value. This indicates that late responding slowly proliferating tissues have a low- and high-dose hypersensitivity indicating a fractionation window around 1.5–2.5 Gy as seen in Fig. 2.11. There are already indications that this low-dose hypersensitivity phenomenon may be clinically relevant since low and high doses per fraction to the skin may cause similar skin reactions even if the total skin doses differ substantially. Clearly, the low-dose hypersensitivity has important consequences for radiation protection issues as well since many low-dose fractions can cause more damage per unit dose.

The newly proposed cell survival model called the RCR model is based on the Poisson statistics. It has some of its basic assumptions in common with the RMR and LPL models, but two distinct classes of repairable damage are introduced. The model distinguishes between two types of damage, namely those that are potentially repairable (PR) but may also be lethal, that is, non-repaired or misrepaired, and those which are conditionally repairable (CR). When the PR damage is repaired, part of the CR ones may simultaneously be correctly repaired. The survival after complete repair is given by:

$$S_{\mathrm{RCR}}(D) = e^{-aD} + bDe^{-cD}, \qquad (2.5)$$

where the term with a describes the cell survival due to avoidance of irreparable cell kill and the b and c terms describe the potentially lethal part which can be repaired if the right environmental conditions prevail for repair. By serial expansion of the LQ equation and Eq. (2.5) at low doses, it is clear that $\alpha \approx a - b$ and $\beta \approx -b(b/2 - a + c)$, whereas the slope at high doses is characterized by c provided $a > c$. The thick solid curve in Fig. 2.11 is from Eq. (2.5) and the dots with error bars are experimental data for lung epithelial cells. By comparing with the dash doted LQ curve, the improved fitting particularly at low but also at high doses mainly outside the figure is obvious.

As shown above in Eq. (2.5), the more severe irreparable damage causes a pure exponential cell survival. If the sublethal damage is allowed to be fully repaired, the repaired fraction will be proportional to the dose at low doses, whereas at high doses it also becomes quasi-exponential, as might be expected based on binomial or Poisson statistics grounds. If the appropriate conditions for repair of sublethal damage prevail, the cell survival may therefore be approximated by a bi-exponential expression, where the first term describes the irreparable cell kill characterized by the radiation sensitivity and the second term describes the increased survival as a result of repairable sublethal damage. As seen in Fig. 2.11, this expression adequately describes the cell survival at low, intermediate and high doses, whereas the classical LQ expression is good only at intermediate doses around a few Gray. Based on Eq. (2.5), it is tempting to think that cell survival is a deterministic process as a function of dose. However, to treat the real cell survival accurately, each cell should be followed separately and the total cumulative effect is due to the effect on all cells. Each cell can either die or survive so that the cell survival is essentially a binomial process where 0 or 1 is the only end point for each cell. The probability distribution of cell survival is thus given by the binomial expression in Fig. 2.11. At zero dose, all cells are alive as given by the sharp peak (barely visible at $\ln S = 0$) at the origin in the figure. When the dose is increased, the probability distribution is spread out as expected due to the laws of statistics. The interesting fact is that at high doses the probability again reaches unity when practically all cells are hit and the probability of having no surviving cells is almost unity or 100%. If the cells were tumor clonogens, this would be identical to total tumor eradication and thus 100% probability of tumor cure or a beneficial treatment.

The present model can basically be derived from the following three simplifying assumptions:

(1) The total amount of initial damage induced is in the first approximation proportional to the absorbed dose D. This is reasonable at low-to-medium doses since the amount of damage should be proportional to the product of the damage cross-section and the fluence or absorbed dose of the associated particles as multiple events are negligible.

(2) The total amount of damage can be subdivided into two distinct Poisson processes, PR and CR damage. The amount of both kinds of initial damage are proportional to the absorbed dose, D but with different coefficients of proportionality.

(3) Only one of the two damaging processes can trigger the repair system, namely the PR damage, which then may also repair some of the CR type of damage.

The rationales for these assumptions are: (1) the observed purely exponential cell survival at high doses for most normal cell lines, (2) the almost perfect exponential response at all dose levels for the largely irreparable high-LET damage, as well as, (3) the purely exponential dose response for repair-deficient cell lines. Figure 2.12

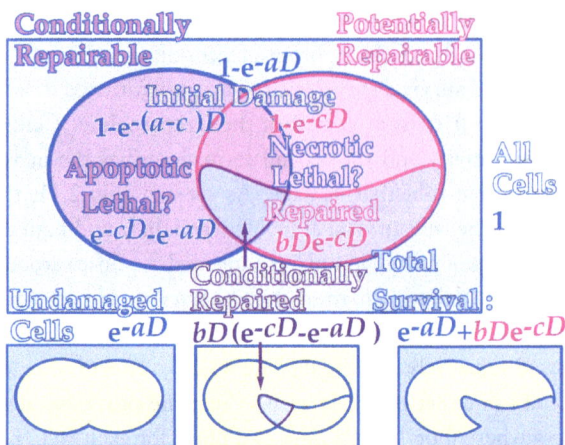

Figure 2.12. Venn diagrams of the RCR model, showing the two main types of events. The different cell fractions corresponding to the different categories of cell deaths or repair are indicated. On the lower panel, it is shown how the different terms of the model account for the different cell fractions and consequently the biological processes that are related to them.

shows a Venn diagram where the sets of events from the two distinct damaging processes are denoted respectively.

First, the limit values at high and low doses give the gross characteristics of a survival model. To describe the slope of the logarithmic survival curve, it is useful to define a quantity $k(D)$ by defining the fractional cell kill per unit dose, which can also be regarded as the relative slope or absolute logarithmic slope of the survival curve according to:

$$k(D) = -\frac{1}{S}\frac{\partial S}{\partial D} = -\frac{\partial \ln S}{\partial D}. \qquad (2.6)$$

A fundamental way to describe low-dose hypersensitivity is to use the fact that the first derivative of the fractional cell kill per unit dose $\kappa'(D)$ is positive in a region where low-dose hypersensitivity is present. This opens up an interesting way to calculate the relation among a, b, and c for the limiting case where there is no low-dose hypersensitivity by solving for an inflection point located at $D \leq 0$. The relation between the parameters then becomes $b \leq 2(a - c)$.

In the high-dose approximation, $\lim_{D\to\infty} \kappa_S(D) = c$, the simplified expression, Eq. (2.5), can thus with advantage be written as

$$S(D) = (e^{-(a-c)D} + bD)e^{-cD} \approx ne^{-D/D_0}, \qquad (2.7)$$

where it is immediately clear that $c \approx 1/D_0$ or that $D_0 \approx 1/c$. Furthermore, the effective extrapolation number n at the dose level D is given by:

$$n = bD + e^{-(a-c)D}. \tag{2.8}$$

From this relation it is seen that the extrapolation number increases steadily with the dose level from which it is extrapolated. This effect is also generally true for experimentally observed cell survival curves, as well as for the LQ relation (Brahme 1984).

Based on linear extrapolation of the high dose behavior, it is possible to express the wasted or quasi-threshold dose D_q according to:

$$D_q = \frac{1}{c} \ln \left(bD + e^{-(a-c)D} \right) \approx D_0 \ln \left(bD + e^{-(a-c)D} \right). \tag{2.9}$$

Like the extrapolation number, the wasted dose increases with the dose level of interest. If we, instead, want to describe the cell survival at low doses, it is according to $\lim_{D \to 0} \kappa_S(D) = a - b$, sometimes more convenient to express S according to

$$S(D) = e^{-(a-b)D}(e^{-bD} + bDe^{-(b+c-a)D}) \approx e^{-(a-b)D} \left(1 - \left(\frac{b}{2} + c - a \right) bD^2 \cdots \right)$$

$$\approx e^{(a-b)D - \left(\frac{b}{2} + c - a \right) bD^2}. \tag{2.10}$$

These expressions show that at low doses the cell survival exponentially decreases with a linear rate $\alpha \approx a - b$. Furthermore, the quadratic term at low doses may be approximated by $\beta \approx b(b/2 + c - a)$. For cell lines with a significant low-dose hypersensitivity, β at low doses is in reality negative ($a \geq c + b/2$). This is also a classical sign of a heterogeneous cell line where the most sensitive cells are lost first at low doses. This expression (Eq. (2.10)) also indicates an interesting possibility for getting almost exponential cell survival curves, for example, for ion beams or repair-deficient cell lines, without having $b \equiv 0$. In the first approximation, $b \approx 2(a - c)$ makes the second and higher order terms as small as possible. Similarly, it is helpful if $b \approx a - c$ so that the low and high dose slopes κ_S become approximately the same. Together these two expressions are obviously best fulfilled by low b and $a - c$ values.

At intermediate doses, the LQ cell survival parameters can be derived since the low-dose hypersensitivity is then largely lost. They can be derived by equating the cell survival and its first derivative with respect to the dose at an arbitrary dose level of interest D.

$$S(D) = e^{-aD} + bDe^{-cD} = e^{-\alpha D - \beta D^2} = S_{LQ}(D) \qquad (2.11)$$

$$S'(D) = -ae^{-aD} + (b - bDc)e^{-cD} = (-\alpha - 2\beta D)S_{LQ}(D) = S'_{LQ}(D).$$

With α and β as unknowns, it is possible to derive their values from these two relations at a given dose level D and given set of survival curve parameters a, b, and c according to:

$$\alpha = \frac{b(1 - cD)e^{-cD} - ae^{-aD}}{e^{-aD} + bDe^{-cD}} - \frac{2}{D} \ln{(e^{-aD} + bDe^{-cD})}, \qquad (2.12)$$

and

$$\beta = \frac{-b(1 - cD)e^{-cD} + ae^{-aD}}{D(e^{-aD} + bDe^{-cD})} + \frac{1}{D^2} \ln{(e^{-aD} + bDe^{-cD})}. \qquad (2.13)$$

Unfortunately the inverse problem, to derive a, b, and c from known α and β values is more complex as the second derivative also has to be considered (Fredriksson 2002):

$$S''(D) = a^2 e^{-aD} + bc(cD - 2)e^{-cD} = ((\alpha + 2\beta D)^2 - 2\beta)S_{LQ} = S''_{LQ}(D). \qquad (2.14)$$

It is still straight forward to solve Eqs. (2.11) and (2.14) for this case, but unfortunately a transcendental equation is obtained, calling for a numerical solution:

$$\frac{S''_{LQ} - a^2 e^{aD}}{S_{LQ} - e^{-aD}} - \left(\frac{S'_{LQ} + ae^{-aD}}{S_{LQ} - e^{-aD}}\right)^2 + D^{-2} = 0. \qquad (2.15)$$

The value of a that satisfies Eq. (2.15) should be inserted in Eq. (2.11) to obtain b and c as well. First, c is given as

$$c = \frac{1}{D} - \frac{S'_{LQ} + ae^{-aD}}{S_{LQ} - e^{aD}}, \qquad (2.16)$$

and then b is given as

$$b = e^{cD}(S_{LQ} - e^{-aD})/D. \qquad (2.17)$$

It is surprising how well this set of equations can produce a first useful set of a, b, and c values even though α and β do not include any information about the low-dose hypersensitivity except possibly through the known property that low α/β ratios and large shoulders are generally linked to a more pronounced low-dose hypersensitivity (Dasu and Denekamp 2000). However, it is important that the fit is done at a dose level D around 1.5–3 Gy where the low-dose hypersensitivity has little influence on the shape of the survival curve and thus α and β are consistent with the true survival curve shape at these doses. It is also interesting to observe that tissues with marked low-dose hypersensitivity are also linked to "high-dose sensitivity" in the form first discussed by Withers (1982) in connection with the severe damage at high doses per fraction to late responding tissues.

2.5.3. *Radiation Biology of Functional Tumor Cells*

The microenvironment of an arbitrary tumor can be quite complex with microscopic variations of tumor clonogenicity, nutrient, and oxygenation pattern as well as cellular density, DNA content, and genetic make-up. The surviving clonogen distribution after irradiation by a dose distribution function $D(r)$ can generally be written as

$$n(r) = \int n_{D_e}(r)S(D(r))dD_e, \qquad (2.18)$$

where $n_{D_e}(r)$ is the spatially dependent clonogen density differential in D_e and D_e is the effective radiation resistance. For a uniform dose or constant dose per fraction D and effective radiation resistance D_e, a purely exponential cell survival is obtained by:

$$S(D) = e^{-D/D_e}. \qquad (2.19)$$

From this expression and Eq. (2.5), D_e can be calculated at an arbitrary dose level or dose per fraction D according to:

$$D_e = -\frac{D}{\ln S(D)} = 1 \bigg/ \left(c - \frac{\ln\left(bD + e^{(c-a)D}\right)}{D} \right). \qquad (2.20)$$

Thus, as derived by Lind *et al.* (2003), $D_e = 1/(a - b)$ at low doses, whereas generally $D_e = 1/c$ at high doses, and there is a continuous transition between these values at intermediate doses. The mean density distribution of surviving clonogens after being exposed to a dose distribution $D(r)$ may be thus rewritten as

$$n(r) = \int n_{D_e}(r)e^{-D(r)/D_e}dD_e. \qquad (2.21)$$

However, at least for milder types of lethal damage, it may take several cell divisions before a damaged cell becomes necrotic or in general non-functional. The functional fraction at time t after a first irradiation at time t_1 to dose D may therefore be approximated by:

$$f_1(t, D) = e^{-aD} + (1 - e^{-aD})e^{-(t-t_1)/\tau}$$
$$+ D(b_1 + b_2 - b_1 e^{-(t-t_1)/\tau_1} - b_2 e^{-(t-t_2)/\tau_2})e^{-cD}, \qquad (2.22)$$

where τ is the time constant of loss of functionality for lethally damaged cells after an irradiation and τ_1 and τ_2 are the fast and slow repair times of sublethally damaged cells (cf Lind *et al.* 2003). The time τ is of the order of a few doubling times of the cells and the repair times of the fast and slow components are of the order of 0.1 and 4 h, respectively. Thus, after a time of about 2 weeks, $f_1(t, D) \approx S(D)$, since all lethally committed cells have been lost and complete repaired and misrepair of sublethally damaged cells have taken place. Since τ is generally much longer than τ_1, τ_2 and 1 day, the τ_1 and τ_2 terms may often be disregarded even when cell cycle blocks are taken into account when studying the functional compartment over time periods longer than 1 day. This implies that we can assume that there is sufficient time to repair sublethal damage and, which is also largely true, that DNA damage repairing cells are still functional hence Eq. (2.22) may be reduced to:

$$f_1(t, D) = S(D) + (1 - S(D))e^{-(t-t_1)/\tau}, \qquad (2.23)$$

where t_1 is the irradiation time. Thus, when the second dose fraction is delivered, some of the lethally damaged cells from the first fraction are still functional, but will eventually be lost a few time constants τ later. However, some of the undamaged cells or fully repaired cells from the first treatment may be lethally hit by the second irradiation at $t = t_2$ so the functional fraction will now be:

$$f_2(t, D) = S^2(D) + S(D)(1 - S(D))e^{-(t-t_2)/\tau} + (1 - S(D))e^{-(t-t_1)/\tau}. \qquad (2.24)$$

As shown in Fig. 2.13, the functional but doomed tumor cell compartment rapidly becomes many times larger than the finally surviving cells and decreases by a rate given by t quite early after the start of the treatment. Already, after 1 week of therapy, the dying but still partly functional cells are dominated by about one order of magnitude. This is clearly seen in Fig. 2.13 where the loss rate of functional cells quite rapidly dominates over the ordinary long-term cell survival. This could also be understood by comparing the time constants τ of the order of several days (typically 2 weeks are used to determine the true long-term clonogen survival), whereas a survival of about 50% per dose fraction of 2 Gy corresponds to a τ value of about 1.3 days during daily fractionation. Not even the hypoxic tumor fraction will influence the situation significantly unless it is very large and totally dominates the early cell kill.

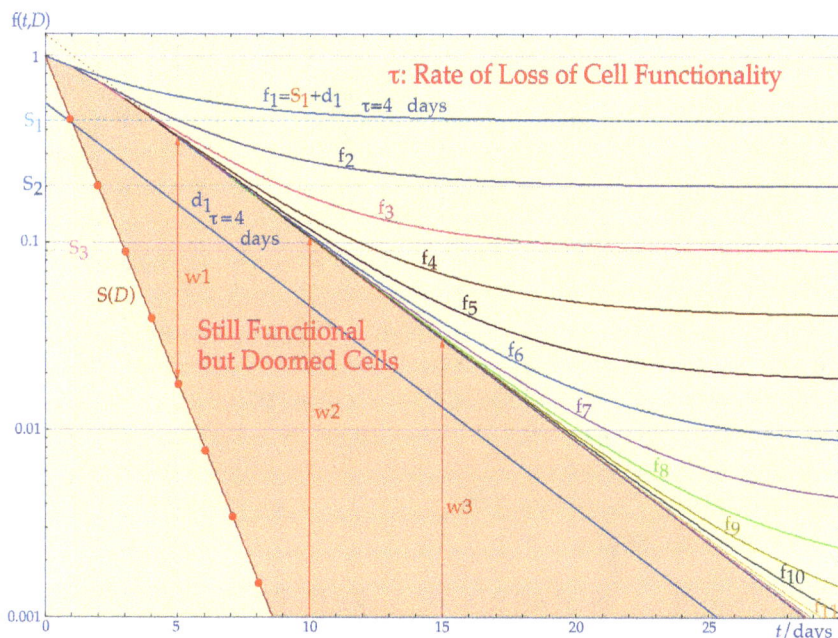

Figure 2.13. The decrease in the finally surviving and the doomed but still functional cell compartments during daily fractionation with 2 Gy is illustrated. Also shown is the decay of the lethally hit compartment assuming a time constant loss of $\tau = 4$ days. The decrease in the functional compartment if the radiation is interrupted on day n ($n = 1 - 11$). It is seen, that the doomed but still functional compartment is more than one order of magnitude larger than the finally surviving fraction per week of therapy (w1, w2 etc). All cells above $S(D)$ are doomed cells that eventually are going to die.

2.5.4. *Mixed Radiation Qualities*

The cell survival after exposures to a multitude of radiation qualities, each with a survival curve $S_i(D_i)$, where D_i is the absorbed dose of radiation quality i, is classically given by:

$$S_{\text{tot}} = \prod_{i=1}^{n} S_i(D_i). \tag{2.25}$$

To be more exact this expression is based on the assumption that $S_i(D_i)$ is the survival after complete repair of possible sublethal damage and that each new radiation quality is given after a time that is longer than the repair time of the previous beam. Obviously, if the repair time is long compared to the cell cycle length, possible cell cycle blocks and subsequent cell growth should also be taken into account.

Since the dominating effect of the cell survival function is exponential as discussed above, Eq. (2.25) can be rewritten based on the high LET or almost repair-free form of the cell survival curve ($P_{CR}(D) = 1 - \exp(-\lambda D)$, where λ is the mean number of CR damage events per Gray):

$$S_{tot} = \prod_{i=1}^{n} S_i(D_i) \approx \prod_{i=1}^{n} e^{-\delta_i D_i} \equiv \prod_{i=1}^{n} \left(e^{-\delta_i D_{tot}}\right)^{\frac{D_i}{D_{tot}}} \approx \prod_{i=1}^{n} S_i(D_{tot})^{\frac{D_i}{D_{tot}}}. \quad (2.26)$$

Even if Eq. (2.26) was derived under assumption of pure exponential cell survival, it is an extremely useful form of the composite cell survival curve since only the survival level at the total dose is needed contrary to Eq. (2.25). Furthermore, it should be very accurate at least for high-LET radiations, where exponential survival dominates. For the simple case of two-component dose delivery, Eq. (2.26) reduces to:

$$S_{tot}(D_{tot}) = S_1(D_{tot})^f \cdot S_2(D_{tot})^{1-f}, \quad (2.27)$$

where we, for simplicity, assume that D_1 and S_1 pertain to the highest LET component and that $f = D_1/D_{tot}$ and $D_{tot} = D_1 + D_2$.

This modified geometrically averaging damage interaction model (Eqs. (2.26) and (2.27)) opens a new way of comparing radiation responses which is more generally applicable in radiation therapy than the classical RBE concept. This is seen by multiplying both nominator and denominator by S_2^f in Eq. (2.27) and rewriting it as

$$S_{tot}(D_{tot}) = \left(\frac{S_1(D_{tot})}{S_2(D_{tot})}\right)^f S_2(D_{tot}) = S_2\sigma_{1,2}^f. \quad (2.28)$$

The survival ratio, $\sigma_{1,2}$ defined as

$$\sigma_{1,2} = \frac{S_1(D_{tot})}{S_2(D_{tot})}, \quad (2.29)$$

is a very convenient way of accounting for the interaction of two radiation modalities. One advantage with this technique is that it like Eq. (2.27) is independent of the exact cell survival model used. The survival ratio also has a considerable advantage in radiation therapy optimization since the RBE is unknown until the survival level is given making it extremely hard to optimize dose delivery before the exact dose or survival level is known. Furthermore, by assuming that the high-LET component,

$S_1(D_1)$, is purely exponential, it is possible to derive a simple relation between the survival ratio and the RBE

$$\sigma_{1,2}(D_2) = \frac{S_1(D_2)}{S_2(D_2)} = \frac{S_1(D_2)}{S_1(D_1)} = S_1(D_1)^{\frac{D_2-D_1}{D_1}}$$

$$= S_1(D_1)^{RBE-1} = S_2(D_2)^{RBE-1}. \tag{2.30}$$

Thus, the logarithm of the survival ratio is approximately proportional to the RBE as might be expected.

Equation (2.27) above can also be used as a definition of the effective dose fraction, f^* since it is possible to solve for f^* provided S_1, S_2, and S_{tot} have been measured separately:

$$f^* = \frac{\ln S_{tot} - \ln S_2}{\ln S_1 - \ln S_2}. \tag{2.31}$$

By plotting this quantity as a function of the real dose fraction $f = D_1/D_{tot}$ for a given total dose D_{tot}, an interesting possibility is obtained to compare different cell survival models in a uniform way. The trivial case of perfect exponential survival would map D_1/D_{tot} linearly on f^* with similar range of variability since f^* by necessity varies from 0 to 1, when D_1 varies from 0 to very high doses or f varies from 0 to 1. The whole data set of Ngo et al. (1981) is thus plotted in Fig. 2.14, illustrating the deviation between the experimental effective dose fraction f^* and the delivered dose fraction of beam quality 1 (neon ions). It is seen that most f^* values are higher than the straight line between 0 and 1, expected based on the assumption of purely exponential survival. This means that the true experimentally observed survival is slightly lower than that expected based on Eq. (2.26), indicating that a true synergistic effect is obtained when simultaneously combining low and high LET. It should also be pointed out that the traditional cell survival model based on Eq. (2.25) corresponds to no synergy or completely independent interactions and is always located below the straight line indicating also that Eq. (2.26) is always a better approximation to the real survival than Eq. (2.25) when using simultaneous irradiation.

Interestingly, there is approximately a 30% increase in f^* when going from sequential to simultaneous irradiation at an f value of around 0.5 as seen in Fig. 2.14. This means that the effective D_1/D_{tot} value increases by about 30%, and thus: $D_{1,eff} \approx 1.3 D_1$, since $f^*/f \approx 1.3$. It is, therefore, very important to deliver the high- and low-LET dose fractions as close in time as possible preferably within a few seconds due to the fast capping of the DNA ends by the Ku molecules of NHEJ. This is so, even if the complete NHEJ repair process may take some 30 min, since

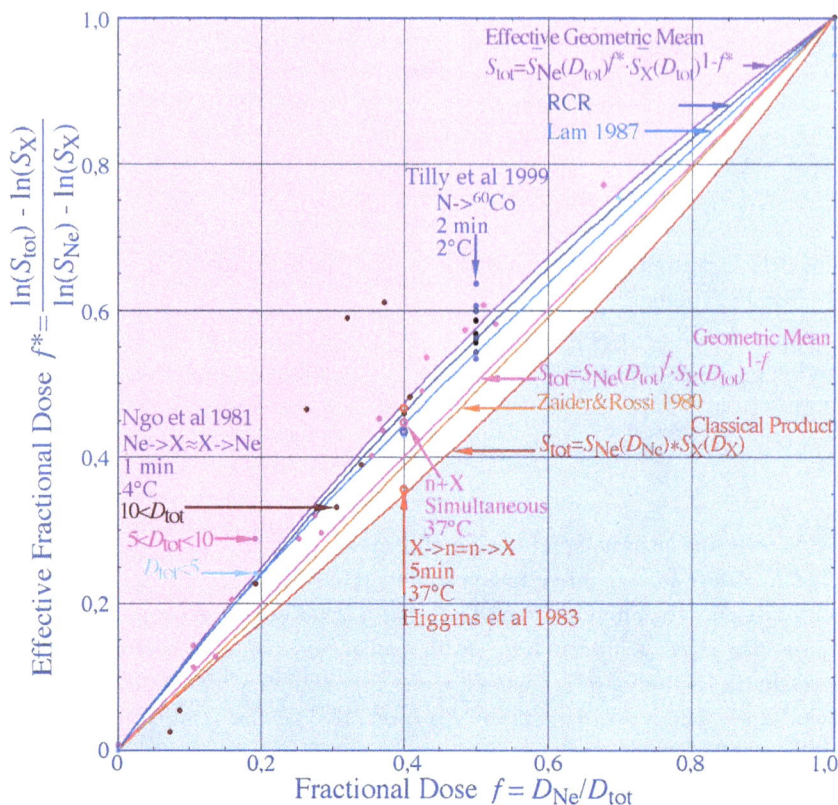

Figure 2.14. Comparison of different methods to calculate the biological effect when combining low- and high-*LET* radiations is shown. The classical product formula Eq. (2.25) is valid with long time interval between irradiations. For simultaneous irradiation, the synergistic effect is significant and largest around $f = 0.5$. The straight diagonal corresponds to the pure geometrical averaging of Eq. (2.26). It is interesting to see that 1 min at 4°C or 2 min at 2°C and simultaneous irradiation at 37°C are almost equivalent, whereas 5 min at 37°C results in an effect which is totally independent of the sequence of irradiations and most low-*LET* sublethal damage is fully repaired with about 20–30% lower biological effect (cf Fig. 8.8e)!

most of the essential early steps of sublethal repair are finished within a fraction of a minute or at least 5 min at 37°C as seen from Higgins *et al.* (1983) data in Fig. 2.14. So, in fact, for optimal radiation therapy outcome, it is essential that the incoming beams have maximum effect in the tumor and that the doses are as high as possible in the tumor and low in normal tissues. Furthermore, a two-field technique or the use of scanned spread out Bragg peaks would substantially benefit by simultaneous dose delivery as described further in Chapter 8 Figs. 8(e) and (f) and 8.38(j)–(l).

Also included in Fig. 2.14 is the cell survival expected according to the lesion interaction model of the theory of dual radiation action (Zaider and Rossi 1980) and the indistinguishable lesion model (Lam 1987). It is seen that these models are just below and above the straight line diagonal, indicating that they are closely approximated by the geometrical average described by Eq. (2.26). The present model describing the interaction of two or more radiation components is indicated in Fig. 2.14 as a solid curve that lies significantly above the straight diagonal. This curve largely coincides with the experimental data of Ngo *et al.* (1981) and also with those of Higgins *et al.* (1983).

2.5.5. *Dose–Response Relation*

When the survival of a certain organ or a tumor is considered, the single-hit multi-target model, often used for cell survival in the past, may still be useful to describe organ reactions as schematically illustrated in Fig. 2.15. The targets are now the N individual voxels making up the FSUs of an organ of an essentially parallel organization or the clonogenic tumor cells of a tumor, instead of the various sub-targets in the cell nucleus previously assumed to cause cell death when being hit. For simplicity, it is also assumed that each clonogenic cell is inactivated when an ionizing particle hits the sensitive volume of the cell as specified by the inactivation cross-section, σ_0, of the cell or FSU, as illustrated in Fig. 2.15.

The incident radiation beam causing cell inactivation is described by the dose D or fluence Φ of ionizing particles. In the case of electron, photon, proton, and ion beams, this is mainly the fluence of secondary electrons or δ rays. This fluence can be obtained from the absorbed dose to the organ by dividing the dose by the mean restricted collision mass stopping power L_Δ/ρ for the electron slowing down spectrum at hand.

When the inactivation cross-section and the initial number of cells are known, the mean cell survival as a function of absorbed dose or fluence can be calculated somewhat in analogy with the nuclear reaction probability in an irradiated medium. Thus, the mean number of inactivated clonogenic cells dN due to a fluence increase $d\Phi$ is given by:

$$dN = -\sigma_0 N d\Phi. \tag{2.32}$$

This is the differential equation resulting in the traditional single hit exponential cell survival:

$$N = N_0 e^{-\sigma_0 \Phi} = N_0 e^{-D/D_0}, \tag{2.33}$$

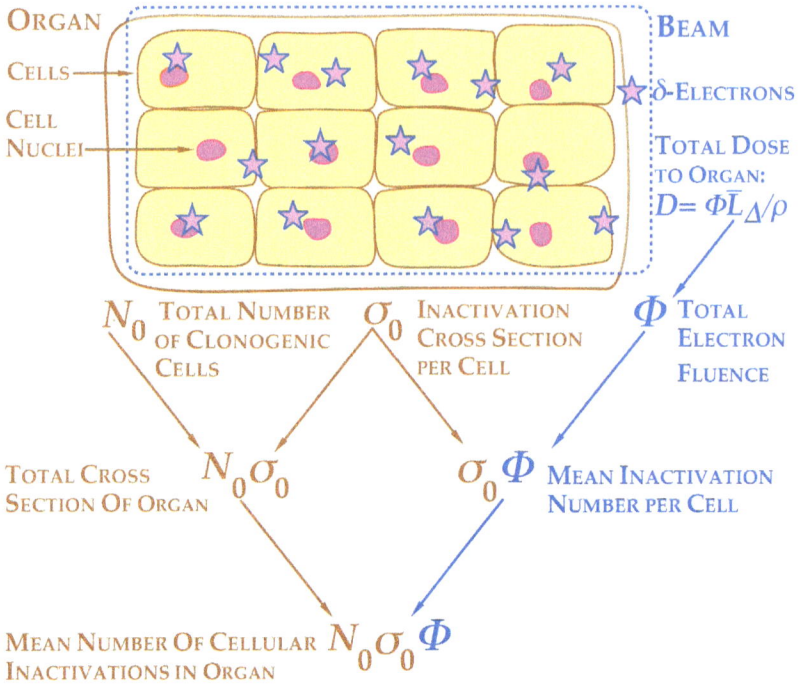

Figure 2.15. Illustration of the model used in calculating the response of an organ consisting of a large number of individual cells (N) each with an inactivation cross-section (σ_0 cm^{-2}/cell), when exposed to a radiation beam as specified by its fluence (Φ/particles cm^{-2}).

where N is the mean number of surviving cells and N_0 is the initial value. The relation between the "radiation resistance" or inactivation dose D_0 and the inactivation cross-section, σ_0, is thus given by. At a dose level of 1 Gy the fluence is therefore typically $3 \cdot 10^9$ electrons per cm^2, assuming a mean stopping power typical for high-energy electrons and photons of about 2 MeV cm^{2}g^{-1} or 0.2 eV/nm of unit density material. Similarly, the density of ionizing events may be calculated based on a mean inactivation event size of about 60 eV in unit density material. At 1 Gy, the density becomes 10^{14} events cm^{-3}. This corresponds to a mean distance between events of about 0.2 μm in approximate agreement with the fluence value just calculated.

If the inactivation cross-section for each cell is very small and the fluence of particles very large, so that the product of the two, that is the mean hit number per cell ($\sigma_0 \Phi$, cf. Fig. 2.15), is finite, Poisson statistics can be applied to estimate the probability for having precisely v hits:

$$P_h(v) = \frac{e^{-\sigma_0\Phi}(\sigma_0\Phi)^v}{v!}. \tag{2.34}$$

Thus, the probability for no hits ($v = 0$) or the probability that a given cell survives is given by Eq. (2.33) above. Exponential cell kill is thus obtained when Poisson statistics is applicable (cf Brahme 2011 and Chapter 8 Eq. (8.7)).

The probability that a single cell is killed (i.e., hit one or more times) is therefore $P_e = 1 - P_h(0)$. Provided the killing of each cell is statistically independent of what happens to every other cell, the probability that a tissue or tumor consisting of N_0 cells is completely eradicated by killing all its N_0 clonogenic cells is given by the binomial type (cf. Eq. (2.39) below) conditional probability:

$$P_e = (1 - e^{-D/D_0})^{N_0}. \tag{2.35}$$

The survival probability for a tissue consisting of N_0 cells is thus $1 - P_e$ which is recognized as being mathematically similar to the traditional single-hit multi-target survival curve equation. Again, applying Poisson statistics a very useful alternative expression for the probability of eradication of a given organ can be derived from the probability of having precisely v surviving cells:

$$P_s(v) = \frac{e^{-N}N^v}{v!}, \tag{2.36}$$

where N is the mean number of surviving cells as given, for example, by Eq. (2.33). From this expression the probability of no survival $P_s(0)$, that is, the eradication probability, P_e, simply becomes:

$$P_e = e^{-N} = e^{-N_0 e^{-D/D_0}} = (e^{-e^{-D/D_0}})^{N_0}. \tag{2.37}$$

This expression is very closely related to Eq. (2.35) particularly at high doses such as over the sigmoidal part of the dose–response curve. Equation (2.37) always gives a slightly larger value than Eq. (2.35) as can be demonstrated by power expansion of Eqs. (2.35) and (2.37) for dose values both larger and smaller than D_0. However, the more basic Eq. (2.35) should, in principle, be more accurate, but the difference is never clinically significant for large values of N_0. Owing to the greater mathematic simplicity, Eq. (2.37) has been extensively used over the years to accurately describe the shape of the dose–response relation. However, when the number of cells or FSUs N_0 is small ($N_0 \ll 100$ or the normalized slope of the dose–response relation $\gamma < 1.5$, see Eq. (2.41) below), the essentially binomial Eq. (2.35) should be used to be accurate.

When Poisson statistics is applied, Eq. (2.37) may be used to derive the isoeffect dose when the cell density or the number of cells in a given volume is reduced or increased beyond N_0 by a factor ρ. The isoeffect dose D_ρ equivalent to the dose D for N_0 cells can be derived directly from Eq. (2.37) and becomes:

$$D_\rho = D + D_0 \ln \rho. \tag{2.38}$$

The isoeffect dose is therefore directly proportional to the logarithm of the cell density factor ρ. Thus, for a D_0 value of 2 Gy and a dose level $D = 50$ Gy, a 10-fold change in the cell density requires $<10\%$ change in dose to get the same response in the tumor or normal tissue.

There are basically two types of response to high dose irradiation that are mathematically modeled. These are cell survival and organ response. Different radiobiological models have consequently been developed trying to describe the pattern of response of the first or the second category.

2.5.5.1. *Binomial and Poisson models*

The radiation response of normal tissues as well as the probability to locally control a tumor can generally be well described by binomial statistics since stem cell damage or eradication of each and every one of the clonogenic tumor cells are needed to induce tissue damage or to control the entire tumor, respectively. If a tumor initially contains N_0 clones, each of which having a mean survival probability $S(D)$ after irradiation to dose D, the probability to have exactly ν surviving clones is according to the binomial theorem given by:

$$P_\nu = \frac{N_0!}{\nu!(N_0 - \nu)!} S^\nu (1 - S)^{N_0 - \nu}, \tag{2.39}$$

as illustrated in Fig. 2.16.

It is clearly seen how the LQ-like cell survival dominates at low doses, whereas at high doses and low survival probabilities, the sigmoidal dose–response relation describes the biological effects very well in comparison with clinical data in Fig. 2.17.

The probability of a beneficial outcome, P_B, with full tumor control and no tumor clones surviving ($\nu = 0$) is thus given by:

$$P_B = P_0 = (1 - S)^{N_0}. \tag{2.40}$$

This expression can, in the limit when N_0 is very large, be well approximated by the Poisson expression as seen by comparison with Eqs. (2.35) and (2.37) above:

$$P_B = e^{-N_0 S}. \tag{2.41}$$

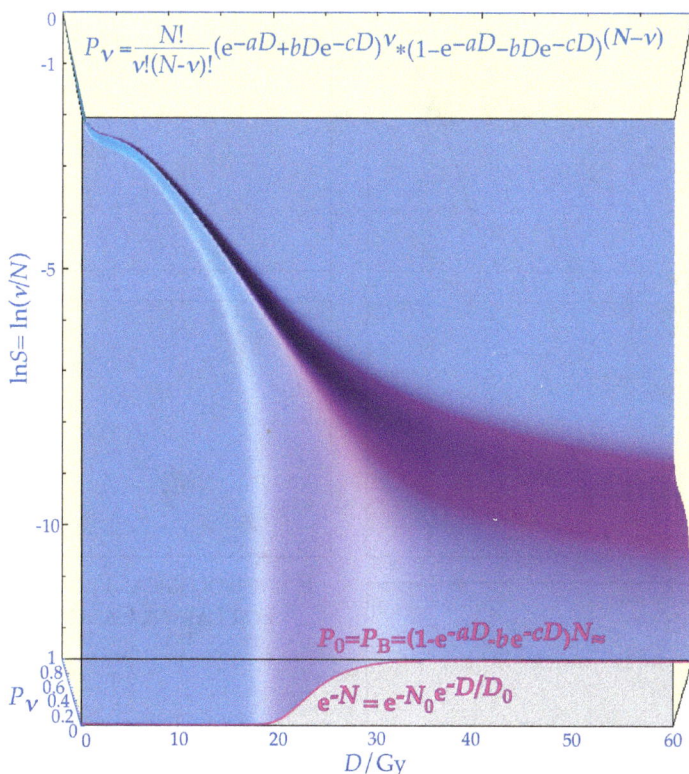

$$P_v = \frac{N!}{v!(N-v)!}(e^{-aD}+bDe^{-cD})^v * (1-e^{-aD}-bDe^{-cD})^{(N-v)}$$

$$P_0 = P_B = (1-e^{-aD}-be^{-cD})N_\infty$$

$$e^{-N} = e^{-N_0}e^{-D/D_0}$$

Figure 2.16. The probability that a certain fraction of the cells survive at a given dose level is shown. At zero dose, 100% of the cells survive, whereas at high doses and low survival probabilities, almost all the cells are eradicated leading to tumor control. Outside these areas the probability mass is spread out over a wider range of survival levels and the resultant probability is lower. At low doses the LQ cell survival dominates, whereas at high doses, the dose–response relation is the most useful concept to describe the radiation effects.

When the survival of the cells is exponential or LQ, the tumor control curve is thus well described by a relation of the type:

$$P_B = e^{-N_0 S(D)} = e^{-N_0 e^{-D/D_0}} = e^{-e^{\ln N_0 - D/D_0}} = e^{-e^{\gamma e\left(1-\frac{D}{D_{37}}\right)}}$$

$$= 2^{-e^{\gamma e\left[1-\frac{\ln(\ln 2)}{e\gamma}-\frac{1}{2(e\gamma)^2}-\frac{1}{6(e\gamma)^3}-\frac{1}{24(e\gamma)^4}\cdots\right]\left(1-\frac{D}{D_{50}}\right)}} \approx 2^{-e^{\gamma e\left(1-\frac{D}{D_{50}}\right)}}, \qquad (2.42)$$

where $\gamma = \ln N_0/e$ is the normalized slope of the dose–response relation and $D_{37} = D_0 \cdot \ln N_0$ and $D_{50} = D_0(\ln N_0 - \ln \ln 2)$ are the doses at which 37% and

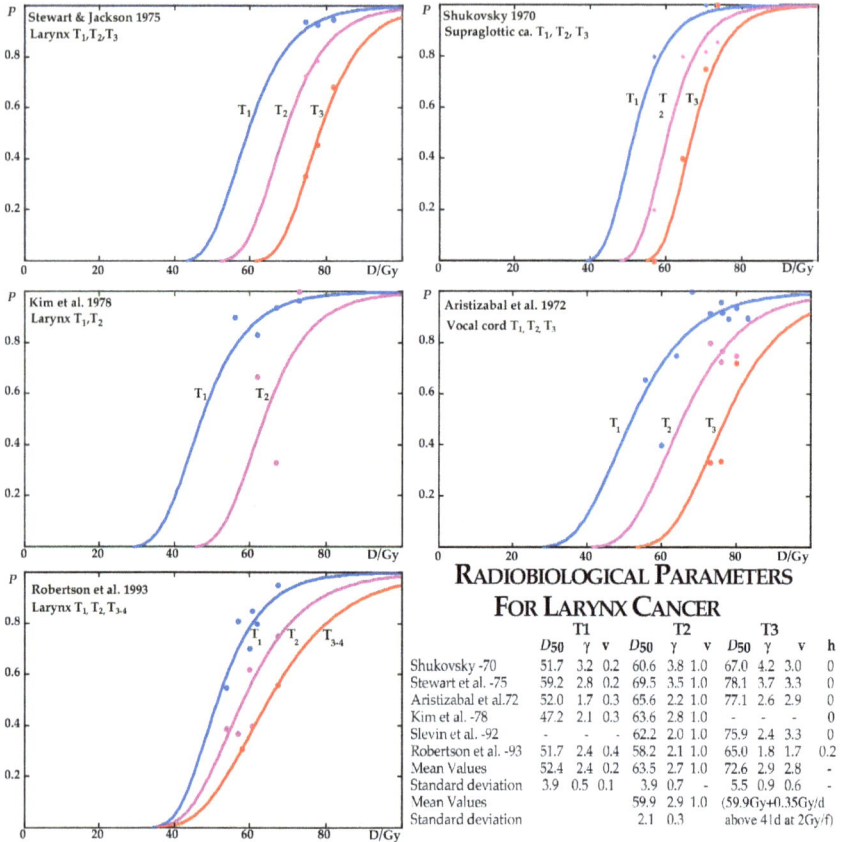

RADIOBIOLOGICAL PARAMETERS FOR LARYNX CANCER

	T1			T2			T3			
	D_{50}	γ	v	D_{50}	γ	v	D_{50}	γ	v	h
Shukovsky -70	51.7	3.2	0.2	60.6	3.8	1.0	67.0	4.2	3.0	0
Stewart et al. -75	59.2	2.8	0.2	69.5	3.5	1.0	78.1	3.7	3.3	0
Aristizabal et al.72	52.0	1.7	0.3	65.6	2.2	1.0	77.1	2.6	2.9	0
Kim et al. -78	47.2	2.1	0.3	63.6	2.8	1.0	-	-	-	0
Slevin et al. -92	-	-	-	62.2	2.0	1.0	75.9	2.4	3.3	0
Robertson et al. -93	51.7	2.4	0.4	58.2	2.1	1.0	65.0	1.8	1.7	0.2
Mean Values	52.4	2.4	0.2	63.5	2.7	1.0	72.6	2.9	2.8	-
Standard deviation	3.9	0.5	0.1	3.9	0.7	-	5.5	0.9	0.6	-
Mean Values				59.9	2.9	1.0	(59.9Gy+0.35Gy/d			
Standard deviation				2.1	0.3		above 41d at 2Gy/f)			

Figure 2.17. Review of the clinically observed dose–response relations for T_1, T_2, T_3, and T_4 larynx tumors. When the clinical data is treated in a uniform way, the agreement between the different data sets is quite good with a relative standard deviation of about 3% in D_{50} and 10% in γ for the T_2 tumors, as seen in the inserted table.

50% of the patients, respectively, have their tumors controlled. At D_{37}, thus, on average one clonogenic tumor cell survives. It can be shown that a similar relation holds also well for the probability of injury (P_I) to normal tissues.

2.5.6. *Volume Effects*

For the normal tissues, the problem is more on the organizational side since the functional arrangement of the subunits of a tissue is fundamental for its response to heterogeneous or partial irradiation, as illustrated in Fig. 2.18. In the upper part of the figure, the different types of functional organization of the subunits are

INFLUENCE OF FUNCTIONAL ORGANIZATION OF TISSUES ON DOSE RESPONSE RELATION

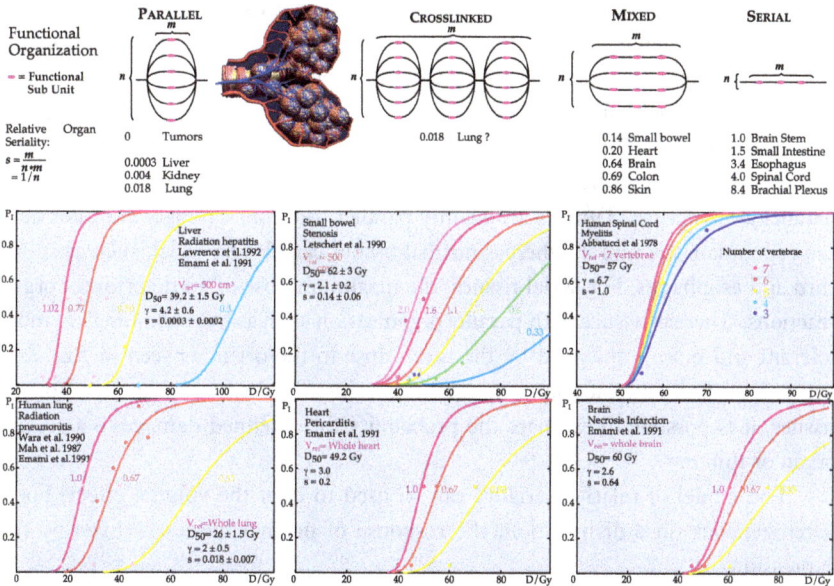

	PARALLEL m	CROSSLINKED m	MIXED m	SERIAL
Functional Organization				
— = Functional Sub Unit	n	n	n	$n \{$ m
Relative Organ Seriality: $s = \frac{m}{n*m} = 1/n$	0 Tumors 0.0003 Liver 0.004 Kidney 0.018 Lung	0.018 Lung ?	0.14 Small bowel 0.20 Heart 0.64 Brain 0.69 Colon 0.86 Skin	1.0 Brain Stem 1.5 Small Intestine 3.4 Esophagus 4.0 Spinal Cord 8.4 Brachial Plexus

P_I

Liver
Radiation hepatitis
Lawrence et al.1992
Emami et al. 1991

V_{ref}= 900 cm³
D_{50}= 39.2 ± 1.5 Gy
γ = 4.2 ± 0.6
s = 0.0003 ± 0.0002

P_I

Small bowel
Stenosis
Letschert et al. 1990

D_{50}= 52 ± 3 Gy
γ = 2.1 ± 0.2
s = 0.14 ± 0.06

P_I

Human Spinal Cord
Myelitis
Abbatucci et al 1978
V_{ref} = 7 vertebrae
D_{50}= 57 Gy
γ = 6.7
s = 1.0

Number of vertebrae
• 7
• 6
• 5
• 4
• 3

P_I

Human lung
Radiation pneumoritis
Wara et al. 1990
Mah et al. 1987
Emami et al.1991

V_{ref}=Whole lung
D_{50}= 26 ± 1.5 Gy
γ = 2 ± 0.5
s = 0.018 ± 0.007

P_I

Heart
Pericarditis
Emami et al. 1991
V_{ref}= Whole heart
D_{50}= 49.2 Gy
γ = 3.0
s = 0.2

P_I

Brain
Necrosis Infarction
Emami et al. 1991
V_{ref}= whole brain
D_{50}= 60 Gy
γ = 2.6
s = 0.64

Figure 2.18. Depending on the functional organization of the subunits of an organ, the dose–response relation will be strongly dependent on the partial volume being irradiated. For tissues with a parallel organization, small volumes can tolerate large doses without severe damage to the organ. Serial tissues, on the other hand, are already severely damaged when a small portion of the organ is irradiated.

illustrated with a continuous change from fully parallel via mixed to fully serial organization. In the parallel tissues, a small part of an organ can accept a high dose, since other parts of the organ can take over the function lost in the irradiated area. The opposite is true for almost perfectly serial tissues such as the spinal cord that are severely damaged if the highest dose is above tolerance (50–54 Gy in 2 Gy fractions over 6 weeks). The mixed group with simultaneously parallel-serial organization of its FSUs has a lesser tolerance to partial irradiation but without a clear threshold level for severe damage. This is the case for small bowel and heart.

Both D_{50} and γ depend on the initial number of FSUs in the tissue. The complications observed in normal tissues following the therapeutic use of radiation have been described in terms of inactivation of FSUs (Ågren 1995). The organization of the FSUs is described in terms of serial, parallel, or more generally a combination of these two structures. Many researchers have provided expressions for estimating the probability of complications using models that account for the volume effect, which stems from the FSU infrastructure of the organs. The volume effect describes how the tolerance dose increases with decreasing partial irradiated volume of normal tissues. Organs with serial infrastructure have small volume dependence since every subunit is vital for organ function. For organs with parallel infrastructure, a strong volume dependence can be expected since the organ can maintain most of its function even when a large portion of its subunits is damaged. For most normal tissues, the FSUs that characterize the tissue are arranged differently depending on whether the tissue is of serial or parallel architecture. Tumors are mainly parallel, since all clonogenic tumor cells can repopulate the tumor, whereas normal tissues can also be serial, such as spinal cord and esophagus. For serial tissues, the maximum dose often determines organ reactions, whereas tissues with parallel organization such as liver and lung are more tolerant and more influenced by the mean dose to the organ as seen in Fig. 2.18 below. So, based on tabulations of γ and D_{50} values for the tumors and the normal tissues, it is possible to estimate the probability for a defined damage to a certain organ or tumor.

The model of relative seriality can be used to treat the volume effect. For a heterogeneous dose distribution, the response of normal tissues is given by the expression:

$$P(\vec{D}) = \left[1 - \prod_{i=1}^{M} [1 - P(D_i)^s]^{\Delta v_i} \right]^{1/s}, \qquad (2.43)$$

where $\Delta v_i (= \Delta V_i / V_{\text{ref}})$ is the fractional sub-volume of an organ that is irradiated compared to the reference volume for which the values of D_{50} and γ were calculated. $P(D_i)$ is the probability of response of an organ having the reference volume and being irradiated to dose D_i, M is the total number of voxels or sub-volumes in the organ, and s is the relative seriality parameter that characterizes the internal organization of the organ. A relative seriality close to zero ($s \approx 0$) corresponds to a completely parallel structure, which becomes nonfunctional when all its FSUs are damaged, whereas $s \approx 1$ corresponds to a completely serial structure which becomes nonfunctional when at least one FSU is damaged. Usually, the whole volume of a healthy organ is used as the reference volume because the volume of an organ is related to the functional needs of the individual human being. In this clinical case, the whole lung constitutes the reference volume to which the model parameters D_{50} and γ are referring.

The individual parameters of the P_B and P_I functions for a given disease are often based on the mean value of γ and D_{50} over a few hundred patients of the relevant tumor stage, age group, sex, etc. However, for a given patient the radiation sensitivity of the tumor and that of the normal tissues may vary considerably from

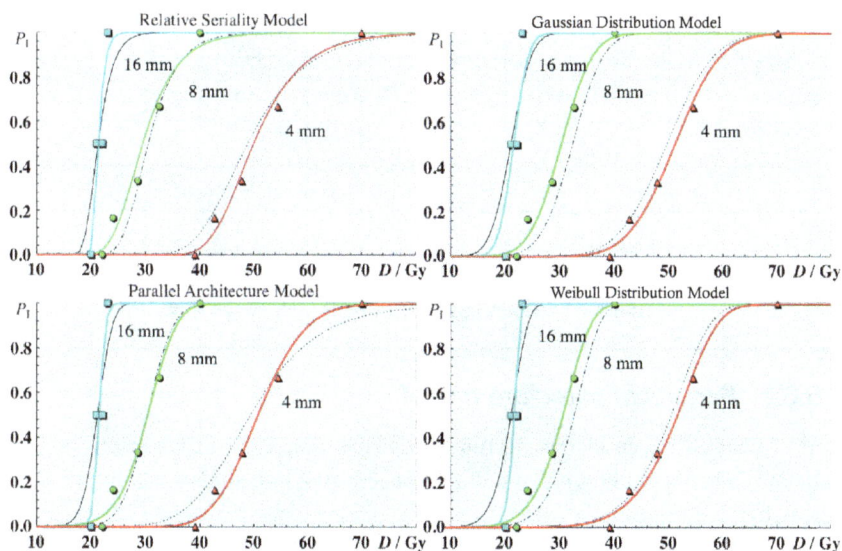

Figure 2.19. Illustration of dose–response relationships for the relative seriality (upper left diagram), Gaussian distribution (upper right diagram), parallel architecture (lower left diagram), and Weibull distribution models (lower right diagram) based on experimental data of white matter necrosis. The symbols correspond to the experimental data, the solid lines have been calculated fitting the parameters of each model to each of the irradiated length of spinal cord and the dotted lines have been calculated fitting the parameters of each model to the whole set of experimental data.

the mean value over the whole patient population and the individual γ value is also likely to be steeper than the mean value. To be on the safe side, it can be shown that it is reasonable to assume that the given patient has a more resistant tumor at the same time as the normal tissues are more sensitive (by about $0.5\sigma_{D_{50}}$) and that the γ value is steeper (cf Ch 6.10). In Fig. 2.19 four different dose response models are compared for the quite serial response of white matter necrosis in spinal cord and the models are described in more detail below (cf Adamus-Górka *et al.* 2011).

2.5.6.1. *Relative seriality model*

The *s* model has been applied and discussed broadly in the literature (Brahme and Ågren 1987; Ågren *et al.* 1990; Källman *et al.* 1992; Karlsson *et al.* 1997; Lind *et al.* 1999; Moiseenko *et al.* 2000). In the relative seriality model, a combination of both serial and parallel organizations of FSUs is modeled. In this model, normal tissue complication probability P_{I} is mathematically expressed by:

$$P_{\text{I}}(D, V) = [1 - (1 - P_{\text{I}}(D)^s)^{V/V_{\text{ref}}}]^{1/s}, \qquad (2.44)$$

where V/V_{ref} is the volume fraction being irradiation to dose D, s is the parameter which expresses the degree of seriality (the value varies from $s = 0$ for a parallel organ to $s = 1$ for a purely serial organ), $P_I(D)$ is given by, for example, the Poisson expression

$$P_1(D, V_{ref}) = \exp\left(-e^{e\gamma - (D/D_{50})(e\gamma - \ln(\ln 2))}\right), \tag{2.45}$$

where D_{50} is the dose associated with the 50% response probability and γ is the maximum of the normalized dose–response gradient, defined as $\gamma = D(dP/dD)_{max}$.

2.5.6.2. *Parallel architecture model*

In this model, P_I is an increasing function of the number of FSUs inactivated by radiation. The probability p that a dose D inactivates an FSU is given by the logit expression:

$$p(D) = \frac{1}{\left(1 + \left(\frac{D_{50}}{D}\right)^k\right)}. \tag{2.46}$$

The above sigmoid dose–response function, $p(D)$, is assumed to describe the probability of damaging a subunit at a given biologically equivalent dose. Apart from the assumption that biologically equivalent doses can be calculated from a LQ formula, no connections of this probability with any underlying vascular mechanism of radiation injury or identification of the subunits involved has been attempted. Instead, it has been chosen to phenomenologically describe the subunit response, using a logistic function of dose parameterized in terms of the dose D_{50} at which 50% of the subunits are damaged and the slope parameter k that determines the rate at which the probability of damaging a subunit increases with dose (k is related to γ through $k = 4\gamma$).

For a given dose volume histogram (DVH), the total fraction of FSUs, being inactivated is given by the sum over all the individual contributions:

$$f = \sum v_i p(D_i), \tag{2.47}$$

where D_i and v_i are the dose and the volume fraction of the ith dose and f is called the fractional damage. To fit the parallel architecture model to clinical data, expressions for both $p(D)$ and the statistical distribution of functional reserves over the patient population are required. Normal tissue complication probability P_I for a general DVH is calculated from the equation:

$$P_{\rm I} = \frac{1}{\sqrt{2\pi\sigma^2}} \int_0^f \exp\left(-\frac{(v - v_{50})^2}{2\sigma^2}\right) dv, \qquad (2.48)$$

in which it is assumed that the cumulative functional reserve distribution can be described as a displayed error function and quantified by the mean value of the functional reserve v_{50} and the width of the functional reserve distribution sigma. In this equation, v is the partial organ volume being irradiated (Jackson *et al.* 1993; Yorke *et al.* 1999).

2.5.6.3. *Gaussian distribution model*

This model has been extensively applied (Burman *et al.* 1991; Kutcher *et al.* 1991; Zaider and Amols 1998; Moiseenko *et al.* 2000) and it is based on the error or probit function form for calculating complication probability:

$$P_I = \frac{1}{\sqrt{2\pi}} \int_{-\infty}^{t} e^{-t^2/2} dt, \qquad (2.49)$$

where the upper limit of the normal probability function is defined as follows:

$$t(D, V) = \frac{D - D_{50}(V/V_{\rm ref})}{m D_{50}(V/V_{\rm ref})} \quad \text{and} \quad D_{50}(V/V_{\rm ref}) = D_{50}(1)\left(\frac{V}{V_{\rm ref}}\right)^{-n}. \qquad (2.50)$$

The model contains four free parameters, namely D_{50}, n, m, and $V_{\rm ref}$; $V_{\rm ref}$ is the reference volume of the organ for D_{50} and $V/V_{\rm ref}$ is the fraction of the organ being irradiated, while $D_{50}(1)$ is the tolerance dose for 50% complications for uniform whole organ irradiation, $D_{50}(V/V_{\rm ref})$ is the 50% tolerance dose for uniform partial organ irradiation. The volume dependence of the complication probability is determined by the parameter n, which quantifies the sensitivity of P_I to the irradiated volume. The slope of the dose response is governed by the value of the parameter m. The slope parameter m is reversely proportional to γ through the relation: $\gamma = \pi/8m$.

2.5.6.4. *Weibull distribution model*

The mathematical expression of normal tissue complication probability, P_I is based in this case on a modified Weibull function:

$$P_I = 1 - \exp\left[-\left(\frac{DV^b}{A_1}\right)^{A_2}\right], \qquad (2.51)$$

where A_1, b, and A_2 are three model parameters that are to be determined from the clinical data (Klepper and Klimanov 2000).

2.5.7. *Dose–Response Relation for Hypoxic and Generally Heterogeneous Tissues*

Clinical data on the radiation effect on tumors and normal tissues have been collected at an increasing rate, during the past decades. Despite the strong development of new response models, two main problems still exist. For the tumors, it is important to be able to quantify the hypoxic or most resistant tumor cell compartment, since it is fundamental for the responsiveness of the tumors even if that compartment is small. Figure 2.20 illustrates the strong influence of a small hypoxic population (1000 cells — dotted curve) even if the bulk of the clonogenic tumor cells is 100,000 times larger (left solid curve). So even if the tumor physically is dominated by the well-oxygenated cells, the small hypoxic compartment totally dominates its radiation response at high doses where the right most solid curve illustrates the total tumor responses (hypoxic and well-oxygenated cells together).

In Figure 2.17, the clinically observed dose–response relation in five different larynx cancer materials is summarized. After converting the different fractionation schemes to a standard 2 Gy/fraction and 0.35 Gy/day for the treatment duration beyond 42 days and as far as possible correcting the data to get actuarial survival, the 50% control dose was found to be 59.9 Gy with a standard deviation of 2.1 Gy or about 3% for the five data sets. The established γ value was 2.9 with a standard deviation of only 10%, which is a much lower uncertainty than commonly observed

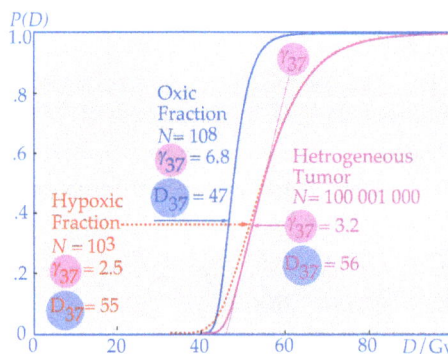

Figure 2.20. Illustration of dose–response curves for a hypoxic tumor and its two constituent cell compartments: the anoxic and the well-oxygenated tumor cells. It is interesting to observe that the small anoxic cell compartment dominates the response at large doses.

in clinical materials. This indicates that the clinical data, when treated in a systematic way, can give us very well defined dose–response relations for use in treatment optimization.

2.5.7.1. *Radiation response with spatially varying dose, clonogen density, and radiation resistance*

The Poisson approximation of the probability of tumor cure or a beneficial treatment, P_B, was generalized to take spatial heterogeneities into account by Brahme and Ågren (Brahme 1984; Brahme and Ågren 1987). They used the elementary exponential model for cell survival together with the Poisson approximation to express the probability of tumor cure according to

$$P_B = \mathrm{e}^{-N} = \exp\left(-\iiint_{V_t} n(\vec{r})d^3r\right) = \exp\left(-\iiint_{V_t} n_0(\vec{r})\mathrm{e}^{-D(\vec{r})/D_0(\vec{r})}d^3r\right),$$

(2.52)

where N is the mean total number of surviving cells, $n_0(\vec{r})$ is the initial clonogen density (i.e., clonogens per unit volume), $n(\vec{r})$ is the density of surviving clonogens, $D_0(\vec{r})$ is the spatial distribution of radiation resistance, and $D(\vec{r})$ is the spatial dose distribution in the organ or tumor of volume V_t. The optimal spatial dose distribution $D_{\mathrm{opt}}(\vec{r})$ giving the lowest possible maximum recurrence probability at a tumor control probability P_B, thus requiring a uniform survival probability, is then simply given by (Brahme and Ågren 1987; Ebert and Hoban 1996; Stavreva *et al.* 1996):

$$D_{\mathrm{opt}}(\vec{r}) = D_0(\vec{r}) \ln\left(\frac{n_0(\vec{r})V_t}{-\ln P_B}\right).$$

(2.53)

The optimal spatial dose distribution with uniform recurrence probability as the end point, neglecting normal tissue complications, should thus be proportional to the radiation resistance $D_0(\vec{r})$ and the logarithm of the initial clonogen density $n_0(\vec{r})$ (cf. Chapter 7, Figs. 7.16 and 7.17 and Chapter 8, Fig. 8.27).

2.5.7.2. *Radiation response with a microscopic distribution of radiation resistance and uniform dose*

Since individual cells are difficult to target by external beam radiation therapy, macroscopic voxels containing a large number of cells are generally considered in treatment planning. Owing to the steepness of intra-tumor oxygen gradients, each voxel on the cellular level will have a microscopic distribution of pO_2 and thus also often a microscopic distribution of radiation resistance D_0. This distribution can be approximated by a continuous clonogen distribution function N_{D_0}, which is differential in D_0 such that the total initial number of clonogens in the voxel is $N_0 = \int N_{D_0} dD_0$. It is convenient to describe the variation in radiation sensitivity by the elementary exponential cell survival model, $S(D) = \exp(-D/D_0)$. For fractionated radiation therapy with a quasi-constant dose per fraction, the effective cell survival is well described by such a model. This could also be regarded as the most significant term of the LQ equation $S_{LQ}(D) = \exp(-\alpha D - \beta D^2)$, where $D_0 \approx \alpha^{-1}$, since β is mostly related to the repair capacity and not so much to the direct radiation sensitivity, particularly at low doses. Obviously the effective D_0, D_e for a given fractionation schedule (Källman *et al.* 1992) can be estimated using the LQ model or other more accurate cell survival models (e.g., the RCR model). For a given cell survival function, $S(D)$, D_e can with a uniform dose per fraction be expressed as $D_e = -D/\ln S(D)$, where D is the total dose. The number of surviving clonogens $N(D)$ in the voxel for a uniform voxel dose D becomes as follows, using the elementary exponential survival model:

$$N(D) = \int_{D_{0,\min}}^{D_{0,\max}} N_{D_0} e^{-D/D_0} dD_0. \tag{2.54}$$

The number of surviving cells as a function of dose will thus decrease quasi-exponentially depending on the differential distribution of cellular radiation resistance N_{D_0}. Cells with low D_0 will initially die off more rapidly. The mean surviving fraction of clonogens $S(D)$ is given by:

$$\overline{S}(D) = \frac{N(D)}{N_0} = \frac{\int N_{D_0} e^{-D/D_0} dD_0}{\int N_{D_0} dD_0}. \tag{2.55}$$

At a given dose level D it is possible to perfectly fit both the number of surviving clonogens according to Eq. (2.54) and its derivative to an exponential expression of cell survival. By introducing the effective initial number of clonogens $N_{0,\text{eff}}(D)$ and the effective radiation resistance $D_{0,\text{eff}}(D)$ at that dose level, we obtain:

$$N(D) = \int N_{D_0} e^{-D/D_0} dD_0 = N_{0,\text{eff}} e^{-D/D_{0,\text{eff}}}. \tag{2.56}$$

$D_{0,\text{eff}}$ is then defined at that dose level according to:

$$D_{0,\text{eff}}(D) = \frac{\int N_{D_0} e^{-D/D_0} dD_0}{\int N_{D_0} D_0^{-1} e^{-D/D_0} dD_0}. \tag{2.57}$$

Similarly $N_{0,\text{eff}}$ can be expressed at this dose level using Eqs. (2.56) and (2.57) as

$$N_{0,\text{eff}}(D) = N(D)e^{-D/D_{0,\text{eff}}} = \left(\int N_{D_0} e^{-D/D_0} dD_0 \right)$$

$$\times \exp \left(\frac{\int N_{D_0} D D_0^{-1} e^{-D/D_0} dD_0}{\int N_{D_0} e^{-D/D_0} dD_0} \right). \qquad (2.58)$$

While varying the dose level D at which the effective values are defined, it is seen from these expressions that $D_{0,\text{eff}}(D)$ increases slowly from the low-dose quasi-well-oxygenated value, whereas $N_{0,\text{eff}}(D)$ decreases steadily as the radiation-sensitive clonogens are rapidly eradicated (Fig. 2.21). At D_{37}, the dose where the tumor cure is 37%, there is on average only one surviving clonogen and thus according to Eq. (2.54) $N(D_{37}) = \int N_{D_0} e^{-D_{37}/D_0} dD_0 = 1$. The effective values defined at D_{37} are thus reduced to:

$$D_{0,\text{eff}}(D_{37}) = 1/\int \frac{N_{D_0}}{D_0} e^{-D_{37}/D_0} dD_0, \qquad (2.59)$$

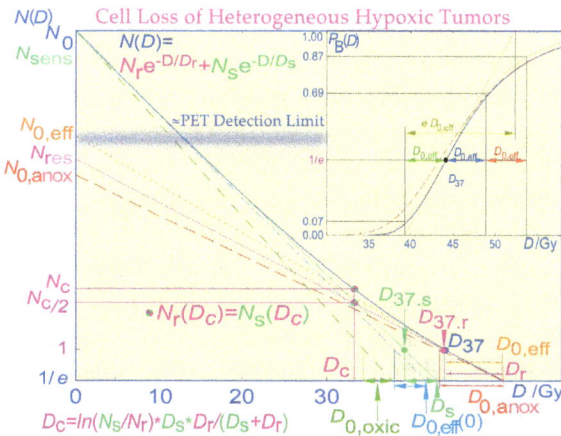

Figure 2.21. Number of surviving clonogens based on the full D_0 distribution (solid curves) and the effective value approximation at D_{37} (dotted curves) (A) and their associated dose–response relations (B) are shown. The dashed curves in both panels are calculated using effective values evaluated in $D_{37} + 2D_{0,\text{eff}}(D_{37})$. In (A), survival curves are also shown corresponding to $D_{0,\text{min}}$ and $D_{0,\text{eff}}(0)$ (dash-dotted curves). In (B), the tangent to the effective $P_B(D)$ at D_{37} (dash-dotted curve) is included. The data correspond to a vascular heterogeneity of 20 μm and a vascular density (relative vessel area) of 3%.

and

$$N_{0,\text{eff}}(D_{37}) = \exp\left(\int N_{D_0}\frac{D_{37}}{D_0}e^{-D_{37}/D_0}\,\mathrm{d}D_0\right) = e^{D_{37}/D_{0,\text{eff}}(D_{37})}. \qquad (2.60)$$

This last expression may also be used to express D_{37} in terms of the effective values as $D_{37} = D_{0,\text{eff}}(D_{37})\ln N_{0,\text{eff}}(D_{37})$. The effective $N_{0,\text{eff}}$ and $D_{0,\text{eff}}$ can be evaluated at any dose level D but are, in the following sections, for simplicity, assumed to be evaluated at D_{37}, which is a dose level of high clinical importance for both tumor cure and normal tissue damage.

When there is no information available about the spatial distribution of the clonogen density or radiation resistance, the optimal dose delivery according to Eq. (2.53) may be assumed to be uniform. In absence of accurate $n_0(\vec{r})$ and $D_0(\vec{r})$ for individual patients, this assumption may also be used in treatment optimization. By using a library of published average distributions of pO_2 for different tumor types and stages or individual patient data together with assumptions of N_0 based on tumor size, it is possible to calculate the expected N_{D_0}, $N_{0,\text{eff}}$, and $D_{0,\text{eff}}$ for individual tumors using the equations presented above. These parameters can then be used for individualized treatment optimization taking into account the expected microscopic distribution of radiation response for each tumor. When the spatial variation is available, for example, through positron emission tomography studies, it is possible to accurately account for this information (as discussed in more detail in Chapter 7).

Diffusion of oxygen in tissue is a complex process due to the nature of cell membranes and vascular walls, oxygen consumption, the effect of oxygen being bound to hemoglobin in red blood cells, extravascular streaming, and so on. If we neglect these factors, we are left with the apparently simple problem of solving the diffusion equation for our geometry of vessels, while treating tissue as a homogeneously diffusive medium. This approach is simplified, but since it gives oxygen distributions very similar to what is being observed in the clinical situation (cf. Fig. 2.22, left panel), it can be considered valid and useful as a model for the gross effects of tissue oxygenation. In the tissue model used by Nilsson *et al.* (2002), the oxygen diffusion is calculated as the sum of the contributions from all vessels. The clinical data presented in Fig. 2.22 comes from pooled Eppendorf measurements presented as histograms (Vaupel *et al.* 1989, 1998). With the Eppendorf polarographic needle, the oxygen profile in tissue is measured as the electrode is stepwise inserted (Kallinowski *et al.* 1990).

In the left panel of Fig. 2.22, average pO_2 frequency distribution for different vascular densities and heterogeneities using 100 random oxygenation calculations for each pair of model parameters is shown. The solid lines represent the model tissue frequency distributions with a bin size of 0.5 mmHg. The histograms with bin size

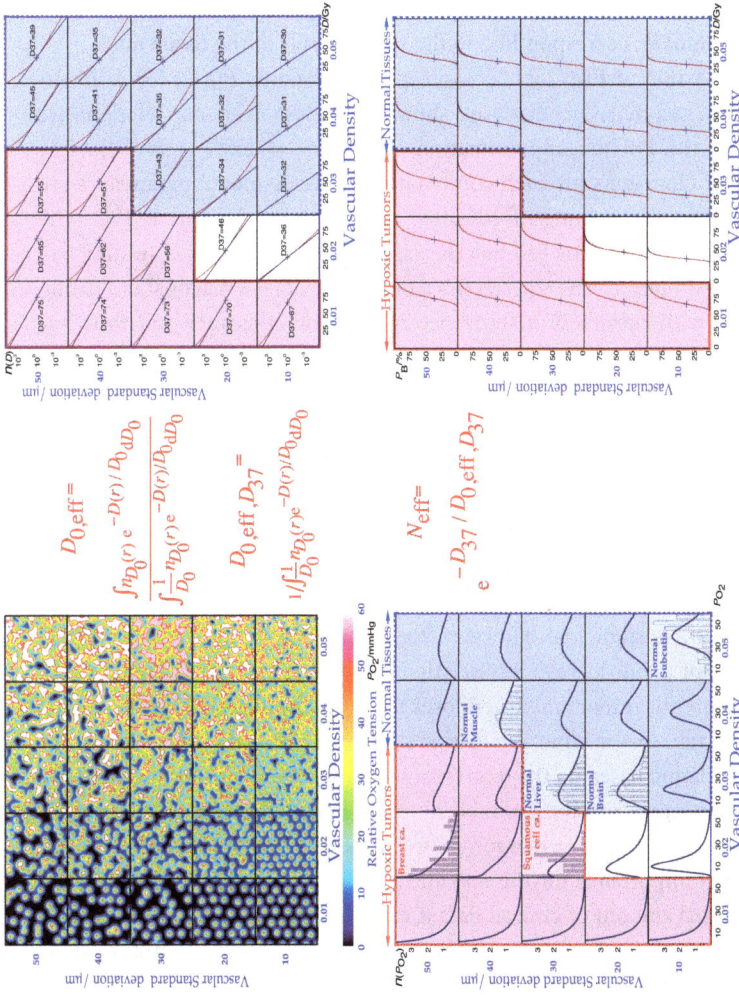

Figure 2.22. Schematic comparison of cellular oxygenation distributions (left panel) and associated dose–response curves (right panel) for different tissue models with increasing vessel density and variation in vascular structure. It is seen that the tissue model agrees fairly well with clinically observable oxygenation distributions both for hypoxic tumors (pink–red shading) and well-oxygenated normal tissues (blue shading). It is clear that the hypoxic tumors are more radiation-resistant than the well-oxygenated normal tissues, as seen by their lower D_{37} values in the right panel. Equally located diagrams correspond to the same tissue structure in these $5 \times 5 = 25$ regions of different vascular characteristics.

of 5 mmHg are pooled data from real tumors (pink histograms) and normal tissues (blue histograms), measured with polarographic needle electrodes and fitted by eye to the tissue model distributions. Clearly, clinical data from tumors end up in the upper left region where the oxygenation is bad, and normal tissues are represented in the more well-oxygenated lower right region. The clinical histograms often extend up to around 100 mmHg, corresponding to arterial pO_2. For spatial reasons, they are cut-off at 60 mmHg, corresponding to the vascular pO_2 in the tissue model.

In the right panel of Fig. 2.22, dose–response relations, $P_B(D)$ for different vascular densities and heterogeneities are shown. The dose–response relations are calculated using the Poisson approximation as: $P_B(D) = \exp(-N(D))$. The initial number of clonogens for each combination of tissue model parameters are 10^6, corresponding to a volume of 100 mm^3 at a clonogen density of 1%. The clonogen number decreases slowly when the vascular density increases, leading to a slightly lower maximum γ. The γ_{37} reaches a minimum along a diagonal indicating cell populations with extremely heterogeneous radiation resistance and thus low effective clonogen numbers.

Since the normalized slope of the dose–response relation is proportional to the logarithm of $N_{0.eff}$ it will be reduced relative to the uniform cell population. Again, this value will be particularly important at the steepest point of the dose–response curve, close to D_{37}. In the right panel of Fig. 2.22, the dose–response relations obtained from the cellular distribution functions in the left panel of Fig. 2.22 are calculated using the more exact first half of Eq. (2.56) as well as the approximate last half. The shape of the resultant dose–response relation both for heterogeneous and uniform tissues is sigmoidal. However, for heterogeneous tumors and normal tissues, the slope of the dose–response relation is lower and therefore higher doses are required to cure the tumor and lower doses are often less tolerated by the normal tissues. An important task in clinical radiation therapy is to quantify the shape of the dose–response relation based on the treatment outcome in terms of both tumor cure and normal tissue injury and the desirable outcome in complication-free cure. It is also essential to quantify the correlation between these outcome measures, since it is of fundamental importance for the maximization of the complication-free cure. Today, a substantial amount of clinical data is rapidly becoming available making it possible to perform a more strict treatment optimization for more complex tumor and normal tissue configurations.

2.5.8. *Influence of Fractionation on the Dose–Response Relation*

When different treatment plans are compared, eventual differences of their fractionation schedules should be taken into account. The dose–response curves in Figs. 2.18 and 2.19 were produced using the values of the D_{50} and γ parameters of the corresponding tissues. However, these parameters depend on whether a fixed uniform dose per fraction, a fixed number of fractions, or an arbitrary nonuniform dose distribution is delivered. Consequently, the shapes of these dose–response curves change according to the fractionation sequence considered.

If a constant dose per fraction is kept during the treatment and total dose increases by increasing the number of fractions, then the normalized steepness, γ becomes largely independent of α, β, the number of fractions, n, and the dose per fraction, d and is determined solely by the effective number of clonogenic cells for the tumors or the normal tissue FSUs N_0 as given by the Eq. (2.33) above:

$$\gamma \approx \frac{\ln(N_0)}{e}. \tag{2.61}$$

However, the $D_{50,d}$ value will depend on α/β ratio and the dose per fraction d, according to:

$$D_{50,d} = D_{50,d_0} \left(\frac{1 + \frac{d_0}{\alpha/\beta}}{1 + \frac{d}{\alpha/\beta}} \right), \tag{2.62}$$

where d_0 is the reference dose per fraction (usually 2 Gy) (Barendsen 1982; Ågren 1995).

When an organ or a tumor is heterogeneously irradiated, the increased dose to some of its parts will automatically be associated with an increased dose per fraction. Because of the nonlinear response of tumors and normal tissues to varying dose per fraction, the first equality of Eq. (2.2) is only a first approximation to the true effects as it is based on the total dose. To get a more accurate description of the total biological effect at varying doses per fraction, the LQ cell survival model of Eq. (2.2) is used instead. To make the transition from the first equality of Eq. (2.2) to the second one, the values of α and β have to be known. If D_{50} and γ are known for a given $\bar{d}$ and α/β ratio, then α and β can be calculated by the following expressions (Ågren 1995; Lind et al. 1999):

$$\alpha = \frac{e\gamma - \ln\ln 2}{D_{50}\left(1 + \frac{\bar{d}}{\alpha/\beta}\right)}, \quad \beta = \frac{e\gamma - \ln\ln 2}{D_{50}\left(\frac{\alpha}{\beta} + \bar{d}\right)}. \tag{2.63}$$

On the other hand, if the number of fractions is kept fixed and the dose increases by increasing the dose per fraction, then the steepness of the dose–response curve will increase. The γ value for fixed fractionation, γ_f can be approximated by the expression:

$$\gamma_f \approx \frac{\alpha/\beta + 2\bar{d}}{\alpha/\beta + \bar{d}}\gamma. \tag{2.64}$$

This simple relation shows that particularly for late-responding tissues with a low $\alpha/\beta = 3\,\text{Gy}$ and at doses per fraction, $\bar{d}$, of around $2\,\text{Gy}$, the normalized dose–response gradient at a fixed number of fractions is about 40% steeper than that at a fixed dose per fraction. For acutely responding normal tissues and tumors, which have α/β values of about $10\,\text{Gy}$, the effect will be reduced but still high enough to be considered in nonuniform dose delivery (Ågren 1995). The D_{50} for a fixed fractionation is given by the following formula:

$$D_{50,f} = D_{50}\left(1 + \frac{n_f\bar{d} - D_{50}}{n_f\bar{d} + D_{50} + n_f\alpha/\beta}\right). \tag{2.65}$$

When different treatment plans are compared, one does not need to convert them to the same fractionation mode. It is important though to use the correct values of the radiobiological parameters of the involved organs according to the fractionation regime, which the treatment plan is intended to be applied with. This way the fractionation effects of each dose plan will be taken into account and their comparison will be compatible.

2.5.9. Generalization of the Normalized Dose–Response Gradient to Nonuniform Dose Delivery and Complex Tissues

2.5.9.1. *Uniform dose delivery*

The concept of a normalized dose–response gradient, γ, was introduced some 30 years ago (Brahme 1984) and has become a convenient tool for the characterization of a tissue and its dose–response relation both for tumors and normal tissues. The normalized response gradient $\gamma(D)$ can be defined at any dose level D according to:

$$\gamma(D) \equiv D\left(\frac{dP(D)}{dD}\right), \tag{2.66}$$

where $P(D)$ is the dose–response relation for the tissue at hand. When $P(D)$ is known, this expression can be used to calculate the normalized steepness of the dose–response relation, that is, the ratio of the percentage increase in response to 1% increase in dose at any dose level D from zero to infinity. Beside the proportionality to the logarithm of the effective clonogen number (cf Eq. 2.60, 2.61 and 2.71) one of the most attractive features of the γ definition above, is that it can be used to predict the change in response from a small change in dose according to:

$$\Delta P(D) \approx \frac{\Delta D}{D}\gamma. \tag{2.67}$$

2.5.9.2. *Nonuniform dose delivery*

A general dose distribution $D(\vec{r})$ will from hereon be denoted by the vector notation, $\vec{D}$. With this notation, a dose–response relation $P(D(\vec{r})) = P(\vec{D})$ can be viewed as a functional or mapping $P : S \subset R^n \to R^1$ that maps the n-dimensional dose vector $\vec{D}$ onto a scalar quantity P describing the probability of response for an arbitrary nonuniform dose distribution. A more general approach to handling arbitrary nonuniform dose distributions is to explicitly define γ as a function of any dose distribution $\gamma(\vec{D})$ (Lind *et al.* 2001). One natural way to generalize γ along this line of thought is to replace the derivative in Eq. (2.66) with a gradient as

$$\gamma(\vec{D}) = \vec{D} \cdot \nabla_D P(\vec{D}) = \sum_{i=1}^{n} D_i \frac{\partial P(\vec{D})}{\partial P_i} \frac{\partial P_i(\vec{D})}{\partial D_i} = \sum_{i=1}^{n} \frac{\partial P(\vec{D})}{\partial P_i} \gamma_i, \qquad (2.68)$$

where D_i is the quasi-uniform dose in voxel i, and γ_i, the local contribution to the normalized dose–response gradient from each voxel is given as

$$\gamma_i = \gamma(D_i) \equiv D_i \frac{\partial P_i(\vec{D})}{\partial D_i}. \qquad (2.69)$$

Thus, the sum of all local normalized response gradients can be viewed as a scalar product between the dose vector and the gradient of the response function with respect to the dose vector. This is a natural generalization since for each voxel the local γ value, γ_i, will be related to the local number of clonogens or FSUs and thus the organ-specific value will be related to the weighted sum over all voxels. The normalized dose–response gradient at the steepest absolute point for nonuniform dose delivery is given as

$$\tilde{\gamma} \equiv \tilde{\vec{D}} \cdot \nabla_D P(\tilde{\vec{D}}), \quad \text{where} \quad \nabla_D P(\tilde{\vec{D}}) = \max_{\vec{D}} \| \nabla_D P(\vec{D}) \|. \qquad (2.70)$$

This equation is also a way to define the effective clonogen number using nonuniform dose delivery and assuming Poisson statistics:

$$N_{\text{eff}} = \exp{(e\tilde{\gamma})}. \qquad (2.71)$$

If there is a heterogeneous distribution of radiation sensitivity within each voxel, the effective number of clonogens N_{eff} can still be calculated by Eq. (2.71) using an appropriate expression for P_B. Similar to Eq. (2.70), an equation may be defined for heterogeneous dose delivery. The highest possible normalized dose gradient is thus given as

$$\hat{\gamma} \equiv \max_{\vec{D}}\{\vec{D} \cdot \nabla_D P(\vec{D})\}. \tag{2.72}$$

This expression can be effectively evaluated using Eq. (2.68).

2.5.9.3. γ Value for multiple target volumes and normal tissues

Tumors: When the control probabilities $P_B^j(\vec{D})$ of n different tumor compartments (j) are all statistically independent, the total probability $P_B(\vec{D})$ of controlling all of the n tumor compartments is given as

$$P_B(\vec{D}) = \prod_{j=1}^{n} P_B^j(\vec{D}). \tag{2.73}$$

Using the definition of Eq. (2.66) and the chain rule, we get:

$$\gamma(\vec{D}) = \sum_{i=1}^{n} \gamma_i(\vec{D}) \prod_{j \neq i}^{n} P_B^j(\vec{D}) = P_B(\vec{D}) \sum_{i=1}^{n} \frac{\gamma_i(\vec{D})}{P_B^i(\vec{D})}, \tag{2.74}$$

where $\gamma_i(\vec{D})$ is the normalized dose–response gradient from Eq. (2.68) for tumor compartment i. Thus, two tumor compartments will have a total normalized dose–response gradient $\gamma = \left(\frac{\gamma_1}{P_B^1} + \frac{\gamma_2}{P_B^2}\right) P_B$.

Normal tissues: First, let us assume that the injury of at least one normal tissue compartment is sufficient to cause a global injury and that the compartment's individual injury probabilities are statistically independent. Then, the total probability of total injury $P_I(\vec{D})$ is given as:

$$P_I(\vec{D}) = 1 - \prod_{j=1}^{n}(1 - P_I^j(\vec{D})), \tag{2.75}$$

where n is the number of tissue compartments. Using the definition in Eq. (2.66) and the chain rule, we similarly get:

$$\gamma(\vec{D}) = (1 - P_I(\vec{D})) \sum_{i=1}^{n} \frac{\gamma_i}{1 - P_I^i(\vec{D})}. \tag{2.76}$$

Thus, two healthy tissues will have a total normalized dose–response gradient

$$\gamma = \left(\frac{\gamma_1}{1 - P_I^1} + \frac{\gamma_2}{1 - P_I^2}\right)(1 - P_I).$$

2.6. Derivation of Dose–Response Relations from Clinical Data

The predictive strength of the different radiobiological models does not only depend on their capability of describing closely the biological mechanisms of the therapeutic irradiation but also depend on the accuracy by which the radiobiological parameters of the models have been determined from clinical data. The appropriate way to estimate those parameters is by using clinical materials with complete treatment and follow-up recording (Eriksson *et al.* 2000; Gagliardi *et al.* 2000; Mavroidis *et al.* 2002a). This is a very complicated task because it depends on many factors that are not standardized among the different radiotherapy centers.

The shape of a dose–response relationship constitutes an association between the dose received by the tissue and the expression of a specified clinical effect (Flickinger 1989; Karlsson *et al.* 1997). The information concerning both radiation treatment and clinical outcome may be available for individual patients or for patient groups. The second case approximates the clinical practice where the clinical data (doses and follow-up results) are described in terms of average values for certain uniform groups of patients. Usually, a mean organ dose or DVH averaged over the patient group is used as the dose reference, whereas the follow-up results are expressed as an incidence rate (number of events over the total patient population). This is the information that is used currently in the clinic where single doses inside the tissue (reference dose) is associated with a certain incidence rate. However, the use of a data reduction procedure always results in a loss in the information structure. In such an analysis, the use of the individual patient dose distributions and treatment outcomes allows the variation of the interpatient radiation sensitivity or the volume effect of the tissue for a certain end point to be identified.

Modeling of normal tissue complication and tumor control requires reliable data from appropriately designed studies of specific clinical cases. This demand is met by studies based on retrospective data collection, which, however, may be subjected to potential methodological inaccuracies. Accurate assessment of the radiation exposure and treatment outcome is critical point in normal tissue complication and tumor control quantification and modeling. Although a considerable number of radiobiological models have been developed to describe the dose–response relations for different tumors and normal tissues, the precision and accuracy of the dosimetric (exposure) and the follow-up (outcome) information of the treatments have not been evaluated systematically. Generally, there is a lack of consistent revision of treatment-related definitions and data measures used by the different models.

In the literature, many studies have determined dose–response relations and tolerance doses for different tissues and clinical end points. However, many of them are based on two-dimensional treatment plans and on approximate end point definitions and follow-up measures. Consequently, the development of models based on these data should be considered as approximate requiring the verification of the obtained results both in terms of the mathematical formalism of the models and of the estimated values of the model parameters. The quantification of dose–response relations from clinical material depends strongly on the accuracy of the clinical data used.

Radiobiological modeling is a complicated process even if the available clinical data are accurate. This is because the available information usually covers only a limited part of the dose–response curve since the clinical data are derived from radiation treatments, which aim at achieving tumor control with minimum normal tissue complications. This means that the part of the dose–response relation outside the region of the clinical data (therapeutic range) is based on the form of the model, which cannot be accurately verified by observations in that particular region. This limitation can be very important if these data are to be applied to other classical or modern intensity-modulated treatment techniques, which may cover another dose range than that covered by the clinical data.

The dosimetric data (DVH or volumetric dose distribution) and the clinical outcome (clinical incidence or individual outcomes) are used as input in the determination of radiobiological parameters. Solely, the dose distribution in the volume of interest is not sufficient to describe the effects of the radiation exposure. Radiobiological modeling should also include the influence of the applied fractionation schedule and the response type of the tissue to radiation (early or late). This information can be taken into account by applying the biologically effective dose concept, which accounts for the isoeffect relationships in radiation therapy.

The final outcome of the fitting process is the determination of the model parameters (i.e., D_{50}, γ, and s for the relative seriality model) for a specific clinical end point. These parameters determine the shape of the corresponding dose–response relation allowing the association of a certain dose distribution with the tissue complication or control probability.

2.6.1. *Tumors*

Contrary to the difficulties in defining the clinical end point of the treatment for normal tissues, the treatment outcome for tumors is more clear cut (control or recurrence). The major difficulty in modeling the response of tumors and AVMs is their volume definition (Friedman *et al.* 1995; Yamamoto *et al.* 1995). Although there are issued guidelines and recommendations (International Commission of Radiation Units and Measurements [ICRU] and NACP reports), the clinical practice differs between different radiotherapy centers. This imposes a significant restriction in the comparison of different studies concerning the radiobiological modeling of similar tumors. Especially for the AVM case, the deviation of the estimated AVM volume from the actual one is large because of the standard two projection imaging technique used (Söderman 2000).

In the upper diagram of Fig. 2.23, the estimation the parameters that characterize the dose–response relation of a tumor are shown. The green region on the top of the diagram shows the combinations of D_{50} and γ in the Poisson model that are best associated with the pattern of clinical outcomes of the patients under consideration. In the lower diagram of Fig. 2.23, the effect of the uncertainties of the estimated radiobiological parameters on the dose–response relation is shown. The determination and demonstration of this effect is very important because it is directly related to the implementation of these parameters into the clinical practice. The range of these uncertainties include the contributions from all the associated sources of errors such as potential deficiencies of the radiobiological model used, dosimetric uncertainties, treatment delivery inaccuracies, etc.

Many studies have been carried out to determine the dose–response relations of different targets (tumors or AVMs) (Ågren 1995; Flickinger *et al.* 1996; Karlsson *et al.* 1997, 1999; Wigg 1999). However, most of these studies suffer from the last of critical information that would allow an even closer estimation of the dose–response relations of different targets and improvement of the mathematical formalism of the different radiobiological models. Such information should be the accurate determination of the hypoxic fraction of many tumors, the factors affecting the radiation sensitivity of different targets (such as the size and location), and the spread and density of the target cells to be eradicated.

2.6.2. *Normal Tissues*

The manifestation of a specific radiation effect (complication for normal tissues) in a given tissue is known as clinical end point. Normal tissue end points are generally grouped in two categories. The first category is related to functional changes in the tissue (e.g., paralysis or death), which take place often over a narrow dose range and can be considered to be binary (presence or absence of the effect). In the second functional end point category, extensive reactions or physiological changes (e.g., a degree of skin damage) are involved. These radiation effects cover a wide dose range since they are associated with nonfatal injuries of increasing, with dose, severity. The development of different radiation end points in the same tissue may depend on the eradication of different cell groups associated to different functionalities.

The complication end points for normal tissues are generally classified as early or late and subjective or objective. Different radiotherapy centers apply different

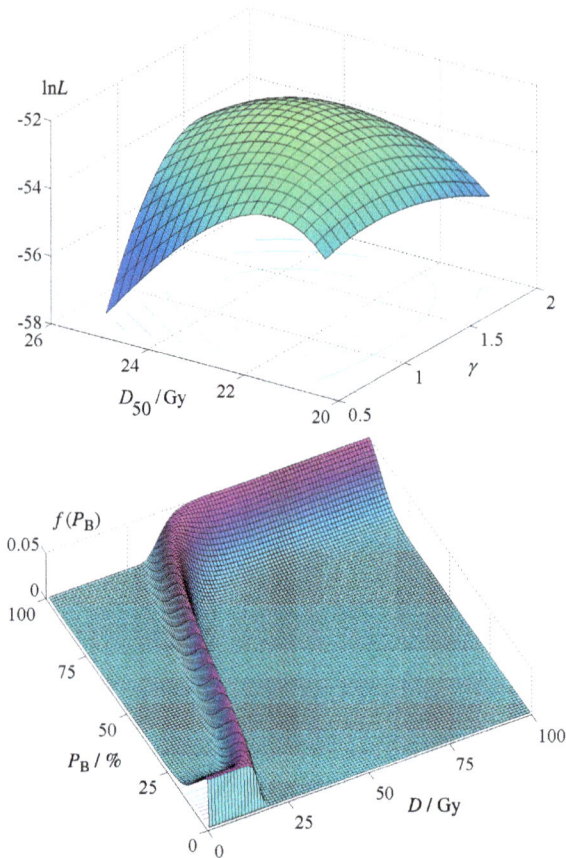

Figure 2.23. *Upper diagram*: The hypersurface of the log-likelihood space, which was used to calculate the uncertainties of the model parameters is shown. The solid line represents the isosurface that corresponds to the 68% probability of deviation from the maximum value of the likelihood. The hypersurface shows how the logarithm of the maximum likelihood function changes in the region around the best estimates of the radiobiological parameters. The maximum point of the diagram corresponds to the best estimate of the parameters (maximum of the log-likelihood function). From the actual hypervolume of the log-likelihood function (it is calculated using all the possible combinations of parameter values), the confidence interval of the estimated radiobiological parameters is calculated. *Lower diagram*: The best estimates of the dose–response curves of the tissue are shown together with their 68% confidence intervals are shown. The range in each case constitutes the confidence interval of the tissue survival curve representing mainly the variation of the interpatient radiation sensitivity and dosimetric discrepancies between the calculated and the delivered dose.

classification systems to record follow-up results. Such classification systems for normal tissue complications are the RTOG/EORTC and the LENT/SOMA (Pavy et al. 1995; Steel 1997). The normal tissue complication end points are usually classified as binary or graded. Binary end points are related to increasing incidence rates with dose and not to increasing response intensity. On the contrary, graded end points consist of a number of data and symptoms (e.g., CT density measurements and diarrhea), which can be translated into graded responses.

Although the proper definition of the clinical end point is important from a clinical perspective, it does not affect the basic process of radiobiological modeling. This is because the estimated dose–response relation for a certain tissue is an association between the dose and the expression of the clinical end point irrespective of the definition and classification of this end point. However, the lack of a protocol based on well-defined and classified end points for the different normal tissues imposes strong limitations to the clinical introduction of the radiobiological models and to the comparison of treatment results between different clinics.

In the derivation of radiobiological parameters from clinical material, a number of important factors have to be taken into account. Such a factor can be the identification of sources related to radiation sensitivity variations. Usually, it is the effective dose–response relations that are estimated, which includes the interpatient and the intrapatient radiation sensitivity variations. The individual patient radiation sensitivity can only be determined with predictive assays. This is a very significant information that has still not been provided and utilized clinically. Intrapatient radiation sensitivity variations are related to non-homogeneous radiation sensitivity distributions within the same tissue or to other clinical factors that may influence the radiation sensitivity of a tissue and the expression of a certain end point.

As it is seen in Fig. 2.24, the volume effect that was discussed above, dramatically affects the shape of the dose–response relations of normal tissues. Consequently, the correct estimation of the parameter that describes the structural organization of the tissue is of major importance. Things become even more complicated when different clinical end points are considered since different groups of cells may be responsible for different end points of the same tissue.

In some cases, it is difficult to determine the source of the complication, especially when graded complications are involved, which consist of a number of symptoms. Such a case is the radiotherapy treatment delivered to a prostate cancer patient is shown. The four-field box technique that is usually applied in such cases delivers high doses to the rectum, which is a sensitive organ at risk. One of the symptoms that are often present after radiotherapy is the fecal leakage (the patient cannot control the defecation of the bowel). However, it is not clear if this symptom stems from radiation-induced functional inhibition of the rectum or

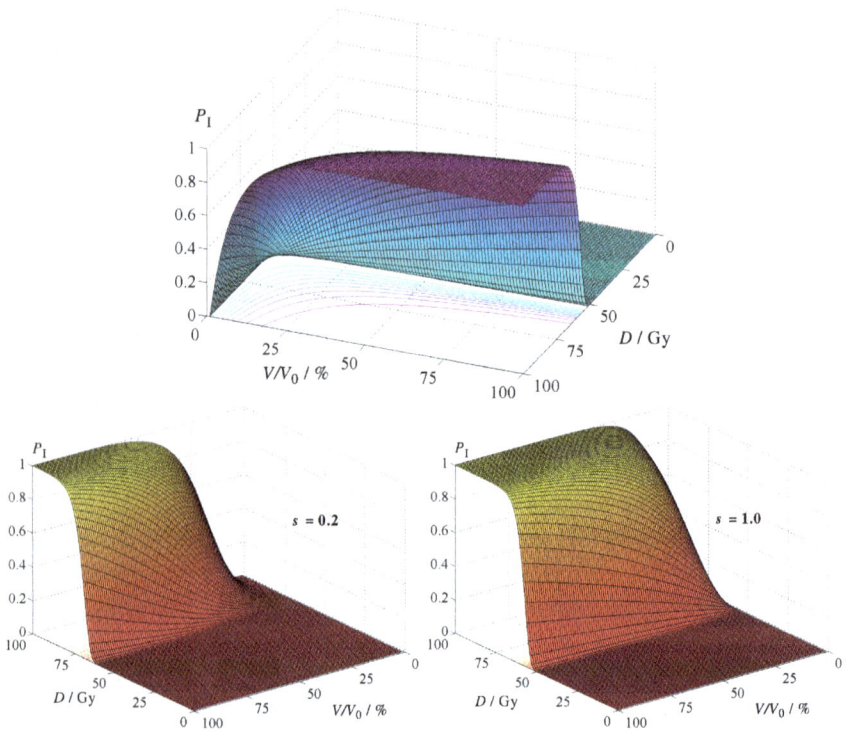

Figure 2.24. *Upper diagram*: The influence of the structural organization of the FSUs in the tissue on its dose–response relation is shown. This is demonstrated by separating the tissue into two compartments, delivering to them the 100% and 10% of the dose shown on the lateral axis and varying the volume proportions of the two compartments ($V/V_0 = 100\%$ corresponds to uniform dose delivery). *Lower diagrams*: Comparison of a fairly parallel (left) with a serial (right) tissue in terms of their behavior to different dose volume combinations. It is seen that for the serial tissue, high levels of complications are expected even if a small portion of the tissue is irradiated with a relatively high dose (the response on the V/V_0 axis raises dramatically). The accurate definition of the structural organization of the tissue is key factor in the correct estimation of its dose–response relation.

injury to the sphincter. Another such symptom is the frequent defecation urgencies of the patients.

Other factors related to the derivation of radiobiological parameters are the imaging information and the resolution of the imaging and dose delivery used. The accuracy by which the tissues are defined and the planned dose is delivered is of primary importance. For example, the exact position and shape of the tissue (e.g., esophagus or rectum) should be known in all parts of the radiotherapy treatment. During treatment planning, the patient is simulated by a voxel representation. In each voxel, a number of clonogenic cells for the tumors or FSUs for normal tissues and a dose value are assigned. The size of the voxels is a significant factor since the actual number of tumor cells or FSUs may significantly vary between different voxels and the dose distribution within each voxel may not be accurately represented by a single value. This detail may affect the accuracy by which estimated associations between dose and clinical effect are derived from clinical trials. Consequently, high-resolution imaging and treatment modalities are desirable for the extraction of clinically valid dose–response relations.

2.6.3. *Statistical Analysis*

The determination of the best estimates of the model parameters is done by fitting the radiobiological model to the clinical data (dose distributions and follow-up results). The shape of a dose–response relation is usually sigmoid, depending on the model used, and it is determined by the model parameters. The fitting method that is most often applied is the maximum likelihood method (denoted here as L) (Herring 1980; Jackson *et al.* 1995; Gagliardi *et al.* 2000). This method is applied on the data of each individual patient. Generally, a model that predicts the induction of a radiation effect consists of one set of parameters (model dependent, denoted as $\vec{X}$) describing the tissue radiation sensitivity and one set of parameters describing the individual treatment effectiveness (dose distribution, denoted as $\vec{\theta}$). Given a set of N patients whose treatment outcome is denoted as r (for simplicity a binary classification of the outcome, that is, $r = 1$, can be assumed, if the patient responds to the treatment and $r = 0$, if the patient does not). The probability that a group of N patients manifests the observed outcome is mathematically expressed by the following formula:

$$L(\vec{X}|\vec{\theta}) = \prod_{i=1}^{N} P(\vec{X}|\vec{\theta_i})^{r_i} \cdot (1 - P(\vec{X}|\vec{\theta_i}))^{1-r_i}. \tag{2.77}$$

The same equation can be expressed in the following way:

$$L(\vec{X}|\vec{\theta}) = \prod_{i=1}^{m} P(\vec{X}|\vec{\theta_i}) \times \prod_{j=1}^{n} (1 - P(\vec{X}|\vec{\theta_j})), \tag{2.78}$$

where m and n represent the fractions of those patients who developed the radiation effect and those who did not, respectively. The best estimates of the parameters are those that maximize the value of the likelihood function. The value of L corresponds to the probability that the model reproduces the observed pattern of treatment outcome.

Several studies, which determined dose–response relation, using individual dose distributions and follow-up data, were based on the maximum likelihood method. In this work, the search for the parameter values was performed by means of a minimization package, MINOS, which has been used in several other applications dealing with optimization problems. MINOS uses a number of minimization strategies in order to converge to the global minimum. However, this is not always guaranteed and a local minimum may instead be found. This can be overcome by using different starting values and investigating different regions in the parameter space. Usually, constrains are imposed on the estimated parameters to keep the parameters within clinically meaningful intervals.

The parameters estimated by modeling of tissue responses can be applied on the dosimetric data of an independent patient group to determine the corresponding response probabilities. The predicted response probability is associated with an uncertainty, which depends both on the clinical data and on the radiobiological model. The determination of the parameter uncertainties can provide information concerning the reliability of the model in describing the clinical data (Schilstra and Meertens 2001). However, their interpretation is complicated due to the inevitable inherent uncertainties characterizing the input data sets and the intentionally limited interval of the dose–response curve that is covered by those data. In practice, a more relevant measure of the quality of the fit of the resulting parameter sets is obtained from the calculation of the uncertainties on the predicted response probability (Hanley and McNeil 1982; Venkatraman and Begg 1996).

As it is shown in Fig. 2.23, the appropriate way of quantifying the uncertainties of the radiobiological parameters is by using the hypervolume of the parameter space. In this figure, the information flow for determining dose–response relations of normal tissues and tumors is illustrated. In the upper diagrams, the treatment data in the form of individual DVHs and the follow-up results in terms of presence or absence of the radiation effect are taken into account (for demonstration purposes, the mean DVHs are used here). Subsequently, the parameter uncertainties are determined from the hypervolume or the hypersurface of the log-likelihood space. This is the most accurate way for pathological and nonlinear functions as in these cases. Finally, the combined influence of the parameter uncertainties on the dose–response relations is applied providing the confidence intervals, which are important for their clinical use.

The validity of the calculated parameters can be examined by applying them to other independent study materials. This way, the sets of parameters determined for a certain model can illustrate the clinical utility of the biological modeling and their clinical accuracy can be estimated. In Fig. 2.25, a statistical verification of estimated dose–response relations is illustrated. It can be seen that the predicted complications and control for the normal tissue and the target, respectively, are very close to the follow-up results for groups of patients receiving similar doses. The normal error distribution, the Pearson's test, and the ROC curves are three statistical methods widely used for evaluating the clinical relevance of such data.

Doses within a certain dose-bin in Fig. 2.25 may refer to a certain classical treatment technique. For such a technique only this part of a dose–response curve is of interest. On the contrary, modern intensity-modulated treatments require the information of the whole curve to evaluate the biological effectiveness of their strongly inhomogeneous dose distributions. Separate dose–response curves can be used for patient subgroups that have special characteristics influencing their

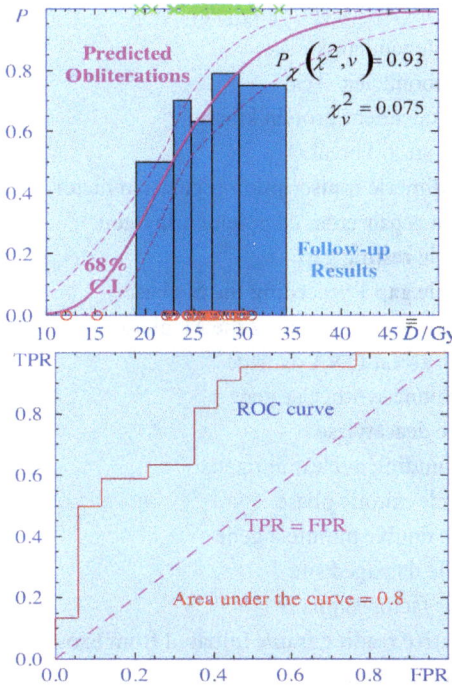

Figure 2.25. *Upper diagram*: The dose–response curve derived using the Poisson model for an AVM case is shown. On the same diagram, the points of the control probability of each patient have been drawn. Those points were calculated using the individual dose distribution delivered to each patient and the model parameters that were calculated. *Lower diagram*: Using the estimated model parameters to calculate the expected complication probabilities of the patients and ordering them as in the upper diagram of the figure, a ROC curve could be constructed using different cut-off thresholds and calculating the corresponding TPR and FPR values. The area under the curve indicates a good agreement between the expected and the observed complication data.

response to dose. Further independent studies using the same tissue delineation method and clinical end point definition can be used to support the validity of the estimated data.

Glossary of Terms

Basic Biological Terms

ATM	Ataxia telangiectasia-mutated gene
BAX	BCL2-associated X protein
BCL-2	B-cell CLL/lymphoma 2
BRCA	Breast cancer susceptibility gene
CDK	Cyclin-dependent kinase
DNA	Deoxyribonucleic Acid
DNA-PK	DNA-dependent protein kinase
DSB	Double-strand breaks
E2F	Heterodimeric transcription regulation factor
ERCC	Excision repair cross complementing gene
G_0	Cell cycle resting gap
G_1	Cell cycle gap 1 preceding the S-phase
G_2	Cell cycle gap 2 preceding the M-phase
GADD45	Growth arrest DNA damage
GS	Glutathione synthetase gene
HDAC	Histone deacetylase
Ku70, 86	DNA-binding nuclear antigens
M	Cell cycle mitotic phase
MDM2	Murine double minute 2 gene
MDS	Multiple damaged site
NF-κB	Nuclear factor-kappa B
p14^{ARF}	Alternative reading frame initiated from exon 1β of the DCKN2A locus
P21	21KD cdk inhibitor protein
RAS	p21^{RAS} protein
RB1	Retinoblastoma susceptibility gene
RCR	Repairable–conditionally repairable
S	Cell cycle DNA synthesis phase
SSB	Single-strand breaks
Topo	Topoisomerase gene
TP53	p53 gene

WT   Wilms' tumor
XPA-G   Xeroderma pigmentosum complementation groups A–G
XRCC   X-radiation cross-complementing gene

Radiobiological and Physical Terms

α	Linear component of the LQ model
β	Quadratic component of the LQ model
a	Low dose irreparable damage coefficient
b	Proportion of reparable damage
c	Repair loss coefficient
δ_{rays}	Low-energy electron tracks
D	Absorbed dose
D_{50}	50% survival dose
LET	Linear energy transfer
OER	Oxygen enhancement ratio
RBE	Relative biological efficiency
f	Real dose fraction
f^*	Effective dose fraction
Φ	Fluence of electrons
γ	Normalized dose–response gradient
L_{Δ}	Restricted stopping power
N_0	Initial number of clonogenic cells
$P_h(\nu)$	Hit number probability
P_s	Survival probability
P_B	Probability of benefit (tumor cure)
P_I	Probability of normal tissue Injury
ρ	Density of the medium
σ_0	Cross-section for a lethal hit
S	Surviving fraction
ν	Number of hits

Bibliography

Adamus-Górka M, Mavroidis P, Lind BK, Brahme A (2011) Comparison of dose response models for predicting normal tissue complications from cancer radiotherapy: Application in Rat Spinal Cord: Cancers 3:2421–2443

Ågren Cronqvist AK (1995) Quantification of the response of heterogeneous tumours and organized normal tissues to fractionated radiotherapy, PhD thesis, Stockholm University, Sweden.

Ågren AK, Brahme A, Turesson I (1990) Optimization of uncomplicated control for head and neck tumors. *Int J Rad Oncol Biol Phys* 19:1077–1085.

Barendsen GW (1968) Response of cultured cells, tumors and normal tissues to radiations of different linear energy transfer. *Curr Topics Rad Res* 4:295–356.

Barendsen GW (1982) Dose fractionation, dose rate and iso-effect relationships for normal tissue responses. *Int J Radiat Oncol Biol Phys* 11:1751–1757.

Brahme A (1984) Dosimetric precision requirements in radiation therapy. *Acta Radiol Oncol* 23:379–391.

Brahme A (1995) Treatment optimization using physical and biological objective functions. In: Smith A (ed.), *Radiation Therapy Physics*, 209–246. Berlin: Springer.

Brahme A, Lind BK (2010) A systems biology approach to radiation therapy optimization. *Radiat Environ Biophys* 49:111–124.

Brahme A. (2011) Accurate description of the cell survival and biological effect at low and high doses and LETs. *J Rad Res* 52:389–407.

Brahme A, Ågren A (1987) Optimal dose distribution for eradication of heterogeneous tumors. *Acta Oncol* 26:377–385.

Burman C, Kutcher GJ, Emami B, Goitein M (1991) Fitting of normal tissue tolerance data to an analytic function. *Int J Radiat Oncol Biol Phys* 21:123–135.

Curtis SB (1996) Lethal and potential lethal lesions induced by radiation: a unified repair model. *Radiat Res* 106:252–270.

Dasu A, Denekamp J (2000) Inducible repair and intrinsic radiosensitivity: a complex but predictable relationship? *Rad Res* 153:279–288.

Ebert M, Hoban P (1996) Some characteristics of tumour control probability for heterogeneous tumours. *Phys Med Biol* 41:2125–2133.

Emami B, Lyman J, Brown A, *et al.* (1991) Tolerance of normal tissue to therapeutic irradiation. *Int J Radiat Oncol Biol Phys* 21:109–122.

Eriksson F, Gagliardi G, Liendberg A, Lax I, Lee C, Levitt S, Lind B, Rutqvist LE (2000) Long-term cardiac mortality following radiation therapy for Hodgkin's disease: analysis with the relative seriality model. *Radiother Oncol* 55:153–162.

Flickinger J (1989) The integrated logistic formula and prediction of complications from radiosurgery. *Int J Radiat Oncol Biol Phys* 17:879–885.

Flickinger JC, Pollock BE, Kondziolka D, *et al.* (1996) A dose-response analysis of arteriovenous malformation obliteration after radiosurgery. *Int J Radiat Oncol Biol Phys* 36:873–879.

Frankenberg D, Frankenberg-Schwager M, Blöcher D, Harbich R (1981) Evidence for DNA double-strand breaks as the critical lesion in yeast cells irradiated with sparsely or densely ionising radiation under oxic or anoxic conditions. *Rad Res* 88:524–532.

Fredriksson M (2002) Biological effects of fractionated radiation therapy, MSc thesis, Stockholm University, Sweden.

Friedman W, Bova F, Mendenhall (1995) Liner accelerator radiosurgery for arteriovenous malformations: the relationships of size to outcome. *J Neurosurg* 82:180–189.

Gagliardi G, Björhle J, Lax I, Ottolenghi A, Eriksson F, Liendberg A, Lind P, Rutqvist LE (2000) Radiation pneumonitis after breast cancer irradiation: analysis of the complication probability using the relative seriality model. *Int J Radiat Oncol Biol Phys* 46:373–381.

Gasinska A, Dubray B, Hill SA, Denekamp J, Thames HD, Fowler JF (1993) Early and late injuries in mouse rectum after fractionated X-ray and neutron irradiation. *Radiother Oncol* 26:244–253.

Gross W, Dicello J, Colvett RD (1974) The dosimetry and microdosimetry of 5.7 GeV neon ions. (Annual report on research project, Radiological Research Laboratory, Columbia University NY, 88-115, COO 3243-3 United States Atomic Energy Commission).

Hamilton CS, Denham JW, O'Brien M, Ostwald P, Kron T, Wright S, Dörr W (1986) Underprediction of human skin erythema at low doses per fraction by the linear-quadratic model. *Radiother Oncol* 40:23–30.

Hamm RN, Wright HA, Katz R, Turner JE, Ritchie RH (1978) Calculated yields and slowing-down spectra for electrons in liquid water: implications for electron and photon RBE. *Phys Med Biol* 23:1149–1161.

Hanley JA, McNeil BJ (1982) The meaning and use of the area under a receiver operating characteristic (ROC) curve. *Radiology* 143:29–36.

Herring DF (1980) Methods for extracting dose-response curves from radiation therapy data, I: A unified approach. *Int J Radiat Oncol Biol Phys* 6:225–232.

Higgins PD, DeLuca Jr PM, Pearson DW (1983) V79 survival following simultaneous or sequential irradiation by 15 MeV neutrons and ^{60}Co photons. *Radiat Res* 95:45–56.

Holley WR, Chatterjee A, Magee JL (1990) Production of DNA strand breaks by direct effects of heavy charged particles. *Rad Res* 121:161–168.

ICRU 35 (1984) Radiation Dosimetry: electron beams with energies between 1–50 MeV. *Int Comm Rad Units and Measurements* (ICRU), Bethesda, Maryland.

Jackson A, Kutcher GJ, Yorke ED (1993) Probability of radiation-induced complications for normal tissues with parallel architecture subject to non-uniform radiation. *Med Phys* 20:613–625.

Jackson A, Ten Haken RK, Robertson JM *et al.* (1995) Analysis of clinical complication data for radiation hepatitis using a parallel architecture model. *Int J Radiat Oncol Biol Phys* 31:883–891.

Jansson T, Inganäs M, Sjögren S *et al.* (1995) p53 status predicts survival in breast cancer patients treated with or without postoperative radiotherapy: a novel hypothesis based on clinical findings. *J Clin Oncol* 13:2745–2751.

Joiner M, Johns H (1988) Renal damage in mouse: the response to very low doses per fraction. *Radiat Res* 114:385–398.

Kallinowski F, Zander R, Höckel M, Vaupel P (1990) Tumor tissue oxygenation as evaluated by computerized pO_2 histography. *Int J Radiat Oncol Biol Phys* 19:953–961.

Källman P, Ågren A, Brahme A (1992) Tumour and normal tissue responses to fractionated non-uniform dose delivery. *Int J Radiat Biol* 62:249–262.

Karlsson B, Lax I, Söderman M (1997) Factors influencing the risk of complications following gamma knife radiosurgery of cerebral arteriovenous malformations. *Radiother Oncol* 43:275–280.

Karlsson B, Lax I, Söderman M (1999) Can the probability for obliteration after radiosurgery for arteriovenous malformations be accurately predicted? *Int J Radiat Oncol Biol Phys* 43:313–319.

Klepper LY, Klimanov A (2000) Calculation of radiation complication probabilities in organs and tissues on the basis of a modified Weibull's function for usage in optimisation algorithms. Manuscript.

Knudson AG (1971) Mutation and cancer: statistical study of retinoblastoma. *Proc Natl Acad Sci USA* 68:820–823.

Kraft G, Krämer, Scholz M (1992) LET, track structure and models. *Rad Env Biophys* 31:161–180.

Lam GKY (1987) The survival response of a biological system to mixed radiations. *Radiat Res* 110:232–243.

Lambin P, Malaise EP, Joiner MC (1994) The effect of very low radiation doses on the human bladder carcinoma cell line RT112. *Radiother Oncol* 32:63–72.

Lambin P, Marples B, Fertil B, Malaise EP, Joiner M (1993) Hypersensitivity of a human tumour cell line to very low radiation doses. *Int J Radiat Biol* 63:639–650.

Lane D (1992) p53, guardian of the genome. *Nature* 358:15–16.

Lind BK (1990) Properties of an algorithm for solving the inverse problem in radiation therapy. *Inverse Problems* 6:415–426.

Lind BK, Mavroidis P, Hyödynmaa S, Kappas C (1999) Optimization of the dose level for a given treatment plan to maximize the complication-free tumor cure. *Acta Oncol* 38:787–798.

Lind BK, Nilsson J, Löf J, Brahme A (2001) Generalization of the normalized dose-response gradient to non-uniform dose delivery. *Acta Oncol* 40:718–724.

Lind BK, Persson LM, Edgren MR, Hedlöf I, Brahme A (2003) Repairable- conditionally repairable damage model based on dual poisson processes. *Radiat Res* 160:366–375.

Marples B, Joiner MC (2000) Modification of survival by DNA repair modifiers: a probable explanation for the phenomenon of increased radioresistance. *Int J Radiat Biol* 76:305–312.

Mavroidis P (2001) Determination and use of radiobiological response parameters in radiation therapy optimization, PhD thesis, Karolinska Institutet.

Mavroidis P, Axelsson S, Hyödynmaa S, Rajala J, Pitkänen MA, Lind BK, Brahme A (2002b) Effects of positioning uncertainty and breathing on dose delivery and radiation pneumonitis prediction in breast cancer. *Acta Oncol* 41:471–485.

Mavroidis P, Laurell G, Kraepelien T, Fernberg JO, Lind BK, Brahme A (2002c) Determination and clinical verification of dose-response parameters for esophageal stricture from head & neck radiotherapy. *Acta Oncol* 42:865–881.

Mavroidis P, Lind BK, Brahme A (2001) Biologically effective uniform dose ($\overline{\overline{D}}$) for specification, report and comparison of dose response relations and treatment plans. *Phys Med Biol* 46:2607–2630.

Mavroidis P, Lind BK, Van Dijk J, Koedooder K, De Neve W, De Wagter C, Planskoy B, Rosenwald JC, Proimos B, Kappas C, Danciu C, Benassi M, Chierego G, Brahme A (2000) Comparison of conformal radiation therapy techniques within the dynamic radiotherapy project "DYNARAD." *Phys Med Biol* 45:2459–2481.

Mavroidis P, Theodorou K, Lefkopoulos D, Nataf F, Schlienger M, Karlsson B, Lax I, Kappas C, Lind BK, Brahme A (2002a) Prediction of AVM obliteration after stereotactic radiotherapy using radiobiological modelling. *Phys Med Biol* 47:2471–2494.

Ngo FQH, Blakely E, Tobias C (1981) Sequential exposures of mammalian cells to low- and high-LET radiations. *Rad Res* 87:59–78.

Nilsson J, Lind BK, Brahme A (2002) Radiation response of hypoxic and generally heterogeneous tissues. *Int J Radiat Biol* 78:389–405.

Paretzke HG (1987) Radiation track structure theory. In: Freeman GR (ed.) *Kinetics of Non-homogeneous Processes*, 89–170. New York: Wiley.

Pavy JJ, Denekamp J, Letschert J, Littbrand B, Mornex F, Bernier J *et al.* (1995) Late effects toxicity scoring: the SOMA scale. *Radiother Oncol* 35:11–15.

Schilstra C, Meertens H (2001) Calculation of the uncertainty in complication probability for various dose-response models, applied to the parotid gland. *Int J Radiat Oncol Biol Phys* 50:147–158.

Söderman M (2000) Volume determination and predictive models in the management of cerebral arteriovenous malformations, PhD thesis, Stockholm University, Sweden.

Stavreva NA, Stavrev PV, Round WH (1996) A mathematical approach to optimizing the radiation dose distribution in heterogeneous tumours. *Acta Oncol* 35:727–732.

Steel G (1997) *Basic Clinical Radiobiology*. The Bath Press, Great Britain.

Svensson H, Brahme A (1981) Fundamentals of electron beam dosimetry. In: Chu FCH, Laughlin JS (eds.) *Proceedings of the Symposium on Electron Beam Therapy*, 17–30. New York: Memorial Sloan-Kettering Cancer Center.

Tilikidis A, Brahme A, Lindborg L (1990) Microdosimetry in the build-up region of gamma ray beams. *Rad Prot Dos* 31:227–233.

Tilly N, Brahme A, Carlsson J, Glimelius B (1999) Comparison of cell survival models for mixed LET radiation. *Int J Radiat Biol* 75:233–243.

Tobias CA (1985) The repair-misrepair model in radiobiology: Comparison to other models. *Radiat Res* 104:77–95.

Tsoulou E, Baggio L, Cherubini R, Kalfas CA (2001) Low-dose hypersensitivity of V79 cells under exposure to γ-rays and 4He-ions of different energies: Survival and chromosome aberrations. *Int J Radiat Biol* 77:1133–1139.

Turesson I, Bernefors R, Book M, Folgegård M, Hermansson I, Johansson KA, Lindh A, Sigurdardottir S, Thunberg U, Nyman J (2001) Normal tissue response to low doses of radiotherapy assessed by molecular markers. *Acta Oncol* 40:941–951.

Turesson I, Joiner MC (1996) Clinical evidence of hypersensitivity to low doses in radiotherapy. *Radiother Oncol* 40:1–3.

Vaupel P, Kallinowski F, Okunieff P (1989) Blood flow, oxygen and nutrient supply, and metabolic microenvironment of human tumors: a review. *Cancer Res* 49:6449–6465.

Vaupel P, Thews O, Kelleher DK, Hoeckel M (1998) Current status of knowledge and critical issues in tumor oxygenation. Results from 25 years research in tumor pathophysiology. *Adv Exp Med Biol* 454:591–602.

Venkatraman ES, Begg C (1996) A distribution-free procedure for comparing receiver operating characteristic curves from a paired experiment. *Biometrika* 83: 835–848.

Ward JF (1988) DNA damage produced by ionizing radiation in mammalian cells: identities, mechanisms and repairability. *Prog Nucleic Acid Res Mol Biol* 35:95–125.

Wigg RD (1999) A radiobiological basis for the treatment of arteriovenous malformations. *Acta Oncol* 38(Suppl 14):3–29.

Withers HR (1982) Biological basis of radiation therapy. In: *Principles and Practice of Radiation Oncology* 67–98. Philadelphia: JB Lippincott.

Withers HR, Peters LJ (1980) Biologic aspects of radiation therapy. In: Fletcher GH (ed.) *Textbook of Radiotherapy*, third edition, 103–180. Philadelphia: Lea & Febiger.

Withers HR, Taylor JMG, Maciejewski B (1988) Treatment volume and tissue tolerance. *Int J Radiat Oncol Biol Phys* 104:751–759.

Yamamoto Y, Coffey RJ, Nichols DA, Shaw EG (1995) Interim report on the radiosurgical treatment of cerebral arteriovenous malformations: the influence of size, dose time and technical factors on obliteration rate. *J Neurosurg* 83:832–837.

Yorke ED, Jackson A, Fox RA, Wessels BW, Gray BN (1999) Can current models explain the lack of liver complications in Y-90 microsphere therapy? *Clin Cancer Res* 5(10 Suppl.) 3024—3030.

Zaider M, Amols HI (1998) A little to a lot or a lot to a little: is NTCP always minimized in multiport therapy? *Int J Radiat Oncol Biol Phys* 41:945–950.

Zaider M, Rossi HH (1980) The synergistic effect of different radiations. *Radiat Res* 83:732–739.

Zaider M, Rossi HH (1986) Microdosimetry and its applications to biological processes. In: Orton CG (ed.) *Radiation Dosimetry*, 171. New York: Plenum.

Zimmer KG (1961) *Studies on Quantitative Radiation Biology*. Oliver & Boyd, London.

The Radiation Biological Basis of Radiation Therapy

3

Anders Brahme, Panayiotis Mavroidis, and Bengt K. Lind

3.1. Introduction

Today surgery and radiation therapy are the major treatment modalities for curing cancer patients. About 20–25% of the diagnosed cancers can be cured by each of these treatment modalities either alone as sole modalities or together in adjuvant treatments. In addition, cytostatic drugs, kinase inhibitors, and other chemotherapeutic agents can help cure about 5–10% of the diagnosed cancers. Thus, over 50% of all cancer patients may be cured and this is also true for that half of the patients who are receiving radiation therapy (cf. Chapter 8, Fig. 8.40).

During the past three decades, radiation therapy has gone through a very dramatic development: during the 1970s and 1980s computed tomography (CT) and magnetic resonance imaging (MRI), respectively, have revolutionized the diagnostic phase of radiation therapy by providing truly three dimensional (3D) diagnostic tools of sub-millimeter accuracy. This phase has matured further during the 1990s and has been amplified by an equally important development of radiation therapy equipment allowing truly 3D dose delivery to the tumor. The time between the first introduction of new advanced medical tools and the full realization of the potential benefits in the clinic is often quite long, especially considering the long follow-up periods (5–10 years) required for many cancer sites. Hopefully, during the first decade of the new millennium we will start to see the benefits of these developments in improved local cure and survival as new 3D intensity-modulated treatment modalities are taken into clinical use.

After a brief review of the whole radiotherapy process, the main goals of radiation therapy will be formulated in radiobiological terms. Furthermore, the treatment results of classical and advanced intensity-modulated radiation therapy will be discussed. Finally, the different treatment modalities employed today for radiation therapy as well as the characteristics of the different treatment units will be discussed and the future development of radiation therapy will be indicated.

During the past 30 years, radiation therapy has developed from classical rectangular beams via conformation therapy with largely uniform dose delivery but irregular field shapes to fully intensity-modulated dose delivery where the total dose distribution in the tumor can be fully controlled in three dimensions. This last step has been developed during the past 15–20 years and has opened up the possibilities for truly optimized radiation therapy. Today it is not only possible to produce almost any desired dose distribution in the tumor volume, but also possible to deliver the dose distribution, which has the highest probability to cure the patients without inducing severe complications in normal tissues. To fully exploit the advantages of intensity-modulated radiation therapy, quality of life or radiobiological objectives have to be used, preferably combined with predictive assay of radiation sensitivity.

The present review of the radiotherapy process starts by explaining the radiobiological basis of radiation therapy and briefly discusses the biological objective functions and the associated advantages in the treatment outcome using new treatment approaches such as intensity-modulated beams. Finally, different possibilities for realizing general 3D intensity-modulated dose delivery are described. Once accurate genetic and/or cell survival-based predictive assays become available, radiation therapy will become an exact science allowing a truly individual optimization considering also the panorama of side effects that the patients are willing to accept.

3.2. Overview of the Radiation Therapy Process

3.2.1. *Diagnostic Work Up*

Dependent on the patient's history, an increasing arsenal of diagnostic procedures will be used, starting from visual inspection and palpation often complemented with different forms of endoscopy, ultrasound (US), and fine needle aspiration biopsy of suspected nodules. In addition, tumor-related host factors, such as prostate-specific antigen, etc., are evaluated.

All modern diagnostic imaging tools may also be useful in the work up of the patient for diagnostic and treatment planning purposes, both as a first step in the evaluation of the stage of the disease and for the delineation of the clinical tissues that will be target and organs at risk during the treatment, as shown in Fig. 3.1. Most commonly CT and MRI are used as the base for treatment planning but also diagnostic X-ray, mammography, positron emission tomography (PET), single-photon emission computed tomography (SPECT), and US may generate useful data for defining the clinical target volume (CTV). The CTV is often the same for the two major treatment modalities: radiation therapy and surgery. Depending on the uncertainty in the diagnostic information gained by all diagnostic modalities

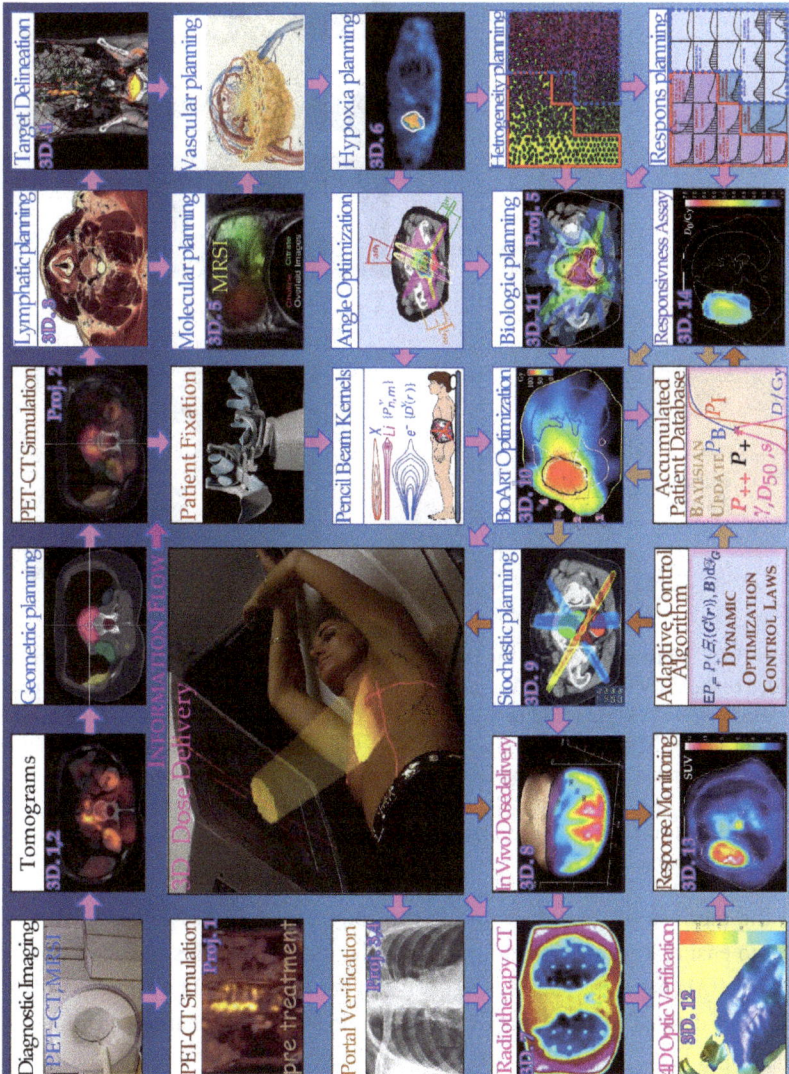

Figure 3.1. Schematic overview of the radiation therapy process describing some of the main activities and their interaction. Two types of imaging, projected 2D and true 3D images, are involved. Portal and simulator imaging and digitally reconstructed radiographs belong to the former projected 2D type. Most modern diagnostic methods produce 3D images including CT, MRI, PET, and SPECT. In addition to diagnostic CT images, the therapeutic beam itself can be used to make truly 3D radiotherapeutic CT images of the patient in treatment position. The some times rather extencive information flow during radiation therapy including the interaction between the different imaging modalities and the adaptive control and correction of the treatment setup are illustrated. The important possibility of continued sequential updating of dose–response relations during treatment follow up are indicated in the lower part of the figure.

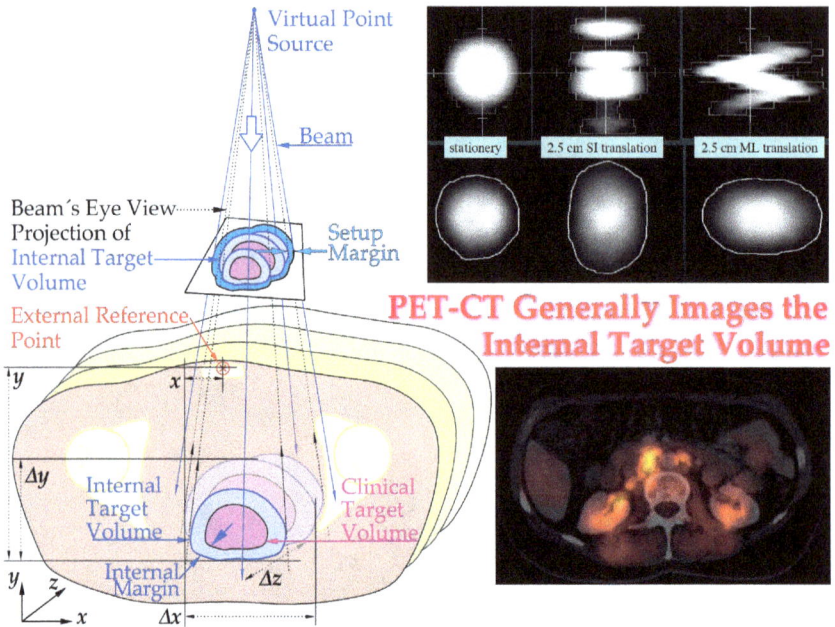

Figure 3.2. Illustration of how the IM due to organ motion and tumor spread uncertainties is added to the CTV to get the ITV accurately defined in relation to extended reference points. The *setup margin* (SM) is added to the beam's eye view projection of the ITV (here drawn on a stack of CT or MRI slices and imaged by PET-CT since the multi minute average image time generally takes target motion into account) to get the cross-section of the therapeutic beam since the ITV is accurately defined in relation to superficial external reference points on the patient's surface; a very accurate beam alignment is possible even if the tumor is not visualized during setup.

and the motions of the tissues inside the body, a margin has to be added to the CTV to ensure a curative treatment. For radiation therapy this so-called "internal margin" (IM) has to be added to the CTV to get the internal target volume (ITV) accurately defined inside the body in relation to superficial anatomic landmarks as indicated in Fig. 3.2. The ITV has to be irradiated to a curative dose of radiation to ensure a beneficial treatment but at the same time the dose to surrounding normal tissues should be low enough to avoid severe normal tissue damage.

3.2.2. *Aim of Treatment*

3.2.2.1. *General*

Radiation therapy is the modality of choice for about 50% of the cancer patients at some stage during their treatment. However, the aim and intent of the treatment can significantly differ from case to case. For a radical treatment with curative intent, radiation therapy may be used alone or in combination with other adjuvant modalities such as surgery or chemotherapy. In the former case, all clonogenic tumor cells should be eradicated by radiation alone, and in the latter case in combination with other modalities. Radiation therapy may also be used with a non-radical or palliative intent in very advanced tumors. In such cases, the target for the irradiation may not necessarily be all the clonogenic tumor cells but those tissues that are giving an immediate clinical problem or are responsible for the symptoms at hand.

In all forms of radiation therapy a balance has to be found between the radiation effects in the target tissues and surrounding normal tissues. Maximum probability of tumor eradication without serious side effects calls for optimal fractionation and precision of the dose to the target volume as well as minimal irradiation of the organs at risk. If the probability of cure is very small or negligible, the treatment will often be considered palliative. Since all these aspects will have a considerable influence on the selection of volume and dose concepts in radiation therapy, clear definitions of the therapeutic intent have to be formulated.

3.2.2.2. Treatment aims

The aim of radiotherapy could be either curative or palliative. In *curative radiation therapy*, the aim is to decrease the number of clonogenic tumor cells to a level that results in permanent tumor control. In *curative radiation therapy*, it is commonly understood that (1) radiation therapy is used as the main local treatment modality, (2) it has a significant probability of tumor eradication, and (3) all tumor cells have a high probability of being included in the defined target volumes. Owing to the random nature of cell kill, the clinical uncertainty in microscopic tumor spread, and the individual differences and variations in radiosensitivity, all patients may not be cured. For a curative treatment, the irradiated volumes have to include all macroscopic tumor tissues (the gross tumor [GT]) and volumes with risk for subclinical spread. To minimize the adverse effects of the irradiation, the GT and subclinical malignant disease should preferably be irradiated to individually selected dose levels (cf. Nordic Association of Clinical Physics [NACP], International Commission of Radiation Units and Measurements [ICRU] 50 reports).

Palliative radiation therapy is given to maintain local tumor control, to relieve a symptom, to prevent or delay an impending symptom, and generally to improve the quality of life but not primarily to increase the probability to eradicate the tumor or improve survival. The target volume(s) for symptom relief does not necessarily need to encompass all tumor tissues in the patient. The doses are generally lower than those used for curative treatments, but they should be sufficient to relieve the symptoms and/or reduce the tumor cell burden in the target volume(s) such that the lifespan or quality of life is increased.

3.2.2.3. Treatment types

In *radical radiation therapy*, the delivered dose should be high enough to decrease the *local* tumor cell burden to a level that results in permanent local tumor control.

In *adjuvant radiation therapy*, radiation therapy is combined with surgery such that the combined modalities have a high probability to eradicate all clonogenic tumor cells to a level that results in permanent local tumor control. To be more specific, the terms pre-, post-, or intraoperative radiotherapy are used, as described below.

Preoperative radiation therapy is the situation where radiation therapy is given before planned surgery and the combined procedure has a high probability to eradicate all clonogenic tumor cells.

Postoperative radiation therapy is the situation where planned radiation therapy has a high probability of eradicating all clonogenic tumor cells left after earlier surgery, and therefore all tumor tissues should be included in the target volume(s).

Intraoperative radiation therapy is the situation where at least one high dose fraction is delivered as the tumor bed is opened for treatment-related surgery.

In *radiotherapy of benign diseases*, such as arteriovenous malformations, eczemas, keloids, and inflammatory processes, all of the affected tissues are not necessarily included in the treatment.

3.2.3. Patient Data for Treatment Planning

The first part of the treatment preparation consists of the acquisition of patient data for the delineation of target volume(s) and organ(s) at risk and a selection of the suitable treatment modality and patient fixation technique. The patient fixation is of importance early in the treatment planning particularly for the large number of advanced tumors where it is important to ensure that the same patient position will be used, both during the anatomic work up, simulation, and treatment execution. In modern radiation therapy, other types of clinically relevant information are also of interest, such as the radiation sensitivity of the patient and the tumor as determined by predictive assays or genetic markers, the tumor growth rate as measured by the tumor doubling time, and the oxygenation status as determined by electrodes, MRI, or nuclear medicine techniques. These latter factors are of great importance particularly when advanced treatment optimization procedures are employed.

In general, it is important that the images are generated in a geometry which is as similar as possible to that during treatment execution. Otherwise, the anatomy can be severely distorted between imaging and dose delivery. Contrast agents may be used to identify the affected tissues. However, contrast-filled organs such as the urinary bladder and rectum may be distended more during the diagnostic procedure than during radiation treatment. This will alter the pixel information in CT images and may complicate density corrections in dose planning. Short image detection times will give sharp images without motion artifacts, while long image detection times may give information about the mean target tissue movements during treatment. Several image processing methods can be used to get more information out of the images, such as regional histogram equalization to improve the contrast range and image matching and fusion to identify different aspects of the target volume in different imaging modalities. This will also make it possible to generate new images such as beam's-eye-views or digitally reconstructed radiographs and to apply 3D surface rendering to improve visualization (cf. Fig. 3.1). For this purpose, it is important to use a local patient coordinate system describing as accurately as possible the location of organs in relation to the target volume.

An efficient image handling system is necessary in defining target volumes and critical organs. Diagnostic systems, dedicated to radiation therapy, are often integrated with the treatment planning system. Especially when 3D information is needed, such systems are mandatory. The system should be able to combine information from all different imaging modalities available in the clinic such that volumes identified by one modality can be transferred to the others. MR images are often distorted, and nuclear medicine images are affected by attenuation and scatter and this has to be taken into account when they are combined with other imaging modalities.

3.2.4. *Treatment Planning*

The treatment planning process consists of several methods for treatment preparation and simulation to achieve a reproducible and optimal treatment plan for the patient. Irrespective of the temporal order, these events include:

- patient fixation, immobilization, and reference point selection;
- dose prescriptions for target volumes and the tolerance level of organ at risk volumes;
- selection and optimization of

    — radiation modality and treatment technique,
    — the number of beam portals,

— the directions of incidence of the beams,
— beam collimation,
— beam intensity profiles,
— fractionation schedule;

• dose distribution calculation considering geometrical, stochastical, and biological factors; and
• treatment simulation.

The position of the patient must be very reproducible throughout the entire treatment course, thus making adequate and reproducible fixation a necessity (cf. Fig. 3.1). There exist several systems, which can be used, such as shells, masks, bite-blocks, etc. Two setup techniques are in common use: the isocentric and the fixed source to surface distance methods. Both techniques have advantages and disadvantages, and one must be aware of these.

For both techniques, laser alignment of the patient will enhance the precision and reproducibility of the patient position. Such equipment should be positioned identically when preparing fixation aids, doing CT or MRI, and performing simulations and treatments. Reference points and reference lines for patient and beam positioning are essential for correct setup.

To find the best dose plan, several parameters of the list above have to be optimized. Filters are often used for shaping the lateral dose distribution across the beam. Both compensation filters and wedge filters are used. The collimation system and/or scanning beam on some accelerators can be used to dynamically vary the dose distribution instead of using static filters or wedge filters. The dose calculation algorithms should be able to handle 3D anatomical information and allow accurate dose calculation in strongly heterogeneous situations. One should be aware of the limitations and inaccuracies of the algorithms employed, since improper use can give rise to unacceptable errors in the delivered dose distribution. The dose calculation system should also be able to utilize all the technical capabilities of existing treatment units and have reliable routines for optimization of most important treatment parameters. For brachytherapy, the optimization should include parameters such as number and shape of applicators and sources, positions of applicators, retractors and sources, treatment time for each source position, and shielding.

The conventional simulator is used both as a localizer like other imaging techniques and as simulator of the treatment setup. A major part of the simulator procedure is to identify suitable anatomical reference points and markings for the beam portals. For these procedures, both CT scanners and ordinary simulators can be used and equipment has been developed that combines the benefits from both. Digitally reconstructed radiographs similar to simulator setup images can be generated by several treatment planning systems and CT scanners, and this will enhance the precision of beam setup and treatment verification.

The performance of each of the methods used should be known, since inaccuracies and variations can severely alter delivered dose distribution. These inaccuracies and variations have to be taken into account when defining beam portals. The estimation of the uncertainties of the different steps in the preparation process can also give rise to mistakes and inaccurate transfer of treatment parameters. A complete information system for radiation therapy, where treatment parameters can be automatically transferred from planning computer to the treatment unit, will reduce this problem, but does not replace the need for a well-trained and educated staff. Also one must be aware that automatic transfer of treatment parameters to the treatment unit, while reducing the risk of data transfer errors in individual fractions, carries the risk of introducing systematic errors which could affect the complete course of treatment.

3.2.5. *Treatment Execution*

During the treatment course, measures have to be taken to ensure that dose delivery corresponds to the established plans. Any deviations have to be recorded and dosimetric consequences have to be evaluated. Equally important are the different methods for treatment verification:

- Patient dose measurements (*in vivo* dose verification)
- Portal imaging
- Isocenter verification by orthogonal X-ray projections ($0° \pm 90°$)
- Verification systems for dose monitor, gantry, treatment couch, light, and radiation beam setup
- Repeated imaging during the course of therapy to verify treatment setup and target volumes

Verification systems for all treatment parameters are valuable and should be mandatory for check of the beam patient setup and verification of monitor settings. However, it is important to use as direct a transfer of treatment parameters as possible, preferably directly from the planning computer to the simulator and treatment unit to minimize errors. Portal imaging and radiotherapeutic CT, if available, will verify the patient setup relative to the therapy beam, and they are the ultimate 3D check of correct beam/patient alignment. Portal imaging, with film or preferably with a real-time imaging device should be used routinely to verify patient setup relative to the therapy beam for external radiation therapy.

These checks are often time-consuming, and the frequency of portal verification should depend on the treatment technique, the immobilization device in use, and the type and magnitude of setup errors that are expected and tolerated. Digital image enhancement techniques can be used to improve the quality of portal images, and where possible computerized techniques should be used to measure setup errors. An important task is to establish tolerance and action levels for initiating corrections to the patient's position, as well as protocols for determining the magnitude of such corrections. The use of internal and external reference points is of utmost importance throughout the whole treatment verification procedure.

3.2.6. *Follow Up*

Standardized codes of practice for patient follow up including registration of tumor response and acute and late normal tissue reactions are as important for the quality of radiation therapy as a precise patient fixation and dosimetric and portal verification methods. This is particularly important for the overall quality assurance of the therapy procedure, particularly when treatment optimization methods are employed and the side effects are expected to be close to the tolerance level. The aim of the follow up of acute and late normal tissue reactions is therefore to continuously improve the accuracy of established dose–response relations and hopefully generate constructive impact on future radiation therapy (cf. Fig. 3.1). In the case of cancer cure, the follow up is thus not restricted to recording of possible recurrences but is used also in meticulously scoring normal tissue damage several years after the treatment. It is extremely important that, in studies of tumor control and adverse reactions in normal tissues, the correlation between benefit (tumor control) and injury (severe normal tissue damage) is reported since they seem to be statistically dependent end points even if this is not generally assumed to be the case. If well planned, the frequency of follow up may still be quite low in most cases (see Sec. 3.2.4).

3.3. Radiobiological Basis of Radiation Therapy

To accurately quantify the relative merits of all different treatment modalities and treatment techniques, clinically relevant radiobiological response models are needed. For that purpose, it is important to adopt biologically sound cell survival models for tumors and normal tissues to clinically observed dose–response data, so that the model at least describes the response of classical radiation therapy well. As discussed in more detail in the associated radiation biology Chapter 2 (cf. also Comprahensive

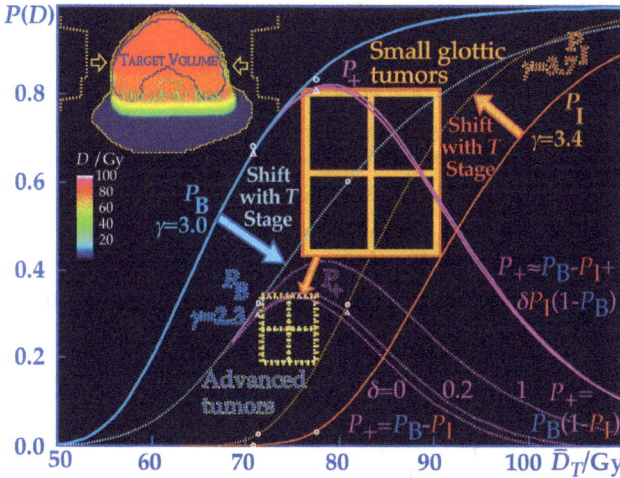

Figure 3.3. The problem of the diminishing therapeutic window in curative radiation therapy of more and more advanced tumors is illustrated. The large arrows illustrate the shift of the tumor control (P_B) and normal tissue complication curves (P_I) as the tumor stage increases from small glottic tumors to larger more advanced tumors. The insert illustrates the rather uniform dose delivery approximately used in obtaining the clinically observed data points.

BioMedical Physics Vol 7 2014 and EOLSS 6.8.4.3: "Radiation Biology") and in a number of recent publications large amounts of clinical dose–response data are continuously being accumulated.

A typical clinical dose–response data set is illustrated in Fig. 3.3, where the dose–response curves for small glottic and advanced head and neck tumors and the associated fatal normal tissue complications are plotted. The mean dose to the tumor, or more precisely the ITV (see Fig. 3.2), is used as independent dose variable. This data set, in a nutshell, illustrates the main problem of radiation therapy of advanced tumors. For the small glottic tumors (solid curves), the therapeutic window between tumor control (P_B and left most sigmoid curve) and fatal normal tissue damage (P_I and right most sigmoid) is very large. For this reason, the probability to achieve complication-free cure (almost 85% at the optimal dose level) is very high. However, for the advanced tumors (dotted curves), the therapeutic window has shrunk to a small fraction of that for the small tumors and the optimal complication-free cure is now as low as about 35%. Figure 3.3 illustrates the case of more or less standard uniform beam radiation treatments where the increased tumor burden for the larger tumors requires substantially larger radiation fields and tumor doses to be eradicated, in accordance with most existing radiobiological dose–response models. Therefore, the tumor control curve (P_B, the probability for beneficial treatment) is shifted to the right to higher doses (left most dotted curve). However, as the larger tumor is irradiated to higher doses, the fatal normal tissue reactions substantially increase and the complication curve (P_I, the probability for fatal injury) moves instead to the left or lower doses. Thus, for advanced disease the therapeutic window is substantially decreased making it very difficult to cure all patients. This is the key problem of curative radiation therapy and we have to solve it in order to cure patients who debut with an advanced form of the disease.

Figure 3.3 also illustrates a few other interesting phenomena that are common in many clinical materials. Not only are the dose–response curves for the larger tumors shifted to larger doses but their dose–response slopes, as expressed by their normalized dose–response gradient γ, are decreased (cf. Chapter 2 and EOLSS 6.8.4.3: "Radiation Biology"). Conversely, the slope is increased for the normal tissues making it even more difficult to achieve a high probability of complication-free cure. This further reduction of the therapeutic window for larger advanced tumors is due to an increased tumor heterogeneity (such as more hypoxic cells, cf. Fig. 2.20) and increased mass of normal tissues being irradiated resulting in more shallow and steeper slopes, respectively.

Furthermore, for the advanced tumor stages, the clinically observed complication-free cure (solid triangles on the middle dashed bell-shaped curve) is lower than what most traditional response models predict (upper dashed bell shaped curve). This is because the traditional models assume that cure and fatal complications are statistically independent processes resulting in a complication-free cure or positive treatment outcome given by the product of the probability for cure and no fatal injury. However, in the clinical material, almost all patients ($\sim$80%) who suffer fatal complications are also cured so the clinically observed P_+ is much closer to the totally correlated response with $P_+ = P_B - P_I$ (lowest dashed bell-shaped curve; see also Sec. 3.4.2 below). The main reason why the tumor control and severe normal tissue complications often are strongly correlated end points in classical clinical trials is that both are exposed to high uniform dose levels, for example, when parallel-opposed beam techniques are being used. However, the correlation may also depend on that most genetic alterations in the normal tissues are preserved in the tumor, such as for example the ataxia telangiectasia-mutated gene, which can make both the tumor and the normal tissues more sensitive than usual. To be able to use clinical dose–response data for treatment optimization it is therefore important to record not only tumor cure and normal tissue damage but also the degree of correlation between these two major end points as it will strongly influence the location of the optimal dose level as seen by the dashed bell-shaped curves in Fig. 3.3.

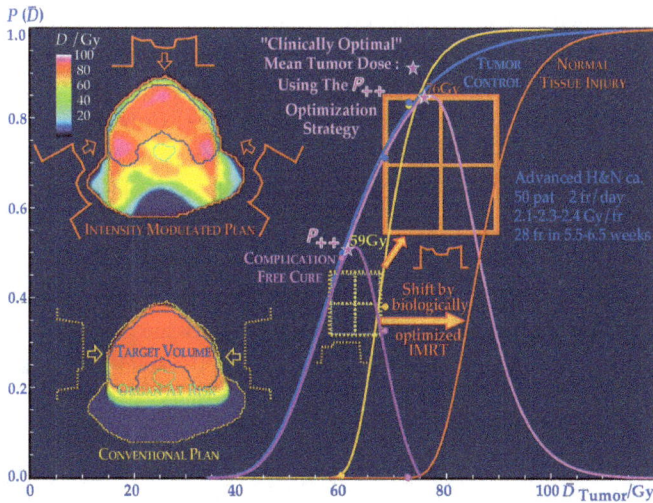

Figure 3.4. Illustration of the possible improvement of the therapeutic window and the clinical outcome for advanced head and neck tumors by using biologically optimized intensity-modulated treatments. By the intensity modulation, the dose to the tumor is increased at the same time as it is reduced to the normal tissues. Therefore, a considerably improved therapeutic window and treatment outcome is obtained by using intensity-modulated dose delivery (solid curves).

We will now illustrate how the merits of the modern biologically based treatment optimization techniques with intensity-modulated beams can be used to solve the problems of advanced tumors, see Fig. 3.4. Again an advanced head and neck tumor similar to that associated with the dotted curves in Fig. 3.3 will be used for comparison. However, it should be pointed out that the response curves in Fig. 3.4 are based on a more uniform subset of the advanced head and neck tumors in Fig. 3.3, since a shorter fixed treatment time was used. Thus, only the dose per fraction was increased always with a fixed total number of fractions. Therefore, the observed steepness of the dose–response curves and the complication-free cure (dotted lines) is higher and the treatment outcome slightly better than in Fig. 3.3. This subset is obviously better suited for use with intensity-modulated beams, which by necessity will give a fixed number of fractions but not necessarily a fixed dose per fraction everywhere, exactly as the case for the data in Fig. 3.4. The intensity-modulated dose distributions have been calculated with one of the most advanced and flexible optimization algorithms available today.

The resultant dose–response relations for intensity-modulated therapy presented in Fig. 3.4 clearly illustrate the considerable therapeutic advantages achieved by using intensity-modulated radiation beams. The inserted isodose diagrams compare in the same slice of a head and neck target: (1) classical and (2) intensity-modulated beams and the associated dose–response curves for the two treatments (dotted and solid curves, respectively). This figure allows a very clear-cut comparison of the two treatment techniques since the common dose axis is in units of the mean dose to the ITV or the tumor. Therefore, the dose–response curves for the tumor will practically be the same in the two cases (left most solid sigmoidal curve). However, the normal tissue complication curves are different since in the intensity-modulated plan the dose to the tumor is significantly increased at the same time as the dose to the critical normal tissues is somewhat reduced. Since all curves are related to the mean tumor dose and has increased over the normal tissue dose by almost 15 Gy compared to the plan with essentially uniform dose, the solid complication curve for the intensity-modulated plan has thus moved about 15 Gy to the right from its location with essentially uniform dose delivery (right most sigmoidal curve).

As a consequence, the therapeutic window has substantially opened up and the bell-shaped curve for complication-free cure (P_+) has increased from almost 50% to 85%. This significant increase in treatment outcome is achieved because biologically optimized intensity-modulated dose delivery results in an increased dose to the tumor at the same time as the normal tissues are spared to a greater extent. By comparison of Figs. 3.3 and 3.4 it is seen that with intensity modulation the treatment outcome for advanced tumors can be brought up to the level of small tumors and uniform dose delivery and even somewhat further, provided the disease is well localized.

This also implies that the classic problem of double trouble in dose escalation (large doses result also in large doses per fraction and thus more severe late normal tissue reactions) is converted to a double or even triple advantage. Not only is the relative dose to the normal tissues reduced but also the dose per fraction and dose rate are also reduced thus substantially reducing the risk for severe late complications. As mentioned, the response curves in Fig. 3.4 were derived with a varying dose per fraction rather than by varying the number of 2 Gy fractions as was done for several of the patients in Fig. 3.3. This data set is therefore ideally suited for use with nonuniform dose delivery where the dose per fraction by definition will vary with the intensity modulation. The comparison in Fig. 3.4 thus presents in a nutshell the advantages that can be gained by biologically optimized intensity-modulated dose delivery.

As a further illustration of the development of radiation therapy, a schematic view of the main steps in the development of treatment techniques that have taken place during the first century of radiation therapy is illustrated in Fig. 3.5. From this figure, it is clear that a lot of things have happened since the dawn of modern radiation therapy in the late 1950s. Obviously, all radiation therapy aims at being conformal: that is to wrap the "therapeutic" isodose as close as possible around the tumor, or

SIX STEPS IN THE DEVELOPMENT OF RADIATION THERAPY

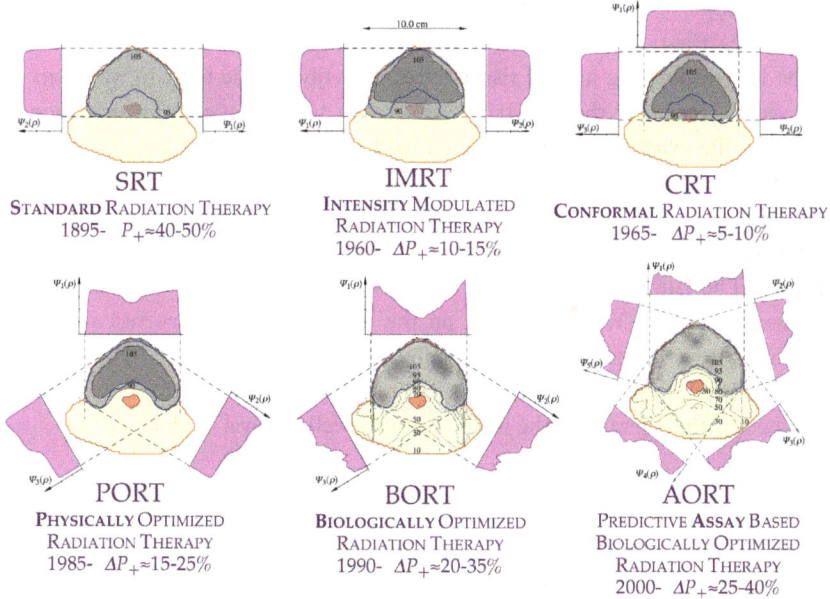

Figure 3.5. Schematic overview of the development of modern radiation therapy techniques. The approximate year of introduction and the resultant increase in complication-free cure, ΔP_+, compared to that of standard radiation therapy are also indicated.

more exactly, around the ITV. In the present review, the term conformal is used to describe this aim similar to the conformation therapy techniques developed in the early 1960s. At the same time, the first optimization algorithms using essentially uniform or wedge-shaped beams were developed by collimation along the projection of the target volume. This is also the method used by most workers today saying that they are doing conformal radiation therapy. However, from Figs. 3.4 and 3.5 it is clear that considerable improvements in treatment outcome are not assured until we enter the area of full intensity-modulation and physically and preferably biologically optimized treatments. It is well-known today that the considerable dose optimization effort that took place from the mid-1960s throughout the 1970s was not so successful due to the low number of free variables used. We really need in the order of thousands of free variables, whereas in the 1960s a few dozen of fields with or without wedges were used. However, when radiobiological dose optimization is employed, it is even possible to reduce the number of free variables, for example, by using few field multi-segmented treatments.

As important as the dose level delivered to the target tissues is the fractionation schedule employed. Considerable advances have taken place here during the past few decades. First, it was understood that large doses per fraction are not tolerated by the late-responding normal tissues that normally set the limit to what can be achieved by radiation therapy. It is well-proven today that this phenomenon is caused by the extended shoulder region of the cell survival curve which is often associated with a large compartment of resting G_0 cells (see Figs. 2.1 and 8.25). Late-responding organs, therefore, have a longer recovery phase after irradiation allowing them to repair a substantial fraction of the sublethal injury induced by each treatment fraction. The late response of such organs may also partly be due to multistep processes where the radiation effect on the microscopic capillary vasculature in addition to direct damage to tissue-specific cells may contribute.

The shape of the shoulder of the cell survival curve at therapeutic doses is often described by the α/β ratio of the linear quadratic cell survival model:

$$s = e^{-\alpha D\left(1 + \frac{D}{\alpha/\beta}\right)}. \tag{3.1}$$

An α/β value of 3 Gy is common for late-responding tissues which means that at 3 Gy the linear cell kill is equal to the quadratic repairable kill associated with the D^2 term. Tumors and acutely responding normal tissues have an α/β of about 10 Gy and large doses per fraction are better tolerated.

It has been known for a very long time that the cell kill is purely exponential at very large doses and not quadratic as Eq. (3.1) indicates. More recently, it has been shown that the linear quadratic model is not so well suited to describe the response at low doses either even though it works quite well in the standard clinical 2 Gy/fraction range. At low doses the cell kill seems to be much steeper than previously realized. The increased sensitivity at low doses as shown in Fig. 3.6 is more pronounced when the shoulder of the cell survival curve is large such as for late-responding tissues with a low α/β value. There are already indications that this low-dose hypersensitivity phenomenon may be clinically relevant since low and high doses per fraction to the skin may cause similar skin reactions even if the total skin doses differ substantially. From the above points of view, a better description of the cell survival curve is obtained by an expression of the type:

$$s = e^{-aD} + bDe^{-cD}, \tag{3.2}$$

where the a term describes the survival of unhit cells and the b term describes the potentially lethal part which can be repaired if the right environmental conditions for repair prevail. The thick solid curve in Fig. 3.6 is from Eq. (3.2) and the dots with error bars are experimental data for lung epithelial cells. By comparing with the dashed linear quadratic curve due to Eq. (3.1), the improved fitting particularly at

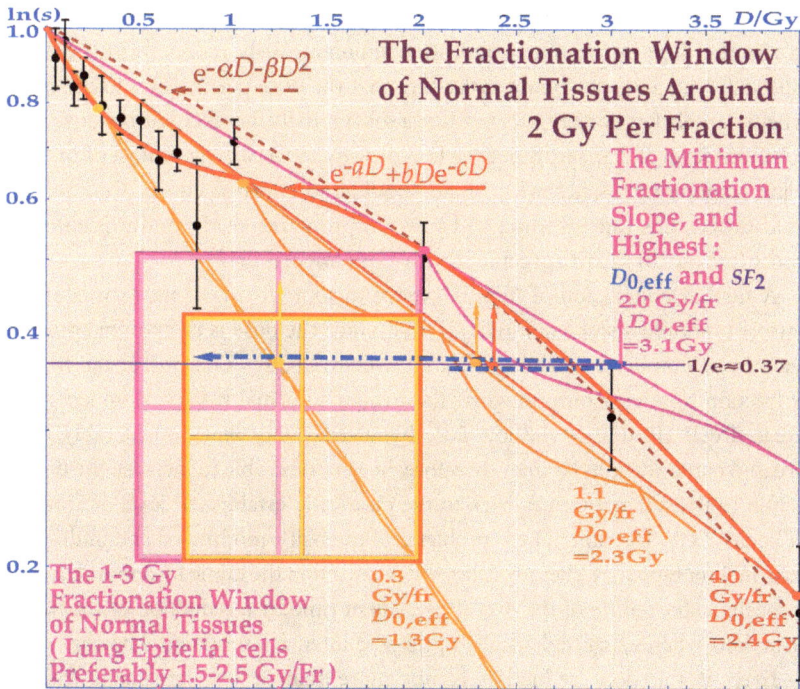

Figure 3.6. Illustration of the fractionation window caused by low–dose hypersensitivity. At low doses and high doses per fraction, the normal tissue damage is increased over that at the standard 2 Gy/fraction level used in classical radiotherapy.

low but also at high doses outside the figure are obvious. A very interesting effect of clinical relevance can be seen in Fig. 3.6 by studying the shallowest possible secant of the survival curve through the origin to find the lowest effective cell kill for a given dose to the tissue in question. For the lung epithelial cells in Fig. 3.6, this is the tangent close to the 2 Gy/fraction dose which means that this dose per fraction is the dose which produces the least possible cell kill per unit dose to this tissue. Thus, the low–dose hypersensitivity not only makes the tissue quite sensitive for low doses per fraction but also forms a fractionation window where the dose level per fraction produces the least possible damage to a given tissue per unit dose when trying to irradiate an underlying tumor. It is very likely that this mechanism is responsible for the observation in most clinical trials that the best possible therapeutic effect is obtained in the dose per fraction range around 2 Gy/fraction.

At higher doses per fraction, the late-responding tissues introduce more complications, whereas at lower doses per fraction the low–dose hypersensitivity may cause more complications as seen in Fig. 3.6.

Thus, when introducing optimized intensity-modulated radiation therapy to increase the dose in the tumor and often simultaneously reduce it in the normal tissues (cf. Fig. 3.4), we should make sure that the dose per fraction to the normal tissues is largely left unchanged to ensure a substantially increased tumor cure and an unchanged level of damage. Since the low-dose hypersensitivity is most pronounced in late-responding tissues with broad shouldered survival curves ($\alpha/\beta \gg 3$) one would expect this phenomenon to be not so pronounced in rapidly proliferating tumors ($\alpha/\beta \approx 10$) in fair agreement with experimental data.

A further conclusion of high relevance to some forms of modern conformal treatments should also be pointed out here, since the dose is then often spread out over a large number of beam portals all over the body. This will make the dose per fraction to a large percentage of the patient's normal tissues quite low and an increased cell kill due to the low-dose hypersensitivity may considerably induce increased complications. It may, therefore, be very desirable to increase the dose per fraction to the normal tissues back to the classically established level of around or just below 2 Gy/fraction. These problems of intensity-modulated and multi-portal conformal therapy may, therefore, largely be solved in the clinic by one and the same approach: a shortening of the overall treatment time by a reduction of the number of fractions when using biologically optimized intensity-modulated beams.

Even if the dose reduction is not more than a few treatment fractions, it represents a new kind of advantage since the effect of accelerated tumor repopulation that often sets in after the first 2–3 weeks of treatment will be less of a problem when the total treatment time is reduced. Often, the delivered dose could be reduced by 0.4–0.6 Gy/day of shortening of the total duration of the treatment compared to the reference situation of 30 fractions in 6 weeks. Alternatively, if the total dose is left unchanged, an equivalent tumor boost of about 0.5 Gy times the number of days of treatment time reduction is obtained.

In conclusion, there are basically the following five kinds of improvement of the treatment outcome possible when using biologically optimized intensity-modulation:

(1) increase in the tumor dose and dose rate;
(2) increase in the tumor dose per fraction;
(3) unchanged or slightly reduced dose per fractionation to the normal tissues;
(4) a reduced overall treatment time and an associated additional gross tumor boost; and
(5) less work in the clinic as fewer treatment fractions need to be delivered.

All of these advantages can simultaneously be effective at a varying degree or all the advantages may be focused on one or two of them making the remaining largely unaffected leaving considerable room for advanced radiobiological treatment optimization such as minimizing the total duration of the treatment.

It is often claimed by physicians that radiation therapy in the future will be replaced by molecular biology approaches such as gene therapy. Surely considerable improvements could be seen in these areas. However, they will also augment the power of radiation therapy as we will be able to use the new genetic markers for radiation responsiveness and radioresistance. Furthermore, gene therapy will probably always be tough on large bulky solid tumors and could gain considerably by combination with radiation therapy. We should, therefore, also expect considerable improvements of radiation therapy when these new tools are coming into clinical use as indicated by the last step of development in Fig. 3.3. It is likely that patient individual predictive assay based on molecular or cellular methods when combined with biologically based therapy optimization can improve the treatment outcome some further 5–15% compared to just using the average value of the response over a population, at least for locally advanced tumors. Furthermore, the simultaneous development of the different adjuvant therapies to radiation, such as surgery, chemotherapy, targeted antibodies, receptors or genetic markers as well as gene therapy hyperthermia, and US etc., will allow further improvements (see Chapter 2.3 and EOLSS 6.8.4.3: "Radiation Biology").

3.4. Radiation Therapy Objectives and Target Definition

3.4.1. *Quality of Life*

The primary objective of radiation oncology, and for all medical care for that matter, is to make the quality of life for the patient already from his or her first contact with the medical system as high as reasonably achievable. In principle, some kind of integral measure of the *quality of life* would be a suitable objective function since a long healthy and comfortable life, accounting also for all steps of medical care, is most desirable. However, a very long life with severe loss of life quality is not generally desirable, so some kind of nonlinear weighted integral measure should, therefore, be most desirable. Obviously, the quality of life of a person has to be a subjective quantity, but from a medical point of view it also has a more strict and objective side as a general health index. This index can range from unity corresponding to perfect health to zero when the person in medical terms is dead or does not want to live anymore. There are several measures which are aimed at quantifying the condition of the patient, such as the Karnofsky status, but all of them are far too complex and difficult to relate to therapeutic actions to be used directly for therapy optimization. We will, therefore, first discuss some simplifications that need to be introduced to get an objective function that is useful for computed radiation therapy optimization.

In practice, two main groups of objective functions, physical and biological, have been used. The most commonly employed objective functions are physical and they often describe the properties of the delivered dose distribution in the *target volume* and affected normal tissues or *organs at risk*. In fact, the definition of these latter concepts is extremely important for how accurate and precise a treatment can be performed as illustrated in Fig. 3.4. It is fundamental that the target volume and the organs at risk are defined with narrowest possible safety margins in relation to the anatomical reference points that are used for setting up the patient for external beam therapy (cf. Fig. 3.2). In principle, the target volume should include all the volumes and margins whose dimensions cannot be affected by changing or developing dose planning or treatment techniques (including fixation of the patient). Therefore, the target volume is generally obtained by adding only an anatomical margin to the oncological volume to account for all internal shape changes of organs and their possible motions relative to the anatomical reference points.

To this target volume, set-up margins may later have to be added depending on the treatment technique and the quality of the treatment unit. It is fundamental that precisely this target volume should be used for dose prescription and reporting, as it is the most relevant one for the treatment outcome. The biological objective functions aim at quantifying precisely this quantity, namely the probability that the patient will have a desirable treatment outcome. From this point of view, the radiobiological objective functions, therefore, do quantify the quality of life of the patient after therapy as will be discussed in more detail now.

3.4.2. *Biological Objective Functions*

There are a large number of factors that by necessity make the biological objective functions statistical quantities. First, the vital end point of killing all clonogenic tumor cells to eradicate the tumor makes the beneficial treatment outcome "tumor control' a stochastic quantity. We can only state the probability, P, and standard deviation, σ, for a certain treatment outcome due to the large uncertainty in hitting the very last tumor clonogen.

In the space of clonogenic tumor cells, the treatment objective can be formulated as: the maximum value of the tumor recurrence probability should be as low as possible throughout the target volume without causing severe damage to normal tissues. This formulation is quite interesting because it solves the inversion problem in a simple way, but it requires information about the density and sensitivity of clonogenic tumor cells. Since tumor eradication in a group of patients is a multidimensional truly binomial process (a given patient can either be cured or not), it is straightforward to calculate the relative standard deviation of the number of cured patients $\sigma_{P_B} = (1 - P_B)P_B$, where P_B is the probability to cure each one of them. The uncertainty has thus its highest value of 25% when P_B is 50% as should be expected.

On top of this uncertainty, there are uncertainties both in defining the target volume, so that all tumor clonogens are really included, and in dose delivery, so that the target volume is fully irradiated. Furthermore, for a given patient and tumor classification (e.g., according to the TNM system, to minimize the influence of the extent of the disease) it is generally not known whether a given tumor is more sensitive or resistant than that of the "mean patient" for which established dose–response relations should be applicable. The same is true for the normal tissues of that patient and both these facts add to the total uncertainty.

For these reasons, it is also generally assumed that the probability to severely injure the patient P_I is statistically independent process from the probability of beneficial treatment outcome P_B, that is, tumor control. The probability of a successful treatment, P_+, can then be expressed since the covariance of P_B and P_I is zero and the product law of statistics can be applied to determine the conditional probability of having tumor control without severe injury:

$$P_+ = P_B - P_B \cap P_I \approx P_B(1 - P_I), \tag{3.3}$$

which is the traditional expression used by most workers — the longer notation NTCP and TCP are sometimes used instead of P_I and P_B, respectively. However, there are a number of factors that can alter this conclusion since it is strongly dependent on the type of dose delivery that is employed and also on whether there might be a true biological correlation between P_B and P_I (cf. Fig. 3.3).

There are several biological mechanisms that can cause such an effect, for example, if the patient happens to have an unusually efficient repair system for double strand. Since most oncogenes are different from the genes responsible for efficient repair of radiation damage, it is probable that both the tumor and the normal tissue will be more resistant to irradiation. Conversely, if the patient happens to have the ataxia telangiectasia gene, which is one of several known genetic defects quite common in cancer patients, there would be an increased risk for normal tissue injury but also an increased probability to control the tumor. Furthermore, some tumors which show an increased risk for metastatic dissemination might be phenotypically distinct and their presence may be used as a predictive indicator which could influence the treatment decision.

However, there are also other genes such as the tumor suppressor gene p53 which has a regulatory effect on how DNA damage is handled by the cell. A mutant p53 gene is one of the most common defects, which is found in tumor cells. This may be an important molecular biological explanation why for such tumor radiation therapy has a more severe effect on the tumor cells than on healthy normal tissues with the wild-type p53 which is better capable of handling radiation-induced DNA damage (see EOLSS 6.8.4.3: "Radiation Biology").

Depending on the type of dose delivery and the proportions of genetic variations for the tumor at hand, deviations from the simple expression (3.3) may, therefore, be seen. It is, therefore, advantageous to base the analysis as far as possible on clinical data. In a recent study on head and neck tumors, a rather uniform dose distribution was used (parallel-opposed beams). It was found that P_B and P_I was totally correlated at low doses, and only for large tumors at high doses, the uncorrelated portion δ reached a value of about 0.2. The clinically observed P_+ value was then accurately described by:

$$P_+ = P_B - P_I + \delta(1 - P_B)P_I, \tag{3.4}$$

for clinically relevant doses as illustrated by the clinical data for large head and neck tumors in Figs. 3.3 and 3.4 above. Figure 3.7 clearly shows how P_+ varies across a tumor normal tissue interface and even if no tumor is seen by PET-CT imaging there may be need for as much as 80% of the GT dose to minimize the recurrence probability at the visible tumor border at least for tumors with an invasive growth pattern. Interestingly, the peak recurrence probability with optimal dose delivery occurs where there is about 30% normal tissue stroma and 70% tumor tissue, assuming the stroma is in an organ at risk.

3.4.3. *Physical Objective Functions*

To allow a strict optimization, the objective function should be a scalar quantity (more precisely a functional) as otherwise two different treatments cannot be compared on the same scale. This property is also true for the radiobiological objective functions presented above. However, to speed up calculations, further simplifications of the objective function are sometimes desirable. This can be achieved by expressing the treatment objectives as simple functions of the dose distribution but also other parameters such as the total treatment time have been used. The success of the dose optimization is directly linked to the ability to define such clinically relevant treatment objectives, and from this point of view the

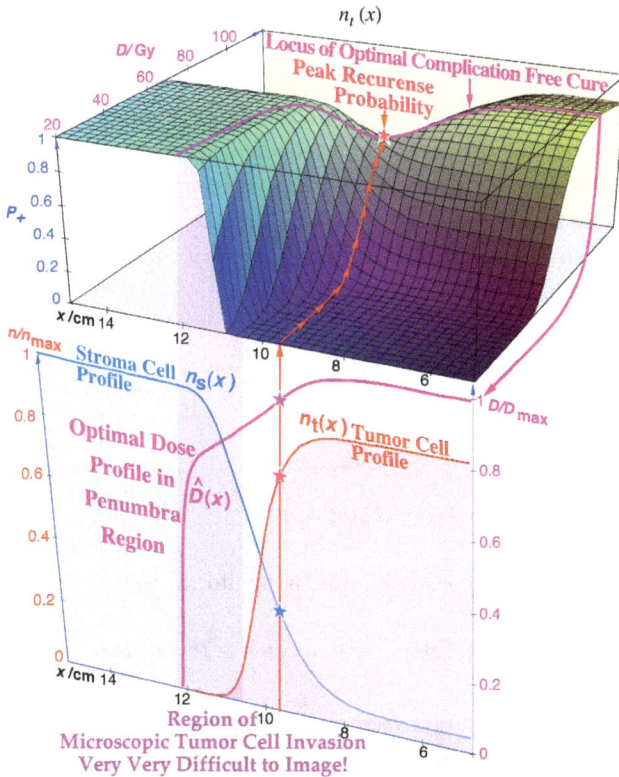

Figure 3.7. Illustration of the P_+ variation for a possible infiltrative pattern of the tumor cell density, $n_t(x)$, and normal tissue stroma cell density, $n_s(x)$, profiles. The dose profile in the penumbra region that maximizes the complication-free cure is indicated. It is seen that even if the tumor cell density is very low, the dose should still be about 80% of that in the GT here for simplicity assuming a well-oxygenated case.

radiobiological objectives are generally most advantageous. The following physical features of importance for tumor control and normal tissue complications are probably the most important alternatives and have been used extensively over the years:

(1) Mean energy imparted to target volume and organs at risk $(\bar{\varepsilon}, \bar{D})$.
(2) Dose variance over target volume (σ_D).
(3) Minimum absorbed dose in target volume (D_{min}).
(4) Peak dose to organs at risk (D_{max}).
(5) Conformity of treatment volume to target volume.

3.4.3.1. *Dose variations and tissue infrastructure*

3.4.3.1.1. Parallel organ

Assume that a parallel organ has the probability P to respond at a dose D. Let us further assume that the organ can be subdivided into two compartments. The response probability, P_p of a single parallel subcompartment is then given by $P + p = \sqrt{P}$ so that $P_p^2 = P$. If, however, a dose $D + \Delta D$ is given to one of the compartments and a dose $D - \Delta D$ is given to the other compartment, the responses of the corresponding compartments are given by:

$$P_p(D + \Delta D) = P_p(D) + \gamma \frac{\Delta D}{D} = \sqrt{P} + \Delta P, \tag{3.5}$$

and

$$P_p(D - \Delta D) = P_p(D) - \gamma \frac{\Delta D}{D} = \sqrt{P} - \Delta P, \tag{3.6}$$

respectively. The total response with this "nonuniform" irradiation is thus:

$$\left(\sqrt{P} + \Delta P \right) \cdot \left(\sqrt{P} - \Delta P \right) = P - (\Delta P)^2.$$

Dose variations will thus always decrease the response for a tumor or an organ with parallel architecture and homogeneous radiation sensitivity over its volume.

3.4.3.1.2. Serial organ

Assume that a serial organ has a response probability P at a dose D. It is again assumed that the organ is subdivided into two compartments. The response probability, P_s of a single subcompartment is then given by $P_s = 1 - \sqrt{1 - P}$ so that $P = 1 - (1 - P_s)^2$. If again one of the compartments receives a dose $D + \Delta D$ and the other one receives a dose $D - \Delta D$, the responses of the corresponding compartments are given by:

$$P_s(D + \Delta D) = P_s(D) + \gamma \frac{\Delta D}{D} = 1 - \sqrt{1 - P} + \Delta P. \tag{3.7}$$

and

$$P_s(D - \Delta D) = P_s(D) - \gamma \frac{\Delta D}{D} = 1 - \sqrt{1 - P} - \Delta P, \tag{3.8}$$

respectively. The total response with this nonuniformly irradiated serial organ then becomes:

$$1 - (\sqrt{1 - P} + \Delta P) \cdot (\sqrt{1 - P} - \Delta P) = P + (\Delta P)^2.$$

Dose variations will, thus, always increase the response of organs with serial architecture.

This means that if a certain dose is delivered to an organ at risk with serial architecture, it should be delivered so that the dose is as low as possible and spread evenly over the organ in order to minimize the complications. On the contrary, if the organ has a parallel architecture, the dose should be delivered through a small part of the organ to eliminate the risk of complications.

For the evaluation of a treatment plan, the mean dose, $\overline{D}$, and standard deviation of the dose distribution delivered to the tumor or more exactly the ITV are often used clinically. However, these data do not really take the biological characteristics of the targets into account. Furthermore, when different plans are compared, the effect of the dose distributions on the tumor is only in the first and second order related to $\overline{D}$ and σ_D, respectively. The effective dose, D_{eff} (Brahme 1994) as suggested in the recent NACP report (Aaltonen *et al.* 1997) is then an alternative. Dose distributions within organs or volumes of interest are never exactly uniform. On the contrary, they can be strongly nonuniform especially for normal tissues. According to D_{eff}, for relatively small dose variations, the effect in the target is well related to the mean target dose, whereas for larger dose inhomogeneities, the minimum target dose is more closely related to the effective dose (Brahme 1994). The equivalent uniform dose (EUD) assumes that any two dose distributions are equivalent if they eradicate the same fraction of clonogenic cells (Niemierko 1997).

Both of these concepts provide a method to account for the biological effects when reporting the absorbed dose. However, they do not apply to all clinical cases since they do not accurately deal with complex targets or organs at risk. Furthermore, they do not provide a common prescription basis for different dose plans. The biologically effective uniform dose, $\overline{\overline{D}}$, introduced by Mavroidis *et al.* (2001) is the uniform dose that causes exactly the same tumor control or normal tissue complication probability as the real dose distribution on a complex target or normal tissue case (Mavroidis *et al.* 2000). In complex patient cases, multiple targets or multiple organs at risk are involved. The biologically effective uniform dose is denoted by $\overline{\overline{D}}$, indicating that it has been averaged both over the dosimetric and the biological information of the patient.

3.4.3.2. *Effective dose, D_{eff} and EUD*

When the desired dose distribution, $D(r)$, in the target volume is known, it may also be possible to quantify the difference between it and the best achievable dose distribution, $\hat{D}(r)$. In principle, one could therefore apply the pth norm on this difference to quantify deviations at all points r_i of interest according to:

$$\Delta p = \left(\sum_i |D(r_i) - \hat{D}(r_i)|^p \right)^{1/p}. \tag{3.9}$$

The elliptic norm may also be of interest in this context since it can be regarded as a generalization of Δ_2 ($p = 2$):

$$\varepsilon_A^2 = d^T A d, \tag{3.10}$$

where A is a positively definite matrix. For the simple case when $A = I$, the unity matrix, the elliptic norm is precisely equal to the second norm. The second norm is of special clinical interest since it can be shown to be related to the probability of achieving local control for a homogeneous tumor. By equating the response probability of a tumor or normal tissue, P_B or P_I (B denotes benefit, whereas I refers to injury) for an arbitrary dose distribution $D(\vec{r})$ with the effect of a fixed effective dose level D_{eff} (Brahme 1984), the following expression can be derived:

$$P_B(D_{eff}) \equiv P_B(D(\vec{r})) \approx P_B(\overline{D}) - \frac{\gamma^2}{2P(\overline{D})} \left(\frac{\sigma_D}{\overline{D}} \right)^2 \tag{3.11}$$

$$P_I(D_{eff}) \equiv P_I(D(\vec{r})) \approx P_I(\overline{D}) + \frac{\gamma^2}{2(1 - P(\overline{D}))} \left(\frac{\sigma_D}{\overline{D}} \right)^2,$$

where

$$\overline{D} = \frac{\int_V D(\vec{r}) dm}{\int_V dm} \quad \text{and} \quad \sigma_D^2 = \frac{\int_V (D(\vec{r}) - \overline{D})^2 dm}{\int_V dm}$$

dose distribution (and thus the second norm) are therefore closely related to the clinical outcome as seen from the expression for the probability to control the tumor. In fact, it can be shown that $\overline{D}$ is also a very suitable quantity for predicting complication-free tumor control and there are also clinical data supporting this. In addition, the use of $\overline{D}$ minimizes the second norm (Δ_2) which is a very valuable property according to Eq. (3.11). For tissues with a steep threshold type response, the infinite norm, Δ_∞, may also be of interest as it is a way of quantifying the maximum deviations.

The above relationships of $P_B(D(\vec{r}))$ and $P_I(D(\vec{r}))$ were used to derive the corresponding effective doses, D_{eff} according to:

$$D_{eff}^B = \overline{D}\left[1 - \frac{\gamma}{2P(\overline{D})}\left(\frac{\sigma}{\overline{D}}\right)^2\right] \tag{3.12}$$

$$D_{eff}^I = \overline{D}\left[1 + \frac{\gamma}{2(1 - P(\overline{D}))}\left(\frac{\sigma}{\overline{D}}\right)^2\right],$$

where $\overline{D}$ is the mean dose delivered to the tumor, γ is the steepest normalized gradient of the dose–response curve, $\sigma_D/\overline{D}$ is the relative standard deviation of the delivered dose distribution, and $P(\overline{D})$ is the probability of local control at the dose level, $\overline{D}$. This expression shows clearly that for parallel tissues such as tumors, the effective dose decreases below $\overline{D}$ as soon as the relative standard deviation of the dose distribution increases. This implies that for relatively small dose nonuniformities (small standard deviation), the dose effectively delivered to the target can be approximated by the mean target dose, though for large dose inhomogeneity the minimum target dose is closer related to the expected clinical effect (Brahme 1984). The mean target dose approach assumes that doses above the mean target dose compensate for doses less than the mean target dose. The minimum target dose approach indirectly impies that a cold spot cannot be compensated by any dose delivered to the rest of the target volume which normally is not true since the extra dose will always reduce the risk for a local relaps.

A related concept that was introduced by Niemierko (1997) is the *EUD* which assumes that any two dose distributions are equivalent if they cause the same radiobiological effect in terms of surviving clonogenic cells. The EUD results in the survival of the same fraction of clonogenic cells as the true delivered dose distribution.

$$S(EUD) \equiv S(D(\vec{r})). \tag{3.13}$$

Under special conditions, the above definition can be expressed in the following form:

$$EUD = D_{ref}\ln\left[\frac{1}{N}\sum_{i=1}^{N}(SF_2)^{D_i/D_{ref}}\right]\Bigg/\ln(SF_2), \tag{3.14}$$

where $D_{ref}(= 2\,\text{Gy})$ is the dose per fraction, which is related to the surviving fraction SF_2 and N is the number of dose calculation points.

All of these suggestions have the goal to provide an appropriate basis for reporting the biological evaluation of a treatment plan. Moreover, they try to provide a common dose scaling base for treatment plan comparison. That is because in a response diagram, the control curve is plotted at the same or almost at the same position for all of the different treatment plans. However, the D_{eff} and EUD concepts were not designed for cases that involve more than one target or organ at risk.

3.4.3.3. *Biologically effective uniform dose, $\overline{\overline{D}}$*

The biologically effective uniform dose, $\overline{\overline{D}}$ is the uniform dose that causes exactly the same total tumor control or normal tissue complication probability as a given nonuniform dose distribution on a complex patient case (Mavroidis *et al.* 2000). The notation $\overline{\overline{D}}$ indicates that the quantity has been averaged over both the dosimetric (dose distribution) and the biological (dose–response relations) information of the complex case. The general expression of $\overline{\overline{D}}$ is defined for a given tumor or tissue from its dose–response relation without dependence on the radiobiological model used and it is then given by:

$$P(\overline{\overline{D}}) \equiv P(D(\vec{r})). \tag{3.15}$$

$\overline{\overline{D}}$ is based on the mean value theorem, which assumes that if one function $f(x)$ is monotonic in the interval $[a, b]$, then there exist a value $a \leq \xi \leq b$ such that

$$\prod_{x=a}^{b} f(x)^{\Delta x} = f(\xi)^{\sum_{x=a}^{b} \Delta x}. \tag{3.16}$$

For normal tissues of parallel infrastructure or tumors, ξ is located in the interval $[a, (a + b)/2]$, whereas for serially organized tissues $\xi \in [(a + b)/2, b]$ (cf. Appendix A, Mavroidis *et al.* 2001). In the case of a target volume consisting of different well-defined gross target volume (GT) and positive lymph node (LN) volumes, $\overline{\overline{D}}$ can be derived from the following expression:

$$P_{\mathrm{B}} \equiv \prod_{i} P_{\mathrm{B}}^{i}(\overline{\overline{D}}) = \prod_{i} P_{\mathrm{B}}^{i}(D(\vec{r}))$$
$$\Rightarrow P_{\mathrm{GT}}(\overline{\overline{D}})P_{\mathrm{LN}}(\overline{\overline{D}}) = P_{\mathrm{GT}}(D(\vec{r})) \cdot P_{\mathrm{LN}}(D(\vec{r})), \tag{3.17}$$

where $D(\vec{r})$ denotes the real dose distribution. The presented expression is more general since it can deal with cases where multiple targets with different biological parameters are involved or even with cases where the radiation sensitivity of the tumor varies over its volume (e.g., in the presence of hypoxic cells).

For example, if the response probability is estimated using the linear-quadratic-Poisson model, $\overline{\overline{D}}$ can be derived from the following expression:

$$\prod_{j=1}^{M_{GT}} \left(e^{-e^{\varepsilon\gamma_{GT} - \alpha_{GT}D_j - \beta_{GT}D_j^2/n}} \right)^{1/M_{GT}} \prod_{j=1}^{M_{LN}} \left(e^{-e^{\varepsilon\gamma_{LN} - \alpha_{LN}D_j - \beta_{LN}D_j^2/n}} \right)^{1/M_{LN}}$$

$$= e^{-e^{\varepsilon\gamma_{GT} - \alpha_{GT}\overline{\overline{D}} - \beta_{GT}\overline{\overline{D^2}}/n}} \cdot e^{-e^{\varepsilon\gamma_{LN} - \alpha_{LN}\overline{\overline{D}} - \beta_{LN}\overline{\overline{D^2}}/n}} \tag{3.18}$$

A better overview of the characteristics of the different concepts is shown in Fig. 3.8 where their behavior is studied for a range of relative standard deviations. The diagrams refer to the case of an organ that is irradiated with a step-wise dose distribution. One half of the organ receives a dose above the mean (80 Gy) and the other half receives a dose below that mean. The difference between the two dose levels follows the range of the relative standard deviations shown. In the case of $\overline{\overline{D}}_2$, a three-step dose distribution was applied. The $\overline{\overline{D}}$ for the relative seriality model, which is more relevant to normal tissues, moves toward the maximum dose of the distribution, whereas for the parallel model, which is suitable for tumors, it

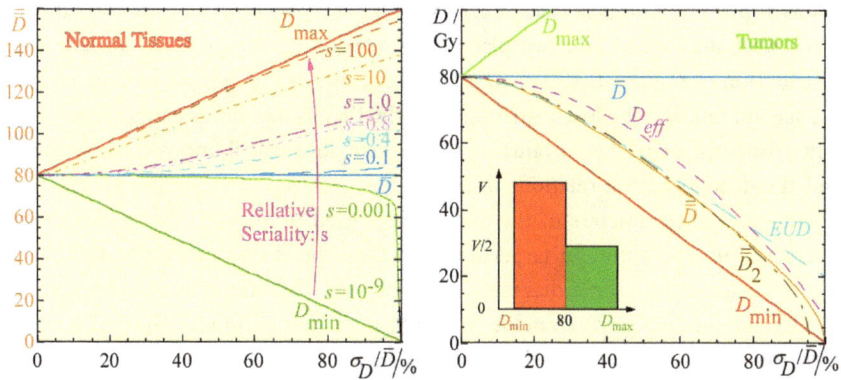

Figure 3.8. The diagrams illustrate the behavior of the different dose report concepts for the case of a target, which is irradiated with a step-wise dose distribution having the same mean but varying relative standard deviation. $\overline{\overline{D}}$ is related to the minimum dose for parallel tissues such as tumors at large dose variations. For organs of high seriality, which is more relevant to normal tissues, $\overline{\overline{D}}$ seems to be better related to the maximum dose. D_{eff} behaves similarly but the curves of $\overline{\overline{D}}$ and D_{eff} differ from each other in their absolute values. All these different dose distributions are characterized by the same mean value implying that $\overline{\overline{D}}$ is not an appropriate unit to describe a dose distribution. The curve of EUD follows closely that of $\overline{\overline{D}}$ apart from the region of low doses where EUD does not give clinically acceptable results.

approximates the minimum dose. Every point of these curves actually refers to a different dose distribution, each of which has a different dose variation. Although all of these distributions have different response probabilities, they have the same mean dose $(\overline{D})$, which means that the mean dose cannot be a good descriptor in biological treatment planning. As it is denoted by the curves of $\overline{\overline{D}}$ and $\overline{\overline{D}}_2$, two different dose distributions can have the same dose variation $(\sigma_D/\overline{D})$ but different response probabilities. This is against the assumption of the D_{eff} concept, which supposes that every dose distribution with the same mean dose and relative standard deviation should have the same response probability. Finally, the EUD concept is not a good biological dose descriptor for low doses. This is because when part of the tumor receives zero dose, then the control probability should be zero. Consequently, the EUD value should also be zero, which is a clinically relevant value, rather than the positive value given by EUD. This problem stems from the fact that Poisson statistics cannot describe well these cases. To come around this problem one should derive the $\overline{\overline{D}}$ expression from, for example, the linear-quadratic-binomial model (Mavroidis *et al.* 2002b). However, it can be proven that $\overline{\overline{D}} \approx D_{\text{eff}} \approx EUD \approx \overline{D}$ for small variation of the dose distribution.

To illustrate the significance of the $\overline{\overline{D}}$ concept in plan reporting and comparison, it was applied to a clinical case of cervix cancer for which two different treatment plans were produced. The two plans were selected on the grounds of their dose distribution characteristics. The first treatment plan is much more conformal than the second one and its dose distribution in the ITV is much more inhomogeneous. The treatment plans were evaluated using their dose distributions to the patients and the dose–response relations of the organs involved in this clinical case. The radiobiological parameters of the organs were used to estimate the normal tissue tolerance and the optimum target dose. Plotting the curves of P_B, P_I, and P_+ of the two plans in the same diagram, it can be seen that the corresponding curves of ITV for the two plans coincide (Fig. 3.9, lower left diagram). Since this is the case, the curves of the P_I are the ones that decide which plan is superior from the radiobiological point of view. This is also shown by the curve of P_+, which confirms that P_+ is an objective that depicts the quality of a treatment plan. It has to be mentioned that the curves of P_B of the two plans coincide because $\overline{\overline{D}}_{\text{ITV}}$ has been used as the unit of the dose axis. That is because $\overline{\overline{D}}$ is derived in a way, which depends only on the radiobiological characteristics of the targets involved in the clinical case. In the lower right diagram of Fig. 3.9, $\overline{\overline{D}}_I$ was used to make the total response curves of the organs at risk coincide for the two plans. Then, for the same complication rate the treatment configuration achieving better total control can be found accordingly. Consequently, it is shown that $\overline{\overline{D}}$ provides the proper dose prescription basis for

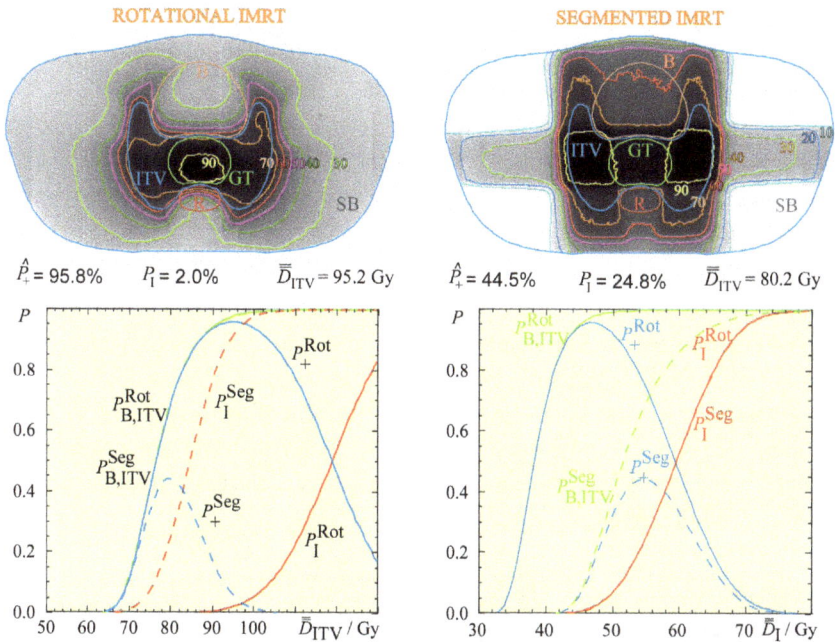

Figure 3.9. The curves derived from the radiobiological evaluation of two treatment plans are plotted in the same diagram using $\overline{\overline{D}}$ on the dose axis. In the lower left diagram where the $\overline{\overline{D}}_{ITV}$ is used as the dose scaling unit, it is shown that the curves corresponding to the response of the ITV ($P_{B,ITV}$) coincide. This way the response curves of the organs at risk determine which plan is superior. In the lower right diagram, $\overline{\overline{D}}_I$ is used as the dose level unit making the response curves, which correspond to the total complication probabilities, lie at the same position for both of the plans. Consequently, by comparing the control curves of the two plans, the superior one can be determined. These diagrams also illustrate the value of the P_+ objective as an evaluator. Such as a dose–volume histogram (DVH) chart is a good illustration of the volumetric dose distribution delivered to the patient, so is the biological evaluation plot of a dose plan a good illustration of the expected clinical outcome. Dose–response diagrams together with the corresponding dosimetric diagrams give a more complete picture of the delivered treatment.

comparing treatment plans through the evaluation of the biological effects of the delivered dose distribution.

Radiobiological evaluation of treatment plans should provide a closer association of the delivered radiation therapy with the clinical outcome. This can be achieved by taking into account the dose–response characteristics of the irradiated targets and normal tissues involved in the clinical case. The simultaneous presentation of the radiobiological evaluation together with the physical data shows their complementary relation in analyzing a dose plan. The use of radiobiological parameters is necessary if a clinically relevant quantification of a plan is needed. The concept of $\overline{\overline{D}}$ was developed to provide a proper dose prescription basis for comparing treatment plans through evaluation of the biological effects of the delivered dose distribution. The concept stems from basic radiobiological principles. It is very simple and easy to use for reporting and comparing different dose plans during treatment planning. The application of the $\overline{\overline{D}}$ concept on the representative treatment plans of cervix cancer revealed its significance in comparing them. $P_+|\overline{D}$ can be a better dose measure of treatment outcome of radiotherapy than several other strictly dosimetric measures commonly used. A comparison of the $\overline{\overline{D}}$ concept with other relevant dose prescription concepts for single and multiple ITV targets revealed its valuable properties. The definition of the biologically effective uniform dose is model-independent and can be used as a suitable dose axis in dose–response diagrams. These diagrams can be considered as the radiobiological version of the extensively used DVH diagrams.

3.4.4. *Degrees of Freedom in Radiation Therapy Dose Delivery*

Beside the selection of radiation modality such as electrons, photons, or protons, the degrees of freedom include dose fractionation schedule, beam energy, beam directions, beam collimation, beam intensity profiles, and the irradiation technique in general as determined by the type of equipment used (see Chapter 5.2.3 for more details!).

Classical conformation therapy as developed in the late 1950s and early 1960s used a continuum of generally uniform or blocked beams conforming to the shape of the target volumes and the organs at risk as seen from the point of view of the beam source (Figs. 3.2 and 3.5). The ultimate step in the therapy development is to allow full freedom in the shape of the delivered beams both with regard to beam energy, beam direction, and beam profiles, as illustrated in the lower right panel in Fig. 3.5. These new degrees of freedom of what might be called generalized conformal or biologically optimized radiation therapy will also allow an accurate shaping of the dose delivery for very complex concave and heterogeneous target volumes. However, such treatments can be quite complicated both to plan and deliver, so a more practically oriented treatment optimization often requires some treatment parameters or degrees of freedom to be locked to make planning and dose delivery practical also at a small clinic. For example, the beam energy or directions of incidence may be preselected to clinically obvious values for the target at hand.

For many simple target volumes, few field techniques with uniform beams are quite sufficient, whereas for more complex shapes nonuniform dose delivery is generally much more advantageous. In fact, it can be shown that the classical conformation therapy method with uniform beams is almost equal to the fully optimized generalized conformal method only for the special case of homogeneous circular symmetric target volumes without organs at risk. For most other target volumes and normal tissue configurations, nonuniform dose delivery will be clearly advantageous.

The number of degrees of freedom and the number of free variables that have to be managed with different irradiation techniques are most easily understood by using the concept of *energy deposition kernels*. The purpose of the energy deposition kernel is to describe, as accurate as possible, the energy deposition by a narrow elementary radiation beam or pencil beam. The kernel should account for all possible radiation effects arising from the interaction of the beam, such as scattering and absorption of secondary electron production. The most elementary kernel and the building block for most composite kernels is the point monoenergetic and monodirectional pencil beam energy deposition kernel which describes the spatial distribution of the mean energy imparted by a narrow particle beam of energy E and direction Ω. By combining an array of such kernels of varying energy and intensity composite energy deposition distribution for intensity-modulated beams can be defined (see: Secs. 3.2.2 and 3.4.1).

3.4.5. *Biologically based Dose Level Optimization in Intensity-modulated Radiation Therapy*

3.4.5.1. *Optimal dose variation in the ITV*

Multileaf collimators (MLCs), pencil beams, and other modern techniques have the ability to deliver very conformal treatments that were not possible a few years ago. There is a continuous progression and evolution of the technological capabilities in delivering better radiation treatments. Simple plans that use only open fields are substituted by plans that use wedges and blocks, which are furthermore substituted by plans that use MLCs, dynamic blocking, and pencil beams. Intensity-modulated external radiation therapy can today produce very conformal distributions that may result in good treatment outcome eliminating the risk for complications in the sensitive normal tissues. The beneficial aspects of the quality margin between the modern intensity-modulated treatments and the conventional practice can be quantified and optimized by the use of the dose–response relations of tumors and normal tissue (see Chapters 5.4 and 5.5 for further details!).

Radiobiological evaluation of treatment plans allows consideration of the radiosensitivity variation within the patient. Generally, parallel presentation of the radiobiological evaluation together with the physical data would help in showing their close association. However, the use of radiobiological parameters is necessary if a clinically relevant quantification of a plan is needed. These parameters incorporate the information that is intrinsically related with the patient into the treatment plan evaluation. The physicians and the medical physicists may benefit from radiobiological modeling and introduce new ways in the clinical way of thinking. This is because radiobiological analysis combines the physical criteria that medical physicists use in their evaluation with the biological and clinical criteria that are employed by the physicians.

Treatment plans generally use many different beam portals to distribute the dose to a large volume of the normal tissue stroma taking advantage of its usually parallel structural organization. When an ITV (usually consisting of a GT and the involved LNs) is assumed to have homogeneous radiosensitivity, treatment plans try to deliver a high uniform dose to the whole ITV. However, when high levels of cellular inhomogeneity in the ITV are present, treatment plans are designed taking the different radiosensitivities of the GT and the microscopic spread volumes into account by delivering a higher dose to the more resistant GT. In these cases, the dose variation inside the ITV follows the changes of the target radiosensitivity, optimizing the effectiveness of the dose distribution. Consequently, the more conformal plans can deliver a higher dose level to the ITV than the less conformal ones (Mavroidis *et al.* 2000, 2001). This is radiobiologically equivalent to achieving increased control rates for the same or even a reduced risk for complications. In practice, this corresponds to a situation of having high doses within the ITV and a steep dose fall-off at the border to organs at risk so that they receive much lower dose than the target.

In the case of an ITV consisting of two targets, to represent the changes of the control probability, P_{ITV} for a range of dose variations is more complicated. If it is assumed that the dose variation within each target is the same and equal to σ_D, and the dose variation between the mean doses to the two targets equal to $\sigma_{\overline{D}}$, then P_{ITV} behaves as shown in Fig. 3.10. The mean dose is 80 Gy in all the cases, which is the D_{50} value of the more radioresistant GT. In the direction of the intra-target dose variation (σ_D), P_{ITV} always decreases with increasing σ_D. This is because the low doses to each target cannot be compensated by corresponding doses above the mean. In the inter-target dose variation ($\sigma_{\overline{D}}$) direction, the change of the control probability increases at the beginning because higher dose is delivered to the more radioresistant target. Although, the less radioresistant target receives less dose than the mean, this dose is higher than its D_{50} value. The maximum control probability is reached when the dose difference between the two targets is equal to the difference of their radiosensitivities as these are expressed by their D_{50} and γ values. As the

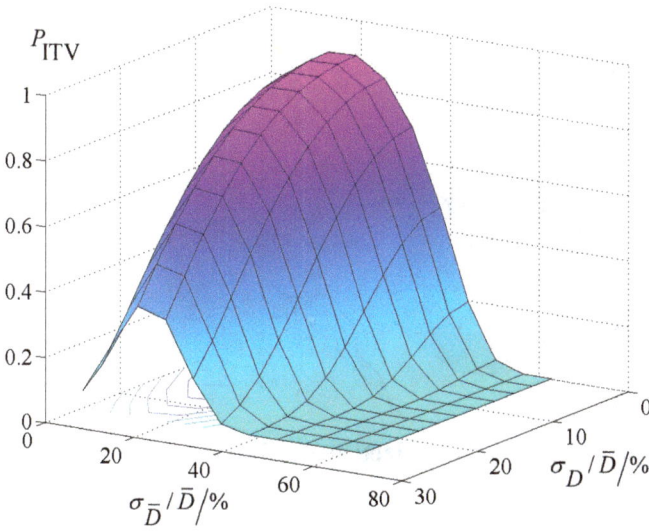

Figure 3.10. The variations of the control probability, P_{ITV}, are demonstrated as a function of dose escalation in the case of an inhomogeneous ITV, which is irradiated with a step-wise dose distribution having the same mean but varying inter-target and intra-target relative standard deviations ($\sigma_{\bar{D}}/\bar{D}$ and $\sigma_D/\bar{D}$ respectively). It is assumed, that the dose variation within the two targets, σ_D, is the same for all the range of relative standard deviations delivered to the two targets ($\sigma_{\bar{D}}$).

inter-target dose variation increases beyond the optimal dose separation, the total control probability decreases because the low doses to the less radioresistant target cannot be compensated any more by the high doses to the rest of the ITV. This has as a consequence the decrease of the P_{ITV} accordingly.

3.4.5.2. *Dose level optimization*

The determination of a treatment configuration that produces a conformal dose distribution to the ITV is the major objective of radiation therapy treatment planning. However, to find the optimum dose, scaling level may also have a significant impact on the effectiveness of the treatment plan. The clinical value of radiobiological objectives in IMRT can be demonstrated by applying it to a clinical case. As an example, the clinical case of a head and neck target with the locally involved LNs of a larynx tumor (GT) is selected. In this patient geometry the local normal tissue stroma, the brain, and the spinal cord are the principal organs at risk (Fig. 3.11). The GT and the involved local LNs are regarded as separate biological structures and therefore they are associated with different radiosensitivities.

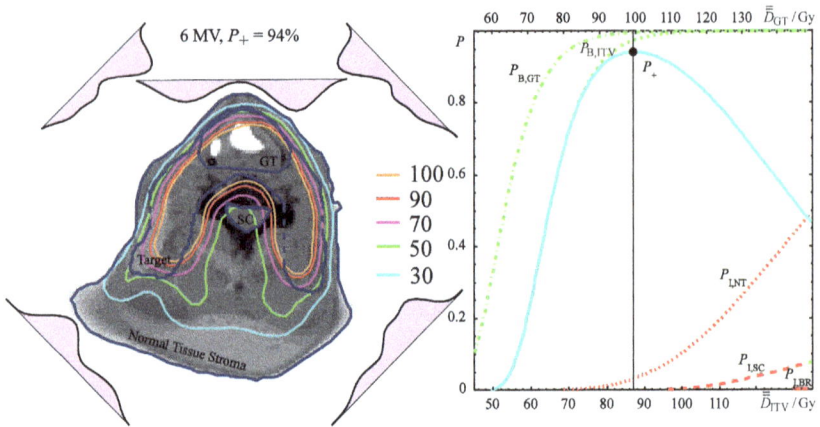

Biologically Optimized Intensity Modulated Beams

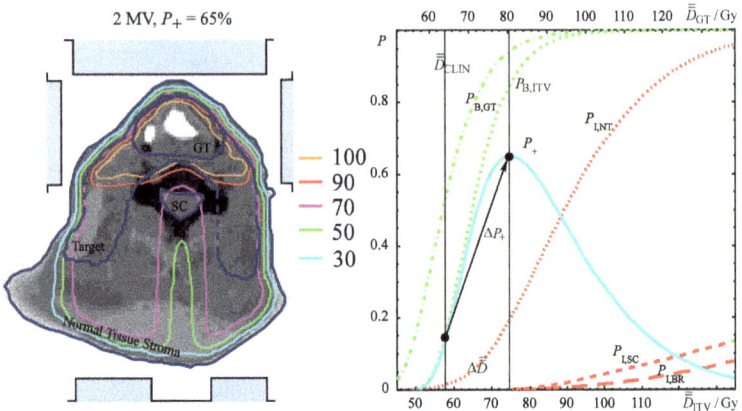

Classical Uniform Beams

Figure 3.11. The dose–response curves of the targets and organs at risk corresponding to two treatment plans of a head and neck cancer case are shown. On the horizontal axis, the biologically effective uniform dose of the target volume, $\overline{D}_{ITV}$, representing the dose prescription is used. $\overline{D}_{ITV}$ was calculated from the control probability of the ITV (using the responses of the GT and the LNs). On the upper axis, the biologically effective uniform dose to the GT, $\overline{\overline{D}}_{GT}$, is given separately. The vertical line, $\overline{\overline{D}}_{CLIN}$, indicates the clinical dose prescription. It can be seen that the clinically prescribed dose (64 Gy to the GT) is far from the optimal one achieved using radiobiological modeling; however, it ensures a lower incidence of complications and so does the P_{++} optimization strategy (cf Figure 3.4).

The presentation of the treatment plans has three purposes: first, to reason the results obtained using radiobiological models; second, to summarize the physical parameters that characterize the plans such as the isodose levels; and third and foremost to show that radiobiological evaluation generally agrees with the results of the physical evaluation, and furthermore it provides more detailed clinical information concerning the expected clinical outcome (Lind *et al.* 1999). This implies that physicians and medical physicists need to move from the deterministic interpretation of the dosimetric parameters to a probabilistic viewpoint, which is more clinically relevant because of the stochastic nature of radiation at the microscopic level.

By analyzing the main dose characteristics of the presented treatment plans, it can be observed that the sparing of the organs at risk is achieved through the very steep dose gradient at the borders of the organs at risk proximal to the ITV. In conformal treatment plans, the dose gradient can be very steep but it is very important to be correctly located toward the inner side of the borders of the ITV otherwise a small underdose of the target can substantially reduce the control probability. The steep gradient should be present at the borders toward all the organs at risk, otherwise the organ that is spared less will be the dominant factor in the treatment outcome. Better conformity means steeper and more accurately located dose fall-off at the borders of the ITV. On the contrary, a decreased dose gradient can lead to increased injury rates.

In this example, it is demonstrated that the complication-free tumor control probability, P_+, can be a valuable parameter in the optimization of the treatment planning process. It is shown, how the complication-free cure can be used to select the best treatment plan but also the best dose level. It can be noticed that the most conformal treatment plan shows a higher value of P_+ (Källman *et al.* 1992; Mavroidis *et al.* 1997; Brahme 2000). In this clinical case, the evaluations of the medical physicists and the physicians agreed with those of the biological models. However, in cases of treatment plans whose conformities do not differ substantially, the clinical personnel may not identify the most effective treatment configuration. This is because the evaluation of the treatment plans by the clinical personnel is mainly based on physical quantities such as the shape of the isodose levels and DVHs, whereas the use of radiobiological data is limited.

In Fig. 3.11, the importance of dose level optimization is clearly demonstrated. An interesting question that may arise in dose level optimization is whether it is possible, by sacrificing some small amount of the complication-free tumor control probability, to achieve a significant reduction of the normal tissue injury. This is a reasonable proposal since at maximum P_+ the rate of increase of P_B and P_I are very similar, whereas in the dose region around the maximum P_+ the gradient of the two curves (P_B and P_I) can differ significantly. This means that by changing the treatment configuration, it may be possible to find a more clinically acceptable combination of tumor cure and fatal complications, while only sacrificing a minimal amount of complication-free cure. This index is mathematically expressed as $P_{++} = ((P_+ - \Delta P_+)^{\max} | (P_I)_{\min})$ that is, minimize the overall normal tissue complication probability under the constraint that the probability for uncomplicated tumor control is $\geq P_+ - \Delta P_+$ (Löf 2000) where $\Delta P_+ \approx 2\%$.

It is not possible for someone to quantify the differences of two dose distributions at a biological level by only looking at their DVHs or isodose charts. In other words, the different radiobiological behavior of each of the different organs involved is affected differently by the examined dose distributions and their uniformity. So, if the planner evaluates a plan based only on the physical criteria, like it is done in the current clinical practice, a less effective plan for the patient may be chosen. According to this analysis, it is obvious that P_+ is a valuable parameter in the optimization of the treatment planning process. It can become the reference point in using the patient special characteristics instead of just optimizing physical functions.

3.4.5.3. *DVH versus DMH*

In practice, the dose delivered to an organ is quantified by isodose charts or DVHs. This is an accurate description when the density of the organ is homogeneously distributed throughout its volume. However, when the organ is heterogeneous in density (such as lungs), the appropriate way to quantify its volumetric dose distribution is by calculating its dose–mass histogram (DMH) (Mavroidis *et al.* 2003b). This can be done by using the density map that is provided by the Hounsfield-numbers of the planning CTs and the volume information from the treatment planning system. DMHs are also more appropriate to be used in calculations concerning the expected complication probability of the lungs (cf. Fig. 3.12). That is because density is related to the number of functional subunits (FSUs), which are responsible for the expression of a certain clinical end point (such as radiation pneumonitis). From a biological point of view it should ultimately be the number of cells that receives a certain dose level so to be accurate the cell size and sensitivity dose also play a role.

3.5. Treatment Techniques

3.5.1. *Target Definitions*

When the tumor and sensitive normal tissues have been localized by some kind of diagnostic technique, such as CT or MRI, it is important to define the target for the

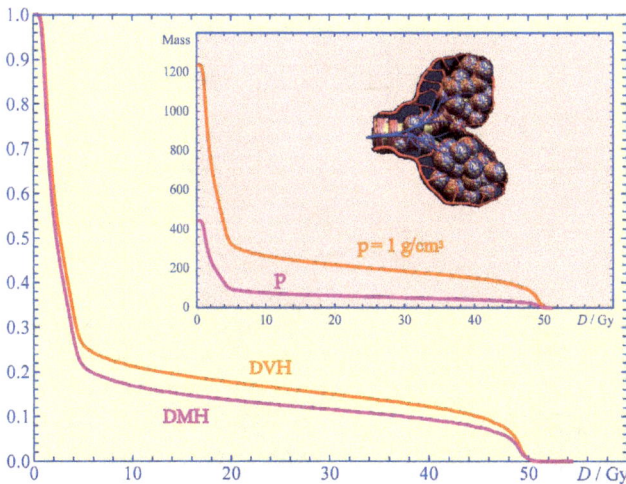

Figure 3.12. Comparison of the DMHs and DVHs of the lung, for a resection-negative node involved breast cancer case. The histograms are presented in absolute and normalized units. For the estimation of lung complications, the use of DMH is better associated with the clinical outcome because it is better related to the density and location of the FSUs, which are responsible for the injury.

treatment. Figure 3.13 schematically illustrates the case of an infiltratively growing tumor surrounded on one side by a sensitive normal tissue.

Together with Figs. 3.2 and 3.14 it illustrates how different margins are needed to ensure a successful treatment. The following definitions are useful to ensure that the complication-free cure will be as high as possible (cf. the NACP and ICRU reports, for further details).

3.5.1.1. *Clinical target volume*

According to ICRU 50: "The *Clinical Target Volume* is a tissue volume that contains a *Gross Tumor Volume*, GTV, which is the gross palpable or visible/demonstrable extent and location of the malignant growth, and/or subclinical microscopic malignant disease, which has to be eliminated. This volume thus has to be treated adequately in order to achieve the aim of therapy: cure or palliation."

One or several *CTVs* can be specified in a patient at the same time. For each target tissue to be treated according to a prescribed dose fractionation schedule, a separate *CTV* should be defined. One *CTV* can include none or several *GTVs*. The *CTV* is defined in a static image without consideration of organ motion.

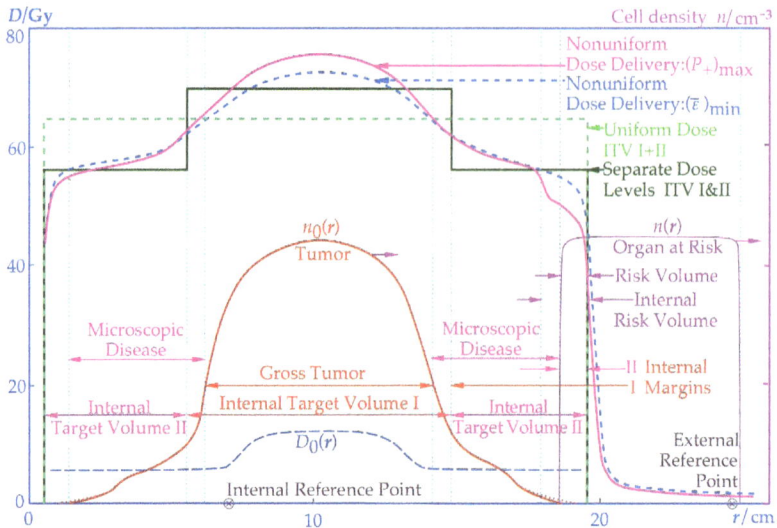

Figure 3.13. Schematic illustration of the lateral dose distribution through a heterogeneous tumor such as the dose distribution through the center of the *gross tumor volume* (GTV) (cf. Fig. 3.2). It is clear that the optimal dose delivery represented by the upper solid curve can be approximated by either a constant uniform dose (dashed line), two uniform dose levels (solid stair-like curve), or the continuous dose distributions, which either maximize complication-free tumor control (solid curve) or minimize dose tissue complications (dashed curve). The tumor cell density (right scale) and associated D_0 distribution (left scale) are also schematically illustrated.

3.5.1.2. *Risk volume*

The *risk volume* (*RV*) is the potential (likely to be irradiated) part of the *organ at risk*. According to ICRU 50, the *organ at risk* is normal tissues whose radiation sensitivity may significantly influence treatment planning and/or prescribed dose. The *RV* is defined in a static image during the treatment planning procedure.

3.5.1.3. *Internal margin*

Owing to physiological changes such as organ motions and/or changes in shape and size, the *CTV* and *RV* are mobile in the *local patient coordinate system*. Therefore, in external beam radiotherapy, *IMs* around the *CTV* and *RV* (see Fig. 2.2) can be used to account for the uncertainty in anatomic information, physiological changes such as expected movements, and/or changes of shape and size of the *CTV* and *RV* in relation to *internal and external reference points*. For each *CTV* and *RV*, the outer boundary of the *IM* defines a fixed volume in the *local patient coordinate system*.

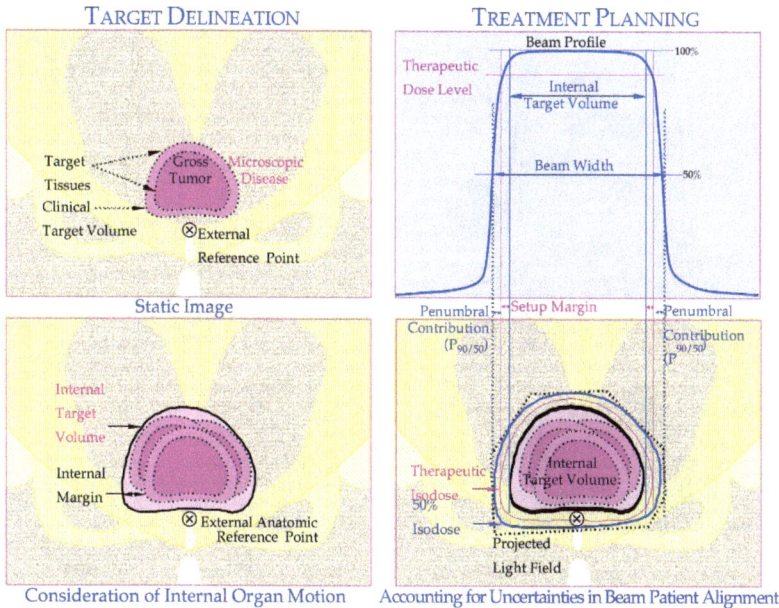

TARGET DELINEATION

Static Image

Consideration of Internal Organ Motion

TREATMENT PLANNING

Accounting for Uncertainties in Beam Patient Alignment

Figure 3.14(a). **Left panels:** Schematic illustration of the handling of internal organ motions and the setup uncertainty during target volume definition and treatment delivery of a prostate tumor as it may appear in relation to the *external reference point* on a series of three simulation or verification films or digitally reconstructed radiographs. During radiotherapy, the target tissues and the *CTV* move particularly due to varying rectal and bladder content so an *internal margin* has to be added to it, in order to always have a well-defined *ITV* fixed in relation to the *external* and *internal reference point*. **Right panels:** The bony structures are never perfectly stationary in the coordinate system of the beam. The skin marks may also move relative to the gantry coordinate setting and the light field. All these motions and uncertainties of the beam in relation to the ITV may be considered by adding a SM to the beam cross-section in beam's eye view during treatment planning (cf. also Fig. 3.2).

Since the *IM* should be anatomically adjusted, it is not recommended to always add a constant margin around the *CTV* and *RV* in a static image. This is because the movements of the *CTV* and *RV* can be strongly varying with direction and their location in a static image may be in an extreme position and not in its mean position of motion.

3.5.1.4. *Internal target volume*

The *ITV* is defined by the outer boundary of the anatomically adjusted *IM* of the *CTV*. The *ITV*, therefore, accounts for the movements of the *CTV* inside the *local patient coordinate system*. This volume contains or has a high probability of containing

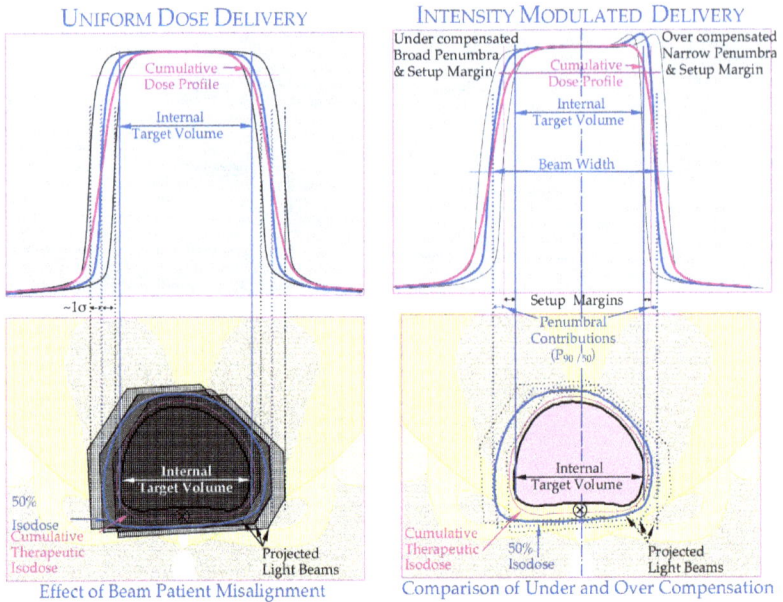

Figure 3.14(b). **Left panels:** During radiation therapy the total cumulative dose profile can be calculated by summing the dose profiles of the quasi randomly positioned beams corresponding to the uncertainty in alignment of the actual dose delivery. The *Set-up Margins* should thus be chosen during the dose planning process such that the cumulative therapeutic isodose delivered during the whole treatment series really encloses the *ITV*. The dashed 50% isodose is approximately coinciding for the most probable single beam and the cumulative dose distribution. **Right panels:** It can be shown, based on radiobiological arguments, that a somewhat better alternative to uniform beam with a broad *SM* (right half panel) is to use an overcompensated beam with a narrow margin (left half panel). As seen from the upper dose profile and the lower isodose image, a closer conformal treatment is possible by using overcompensated beams with a reduced *SM*.

the *CTV*, and thus the target tissues, throughout all treatment sessions. Therefore, the *ITV* is the volume in those patients who should receive the prescribed dose with a high degree of probability. Each *CTV* will have its own *ITV*.

The *ITV* is a geometrically defined volume fixed in the *local patient coordinate system* and it is specified in relation to *internal and external reference points* which preferably should be rigidly related to each other through bony structures.

3.5.1.5. *Internal risk volume*

The *internal risk volume* (*IRV*) is defined by the outer boundary of the anatomically adjusted *IM* of the *RV* and it therefore accounts for the movements of the *RV* in the *local patient coordinate system*. The *IRV* is a geometrically defined volume fixed in the *local patient coordinate system* and it is specified in relation to *internal and external reference points*. Note, that the true *IRV* can only be specified after the *cumulative dose distribution* has been calculated. The true *IRV* is then that volume of the *organ at risk* which receives a significant dose with regard to the tolerance of the organ.

3.5.1.6. *Setup margin*

To account for uncertainties in the positioning of the patient and inaccuracies in the alignment of the therapeutic beams during dose planning and patient irradiation through all treatment sessions, margins for the setting of the beams are needed. These so-called, *SMs* are external margins which account for relative movements of the *local patient coordinate system* and thus the *ITV* or *IRV* in relation to the radiation beams.

The *SMs* are considered during treatment planning to ensure that the prescribed dose is really delivered to the whole *ITV* and the dose to healthy normal tissues is below the agreed tolerance levels (cf. Figs. 3.2 and 3.14). The *SMs* for different beams have to be defined separately and in general they depend on the treatment technique. They are accounted for when choosing treatment beam sizes in the beam's eye view (cf. Fig. 3.2) or beam energy when the setup uncertainty along the beam axis has to be considered (e.g., for treatments where the distal fall-off in an electron or proton beam or the width of the photon build-up region is used to increase treatment conformality).

The *SM* has to account for uncertainties in (1) patient positioning (interfractional movements), (2) movements of the patient during each treatment fraction (intrafractional movements), (3) dose planning and treatment technique in general through the resolution of the treatment planning system, and (4) treatment unit performance characteristics. The latter are determined by the quality of the treatment unit, for example, with regard to uncertainties in the shape of the depth–dose curve, the shoulder of the penumbra region, and the light field and radiation field alignment as well as the size of isocentric deviations. The size of the *SM* should ideally be chosen iteratively such that the cumulative therapeutic isodose encloses the *ITV* (Figs. 3.14(a) and (b)).

Note that *SMs* are not included in the ITV. The *SMs* are defined in the *treatment unit coordinate system* as margins to be added to the beams to account for the uncertainty in beam patient alignment (cf. Figs. 3.2 and 3.14). The setup uncertainty may be clinically determined by repeated portal verification techniques (cf. Figs. 3.1 and 3.2).

3.5.2. *Setup and Organ Motion Uncertainties*

The more conformal a treatment technique is, the more precise and accurate the setup process should be. In these techniques, the dose distribution is so well matched with the radiation sensitivity map of the clinical case that a small misalignment in the setup can very much reduce the effectiveness of the therapy. If a reliable positioning and dose delivery procedure is not available, a less conformal technique could be more effective and trustworthy. The quality of radiation therapy does not only depend only on the conformity of the applied treatment technique but also on the quality of the supporting services.

Although ITV is the volume that should be irradiated with the prescribed dose, SM are also added to account for uncertainties in the positioning of the patient and inaccuracies in the alignment of the therapeutic beams between dose planning and patient irradiation through all treatment sessions. These SMs are external margins, which account for the relative movement of the local patient coordinate system and thus the ITV in relation to the radiation beams (Aaltonen *et al.* 1997; ICRU 62 1999). SMs are considered during treatment planning to ensure that the prescribed dose is really delivered to the whole ITV. For different beams, SMs have to be defined separately and in general they depend on the geometry of the treatment configuration (Ekberg *et al.* 1998; Mavroidis *et al.* 2002a). SMs have to be accounted mainly for uncertainties in (i) patient positioning (interfractional movements), (ii) movements of the patient during each treatment fraction (intrafractional movements), and (iii) dose planning and treatment delivery through the geometrical resolution of the treatment planning system. SMs are defined in the treatment unit coordinate system as margins to be added to the beams to account for the uncertainty in beam patient alignment. The setup uncertainty may be determined clinically by portal verification techniques and PET-CT *in vivo* dosimetry.

The relations between the different clinical volumes are demonstrated in Fig. 3.15 where the treatment techniques of two different clinical cases related to breast cancer are shown. The CTVs and consequently the corresponding ITVs are different between the two cases. However, for each on them, a number of different treatment plans have to be produced before the best one is selected. In the examined plans of each case separately, the ITVs are the same (since ITV is defined in the local patient coordinate system), whereas the PTVs are different, since different treatment configurations are generally associated with different SMs (Brahme 1997). In Fig. 3.15, even though the ITVs of the two cases are different, the influence of the beam geometry on the PTV definition is well demonstrated. According to this analysis, it is apparent that the definition of the different clinical volumes before

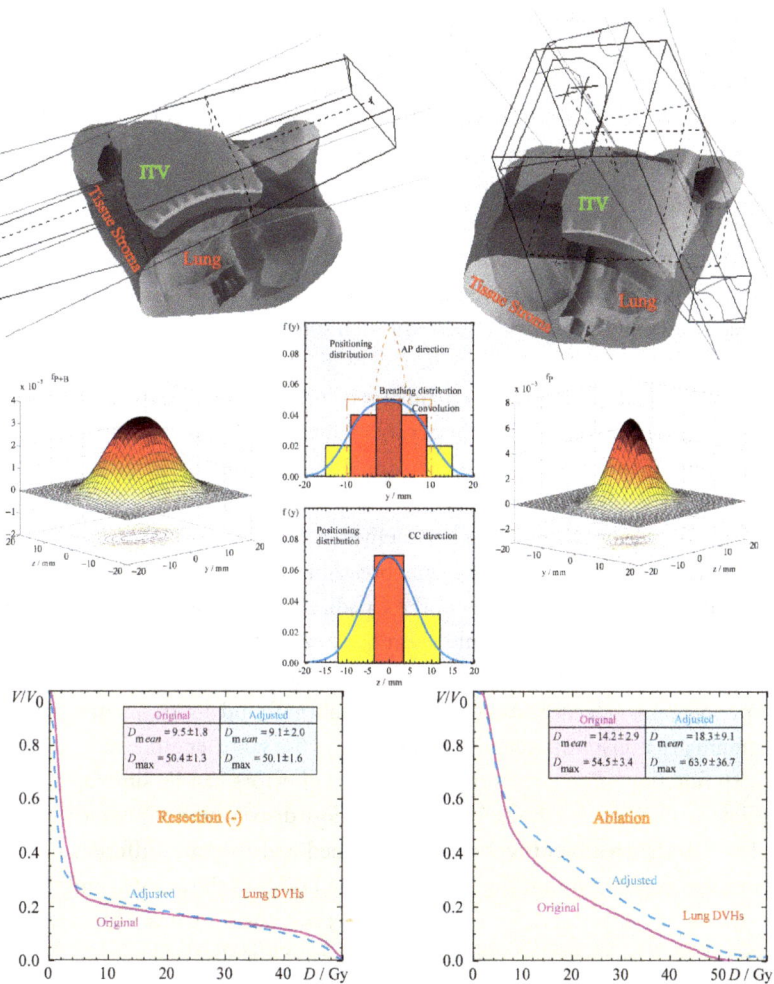

Figure 3.15. The treatment techniques that are usually applied in the breast cancer cases of resection-negative node involvement (upper left graph) and ablation (upper right graph) are demonstrated in a 3D view. In these graphs, the geometrical relation between the irradiating beams and the target is shown. This geometrical relation strongly influences the effect of the patient setup errors and the breathing effects on the delivered dose distribution. The joint 3D uncertainty distributions imposed by patient setup errors and breathing effects are composed by the separate error distributions in the anteroposterior (AP) and the craniocaudal (CC) directions. The distribution in the AP direction is the result of a convolution of the patient positioning and breathing uncertainties. In the lower diagrams, the DVHs of the original and adjusted dose distributions are presented together with quantifications of their differences.

treatment planning is not a simple task. It seems more appropriate that the ITVs and IRVs be drawn from the oncologists since they have a better insight of the CTV and the involved organ motions, whereas the corresponding PTVs and PRVs be drawn from the medical physicists since they are more involved with the applied treatment configuration and the related setup errors.

Quality control is of outmost importance in radiation therapy because of the existence of many potential sources of errors. Such errors, which take place during the delivery of the treatment to the patient, have as a result the degradation of the curative power and effectiveness of the treatment (Löf *et al.* 1995). Positioning uncertainties and breathing effects are such sources of errors because they lead to a dose delivery that is different to the one originally intended to be given (Mavroidis *et al.* 2002). Lung is usually the main organ at risk in radiotherapy for breast cancer, which is usually applied after conservative surgery or radical mastectomy. Therefore, restriction of the dose to the lung to a minimum is among the guidelines followed by clinically applied treatments.

However, the deviation of the delivered from the planned dose distribution to the lung due to positioning and breathing uncertainties can be significantly large. Positioning uncertainties in the CC direction are generally larger than in the AP direction. These uncertainties can be approximated by Gaussian distributions based on the fact that the setup errors are random. On the other hand, breathing is periodical (such as a saw-toothed distribution) and its frequency distribution, assuming that inspiration and expiration have the same length, can be described by a step function. The effects of both sources of errors can be simulated using very realistic values and methods. However, the dose distribution adjustments are closely related to the treatment techniques examined and they are different for different treatment configurations. For the R case, the DVHs of the original and the adjusted treatment plans do not differ much partly because their integral doses are almost the same (Fig. 3.15, lower left diagram). In this case, the displacement of the beams in respect to the lung results in only a small volume of the lung lying always in the high dose region and a large volume lying in the intermediate dose region. In the case of the ablation, the dose of the adjusted DVHs is much higher than that of the original DVHs because the contribution of the frontal supraclavicular electron field becomes larger leading to a significant increase in the integral dose to the lung (Fig. 3.15, lower right diagram).

The simulation process can be very accurate by using a larger number of fields approximating even more the frequency distribution functions of the different error sources (Fig. 3.15, middle diagrams). The adjusted dose distributions shown are not the true delivered dose distributions but a reasonably good approximation of them. For other beam configurations, a similar study has to be carried out to estimate the influence of the positioning uncertainties and breathing effects on the delivered dose distribution because this depends on the geometrical relationship of the treatment fields and the patient. The most appropriate way to carry out this study is to estimate these uncertainty values from the same patient material using portal imaging or other treatment verification means.

The radiobiological parameters applied to a certain patient material should be compatible with it. These parameters are generally derived from patient materials where the dose delivered to each patient and the follow-up records are available. Because the treatment methodologies among different institutions are likely to differ to some extent and the clinical information is still limited (imaging at a cellular level, accurate determination of the dose delivered to the patient, radiation sensitivity of the individual patient, etc.), the derived parameters from such studies are subjected to those factors. To estimate the expected response for adjusted treatment plans, another set of parameters should be used (derived from a patient material where positioning uncertainties and breathing effects have been taken into account).

3.5.3. *Dose Delivery Methods*

Basically there are four different methods to increase the flexibility in dose delivery in external beam radiation therapy as illustrated in Fig. 3.16. Starting from narrow pencil beams via elongated intensity-modulated fan beams and classical block collimated beams with wedge filter to general nonuniform beams of irregular cross-section, for example, generated by dynamic MLC. We will start off by first presenting the methods that are available today for nonuniform dose delivery and then describe the more "differential" approaches.

3.5.3.1. *General methods for nonuniform dose delivery*

The main methods for nonuniform dose delivery are reviewed in Fig. 3.17. From the figure, it is seen that if full dynamic flexibility and reasonable treatment times are required in the clinical application of the new multidimensional optimization techniques, the fastest methods for nonuniform dose delivery are dynamic MLC and scanned elementary beams.

The dual dynamic jaw collimation method in Fig. 3.17 also allows in principle full modulation of the incident beam but at the cost of very extended treatment times as the mechanically moved narrow elementary beams in Fig. 3.16. Furthermore, it requires that both the upper and lower jaw pairs are fully asymmetric so that a narrow rectangular beam spot could be scanned arbitrarily across the entire target volume. If very high dose rates were available and the speed of motion of the collimator

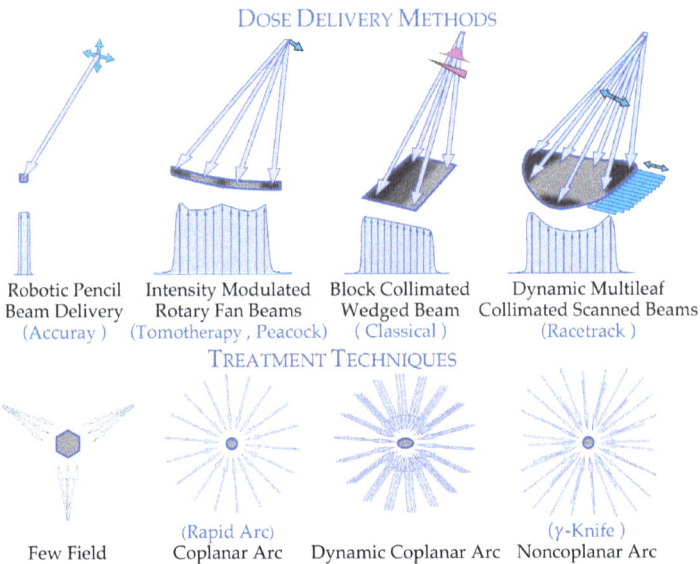

Figure 3.16. The four major groups of dose delivery systems for external beam radiation therapy are illustrated. The most differential are the pencil beam methods either using electromagnetically scanned beams or mechanically moved beams. To the fan beam group belongs the Peacock device and the tomotherapy method. Classical external beam therapy mainly employs uniform rectangular block collimated beams, sometimes with a wedge filter. Discrete nonuniform beams can be delivered with any of the methods in Fig. 3.17 and the resultant treatment techniques are generally quite fast, simple and reliable, and easy to verify. Continuously moving beam techniques are using the degrees of freedom in dose delivery more completely but are complex to verify when the intensity modulation is rapidly changing.

jaws was very fast the time required could be reduced but this is not a very realistic method at least with presently available accelerator systems.

The classical filter and block techniques have the flexibility but are quite time-consuming, so they are probably not realistic for more than some three portals per patient. They could be used either by manual change or with a filter revolver on the front end of the treatment head carrying three to five filters. In recent years, several compensator optimization techniques have been developed which are quite useful to handle few field techniques provided suitable beam directions can be identified. In reality, the optimal choice of beam direction is one of the most difficult problems of treatment optimization since it involves a restriction on the whole phase space of feasible beam combinations. This cannot be achieved without having located all beam combinations corresponding to local optima, which in practice is equal to a global optimization. It also accentuates a difficult radiobiological problem, in a

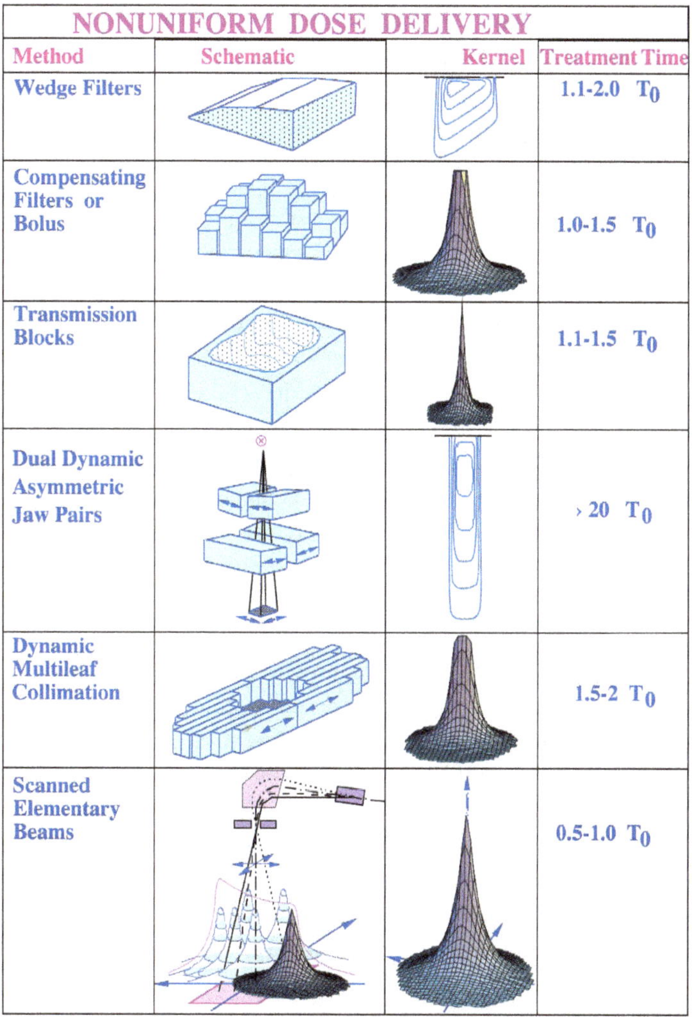

NONUNIFORM DOSE DELIVERY

Method	Schematic	Kernel	Treatment Time
Wedge Filters			$1.1\text{-}2.0\ T_0$
Compensating Filters or Bolus			$1.0\text{-}1.5\ T_0$
Transmission Blocks			$1.1\text{-}1.5\ T_0$
Dual Dynamic Asymmetric Jaw Pairs			$> 20\ T_0$
Dynamic Multileaf Collimation			$1.5\text{-}2\ T_0$
Scanned Elementary Beams			$0.5\text{-}1.0\ T_0$

Figure 3.17. Comparison of six different methods available for delivering of nonuniform therapeutic beams is shown. T_0 is the standard treatment time of about 1 to 2 min for uniform dose delivery to target volume. Only the lower three methods allow dynamic beam shaping but at greatly varying treatment times.

way the Scylla and Charybdis of radiation therapy: With a single beam, the small volumes of normal tissues in the entrance region receive a rather high local dose, whereas on the other extreme, with a continuum of arc beams, large volumes receive rather low doses. To allow a strict optimization, accurate radiobiological objective functions capable of distinguishing between these extreme situations are needed.

With present dose–response models, this steering between Scylla and Charybdis can only be made in a rather approximate manner.

3.5.3.2. *Scanning beam therapy*

Radiation therapy is traditionally performed with stationary beams and flattening filters to make the beam uniform. The fastest and probably safest way to deliver nonuniform beams in real time today is by moving a small elementary electron, photon, or proton beam over the patient similar to the electron beam in a TV monitor. Such beams are since almost 30 years available on a 5–50 MeV racetrack accelerator. Because the elementary essentially Gaussian electron beams and the bremsstrahlung beams have fairly wide half-widths (≥ 12 and ≥ 40 mm, respectively, at isocenter), the MLC has to be used for spatial modulation when higher geometric precision is required. Despite this shortcoming, the scanned beams are very useful and many times sufficient at least for beam compensation. In combination with dynamic MLC, a very fast and flexible dose delivery is possible and ideal for few field nonuniform generalized conformal therapy with treatment times of the order of a few minutes in most cases. An exploded view through the treatment head of an accelerator combining these two capabilities is illustrated in Fig. 3.18. The scanning system is based on the beam optical property of the last bending magnet imaging the scanning center of a first scanning magnet on a second magnet such that the first scanning magnet deflects the beam in the bending plane of the rotary gantry and the second magnet on top of this deflects the beam in and out of the bending plane. This is described in more detail in Chapters 4.3.3.7, 4.4.3 and 4.5 below.

Dynamically scanned proton or light ion beams will probably be the ultimate radiation therapy modality when high geometrical precision is required since the pencil beam penumbra is so narrow that additional collimation is not required and their finite range also protects tissues beyond the tumor volume.

3.5.3.3. *Fan beam therapy*

Today, there are a large number of projects centered around the use of uniform and nonuniform fan beams (cf. Fig. 3.16). The earliest was probably in the computer-controlled therapy in Chicago and Boston (cf. 3.5.3.4 and Chapter 4) where the length of a narrow elongated slit beam through the isocenter (the fan beam) was varied as the gantry rotated and the patient was slowly moved through the beam. The treatment time was often long, of the order of 20 min, and the setup time was also considerable. This problem is shared with all small volume irradiation techniques, unless the dose rate and speed of rotation is increased by at least one order of magnitude.

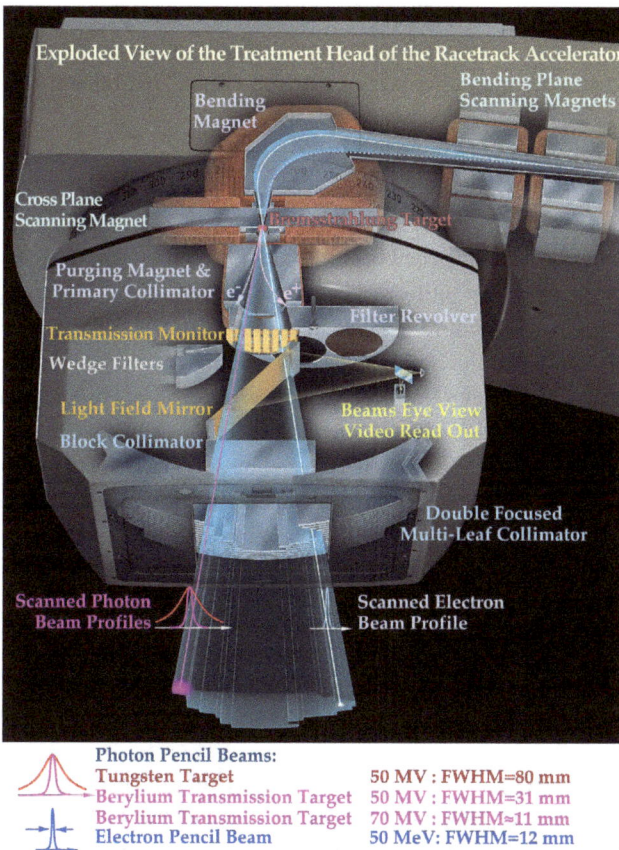

Figure 3.18. Exploded view through the treatment head of scanning beam treatment unit combining electron and photon beam dose delivery with the same double focused MLC, since the electron beam impinging on the target is scanned in two orthogonal directions parallel and perpendicular to the bending plane of the 96° bending magnet. The beam spot on the patient is illustrated for both electron beam (to the right) and photon beam delivery (to the left; this treatment unit is available at TopGrade Healthcare, Beijing China).

A similar technique has also been developed keeping the patient fixed and moving an oblique fan beam by using a dynamic pair of asymmetric collimator jaws. More recently, a special modulated fan beam collimator has been developed for this purpose. This device allows temporal modulation of the treatment time along the fan beam and thus allows nonuniform dose delivery. The latest development along these lines is helical tomotherapy. The idea is then to use a fan beam modulating collimator for "spiral irradiation" with a longitudinally moving patient much in the same way as used in spiral CT. Unless the accelerator output is very high, all the fan beam approaches described here will require fairly long treatment times. Furthermore, they will basically be limited to coplanar irradiations, unless the patient is moved in more complex patterns or when wide field "cone beam" tomotherapy is employed.

3.5.3.4. *Pencil beam therapy*

At the cost of a further increase in treatment time, it is possible to use a moving narrowly collimated beam (pencil beam) to deliver nonuniform dose distributions. The pioneering work for uniform beam delivery was done in Chicago using a mechanically moving bending magnet in a rotary gantry. Since the dose rate in the electron beam was quite high, the treatment time was not too much increased.

More recently, several robot-mounted linear accelerators have been developed. Such devices have the advantage of a high degree of freedom since the computer-controlled dynamic dose delivery is performed by a robot. However, for large target volumes, this device requires very long irradiation times since the beam is narrow (<4 cm) and the dose rate is normal (3 Gy/min). An ideal algorithm for planning and biological optimization of pencil beam and also more general types of treatment techniques employing MLC have recently been developed (cf. Figs. 3.21 and 3.22 and Chapters 5.4–5.5 for more details).

3.5.3.5. *Design characteristics of modern treatment units*

To get a feel of the components involved in forming high-quality photon and electron beams, an "exploded" view of the internal components of a radiation therapy treatment unit are shown in Fig. 3.18. Most treatment units for intermediate-to-high energies (8–50 MeV) have a bending magnet that bends the electron beam toward the patient. Low-energy photon machines (4–6 MV) do not need this bending magnet and use a short linear accelerator directed onto the patient. In the machine, in Fig. 3.18, the beam optics of the bending magnet is used to image the scanning motion of a first scanning magnet deflecting the beam in the plane of the bending magnet onto a second scanning magnet placed just after the bending magnet. Since this magnet deflects the beam in and out of the bending plane, the emerging electron beam can be scanned in two perpendicular directions over the patient surface. The scanning motion can be used to make uniform or intensity-modulated electron and photon beams as seen in Fig. 3.18. In machines without scanned beams, uniform beams are generally produced by photon absorption in a flattening filter.

After the beam has been collimated, generally some kind of transmission monitor is used to detect if the beam is shaped uniform or intensity-modulated to ensure correct dose delivery to the patient. Below the transmission monitor is generally a thin foil mirror allowing an optical light field to simulate the path of the therapeutic beam. Finally, some kind of collimation system is used to ensure that the normal tissues around the tumor are not irradiated more than necessary. The treatment unit in Fig. 3.18 is equipped with a so-called MLC allowing the fine adjustment of the beam shape so only tumor tissues get a high therapeutic dose. By combining scanned electron and photon beams with dynamic MLC, beams with almost arbitrary dose profiles can be delivered, for example, to heterogeneous target volumes (cf. Fig. 3.13). At the cost of a longer treatment time, this could be done simply by using dynamic MLC in uniformly filtered beams.

3.5.4. *Development of New Treatment Techniques*

Irrespective of the method of dose delivery, a large number of treatment techniques are possible as illustrated in the lower half of Figs. 3.16 and 3.19 and discussed in further details in Chapter 4. Most classical treatment techniques are coplanar with all the beams incident in one plane by the rotation of the radiation source mounted in an isocentric rotary gantry. Since there is a significant exponential dose fall-off in most photon beams, despite their high energy, the most common treatment techniques employ symmetric beam configurations to largely eliminate dose gradients due to photon absorption over the tumor. Thus, parallel-opposed and four-field box techniques are quite common (Fig. 3.19 upper row). However, the symmetrical or nearly symmetrical three-field technique is generally more advantageous since entrance and exit volumes are almost fully separated and the high dose volume in normal tissues surrounding the tumor is significantly reduced. A two-field technique which is more advantageous than the standard parallel-opposed configuration is therefore to use more oblique nonuniform fields at ~100° to 120° from each other (Fig. 3.19 middle row, even if strictly speaking two fields always are coplanar). Similarly, with nonuniform dose delivery, it is sometimes better to turn the three fields out of the plane of rotation to better separate the incident and existing beams from each other (first row of noncoplanar techniques in Fig. 3.19).

   A much better four-field technique than the standard four-field box is obtained by letting the beams enter through the main tetrahedral directions again reducing the normal tissue dose by non-coinciding entrance and exit regions. There are also five- and six-field techniques which largely keep the symmetry of the non-coplanar three- and four-field techniques. All the treatment techniques in the last row of Fig. 3.19 has the interesting property of generating almost uniform target doses independent of using wedge filters, at least on a quasi-spherical patient. This is a very valuable property of many semi-optimal few field techniques. No exact symmetric solution is known for five or six fields like the tetrahedral solution for four fields since the

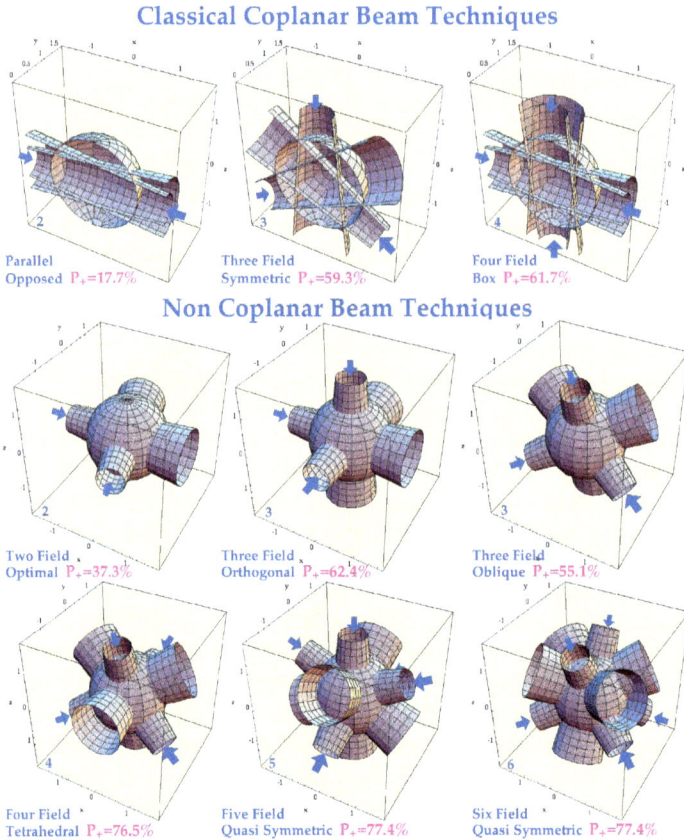

Figure 3.19. Comparison of classical coplanar and more general non-coplanar treatment techniques is shown. The symmetric planar configurations allow, in most cases, a fairly uniform dose to the tumor. The non-coplanar techniques are more advantageous with regard to the dose to normal tissues. For the new field techniques to be effective it is important to use nonuniform dose delivery. The optimal angle between two neighboring fields is then often around 100° to 120° to maximize the effect on the tumor and minimize the dose to normal tissues. Typical achievable complication free cure values P_+ are also indicated.

close packing problem of circles on a sphere is unsolved for arbitrary numbers of circles.

The theoretical limits for irradiation with external photon beams for different treatment geometries are compared by their dose distribution kernels in Fig. 3.20. Three external beam kernels are shown and one isotropic internal source, all for an monochromatic energy of 1.25 MeV corresponding to ^{60}Co γ rays. The basic building block for external beam therapy is the point monodirectional energy

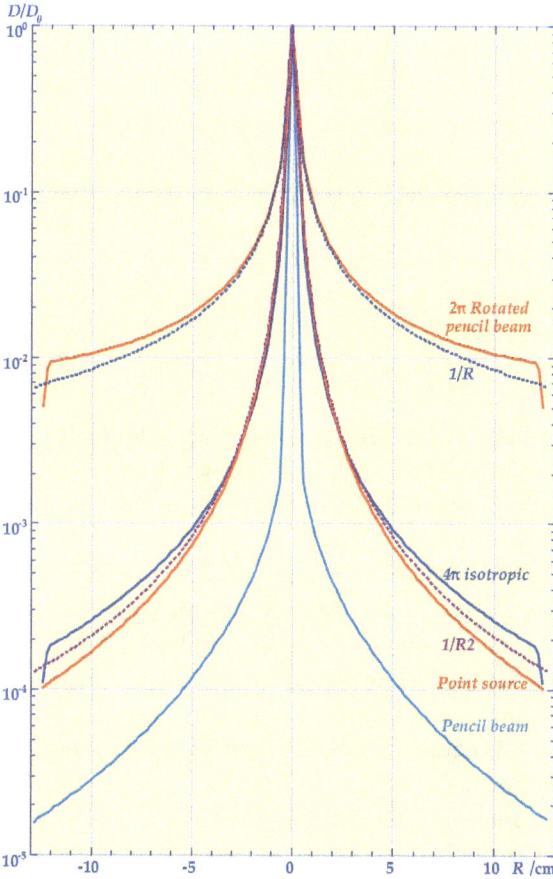

Figure 3.20. The radial dose profiles through four different ^{60}Co beam kernels: point monodirectional pencil beams (lower solid line), planar rotated pencil beams (upper solid line), 4π isotropically incident pencil beams (middle solid line), and isotropic internally emitting nuclides (middle thin line without build up at the surface) are shown. For comparison, the dotted lines correspond to $1/R$ and $1/R^2$ radial dependencies truncated at small radii. For the 2π and 4π curves, the deviations from the $1/R$ and $1/R^2$ dependencies are due to photon absorption in the patient. In principle, tomotherapy would allow very high resolution along the axis of rotation as seen from the lower pencil beam profile but unfortunately this would require very long treatment times and generally a few centimeter axial width is used instead. Across the rotational axis, the resolution is less good (upper profile) and is used most of the time. The approximate 4π isotropic γ-knife has then generally a higher geometric resolution but not really in the axial direction!

deposition kernel or pencil beam, here represented by the lowest solid line radial dose profile in Fig. 3.20. If the pencil beam is rotated 2π in the plane, the resultant dose distribution in the plane corresponds to a full arc treatment of a point target (upper solid line). For symmetry reasons, the dose distribution across the plane of rotation of a fixed intensity pencil beam is the same as the radial distribution through the pencil beam (lowest solid line). For comparison, the limiting case of a beam without attenuation is also shown with a fall-off proportional to $1/R$ (truncated at the central axis and normalized at a large radii, upper dotted line).

The ultimate dose kernel is obtained by superposing the contributions from pencil beams from every angle over 4π and the resultant dose distribution corresponds to a true isotropic spherical irradiation (middle solid line). For comparison, the limiting case without attenuation is also shown for this case with a fall-off proportional to $1/R^2$ (lower dotted line). With the same photon energy but using an internal point source, a dose distribution very much like the isotropic external irradiation is obtained. The main difference can be seen in the small dose build up near the surface of the focused external beams. It is interesting that for fully isotropic dose delivery it does not matter very much whether internal or external radiation sources are used since the dose distribution is mainly determined by the convergent geometry and not by beam attenuation. Thus, by using external convergent beams it is possible to make dose distributions which are almost as good as with internal intracavitary sources. This is basically the way the "γ knife" works with convergent ^{60}Co beams for radiotherapy of brain tumors but the principle may equally well work for a whole body device. This has the considerable dose distributional advantage of a dose fall off close to $1/R^2$ compared to nonuniform arc therapy with the fan beam devices discussed above, basically limited to a $1/R$ dose fall-off as illustrated in Fig. 3.19. Of course, the axial modulation is theoretically better with classical arc therapy as shown by the lowest profile across the plane of rotation. However, a multiple beam device such as the γ knife has the advantage of delivering all beams simultaneously with a high dose rate at the focus and considerably lower dose rate everywhere else. The low dose rates to normal tissues imply an additional radiobiological spearing of healthy tissues around the tumor. This effect is not obtained by ordinary arc- or tomotherapy techniques, where the beams are delivered at an almost constant dose rate (see Sec. 3.5).

3.6. Treatment Delivery

To illustrate the development of radiation therapy in recent years, a classical "four-field box" technique is compared to radiobiologically optimized intensity-modulated radiation therapy of a prostate tumor in Fig. 3.21.

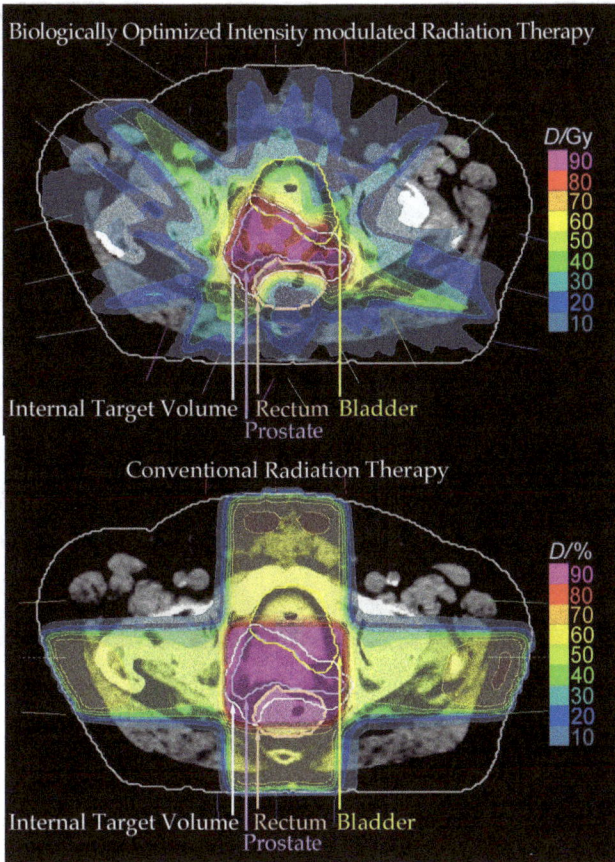

Figure 3.21. Comparison of intensity-modulated dose delivery with the standard uniform or wedged four-field box treatment technique is shown. The high degree of conformity of the delivered dose distribution using intensity modulation with the prostate ITV is striking.

This case was chosen as it is typical for a large group of pelvic tumors, such as cervix, bladder, and rectal tumors, and therefore of general interest. It is clearly seen that with the standard four-field box technique quite high doses are also given to bladder and rectum when treating this prostate cancer. The intensity-modulated plan, on the other hand, spares both bladder and rectum quite well even though it is almost impossible to completely eliminate rectal irradiation since the margin between prostate and rectum is only some 3–4 mm as discussed in further details in Chapter 5.5. Since the organ motion can be quite substantial, it is seen that the ITV will include part of the rectal wall due to the influence of varying rectal filling. With a curative intent, it is thus very difficult to avoid slight rectal complications.

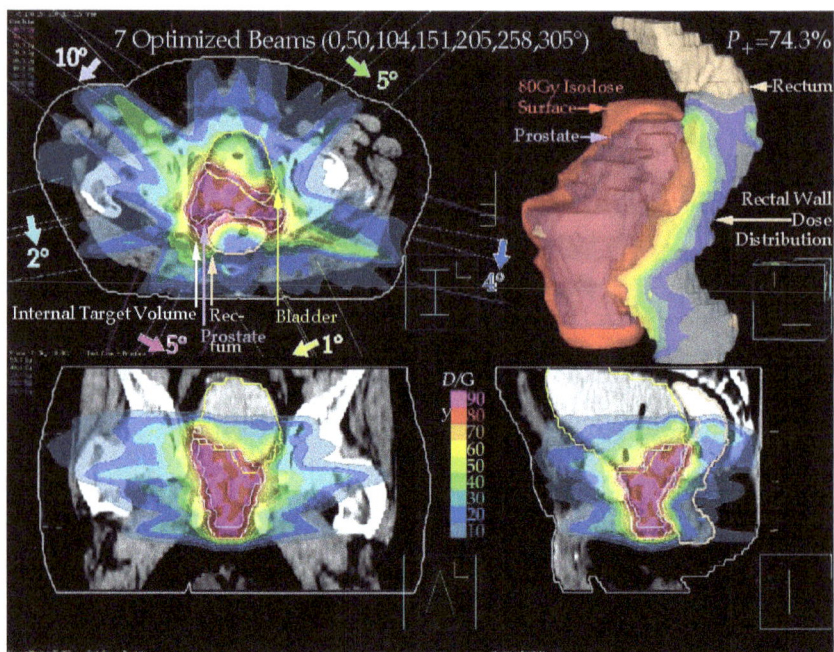

Figure 3.22. The dose distribution in a patient with a prostate cancer exposed to seven directionally and intensity-modulated photon beams is shown. It is seen that in all views of the tumor, the dose distribution is decreasing very rapidly outside the tumor (see Chapter 5.5 for further details).

To illustrate the power of modern radiobiological treatment optimization, the angles of incidence have been optimized at the same time as the intensity modulation as shown in Fig. 3.22.

This is a seven-field plan for a prostate tumor where the initial suggestion was seven nearly equally spaced beam portals. An almost 6% improvement in the complication-free cure could be achieved by fine adjusting the angles of incidence so that they better separate the dose delivery to the prostate from the "spill-over" to rectum and bladder. The optimized beam directions are better aligned to the shape of the tumor at the same time as they reduce unnecessary irradiation of organs at risk. This ability to biologically optimize the directions of incidence is very useful clinically as it performs a fine adjustment of the angle of incidence to a degree, which is hard to achieve by ocular inspection even by a very experienced clinician. The treatment plan in Figs. 3.21 and 3.22 was made by one of the most advanced and flexible radiobiological optimization algorithms that has been developed to date, called ORBIT (optimization of radiation therapy beams by iterative techniques), here shown fully integrated in a regular treatment planning system. It allows optimum combination of different radiation modalities or beam energies, but also optimization of the fractionation schedule considering low-dose hypersensitivity such that the optimal dose level per treatment fraction is achieved in the late-responding organs at risk. It is obvious that all patients do not really need the full power of efficient intensity modulation. For example, many of the early diagnosed small T1–T2 tumors are already treated very well with ordinary uniform beams of rectangular cross-section.

3.6.1. *Radiobiological Quantification of IMRT Dose Delivery*

Most of the recent studies concerning IMRT delivery perform a physical verification of the planned treatment by examining dosimetric quantities (Oldham and Webb 1997; Partridge *et al.* 2000; Ting and Davis 2001; Ploeger *et al.* 2002). Few authors had referred, in the past, to the need for a radiobiological approach to the quantification of IMRT delivery, pointing out that such information would be more useful in the clinical practice (Brahme *et al.* 1988). According to this approach, it is important to use radiobiological parameters for a more complete evaluation of the quality of a treatment plan, closer to the clinical outcome and to show the necessity of using these radiobiological measures in treatment verification and quality assurance of treatment planning.

In Fig. 3.23, the deviation between the planned and delivered dose distributions is shown quantitatively and qualitatively. In the dose distributions presented, a very steep dose gradient can be observed at the borders of the organs at risk proximal to the ITV causing the sparing of large volumes of mainly the bladder and the rectum. The presented treatment plan utilizes five different beam portals to distribute the dose to a large volume of the small bowel, taking advantage of its low relative seriality ($s \approx 0.14$). Furthermore, it tries to deliver a high nonuniform dose to the ITV depending on its underlying radiation sensitivity variation. The treatment plan was designed taking into account the different radiosensitivities of the GT and LNs, by delivering a boost dose to the more resistant GT. It is illustrated that the dose variation inside the ITV follows the changes of the target radiosensitivity, optimizing the effectiveness of the dose distribution. This implies that an increased control rate for the same or a reduced risk for complications should be expected for the increased conformity of the treatment plans.

From the isodose graphs of Fig. 3.23, it can be seen that the two dose distributions are very close but not identical. Although, the two dose distributions seem to be closely associated, their maximum values of P_+ and $\bar{\bar{D}}$ indicate that their

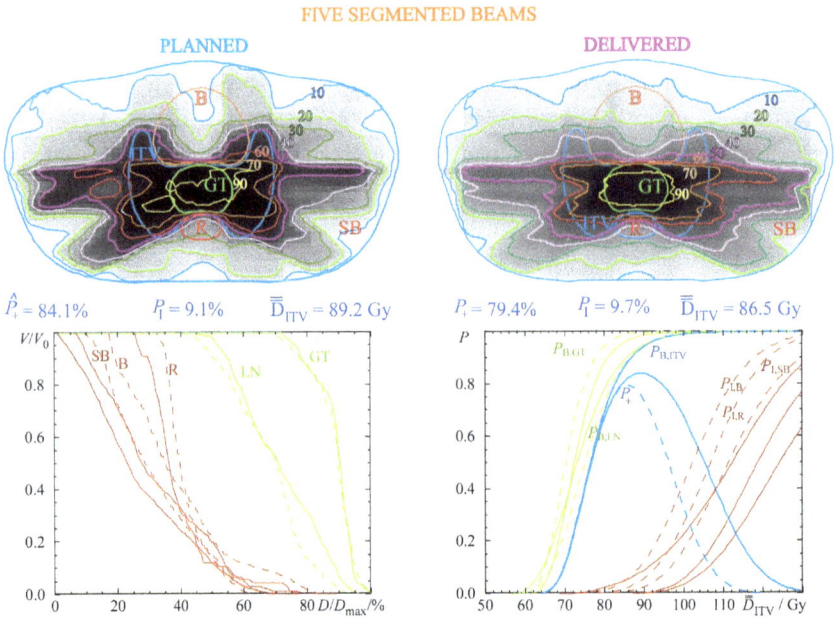

Figure 3.23. *Upper panels*: The planned and delivered dose distributions are presented simultaneously, in the form of dose matrix and isodose levels with the targets and organs at risk in position. A 50 MV racetrack supported by MLCs was used, in a beam configuration of five directions. In each direction, four beam segments were used, with different weighting factors, entry points, and angles. *Lower left diagram*: The DVHs of the targets and organs at risk as they were extracted from the planned and delivered dose distributions are shown. The solid lines are for the planned, whereas the dashed lines are for the delivered dose distribution. *Lower right diagram*: The response curves of the different organs are presented for the planned and delivered treatments. The shift of the normal tissue curves toward those of the targets for the delivered dose distribution (dashed lines) imply a degradation of the effectiveness of the treatment plan, which is also illustrated by the corresponding narrowing and decrease of the P_+ curve.

effectiveness differ significantly. For this reason, it would be more clinically relevant to examine the dose distributions delivered to each of the targets and organs at risk separately, to really quantify the effectiveness of the two distributions. Although the DVHs of the whole dose distributions are close, the DVHs of the individual organs show more extended differences (Fig. 3.23, lower left diagram). Especially for the small bowel, the bladder, and the rectum, these differences seem to be quite significant. The impact of this deviation is better illustrated in the lower right diagram of Fig. 3.23, where the expected responses of the different organs are presented for a range of prescribed doses. The use of $\overline{\overline{D}}$ in the scaling axis provides a common prescription basis for both of the dose distributions, since it forces their $P_{B,ITV}$ curves to coincide. In comparison to the planned treatment, the delivered dose distribution is less conformal to the ITV, which makes the response curves of the organs at risk move closer to those of the targets (Mavroidis *et al.* 2003). This shift has as a consequence the narrowing of the therapeutic window, which is expressed by the width and the maximum value of the curve. The response curves of the targets also show a small shift, with the GT being more effectively irradiated at the expense of a worse control rate for the LNs. It can be observed that the pattern of relations between the response curves depicts almost the pattern of their DVHs.

The discrepancy between the two dose distributions stems from a number of different sources of errors with different weights in their contributions. Among those sources, the most important was the uncertainties associated with the dose delivered by small field segments. Furthermore, the spatial resolution that is used during treatment planning may play a significant role in knowing the effectiveness of the produced dose distribution with high accuracy. At the same time, the resolution in the dose distribution derived from a film can be quite high. This means that a film can provide more fine details of the dose distribution. Another two classical sources of error are the setup uncertainties mentioned before and the accuracy of the algorithms used by the treatment planning system, which are always present during treatment delivery. Finally, a small part in the deviation observed between the dose distributions should be attributed to uncertainties during the film dosimetry even though their impact on the overall comparison was minimal. All these factors can dramatically affect the effectiveness of the intended treatment delivery, influencing not only the radiobiological estimation of the expected clinical outcome but even more the initial derivation of the dose–response parameters used in radiobiological modeling and BioIMRT (biologically intensity modulated radiation therapy) optimization (Schilstra and Meertens 2001; Mavroidis *et al.* 2002a).

Today, a method for a 3D *in vivo* treatment plan verification and dose control during external radiation therapy using PET is investigated (Janek 2003, 2013). This technique is based on the activation of tissue by photons from a 50 MV photon beam, resulting in the production of ^{11}C, ^{13}N, and ^{15}O, β^+-emitting isotopes. PET is a coincidence counting technique for detecting pairs of annihilation γ-rays and it is used to monitor the positron activity induced in the tissue. Therapy monitoring can be achieved by comparing the measured β^+ activity distribution with an expected pattern calculated from the treatment plan as well as the anatomical information of the patient. This results in an appropriate method of treatment plan verification, aiming at the minimization of the damage to sensitive healthy tissues. Such a dose distribution can be more accurately compared with the intended treatment plan and associated with the clinical outcome. These images can be taken after each fraction during the course of radiation therapy allowing the correction of eventual deviations.

Although the treatment configuration that was applied is strictly related to the clinical case under consideration, the general results and conclusions about the need of using radiobiological measures in treatment plan verification did not change. Solely, the complete 3D dose distribution, together with the fractionation schedule, do not suffice to describe the quality of a treatment plan. Biological objective functions are needed to evaluate the curative power of the delivered radiation therapy. In the clinical routine, physicians use biological data in their evaluation criteria (e.g., tolerance doses, fractionation protocols, tumor stage, etc.). Presently, their practice is often governed by a deterministic interpretation of the dosimetric data, namely certain prescribed doses correspond to tumor control (control certainty), etc. However, this is not the case since the radiation beams by necessity have a stochastic effect on tumor control at the microscopic level. For this reason, a more accurate radiobiological and stochastic consideration of the dosimetric data, meaning that certain prescribed doses correspond to certain control probabilities, is more clinically relevant.

In Fig. 3.23, for the planned dose distribution, the maximum value of P_+ was 84.1% for a $\bar{\bar{D}}$ to the ITV of 89.2 Gy. At this dose level ($\overline{D}_{ITV} = 93.3$ Gy), the control rate for the whole ITV, P_B, was 93.2%, whereas the total complication probability, P_I was 9.1%. For the delivered dose distribution, the value of P_+ after the dose escalation ($\overline{D}_{ITV} = 97.9$ Gy) was 79.4% with a $\bar{\bar{D}}$ to the ITV of 86.5 Gy. The associated values of P_B and P_I were 89.1% and 9.7%, respectively. In the lower left diagram, it can be seen that the curves of the organs at risk are well separated from the ones of the targets, meaning that normal tissues are spared well and this technique provides a good conformation. Contrary to the upper panel, where the total dose distributions are examined, a significant deviation is observed between the two distributions in the dose delivered to the different clinical structures separately (mainly to the normal tissues). It is noticed, that the pattern of the response curves depicts the pattern of the corresponding DVHs.

The radiobiological evaluation that is used in this example provides a better description and more complete utilization of the biological data already used clinically. These radiobiological parameters are based on data derived from patients, who have been treated in the past and whose follow-up results had been registered (Emami *et al.* 1991; Ågren 1995). These data are to a large extent reliable for a large number of organs and soon a complete library of them will be available for this kind of evaluations, taking into account the individuality of the patients. One needs both the complete dosimetric data and the biological description of the individual case to evaluate the quality of a treatment plan and the accuracy by which this plan is delivered. In this way, the increase of the risk of a tumor control failure or a severe normal tissues complication, because of the deficiency of the clinical methodology to deliver exactly the planned radiation therapy, can be estimated and taken into account accordingly.

From the analysis of the treatment plans shown in Figs. 3.11 and 3.23, it is apparent that the more conformal ones can deliver a higher dose level to the target than the less conformal ones. The more conformal treatment plans thereby allow increased control rates for the same or a reduced risk for complications. This is radiobiologically equivalent to having a high dose within the target and a steep dose gradient at the border of the ITV proximal to the organs at risk. Figure 3.24 illustrates a study of how the effectiveness of the applied dose distribution drops when the target is not aligned correctly with the beam. The study was carried out for a highly conformal treatment plan (upper diagram) and a plan of reduced conformation (lower diagram). In this diagram, the dose distribution has been shifted within a 8 cm range in the AP and the sinister-dexter directions. The values of P_+, which expresses the quality of the treatment plan, drop much more dramatically if the dose distribution is shifted from its optimum position in the vertical direction where the primary organs at risk, the bladder and the rectum, have their proximal borders in direct contact with the ITV. Generally, such a degradation of the effectiveness is more pronounced in conformal treatment plans, where the setup of the phantom is more critical. A method that can make such dose plans more robust against misalignments between the irradiating beam and the patient has been proposed by Lind and Löf (Lind 1991; Lind et al. 1993; Löf et al. 1995).

Although the techniques that were applied depend on the clinical case at hand, the general results and subsequent conclusions about the current status of the technological capabilities in radiation therapy do not change; neither does the need for a biological evaluation of treatment plans. It is claimed that biological objective functions allow much higher dose conformity. However, the dose distribution becomes more conformal only through the use of technological capabilities. The biological objective functions help the dose distribution to conform to the radiation sensitivity map of the case. The complete dose distribution (in 3D space), together with the fractionation schedule applied, contains most of the information of a treatment plan. Radiobiological evaluation provides just better and more complete description of the biological data already used clinically. Both the complete dosimetric data and the biological description of the individual case are needed to evaluate the quality of a treatment plan.

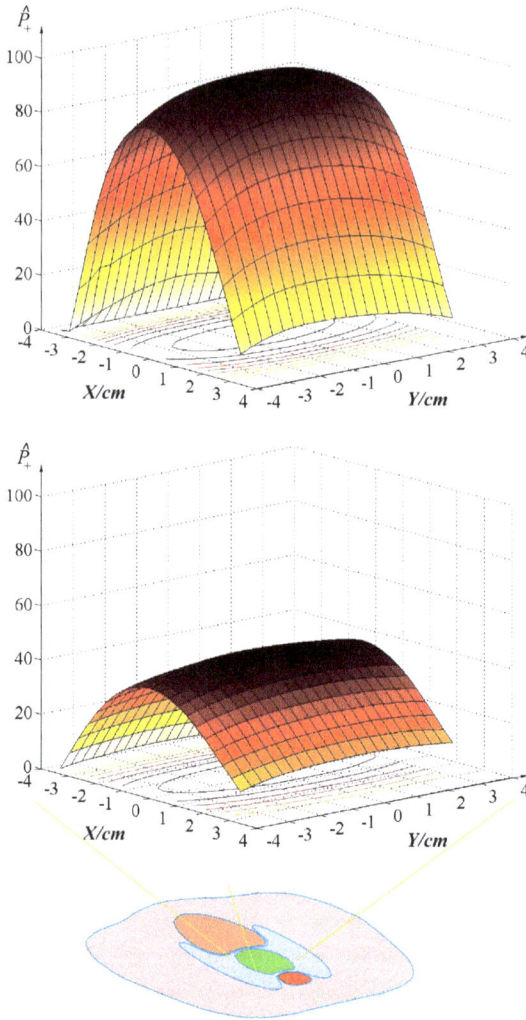

Figure 3.24.   Demonstration of the effects of patient setup on the expected clinical outcome. It is observed that a setup misalignment in the direction, where the bladder and the rectum are located, reduces rapidly showing that conformal plans are very sensitive to positioning errors. If a reliable setup procedure is not available, a less conformal treatment technique could be more effective and secure. The gross target volume (cf. Fig. 3.9) in the illustrated anatomical structure depicts the range of misalignments between the delivered dose distribution and the patients who were studied.

3.7. Modern Developments in Radiation Therapy

Many of the new developments in radiation therapy have the potential to considerably improve the treatment outcome for most target sites. This does not mean
that we need to implement the most advanced form of IMRT on all treatment
units in every clinic. Most small tumors can be treated very effectively by standard
treatment techniques even though biologically optimization still may reduce late
complications in most tumors. However, when the new optimized treatment
techniques have been fully integrated in modern equipment with full computer
control and biologically based treatment planning, it will not be more complex to
make advanced optimized intensity-modulated treatment plans for all patients and
deliver them using computer-controlled intensity-modulated treatment units. In
fact, it will be simpler since most of the new methods are ideally suited for full
automation today when we are mainly using computer-controlled equipment in the
clinic. In the transition period that we are living in, costs will temporarily go up,
since we have to improvise until most existing treatment units and planning systems
are fully developed and well integrated. This obstacle should be gradually resolved,
hopefully some five years from now. After that time it should be much easier to get
fully integrated equipment and their costs may also start to reduce slowly.

In a 5- to 10-year perspective, we will not only have well-functioning and
integrated treatment units and planning systems but also many rather accurate
genetically based assay techniques to better predict optimal treatment strategies for
individual patients. In fact, the new biological optimization methods will become a
central tool for the development of improved dose–response models (cf. Fig. 8.29(b))
and predictive assays. This is because the new biological models and optimization
technique are the ideal tools for radiation biologists, radiation oncologists, and
radiation physicists, to work together for the development of the profession. Not only
will they help in optimizing the outcome of a given treatment, they will also indicate
those organs and sites where the treatment is most likely to produce adverse reactions
to simplify follow up and improve feedback of response data to the models. In fact,
it is likely that all the new optimization methods that are now becoming available
may even alter the way randomized clinical trials are performed. The improved
knowledge about the patient and the substantially improved therapeutic modalities
and methods of follow up will make the treatment of every patient with advanced
disease a radiobiological optimization challenge. The treatment outcome will then
be compared with the accumulated historical experience, which is contained in the
predictions of the biological models, and fewer and fewer patients are likely to be
randomized to suboptimal treatment arms. The large historical data sets built into
the biological models will ensure a minimal contribution from patient individual

variations, which cannot be avoided in classical randomized trials. Furthermore, biological optimization model-based techniques will not only result in optimized treatment plans for each patient but also ensure the most efficient feedback for the development of the radiobiological models and to the fastest possible benefit of future patients.

It is becoming increasingly clear that we have essentially one real chance to cure a patient by radiation therapy, namely at the time of the first treatment. Many tumors, where the initial treatment was not radical enough, seem, when they reoccur locally several years after the treatment, to have progressed in severity and often metastasized. It is therefore important, not least from a cost–effectiveness point of view, to be really radical at the first treatment and make use of the tolerance of the patient and his own judgments of acceptable side effects to really maximize the complication-free cure and to some extent the quality of life. Such arguments will increase the use of advanced treatment techniques and will be our best way to improve the efficacy of radiation therapy in the future.

Glossary of Terms

$\overline{\overline{D}}$ Biologically effective uniform dose
D_{eff} Effective dose
EUD   Equivalent uniform dose
P_+ Complication-free tumor control probability
P_B Probability of benefit (tumor cure)
P_I Probability of normal tissue injury

Bibliography

Aaltonen P, Brahme A, Lax I, Levernes S, Näslund I, Reitan JB, Turesson I (1997) Specification of dose delivery in radiation therapy. Recommendations by the Nordic Association of Clinical Physics (NACP). *Acta Oncol* 36(Suppl 10):1–32.

Ågren A, Brahme A, Turesson I (1990) Optimization of uncomplicated control for head and neck tumors. *Int J Rad Oncol Biol Phys* 19:1077–1085.

Ågren AK (1995) Quantification of the response of heterogeneous tumors and organized normal tissues to fractionated radiotherapy, PhD thesis, Stockholm University, Sweden.

Åsell M, Hyödynmaa S, Söderström S, *et al.* (1999) Optimal electron and combined electron and photon therapy in the phase space of complication free cure. *Phys Med Biol* 44:235–252.

Brahme A (1982) Physical and biologic aspects on the optimum choice of radiation modality. *Acta Radiol Oncol* 21:469–479.

Brahme A (1984) Dosimetric precision requirements in radiation therapy. *Acta Radiol Oncol* 23:379–391.

Brahme A (1987) Design principles, beam properties and clinical possibilities of a new generation of radiation therapy equipment. *Acta Oncol* 26:403–412.

Brahme A (1994) Which Parameters of the Dose Distribution are Best Related to the Radiation Response of Tumours and Normal Tissues. *Proc Interregional Semin for Europe, the Middle East and Africa*, Organized by the IAEA, Leuven, pp. 37–58.

Brahme A (1995) Treatment optimization using physical and biological objective functions. In: Smith A (ed.), *Radiation Therapy Physics*, 209–246. Berlin: Springer.

Brahme A (1997) The need for accurate target and dose specifications in conventional and conformal radiation therapy — an introduction. *Acta Oncol* 36:789–792.

Brahme A (2000) Development of radiation therapy optimization. *Acta Oncol* 39:579–595.

Brahme A (2011) Accurate description of the cell survival and biological effect at low and high doses and LETs. *J Rad Res* 52:389–407.

Brahme A, Lind BK (2010) A systems biology approach to radiation therapy optimization. *Radiat Environ Biophys* 49:111–124.

Brahme A, Ågren A (1987) On the optimal dose distribution for eradication of heterogeneous tumors. *Acta Oncol* 26:377–385.

Brahme A, Chavaudra J, Landberg T, McCullough EC, Nüsslin F, Rawlinson JA, Svensson G, Svensson H (1988) Accuracy requirements and quality assurance of external beam therapy with photons and electrons. *Acta Oncol* 27(Suppl 1):1–76.

Brahme A Ed. (2014) In Chief: Comprehensive Comprahensive BioMedical Physics, Major Reference Work Vol 8 Eds Hendry and Brahme Elsevier Oxford.

CROS (Committee for Radiation Oncology Studies) (1978) Proposal for a program in particle-beam radiation therapy in the United States. *Cancer Clin Trials* 1:153–208.

Ekberg L, Holmberg O, Wittgren L, Bjelkengren G, Landberg T (1998) What margins should be added to the clinical target volume in radiotherapy treatment planning for lung cancer? *Radiother Oncol* 48:71–77.

Emami B, Lyman J, Brown A, Coia L, Goitein M, Munzenrider JE, Shank B, Solin LJ, Wesson AM (1991) Tolerance of normal tissue to therapeutic radiation. *Int J Radiat Oncol Biol Phys* 21:109–122.

EOLSS http://www.eolss.net/outlinecomponents/physical-methods-instruments-measurements.aspx

Fletcher GH (1980) *Textbook of Radiotherapy*. Lea & Febiger, Philadelphia, USA.

Gustafsson A, Lind BK, Brahme A, *et al.* (1994) A generalised pencil beam algorithm for optimization of radiation therapy. *Med Phys* 21:343–356.

Hall EJ (2000) *Radiobiology for the Radiologist*. Lippincott, Philadelphia, USA.

Hope CS, Orr HS (1965) Computer optimization of 4 MeV treatment planning. *Phys Med Biol* 10:365–373.

ICRU 50 () Prescribing, recording and reporting photon beam therapy. *Int Commission of Radiation Units and Measurements* (ICRU), Bethesda, Maryland, USA.

ICRU 62 (1999) Prescribing, recording and reporting photon beam therapy (supplement to ICRU 50). *Int Commission of Radiation Units and Measurements* (ICRU), Bethesda, Maryland, USA.

Janek S (2003) 3-dimensional patient dose delivery verification based on PET-CT imaging of photonuclear reactions in 50 MV scanned photon beams, MSc thesis, Stockholm University, Sweden.

Janek Strååt S, Andreassen B, Jonsson C, Noz ME, Maguire GQ, Näfstadius P, Näslund I, Schoenahl F, Brahme A (2013) Clinical application of in vivo treatment delivery verification based on PET/CT imaging of positron activity induced at high energy photon therapy *Phys Med Biol* 58: 5541–5553

Källman P, Ågren A, Brahme A (1992) Tumour and normal tissue responses to fractionated non-uniform dose delivery. *Int J Radiat Biol* 62:249–262.

Korevaar EW, Heijmen BJM, Woudstra E, *et al* (1999) Mixing intensity modulated electron and photon beams: combining a steep dose fall-off at depth with sharp and depth-independent penumbras and flat beam profiles. *Phys Med Biol* 44:2171–2181.

Lind BK (1991) Radiation therapy planning and optimization studied as inverse problems, PhD thesis, Stockholm University, Sweden.

Lind B, Brahme A (1995) Development of treatment techniques for radiotherapy optimization. *Int J Imag Syst Technol* 6:33–42.

Lind BK, Källman P, Sundelin B, Brahme A (1993) Optimal radiation beam profiles considering uncertainties in beam patient alignment. *Acta Oncol* 32:331–342.

Lind BK, Mavroidis P, Hyödynmaa S, Kappas C (1999) Optimization of the dose level for a given treatment plan to maximize the complication-free tumor cure. *Acta Oncol* 38:787–798.

Löf J (2000) Development of a general framework for optimization of radiation therapy, PhD thesis, Stockholm University, Sweden.

Löf J, Lind BK, Brahme A (1995) Optimal radiation beam profiles considering the stochastic process of patient positioning in fractionated radiation therapy. *Inverse Problems* 11:1189–1209.

Löf J, Lind BK, Brahme A (1998) An adaptive control algorithm for optimization of intensity modulated radiotherapy considering uncertainties in beam profiles, patient set-up, and internal organ motion. *Phys Med Biol* 43:1605–1628.

Löf J, Lind BK, Liander A, *et al.* (1999) Simultaneous beam orientation and intensity modulation optimization using the new universal radiotherapy optimization code ORBIT. *Int J Radiat Oncol* (Submitted).

Lyman JT (1985) Complication probability as assessed from dose-volume histograms. *Radiat Res* 104:S13–S19.

Maehle-Schmidt M, Palmgren J, Lind B, *et al.* (1999) A Bayesian Sequential Model for Updating Radiobiological Parameters in Radiation Therapy. *Second European Conf on Highly Structured Stochastic Systems*, Pavia, Book of Abstracts, pp. 180–181.

Maor MH, Hussey DH, Fletcher GH, *et al.* (1981) Fast neutron therapy for locally advanced head and neck tumors. *Int J Rad Oncol Biol Phys* 7:155.

Mavroidis P, Axelsson S, Hyödynmaa S, Rajala J, Pitkänen MA, Lind BK, Brahme A (2002a) Effects of positioning uncertainty and breathing on dose delivery and radiation pneumonitis prediction in breast cancer. *Acta Oncol* 41:471–485.

Mavroidis P, Ferreira BC, Papanikolaou N, Svensson R, Kappas C, Lind BK, Brahme A (2006a) Assessing the difference between planned and delivered intensity-modulated radiotherapy dose distributions based on radiobiological measures. *Clin Oncol* 18: 529–538.

Mavroidis P, Kappas C, Lind BK (1997) A computer program for evaluating the probability of complication-free tumor control incorporated in a commercial treatment planning system. *J Balcan Union Oncol* 3:257–264.

Mavroidis P, Lind BK, Brahme A (2001) Biologically effective uniform dose ($\overline{\overline{D}}$) for specification, report and comparison of dose response relations and treatment plans. *Phys Med Biol* 46:2607–2630.

Mavroidis P, Lind BK, Brahme A (2002b) $\overline{\overline{D}}$, an effective uniform dose linked to the probability of response. *Phys Med Biol* 47:L3–L9.

Mavroidis P, Lind BK, Van Dijk J, Koedooder K, De Neve W, De Wagter C, Planskoy B, Rosenwald JC, Proimos B, Kappas C, Danciu C, Benassi M, Chierego G, Brahme A (2000) Comparison of conformal radiation therapy techniques within the dynamic radiotherapy project "DYNARAD." *Phys Med Biol* 45:2459–2481.

Mavroidis P, Plataniotis GA, Adamus-Gorka M, Lind BK (2006b) Comments on "reconsidering the definition of a dose–volume histogram"—dose–mass histogram (DMH) versus dose–volume histogram (DVH) for predicting radiation-induced pneumonitis. *Phys Med Biol* 51:L43–L50.

Mijnheer B (1997) Current clinical practice versus new developments in target volume and dose specification procedures: a contradiction. *Acta Oncol* 36:785–788.

Montelius A, Blomquist E, Naeser P, et al. (1991) The narrow proton beam therapy unit at the Svedberg Laboratory in Uppsala. *Acta Oncol* 30:739–745.

Niemierko A (1997) Reporting and analyzing dose distributions: a concept of equivalent uniform dose. *Med Phys* 24:103–110.

Oldham M, Webb S (1997) Intensity-modulated radiotherapy by means of static tomotherapy: a planning and verification study. *Med Phys* 24:827–836.

Partridge M, Symods-Tayler JRN, Evans PM (2000) IMRT verification with a camera-based electronic portal imaging system. *Phys Med Biol* 45:N183–N196.

Ploeger LS, Smitsmans MHP, Gilhuijs KGA, van Herk M (2002) A method for geometrical verification of dynamic intensity modulated radiotherapy using scanning electronic portal imaging device. *Med Phys* 29:1071–1079.

Schilstra C, Meertens H (2001) Calculation of the uncertainty in complication probability for various dose-response models, applied to the parotid gland. *Int J Radiat Oncol Biol Phys* 50:147–158.

Söderström S, Brahme A (1993) Optimization of the dose delivery in few field techniques using radiobiological objective functions. *Med Phys* 20:1201–1210.

Söderström S, Eklöf A, Brahme A (1999) Aspects on the optimal photon beam energy for radiation therapy. *Acta Oncol* 38:179–187.

Sterling TD, Perry H, Weinkam JJ (1965) Automation of radiation treatment planning V. Calculation and visualisation of the total treatment volume. *Br J Radiol* 38:906–913.

Svensson R, Lind BK, Brahme A (1998) Beam characteristics and possibilities of a new compact treatment unit design combining narrow pencil beam scanning and segmental multileaf collimation. *Med Phys* 25:2358–2369.

Takahashi S (1965) Conformation radiotherapy, rotation techniques as applied to radiography and radiotherapy. *Acta Radiol* (Suppl 242).

Ting JY, Davis LW (2001) Dose verification for patients undergoing IMRT. *Med Dos* 26:205–213.

Withers HR, Taylor JMG, Maciejewski B (1988) Treatment volume and tissue tolerance. *Int J Radiat Oncol Biol Phys* 14:751–759.

Wolbarst AB (1984) Optimization of radiation therapy II: the critical-voxel model. *Int J Radiat Oncol Biol Phys* 10:741–745.

Wright KA, Proimos BS, Trump JG (1959) Physical aspects of two million volt X-ray therapy. *Surg Clin North Am* 39:567–578.

Development of High Quality Beams for Uniform and Intensity-Modulated Radiation Therapy

4

Anders Brahme, Roger Svensson, and Bo Nilsson

4.1. Toward Advanced Dose Delivery Techniques

Truly, 3D and 4D conformal dose delivery is currently the goal in radiation therapy, aiming at highest possible treatment outcome, and will be brought to routine clinical use. The treatment plan will be individually optimized for each patient and automatically and safely executed. Random setup errors and the motion of the patient, internal organs, and the tumor are other possible dynamic effects during the treatment course that should be taken into account by different gating and breath hold techniques (cf Figs. 8.28, 8.38e, f, and 8.39 and Brahme *et al.* 2008). The principal goal of curative therapy is to eradicate all tumor clonogens without causing severe injury in normal tissues or more generally to maximize the quality of life of the patient. In general, a high curative dose should be delivered to the target volume and as a low dose as possible to the healthy tissues and to radiation-sensitive organs. This requires that the quality of the beams be well defined, stable, and accurately setup during the whole fractionated treatment schedule.

During the past 50 years, since the first high-energy therapy machines where introduced in the early 1950s, the development of therapy equipment and treatment techniques for advanced cancers has been fantastic. The widely used ^{60}Co-sources mounted for isocentric photon therapy treatments has almost disappeared and been replaced by sharper and more flexible therapy sources such as traveling- and standing-wave linear accelerators and high power recirculated accelerators such as circular microtrons and racetrack microtrons. The high-energy electron accelerators generally have the possibility to deliver both photon and electron beams of a broad range of energies. This is the case not least for the racetrack microtron delivering energies from 2.5–50 MeV to 10–50 MV (Brahme *et al.* 1980, 1987, 1988). Light ion beams with their special dose distributional and radiobiological properties are probably the ultimate radiation treatment modality and are slowly being established on a global scale.

Today, most treatments are delivered using uniform electron beams or block or wedge-filtered photon beams, although the use of intensity -modulated photon beams are steadily increasing, but not least in the United States where the improved methods are stimulated by a rather generous reimbursement systems due to private insurance companies. It is well known from treatment planning using radiobiological optimization algorithms that the treatment outcome could be significantly increased if the dose delivery could be improved by intensity modulation of the incoming beams so the tumor dose is in average increased and sensitive normal tissues spared. The technical development is mainly concerned with how such a beam modulation is most effectively done. Numerous laborious and time-consuming methods has been proposed to spatially modulate the incoming beams, such as mechanical compensators (cf. Ulsø and Brahme 1988) and conventional molding techniques and dynamic asymmetric jaws (cf. Kijewski *et al.* 1978).

However, these irradiation techniques are now largely replaced by more generalized and flexible dynamic intensity-modulating techniques such as real-time dynamic multileaf collimation and/or scanned elementary beams (Brahme 1988). The advantage of dynamic techniques over, for instance, molding techniques is not only the obvious increased flexibility in dose delivery but more important the inherent potential of quickly changing the incoming beam both between treatment fractions and beam portals and even instantaneously during the treatment by dynamic online tailoring the dose delivery depending on organ motions. The more knowledge we have about the state of the patient, such as the location of gas pockets in rectum when treating a prostate cancer or breathing movements when treating a breast or lung tumor, the more valuable it is to perform the real-time dynamic dose delivery. If the dynamics during the course of the treatment is not properly considered by gated breath hold or dynamic therapy, the requirements for intensity modulated are not adequately met.

Since a large amount of treatment units using different therapy sources and treatments techniques have been developed over the years, the development is here briefly reviewed and the main design features of modern treatment units for high precision conformal intensity-modulated radiation therapy (IMRT) are discussed in some detail.

4.2. Development of Radiation Sources for External Beam Treatments

Less than a year after the discovery of X-rays in 1895, clinical machines were built and used for eradication of superficial malformations and tumors. The clinical need to treat deep and large tumors and also to perform the treatment in a reasonable time stimulated the engineers to develop therapy machines toward photon and electron beams of higher energies and beam intensity. The radiotherapy community has since then witnessed an enormous development during this past century and many different types of therapy sources and dose delivery techniques have been used and proven to be successful for radiation therapy. The machine types that has been used for therapy falls mainly into three large groups: *recirculated* accelerators, such as betatrons, microtrons, racetrack microtrons synchrotrons, and cyclotrons for electrons, protons, or light ions, and *linear* accelerators such as X-ray tubes, Van de Graaff generators, and traveling- or standing-wave linacs, today the most widely used accelerator, and finally therapy machines where the source is a *radionuclide* such as ^{60}Co and ^{252}Cf. All these groups have been developed in parallel during the past 60 years. The racetrack microtron has a special role as it can be regarded as a hybrid with both linear and recirculation features. The main principles of the therapy machines regarding injection, acceleration process, and extraction of the beam and beam quality will be briefly discussed here.

^{60}**Co-based therapy machines.** Shortly after Röntgen discovered X-rays, a great interest was devoted to their use for treatment of cancer. After World War I, radium-226 was used as an external or intracavitary source with some success although it was costly and had a low dose rate. As a spin-off from the interest in nuclear energy after World War II, ^{60}Co that possess "megavoltage" mean energy of 1.25 MeV proved to be a cost-effective source for radiation treatment, and the first collimated ^{60}Co teletherapy machine came into use in the early 1950s. These units were quite simple and easy to maintain compared to the more complex machines based on linear or recirculated accelerators, so during the 1960s, most treatments were performed with isocentric ^{60}Co units. Today, the more flexible gantry-mounted linear accelerators that have a wider range of energies from 4 to 20+ MeV both for electrons and photons have practically replaced the ^{60}Co units. The development of the γ-knife, where 201 collimated ^{60}Co-sources are uniformly distributed over a hemispherical area and focused on the same point, has proven to work well for small intracrainal tumors. This machine is dedicated to treat lesions in the head and is capable of delivering a single fraction of high dose with sub-millimeter precision (Larsson *et al.* 1974). A similar multi-cobalt technique has been developed for the whole body (cf. Brahme and Lind 1995). This multi-cobalt unit looks a little bit like a computerized tomography unit and around 100 sources are uniformly distributed around the patient. Two concentric collimator rings can be rotated and moved axially so that arbitrary cylindrical target volumes can be varied from zero up to a diameter × height of $7 \times 7 \, \text{cm}^2$. The machine can in real time deliver intensity-modulated fields by moving the gantry during the treatment. The advantage with these multi cobalt devices is that the common large source is replaced by many small sources making the penumbra very sharp.

Circular accelerators. *Betatrons*: In the beginning of the century it was well known from Maxwell's equations that the azimuthal electrical field associated with a time-varying axial magnetic field could accelerate charged particles. Many attempts were made to control the acceleration process of the electrons, and Slepian (1922) was probably the first to suggest the basic principles regarding acceleration of electrons to high energies. He proposed an induction accelerator similar to a transformer where the induction forces from the time-varying magnetic field in the yoke could act on freely moving electrons around the central leg of the transformer. The circular bending of the electron orbits was done with a permanent magnet within the transformer. The magnet was designed intuitively correct in such a way that when the electrons were accelerated toward higher energies, they increasingly spiraled outward the magnetic field so they could remain as long as possible in the orbit. However, this will cause poor axial stability well known from cyclotron theory, and probably such a design could in reality never work. The problem with axial focusing of the electrons were earlier recognized and in principal solved by Tuve and Breit (1927) at the Bureau of Terrestrial Magnetism in Washington when they worked on discharging methods through air solenoids to generate fast time-varying magnetic fields. The most important contribution to the development of the working principle for the betatron was made by Wideröe (1928) as he, without knowledge about Slepians design, developed a transformer-based accelerator. Instead of using a permanent magnet within the transformer to bend the electrons in a circular orbit, he used the same magnetic flux coming from the same coil to couple the bending of the electrons with the induced acceleration. Thus, when the electrons gained higher energies from the magnetic field increasing with time, they stayed in orbit at the same radius due to the higher bending magnetic field. This principle works as long as the Wideröe condition is fulfilled, that is, that the magnetic flux at the orbit should be equal to half the average magnetic flux over the mean orbit area. Wideröe also made a complete design for both injection of electrons from an external cathode or ray gun and extraction of the high-energy electrons by deflecting coils. Probably his injection of electrons from outside the magnetic field was not so effective since only a small part of the electrons if any would be captured by its rather slowly rising magnetic field and consequently no useful betatron with external injection was ever built. At the end of the 1930s, most of the problems about keeping the electrons in the equilibrium orbit such as radial and axial focusing and damped oscillation were worked out. The problem of injecting electrons was first solved by Kerst (1941). He placed the injector within the magnetic field close to the equilibrium orbit and succeeded in building the first operational small betatron exceeding 1 MV (2.3 MeV) even though of low intensity. Shortly after, in Germany, a 6 MeV betatron was built in 1944 by Siemens in Erlangen (cf. Gund and Paul 1950). In the mid-1940 s, in United States, the electron energy was raised to 22 MeV and the intensity was high. Only a few years later, Kerst together with, among others, Skaggs, Laughlin, and Lanzel at the University of Illinois, worked on a therapeutic unit and in the late 1940s they treated their first patient (Kerst 1975). Their machine also became the prototype for Alice-Chalmers 25 MeV betatrons for industrial radiography, radiation therapy, and research. Simultaneously, Siemens compact 18 MeV and later their large 42 MeV betatron was developed, whereas Wideröe et al. developed Brown Boveries 35 and 45 MeV machines. The fan beam, small angel pendulum technique with the betatron (cf. Fig. 4.43(a) below) allowed thin scattering foils and rather good depth-dose distributions at high energies (Rassow 1970).

Synchrotrons: Only a couple of years after Kerst's first working betatron, Veksler (1944) and McMillan (1945) proposed that the energy of high-energy electrons in the equilibrium orbit of a betatron could additionally be accelerated by synchronous action of a high-frequency electromagnetic field tangential to the orbit. The synchronous acceleration is possible since a high -energy electron bunch travels in the same orbit with almost constant speed very close to the speed of light and additional energy gain only raises the electron mass. When the electron energy is raised, the magnetic field, thus, has to be increased to keep the electrons in the same orbit. Goward and Barnes (1946) succeeded to convert a 4 MeV Kerst betatron into a 8 MeV and later to 14 MeV synchrotron. A dedicated 30 MeV synchrotron was installed at Addenbrooke's Hospital, Cambridge 1949 (Mitchell et al. 1953). The magnet weight was about 4 t. As injection gun, Kerst original design was used, and the electrons was pre-accelerated by betatron, action up to 4 MeV before synchrotron action took over. At full energy, the resonator energy was switched off with a still raising magnetic field such that the electrons were bent inward hitting a 2 mm tungsten target and flattened by a conical copper filter. Even though they treated a few patients with this machine, it suffered from rather low dose rate and instability and this is probably the reason why synchrotrons was never used more often, except possibly the 70 MeV synchrotron of General Electric.

Microtrons: The principle of resonant electron acceleration was first published by Veksler (1944), although the idea had been around for a couple of years, and it did not come into clinical use until the mid-1970s. The electrical field in a single small resonant cavity fed by microwaves accelerates the electron bunches. After acceleration, the electrons travel in a circular orbit forced by the surrounding uniform magnetic field in the gap between two circular cylindrical iron disks. For each turn, the electron energy is raised and the orbital radius for the electron bunch becomes larger. This will only work if the return of the electrons to the resonator cavity is in phase with the accelerating field such that the time spent for the electron bunch in each orbit must be an integral number of full periods in the microwave field. When this microtron resonance condition is fulfilled, the electron bunch can be repeatedly accelerated by the cavity to achieve very high energies. The extraction for a specific electron energy is simply done by moving an iron tube to the corresponding electron orbit. The first microtron was used as early as 1948 (cf. Henderson *et al.* 1948). That machine and the other following it in the 1950s suffered from a low beam-intensity since the injection system was based on the field emission in the resonator and thus most of the electrons was out of phase for acceleration although the injected current could be high. A practical microtron delivering high intensity could not be build until an external small electron gun was proposed (Sedlacek 1955), so that the injection could be controlled (Kapitza *et al.* 1960; Wernholm *et al.* 1974). The microtron then became widely used due to the high output current and low energy spread so it was often used for activation analyses and as an injector for synchrotrons and storage rings. The high beam quality, though with low-energy spread, and the low emittance of the intrinsic beam was very useful for radiation therapy and radiography. The dual scattering foil technique was developed and optimized for the microtron to keep the high quality in the electron beams (cf. Fig. 4.43(f) and Brahme 1972; Grusell *et al.* 1994).

Racetrack microtrons: By splitting the magnet of the microtron in two halves or more and separating them with a linear accelerator in between, the energy gain for each turn could be considerably increased (Schwinger 1946). With this method, it is possible to reach the high-energy therapy range (Wiik and Wilson 1970; Wernholm *et al.* 1979) and even much higher energy therapy range $>500\,\text{MeV}$ (Wiik and Wilson 1967). For this accelerator, the scanning beam technique was developed both for electrons and photons (cf. Figs. 4.10, 4.40, and 4.43(d) and Brahme 1987).

Linear accelerators. *X-ray tubes*: The principle of the X-ray tubes is to emit electrons from a hot filament and accelerate them from the cathode in a static electrical field to the anode where characteristics X-rays and bremsstrahlung are produced. Tube potentials up to around $300\,\text{kV}$ were used for therapy and therefore only superficial lesions could be treated well. Special units reaching the MV region were tested, such as Cockcroft–Walton and resonant transformers, but these did not reach high clinical interest.

Van de Graaff accelerators: The first linear accelerator for energies $>300\,\text{keV}$ was based on accelerating charged particles in a high static electric field such as from a Van de Graaff accelerator. The high electric field strength required was difficult to maintain. High-energy Van de Graaff accelerators up to about 4 MeV were used for therapy as late as in the 1980s.

Traveling- and standing-wave type linear accelerators: The basic principle of the linac is to feed a series cylindrical copper cavities with microwaves and inject the electrons along the central axis to accelerate them. To achieve a continuous acceleration of the electrons, the phase velocity of the field must be slowed down below the velocity of light by dividing the waveguide into a series of cavities. In traveling-wave machines, the first few cavities where the low-energy electrons enter are usually longer and the electrons are bunched. The electrons then travel with approximately the speed of light but only their mass are increased. Two main types of waveguides have been used: the *standing-* and *traveling-* type linear accelerator. The first working linear accelerator for ions was actually built by Wideröe. The working principles and technical development has been reviewed, for instance, by Karzmark (1984) and Karzmark *et al.* (1993).

In the traveling-wave-type accelerator, the microwave power is fed in one end of the structure and the electrons are accelerated in phase with the traveling wave through the whole structure. For a given field strength in the cavity, the length of the structure determines the output energy. In the beginning of the 1960s, Lovinger *et al.* used a 70 MeV traveling-wave linac of several meters length that was powered by two large klystrons, at Stanford. It was placed in a room next to the treatment room and the electron beam from the accelerator was guided into the treatment room with scattering foils and collimators. A similar system with a gantry was also used at Argonne Cancer Research Center in Chicago with a unique mechanical pencil beam displacement system to generate broad 50 MeV electron beams (cf. Fig. 4.43(b) and Carpender *et al.* 1963). Previously, in the mid-1950s, Hsieh and Uhlmann (1956) also evaluated and used a 45 MeV traveling-wave linac as a source to generate broad flattened electron beams at Michael Reese hospital also in Chicago by defocusing the narrow electron beam by the variable pole tip of the bending magnet (Fig. 4.43(e)).

In the standing-wave type accelerator, the microwave power is reflected at the ends of the guide to build up a standing wave system. For equidistantly spaced cavities, the resultant electrical field would always be zero in every second cavity and will not contribute to electron acceleration. Hence, those cavities can be made short or moved to side from the accelerator axis to serve as coupling cavities. By interlacing two such structures, it is possible to get acceleration in every cavity (the interlaced linac developed by Victor Vagin). Thus, the microwave power can be inserted anywhere along the whole structure at such places. In this manner, the whole structure can be made short but it will need a very high microwave power. Most of the medical linacs works in the 3 GHz frequency range (S-band) corresponding to cavities around 10 cm in diameter and 2.5–5 cm length. A gantry-mounted side-coupled standing wave linac such as in the Clinac 2500 could use either a 6 or a 24 MV beam. The length of the structure is about 2 m. With a much shorter accelerator length, the same energy can be achieved by multiple passes of the electron beam. In the Therac 25 machine, the electron beam was allowed to pass two times through the same structure using a achromatic butterfly-like reflection magnet. Today, at the Stanford Linear Accelerator Center, they are developing a X-band 11.4 GHz linac of the traveling wave working at energies up to 50–100 MeV/m which may soon be possible with high powered klystrons (50–100 MW in the pulse). Since the wavelength is only one-third of a conventional S-band linac, the tight geometrical tolerances will be more difficult to meet. The 32–40 MeV stationary traveling wave linac developed in France used a compact quadrupole to scan the electron beam in Lissajou-type figure for beam flattening (cf. Fig. 4.43(c)).

For all the above accelerator designs, there is a common problem of producing therapeutically useful clinical beams, which will be discussed in some detail in the following sections.

4.3. Traditional Uniform Beam Dose Delivery

4.3.1. *Treatment Head Design for Broad Photon and Electron Beams*

Depending on tumor location, different quality parameters are important. For deep-seated tumors, a photon beam with a deep penetration, low dose to the skin and superficial tissues, and a narrow penumbra is to be preferred. More superficial targets are best treated with electron beams with a high ratio of the therapeutic range to the practical range and therefore a steep dose fall-off gradient. To produce radiation beams with desired quality characteristics, electron accelerators, such as linear accelerators or microtrons, are normally used. In some situations, photon beams from radioactive sources, such as ^{60}Co, may be an acceptable alternative due to their high reliability.

Protons and, in particular, heavier light ions have special advantages due to their small range straggling and sharp penumbra and high therapeutic effects at the end of the range and thus giving a better dose and more importantly a better therapeutic effect distribution in the patients (see Chapter 8).

Electron beams are normally more dependent on the design parameters of the machine than the photon beams. The reason is that electrons are much more easily influenced by scattering and energy loss interactions with the materials in and along the beam. The electrons will on its way pass through a vacuum window, scattering foils, transmission monitors, mirror protection foils, and air. The electrons will also be scattered by the photon collimators. All these materials will act as scatterers and absorbers and will influence the position, size, and energy distribution of the effective electron source. Also, the production of contaminating bremsstrahlung may influence the quality of the electron beam.

Photon beams, on the other hand, are not as much influenced by photon scatter and absorption due to their lower interaction cross-sections. However, the influence of contaminating radiation such as secondary electrons may have a major impact on the dose distribution mainly in the build-up region but also to some degree at larger depths. Obviously, the thick flattening filters in high-energy photon machines may also influence the beam quality and dose rate substantially.

Conventional treatment head. The basic components of the treatment head are the following: the *bremsstrahlung target* where the intrinsic accelerator electron beam impinges and is converted into a bremsstrahlung beam (see Fig. 4.12 below). Right below the primary collimators, *the beam-flattening filter*, which is conically shaped, flattens the beam into a broad uniform photon beam. The monitor chamber for monitoring the dose, dose rate, and beam uniformity is positioned right below the flattening filter. The beam shaping is done with orthogonal block or multileaf collimators (MLCs) or a combination of both. The collimated beam is simulated by a field light that in turn simulate the therapeutic beam by a light source reflected through a thin mirror so it s virtual position coincides with the radiation source upstream of the collimator. A tray holder for wedges is often positioned before the collimator. Today, all vendors are focusing on more effective IMRT capabilities of the treatment head that is, high-resolution MLCs of high speed (see below).

Scanned beam treatment head. The major difference between a conventional treatment head, where the raw electron beam is scattered out and the elementary photon beam is flattened by a flattening filter, and a scanning beam head is that the elementary narrow primary electron or photon beam is electromagnetically scanned over the whole patient surface. In this way, intensity-modulated high-quality broad electron and photon beams can be produced.

The development today of the treatment heads for scanned beam is as for conventional treatment heads focused on improving the efficiency of intensity - modulated dose delivery. Narrow scanned beams are inherently suitable for IMRT. The ultimate treatment head for scanned beam consist of minimal collimating device or even none at all when the scanned pencil beam is sufficiently small (≤ 1 mm). All intensity modulation can then be carried out by scanning the very narrow photon pencil beam of high energy and dose rate (Brahme 1987; Svensson *et al.* 2002; Fig. 4.10). All the individual pencil beam pulses should preferably be monitored by a high-resolution segmented transmission ion chamber, preferably based on the high sensitivity gas electron multiplier (GEM) technology to improve speed and accuracy. An advanced treatment should preferably be executed in a small amount of time and the intensity-modulated dose delivery. The treatment head may also be integrated with a positron emission tomography-computed tomography (PET-CT) camera for combined diagnostics and treatment verification in the same treatment position (cf Fig. 8.39).

4.3.2. *Electron Beams*

4.3.2.1. *Fundamentals of electron beams design*

Two fundamentally different processes complicate the design of broad uniform electron beams. First, due to the high scattering power of the air and other materials in and along the beam, electron beam collimation is difficult at low energies ($<10\,\text{MeV}$). Second, for high energies ($>20\,\text{MeV}$) the combined effect of a high radiation stopping power and low scattering power in most materials makes it difficult to obtain a flattened electron beam at least with scattering foils without excessive bremsstrahlung production and beam quality degradation (Brahme and Svensson 1979). Figure 4.1 shows a typical treatment head in electron mode for a high-energy dual modality accelerator where both electron and photon beams are available (third-generation machine, Brahme 1987).

4.3.2.2. *Electron beam quality parameters*

To characterize the therapeutic, dosimetric, and physical properties of the electron dose distribution, various quality parameters are needed. The most common parameter is the practical range, R_p, which can be used to determine the most probable electron energy at the phantom surface. The half-value depth, R_{50}, may be used to obtain the mean electron energy at the surface and this is an essential quantity for determining dosimetric factors, such as stopping power ratios. The relation between R_p, R_{50}, E_p, $\overline{E}$ and other energy dependent parameters are described in several dosimetry protocols (e.g., NACP 1980; AAPM 1983; ICRU 35, 1984; IAEA 1997, 2000; Sorcini *et al.*, 1995). The clinical usefulness of the beam may be characterized by its therapeutic range, which is related to a certain absorbed dose level, usually 85–95%. The dose fall-off behind the therapeutic range may be specified by the normalized dose gradient, D^*, as defined in Fig. 4.2 (Brahme and Svensson 1976). The shape of the dose distribution significantly vary with the energy and angular distribution of the electrons in the beam and depends on the bremsstrahlung contamination, and thus on the design of the treatment head. It is, therefore, important to fully understand how these parameters influence the dose distribution when comparing or taking a new beam into clinical use.

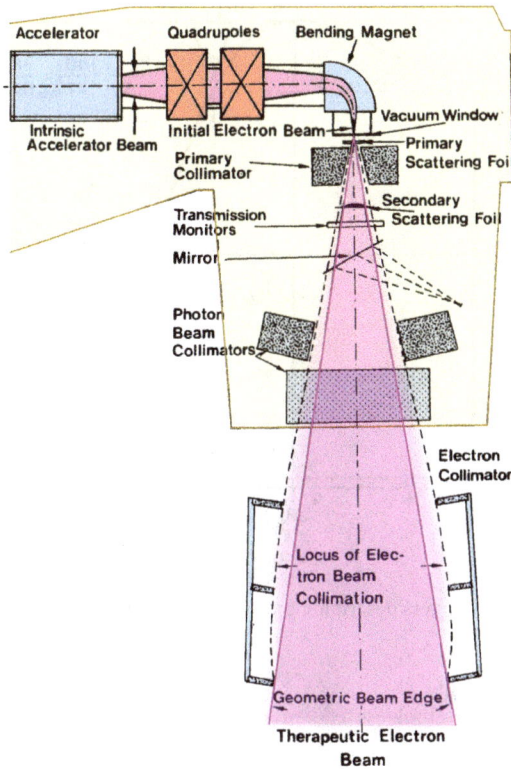

Figure 4.1.   Schematic components of the gantry and treatment head of a treatment unit in electron therapy mode.

4.3.2.3. *Electron beam flattening*

Several different methods of beam flattening are in use. The most common and straight forward method is to use high atomic number scattering foils. The choice of the material in the foil is based on the relative importance of the three different interaction processes in the foil: coulomb collisions, elastic scattering, and inelastic scattering (Bremsstrahlung production). The aim is to scatter the electrons without losing energy and producing bremsstrahlung. With a low atomic number material, the radiation stopping power is low and thus a low production of bremsstrahlung is obtained. On the other hand, there is a low scattering power resulting in the need of a thick scattering foil in order to be able to scatter out the electrons. The collision stopping power is, however, high for low number atomic materials. With a high atomic number material, there is, on the other hand, a high radiation stopping power and high scattering power but a low collision stopping power. Combining all three processes shows (Fig. 4.3) that in order to scatter out the electrons to a certain

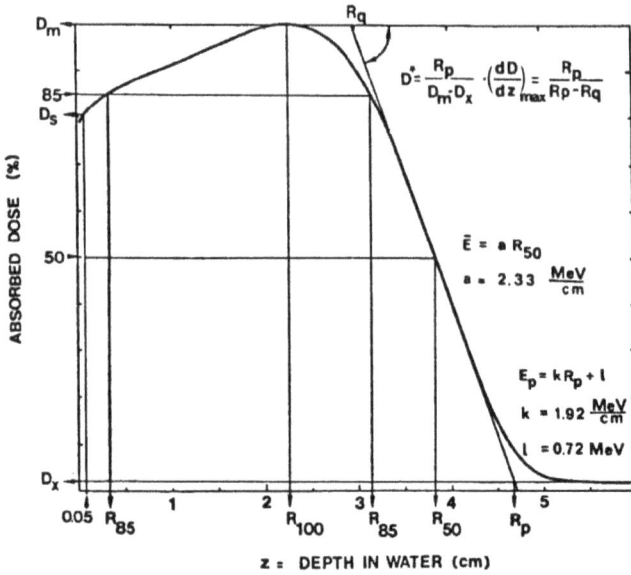

$$D^* = \frac{R_p}{D_m \cdot D_x} \cdot \left(\frac{dD}{dz}\right)_{max} = \frac{R_p}{R_p - R_q}$$

$$\bar{E} = a \, R_{50}$$

$$a = 2.33 \, \frac{MeV}{cm}$$

$$E_p = k R_p + l$$

$$k = 1.92 \, \frac{MeV}{cm}$$

$$l = 0.72 \, MeV$$

Figure 4.2. Key parameters describing the quality of an electron depth-dose distribution.

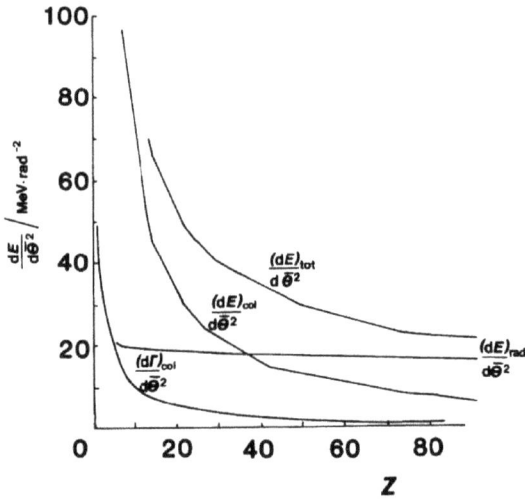

Figure 4.3. The mean, the most probable and the radiation energy losses and the energy straggling per unit mean square angle of scattering for 20 MeV electrons as a function of the atomic number of the scatterers are shown.

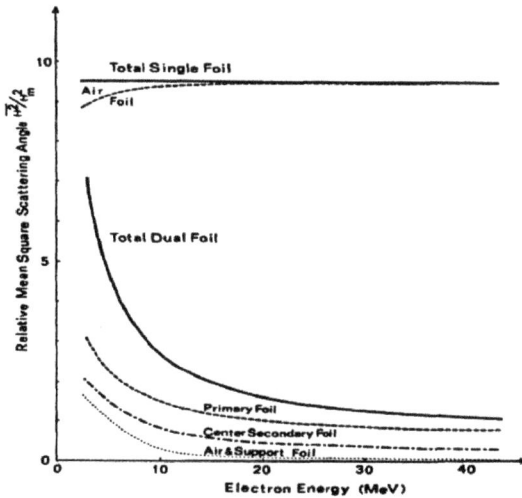

Figure 4.4. The relative mean square scattering angle and thus the effective foil thickness needed for different electron energies with single and dual foil scattering systems (Fig. 4.43(f)) are illustrated.

mean scattering angle the production of radiation is constant with atomic number but the energy loss decreases with increasing atomic number and thus a high atomic number scattering foil shall be used.

Improved beam quality may be further obtained if two separate principal scattering foils are used. The shape and thickness of the scattering foils may be optimized by elementary calculations using multiple scattering theories (Brahme 1972; Brahme 1977; Brahme and Svensson 1979; Brahme and Kempe 2014). Figure 4.4 shows a comparison between the relative foil thickness required with a single foil and the dual foil techniques. The vertical scale gives the relative mean square scattering angle or foil thickness needed to make a flat beam. For low energies, <5 MeV, little improvement is obtained with two foils. At higher energies, however, a significant reduction in the necessary scattering angle is obtained and thus a reduction in total foil thickness and energy degradation. For energies >30 MeV, the increased bremsstrahlung production and the decreased scattering power makes beam flattening by scattering foils impractical, and different methods to modulate the flatness of the beam through scanning the beam is to be preferred.

The effects of the scattering in the foils and other materials in the beam are a decrease of the effective Source Surface Distance (SSD) and a broadening of the effective source size. The decrease in SSD will result in depth-dose curves of decreased therapeutic range and the increased source size will make the collimation

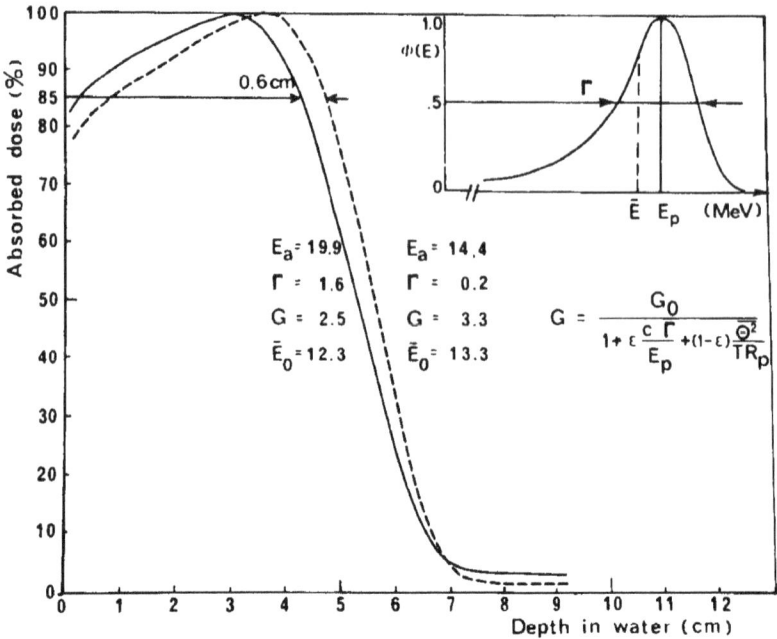

Figure 4.5. Influence of the energy spread on electron depth–dose distribution (Brahme 1977, 1979) is shown.

more complicated and introduce more scattered electrons in the beam and thus further increasing the surface dose. The materials along the path of the electrons will not only change the angular distribution but also degrade the energy of the electrons resulting in a lower mean energy and broader energy distribution. In Fig. 4.5 the dashed curve is taken from a normal beam (energy 14.4 MeV) with small energy spread. The solid curve is produced by a 19.9 MeV electron beam degraded by carbon, to obtain the same practical range as for the clinical beam. The influence of the broad energy distribution, in particular, on the dose gradient is evident.

An important positive phenomenon is the scatter filtering effect of the air which predominately removes low-energy electrons and thereby decreases the energy spread of the beam. Owing to the rapid increase of the mass scattering power with decreasing electron energy, a large fraction of the low-energy electrons will, therefore, be scattered out of the useful beam and generally be removed by the collimating system.

The influence of this process is illustrated in Fig. 4.6 where the fine dashed curve is for a normal clinical beam of about 22 MeV. The two thick curves are obtained by degrading the normal beam by 1 cm of PMMA. For the solid curve, the PMMA is placed as bolus at the surface of the phantom, and for the thick dashed curve,

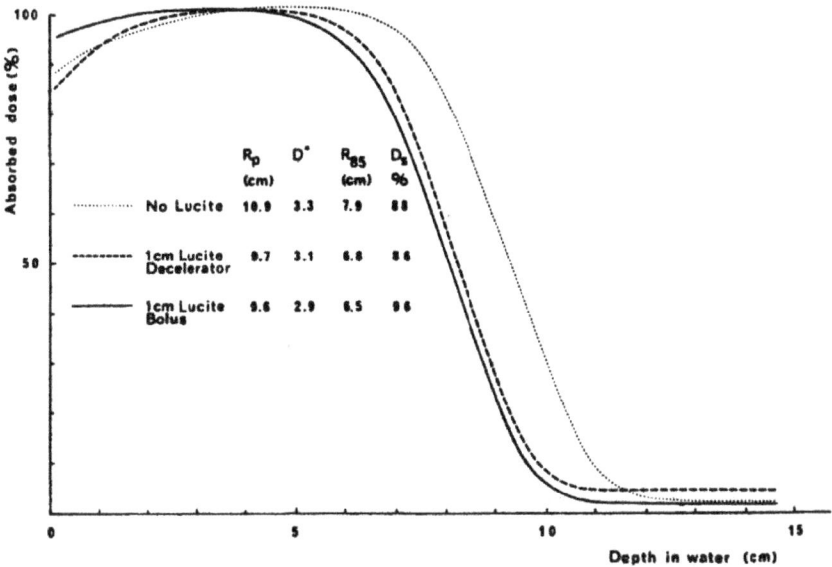

Figure 4.6. Influence of scatter filtering on electron depth-dose distribution is shown (Brahme 1977).

it is placed as a decelerator after the scattering foils. The most obvious difference is that the surface dose decreased from 96% to 86% when the PMMA is moved from the phantom up into the treatment head. This effect is mainly attributed to the scatter filtering in the air of the low-energy electrons from the PMMA. The consequent slight increase in range is due to the smaller energy and angular spread of the electrons reaching the phantom.

4.3.2.4. *Electron beam collimation*

The high scattering power of low-energy electrons make collimation difficult and the collimation of an electron beam cannot be treated purely as a beam geometrical problem due to the considerable influence of air scattering at energies <15 MeV. A broad electron beam must not only be collimated close to the irradiated surface but the upstream beam cross-section must also be significantly wider than the geometric beam in order to obtain a balance between the electrons scattered into and out of the beam (Fig. 4.7, Brahme 1977; see also the collimation system of the schematic treatment head in Fig. 4.1). The collimator shall not be made of a tube or cone because this will produce a lot of scattered electrons that will reach the patient with lower energy and under high incident angles. This will result in a high surface dose and a less steep dose gradient.

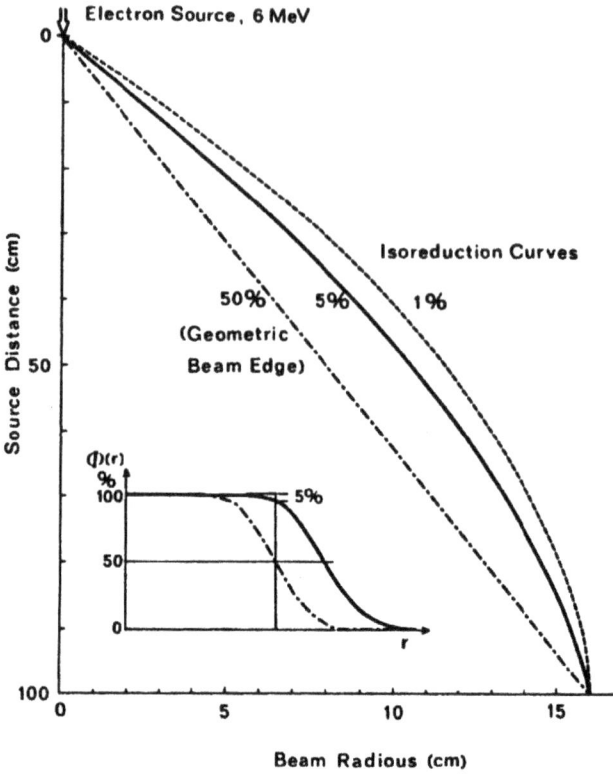

Figure 4.7. Illustration of how the electron air scatter affects the position of nearest possible electron collimator location (cf. Fig. 4.1).

The choice of the collimating material will influence the lateral electron scatter and, to a smaller extent, the bremsstrahlung production. A high atomic number collimation material will contaminate the beam more by bremsstrahlung. On the other hand, if a low-density material is used, a considerable contamination of low-energy electrons is obtained. This is because a much larger volume of the collimator edge will produce low-energy secondary electrons that are scattered into the beam. These electrons may produce hot spots along the edges of the field. These effects are significantly reduced, if the low atomic number material in the collimator is lined with a high density material on the collimator edge (Lax and Brahme 1980). Figure 4.8 shows the isodose distribution of a 22 MeV electron beam that has been collimated by this method. The sharp penumbra and flat isodoses even close to the beam edge can be clearly seen in the figure.

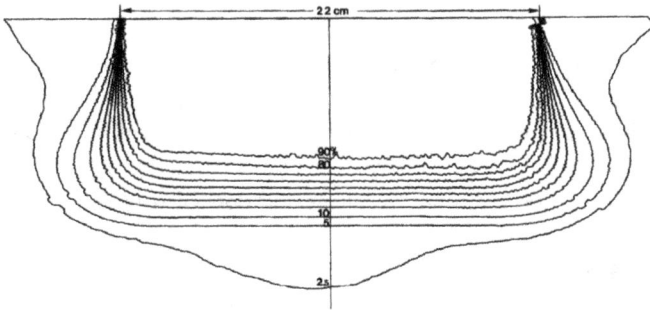

Figure 4.8. Illustration of the isodose distribution of a 21 MeV electron beam from a 22 MeV medical microtron.

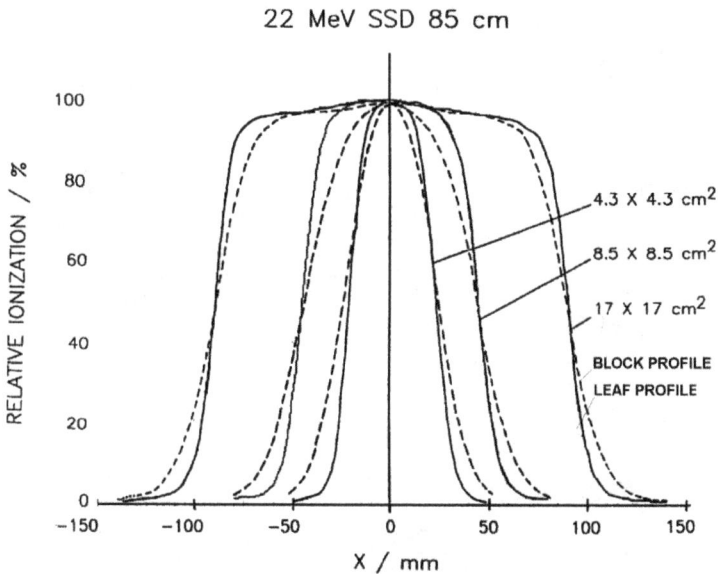

Figure 4.9. Illustration of beam profiles obtained with photon beam collimators in a microtron with 22 MeV electrons at an SSD of 85 cm (cf Figs. 4.1 and 4.7 and Montelius and Rikner 1992).

For high-energy electron beams, the scattering effect is much smaller and it is possible to use only the photon collimators even for electron treatments which have several practical advantages. Figure 4.9 shows measured beam profiles illustrating the loss of electron penumbra when using only the photon collimators for different field sizes as should be expected from Fig. 4.7.

4.3.2.5. *Scanned beams*

At high electron energies ≥ 25 MeV, beam flattening with scattering foils is not so suitable due to the introduction of a significant energy and angular spread and bremsstrahlung contamination. At these high energies, the best method is to scan a narrow unflattened beam over the area to be irradiated. In this way it is also possible to form the field both regarding the shape and dose distribution without special field-shaping devices.

Figure 4.10 shows an example of a treatment head for scanned beams (Brahme 1987). With such scanned beams, it is possible to obtain practically point

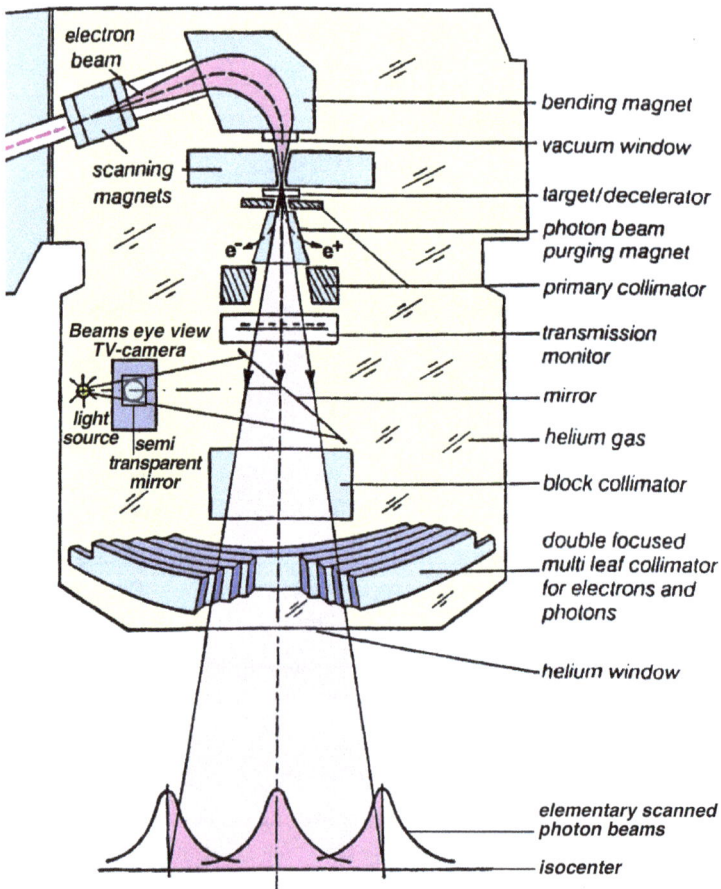

Figure 4.10. Principal design of the treatment head of an accelerator with scanned photon and electron beams (cf. Figs. 4.40, 4.45, and 4.75 below and Figs. 3.18 and 8.39).

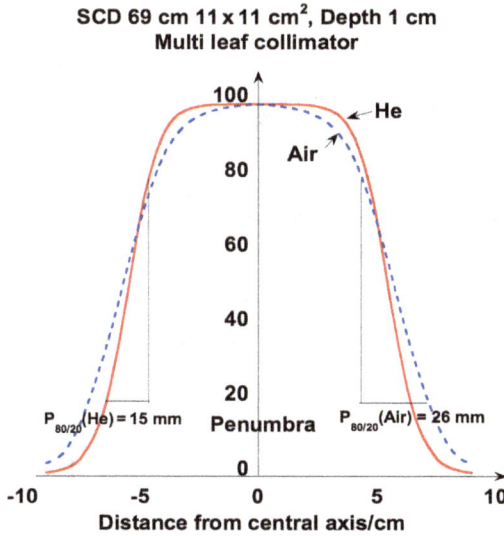

Figure 4.11. Electron beam profiles obtained with air and helium in the treatment head (cf. Fig. 3.1 and Svensson 1985).

isotropic beams of high-energy electrons with the same energy spectrum and fluence distribution over a large cone in the forward direction.

In treatment heads that use scanned electron beams for beam shaping, it is important to reduce, in particular, the scattering effect of the air. This may be achieved by introducing helium in the treatment head instead of air as shown in Fig. 4.10. Figure 4.11 shows two beam profiles obtained with air and helium in the treatment head, respectively. A clear reduction in penumbra is seen. At energies $\geq 10\,\text{MeV}$, the same electron and photon collimation system can then be used which have several practical advantages.

4.3.3. *Photon Beams*

More than 95 % of the different tumor sites are treated with high-energy photon beams. Therefore, a lot of effort has been spent on designing the treatment head that will deliver a therapeutically accurate and safe photon beam. Up to the 1980s, the development was mainly on the dosimetric properties and characterization of the beam quality and a lot of the work was spent on improving the beam quality. With the advent of IMRT in the mid-1980s, more attention has been turned to more effective ways to deliver intensity-modulated beams and verification of the delivered beams. In fact scanning beam IMRT was developed during the development of the

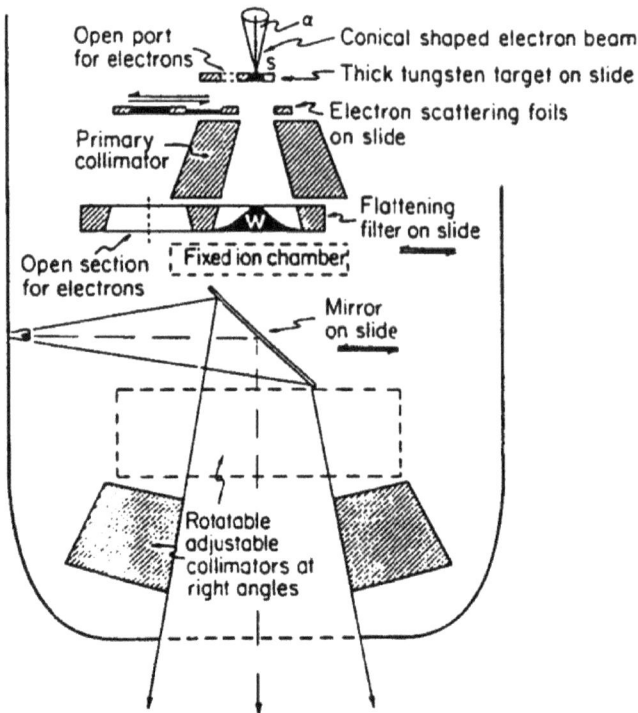

Figure 4.12. Illustration of treatment head proposal for flattening filtered-based photon beam therapy.

Racetrack Microtron during the late 1970s since ordinary scattering and filtering techniques are not feasible at high energies (Brahme *et al.* 1979). This, by necessity, requires much more sophisticated delivery systems such as dynamic high-resolution multileaf collimation or ultrafast electromagnetic scanning of narrow photon pencil beams. Although, scanned beam is the most flexible way to deliver IMRT beams, the system design today is too expensive for the average clinic size and therefore more than 99% of the installed treatment units have more or less conventional design of the treatment head. It should be noted that before the 1990s there were almost no MLCs in use and the focus was on the delivery of large flattened square field of high mean energy for deep penetration. Today, all radiation therapy departments require MLCs that also could be run in a dynamic mode. In fact the first gantry mounted multileaf collimators was produced during the development of Neutron therapy and the Racetrack microtron during the late 1970s (Brahme 1979, 1980, 1985).

A narrow beam increases the possibility to irradiate the geometrical fine structure in the tumor and will minimize irradiation of sensitive normal tissue

(Brahme *et al.* 1990; Lind *et al.* 1991; Söderström *et al.* 1993). Furthermore, when beam scanning and dynamic multileaf collimation are concomitantly used (Svensson *et al.* 1994; Gustafsson *et al.* 1995), the narrow scanned bremsstrahlung beam considerably reduces the treatment time and at the same time the highest possible calculated therapeutic outcome is achieved.

4.3.3.1. *Bremsstrahlung production*

When high-energy electrons are stopped, particularly in high Z materials, intense bremsstrahlung beams that are strongly peaked in the forward direction are produced. There has been a wide spread interest in the properties of bremsstrahlung production ever since the early 1950s (Schiff 1951; Hisdal 1956; Sirlin 1957; Hall *et al.* 1963; Ferdinande *et al.* 1971; Brahme and Svensson 1976; Nordell and Brahme 1984; Svensson and Brahme 1996). To have a high total production of bremsstrahlung, a high atomic number material shall be used (see Fig. 4.13 that illustrates the relative radiation stopping power for some materials). The angular distribution will depend on the thickness and the atomic number material of the target because a thicker target will scatter the electrons more before they produce bremsstrahlung and a higher atomic number has a higher scattering power. Figure 4.14 presents the relative energy fluence profile for different target thicknesses and materials (Nordell and Brahme 1984). The choice of target thickness and target

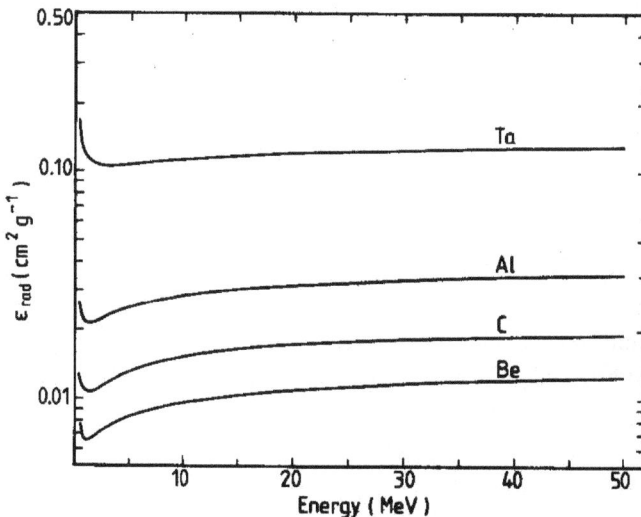

Figure 4.13.   Variation of the specific radiation mass stopping power with energy and atomic number.

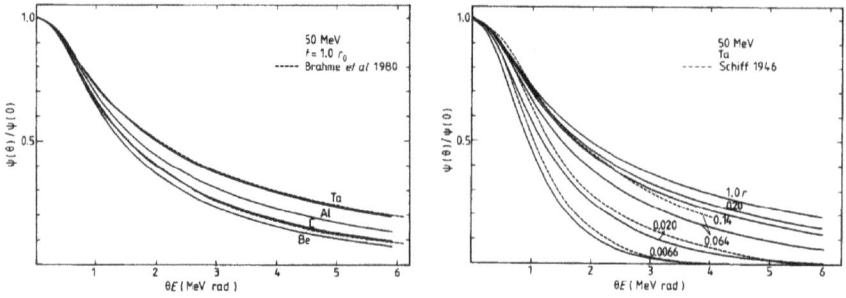

Figure 4.14. Relative angular distribution of the energy fluence as a function of target thickness and atomic number is shown.

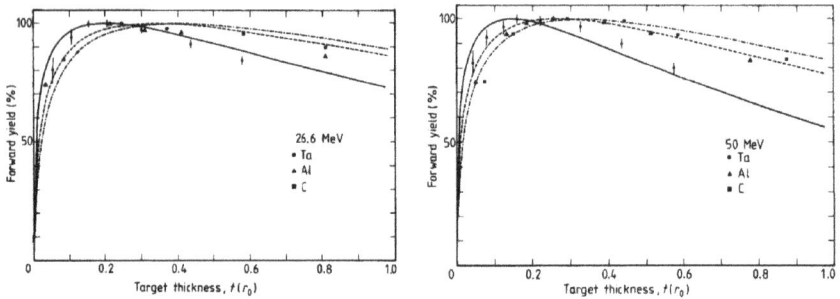

Figure 4.15. Illustration of bremsstrahlung output as a function of target thickness for three different target materials for 26.6 and 50 MeV electron energy, respectively.

material for radiation therapy is thus dependent on the desired characteristics of the bremsstrahlung beam profile and the method of beam flattening used.

The thickness of the target to obtain maximum fluence is illustrated in Fig. 4.15. For high atomic number only a thickness of 0.1 to 0.2 of the electron range is needed, while for lower atomic number 0.3–0.4 of the range is optimal. For an optimal target thickness, then, there will be several electrons that will pass through the target and thus may reach the patient. This can be handled in two ways. Either there is an electron stop behind the target made of a low atomic number that stops the electrons with a small production of bremsstrahlung or purging magnets that sweep the electrons out from the beam.

Classical flattening filters. For clinical photon beams shaped by classical flattening filters both widest possible high-energy bremsstrahlung beam profile and a high-energy fluence rate is desirable such that the flattened energy fluence rate is as high as possible for a given beam current. A small effective source size is desirable to keep the penumbra as small as possible. The downstream end of the target should not generate too many contamination electrons. A medium thickness target of a high Z material followed by a low atomic number electron stopper results in a broad beam with low electron contamination and a high integral yield which is ideal for stationary filtered beams (cf. Brahme 1982; Lambert *et al.* 1982; see also Sec. 4.4). To simultaneously obtain uniform intensity and the same photon spectra in all directions, a higher atomic number of the filter should be used in the forward direction, whereas low atomic number should be used at large radii (cf. Figs. 4.23 and 4.24 and Brahme and Svensson 1979). Unfortunately, inherently flattening filters remove many of the high-energy photons and add secondary electrons and positrons that will increase the surface dose and deteriorate the depth of dose maximum moving it to shallower depths.

Flattening with scanned beams. For beams of high energy and dose rate, the flattening should preferably be done with scanned beam technique to preserve high penetration and a deep dose maximum (Brahme 1987). With scanned photon beams, a narrow forward concentration of the high-energy bremsstrahlung photons with a minimal large angle component is desirable, that is, the conversion of electron energy to bremsstrahlung photon energy should be as efficient as possible only in the forward direction (Brahme *et al.* 1980). A low atomic number target gives a narrow beam due to less multiple scattering of the electrons but a lower integral yield in the bremsstrahlung conversion, at low energies (ICRU–35 1984). To overcome the inherent limitations of a single target material, various types of unorthodox sandwich targets built up of uniform layers of different materials can be used. By combining the properties of low- and high-Z materials, the characteristics of the therapeutic beam can be significantly improved.

Fast high resolution IMRT with scanned beams. For efficient IMRT, scanned beam should preferably be generated with the newly developed Karolinska Institutet's thin transmission target technique (Svensson *et al.* 1998). The target is made of a thin beryllium layer (1–6 mm) that will keep the electron expansion low laterally and with low electron scattering. In this way, the bremsstrahlung is mainly generated in a forward direction. Instead of stopping the electrons in a thick, high atomic number layer that will broaden the photon beam, the electrons is instead transmitted with a strong purging magnet deflecting the transmitted electrons to a dedicated electron collector. Figure 4.14 shows some beam profiles obtained with both more standardized W-Cu targets and thin Be-targets for scanning beams (Svensson 1998).

To keep the intensity high and the penumbra in the collimated beam as small as possible, the collimator edges should be sharp with high atomic number focused at the periphery of the effective photon source on the same side as the collimator is located. The penumbra is then mainly influenced by the size of the effective photon source. Fortunately, most photons are generated at shallow depths in the target, where the forward-directed photons dominate, so that the effective photon source is still fairly small with low atomic number targets. The concept of position and size of the effective source in an electron beam is well known (ICRU–35 1984) and has been investigated experimentally by Jamshidi *et al.* (1986). The shape and location of the effective photon source for clinical accelerators have been measured by Jaffray and Battista (1993). A general model supporting their work and useful for calculation of the geometrical penumbra, particularly for scanned beams, has been developed by Svensson and Brahme (1996).

4.3.3.2. *Elementary bremsstrahlung pencil beam production*

When electrons of high energy are slowed down in a material, bremsstrahlung is produced at all depths as shown in Fig. 4.16. The total photon energy fluence, Ψ_Ω, in direction Ω for a uniform target, differential in angle was described by Nordell and Brahme (1984) and later extended as an integral over all layers for multilayered targets by Svensson *et al.* (1996):

$$\Psi_\Omega = C \int_0^t \Phi_0 \left(1 + \frac{\overline{\theta^2}(z)}{2}\right) \overline{E}(z)\varepsilon_{\mathrm{rad}}(0) \frac{\exp\left[-\theta^2 \sqrt{\theta_X^2(z)}\right]}{\pi \theta_X^2(z)} \eta_i(z,t)\mathrm{d}z, \quad (4.1)$$

where C is a normalization factor, $\overline{E}(z)$ is the mean electron energy (ICRU–35 1984), $\varepsilon_{\mathrm{rad}}(0)$ is the specific radiation stopping power; $\eta_i(z,t)$ is the attenuation factor for

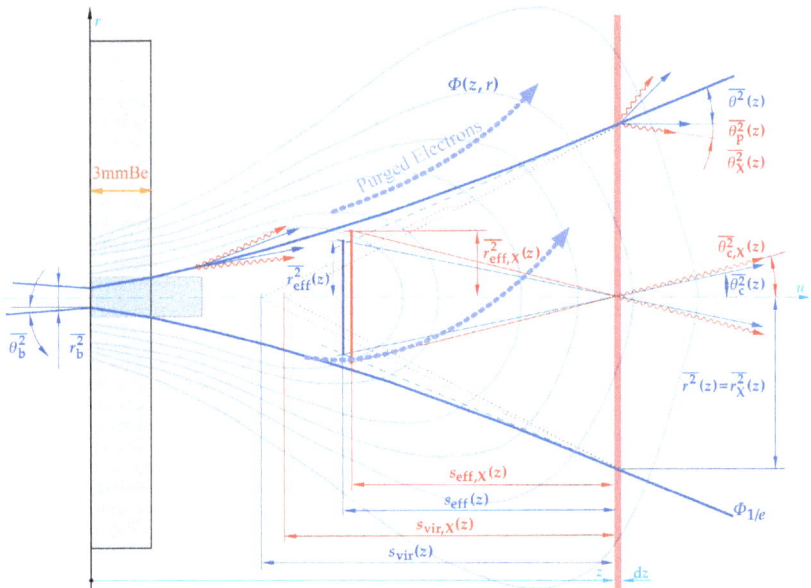

Figure 4.16(a). Illustration of the key factors affecting bremsstrahlung beam production.

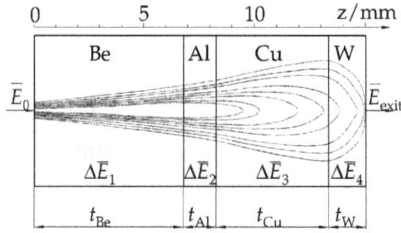

Figure 4.16(b). Electron fluence development in a compact composite target of high efficiency is depicted.

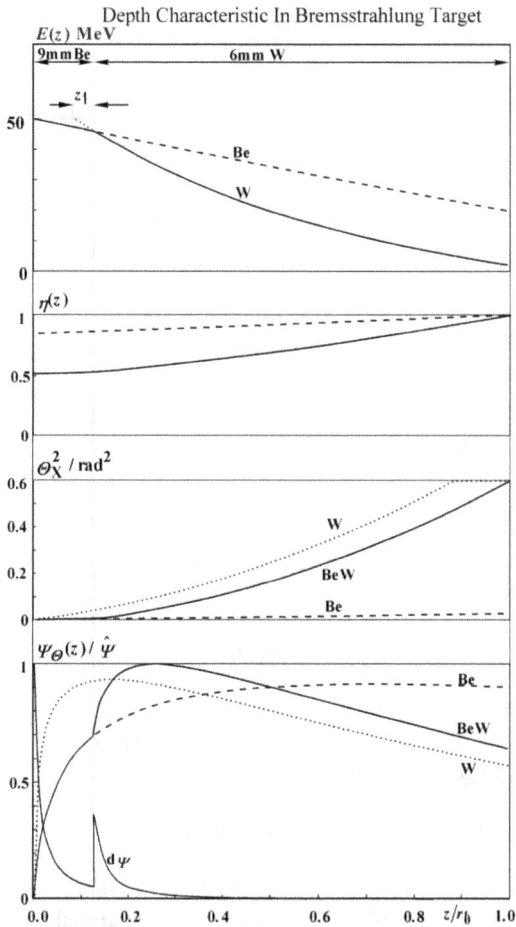

Figure 4.16(c). Depth distribution of key quantities for bremsstrahlung beam production is depicted.

photons generated at depth z which are attenuated by a layer of thickness $t - z$ in the target (cf. Fig. 4.16(c)). The primary electron fluence incident on the target and its build up is described by Φ_0 and $(1 + \overline{\theta^2}(z)/2)$, respectively. The angular distribution of the photons generated at depth z is given by the Gaussian term, where the total mean square angular distribution of the photons generated in a layer in the target $\overline{\theta_X^2}(z)$ can be expressed as the sum of the mean square angular spread of the intrinsic bremsstrahlung process $\overline{\theta_p^2}$ and that of the incident electron beam $\overline{\theta_b^2}$ as well as the multiple scatter in the target $\overline{\theta_m^2}$.

$$\overline{\theta_X^2}(z) = \overline{\theta_p^2}(z) + \overline{\theta_m^2}(z) + \overline{\theta_b^2}(z). \tag{4.2}$$

The bremsstrahlung energy fluence distribution differential in angle generated in a layer, dz, at depth z in the target can be separated in a spatially dependent part depending on the lateral coordinate x and a part, describing the variation of the angular distribution. At the patient surface ($z = d$), this distribution becomes:

$$d\Psi_\Omega(\theta_x, x, z, d) = C\Phi_0 \left(1 + \frac{\overline{\theta^2}(z)}{2}\right) \overline{E}(z)\varepsilon_{\text{rad}}(0) \frac{\exp[-x^2/\overline{r_X^2}(z, d)]}{(\pi \overline{r_X^2}(z, d))^{1/2}}$$

$$\times \frac{\exp\left[-\left(\theta_x - \frac{\overline{r\theta_X}(z,d)x}{\overline{r_X^2}(z,d)}\right)^2 \middle/ \overline{\theta_{X,c}^2}(z, d)\right]}{(\pi \overline{\theta_{X,c}^2}(z, d))^{1/2}} h(z, t)dz, \tag{4.3}$$

where

$$\overline{\theta_{X,c}^2}(z, d) = \overline{\theta_X^2}(z, d) - \{\overline{r\theta_X}(z, d)\}^2/\overline{r_X^2}(z, d) \tag{4.4}$$

is the mean square angular distribution of the photons at the patient surface for the bremsstrahlung that was generated at depth z, and

$$\overline{r_X^2}(z, d) = \overline{r^2}(z) + 2\overline{r\theta}(z)(d - z) + \overline{\theta_X^2}(z)(d - z)^2. \tag{4.5}$$

The total energy fluence differential in angle in the patient plane for a stationary beam is obtained by integration of Eq. (4.3) over all layers in the target. The total distribution in position and angle is given by:

$$\Psi_\Omega(\theta_x, \theta_y, x, y, d) = \Psi_\Omega(\theta_x, x, d)\Psi_\Omega(\theta_y, y, d). \tag{4.6}$$

When using scanning beams, a narrow beam is often more advantageous and therefore a thin target made of a low atomic number material or a ultra thin target made of a high atomic number material is most useful. Figure 4.17 shows some beam profiles obtained with both more standardized W–Cu targets and thin Be-targets for scanning beams (Svensson 1998).

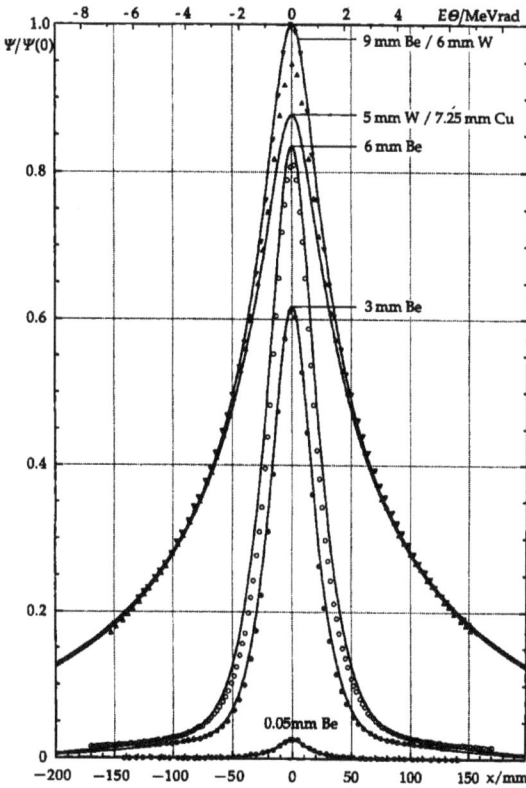

Figure 4.17.   Illustration of experimental beam profiles obtained for different target material combinations.

The energy distribution obtained will depend on target thickness with a higher mean energy for thinner targets. The mean energy also depends on the maximum energy. Figure 4.18 shows the variation of the mean energy with maximum energy, ranging from about 0.4 at low energies to 0.25 at high energies. Also, the energy distribution will change with emission angle with a higher energetic spectrum in the forward direction. Figure 4.19 shows Monte Carlo calculations showing variation of mean energy with distance from central axis. This variation will of course affect the depth-dose dependence.

4.3.3.3. *Effective source and penumbra*

To keep the intensity high and the penumbra in the collimated beam as small as possible, the collimator edges should be sharp with high atomic number focused at the periphery of the effective photon source on the same side as the collimator is

■ Brahme *et al.* ■

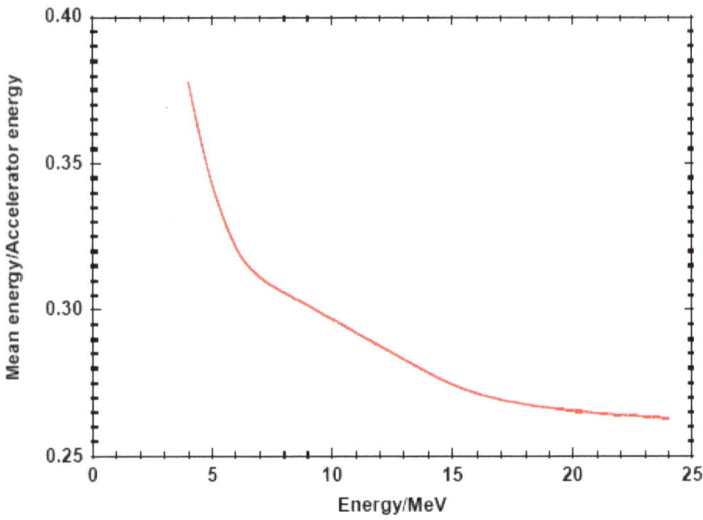

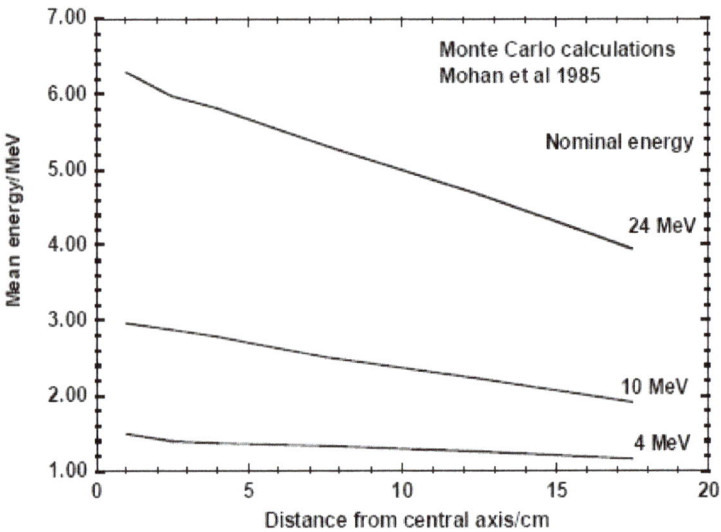

Figures 4.18 and 4.19. Variation of mean photon energy with electron energy and distance from central axis (cf Fig. 4.20 and Mohan *et al.* 1985).

located. The penumbra is then mainly influenced by the size of the effective photon source. Fortunately, most photons are generated at shallow depths in the target, where the forward -directed photons dominate so the effective photon source is still fairly small with low atomic number targets. The concept of position and size

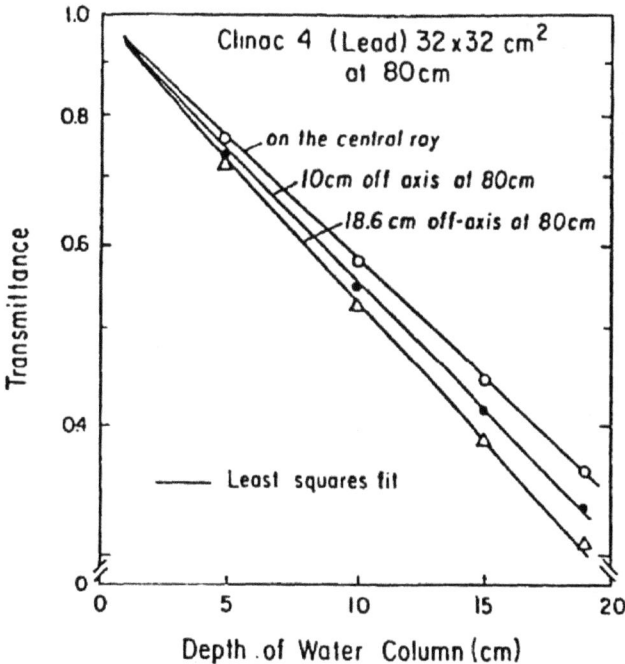

Figure 4.20. Variation of depth-dose distribution with distance from central axis of the beam (cf. Figs. 4.21 and 4.23).

of the effective source in an electron beam has been well described (Brahme 1982; ICRU-35 1984) and has been investigated experimentally by Mandour *et al.* (1982) and Jamshidi *et al.* (1986). The shape and location of the effective photon source have been measured by Jaffray and Battista (1993) for clinical accelerators. A model has been built, which is briefly discussed here, that is suitable for calculation of the geometrical penumbra, not least for scanned photon beams (Svensson *et al.* 1996).

The total dosimetric penumbra depends on (cf. Fig. 3.21):

(1) effective electron source;
(2) associated effective photon source;
(3) photon scatter in filters and collimators;
(4) photon transmission (through leaf collimator); and
(5) phantom scatter (electrons and photons).

As seen in Fig. 4.21, the electron source penumbra (1) results from the shape and the movement of the electron beam during scanning and the multiple scatter in the target. The location of the virtual photon point source thus depends on the

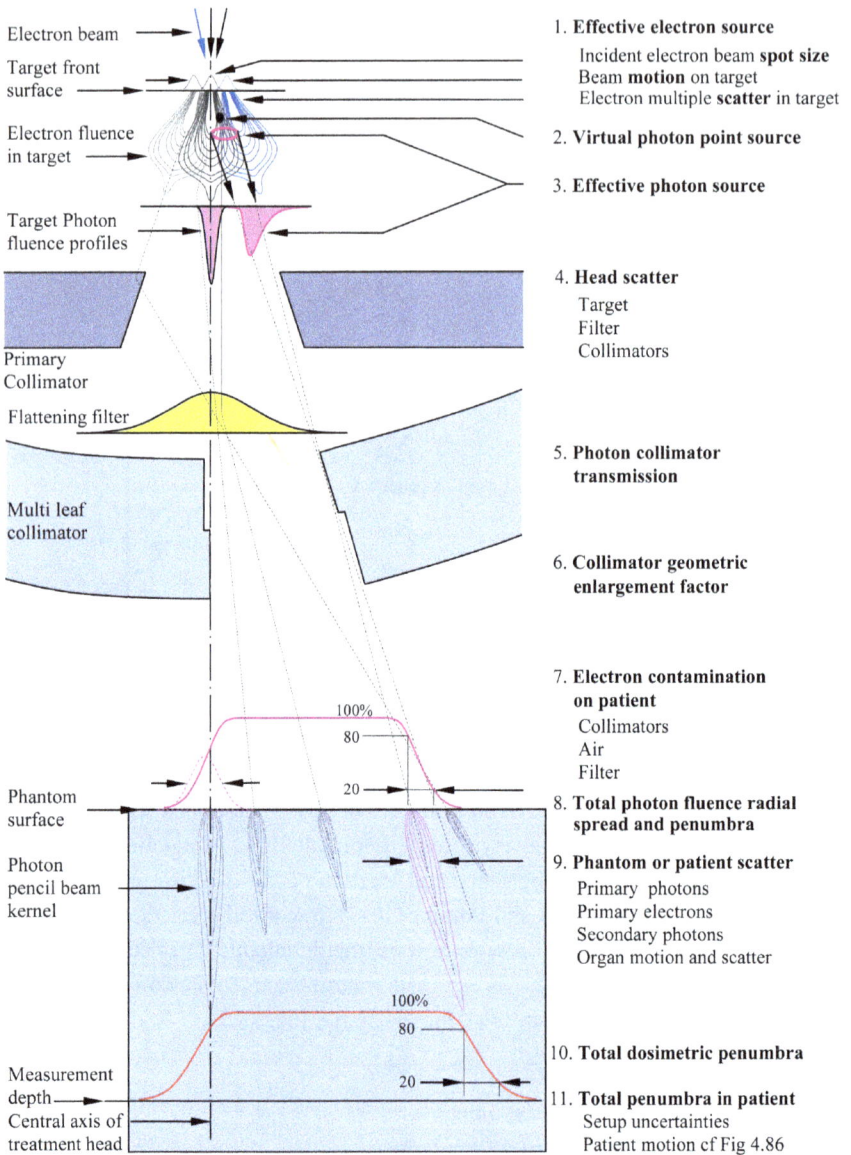

Electron beam

Target front surface

Electron fluence in target

Target Photon fluence profiles

Primary Collimator

Flattening filter

Multi leaf collimator

Phantom surface

Photon pencil beam kernel

Measurement depth
Central axis of treatment head

1. **Effective electron source**
   Incident electron beam **spot size**
   Beam **motion** on target
   Electron multiple **scatter** in target

2. **Virtual photon point source**

3. **Effective photon source**

4. **Head scatter**
   Target
   Filter
   Collimators

5. **Photon collimator transmission**

6. **Collimator geometric enlargement factor**

7. **Electron contamination on patient**
   Collimators
   Air
   Filter

8. **Total photon fluence radial spread and penumbra**

9. **Phantom or patient scatter**
   Primary photons
   Primary electrons
   Secondary photons
   Organ motion and scatter

10. **Total dosimetric penumbra**

11. **Total penumbra in patient**
    Setup uncertainties
    Patient motion cf Fig 4.86

Figure 4.21. Illustration of the major processes contributing to the photon beam penumbra in scanned and filtered beam treatment heads.

location of the observer. The photon production penumbra (3) results from the fact that the real effective photon source cannot be described by a point source at least when the effective electron source size is significant, especially in low atomic number targets (4). The contributions to the head scatter penumbra (4) are due to secondary photons created in the target (4.1), in filters (4.2), or in collimators (4.3). The relative fluence of scattered photons is below 1% of the total photon fluence, at high-energy machines without filter, but may be much larger at medium energies (Nilsson and Brahme 1981). The photon collimator transmission penumbra (5) is due to transmission of primary photons through the edges of the collimator leaves and is particularly large for machines with rounded leaf edges. The photon and electron phantom scatter penumbra (9) is caused by the transport of electrons (9.2) and primary (9.1) and secondary (9.3) photons created in the phantom and is the main contributor to the total dosimetric penumbra at high energies.

The contribution to the penumbra along different beam shaping components and field edges for a double-focused MLC are shown in Fig. 4.21. Another source of radiation outside the defined field edge is the transmission through and leakage between the leaves both for a scanned and filtered beam.

Since the photon production is extended over all depths in the target, particularly for low atomic number targets, the photon source described by the above equation may become large if the distance from the point of observation to the central axis ($x = y = 0$) is large. This is especially true for scanned beams, where the angle of observation may be far away from the central axis of the scanned beam.

However, the angular distribution at a point, (x_0, y_0), on the patient surface generated from a single beam in direction Ω_s, hitting the surface at (x_s, y_s), is only dependent on the distance r between the points. The angular distribution is a matter, of course, also dependent on the direction, the angle φ_s between the x-axis and the line through (x_0, y_0) and (x_s, y_s) (cf. Fig. 4.16(a)). This is taken into account by rotation of an angle independent of angular distribution $\Psi_\Omega(r_s, d)$. The total distribution for a whole field is the contribution from each scan point which is given by:

$$\Psi(x, y, x_0, y_0, d) = \sum_s \text{Re}(\exp(-i\varphi_s))\Psi_\Omega(r_s, d), \qquad (4.7)$$

where $\exp(-i\varphi_s)$ formally describes the rotation of the angular distribution with angle φ_s relative to the x-axis from each scan point.

The distance r from each scan point, with coordinates (x_s, y_s), to the point of observation, (x_0, y_0) (cf. Fig. 4.16(a)) is given as:

$$r = \sqrt{(x_0 - x_s)^2 + (y_s - y_0)^2}. \qquad (4.8)$$

The law of cosines gives us an expression for the angle φ_s:

$$\varphi_s = \arccos\left(\frac{r^2 + (x_0 - x_s)^2 - (y_s - y_0)^2}{2r(x_0 - x_s)}\right). \tag{4.9}$$

Photon collimator transmission penumbra: Knowing the total energy fluence differential in angle on the patient plane for a stationary beam, one can calculate the photon collimator transmission penumbra contribution to the total penumbra. This arises because the photon production is extended in the target (especially for a BeW-target) and consequently the collimator edges will not in reality be perfectly "double-focused."

The amount of transmitted bremsstrahlung $T(l)$ is of course dependent on the distance l that the beam travels through the collimator, and can be expressed as:

$$T(l) = 1 - \frac{\int H(\theta_x - \theta_c)\exp(-\mu l)\Psi(\theta_x, \theta_y, x, y, d)d\theta_x}{\int \Psi(\theta_x, \theta_y, x, y, d)d\theta_x}, \tag{4.10}$$

where $H(\theta_x - \theta_c)$ is Heaviside's step function, which gives the value one if $\theta_x - \theta_c > 0$ and 0 otherwise.

The mean square radial spread at the projected collimator edge on the patient surface: The mean square radial spread of the total photon fluence $r^2_{t,X}$ including all secondary and scattered photons can be separated into the contribution from the primaries $r^2_{p,X}$, with a fairly small mean square radial spread and the total head scatter $r^2_{hs,X}$ that has a large mean square radial spread but small intensity. The mean square radial spread of the primary photons is a magnification of the effective photon source:

$$\overline{r^2_{p,X}}(d) = \overline{r^2_{eff,X}}(d). \tag{4.11}$$

The mean square radial spread on the patient surface at a point x_0 tangential to the collimator edge of the secondary photons created in the treatment head can be expressed as:

$$\overline{r^2_{hs,X}}(x_0) = \frac{\psi_{hs,t}\overline{\theta^2_{hs,t,X}}f^2 + \psi_{hs,f}\overline{\theta^2_{hs,f,X}}(f - ff)^2 + \psi_{hs,c}\overline{\theta^2_{hs,c,X}}(f - pc)^2}{\psi_{hs}}, \tag{4.12}$$

where the different parts in the sum are contributions from scattering in the target (t), the flattening filter (f), the primary collimator (c), and the energy fluence of the scattered photons $\psi_{hs} = \psi_{hs,t} + \psi_{hs,f} + \psi_{hs,c}$. The distances in the equation are illustrated in Fig. 4.16(a).

The total mean square radial spread can, assuming that the transmission of primary and secondary photons is approximately the same, be described as

$$\overline{r_{t,X}^2}(x_0) = \frac{\psi_p \overline{r_{p,X}^2} + \psi_{hs} \overline{r_{hs,X}^2} + \psi_t \overline{r_{ct}^2}}{\psi_t}, \tag{4.13}$$

where $\overline{r_{ct}^2}$ is the radial spread of the photons transmitted through the collimator.

Total photon fluence penumbra: The total radial mean spread of the photon fluence is assumed to be Gaussian-like, so the extrapolated penumbra P_e, given by the distance between the two points where the tangent in the inflection point of the edge of the total photon fluence on the patient surface reaches the zero, and the top level (cf. figure 1b), can be expressed as

$$P_e \approx \sqrt{\pi \overline{r_{t,X}^2}}. \tag{4.14}$$

This expression matches almost the $P_{90/10}$ penumbra, so:

$$P_{90/10} \approx \sqrt{\pi \overline{r_{t,X}^2}}. \tag{4.15}$$

The total photon fluence penumbra $P_{80/20}$ will then be given by:

$$P_{80/20} \approx 0.6 P_e. \tag{4.16}$$

4.3.3.4. *Beam-flattening filter*

To generate a uniform beam flattening filters are introduced in the beam. If a low atomic number material is used for beam flattening, the variation in energy over the field will be increased resulting in a flat field at a particular depth but hot spots at smaller depths and round corners at larger depths. This is an effect of the higher attenuation coefficient of low-energy photons. With a high atomic number material, the mean energy may be lowered. Due to the increasing pair production with increasing energy, high-energy photons are attenuated more. This is illustrated in Fig. 4.22 that shows photon spectra for 21 MV X-rays before and after an Al or a Pb filter. The spectra are normalized to 5 MeV. To increase the uniformity of the energy distribution over the beam, one may use a filter with a high atomic number material in the center and a low atomic number close to the periphery of the beam. In this way, a flattening filter of two materials and a nearly constant total thickness may be made (Brahme and Svensson 1979).

Figure 4.23 shows the variation of the HVL across the beam measured for different accelerators, indicating that the energy compensating filters will give a more homogeneous beam. With scanning beams without beam-flattening filters, the

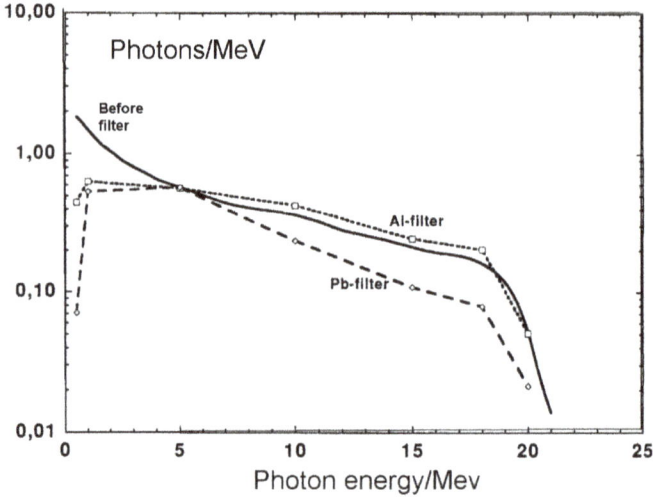

Figure 4.22.   Photon spectra after Al and Pb beam-flattening filters (cf. Fig. 20).

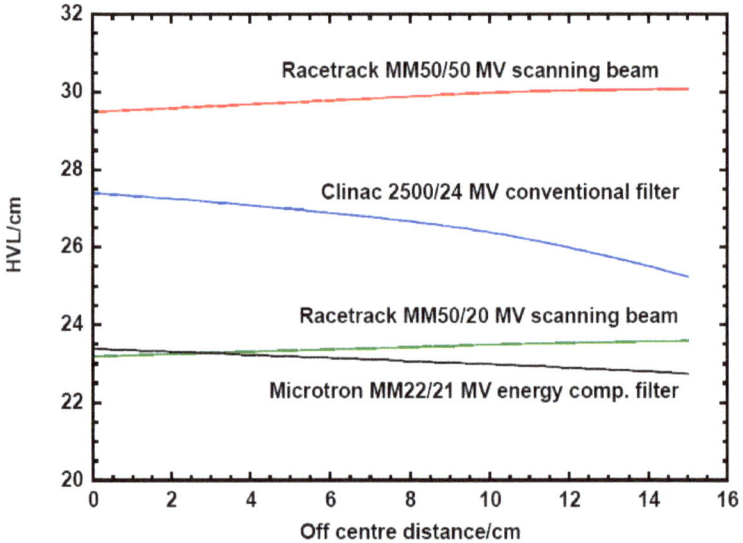

Figure 4.23.   Illustration of the variation of the effective photon energy (as described by the HVL) with the distance from central axis of the beam with different accelerators and beam flattening systems (Modified from Staël von Holstein *et al.* 1992). The peripherally harder beams generate more uniform beams over a wide range of depth intervals as seen in Fig. 4.24.

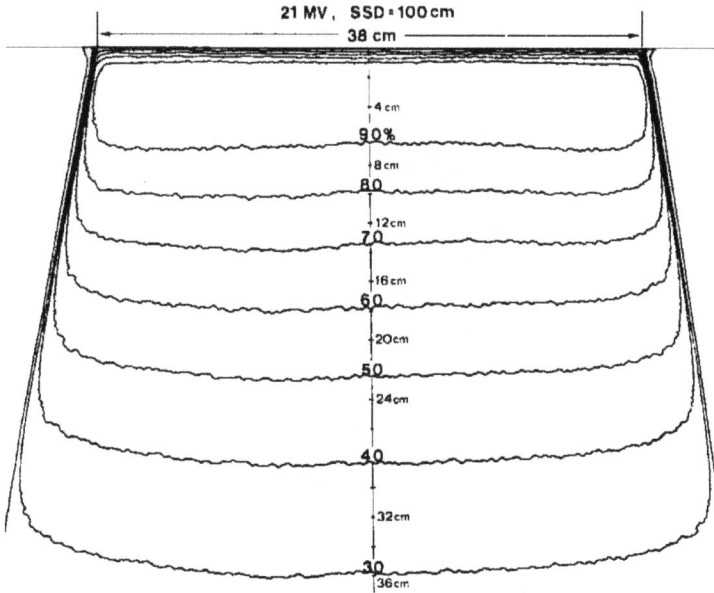

Figure 4.24. Illustration of broad photon beam flattened by lead in the center and aluminum at the periphery to obtain a uniform dose and radiation quality at all depths without over-flattening at the surface (cf. Fig. 4.23 and the text).

variation with energy over the field is very small, especially with thin target narrow photon beam scanning. Also with thick targets this is the case as seen in Fig. 4.23 for the racetrack microtron where the energy even increases outside the central axis to compensate for the longer path through tissue by mainly directing the pencil beam axis at large angels.

4.3.3.5. *Secondary photons*

The bremsstrahlung beam produced in the target and flattened by the filter will be deteriorated by secondary radiation produced between the target and the patient. Figure 4.25 shows different components contributing to the secondary radiation contaminating high-energy photon beams. The main sources are the primary collimator, the beam-flattening filter, the secondary collimators, beam-shaping devices, and the air volume between the treatment head and the patient. The patient, of course, is also a source of secondary radiation.

The photoelectric effect, incoherent scatter, and pair production produce secondary photons. Coherent scatter may also in some situations be of importance. In the energy region commonly used today, the incoherent scatter will

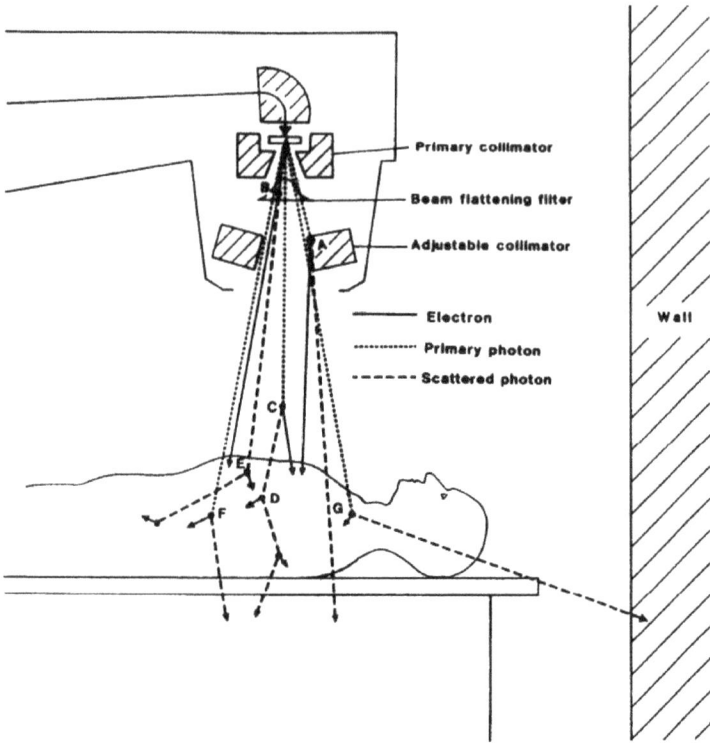

Figure 4.25. Illustration of the components contributing to the contaminating radiation in a photon treatment.

dominate both due to the high cross-section and to the small scattering angle at therapeutic energies. At high energies, the cross-section for pair production will be important but the isotropic emission of the annihilation photons and their relatively low energy make the contribution small.

Figure 4.26 shows the relative absorbed dose distribution due to scattered photons for 6 and 21 MV X-rays with beam radius of 10 cm as obtained from Monte Carlo calculations (Nilsson and Brahme 1981). Secondary photons produced by incoherent scatter and annihilation photons produced by pair production are included in the calculations. Data are normalized to the dose at dose maximum. The main sources of scattered photons are the primary collimator for 6 MV X-rays and the beam-flattening filter for 21 MV X-rays. The total contribution for a 20×20 cm^2 field to the dose at the central axis from scattered photons is, in these calculations, <3% for both energies. For larger fields, the scattered photon dose may reach higher values. Outside the primary beam, the dose decreases to only 0.5%, 5 cm

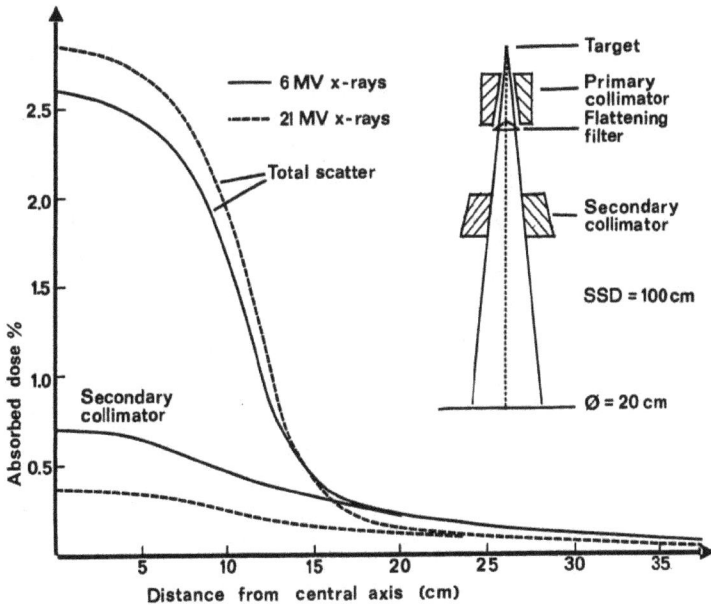

Figure 4.26. Relative absorbed dose from scattered photons at dose maximum for 6 and 21 MV X-rays.

from the field edge. Here, the main source is the adjustable collimator. The energy distributions of the scattered photons are shown in Figs. 4.27 and 4.28 for 6 and 21 MV X-rays. The spectra show that inside the useful beam the secondary photons are mainly single-scattered photons scattered at small angles and thus with a small energy degradation. Outside the beam, the scattering angles and the probability of multiple scatter is larger and thus the energy is lower. For 21 MV X-rays, there is also a contribution from annihilation photons. The contribution to the absorbed dose is, however, small due to the isotropic emission of the annihilation photons. At the central axis, <1% of the secondary photon dose is due to annihilation photons and <5% outside the beam, 25 cm from the central axis. The results obtained indicate that the secondary photons will not affect the depth–dose distribution in a significant way due to the small angle scatter of the photons, the small separation between the scatter source and the target, and finally due to the high mean energy of the scattered photons.

Calculations also show that outside the primary beam the dose to the patient is dominated by photons scattered in the patient, and contaminating photons from the head will have a very small influence. This is shown in Figs. 4.29 and 4.30 where beam profiles in water for 6 MV X-rays at 10 and 20 cm depth and field sizes

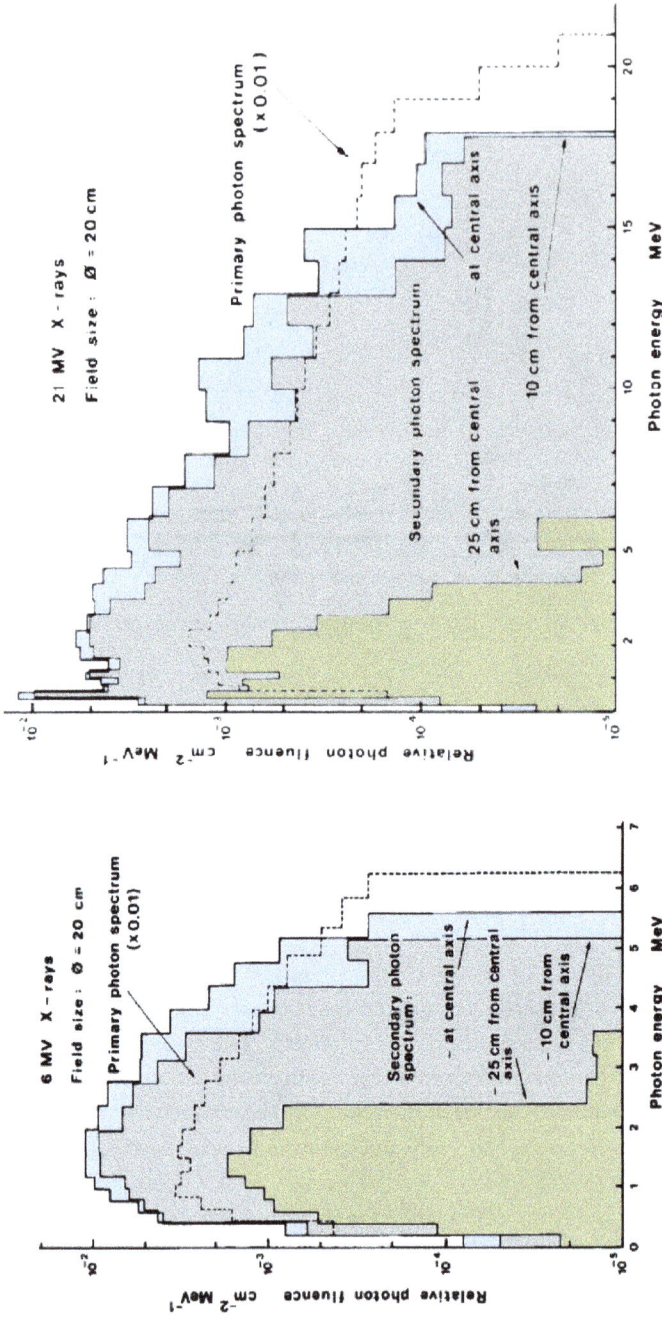

Figures 4.27 and 4.28.　Illustration of photon spectral distribution of secondary photons at central axis and outside beam for 6 and 21 MV X–rays.

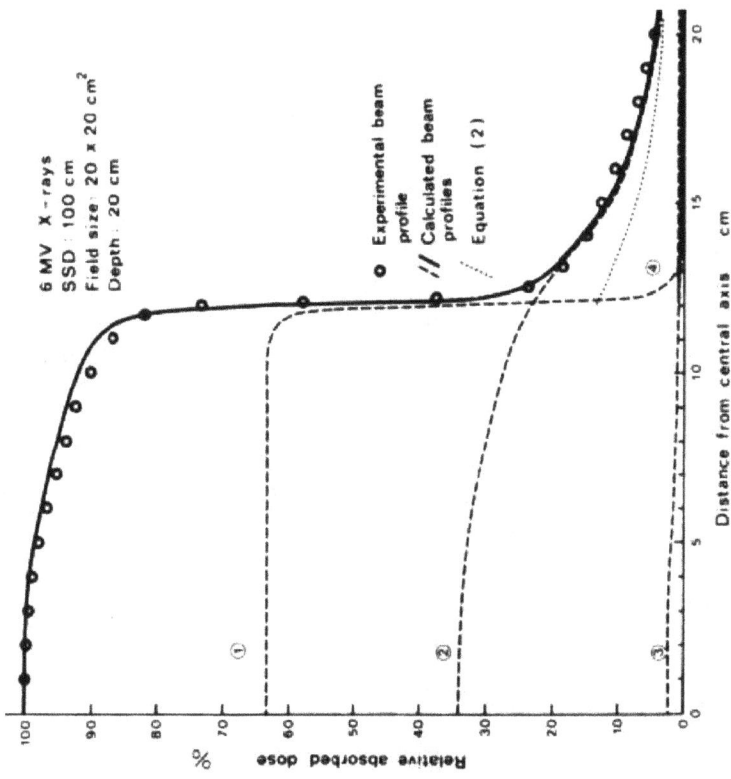

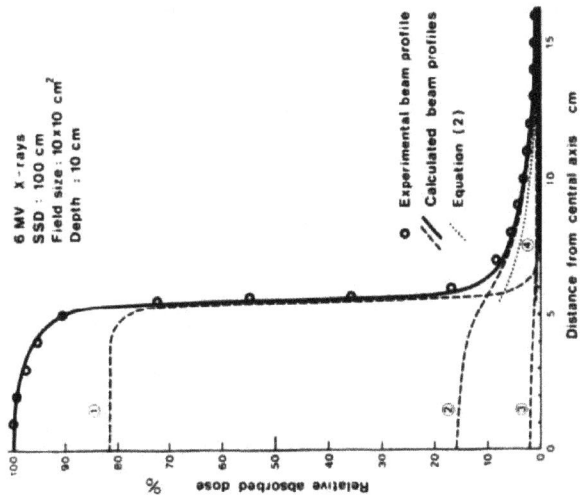

Figures 4.29 and 4.30. Illustrations of beam profiles for a 6 MV X-ray beam for different field sizes and depths and the considerable phantom scatter in broad low-energy beams.

$10 \times 10\,\mathrm{cm}^2$ and $20 \times 20\,\mathrm{cm}^2$, respectively, are presented. The secondary photons from the treatment head are a small contribution compared to the phantom scattered photons. In these figures, the photons transmitted through the secondary collimator are also included which in the geometry used was just a few tenths of a percentage of the total dose.

However the calculations hold only for the geometry used and with larger field sizes and thicker beam-flattening filters there may be a larger dose due to secondary photons. Also, in the calculations, coherent scatter were not included. The coherent scattered photons do not lose any energy and are scattered over small angles so they often are not included in this type of calculations. However, they may be of importance, particularly when measured depth-dose data are used for estimating beam parameters to be used as input data for the dose planning systems.

Also when blocking material or especially when wedges are inserted in the beam close to the patient, the contribution of secondary photons may increase and also influence the effective source–skin distance. Some measurements indicate a change in effective SSD of 4–5 cm when introducing a wedge in the beam.

4.3.3.6. *Secondary electrons*

Secondary electrons are produced in the same interaction processes as the secondary photons. The main electron sources in a treatment head are target, beam-flattening filter if any, collimators, air, and often beam blocks or electron filters.

The electron contamination will mainly influence the surface dose and the build-up region and it is important to know the different sources of this contamination in order to be able to correctly calculate the dose in the build-up region. Tests with the Helax TMS dose planning system show that the dose at dose maximum may be in error with a few percentage due to the simplified assumption that the only source for electron contamination is the flattening filter (Montelius *et al.* 1992). Owing to the relatively large range of the electrons at high energies even the dose distribution at large depths may be affected and it is important that the reference dose is determined at a depth where the contribution from contaminating electrons is negligible. The reference depths have so far been 5 or $10\,\mathrm{g/cm}^2$ for low photon energies ($\mathrm{TPR}_{10}^{20} < 0.70$) and $10\,\mathrm{g/cm}^2$ for higher energies (IAEA 2000). With increasing use of higher energies and scanning beams, it may be necessary to further increase the reference depths.

4.3.3.6.1. Air scatter

The air volume between the source and the patient is important both as a source of electrons and as a scattering material for electrons produced in the treatment head.

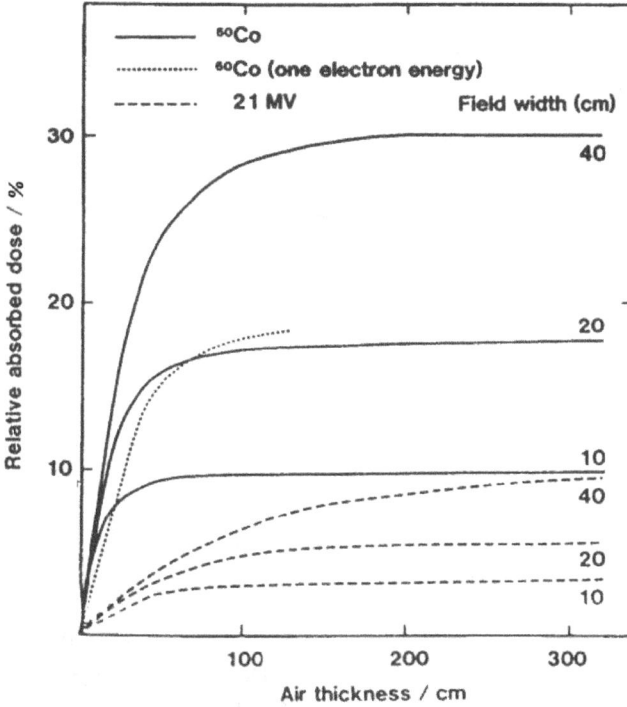

Figure 4.31. Illustration of relative surface absorbed dose from electrons produced in air for ^{60}Co-γ-rays and 21 MV X-rays.

Figure 4.31 shows the calculated surface dose produced in air for ^{60}Co and 21 MV X-rays (Nilsson 1985).

The surface dose due to electrons from the air is dependent on field size, thickness of the air volume, and photon energy. As shown in the figure, there is a fast increase in dose with the air volume for low photon energies. However, for a larger air density, due to the high scattering power of low-energy electrons they will be scattered out of the beam and thus decrease the contribution from layers at large distances from the phantom. With increasing field size, more of the out-scattered electrons will be compensated by in-scatter. However, for ^{60}Co, it is necessary with a field width of more than 120 cm to compensate for the out-scattered electrons. Thus, for photons in the lower energy region (1–2 MeV), most of the contaminating electrons reaching the phantom are produced rather close to the phantom. This implies that a magnet placed just below the treatment head to sweep out the electrons will not be very effective because the air volume below the magnet will be the main source of contaminating electrons.

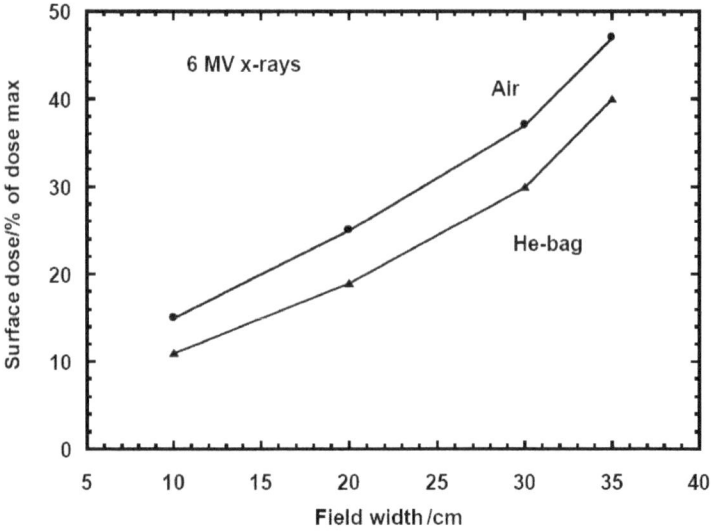

Figure 4.32.   Effect on electron surface dose with air or helium as medium between the vacuum window and the patient.

For higher energies, the interaction coefficients are lower and thus the emission of leptons will be lower for small air density. However, the longer ranges, smaller emission angles, and lower scattering power of the high-energy electrons imply that a larger air density will increase the surface dose. In whole body treatments, the source to skin distance may be several meters and in this situation electron contamination from the air may be large at high photon energies.

One possibility to reduce the electron contamination at the surface is to replace the air between the source and the patient with helium (Nilsson and Brahme 1979). Owing to the smaller number of emitted electrons, the surface dose from electrons emitted in the gas volume may be decreased as shown in experiments in Fig. 4.32 (e.g., Yorke *et al.* 1985). Monte Carlo calculations show that for 50 MV X-rays from a racetrack microtron, the electron energy fluence is reduced with about 10% when the treatment head is filled with helium instead of air (Staël von Holstein *et al.* 1992).

4.3.3.6.2. Filters

In most treatments, some material is placed between the radiation source and the patient. Examples are beam-flattening filters, wedges, shadow trays with beam-shaping devices, and/or compensators. In all these materials secondary electrons will be produced which contribute to the surface dose. This contamination depends on atomic number, photon energy, filter-phantom distance, filter thickness, and

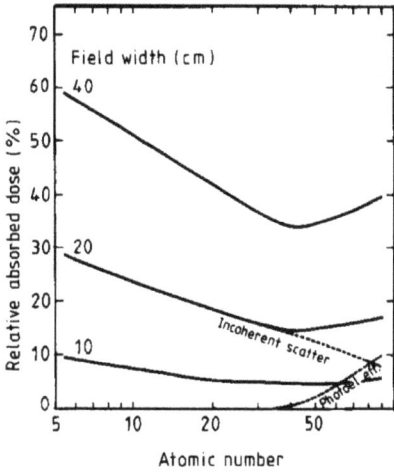

Figure 4.33. Illustration of surface dose from filter produced electrons in a ^{60}Co beam (Nilsson 1985).

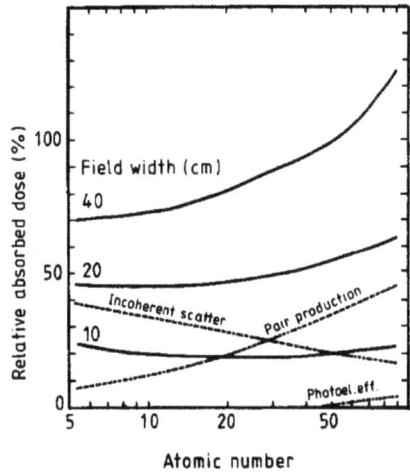

Figure 4.34. Surface dose from filter produced electrons in a 21 MV X-ray beam (Nilsson 1985).

field size, and these parameters must be studied in order to be able to construct a treatment head with minimal electron contamination.

Figure 4.33 shows the variation of electron contamination with atomic number for a ^{60}Co source. In the figure, the different contributions to the contamination is illustrated showing the decrease of incoherent scatter with increasing atomic number and the increase of photoelectric effect at high atomic numbers. There is a minimum in electron contamination around medium atomic numbers, for example, Fe and Cu.

Figure 4.34 shows the corresponding variation of electron contamination with atomic number for 21 MV X-rays with a filter plate located 20 cm above the phantom surface showing that in addition to incoherent scatter and photoelectric effect, pair production is also of importance (Nilsson 1985). From this energy on, there will be an increase in electron contamination in high atomic number materials due to their high pair production cross-section. For large field sizes and filter thickness, this will give a linear increase in lepton contamination with atomic number. For small field sizes, the higher scattering power in high atomic materials will scatter electrons more which mean that a larger proportion is lost from the beam by scattering out. Thus, for small field sizes, the variation with atomic number is rather small. For broad beams and very high-energy photons a low-to-medium atomic number material will therefore give the smallest electron contamination.

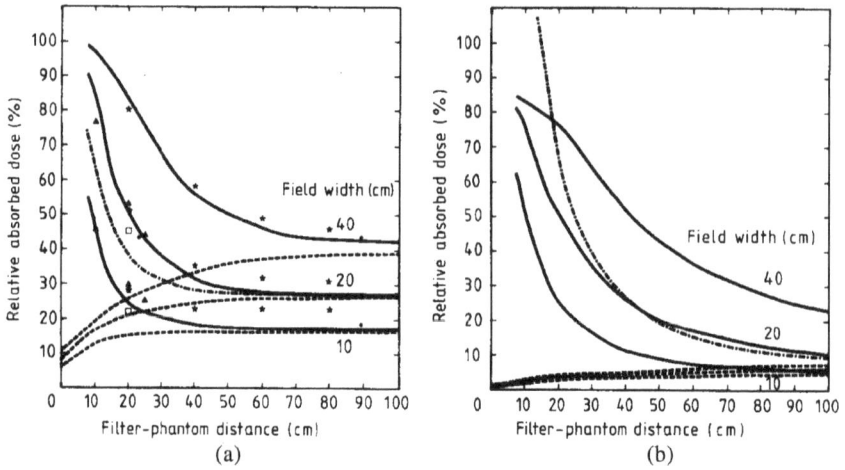

Figure 4.35. Variation of surface dose with filter-phantom distance for (a) ^{60}Co, (b) 21 MV X-rays (Nilsson 1985).

The filter phantom distance is, of course, also a very important parameter. The variation of the surface dose for ^{60}Co γ-rays and 21 MV X-rays with filter-phantom distance for filters made of PMMA and lead is shown in Figure 4.35. For PMMA filters (solid lines), calculations for different field sizes are included showing the influence of the scattering effect decreasing the electron contamination for small fields. The larger variation with distance for lead is due to the larger emission angle of the electrons emitted from lead (dash-dotted line). Dashed lines show the contribution to the dose from backscattered radiation and air produced electrons.

4.3.3.6.3. Collimators

Another source of electrons is the collimators, and primarily the adjustable collimator is situated downstream of the beam-flattening filter. Because of the small lateral travel of the electrons in dense materials, it is only a small layer close to the collimator surface that can contribute to the electron fluence. This layer is thinnest for high density and high atomic number materials. A lower lepton contamination will thus be obtained if the collimator is made of a high density material. It is even not necessary that the whole collimator is made of such a material. Most important is the first photon path tenth value layer along the beam and one lateral electron range perpendicular to the beam.

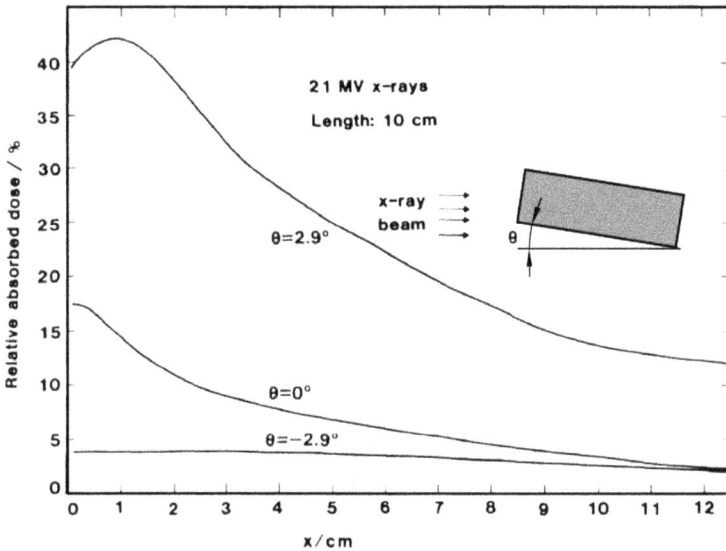

Figure 4.36. Illustration of surface electron dose due to electrons emitted through the sidewall of a lead collimator irradiated with 21 MV X-rays and a field width of 20 cm. With $\Theta = 0$, the collimator side is perfectly aligned with the beam edge. With $\Theta = 2.9°$ the photons can hit the collimator sidewall and for $\Theta = -2.9°$ no photons are assumed to hit the side wall (Nilsson and Brahme 1986, with permission).

Measurements and calculations show that the contribution by collimator-generated electrons is fairly small in most cases and it was estimated that the surface dose for a 40×40 cm² field was <5% for 21 MV X-rays (Nilsson and Brahme 1986). It is, however, important that the collimators are correctly aligned. If the collimating surfaces are hit directly by the primary photons, a much higher electron dose may be obtained as shown in Fig. 4.36 where the lateral distribution is plotted for three different alignment geometries.

This is unfortunate as the optimal alignment of the collimators for reducing the geometrical penumbra requires peripheral source alignment. However, for very high photon energies, the geometrical penumbra is small and the electron contamination is high, so it may be more important to align the collimators to minimize the lepton contribution rather than minimizing geometrical penumbra.

4.3.3.6.4. Surface dose beam profiles

The secondary electrons will contribute to dose outside the primary beam and increase the risk for superficial organs, such as inducing cataracts in the lens of the

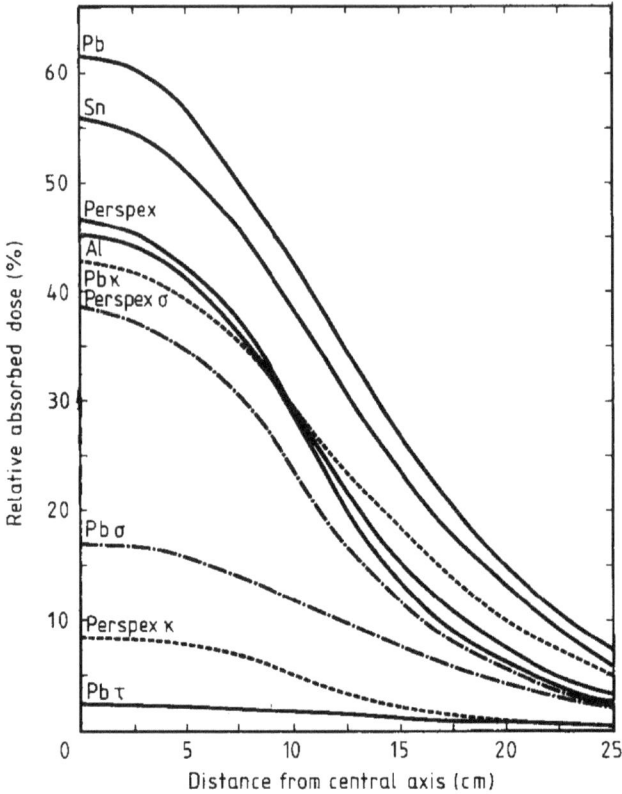

Figure 4.37. Variation of absorbed dose from filters of different materials situated 20 cm from the phantom surface irradiated with 21 MV X-rays is illustrated. Field size: 20×20 cm^2. The separate contributions from photoelectric effect (τ), incoherent scatter (σ), and pair production (κ) are included (Nilsson 1985, with permission).

eye or genetic effects in the testes, when these organs are close to the primary beam. Figure 4.37 shows beam profiles obtained for 21 MV X-rays with filter materials of different atomic numbers placed 20 cm above the patient (Nilsson 1985); 5 cm from the field edge the surface dose may be as high as 25% of the dose at maximum photon dose when a high atomic number material is used as filter material.

The dose of interest is, of course, the dose to the sensitive tissue in the organ and, for example, the lens is situated at a depth of about 3 mm. The depth-dose distribution at a point 3 cm outside the field edge as measured with an extrapolation chamber is plotted in Fig. 4.38 (Nilsson and Sorcini 1989). At a depth of 3 mm, the dose has decreased <10% for 21 MV X-rays. Thus, it may be necessary to insert an electron shield outside the primary beam just above the patient

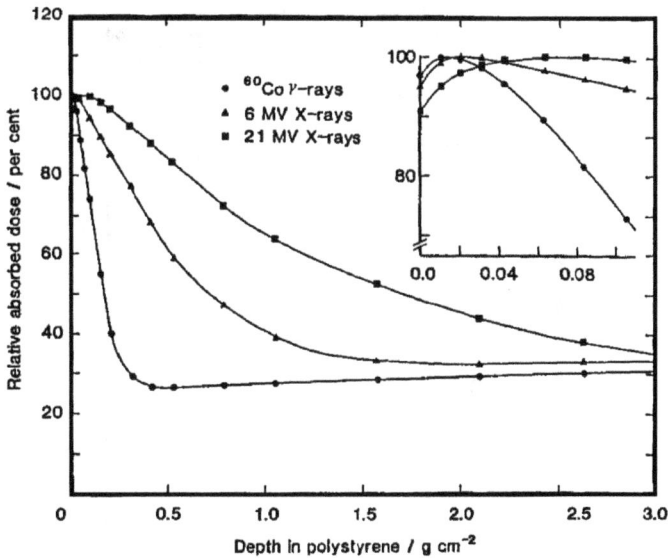

Figure 4.38. Depth–dose distribution for contaminating electrons measured 3 cm outside the field edge is illustrated. The insert shows the dose distribution over the first $0.1 \, \mathrm{gcm}^{-2}$. Field width and filter phantom distance: 20 cm.

to protect sensitive tissues as the lens of the eye or the testes. This shield should have a thickness of about $1 \, \mathrm{g/cm^2}$ at 6 MV X-rays and $3 \, \mathrm{g/cm^2}$ for 21 MV X-rays. The electron contamination may also be a large problem when using irregular fields where tissues shadowed by the beam blocks may still receive a large surface dose due to electron scatter in the air.

4.3.3.6.5. Total contamination

The different sources of contamination can now be summarized. In Fig. 4.39, the different contributions to the electron contamination are plotted as a function of energy for a large and a small field size. It is assumed that a beam–flattening filter of lead is placed at 80 cm from the phantom surface and that the upper surface of the collimators is placed at 60 cm from the phantom surface. The figure shows a minimum in surface dose for monoenergetic photon energies around 3 MeV corresponding to about 10 MV X-rays and a high increase for high energies and large fields. The figure also shows the decrease of the air contribution and increase of the contribution from beam–flattening filter with increasing energy. In clinical situations, this is further stressed due to the increasing thickness of the beam-flattening filter with increasing photon energy. For large field sizes, the air is the

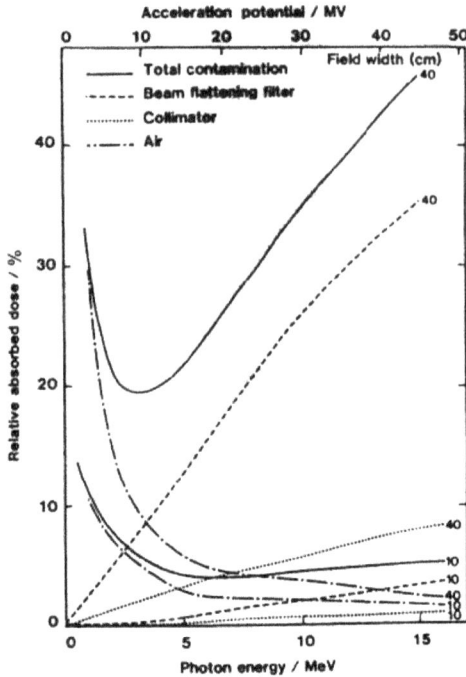

Figure 4.39.   Variation of the different electron contamination sources with photon energy is illustrated.

dominating source up to ~3 MeV or 10 MV, while for small field sizes, air produced electrons dominate up to over 15–20 MV due to the larger influence of scatter filtering in the air at small field sizes. When well aligned, the contribution from the collimators is a small fraction of the total surface dose.

4.3.3.7. *Scanned beams*

The thick flattening filters needed at high photon energies degrade the energy distribution by attenuating the high-energy photons. They also decrease the efficiency of bremsstrahlung production by reducing the photon fluence to as low as 20% or even less at high photon energies. By, instead, using scanned beams, no beam-flattening filter is needed and a more flexible beam shaping and better depth-dose distribution may be obtained as shown in Figs. 4.40 and 4.42.

With a scanned beam, it is also possible to generate compensated or wedged fields without material filters. Figure 4.54 shows a dose distribution for a 50 MV photon beam obtained from six elementary beams as indicated in the figure

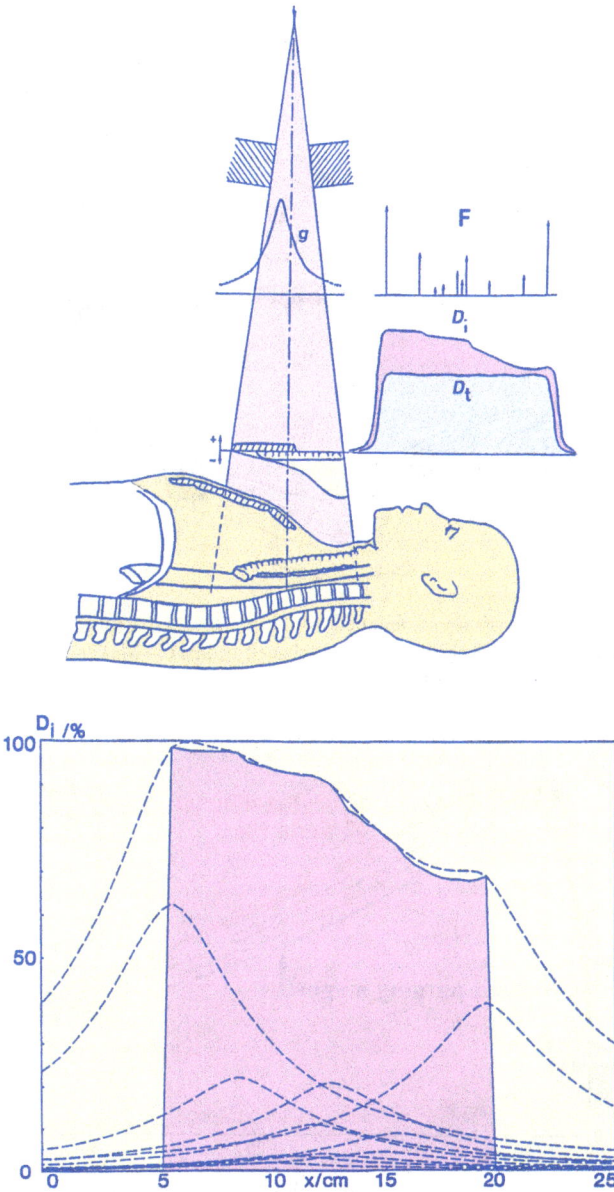

Figure 4.40. Irregular wedge–like beam (upper dashed line) obtained by adding physically optimized collimated scanned pencil beams, g (dashed bremsstrahlung lobes, F is the optimal scanning density, cf. Fig. 4.10). The lower pink dose distribution is desired and the upper dashed curve is the optimal result using the minimum overdosage algorithm (cf Lind 1990).

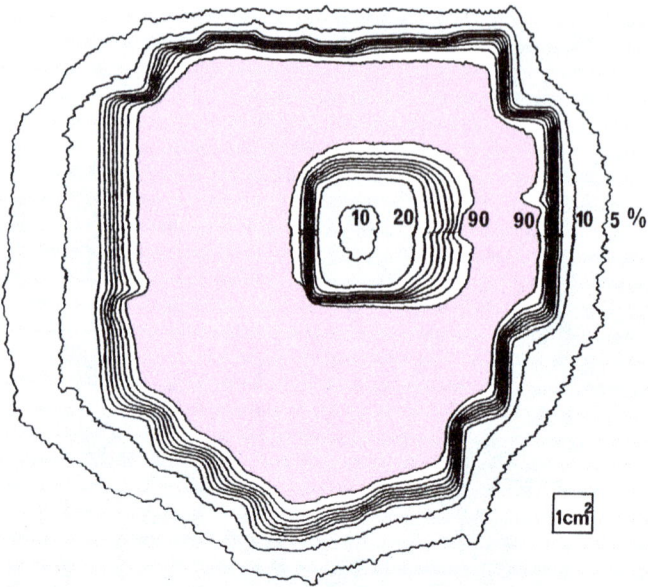

Figure 4.41. Dose distribution obtained with the very first gantry mounted MLC by blocking an island using two abutted crescent-shaped beams (50 MeVp->Be neutron beam, modified from Montelius *et al.* 1984).

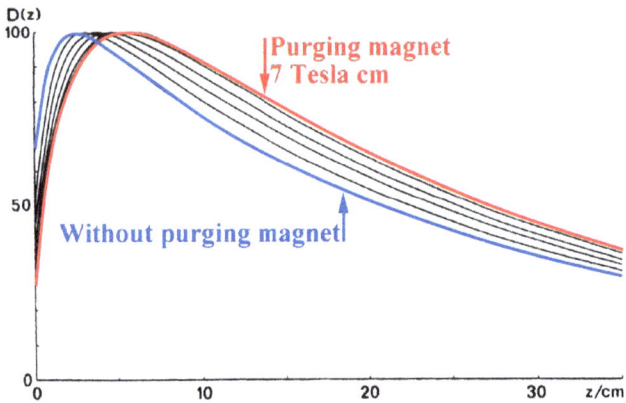

Figure 4.42. Illustration of the effect of high-energy electron contamination on depth-dose distribution of a 50 MV scanned photon beam and how it can be removed with a strong purging magnet (cf Figs. 4.10, 4.48 and 4.53).

(Näfstadius *et al.* 1984). Owing to the larger full width at half maximum (FWHM) for photon beams, additional collimating is necessary. To form the beam in any shape, a double focused MLC together with a scanned beam will be the ideal combination. Figure 4.41 shows the resultant dose distribution obtained with a MLC with a 20 MV photon beam (Brahme 1987). It is clear that the smooth shape of the therapeutic 90% isodose is ideal for beam collimation.

As discussed above, the filters are severe sources of electron contamination and they may be a serious detriment for the beam quality. With scanned photon beams, no flattening filter is needed and it is possible to install an effective purging magnet inside the treatment head directly below the target where the cross-section of the beam is small (cf. Fig. 4.10 and Sec. 4.4.3.5 below). Figure 4.42 shows depth-dose curves for a 50 MV photon beam with a field size of $30 \times 30 \, \text{cm}^2$ obtained with different magnetic field integrals from 0 to 7 Tcm. Without a purging magnet the electron contamination is high, but with full purging magnet excitation, a large lepton removing effect is possible (both electrons and positrons are removed). Monte Carlo calculations show that with a purging magnet a reduction of the electron contamination with a factor of about 10 may be obtained (Staël von Holstein *et al.* 1992) largely depending on air or helium, which is used in the treatment head.

4.3.3.8. *Photon beam quality parameters*

The variation of secondary radiation with the treatment situation shows the importance of having beam quality descriptors that takes into consideration not only the depth-dose distribution but also other factors relevant for dosimetry and dose planning. Quality descriptors should also include parameters relevant for shielding calculations both with respect to the patient and to the personal.

One clinically important quality parameter is for selection of stopping power ratios in order to determine the dose to water from ionization chamber measurements. One early parameter used was just the accelerating potential or the maximum energy of the photons. However, this parameter was found to be an unsuitable parameter for dosimetry. Different parameters using the transmission properties were instead used, for example, $d_{80\%}$ (*Brit. Inst. Radiol.* 1983). However, since the dose at dose maximum may be contaminated with electrons, this parameter was considered unsuitable and instead dose ratios at larger depths have been suggested ($\mathcal{J}_{15}/\mathcal{J}_5$, $\mathcal{J}_{10}/\mathcal{J}_{20}$, D_{10}/D_{20}, $_{PP}D_{10}/D_{20}$, TPR_{10}^{20}) by different organizations (e.g., NACP 1980; AAPM 1983; IAEA 1987). With the new high-energy accelerators, even these ratios may be dependent on electron contamination and perhaps TPR_{15}^{25} or even TPR_{20}^{30} should be used at these high energies, as mentioned above. The AAPM protocol TG 51 introduces another concept DD(10), that is, the depth dose at 10 cm depth. AAPM claims that this quality factor is better than the TPR^{20}_{10} for determining the stopping power ratio. However, IAEA still believes that the TPR^{20}_{10} is better (cf. Andreo and Brahme 1986).

Other parameters that should be included in a description of a photon beam are the surface dose and the dose at small depths due to the electron contamination. The dose due to secondary photons and the origin of these photons are also important parameters affecting the depth-dose distribution. These parameters should also be included in dose planning programs.

4.4. Intensity-modulated Beams

Techniques for the realization of flexible nonuniform dose delivery are now being available with the equipment from most accelerator manufactures with the introduction of MLC. Optimized nonuniform photon beams can be generated by classical compensator filters and transmission blocks techniques (Ulsö and Brahme 1988). One of the earliest attempts to achieve a conformal dose distribution was by the use of gravitationally controlled compensators (Proimos 1966). Fast and accurate computerized molding techniques and improved materials for casting such as stainless steel granulates (van Santvoort *et al.* 1995; Mejaddem *et al.* 1997) and low melting point heavy metal alloys (Walz *et al.* 1973) together with clever filter revolver techniques (Mejaddem and Söderström 1998) mounted in the treatment head or below its exit window may improve the flexibility, at least for optimized few field techniques (Söderström *et al.* 1995).

However, such filter molding techniques may be superseded by more generalized treatments and can be regarded as an intermediate step toward more cost-effective and dedicated dose delivery techniques for conformal treatments required by most modern therapy centers. Dynamic dual asymmetric jaws can also be used to arbitrarily shape the incoming beam if all four jaws can move independently and close in anywhere in the field. Unless the dose rate of the accelerator can be heavily increased, the dual jaws should in practice only be used where very smooth shaping is needed and for instance using a close in technique in order to keep down the treatment time. The asymmetric dual jaws can be further developed by adding an extra 45° rotated dual jaw pair on top of the first one so that an arbitrary field of octahedral shape can be produced that may be of interest in combination with narrow scanned beams. Today, the most flexible method to deliver nonuniform beams is by dynamic multileaf collimation and scanned beams.

Scanned narrow photon beams: It was earlier recognized that scanned beams could not only be used for synthesizing large uniform photon and electron beams of high therapeutic quality (Brahme *et al. 1980*) also could be used to intensity modulate the incoming beams to improve the dose uniformity in the target volume or avoid organs at risk. By increasing the intensity at the edges of the field with the elementary beam, the flatness of the dose distribution could be exactly controlled and fine-tuned at any depth. This was one of the starting points for the extensive research of how the incident beam should optimally be shaped and how the modern techniques of *inverse treatment planning* and Intensity Modulated Radiation Therapy (IMRT) were born (Brahme 1982, 1985, 1988; Lind and Brahme 1985). At that time the elementary bremsstrahlung beams that could be generated was quite broad (FWHM of 5–10 cm) with large tails and therefore such beams had limited possibility to shape steep gradients within the photon beams. Primarily the bremsstrahlung targets were designed to give a high forward and integral yield. Today, narrow elementary photon beams, suitable for advanced pencil beam scanning are possible to generate. The necessary tools are: a high-energy electron beam of low emittance, optimal target design, and advanced purging magnet techniques.

4.4.1. *Modern Advanced Dose Delivery Techniques*

Beside the classical uniform filtered beam techniques, it is possible to distinguish between two major groups of dose delivery: fan beam and pencil beam scanning techniques. All of them have the possibility to deliver intensity-modulated beams to complex target volumes. However, the fan beam and pencil beam methods for photon therapy may require long treatment times for extensive disease due to the small portion of the target that could simultaneously be irradiated by an incident fan beam or pencil beam.

4.4.1.1. *Pencil beam therapy*

The first working pencil beam scanning machine was developed in Chicago in the early 1960s (Carpender *et al.* 1963; Fig. 4.43(b)) and was the first working therapy machine that could scan a high-energy electron pencil beam over the patient. The beam source, which took up a rather large room, was a linear accelerator of the traveling wave type with the capacity of delivering electrons in the range of 5–50 MeV. The electron was guided into a rotary gantry for full isocentric treatments. The last bending magnet could be moved mechanically forward and backward in its own bending plane and it could also be rotated to perform concentric pendulum motions and transversal scans. In this machine, the elementary electron beam width was around 5 mm and of high dose rate, so it could rather rapidly buildup an electron field of about $20 \times 20 \, \text{cm}^2$ on the patient surface.

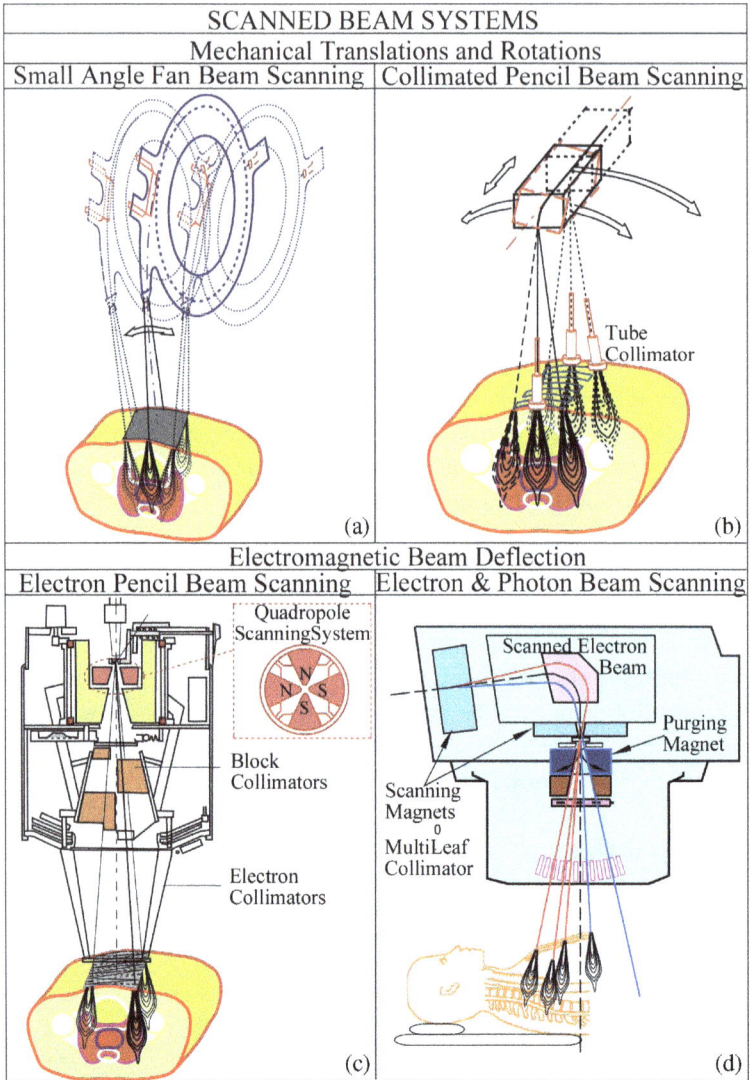

SCANNED BEAM SYSTEMS

Mechanical Translations and Rotations

Small Angle Fan Beam Scanning | Collimated Pencil Beam Scanning

Tube Collimator

(a) (b)

Electromagnetic Beam Deflection

Electron Pencil Beam Scanning | Electron & Photon Beam Scanning

Quadropole ScanningSystem

Block Collimators

Electron Collimators

Scanned Electron Beam

Purging Magnet

Scanning Magnets 0 MultiLeaf Collimator

(c) (d)

Figure 4.43 (a)–(d). Four different advanced beam-shaping systems for broad beams. The upper left panel (a) (Rassow 1970) shows a high-energy betatron that can move the natural electron fan beam by small angle pendular movements. The upper right panel (b) shows a collimated pencil beam scanning system where the bending magnet and collimator can be moved and tilted as indicated (Carpender *et al.* 1963). The high-energy linac in the lower left panel (c) (Aucouturier 1970) scans the electron beam using a quadrupole. The racetrack microtron in the lower left panel (d) (Brahme *et al.* 1987) can scan both electron and photon pencil beams due to the well-defined focus after the bending magnet.

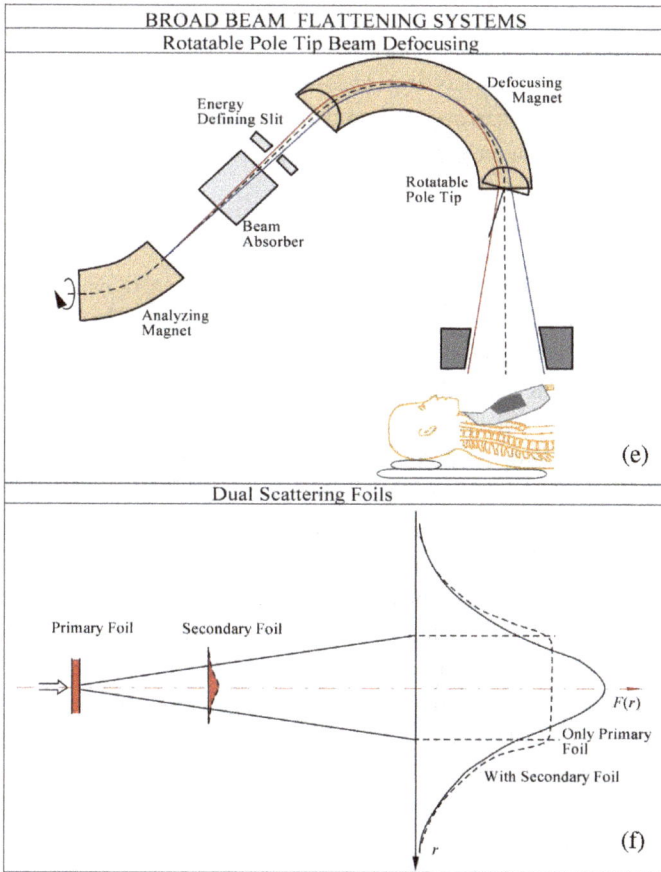

Figure 4.43 (e)–(f). Two different beam flattening systems for broad beams of electrons. The upper panel shows how high-quality high-energy broad electron beam of varying field sizes can be generated by defocusing the initial electron beam in the fringing field of a rotatable pole tip in the bending magnet or using a quadrupole. In principle, a shrinking field technique is possible to carry out but was never explored. The lower panel shows the dual scattering foil technique that can generate high-quality electron beams up to about 25 MeV.

Dynamic asymmetric dual jaws have the possibility to mechanically scan a narrow rectangular photon beam over the patient. A nice feature with this scanning technique is that the rectangular beam "spot" can be varied from very small to large sizes anywhere in the field so that large areas that may require uniform dose can be delivered relatively fast and the jaws can close in or open up on more complex portions. However, for more generalized treatments, the dual jaws technique will become very time-consuming.

Recently, a small linear accelerator has been mounted in a robot arm with a high degree of freedom. The narrow beams and low dose rate gives very long treatments time at least for large target volumes.

4.4.1.2. *Fan beam therapy*

A type of fan beam treatment was tested early in Boston (Kijewski *et al.* 1978) where an elongated fan beam slit in a rotary gantry were allowed to be varied simultaneously as the patient were slowly moved through the beam. Since most of the useful photons are collimated during the treatment, the irradiation time becomes quite long particularly for larger target volumes similar to tomotherapy. The intensity-modulated fan beam collimator (Carol 1992, 1995) has the possibility to deliver advanced intensity-modulated beams. The fan beam collimator can be mounted on an ordinary rotary gantry and have two opposed collimator banks each containing 20 stubby leaves. In each bank, every leaf can attain either of two states, fully open or closed. They are thus a particularly simple case of a MLC that does not rotate and only move a small increment. The patient moves through the gantry as it is rotated like in a CT-scanner. This type of collimator using a fan beam helical spiral technique is used in the tomotherapy machine and it is now installed at several centers (Mackie *et al.* 1993). This machine is promising since it is as robust as a CT and it can take spiral CT images during the treatment.

4.4.1.3. *Scan beam therapy*

The most general and least time-consuming method to perform flexible nonuniform dose delivery is to use electromagnetically scanned electron, photon, or light ion pencil beams. During the past 10–15 years, large electron or photon beams of high quality has been available from racetrack microtrons at several centers around the world. Even though the elementary photon beams have been rather wide ($\sim$9 cm FWHM), some fruitful compensation can still be made. Largely due to the absence of scanning beam models in commercial treatment planning systems this capability has not yet been really exploited. Spatial modulation of higher geometric precision is, therefore instead, done with a double-focused MLC. The original concept was to improve the beam quality and at the same time be able to treat large volumes and deep-seated target in a short time. The racetrack accelerator was a natural choice because of its high current capacity, its broad energy range up to 50 MeV and its low emittance and low-energy spread which is ideal for beam scanning. Since no flattening filter is needed for the high energies as in common machines, the scanned broad beam becomes exceptionally clean from scattered irradiation not least since the gantry head is filled with helium. It was earlier recognized that scanned beams could not only be used for synthesizing large uniform photon and electron beams of high therapeutic quality (Brahme *et al.* 1980), but also could be used to intensity modulate the incoming beams to improve the dose uniformity in the target volume. By increasing the intensity at the edges of the field with the elementary beam, the flattening of the dose distribution could be exactly controlled and fine-tuned at any depth. Today, with elementary photon beams of widths <3 cm, together with a few multileaf segments, such techniques can perform an almost perfect conformal therapy within minutes. The technique is still under development and not yet commercially available.

Dynamically scanned light ion beams will probably be the ultimate radiation therapy modality when high geometrical precision and high effect in resistant tumors is required since the pencil beam penumbra is so narrow that additional collimation is not generally required and the finite light ion range protects tissues also beyond the tumor volume (Graffman *et al* 1985; Carlsson *et al.* 1997; Brahme 2003, 2004; and Chapter 8).

4.4.2. Dynamic Multileaf Collimation

Since the widespread introduction of MLCs at many centers, several research centers have been focusing on how to utilize the dynamic capability of MLCs as a technique for flexible nonuniform dose delivery. Dynamic multileaf collimation is often used as a general term for different types of irradiation techniques where the MLC is used to shape the incoming beams dynamically or segmentally to arbitrary lateral profiles (Källman *et al.* 1988). The techniques could thus be divided into two major groups: continuous velocity modulation techniques and segmental methods such as stop and shoot techniques. The latter technique was earlier implemented at some places due to more simplistic dosimetric quality assurance (Bortfeld *et al.* 1994). Today, several advanced radiation therapy department have implemented velocity modulation techniques into real clinical systems and they are in routine use at several centers around the world. A prerequisite for the velocity modulation technique is that the collimator leaves can move independent of each other and that they can travel over the whole field. Implementation of velocity-modulated collimation techniques had some natural resistance among physicists and oncologists due to the dynamics of the treatment combined with lack of quality assurance programs for IMRT. Physicists and clinicians also believed that the important matter was to use IMRT, but minimizing the treatment time was not yet a primary goal. Today, we know that fast treatments are of outermost importance due to the increasing need for radiation therapy. Fast IMRT by itself is important due to less integral patient motion. An even more flexible method to deliver large intensity-modulated fields is by beam scanning in combination with general or segmental multileaf collimation and this method is still under development. However, the most important tools such as treatment planning using generalized pencil beam algorithms for the optimization of the scanning patterns and multileaf motion as well as equipment for fast dose delivery by scanning beam techniques were earlier developed in a research code at Karolinska.

4.4.2.1. *MLC development*

Nonuniform dose delivery is essential for optimization of dose distributions. Traditionally, simple nonuniform dose delivery has been produced by wedge filters or compensators. This approach is only suitable for static fields and may, thus, only with difficulty be used for moving fields, that is, arc and rotation therapy. While moving the gantry around the patient, it is desirable to continuously conform the field shape to the target volume. This can be achieved in real time with a MLC, where the positions of the individual leaves can be adjusted as the gantry rotates around the phantom.

The first MLCs were motor-controlled and manufactured in Japan in the mid-1960s. The drawbacks of these MLCs were the lack of computer control and their large leaf width (30 and 48 mm), which made close conformation to complex targets impossible. Modern state-of-the-art MLC systems have projected leaf widths between 5 and 12.5 mm and are equipped with computer-controlled systems for both stationary and dynamic use. Today all manufacturers produce leaf collimators.

The main applications of MLCs are:

- for irregular collimation to replace simple lead and cerrobend blocks and
- for intensity modulation by moving the leaves of the MLC during the irradiation (dynamic multileaf collimation).

The MLC has a number of other advantages compared to conventional field shaping with lead or cerrobend blocks. The beam setup time is reduced, leading to a reduction of the total treatment time. The cost of production and storage of beam blocks as well as the production time of the blocks is eliminated. Furthermore, lifting of the heavy blocks and the possibility of making errors while mounting them are eliminated. Inhalation of toxic vapors during the casting of the blocks is also avoided.

However, there are disadvantages with some early MLC designs that need to be mentioned. The independently moving collimator leaves may cause some radiation leakage between the leaves. This may cause hotspots and may potentially cause normal tissue injury to sensitive organs, particularly this might be a problem for long irradiation time. A second disadvantage is that the jagged edges of the treatment field may not properly conform to the target volume causing unnecessary injury to surrounding normal tissue or lack of tumor control at the periphery of the target.

The very first commercially available double-focused MLC on the Scanditronix Racetrack accelerator is shown schematically in Fig. 4.10. It has 32 pairs of independently moving tungsten leaves. The leaves have an effective thickness of 7.5 cm, corresponding to $\sim 145\,g/cm^2$, which is equal to more than 10 half-value layers (the transmission is $<10^{-3}$) for the 50 MV bremsstrahlung beam produced by electrons with an energy of 50 MeV in a beryllium-tungsten target.

The projected leaf width at isocenter (100 cm) is 1.25 cm. Each leaf can be moved 16 cm out from the central axis. The maximum over-travel, that is, the possible movement across the central axis, is 14 cm. This allows a maximum field size of $40 \times 32\,cm^2$ and the maximum area over which opposed leaf pairs can close is $40 \times 28\,cm^2$. The highest leaf speed is 1.5 cm/s. The treatment head is also equipped with a block collimator, moving perpendicular to the leaves of the MLC. Its main function is to allow arbitrary field widths perpendicular to the direction of the leaf motion, but it can also be used to reduce radiation leakage between the leaves. It is made of lead with steel edges and has a thickness of 11 cm. They are, when used, automatically set to the position of the field edges, where the left end and the right end penumbras are measured (cf. Fig. 4.10).

The leaves of the MLC are curved. Both the upper and the lower leaf surface form parts of a sphere with the center at the target. The leaf end and leaf sides are wedge-shaped to be tangential to ray lines converging at the virtual photon point source near the target. These characteristics are particular for the type of collimator that are called a "double-focused" to make sharpest possible penumbra in the beam.

Each leaf is equipped with a step on the sides of the leaf and at the leaf front end, to minimize the leakage of radiation between closed leaf pairs (cf. Fig. 4.63 below). Interestingly this MLC is equally optimized for photon and electron beam therapy!

4.4.2.2. Dynamic multileaf collimation

The first method for effective optimization of dynamic multileaf collimation was developed at Karolinska (Brahme 1985, 1987, 1988; Källman *et al.* 1990). The researchers suggested that the leaf edge should move independently of each other to dynamically produce an optimal opening density and gave universal solutions that were independent of the how the motion was produced. A few years later, a *numerical* method for the leaf motion was developed by Rosenbloom *et al.* (1992) using the sliding window approach. The same year at the *annual medical meeting in Sweden*, Svensson *et al.* (1994) presented the analytical equations of motion for dynamic multileaf collimation. They outlined theoretically a very simple method as to how the leaves should move over the radiation field to shape arbitrary energy fluence distributions on the surface of the patient as quickly as possible taking even mechanical restrictions such as maximum speed and acceleration of the leaves into account. In addition, the beam-shaped incident on the collimator was taken into account and optimized by elementary beam scanning to as far as possible reduce treatment time. This method was published (Svensson *et al.* 1994) and the same year also two other groups (Spirou *et al.* 1994; Stein *et al.* 1994) presented similar equations. They all showed that the treatment time could be considerably reduced, if opposed leaves in the collimator are allowed to move independent of each another such that the size of the aperture varies during the travel across the field. This will significantly reduce the ratio of the incoming unmodulated and the transmitted beam. The optimal collimator motion also depends on the shape of the initial energy fluence distribution upstream of the MLC as well as which mode the collimator will be operated in. A shrinking field technique may be optimal for essentially single-peaked distributions for reducing collimator movements across the field. Such a method is also preferable, considering a full light field simulation for the safety of patient set up or recording a portal image before the actual treatment starts. For more complex distributions having several peaks, a sliding window technique may be preferable. If a beam scanning system generates most of the desired energy fluence profile, the residual modulation with the leaves becomes small.

Generally, the initial energy fluence rate upstream the collimator, $\psi(x, y, z)$, is temporally and spatially varying. A straight forward but time-consuming method to arbitrarily shape an incoming uniform photon beam where $\psi(x, t) \equiv 1$ is to move a very small collimator opening across the field, for instance an almost closed block or multileaf opening, that is, mechanical pencil beam scanning, with a speed $v(x, y)$ at each point that is inversely proportional to the desired intensity below the collimator according to $\Psi(x, y) = \int_{\Delta t} \psi(x, y, t)/v(x)\mathrm{d}t$. However, using such a small aperture will require very extensive irradiation times because of the ineffective use of the incoming fluence.

4.4.2.3. *Equation of motion during dynamic multileaf collimation*

In the definition of the camera shutter technique it is assumed that the leaves start closed on side and in a single sweep with varying aperture end up closed on the other side. The leading collimator leaf is indexed with 1, and the lagging leaf with 2. Only one pair of leaves having zero penumbra width is here analyzed to understand the principles. The leaves begin at one side at position x_b and move to the other side end at position x_e. The time spent by the leading leaf to reach position x is the time $t_1(x)$ at position x. When the leading leaf passes position x, this point is no longer blocked and the energy fluence is transmitted until the lagging leaf reaches this point at time $t_2(x)$ and closes it from further irradiation. The total time that this position has been open for transmission is simply $t(x) = t_2(x) - t_1(x)$. The transmitted energy fluence that reaches the patient through the collimation system is just the time integration of the dose rate for local opening density according to

$$\Psi(x) = \int_{t_1(x)}^{t_2(x)} \psi(x, t)\mathrm{d}t. \tag{4.17}$$

The problem is now to find the optimal motion of each collimator leaf, for a given energy fluence rate $\psi(x, t)$ upstream the collimator, that will minimize the time required to shape a desired energy fluence profile $\Psi(x)$ at the patient surface. The desired energy fluence profile is not the same for each leaf pair and it should be noted that different leaf pairs end up with different minimum treatment time. The minimum total treatment time is then determined by that leaf pair which have the longest minimum treatment time.

Since the temporal variation of the initial energy fluence rate distribution upstream the collimator is often due to stochastic processes in the electron source or the beam transport system, it will influence the whole beam in a similar way. The energy fluence rate distribution can then generally be separated into a spatial and a temporal part $\psi(x, t) = \psi(x) \cdot \psi_t(t)$. This is basically also possible for scanned beams, since in general the average ψ for a whole beam scanning cycle is quite stable. The equation then reduces to

$$\Psi(x) = \int_{t_1(x)}^{t_2(x)} \psi(x) \cdot \psi_t(t)\mathrm{d}t = \psi(x) \int_{t_1(x)}^{t_2(x)} \psi_t(t)\mathrm{d}t. \tag{4.18}$$

The time-varying component of the energy fluence rate is generally not well known. As a first approximation, we will therefore neglect the temporal variation of ψ, which correspond to assuming that the integrated delivered energy fluence or dose, sometimes measured in monitor units, is used as reference scale. The equation then reduces to

$$\Psi(x) = \psi(x)(t_2(x) - t_1(x)). \tag{4.19}$$

Thus, when the desired energy fluence $\Psi(x)$ should be generated with time-independent energy fluence rate distribution $\psi(x)$, the distribution of interest is the local opening density or local irradiation time $t(x)$, which is given by

$$t(x) = \frac{\Psi(x)}{\psi(x)} = t_2(x) - t_1(x). \tag{4.20}$$

The gradient of the local opening density is given by

$$\frac{dt}{dx} \equiv \frac{\Psi'(x)\psi(x) - \Psi(x)\psi'(x)}{\psi^2(x)} = \frac{dt_2}{dx} - \frac{dt_1}{dx} = \frac{1}{v_2(x)} - \frac{1}{v_1(x)}, \tag{4.21}$$

where

$$v_2(x) = \frac{dx}{dt_2} \tag{4.22}$$

and

$$v_1(x) = \frac{dx}{dt_1}. \tag{4.23}$$

This expression illustrates that the higher the maximum collimator speed the faster the desired beam profile will be generated. Obviously, the gradient of the opening density vanishes if the incident energy fluence has approximately the same shape as the desired energy fluence and the leaves should practically be left wide open during the treatment. The total treatment time will in this formulation by definition be equal to $t_2(x_e)$ or $t_1(x_e)$.

The minimization problem now becomes quite simple:

$$\min(t_2(x), x = x_e) = \min(t(x) + t_1(x), x = x_e) \tag{4.24}$$

or

$$\min(dt_2(x)/dt, x = x_e) = \min(dt(x)/dt + 1/v_1(x), x = x_e). \tag{4.25}$$

Thus, when $t'(x) > 0$, the gradient $t'_2(x)$ is always as low as possible when $v_1(x) = \hat{v}$ and vice versa when $t'(x) < 0$; the lowest possible value for the gradient of $t_2(x)$ to satisfy the equation is when instead the lagging leaf is moving with maximum velocity. The equation of motion then becomes:

$$
\begin{cases}
v_1(x) = \hat{v} \\
v_2(x) = \dfrac{\hat{v}}{1 + \hat{v}\frac{dt}{dx}}
\end{cases}
\qquad \frac{dt}{dx} \geq 0,
\tag{4.26}
$$

$$
\begin{cases}
v_1(x) = \dfrac{\hat{v}}{1 - \hat{v}\frac{dt}{dx}} \\
v_2(x) = \hat{v}
\end{cases}
\qquad \frac{dt}{dx} < 0
.
\tag{4.27}
$$

In words, if the spatial gradient of the desired fluence distribution is positive, the leading leaf should move with maximum speed and the lagging leaf that is really shaping the field should move with a velocity, which is inversely proportional to the gradient of the desired dose distribution. When the leaves reach a peak or valley where the gradient is zero, both leaves have the same maximum speed. In regions where the gradient is negative, it is instead the leading field that actually will intensity-modulate the field and the lagging leaf moves with maximum speed. Since there are mechanical constraints on the collimator such as maximum acceleration, fast velocity changes that may be needed near any peak or valley region must be accounted for.

The minimum treatment time $T_{\min}$ to shape any given continuously varying irradiation time distribution is now easily calculated by integrating the spatial gradient of the lagging leaf over the region of interest.

$$
\begin{aligned}
T_{\min} &= \int_{x_b}^{x_e} \left(\frac{dt_2}{dx} \right) dx = \int_{\frac{dt}{dx} < 0} \frac{dx}{\hat{v}} + \int_{\frac{dt}{dx} \geq 0} \left(\frac{1}{\hat{v}} + \frac{dt}{dx} \right) dx \\
&= \int_{x_b}^{x_e} \frac{dx}{\hat{v}} + \int_{\frac{dt}{dx} \geq 0} \left(\frac{dt}{dx} \right) dx = \frac{x_e - x_b}{\hat{v}} + \sum_{i=1}^{n} \Delta t_i.
\end{aligned}
\tag{4.28}
$$

The sum is taken over all regions where $dt/dx \geq 0$. The time required to modulate each section with positive slope $\Delta t_i = t_{\max,i} - t_{\min,i}$ is equal to the increment of the desired opening time distribution between a local minimum and the proceeding local maximum value. The total treatment time given by Eq. (x) is made up of two different terms. The first term in Eq. (x) is simply the time it would take for the left leaf to cross the field with maximum speed. The second term is the additional time required to actually shape the desired irradiation time distribution. It should be noted that with the camera shutter technique the minimum treatment time depends only on the sum of differences between consecutive minima and maxima of the desired opening time distribution and not on the detailed shape of the peaks.

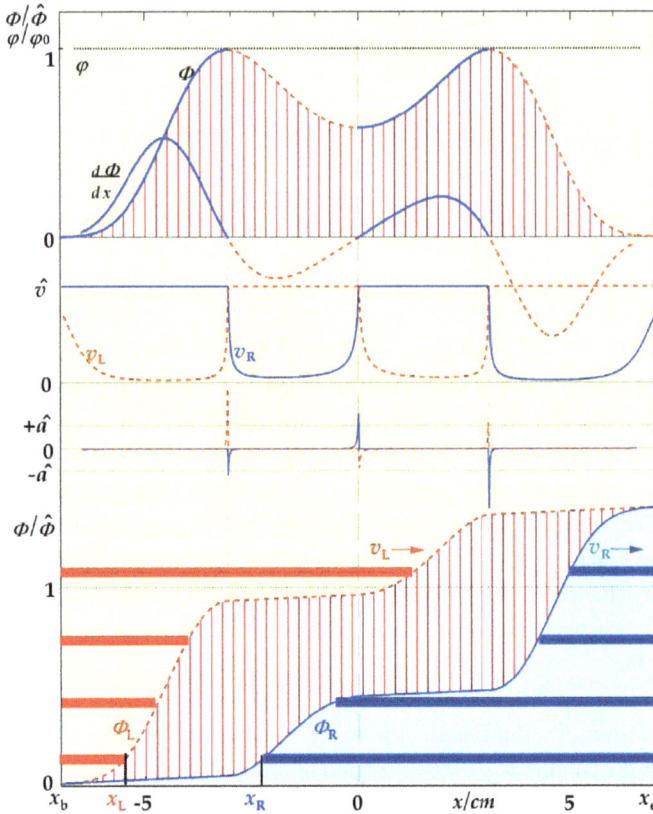

Figure 4.44. Illustration of over-flattened beam (upper vertical lines) obtained by dynamic multileaf collimation (motion indicated by red and blue collimator bars in the lower panel).

Therefore, distributions with varying complexity can be generated in the same time when the sum in Eq. (4.28) is equal for the distributions.

Optimal start and stop positions: With the camera shutter technique, it is assumed that the leaves of the collimator start at the left edge of the radiation field. In principle, the leading right leaf could start close to the first peak and the left leaf alone could modulate the initial part of the desired energy fluence distribution, as shown by the dashed dotted lines in Fig. 4.44. The irradiation time will then be reduced by exactly the time it would take for the right leaf to travel with full speed from the left edge to the first peak. Similarly, when the lagging left leaf reaches the last peak, it should stop there and the right leaf alone would modulate the final part of the desired energy fluence distribution. If the coordinates of the first and last peak values are

x_1 and x_n, respectively, then the distance where at least one of the leaves is moving with maximum speed is the distance between these two outermost peaks $x_n - x_1$. Thus, the minimum irradiation time according to Eq. (4.28) reduces to

$$T_{\min} = \frac{x_n - x_1}{\hat{v}} + \sum_{i=1}^{n} \Delta t_i. \tag{4.29}$$

Thus, the transit time in Eq. (4.29) reduces to $(x_n - x_1)/\hat{v}$, and if the two outermost peaks in the desired distribution are close to each other, then the transit time is small. However, since the left leaf alone modulates the initial part of the first peak, its velocity has to approach infinity as the gradient approaches zero at the peak. The optimal starting point for the right leaf is, therefore, not precisely at x_1 but a short distance in front of the peak. To be exact, a similar margin is needed at x_n but on the opposite side of the peak.

4.4.3. Narrow Scanned Photon and Electron Beams

4.4.3.1. Introduction

More than 95% of all tumors are well treated with high-energy photon beams than with high-energy electron beams. Therefore, a lot of efforts have been devoted to the design of a treatment head that will deliver an accurate high-quality therapeutic photon beam. Up to the 1980s, the development was mainly on the dosimetric properties and the characterization of the beam quality and a lot of the work was spent on improving the beam quality. With the birth of IMRT in the early 1980s, more attention has been turned to more effective ways to rapidly deliver intensity-modulated beams and verification of the delivered dose distributions. This by necessity requires much more sophisticated delivery systems such as high-resolution double-focused dynamic multileaf collimation or ultrafast electromagnetic scanning of narrow photon pencil beams. Although, scanned beams are the most flexible way to deliver IMRT, the system design available today is still too expensive for the average clinic and, therefore, more than 90% of the installed treatment units have more or less conventional design of the treatment head even if multileaf collimation is standard. It should be noted that the first MLCs came into use in the mid-1980s and the focus was on the delivery of large flattened square field of high mean energy for deep penetration. Today, the radiotherapy departments require MLCs that could also be run in a dynamic mode. The potential advantages of using scanned photon beams of high energy are large both considering total treatment time and dose distribution particularly for broad heterogeneous beams of high energy (Brahme et al. 1990; Gustafsson et al. 1995).

To irradiate the geometrical fine structure in the tumor and to avoid damaging the surrounding sensitive normal tissue, a narrow fast scanned photon beam is the ideal modality (Brahme 1987; Brahme *et al.* 1990; Lind *et al.* 1991; Söderström *et al.* 1993). Furthermore, when beam scanning and dynamic multileaf collimation are used concomitantly, it was shown that the narrow scanned bremsstrahlung beam considerably reduces the treatment time and at the same time the highest possible calculated therapeutic outcome is achieved(Svensson *et al.* 1994; Gustafsson *et al.* 1995).

4.4.3.2. *Treatment head design for scanned beams*

An interesting compact design was proposed by Svensson *et al.* (1998) for an efficient treatment unit using narrow scanned photon and electron beams. The treatment unit may also be used for *in vivo* monitoring of the delivered 3D dose distribution using PET-CT verification. The unit is using a newly developed transmission target irradiation technique. An important prerequisite to generate narrow scanned photon beams is to use an high current electron source of low emittance and high energy, that is, the electron spot size on the target should be in the order of 1 mm simultaneously as the intrinsic electron angular spread should not cause the expansion of the beam in vacuum more than around 1 mm.

The development of the treatment heads for scanned beams as well as for conventional treatment heads is focused on improving the accuracy and efficiency of the dose delivery, particularly for intensity-modulated beams. Narrow scanned beam is inherently the most efficient way for IMRT and a flattened beam is just a special case of an intensity-modulated beam. The key components of the treatment head for scanned beam become quite different compared to components of a conventional treatment head and for instance components such as beam blocks and MLCs and some components may even be redundant and can thus be removed to save space and costs. An exploded view of a modern treatment head and its internal components are described in Figs. 4.10 and 4.45.

In this design, the isocenter to the patient has been shortened to 65 cm which considerably decreases the pencil beam width for more effective intensity modulation of the dose delivery. The clearance to the patient is 35 cm. This is possible since the block collimator has been removed in the treatment head and the MLC has been moved to the target.

The ultimate treatment head for effective narrow scanned photon beam may contain a very thin collimating device (Svensson *et al.* 2007). All intensity modulation may be carried out by narrow photon pencil beam scanning by a beam of high energy and dose rate (cf. Figs. 4.40, 4.41, 4.47, and 4.50). Each

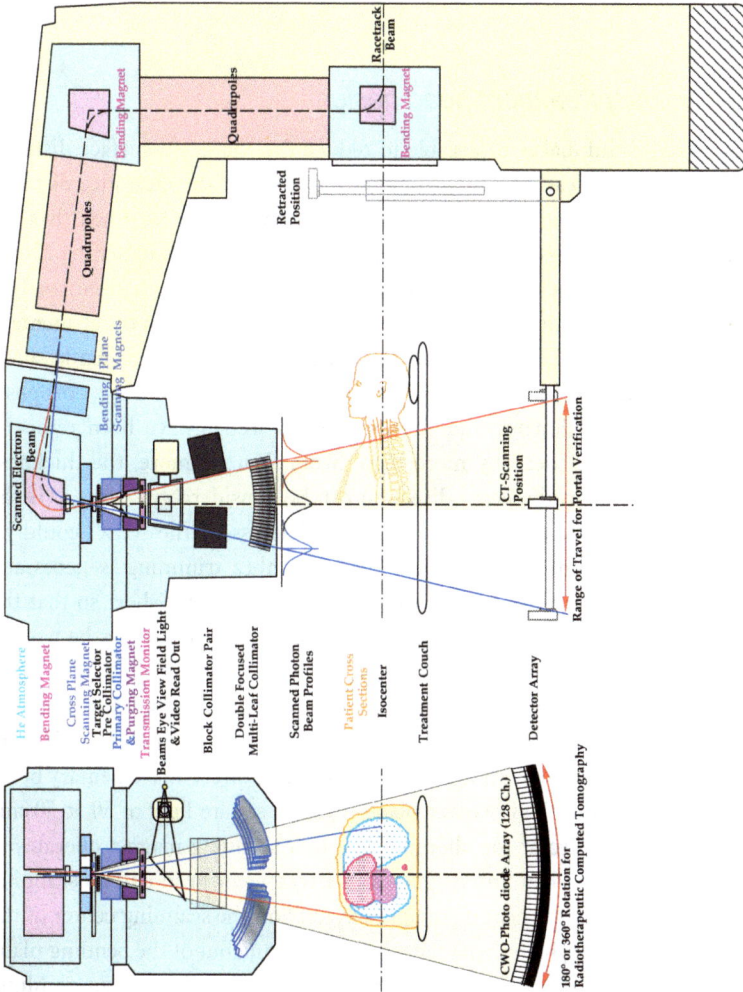

Figure 4.45. Cross-section through the treatment unit based on the Racetrack Microtron.

individual beam is monitored by a high-resolution segmented transmission ion chamber, or by position sensitive GEM detectors. Advanced treatments may be executed in a few minutes. The treatment head will in the future be integrated with a PET-CT camera for combined patient position diagnostics, dose delivery verification and treatment verification with the patient on the treatment couch (Fig. 4.83).

4.4.3.3. *Compact treatment head design*

A small treatment head makes it possible to reduce the source-to-surface distance between the treatment head and the patient without losing the clearance distance between the treatment head and the patient. In a conventional treatment head, a lot of space is necessarily used by beam-shaping devices, such as flattening filters, compensators, wedges, blocks, and MLCs. When scanned beams are used, wedges and beam compensation is done by the elementary beam itself. When the elementary beam is around 1 cm (FWHM) even the beam penumbra will be sufficiently sharp that the thickness of the collimator can be reduced. In the new design proposal (Fig. 4.46) the block collimator and the mechanical wedge have been removed shortening the treatment head by more than 20 cm. Furthermore, the thickness of the collimator leaves in the beam direction can be considerably reduced. If the machine would be dedicated for ≥ 50 MV, the thickness of the MLC could be reduced or even be removed altogether if no penumbra trimming is necessary. In this way, the source to axis distance can be made sufficiently short so that the beam half-width is around 1 cm and indeed no collimating devices may be needed to maximize the treatment outcome.

Scanning procedure, efficiency, and verification: The scanning system operating in the racetrack MM50 is fully computer-controlled (cf Figure 4.76a below) and it can, by using magnetic deflection, within a few microseconds, direct an elementary beam at an arbitrary position in the isocenter plane within a square field of 50×50 cm^2 for electrons and 50×40 cm^2 for photons. The first scanning magnet operates in the same plane as the last bending magnet in the treatment head and its scanning center is imaged through the bending magnet exactly on the scanning center of the second scanning magnet which then scans the beam in and out of the bending plane (Figs. 4.10 and 4.46). Immediately below the last scanning magnet, a bremsstrahlung target can be inserted. The upstream side of the target is placed inside the poles of the scanning magnet to achieve a well-defined focus for all scan directions. A target selector makes it easy and fast to switch between different target designs for various purposes. A user-defined scan matrix calculated by an optimization algorithm for intensity-modulated fields as well as standard circular scans for uniform fields can

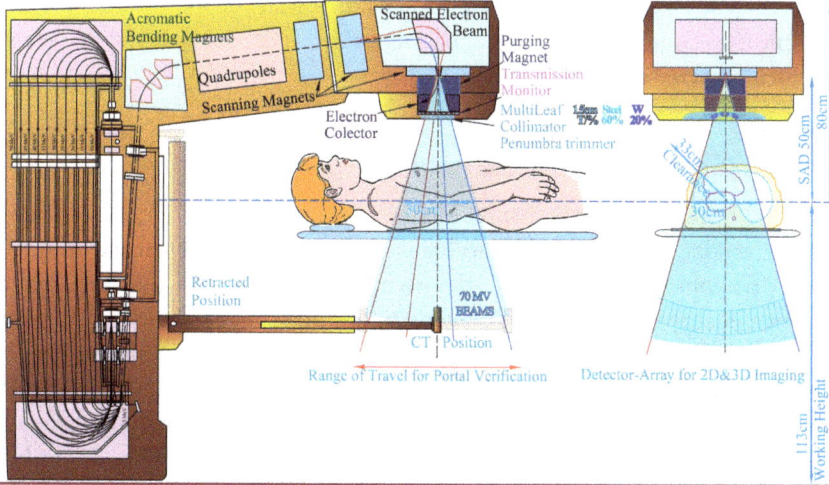

FLEXIBLE DOSE DELIVERY WITH A VERY COMPACT GANTRY MOUNTED RACETRACK (5–70MeV) USING NARROW SCANNED ELECTRON AND PHOTON PENCIL BEAMS

Figure 4.46. Cross-section through a proposed ultra-compact treatment unit based on the Racetrack Microtron.

easily be fed to the therapy system. When the machine is operated at pulse repetition frequency of 200 Hz (5–300 Hz), it takes 1 s to complete a scan with a scan matrix containing 200 positions. The duration of each scan pulse in the scan matrix can be varied within $\sim$1–6 μs, so pulse length intensity modulation can be combined with spatial modulation. Today, the dose regulation system utilizes the possibility to vary the pulse length for the last few scans at the end of a treatment in order to fine-tune the absolute dose level that may otherwise limit the degree of freedom for intensity modulation. When the electron injection system has been replaced by a gridded fast electron gun, the current in each pulse can instead be varied which is more convenient for intensity modulation. For heavy irregular scan matrixes, the path through the scanning matrix will also be optimized to minimize the load on the scanning magnets and reduce magnetic remanence as much as possible and speed up the scanning procedure. A segmented transmission ion chamber is placed below the purging magnet and it monitors the dose delivery. The size of each of the 16 segments is about 4 cm^2 projected on the isocenter and the position and dose contents of each delivered elementary beam can be determined by real-time calculations based on the chamber signal. The check and confirm system can immediately interrupt a treatment if something goes wrong with the positioning or if the dose content in a pulse is erroneous or it can decide to continue and if possible compensate for earlier errors. Portal imaging may be used utilizing a CWO detector array placed in the rotary gantry to monitor patient setup and movements. The portal imaging device together with the transmission chamber can exactly in time record the whole treatment execution and in real time or at consecutive treatment fractions fine adjust the total delivered dose distribution.

4.4.3.4. *Scanned bremsstrahlung beams*

Flattening with scanned beam. For the highest energy beams >20 MV, the flattening should be done by scanning beam techniques since the filter will remove the high-energy photons. With scanned photon beams a narrow forward concentration of the high-energy bremsstrahlung photons with a minimal large angle component is desirable, that is, the conversion of electron energy to bremsstrahlung photon energy should be as efficient as possible only in the forward direction (Brahme *et al.* 1986). If the target is made of a low atomic number, the electron multiple scattering becomes low and consequently gives a narrower beam and higher forward yield but a lower integral yield in the bremsstrahlung conversion. When the electrons are slowed down for deeper layers, the multiple scattering increases, and at the same time the width of the generated bremsstrahlung from deeper layers also increases. The width measured as the FWHM of such elementary bremsstrahlung beams is of the order 6–9 cm (Nordell and Brahme 1984). If the target space in the treatment head is limited, the electrons must be stopped in a shorter distance but at the same time the conversion rate should be as high as possible in the forward direction and the width of the beam should be kept small. Various types of unorthodox sandwich targets build with uniform layers of different materials can generate such beams. For instance, with a target made of 9 mm beryllium followed by 6 mm of tungsten, the width of the 50 MV photon beam will be around 7.5 cm at a SSD of 100 cm. For the same available target space, the target could be further optimized by optimally exchanging the materials with depth using intermediate atomic numbers. If the target is made up of materials of optimal thickness with increasingly higher atomic numbers, the electrons will be stopped in a "soft" manner resulting in a narrower and more intense beam. Unfortunately, the gain by using a target made of three or four layers is not so much compared with a target made of two layers. However, there still may be some beam quality advantages of using optimized targets made of more layers than two.

Fast high resolution IMRT with scanned beams. Most parts, more than the last two-third of the electron range of such composite targets where the electron energy is ≤15 MeV contributes to the wide tails of the elementary beam and such photons also have a lower mean energy. This part can be removed and the transmitted electron beam should instead be deflected out of the therapeutic beam.

Probably the best way to generate really narrow elementary photon beams is to use a thin target of low atomic number such as beryllium or even liquid hydrogen may be used in order to keep down the multiple scattering in the target as much as possible.

For efficient IMRT, scanned beam should preferably be generated with the newly developed thin transmission target technique (Svensson *et al.* 1998). The target is made of a thin beryllium layer (1–6 mm) that will keep the laterally electron expansion and angular scattering low. In this type of target, relatively more bremsstrahlung of high energy is generated in a forward direction. Instead of stopping the electrons in a high atomic number layer that will broaden the photon beam, the electrons are instead transmitted right through, where a strong purging magnet deflects the beam to an electron collector. The bremsstrahlung conversion efficiency in the forward direction for the very thin targets is still very high and of the same order as in a full range target, for instance, the conversion efficiency in the forward direction as high as 70% of that in a full range target made of tungsten. This is due to the very low almost negligible angular spread of the initial electron beam (<1 mrad), and because the electron beam will suffer a very small degradation by multiple scatter in a thin low atomic number target (<30 mrad rms radii at the exit side of the target) resulting in a very narrow forward peaked intrinsic bremsstrahlung distribution (around 15 mrad at FWHM, cf. Fig. 4.47).

To have a high total production of bremsstrahlung, a high atomic number material shall be used (see Fig. 4.13 that illustrates the relative radiation stopping power for some materials). The angular distribution will depend on the thickness and the atomic number material of the target because a thicker target will scatter the electrons more before they produce bremsstrahlung and a higher atomic number has a higher scattering power. Figures 4.14 and 4.15 present the relative energy fluence for different target thickness and materials (Nordell and Brahme 1984). To obtain a large angular distribution, one should thus use a thick target with a high atomic number.

The FWHM of 50 MV scanned elementary photon beams from well-designed high-energy photon targets may be of the order 2–4 cm for very thin targets made of beryllium of thicknesses around 3 mm, but as large as 7–12 cm for thick targets (~5–15 mm) that are fully stopping the electrons for typical SSDs.

Different more or less effective design approaches and combination of them can be used to make the scanned beam as narrow as possible and still keep the intensity high (Brahme 1985; Svensson 1998):

- Increasing the mean energy of the incident electron beam will reduce the width of the elementary photon beam since both the scattering of photons and the

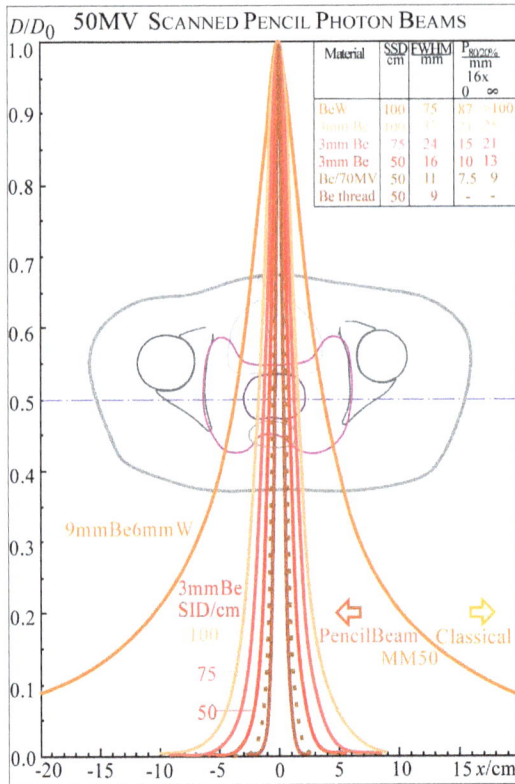

Figure 4.47. Narrow bremsstrahlung beam profiles compared with a typical patient size.

intrinsic bremsstrahlung process and the electron multiple scattering in the target are in principle proportional to the mean energy of the electrons.

- A thinner transmission target will reduce the beam width at the cost of a lower forward yield.
- A transmission target of lower atomic number than beryllium such as lithium, liquid helium, or hydrogen. Lithium is easily vaporized in open air and must be safely contained but could be a useful alternative target. Helium that has a boiling point around 2 K is easy to handle and is readily used in most laboratories and can be used to cool hydrogen that has a much higher boiling point. The density of liquid hydrogen is ten times lower than beryllium and a target space of at least 20 mm is required to get the same forward yield as for beryllium even though the width is reduced by about 5 mm. The beam current on the target is not really a limiting factor when using a microtron and in principle the reduced forward

intensity for the thinnest target is easily compensated for. However, the photon background at isocenter will eventually be largely increased compared to the peak value in the elementary beam if the target is too thin.

- Shortening the distance from the source to the patient surface since the beam width is proportional to this distance. At the same time the dose rate in the beam is increased quadratic.
- Optimizing the target geometrically in such a way that only forwardly directed electrons will produce photons and multiple scattered electrons with too large angles are forced to slip out of the target and are removed by the purging magnet.
- Producing a broader incoming electron beam of controlled angular distribution such that the photon fluence on the patient is sharply focused.
- When other photon generation processes than bremsstrahlung become available, still narrower beams can be produced such as from free electron lasers, inverse Compton effect or X-ray or femtosecond petawatt lasers.

Several of the above methods can be combined either with a uniform beam or together with the pencil beam model in order to give the narrowest possible intense elementary photon beam of high energy.

4.4.3.5. *Purging magnet and electron collector design*

4.4.3.5.1. Purging magnet design for full range target

The purging magnets installed at the MM50 racetracks in China are designed to mainly remove pair-produced leptons (electrons and positrons) of mean energies up to about 25 MeV. These are generated mainly by bremsstrahlung photons that interact near the exit of the target. The purging magnet of the racetrack accelerator is resistive and quite weak to dynamically fully deflect the 50 MeV electron beams. The magnetic field integral is around 5 Tcm with a peak value of 0.6 T. The sidewalls of the pole tip are toothed, which works as an effective distributed lepton collector as long as the lepton path is aligned with the inner side of the tooth. Simultaneously it works as a primary photon collimator with multiple 5 mm electron traps.

4.4.3.5.2. Purging magnet design for transmission target technology

The dose content in one pulse of the transmitted electron beam is a few hundred times higher than that of the useful photon beam. Thus, it is very important that the transmitted beam is bent out from the therapeutic photon beam and safely collected in an optimized electron stopper generating a minimal contamination

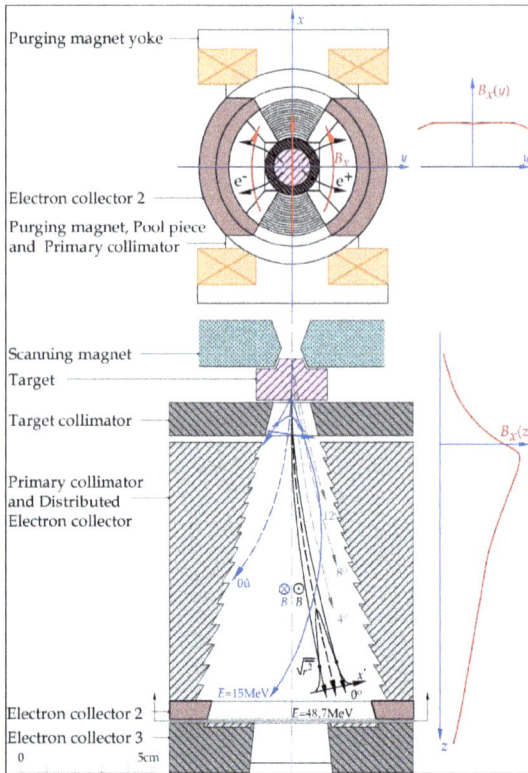

Figure 4.48. Cross-section through the cylindrical purging magnet of the racetrack microtron. The illustration shows the components along the beam from the last scanning magnet, target, target collimator, purging magnet, and integrated electron collector. At the end of the purging magnet and the distributed electron collector, an extra revolver mounted collimator may be inserted. Four calculated trajectories for electrons with 48.7 MeV exit mean electron energy (solid lines, right panel) of four different incident scan angles ($0° - 12°$) influenced by the magnetic field represented by the B_x and B_y components are shown (dashed dotted lines). The rms radii of the transmitted electrons is indicated as solid lines. A calculated trajectory (dashed dotted line, left panel) for an optimized four-layered target with 15 MeV exit mean electron energy is shown for a forward-directed photon beam. The incident mean electron energy is 50 MeV. The angle of the electron absorbers should be equal to the angle the electrons have been deflected.

in the photon beam. An extended analyze of the electron contribution to the therapeutically photon beam for different transmission targets made of different thicknesses of beryllium is shown in Figures 4.48, 4.49 and 4.58 below (cf. Svensson *et al.*). The deflection of the transmitted beam was done with the same purging magnet strength. A thicker target broadens the transmitted electron beam by more

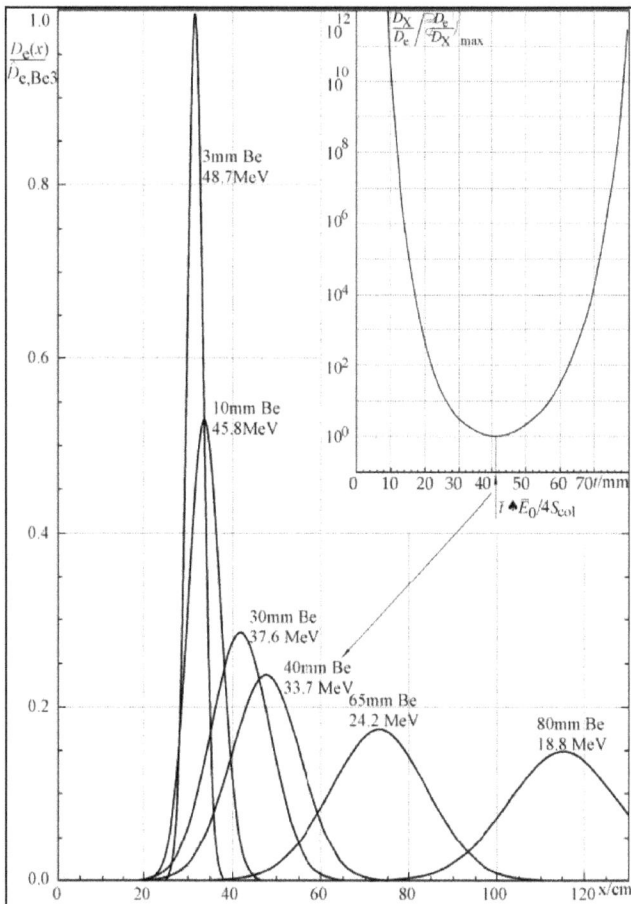

Figure 4.49. The lateral deflection by the purging magnet for different transmitted electron beam energies with different beryllium targets thicknesses is shown. The lateral spread of the transmitted electron beam is proportional to the target thickness and the integral electron fluence is conserved. The photon dose relative to the electron dose at the central axis for a forward-directed photon beam is high for very thin target as well as for thick target (insertion).

electron multiple scattering. However, at the same time transmitted electron energy that is lower increases the deflection angle of the electron beam. On the other hand, a very thin target transmits a high-energy beam that is harder to deflect. However, the transmitted beam is now much narrower due to the low scattering contribution in the thin target, and contamination tail becomes low. Intermediate thicknesses give the highest contamination tail due to a rather broad transmitted beam and high energy. Conclusively, a narrow beam is still preferred since it is easier to control and

The purging magnet should deflect the transmitted primary electron beam of almost the same energy as the incident electron beams, that is, 50–70 MeV. Consequently, the design of the new magnet will differ quite a lot from the previous design. The purging magnet should be as short as possible in order to fit into a compact treatment head and it must be strong, fast, and reliable. Since narrow scanned photon beam makes a rotational treatment head unnecessary, the magnet can have planar pole edge instead of rounded as in the previous design. The magnet can be made much stronger since much more magnetic flux can be transported over the pole gap before the iron is saturated. The magnet should be strong enough such that it manages to deflect a 50–70 MeV electron beam more than 20° out of the therapeutic beam for all beam directions. It needs to deflect the primary electrons out of the photon beam such that a dedicated electron collector placed downstream of the purging magnet could safely collect the electron beam and minimize bremsstrahlung leakage well below 0.5% of the peak value in the forward photon beam. Figure 4.50 shows how

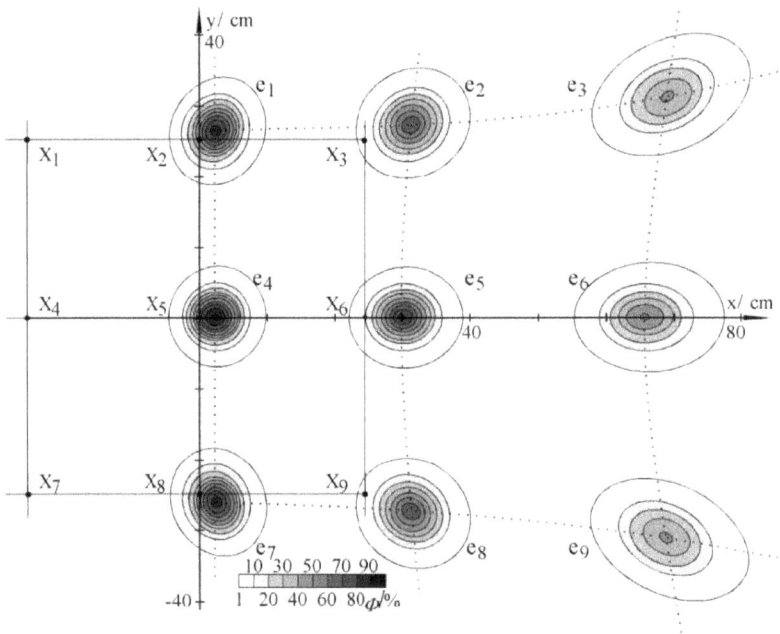

Figure 4.50. Illustration of deformation of the electron pencil beam when deflected over large angels. Calculated transmitted electron fluence distributions normalized to one incident electron on a 3 mm beryllium target for different scanning angles (e_i). All coordinates are projected on the isocenter plane 100 cm from the target. The associated peak photon fluence distributions (X_i) is also indicated. The electron isofluence distributions are normalized to e_4 and include 1%, and 10% to 90% in 10% steps.

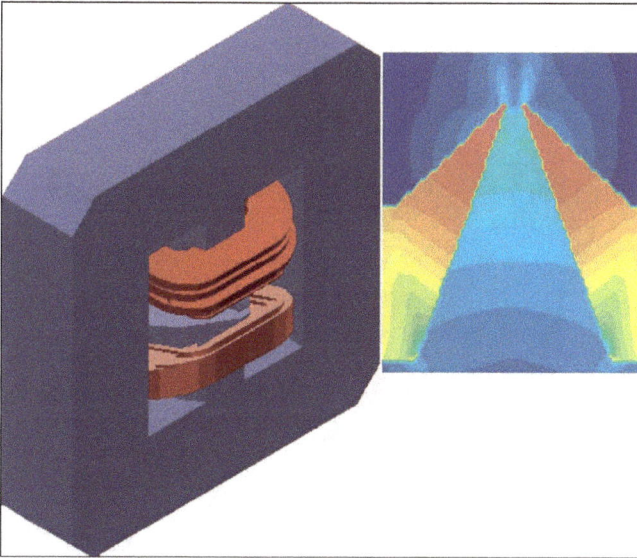

Figure 4.51. Magnet design of the photon beam purging magnet showing the field strength in pool tips and pool gap, see also Figs. 4.52–4.54.

the incident electron beam with different scan positions (X_1–X_9) is transformed by the purging magnet into the positions e1 to e2 at isocenter.

Today, the bending plane of the purging magnet is orthogonal to the bending plane of the last scanning beam so that the influence of the leakage field from the purging magnet is minimized (cf. Figs. 4.10 and 4.45). In most treatment situations of large volumes, the target is elongated in the axial direction of the patient, which is the same as the present bending plane. By rotating the purging magnet 90°, the bending plane of the magnet becomes across the patient and therefore the necessary deflecting capacity might be lowered. For a high degree of freedom in the scanning technique, the amplitude of the magnet field in the purging magnet should be inversely proportional to the amplitude in the scanning magnet. However, when the photon beam is directed forward, the field of the purging magnet should be strong enough, such as the transmitted electron beam is deflected to the electron stopper. When the photon beam is directed on to the field edge, the purging magnetic field should be released almost to zero. In this way, the transmitted electron beam always hits the same spot where the dedicated electron stopper is placed and prevents the electrons from being deflected back into the field by the purging magnet. Since the photon beam can be positioned on either side of the plane orthogonal to the bending plane, the magnetic field should turn direction exactly when the photon beam is

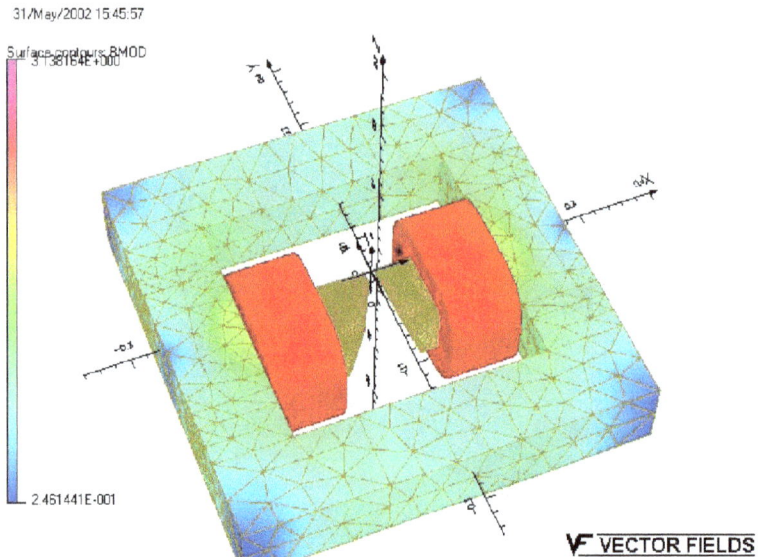

Figure 4.52.   Magnet design of the photon beam purging magnet showing the field strength in the yoke.

directed forward always bending at the same direction as the photon beam. Since a laminated magnet with the same size as a resistant one may lower the field strength by 10–20%, a slower magnet may be used. For instance, the beam system can deliver a scan matrix row-wise that will decrease the necessary speed of the magnet field strength by a factor proportional to the square root of the pulse repetition frequencies.

4.4.3.5.3. Design and magnetic field simulation in OPERA 3D

A commercial simulation program OPERA 3D from Vectorfield™ was used as the main tool for designing the new magnet. The program is made up of three modules: a modeling part (post-processor) similar to a modern cad program; a simulation part called TOSCA for calculating the magnetic fields; and an interactive analyzing part, where the results can be used and displayed in various ways, such as simulation of particle transport in three spatial dimensions in an magnetic, and external electric field and also taking into account the beam space charge. TOSCA uses a discrete finite element model to solve the partial differential equations governing the behavior of electromagnetic fields. The TOSCA method computes the total potential in the regions where source currents have been specified. The reduced potential represents only that portion of the field produced by magnetization, the remainder of the field being computed directly from source currents.

A strong compact purging magnet was modeled in OPERA 3D with a size of $40 \times 40 \, cm^2$ in width and 11.5 cm thick in the beam direction. The first part of the magnet, the pole tip edge, which is closest to the scanning magnet is about 3 mm thin, to minimize the field leakage and the opening is around 7 mm.

The inserted panel in Fig. 4.53 illustrates the integral field over the pole gap with different current density, from 2.5 to 50 A/mm² (15 A/mm² for the field picture is shown, cf. Fig. 4.51). The integral field in the beginning of the purging magnet is very high and can reach up to 17 Tcm or even more depending on the current and the cooling possibilities, which is around three times higher than the existing design.

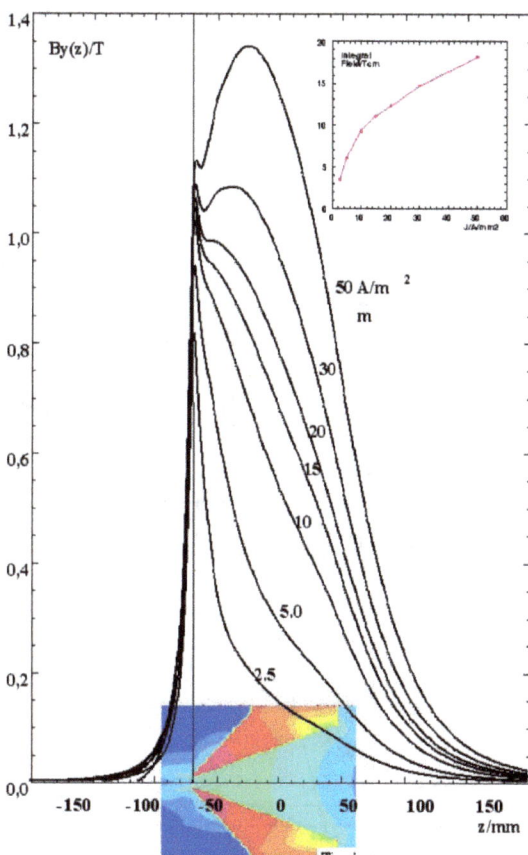

Figure 4.53. Illustration of central axis magnetic field strength of the photon beam purging magnet at increasing current densities.

The saturation field in the pole edge is around 1 T. Deeper in the magnet, there is more space for the field lines such that it can reach up to 1.3 T. A current density around 30 A/mm^2 is quite in order compared to the existing coils in the MM50 system. The peak value is around 1.2 T, depending on magnetic field saturation of the iron used.

4.4.3.5.4. Electron collector

A dedicated electron collector placed downstream of the purging magnet could more effectively stop the primary transmitted electrons possibly in combination with the MLC (cf. Fig. 4.54(a)). None of the incoming electrons should be reflected and the produced bremsstrahlung should be effectively absorbed. The design of the collector is critically influenced by the incoming electron energy, beam size, and angle

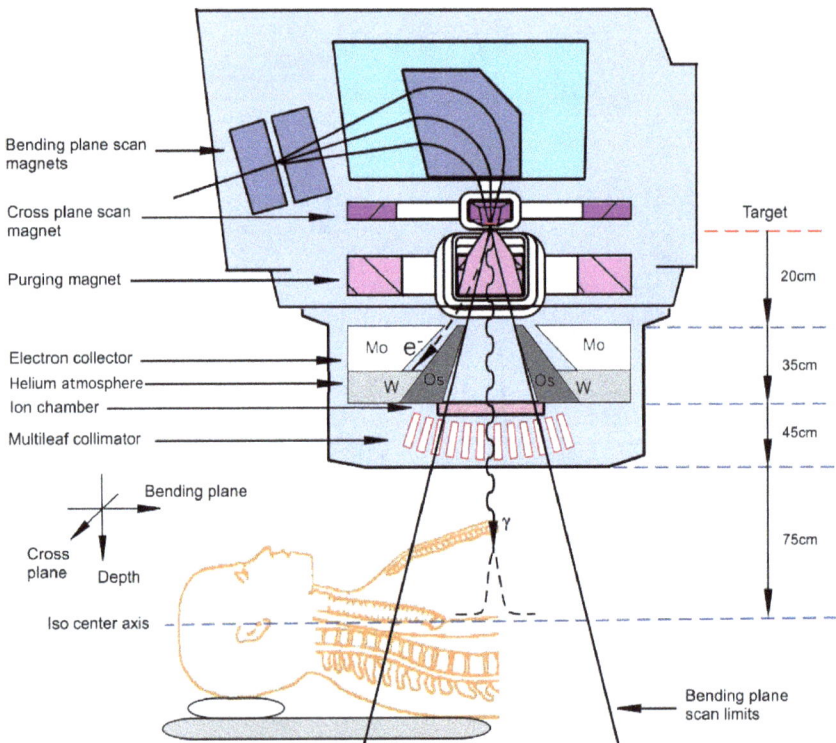

Figure 4.54(a). Cross-section through an accurately designed scanning magnet, purging magnet, and electron collector system (cf. Andreassen *et al.* 2009, 2011; this type of treatment unit is available at TopGrade Healthcare, China).

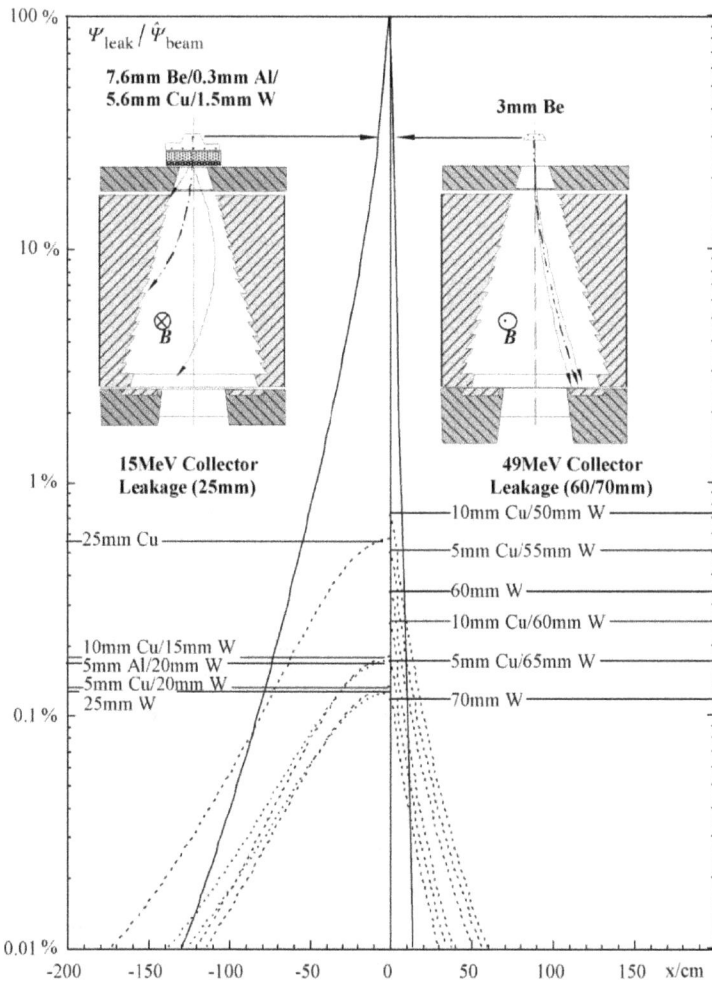

Figure 4.54(b). Calculated photon energy fluencies at isocenter (100 cm) for two different optimized targets (50 MV) and collector designs (left and right panels) are shown. The elementary bremsstrahlung beams generated from the target and the transmitted primary fluence from the collectors are shown as solid and dashed lines, respectively. The contaminating collector-generated photon fluence is generated from transmitted electrons of mean electron energy 15 MeV (left panel) and 49 MeV (right panel).

of incident of the electron beam, which in turn is dependent of the scanned beam dose delivery technique used. As was previously shown, a thick transmission target transmits electrons of lower energy that is easier to deflect and requires a thinner electron collector to absorb generated bremsstrahlung in the collector. However, the transmitted electron beam now acting as contaminating source becomes large and less controllable. The electrons are most efficiently collected if they are kept closely together, by minimizing the multiple scatter of the electrons and consequently the half-width of the generated bremsstrahlung photon beam. This can be achieved by using a thin low atomic number target. A promising technique would be to design a laminated purging magnet, which is as fast as the scanning magnet. It is then possible to counteract all scan angles with the purging magnet so the transmitted electrons are deflected out of the therapeutic beam always along a fixed path to be totally absorbed by an efficient electron stopper. The collector could then be more compact and optimized with regard to its bremsstrahlung absorption. Such a technique has the advantage of collecting the electrons higher up in the treatment head and reducing the scattering of primary electrons on components in the head. It is also possible to combine this technique with the MLC downstream, particularly with dynamic multileaf collimation. The photons for scanned beam are effectively used when the scanned beams are directed to the opening of the collimator. With such a technique the purging magnet does not need to deflect the electrons as much as when the electron collector is integrated with the primary collimator and the purging magnet. Of course, such a technique may link the collimator motion to the electromagnetic scanning pattern, which ideally should be avoided to have maximum flexibility in dose delivery.

It is clear that a tungsten shield of at least the same thickness as the standard collimator (7 cm) should be used in order to keep the bremsstrahlung leakage below 0.5% of the therapeutic beam intensity. The collector is integrated with the purging magnet, that is, right below where the magnetic field is zero and as close as possible to the target. In this way the electron source is small and easier to collect. The deflection of the purge electron beam always goes in the same direction of the central axis as the scanned photon beam so in fact the collector is made up of two, one at each field edge and elongated in the orthogonal scan direction. An analytical calculation that included only primary photons was done using the same equation as was used for bremsstrahlung production in sandwich target.

4.4.3.5.5. Collector material

The same analytical expression that was used to calculate the bremsstrahlung from optimized multilayered targets was used to calculate the transmitted bremsstrahlung contamination produced in the electron collector that may reach the patient. With a fixed spacing for the electron collector, the problem is to minimize the transmitted photon contamination for a given maximum collector thickness. The transmitted electrons from different target compositions were measured right below the purging magnet by a collector of different compositions. The bremsstrahlung production and photon attenuation in the collector was investigated. The bremsstrahlung was transported to the isocenter at 1 m distance behind different collector materials — both layered and homogenous materials. It was shown that even if the bremsstrahlung production is higher in a high atomic number material the photon attenuation per unit length is relatively much higher in high-density material than in a low-density material resulting in less bremsstrahlung contamination. For instance, the maximum photon energy fluence after the copper collector of 25 mm thickness is 4.5 times higher than from the tungsten collector with the same thickness with an electron energy incident on the collector equal to 15 MeV. Thus, the optimal photon attenuator is the most important criterion indicating, if only the density is considered, osmium and rhenium as optimal material if the collector space is limited (cf. Figs. 4.54(a) and 4.54(b)). Of course, this is only true for a collector design that does not reflect some part of the incident electrons or have side leakage of electrons or secondaries. Thus, the ultimate collector design may very well be multilayered to minimize electron reflection.

4.4.3.5.6. MLC as electron collector

It is possible to use the MLC as an electron collector. The MLC should then be positioned so that it fully stops the deflected electron beam. This may mean that the

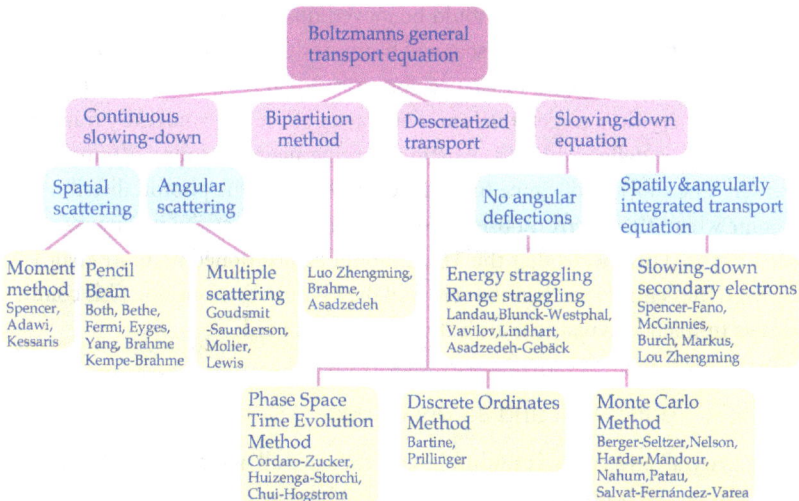

Figure 4.55. Illustration of the different steps in the development of particle transport calculations.

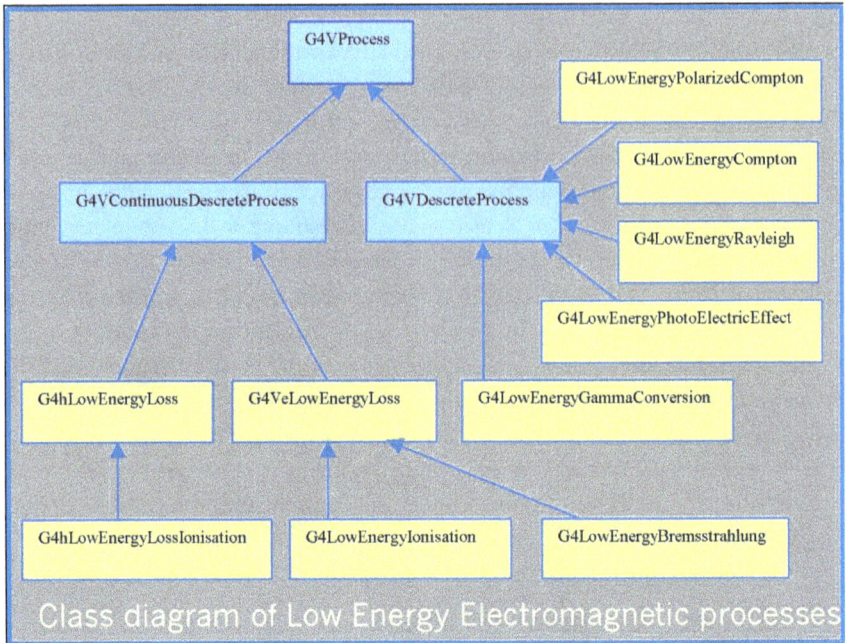

Figure 4.56. Particle transport simulation using the Mote Carlo code GEANT-4 cf Figs. 4.55 and 4.57.

leaves in certain configurations should be moved during the irradiation, somewhat depending on the position of the scanned beam. If dynamic multileaf collimation is used, this is often the case. For instance, for a forward-directed 50 MV photon beam from a 3 mm beryllium target, the center of gravity of the concomitantly transmitted electron beam is deflected by the existing purging magnet around 30 cm from the isocenter. The shortest distance from the central part of the photon distribution to the point where the electron dose has decreased to <1% of the maximum photon dose is 18 cm. This means that the MLC should be positioned no more than 18 cm away from the central axis than the center of the photon distribution. This condition is always fulfilled if dynamic multileaf collimation is used.

4.4.3.5.7. Verification and safety issues

The security arrangement is multifold and they are all associated with check and confirm about the transmitted electron beam. The dose content in one pulse of the transmitted electron beam is several hundred times higher than in the therapeutic photon beam and electron pulse is now allowed to hit the patient.

A tight electrical coupling between the scanning magnets bending in the same plane is preferred, as the purging magnet will be a safe arrangement. In this way the magnetic energy is pending between the scanning magnet and the purging magnet so that when the therapeutically beam is directed forward, there is no deflection power in the scanning magnet. All the magnetic is used by the purging magnet, which fully deflects the transmitted electron beam. On the other hand, when the therapeutic beam is directed with high angles, the excitation of the purging magnet can be relaxed. The total integral magnetic field that the electron beam experience in either case will approximately be the same.

The pulse repetition frequency can be varied from 50 to 500 Hz and a realistic frequency is around 300 Hz. Thus, the time between each pulse is 5 to 20 ms. The lowest time between the pulses is still enough to register with Hall element that magnetic field of the scanning magnet and the purging magnet are carefully set.

The current and voltages of each bending element could easily be measured.

The segmented ion chamber placed immediately downstream the purging magnet will directly sense if any of the photon beam or the transmitted electron beam are wrong according to a preset value. If so, the beam will immediately be turned off.

4.4.3.6. *Clinical characterization of scanned photon beams*

A very clean beam of high photon mean energy with high penetrating properties and uniform energy distribution across the beam can be generated from a thin target as shown in Figures 4.57–4.59.

From a dosimetric point of view, the scanned beam is superior to dynamic collimation since the photon beam transport is only disturbed by helium between the target and the patient. The collimator should be used as little as possible since it has interleaf effects such as leakage and collimator scattering, etc. The integral leakage outside the field will also be reduced.

When using dynamic multileaf collimation, the resolution in the direction of motion is only limited by the penumbra from the leaf edge, but in the orthogonal direction, it is limited by the leaf width. The resolution in a scanned beam is mainly limited by the shape of the elementary photon beam. Thus, a scanning system benefits from both a dosimetric and treatment time point of view. What about the costs? The price for a scanning system is of the same order as the price for a very advanced double-focusing collimator. A scanning system could be combined with a more simple leaf collimator in order to reduce the total costs. Dynamic mechanical systems often are troublesome with wear, which does not exist for a scanning system. When a narrow beam is scanned to form broad beams, the therapeutic beam

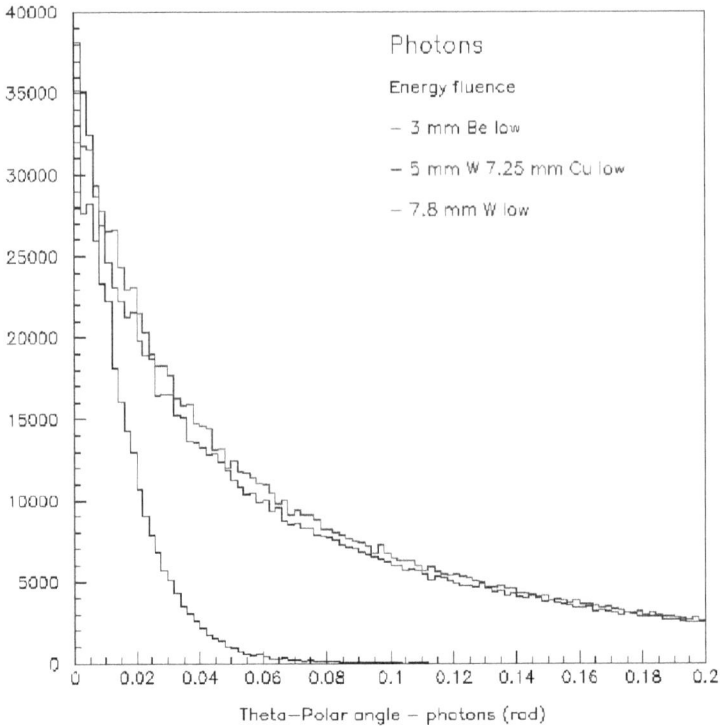

Figure 4.57. Photon energy fluence profiles for thick and a thin target designs calculated by the Mote Carlo method (cf Fig 4.56).

becomes almost perfect, with a well-defined beam energy spectrum and negligible contribution from scattered secondary particles. Consequently, the depth of dose maximum is deep and the surface dose low.

A scanned beam generated with a thin target irradiation technique where all the transmitted electrons have been removed will become particularly clean from secondary particles and have a rather high mean energy. This is indicated in the upper panel of Fig. 4.60, where the dose fall-off measured by the ratio of doses at depths of 10 and 20 cm for a broad beam at an SSD 100 cm corresponds approximately to an equivalent accelerator potential of 80 MV when using a thick tungsten target. The surface dose is <20% of the dose at dose maximum located at a depth of 7.3 cm. Owing to the high energy, the exit dose at 25 cm is as high as 56% at an SSD of 100 cm. At a shorter source to isocenter distances, the beam becomes more divergent and the dose fall at isocenter becomes steeper (upper right panel, Fig. 4.2). Consequently, the equivalent accelerator potential as judged from the depth-dose

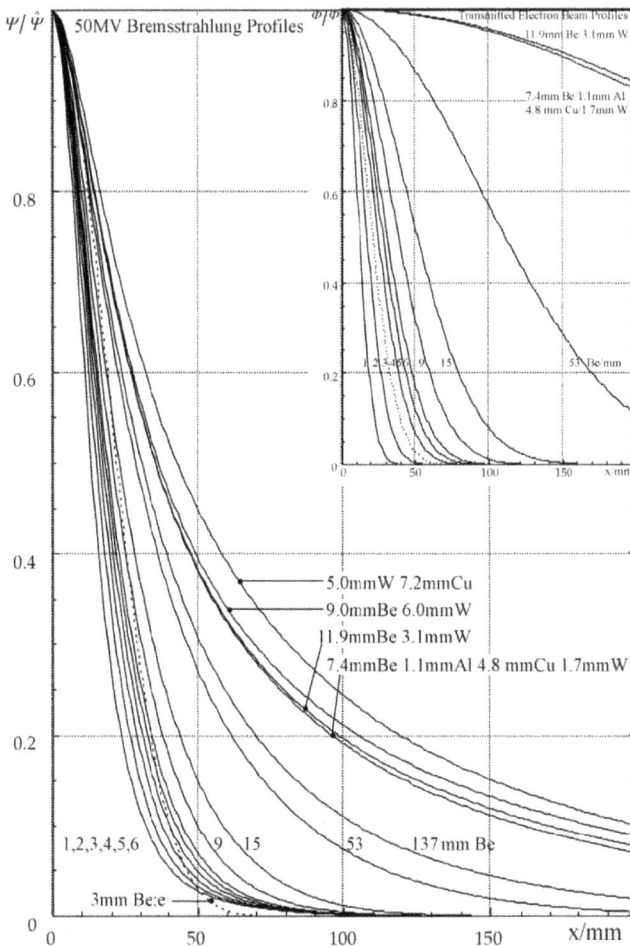

Figure 4.58. Close up of the bremsstrahlung beam profiles. The calculated lateral bremsstrahlung profiles for four different thin multilayered targets (thick solid lines) and for 10 extreme and thin homogenous targets made of 1–137 mm beryllium (thin solid lines) at 1 m distance with initially 50 MeV electrons. The fluence profiles for the transmitted primary electrons (dashed lines and inserted panel) for the thin beryllium targets are basically the same as the photon beam, particularly for the thinner targets.

characteristics is decreased to 35 MV at an SSD of 75 cm and to around 10 MV at an SSD of 50 cm. With decreasing SSD, the exit dose is significantly lowered but the surface dose is almost constant. The depth of dose maximum is only slightly moved to the surface when the SSD is shortened. The depth dose characteristics

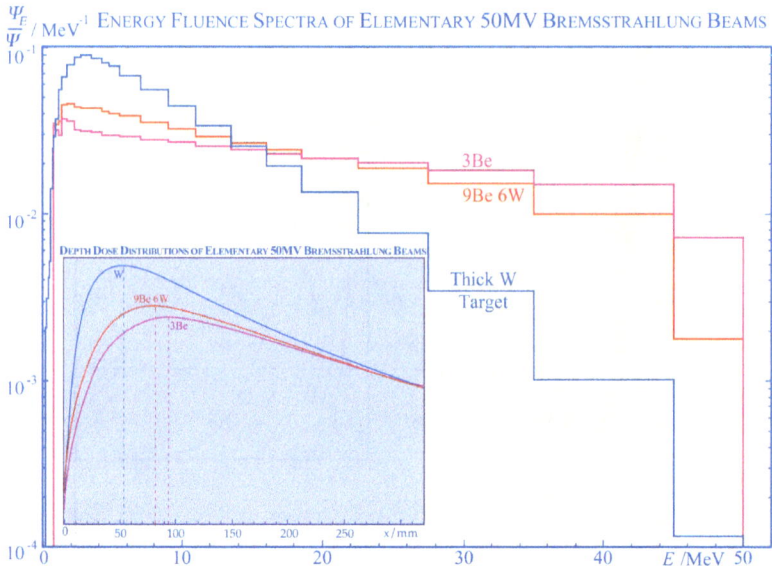

Figure 4.59. High-energy bremsstrahlung beam spectra and their associated depth-dose distributions.

of narrow scanned high-energy beams of a shortened SSD are clinically very useful due to the low entrance and exit doses.

4.4.3.6.1. Lateral beam properties

Laterally, the width of the penumbra region and the shape of the beam shoulder region characterize the photon beam. The width of the elementary scanned beam that is generated in the bremsstrahlung targets used today is in the range of 7–20 cm at isocenter and if such a beam is used without collimation, the dose contribution to the normal tissues outside the target volume would by far be too high. By using the proposed thin target irradiation technique, the elementary photon beam width at SSD 100 cm is 3 cm and as low as 1.6 cm at SSD 50 cm (Figs. 4.17, 4.21 lower right panel, 4.47 and 4.58). The corresponding 80–20% penumbra for the elementary beam varies from 20 mm for a SSD of 100 cm down to 10 mm for a SSD of 50 cm. The central panels of Fig. 4.60 show the edge in a broad uniform beam produced by such a narrow beam optimally scanned at different SSDs. The optimization uses a constrained iterative optimization algorithm (Lind and Brahme 1985, 1987; Lind 1990) that minimizes the overdosage in the target volume at the same time as the underdosage is avoided. The desired uniform dose distribution was 15 cm in one direction and was varied between zero and infinity in the orthogonal direction.

50MV PHOTON BEAMS FROM A 3mm Be TARGET			
SSD	100cm	75cm	50cm
BROAD BEAM LONGITUDINAL DATA			
μ_p/m^{-1}	4.1	4.8	5.6
μ_e/cm^{-1}	0.27	0.27	0.32
v	0.79	0.80	0.84
D_{10}/D_{20}	1.40	1.49	1.66
$D_{surface}/D_{max}$	19%	20%	21%
D_{exit}/D_{max}(25cm)	56%	51%	42%
R_{100}	7.3cm	6.8cm	6.0cm
Equiv.acc. μ_p	55MV	13MV	6MV
potential $D_{10/20}$ SSD=100cm,W-target	80MV	35MV	10MV
Central axis Depthdose:			
FieldSize			
BROAD BEAM LATERAL DATA			
$P_{80/20\%}$ 15-∞	2.0-2.5cm	1.5-2.1cm	1.0-1.3cm
Penumbra in an uncollimated scanned beam:			
ELEMENTARY BEAM LONGITUDINAL DATA			
D_{10}/D_{20}	1.55	1.64	1.85
$D_{surface}/D_{max}$	11%	10%	9%
D_{exit}/D_{max}(25cm)	42%	37%	25%
R_{100}	5.8cm	4.8cm	3.8cm
Central axis Depthdose:			
ELEMENTARY BEAM LATERAL DATA			
FWHM	3.3cm	2.4cm	1.6cm
Lateral Bremsstrahlung Profile :			
$P_{80/20\%}$ 0-∞	2.0cm	1.5cm	1.0cm

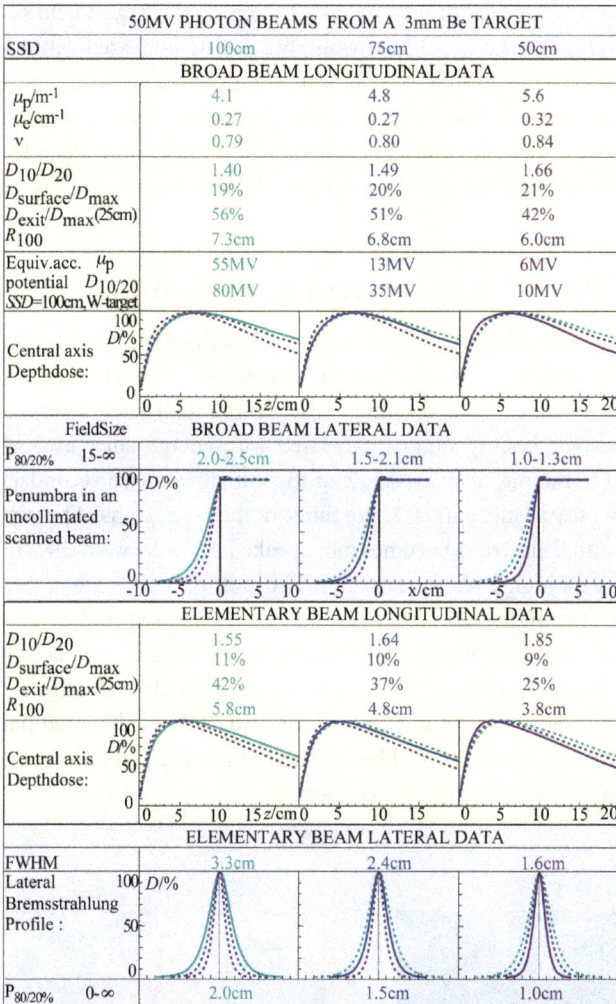

Figure 4.60.   High-energy bremsstrahlung beam profiles and their associated depth-dose distributions using a scanned 3 mm thick beryllium target for different SSD are shown. For a short SSD, the beams become more divergent and consequently the dose fall-off becomes steeper and the attenuation equivalent acceleration becomes much lower and results in a lower exit dose as shown in the right column. The surface dose and depth of dose maximum is only slightly changed for shorter SSD. The width and the penumbra of the elementary beam profile are inversely proportional to the SSD. Owing to the Gaussian shape of the elementary beam, the penumbra remains small even for large irradiated volumes as shown in the lowest panel.

A transversal cut in the resulting dose distribution for the largest field size for all the different SSDs is shown here. The penumbra is only increased 2 mm for medium field sizes up to about 5 mm for broad beams. This is because the elementary beam is not exactly Gaussian and the tails of the elementary beam will add to the total field penumbra. Thus, the penumbra at SSD 75 cm will vary between 15 and 20 mm depending on the field size which is comparable with that of old ^{60}Co units.

4.4.3.6.2. Penumbra and effective source size

As discussed above (Sec. 4.3.3.2 and Fig. 4.21) high-energy photon beams need a well-designed collimator to minimize the lateral secondary electron transport and the penumbra (Nilsson and Brahme 1981). In the high-energy range, the lateral electron transport is only slightly increased with energy since at first order it is proportional to the mean square angle of the initially emitted secondary electrons, their multiple scattering, and effective range of these electrons. The distribution of the initial emitted electrons becomes more peaked in the forward direction and is of higher energy for higher photon energies. The multiple scattering of these electrons becomes less for higher energy. Thus, the total lateral spread of the secondary electrons decreases with energy, whereas the range linearly increases with energy such that the contribution to the penumbra from secondary electrons stays almost constant with energy as seen in Fig. 4.61 for a 50 MV uncollimated photon pencil beam (cf. also Fig. 4.66 below). The contribution to the penumbra from scattered photons is rapidly decreasing with energy.

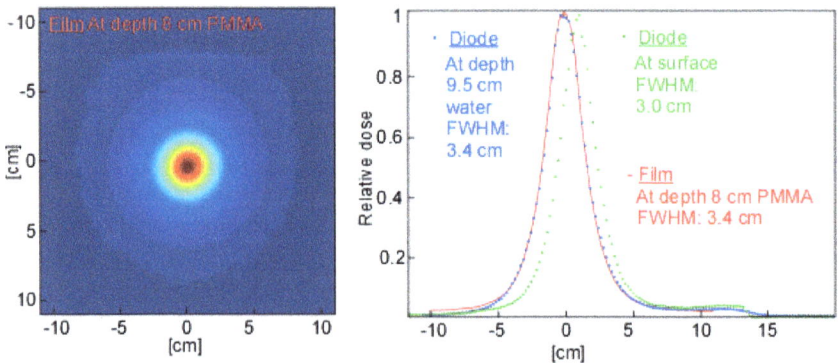

Figure 4.61. Cross-section through a 50 MV bremsstrahlung pencil beam at a depth of about 10 cm in water (cf. also Figs. 4.66 and 4.70 below).

4.4.3.7. *Collimator design for scanned beams*

Modern clinics today use multileaf collimated uniform beams instead of beam blocks. A leaf resolution of around 1 cm provides a good lateral protection of normal tissue almost as good as perfectly aligned beam blocks. However, for such treatments, the longitudinal protection of the normal tissue becomes poor. For advanced treatments where the intensity modulation is performed by narrow scanned pencil beams, all the longitudinal protection is performed by the scanned beam itself. For sufficiently narrow elementary beam, a lot of the lateral protection is also performed by the scanned beam due to the rather steep lateral dose fall-off with a narrow elementary beam and thus the requirements on the leaf width becomes less strict. The influence of treatment outcome on different aspect of the collimator design as well as the scanned beam that is most clinically relevant is best caught by performing radiobiological optimization. Thus, the clinical outcome for different collimator design collimating different scanned beam was investigated using our treatment planning optimization algorithm (Figs. 5.4(b) and 4.60). To somewhat stress the algorithm and enhance some aspect of delivery technique, a cervix cancer stage IV was used as target volume.

The uncomplicated tumor control P_+ is used to score the treatment outcome by the optimization algorithm (Ågren *et al.* 1990; Söderström and Brahme 1993). The algorithm simultaneously optimizes the MLC settings and the scanning pattern (Gustavsson *et al.* 1995; Löf *et al.* 2001).

4.4.3.7.1. Influence of the collimator leaf width on treatment outcome

The treatment outcome for this cervix case was studied by Svensson *et al.* (1998) as a function of leaf size both in combination with uniform beams and for scanned elementary bremsstrahlung beams of different half-widths as shown in Fig. 4.62.

For uniform beams, that is, collimating a broad beam (FWHM $= \infty$, in Figs. 4.64 and 4.65) and a leaf width of 4 cm, the treatment outcome is drastically lowered due to both low leaf resolution within the field and poor matching at the sides for the last leaf pair at the border to the normal tissues. The possibility of the MLC to laterally protect the normal tissue around the target volume rapidly increases with decreasing leaf width. For this particular treatment volume, the dose escalation to the gross tumor can be enhanced further, since decreasing the leaf width better protects the normal tissues. The saturation is due to the lateral electron transport and photon scatter, which reduces the need for very fine collimator resolution when

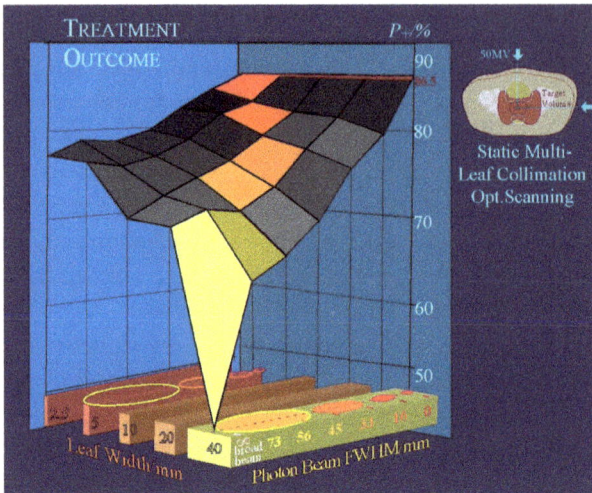

Figure 4.62. Treatment outcome with different scanned beams and multileaf collimation designs. The increase in treatment outcome with decreasing pencil beam width and leaf width of the MLC is clear. Narrow scanned beams improve the treatment outcome faster than a finer collimator resolution.

optimized intensity modulation is used. Now, turning to scanned beam optimization in combination with multileaf collimation using collimators of different leaf widths, the treatment outcome increases as expected with decreasing half-width of the elementary beam and this increase is faster with thicker than with narrow leaves due to the influence of electron scatter. Already, the treatment outcome is drastically increased for quite broad elementary beams such as. For the narrowest elementary beams around 15 mm, the treatment outcome is almost maximized for broad leaves and only a small therapeutic gain can be achieved by increasing the resolution of the MLC.

4.4.3.7.2. Fast and efficient MLC

Obviously, the design of the MLC could be optimized for use in combination with narrow scanned beams. Simultaneous optimization of the scanning pattern and the multileaf settings automatically drives the elementary beam direction toward the opening of the collimator for most effective use of the beam. Thus, the role of the collimator to longitudinally modulate the beam by dynamic collimation is reduced by the scanning system. The task of the collimator instead is to laterally protect the patient and thus sharpen and cut the tails of the elementary scanned beam. With this new role for the collimator, the thickness can then be reduced to

Flat and Thin Collimator for Scanned Narrow Photon Beams

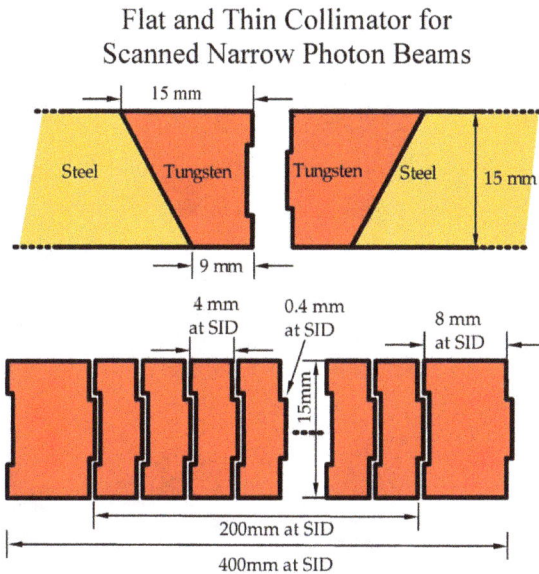

Figure 4.63. Schematic diagram of the new compact penumbra trimming MLC for scanned photon electron and light ion beams. The collimator leaf is flat and only 15 mm in the beam direction. The weight of such a collimator is only a fraction of that of conventional 60–80 mm thick MLCs made of tungsten.

only a few centimeters of tungsten so it does not need to use focused edges. In this way, the source to axis distance can be made sufficiently short such that the beam half-width is around 1 cm and indeed a much simpler collimating device is needed to maximize the treatment outcome. Such a MLC can be ultralight and ultrafast compare to the existing design from vendors such as Varian, Elekta, Siemens, and Scanditronix. Compared to the MM50 collimator, the collimator plane distance from the source is approximately reduced by half which reduces the physical collimator surface by one-fourth, the absorbing material thickness of the collimator is only 20% and the density of steel is only 40% of tungsten. The total weight of the proposed collimator compared to the Scanditronix collimator and covering the same field size thus become only 2% corresponding to a weight well <10 kg. For such a small collimator, the leaf width is small and the interleaf leakage is higher compared to a conventional MLC. However, considering the simultaneously use of narrow scanned beam, only a very small fraction of the primary scanned beam will actually hit the collimator. Consequently, the interleaf effect has not been considered in the following calculations.

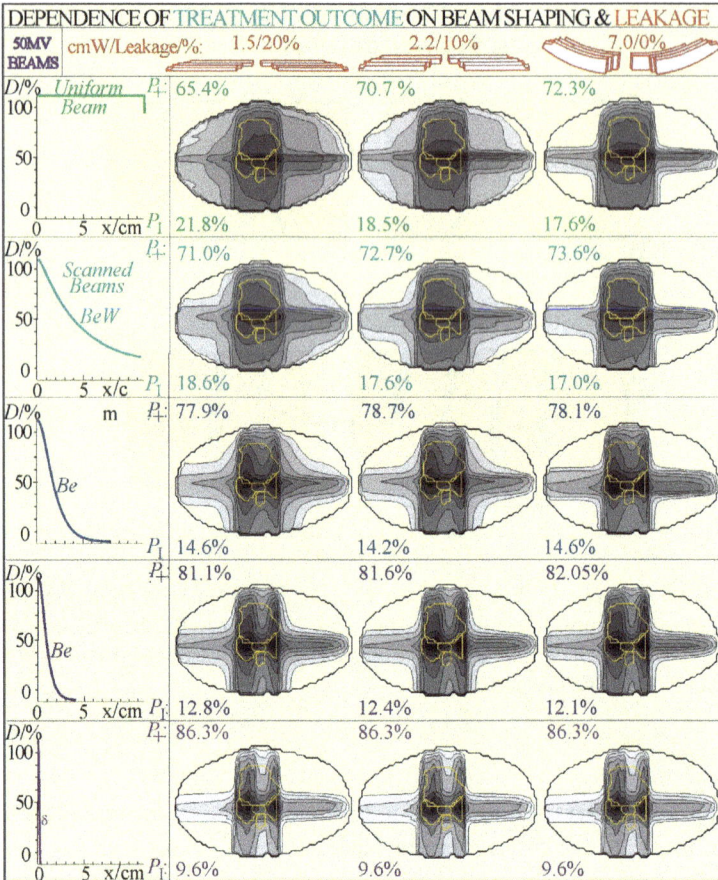

Figure 4.64. Treatment outcome optimization for different primary beam qualities, uniform and scanned beam simultaneously optimized with one MLC settings for different thicknesses of the collimator is illustrated. Note that the leakage in percentage is defined as the transmitted part of the primary incident energy fluence.

4.4.3.7.3. Optimization of the leaf thickness and edge shape

The influence on the treatment outcome for the previous cervix case on different collimator design collimating different scanned beam was investigated. Here, only one optimal multileaf setting was investigated. For all different collimator designs, different beam qualities and delivery techniques was investigated, namely uniform beams, scanned beams using a full range target made of BeW (FWHM = 75 mm), a transmission target technique using 3 mm Be (FWHM = 30 mm) and finally the beam size was reduced by using a thinner target, higher energy, and shorter distance

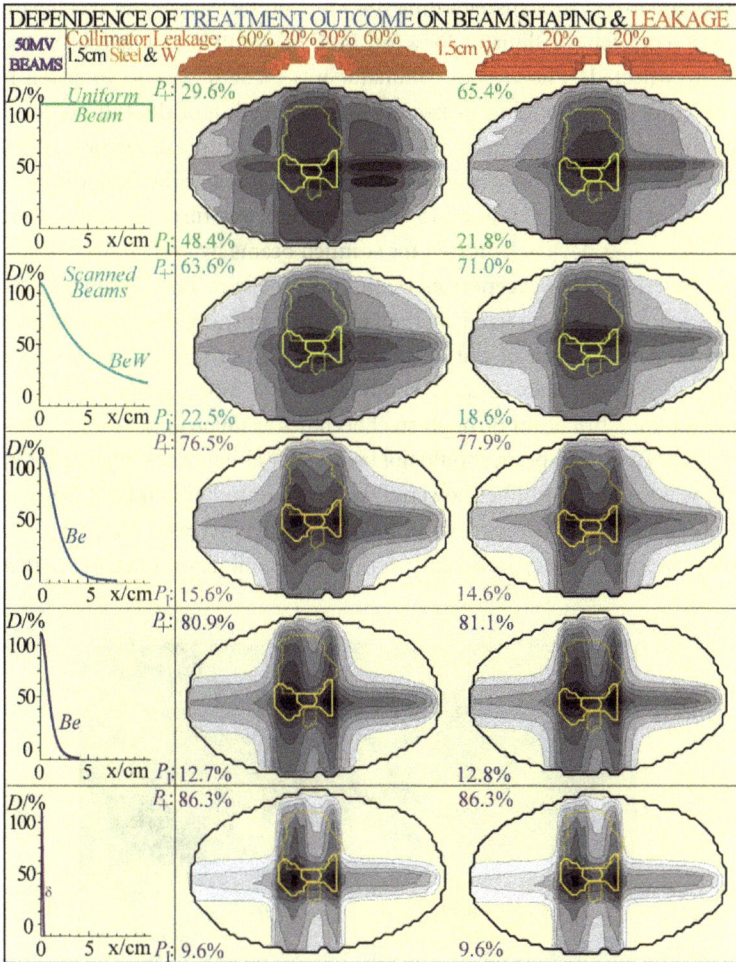

Figure 4.65. Treatment outcome optimization for different primary beam qualities, uniform and scanned beam simultaneously optimized with one MLC settings for two different collimator design of the same thicknesses is illustrated. The collimator to the left is made of steel with tungsten edges and the collimator to the right is made of only tungsten.

so that the FWHM was reduced to 15 mm. As an upper benchmark, the size of the beam was reduced to a Dirac pulse.

First, the treatment outcome was optimized for collimator designs made of tungsten of different thicknesses ranging from 15 to 70 mm (Fig. 4.64 below). The thickest collimator and also double focusing is today used in the MM50 system and for 20 years at Karolinska and will here work as an upper benchmark for an almost perfect collimating system disregarding the width of the leaves. The average photon transmission of such a collimator for a 50 MV BeW-beam is well below 1% and has here been neglected. The second collimator was made of 20 mm tungsten in the beam direction corresponding to around 90% absorption of the full photon beam and thus 10% of the photon beam is transmitted. The third collimator was made of only 15 mm tungsten corresponding to 20% transmission of the full beam.

The third collimator design (15 mm tungsten) was further pushed toward a lighter and simpler device dedicated for scanned beams by exchanging most of the tungsten material to steel except from the collimator edges. This is possible since the narrow beam is mostly directed in the collimator opening. For simplification of the transmission function of the optimization code the tungsten edge was made 15 mm wide and not slanted as it should be for a more optimized edge. The transmission for a full beam through steel becomes as much as 60%.

Obviously, uniform beam should not be used for either of this type of collimator since the whole patient will be showered with transmitted photons. When scanned beams are used, as the treatment outcome heavily increases the narrower the elementary beams become, particularly for the steel collimator, since the elementary

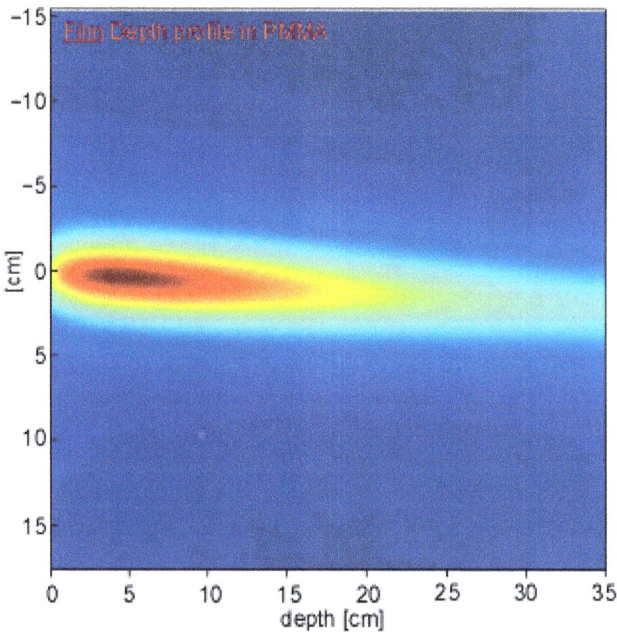

Figure 4.66. The narrow pencil beam kernel from a 3 mm beryllium target measured by film dosimetry.

beam automatically by the algorithm will be directed to the collimator opening. A combination of beams from a full range target and either of the thin collimators is not enough to sufficiently increase the treatment outcome. For the narrowest beam, the treatment outcome is almost saturated and will almost have the same treatment outcome as if the collimator had full thickness for the beam. For the narrowest beam the difference between a collimator made of steel with tungsten edges and a collimator made of only tungsten is negligible since the scanned beam itself modulates the incoming beam both longitudinally and laterally. Still the tungsten edge is needed for collimating the bremsstrahlung tail and laterally protecting organs at risk. Narrow scanned beam opens up new possibility for ultrafast IMRT treatments.

Thinking in the terms fast beam scanning, it is even possible for real orthodox collimating devices, such as rotating collimator discs, where a small opening or several for that matter in a spiral manner rapidly scans over the patient surface simultaneously as the elementary beam always is directed in the moving aperture. In this way, almost all of the generated primary photon will be effectively used in the therapeutic beam of sharp penumbra.

4.5. Clinical Possibilities Using High-energy Photon Pencil Beam Scanning

4.5.1. *Ultrafast IMRT using Scanned Beams Combined with Multileaf Collimation*

It is shown that in most cases a few collimator settings are sufficient for each beam portal when narrow scanned photon beams are available and they can be delivered sequentially with almost negligible delay due to repositioning of the collimator. Furthermore, with narrow scanned beam, there is a minimal loss of therapeutic photons as there is no need to position the pencil beam where the leaf collimator is closed. Consequently, the integral collimator leakage and photoneutron production will be considerably reduced.

Since the electromagnetic scanning of a narrow electron or photon beam is ultrafast, more than 1000 times faster than the leaf motion of most heavy leaf collimators, the intensity modulation should as far as possible be done with the scanning system.

If the incident electron fluence is the same for different target designs, the integral photon yield will naturally be higher for thicker targets of the same atomic number. For this reason, the treatment time will be lowered depending on the intensity modulation. At the same time, a more complex residual dose modulation must be made by the dynamic collimator, which increases the treatment time. For instance, when the intensity modulation is done with thin target irradiation technique, more than 90% of the target-generated photons are used therapeutically, whereas <10% of the target-generated photons are therapeutically used when the beam is modulated by 10 collimated segments in a step and shoot fashion, although the elementary scanned beam is directed toward the opening of each segment. Effective use of target-generated photons is even worse with flattening filters, since almost 70% of the generated photons are absorbed in the filter. The generated bremsstrahlung from the target should be used as effectively as possible. A traditional treatment unit generates a rather broad Gaussian-shaped initial bremsstrahlung distribution that is flattened by a flattening filter thereby removing more than two-third of the incident beam. In the next step, the uniform beam is collimated preferably with a MLC. For the largest field sizes, more than half of the incoming energy fluence is collimated and for small field sizes around $5 \times 5 \, cm^2$ additionally $1/64$ of the photons are wasted by collimation. When the intensity-modulated field is made up by 10 segments and each segment on the average should deliver 0.5 Gy for a four-field technique. This translates into a 10% use of the flattened beam or only 3% ($10 \times 30\%$) use of the generated photons. For scanned beams and the most narrowest 10 mm beams, almost the whole field is modulated by only using the elementary beam. To be fair when comparing in this way, one should keep in mind that the integral photon yield per incident electron on the target for a ultra-narrow photon beam is only a couple of percentage of a beam produced with a full range target. Most of the photons are actually produced and also absorbed in the downstream electron collector.

The effectiveness of the scanning system to deliver strongly heterogeneous dose distribution by scanning elementary bremsstrahlung beam (cf. Brahme *et al.* 1991) is demonstrated in Fig. 4.68 for two different target volumes and for different elementary bremsstrahlung beam shapes. The optimal scanning pattern for the different bremsstrahlung profiles (see Fig. 4.3 and Svensson and Brahme 1996), under the constraint that underdosage should be avoided, was calculated using the iterative optimization algorithm developed by Lind (1990).

As expected, the narrower the bremsstrahlung profiles, the better the delivered field conforms to the desired field. The optimized four-layered target composed by beryllium, aluminum, copper, and tungsten only slightly decreased the width of the elementary beam and consequently the improved conformity to the desired dose distribution became low compared to the full range target composed by only beryllium and tungsten (Fig. 4.68). If the beryllium–tungsten target is allowed to transmit the same mean electron energy as the optimized four-layered target (15 MeV), then the improvement is quite low. It should be pointed out that with

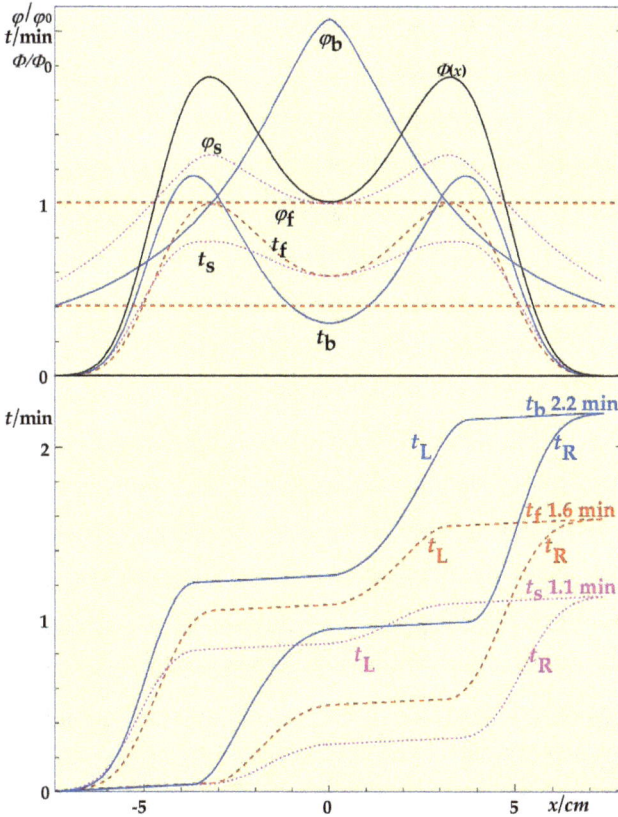

Figure 4.67. Pencil beam scanning pattern and leaf collimator motion (lower panel) are illustrated. The optimal collimator motion for three different initial fluence distributions, φs, φf, and φb due to a scanned, flattened, and a stationary photon beam, respectively, are shown. The required opening time distributions are, ts, tf, and tb and total treatment times are 1.1, 1.6, and 2.2 min, respectively. $\Phi(x)$ is the desired dose distribution in all cases.

the existing accelerator (MM50), a 3 mm beryllium target for 50 MV, the scanning system without collimation almost perfectly conforms with the complex desired dose distribution at least in the interior part of the field. It is interesting to see that a 100 MV photon beam having a FWHM of <20 mm only slightly increases the conformity with the desired dose distribution. However, such high-energy photon beams suitable for large deep-seated tumors could be advantageous when delivering even steeper optimized dose distributions. The penumbra of such narrow elementary beam, as described above, is only 12 mm for a 100 MV beam and 20 mm for a 50 MV

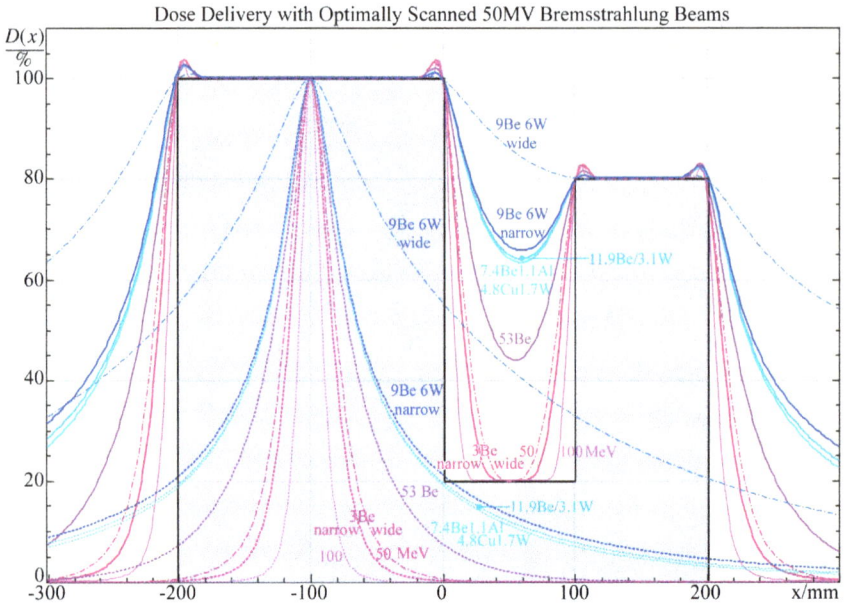

Figure 4.68. Possible dose delivery using scanned 50 MV bremsstrahlung beams. The importance of a narrow bremsstrahlung width in order to achieve a tight dose distribution to a complex target volume is illustrated. A multiple segmented uniform target is irradiated by optimally scanning an elementary beam (short dashed). The resultant dose distribution (solid lines) conforms quite well with the box target when using thin targets and high energy.

beam which is comparable to the 6–8 mm penumbra in a collimated field from a 50 MV microtron using a W-Cu target.

To simulate a more realistic case and to investigate the influence from the large tails of the thick targets, the narrow target volume (Fig. 4.68) was also extended to different broad beams (Table 4.1). The non–projected (dotted lines) and projected (dashed dotted lines) bremsstrahlung kernels from the 3 mm beryllium target and the beryllium–tungsten target are shown in Fig. 4.68. For the almost Gaussian 3 mm Be target, the kernel becomes only slightly broader, which is clearly reflected by the small change in their optimized dose distributions (dashed dotted lines). Unfortunately for the broad BeW kernel of non-Gaussian-like shape, the contribution from the wide tails is large and consequently the optimized broad beam dose distributions become rather shallow. It is also interesting to note that when the scanned beam by itself almost perfectly conforms to the desired dose distribution (3 mm Be), the required residual modulation by the MLC is small.

Table 4.1. Deposited dose from 50 MV photon beam for different targets and related dose from transmitted beam of primary electrons deflected by the purging magnet.

Target				Transmitted Electron Beam					$D_X(0)/$ $D_e(0)$
					$\sqrt{\overline{r^2}}$		Phantom	ξ Purging	Relative
			$\sqrt{\overline{\theta_t^2}}$	Target	Fixed (b)	MLC	Surface	Magnet	Photon
Quantity	t	$\overline{E}_t$		Exit	Collector	(c)	(d)	Deflection	Dose
Element	[mm]	[MeV]	[mrad]	[mm]	[mm]	[mm]	[mm]	[m]	[%]
Be	3.0	48.7	27	0.046	3.4	17	28	0.31	0.6
	6.0	47.5	38	0.13	4.9	24	39	0.32	1.5
	9.0	46.3	48	0.24	6.0	30	49	0.33	2.7
	15.0	43.8	63	0.53	7.7	39	64	0.35	5.3
	52.6	28.9	130	3.7	14	80	133	0.57	28
	136.7	0.0	—	—	—	—	—	—	—
Be/Al/ Cu/W	7.4/1.1/ 4.8/1.7	15.0	450	0.50	52	276	454	3.0	310
Be/W	11.9/3.1	15.0	470	0.76	55	291	479	3.0	340
Be/W	9.0/6.0	2.1	770	2.2	69	475	782	—	750
W/Cu(a)	5.0/7.25	0.0	—	—	—	—	—	—	—

[a] Standard target for MM50.
[b, c, d] AT 13, 63 and 103 cm from target entrance.
Characteristic of transmitted primary electron beam for different targets designs and relative dose from 50 MV photon beam.

The influence of the photon beam profile on the total treatment time for a given target dose distribution and incident radiant energy using a fast scanning system and dynamic multileaf collimation is shown in Fig. 4.69 and in Table 4.1. When the dynamic collimator is introduced a uniform open field is used as a reference. Since the bremsstrahlung beam is of finite width even it is narrow, large dose gradients in the interior of the target volume and its near edges of the field can only be exactly shaped by combining beam scanning and dynamic multileaf collimation. The major part of the desired field is thus irradiated by the fast scanning system and the residuals are taken care of by dynamic collimation (Källman et al. 1988; Gustafsson et al. 1995), for instance, using the shrinking field or camera shutter irradiation technique. The optimal scanning pattern for the actual elementary bremsstrahlung beam is first calculated using the same iterative algorithm as above. The required collimator opening density distribution (Gustafsson et al. 1995) is simply calculated as the quotient between the desired energy fluence distribution and the optimally scanned energy fluence rate distribution incident upstream the collimator. The narrow target volume as seen in the lower panel in Fig. 4.69 (cf also Table 4.2)

Table 4.2. Relative treatment time for combined scanned beam and dynamic MLC for a given nonuniform dose (cf. Fig. 4.69) and incident radiant energy at 50 MV.

Target		Profile	Forward		Yield Integral				Relative Treatment Time %								
					°×0		°×°		Optimal Scanning Pattern			Dynamic MLC Increase			Total + Transit Time		
		FWHM Phantom	MeV		GeV												
Quantity Element	t [mm]	Surface [mm]	cm²	[%]	cm	[%]	GeV	[%]	15×0	15×10	15×°	15×0	15×10	15×°	15×0	15×10	15×°
Be	3.0	32.8	178	72	0.71	20	1.39	3	216	405	860	0.0	0.0	0.0	144	247	523ᵇ
	6.0	37.9	234	94	1.08	30	2.41	5	149	256	521	0.3	0.3	0.7	114	171	334
	9.0	41.3	265	107	1.35	38	3.33	7	124	202	409	0.4	12	12	98	153	292
	15.0	46.0	303	122	1.76	49	5.08	10	102	155	299	4.6	26	26	86	133	243
	52.6	57.3	351	142	2.83	79	13.2	27	79	97	157	21	38	38	78	96	146
	136.7	61.3	304	123	2.85	80	17.5	35	88	102	143	23	39	39	87	101	136
Be/Al/ Cu/W	7.4/1.1/ 4.8/1.7	71.0	334	135	4.36	123	49.4	100	75	77	84	27	40	40	78	80	86
Be/W	11.9/3.1	71.1	330	133	4.43	125	51.8	105	76	77	81	27	40	40	79	80	83
Be/W	9.0/6.0	73.2	248	100	3.55	100	49.4	100	100	100	100	28	40	40	100	100	100
W/Cu(a)	5.0/7.25	85.8	214	86	3.37	95	47.9	97	112	106	104	32	41	41	113	106	104

ᵃStandard target for MM50. ᵇOpen field 3 mm Be

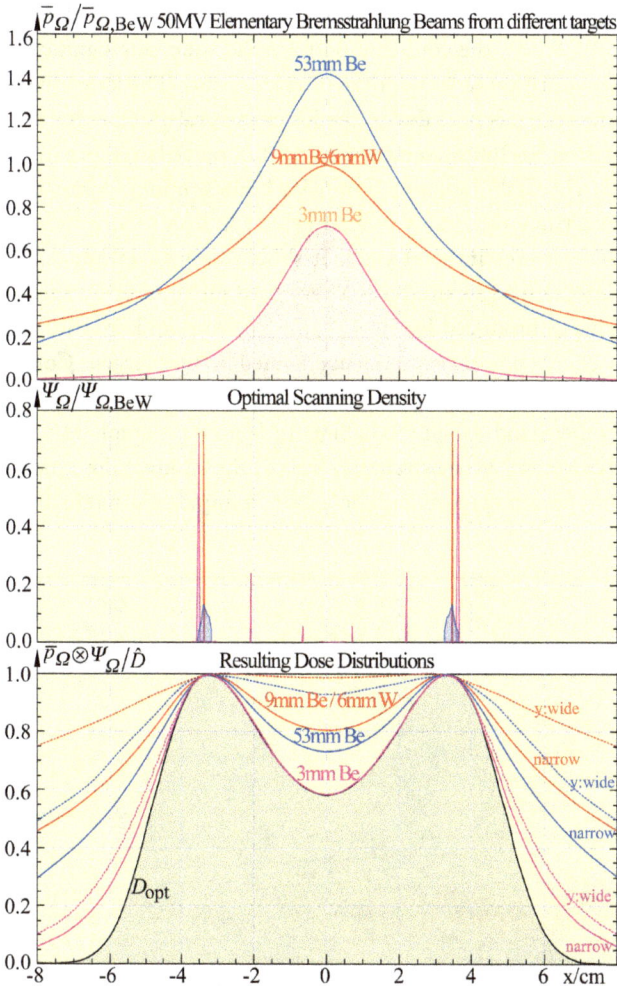

Figure 4.69. The relative merits of three different bremsstrahlung pencil beams (upper panel) are compared for optimal scanning patterns (middle panel) when irradiating a complicated but smooth target volume (gray in lower panel). As seen in the lower panel, only the narrowest bremsstrahlung beam perfectly modulates the interior of the desired distribution Cf Table 2 for details.

was also extended in width to 10 cm and to infinity. To be strict in the comparison between the different target designs, the beam current or more precisely the incident electron fluence on the target is assumed to be fixed. The optimal scanning of an elementary beam from thin targets made of 3–6 mm beryllium will conform very well to the desired distribution also in the interior part of the field. This is also reflected in

Table 4.1 as negligible residual modulation by the MLC is required. For larger target thicknesses of ≥9 mm, the contribution from the wide tails significantly increases the residual modulation as seen for the broader beams. The longer irradiation time for the narrow beams is here due to the lower integral yield, although the forward yield is as high as 70–100% compared to the full range beryllium–tungsten target. Consequently, the total irradiation time using beam scanning alone with a thin target will take about twice the time for small fields and up to four times for very large fields due to the lower integral yield. It should be kept in mind that these thin low atomic number targets produces a very hard photon energy spectrum because bremsstrahlung is produced by 50–49 MeV electrons and are thus not associated with the low-energy photon component obtained in thick targets. Consequently, the irradiation time may further be marginally reduced for deep-seated tumors.

On the other hand, when the collimator is used concomitantly with scanning of broader elementary bremsstrahlung beams, the residual dose modulation with the collimator becomes more complex. For example, the modulation time with

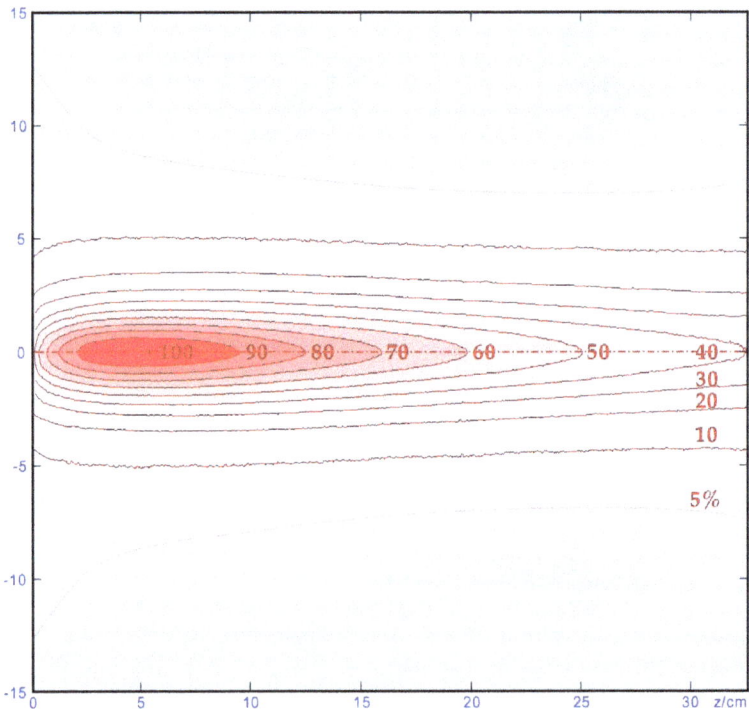

Figure 4.70(a). The experimental narrow pencil beam dose distribution from a 3 mm thick Be target at 50 MeV at an SSD of 100 cm is shown.

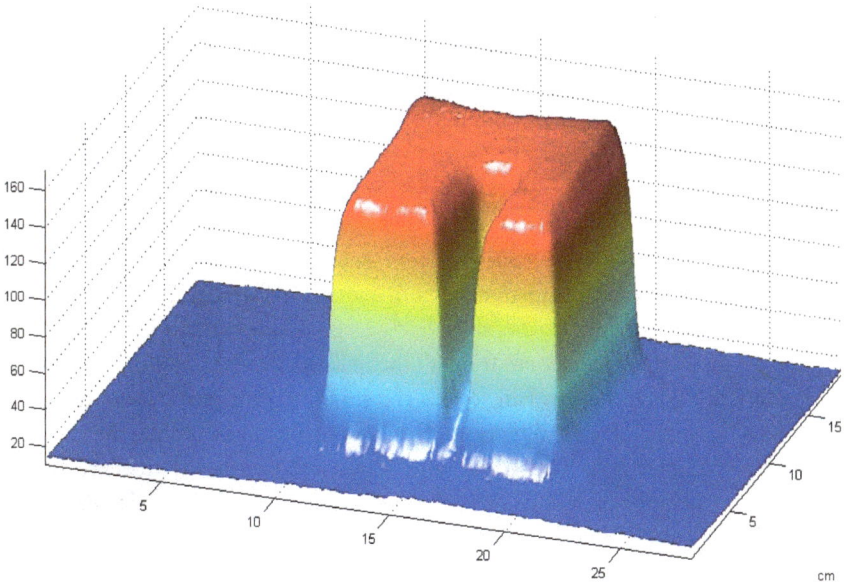

Figure 4.70(b). The lower experimental scanned dose distribution is generated in the geometry of Fig. 4.45 (cf. Figs. 3.15 and 8.39).

the collimator will be almost 40% longer using a standard MM50 sandwiched tungsten–copper target compared to a 3 mm beryllium target. To perfectly compensate for the necessary extra modulation time the target can be made slightly thicker to increase the forward yield (which is naturally field size-dependent). For instance, considering the narrow case, the time it takes to optimally scan the field with an elementary beam from a 9 mm beryllium target that has almost the same forward yield as the full range beryllium–tungsten target is only around 25% longer than that for a full range Be-W target. However, the time it takes for the collimator to modulate the residual dose is instead around 25% shorter. Thus, the total treatment time will consequently be the same for those two target designs, but the negative dosimetric influence from the collimator, such as penumbra, integral collimator transmission, and wear on the collimator, is considerably lowered for the thin target design. When only the lowest possible treatment time is considered when irradiating narrow fields, a beryllium target made of around 50 mm beryllium should be used only for very small field sizes, since the integral yield is most important. With that thickness compared to the Be-W target the forward yield is 40% higher and FWHM is around 30% lower and a 20% decrease in the scanning time is obtained. The integral yield for that beryllium target is of the same order as the integral yield of Be-W target due to the low attenuation of the generated photons in the beryllium target. Consequently, since the residual modulation with the collimator is almost the same as the residual modulation for BeW the total irradiation time is also decreased by around 20%. However, the transmitted electrons from such a target of high integral photon yield have a high electron mean energy, around 30 MeV, and are very diffuse so that the complexity of the design of the purging magnet and the electron collector is increased (cf. Table 4.1 and Fig. 4.49). Another drawback of such a thick target of low atomic number made for scanning a beam is that the size of the effective photon source, particularly at large angles of incidence, will be considerably extended both laterally and axially and consequently the geometrical penumbra becomes relatively large (Svensson and Brahme 1996). Obviously, the high integral yield from all the half-thick targets and the optimized two- and four-layered targets will reduce the scanning time particularly for large fields. Owing to the small difference in the shape of the beam profile, the residual modulation by the MLC will be the same and consequently the total treatment time for all such target will also be the lowest. For very large field sizes, the slightly larger tail of the transmission target made of BeW somewhat lowers the treatment time due to the higher integral yield compared to the optimized four-layered target.

It is interesting to note that when the scanning system by itself cannot shape the desired dose distribution as recommended by modern dosimetry protocols (NACP 1997), the increased modulation time by the dynamic collimation is necessary even for rather small adjustments (cf. Gustafsson *et al.* 1995). For instance, if the camera shutter technique is used, the leaves must always travel across the whole field. This travel time, although it is constant compared to the time when the leaves are actually modulating the field, becomes large for fields that are almost perfectly delivered by the beam scanning alone. If the speed of the leaves is around 1 cm/s the travel time becomes much larger than the actual modulation time of the field.

Finally, it should be emphasized that the potential for the MM50 to increase the beam current on the target to compensate if necessary for the lack of integral yield when using a thinner target is of no problem at all in this high-energy range, since the machine is designed to deliver sufficiently high dose rate even for photon beams of very low energy. In fact, this is naturally already done for calibration purposes when switching between different bremsstrahlung targets and energies such that the total irradiation time for a given dose is fixed.

4.5.2. *Treatment Outcome with Bremsstrahlung Beams*

The optimization of a fixed MLC setting in combination with a narrow scanned bremsstrahlung beam often allows a very high treatment outcome. The dose delivery

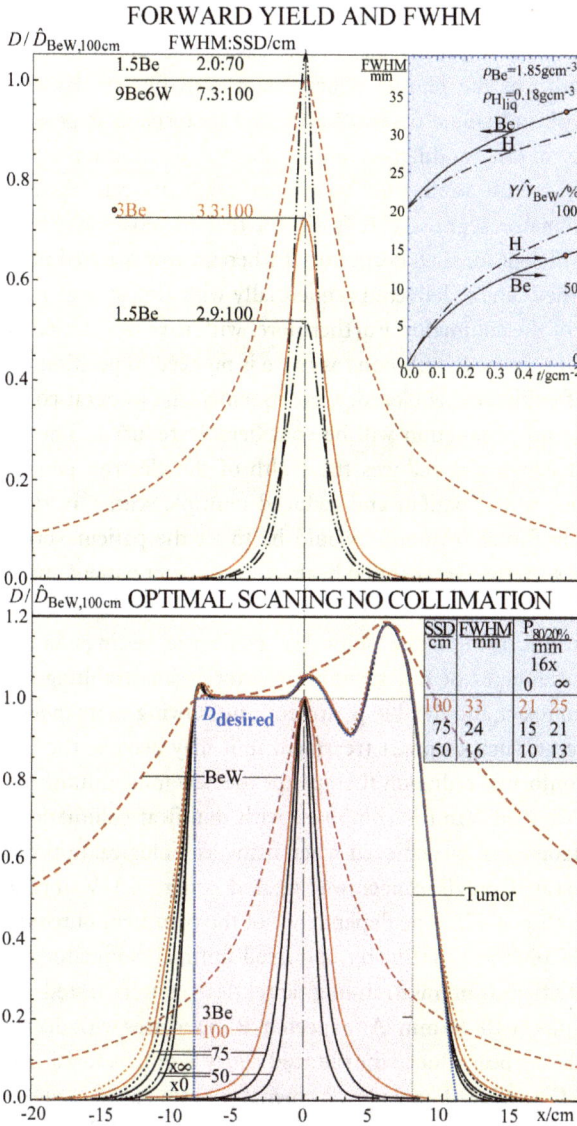

Figure 4.71. The upper panel shows the elementary bremsstrahlung beams for different target designs. The bremsstrahlung from the transmission targets is almost of Gaussian shape. Note that the forward dose rate from the narrowest beam produced by a 1.5 mm Be target at SSD 70 cm is the same as for the BeW target at 100 cm. A lower atomic number target such as liquid hydrogen may also be used (insert). The narrowest beam associated with the narrowest penumbra almost perfectly shapes the target volume for all field sizes, particularly when the optimal dose edge should be unsharp to account for patient motions (cf. Löf *et al.* 1995).

can be improved further by reducing the source to isocenter distance and the size of the elementary photon beam. Increasing the initial electron energy can reduce the width of the photon pencil beam even further. By delivering a few static fields with individual narrow beam scan patterns, it is possible to combine the advantage of fast modulation of the fluence distribution by narrow pencil beam scanning, at the same time as steeper gradients can be delivered using a few static collimator segments. It is shown that in most cases a few collimator settings are sufficient for each beam portal when narrow scanned photon beams are available and they can be delivered sequentially with almost negligible delay due to repositioning of the collimator. Furthermore, with narrow scanned beam there is a minimal loss of therapeutic photons as there is no need to position the pencil beam where the leaf collimator is closed. Consequently, the integral collimator leakage and photoneutron production will be considerably reduced. The reduced source to isocenter distance also reduces the width of the electron pencil beam due to shorter distance to the patient and reduced multiple scatter in the air. Finally, a very interesting future technique would be to let the patient verification system work together with the fast scanned beam to counteract patient and internal organ motions.

In conclusion, using a few static leaf collimator settings and the significant engineering advantages of fast scanning narrow bremsstrahlung beams can thus make a very compact and flexible treatment unit. Owing to its increased efficiency and reduced cost, such compact treatment unit may become the ideal equipment for accurate conformal radiation therapy for the new millennium.

When a scanned beam is combined with multileaf collimation, an increased treatment outcome can be achieved when using as biological optimization models. First a deep-seated cervix cancer was treated using 50 MV anterior and lateral photon beams (Fig. 4.72). The dependence of the treatment outcome on the width of the scanned photon beam using one fixed optimal collimator setting was also investigated. Different bremsstrahlung target designs were tested with elementary beam half-widths of 10–85 mm. As expected, P_+ increases with decreasing width of the bremsstrahlung beam due to the reduced injury in organs at risk and a higher dose to the tumor (Fig. 4.72). The widest elementary beam tested, generated by a WCu target, P_+ has its lowest value due to a combined effect of decreased flexibility to shape the incoming fluence and a lower mean photon energy. For sufficiently narrow elementary bremsstrahlung beams, in combination with fixed block or MLCs, the scanned beam by itself can modulate most of the fine structure of the required dose distribution. This is clearly illustrated in Fig. 4.72 for the thinnest target design, which within about 5% can realize the best possible treatment. For the advanced cervix cancer with infiltration into surrounding lymphatic tissues, Figs. 4.73a, b and

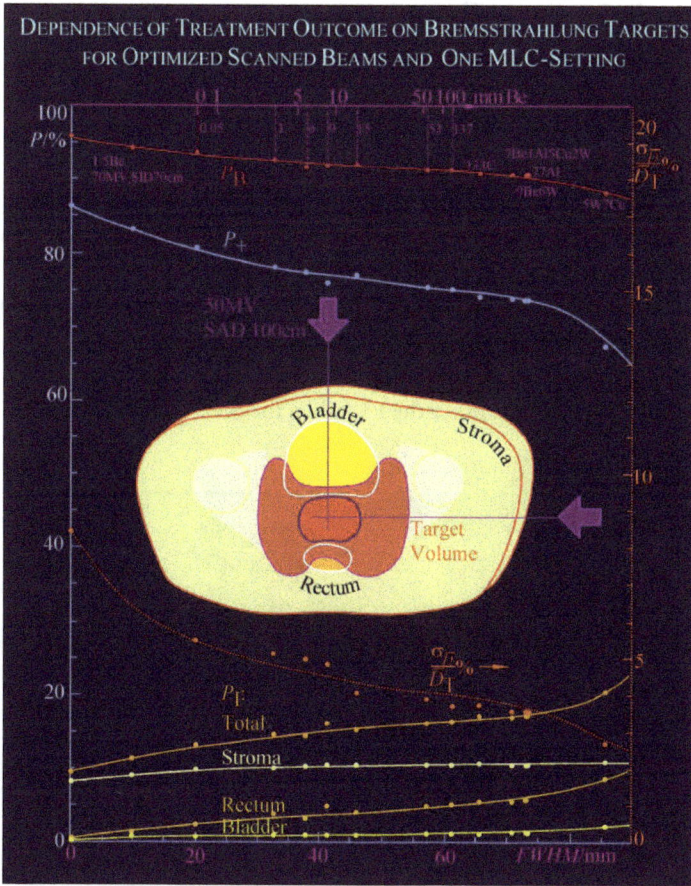

Figure 4.72. The probability of complication-free tumor control for an advanced cervix tumor treated with optimally scanned photon beams generated for different bremsstrahlung targets and one fixed optimal collimator setting is illustrated. The treatment outcome is decreased in proportion to the width of the elementary bremsstrahlung beam. Although the half-width of the BeW target is only slightly smaller than the WCu target, the deep depth dose maximum for the BeW beam is mostly responsible for the significantly higher treatment outcome.

4.75 illustrate the effect of the SAD on the treatment outcome for different dose delivery techniques in full 3D geometry. First, optimal static block and multileaf collimation of uniform incident beams were tested. Second, the incoming beam was also improved by optimally scanning the narrow photon beam. For all dose delivery optimization techniques, the highly penetrative 50 MV 3 mm beryllium bremsstrahlung dose distribution kernel was used (cf Fig. 3.73a).

4.5.3. *Uniform Beam Optimization*

The dependence on SAD and collimation technique is small for uniform beams where P_+ increases from 67% at an SAD of 50 cm and block collimation to 72% at an SAD of 100 cm using multileaf collimation. A systematic reduction of P_+ by 2% going from an SAD of 100–50 cm is also seen independent of collimation technique largely due to the loss in depth dose for deep target volumes. Interestingly, the dose to the gross tumor can be slightly increased at the shorter SAD at the same time as the dose to the rectum is lowered due to the steeper dose fall. The probability for injury to the rectum is decreased by around 1% and the gross tumor is better controlled by around 0.5%. Unfortunately, this is balanced by the increased probability of inducing injury to the bladder by about 1.5%. This and the reduced control of the nodes and increased stroma lower the P_+, when the SAD is shortened. As expected

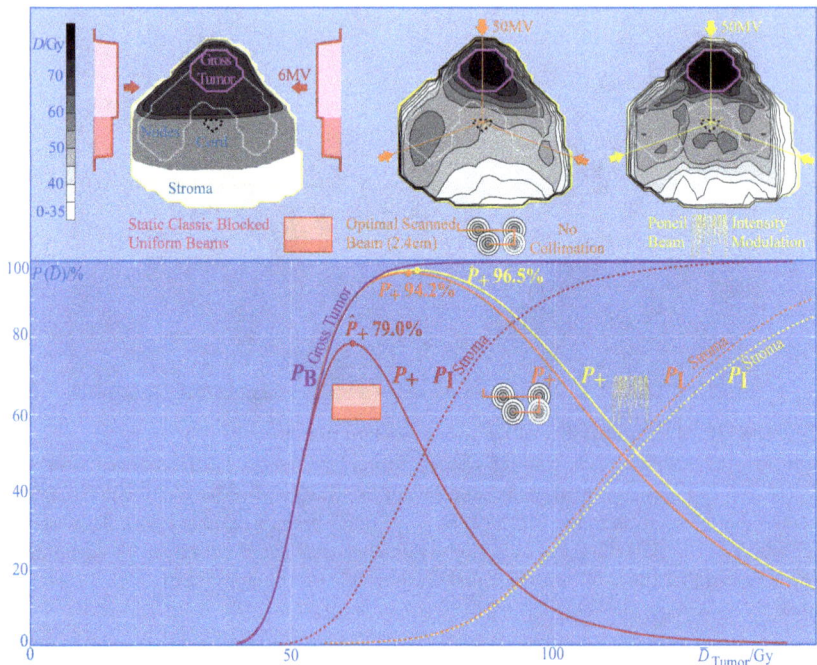

Figure 4.73(a). The treatment outcome for a head and neck cancer with lymphatic spread using a Static Classic Blocked Uniform Beam two-field treatment technique as well as two Optimal Scanned Beam treatment techniques without collimation with Macroscopic and narrow point monodirectional pencil beams. It is seen that reducing the dose to organs at risk substantially increases the therapeutic window and thus pushes the normal tissue injury curves to higher mean tumor doses (dashed curves for spinal cord and normal tissue stroma).

DEPENDENCE OF TREATMENT OUTCOME ON BEAM SHAPING & SAD

SAD:	100cm	75cm	50cm	$\Delta P_+ \approx$
BEAM SHAPING	**OPTIMAL STATIC UNIFORM BEAMS**			
Static Block Collimation P_+	69.3%	68.6%	67.0%	
P_1	18.1%	18.4%	18.9%	$\Delta P_1 \approx 3\%$
Static MultiLeaf Collimation P_+	72.3%	70.9%	70.2%	
P_1	17.6%	18.8%	18.8%	
50 MV 3mm Be	**OPTIMAL SCANNED PENCIL BEAMS**			$\Delta P_+ \approx 5\%$
No Collimation P_+	67.4%	73.5%	77.6%	
P_1	?%	17.0%	14.5%	$\Delta P_1 \approx 3\%$
Static Block Collimation P_+	74.0%	76.1%	77.8%	
P_1	16.5%	15.6%	14.0%	$\Delta P_+ \approx 3\%$
Static MultiLeaf Collimation P_+	78.1%	79.0%	80.3%	
P_1	14.6%	13.9%	13.0%	$\Delta P_+ \approx 3\%$
Pencil Beams P_+	86.3%	86.1%	85.4%	
P_1	9.6%	9.7%	9.9%	

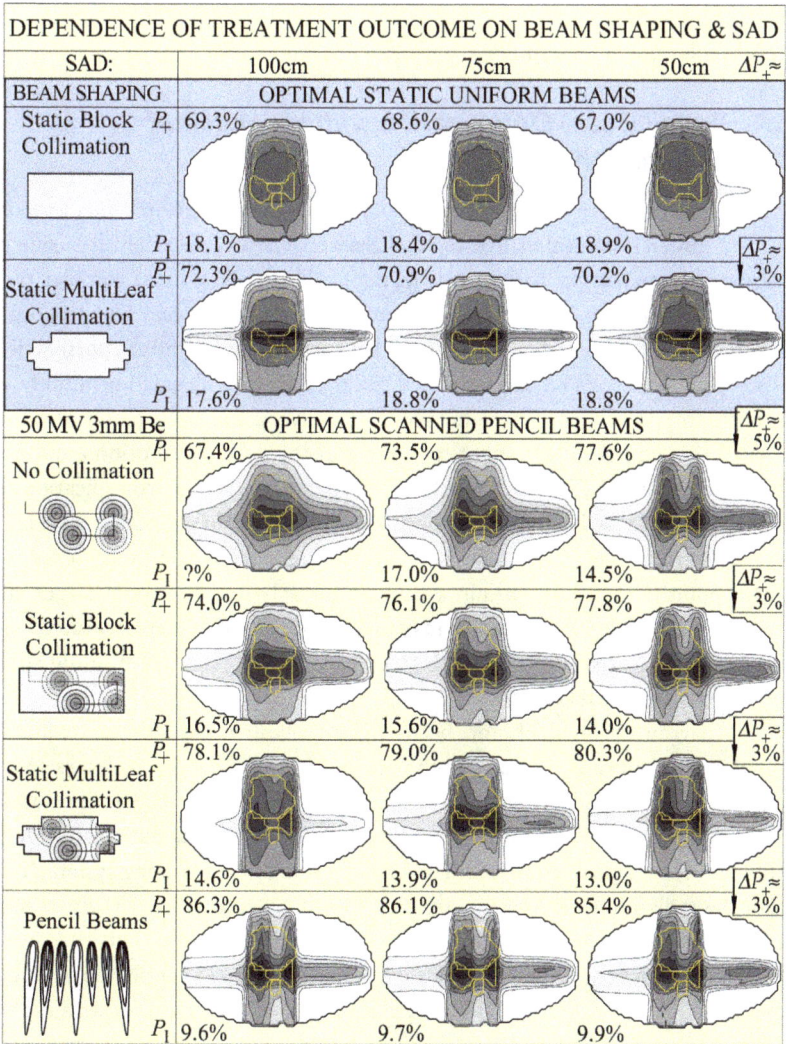

Figure 4.73(b). The treatment outcome for an advanced stage of cervix cancer with lymphatic spread using a two-field treatment technique and different SIDs and different degrees of freedom in the dose delivery starting from classical collimation to advanced pencil beam scanning combined with multileaf collimation are illustrated. For uniform beams, a lower treatment outcome is seen when the SID is shorter due to the shallower depth of dose maximum. The increased flexibility in dose delivery starting with optimal scanning without collimation (upper panel) improves the dose plans which are reflected in an increased treatment outcome. The leaf width of the multileaf collimator was 1 cm.

a small systematic increase in P_+ of about 3% is achieved, going from block to multileaf collimation with uniform beams.

4.5.4. *Pencil Beam Optimization without Multileaf Collimation*

Owing to the very narrow photon beam profile available with the presently proposed treatment unit, it is natural to investigate the effectiveness of dose delivery using just the scanned beam, that is, without any form of collimation. When a complex target volume is treated, such as a complicated tumor surrounded by very sensitive organs at risk, it is vital that the lateral dose fall-off between the tumor and the normal tissue is as steep as possible. The narrow pencil beam width with its small penumbra will drastically improve intensity-modulated dose delivery due to increased flexibility in modulation and the consequent tightening of the dose distribution around the target volume (Figs. 4.73a, b and 4.75). The probability to achieve complication-free cure P_+ is as high as 74% to 78% for a scanned photon beam with a FWHM of 2.4 and 1.6 cm, respectively. This is about 5% higher than can be achieved with a single optimal multileaf setting with a uniform beam. For some target sites, uncollimated narrow scanned beams may even suffice as will be demonstrated below for an advanced larynx tumor, though block or multileaf collimation would still be desirable as a safety backup (cf Fig. 3.73a).

The narrow scanned photon beams may thus alter the way the collimator should be designed and optimized both with regard to the number of leaves and the thickness of each leaf. The lateral protection is still largely performed by the multileaf collimation system, whereas the longitudinal protection of organs at risk in front of, behind, or even inside the tumor will mainly be made by the scanning system. Thus, part of the radiation shielding function has been moved from the dynamics of the MLC to the ultrafast motion of the narrow scanned photon beam. However, even if the collimator can be simplified it is probably better to keep the improved penumbra due to pencil beam scanning and improve it further by combining it with the best possible leaf collimation technique. The ultimate technique would be to scan with a point monodirectional pencil beam (cf. Fig. 4.73b lowest panel). In practice, such a technique can today only be approximated in the clinic by using absorbers of high resolution (about 2 mm) or dynamic multileaf collimation of high quality, that is, using a double-focused collimator of minimal interleaf leakage and transmission.

4.5.5. *Scanned Beam Optimization with a Static Collimator Setting*

When the scanned beam is collimated either by a block or a MLC, the P_+ increases several percentages at least for elementary beams that are broader than 2.5 cm (see

Fig. 4.73). At an SAD of 100 cm, P_+ increases from 67% without collimation to 74% with optimally position of block collimator and 78% with a fixed optimal MLC setting. Note also that at an SAD of 50 cm, the optimal scanning pattern without collimation is almost as advantageous as optimal scanning combined with a fixed multileaf setting at an SAD of 100 cm. The role of the static collimation of the incoming beam is reduced when the width of the elementary beam is reduced to about 2 cm. For instance, at an SAD of 50 cm, the treatment outcome is not increased much when adding block collimation and increased by only about 3% when using a fixed multileaf setting. This is due to the narrow penumbra associated with the elementary beam ($\sim$1 cm).

4.5.6. *Scanned Beam Optimization with a Number of Collimator Segments*

To improve the ability to deliver steep dose gradients inside the beam, the dose delivery can be augmented by allowing the algorithm to divide each beam portal into a number of optimized MLC segments as shown in Figs. 4.74 and 4.75. Each multileaf setting was simultaneously optimized with its individual scanning pattern. With two optimized segments, P_+ increased from 78.1% to 80.5%. However, when three or four or even more segments are allowed, P_+ will increase more slowly and it will saturate at 82.5% for more than 10 segments. This indicates that a few optimized segments are usually sufficient to get a high treatment outcome when combined with narrow pencil beam scanning.

4.5.6.1. *Uniform beams*

For uniform beams and a fixed optimal multileaf or block collimator setting, the treatment outcome is inferior to the situation where the incoming beam on the collimator was first optimized with the scanning system (see Fig. 4.73). This is illustrated in Fig. 4.75 for uniform beams and a leaf width of 4 cm where P_+ dropped to 50% due to both low leaf resolution within the field and poor matching at the sides for the last leaf pair at the border to the normal tissues. The possibility of the MLC to laterally protect the normal tissue around the target volume increases rapidly with decreasing leaf width so that P_+ increases to around 70% when the leaf width is around 2 cm which is comparable with an optimized rectangular beam blocking. The dose escalation to the gross tumor can be enhanced further since the normal tissues are better protected by decreasing the leaf width, and the treatment outcome increases further by more than 5% for the finest collimator resolutions when collimating a broad beam (FWHM $= \infty$ in Fig. 4.62). The saturation is due to the lateral electron transport and photon scatter, which reduces the need for very fine collimator resolution when optimized intensity modulation is used.

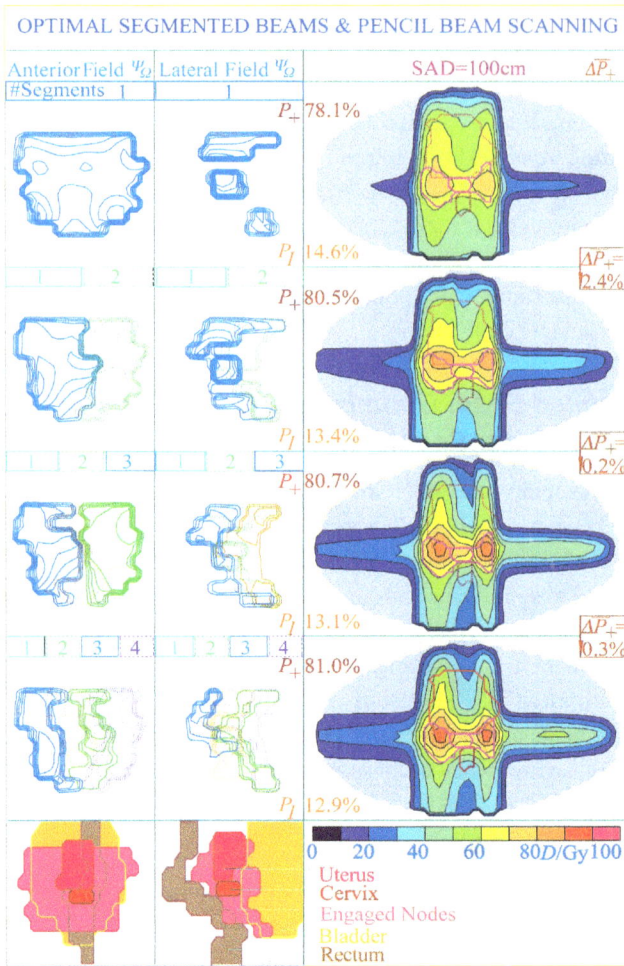

Figure 4.74. The treatment outcome is considerably improved when optimal pencil beam scanning is combined with a few optimally selected collimator settings and is only slightly improved for a large number of settings. The two left panels show the incident energy fluence for the anterior and lateral fields. The boost on the gross tumor from the lateral fields is clearly seen when two or more collimator settings are allowed.

4.5.6.2. *Scanned beams*

Already for a fairly broad elementary bremsstrahlung beam such as that coming from the BeW target, P_+ is raised as much as 15% compared to uniform beams just by adding beam scanning and moving to the right in Fig. 4.62. The treatment outcome increases as expected with decreasing half-width of the elementary beam

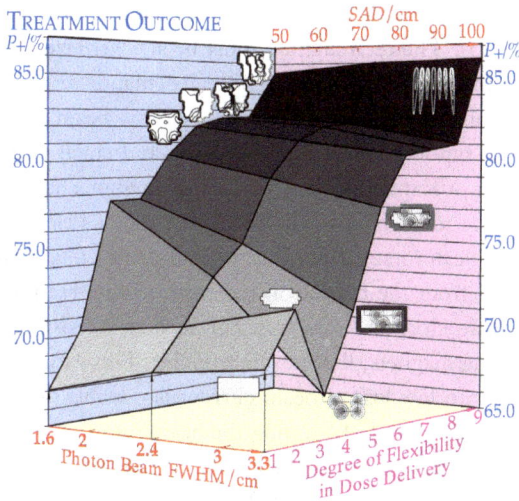

Figure 4.75. The quantitative increase in treatment outcome with decreasing pencil beam width and increasing flexibility in intensity modulation as showed in Figs. 4.73 and 4.74.

and this increase is faster with thicker than with narrow leaves due to the influence of electron scatter. Interestingly enough, for the narrowest elementary beams around 15 mm or so, P_+ is over 80% also for rather broad leaves and increases slowly up to some 85% for the finest leaf widths. Thus, for intensity-modulated fields only a small therapeutic gain can be achieved by increasing the resolution of the MLC. The highest gain is obtained by decreasing the elementary beam width <20 mm.

4.6. Compact High-energy Treatment Units based on Electron Recirculation

High-energy compact electron sources. A recirculated high-energy electron accelerator such as the racetrack microtron is a well-known reliable accelerator of high efficiency considering the microwave power needed. The linear accelerator, linac, and the microwave power package in the S-band frequency domain needed for the racetrack can be made of standard components, power supply, modulator, klystron, and linac structure, available from several vendors. The microwave peak pulse power fed to the linac structure in the racetrack around 2–5 MW during a pulse of length 1–5 us is the same as for other conventional linac-based therapy units. Unfortunately, due to the rather complex construction of the racetrack, the nominal price is quite high and only the largest radiotherapy centers could afford to invest in such a sophisticated system. Such a unit for radiation therapy is shown in Figs. 4.76a and b that is installed at Radiumhemmet, Karolinska Hospital in the beginning of 1990 (Brahme 1982; Andeberg *et al.* 1985). The recirculation must be synchronized in phase with the microwave to keep the system in resonance, which demands very accurate alignment of the racetrack main magnets. The extracted electron beam from a microtron is characterized by a combination of high intensity and small energy spread. The racetrack accelerator of Karolinska Hospital is mounted in a horizontal position on a bench in a separate room next to the treatment room. Placed in the treatment room is a gantry that can rotate around the patient during treatment. The electron beam is transported from the microtron racetrack to the treatment gantry in the treatment room by an electron beam transport system consisting several bending magnets and focusing magnets, that is, quadrupoles. This rather long transport of the electron beam with low loss of beam current is possible due to low emittance of the extracted beam. Instead of bending the beam into the treatment room, it is possible to bend it into another path where, for example, an experimental line could be set up as has been done at Karolinska Hospital or to a second treatment room. Mounting the racetrack in the gantry would shorten the path and make the therapy unit easier to install and maintain. The therapy unit will also take up less space. Three installations have recently been ordered by newly built hospitals in China where the racetrack will be placed horizontally directly outside the treatment room in such a way that the electron transport to the treatment gantry follows a short straight line. Thus the trend is to make the installation as simple as possible and the tuning of the electron transport as fast as possible. Naturally, the drawback of such a system is that the cost per treatment unit is more expensive since one racetrack is installed per unit and the accelerator is the most expensive part of the system.

Ergonomically considerations. Ideally, a treatment unit should be ergonomically to work around. The tabletop height should preferably be lower than 1 meter. The design presented here has a source to isocenter distance that is less than 70 cm. This together with the compact gantry mounted x-band linac allows a small unit with a height of around 185 cm. Thus, properly balanced the isocenter height can easily be made less than 90 cm. Also, the overall compact therapy unit size saves space in the treatment room that also can be built with normal roof height.

4.6.1. *Gantry-mounted Racetrack*

It was proposed to mount the racetrack in the gantry as a solution to get a compact treatment unit for advanced dose delivery using scan narrow beams (Fig. 4.77, Svensson *et al.* 1998). This has several advantages compared to existing racetrack installations namely: (1) the whole system will be more compact and easier to install which will lead to faster installations at lower costs; (2) considerably shorter beam

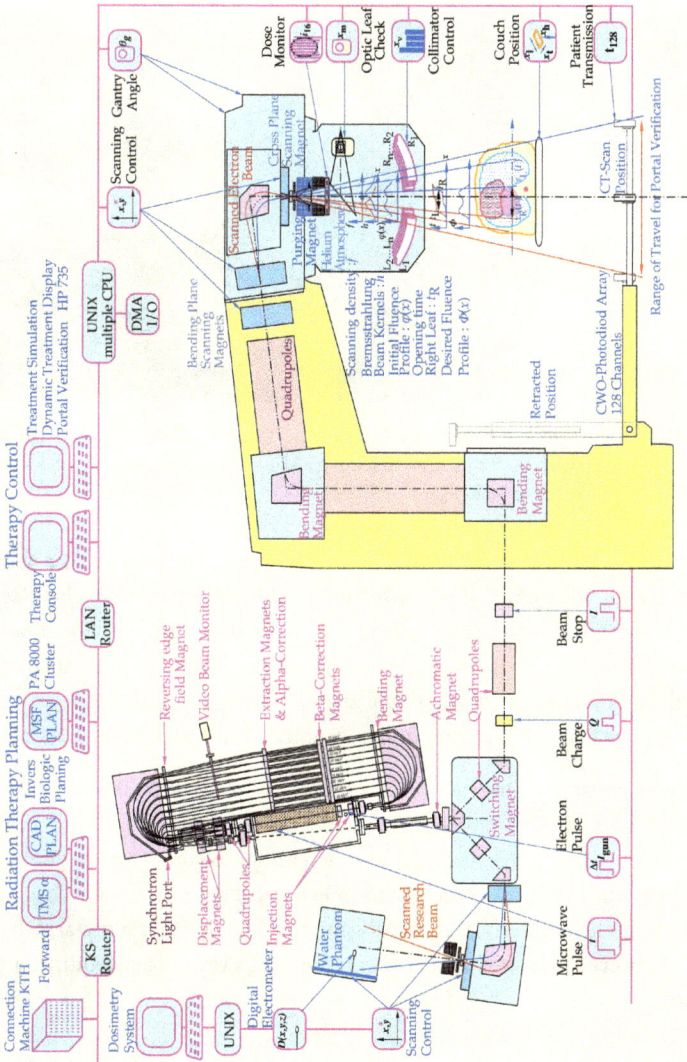

Figure 4.76(a). The Racetrack Microtron at Karolinska with the computer controlled system for the accelerator, beam transport to the Gantry and the research beam line using scanning beam dose delivery: The research behind the development of this treatment unit in the late 1970ies started the field of external beam inverse treatment planning, static and dynamic multileaf collimation and IMRT in general!

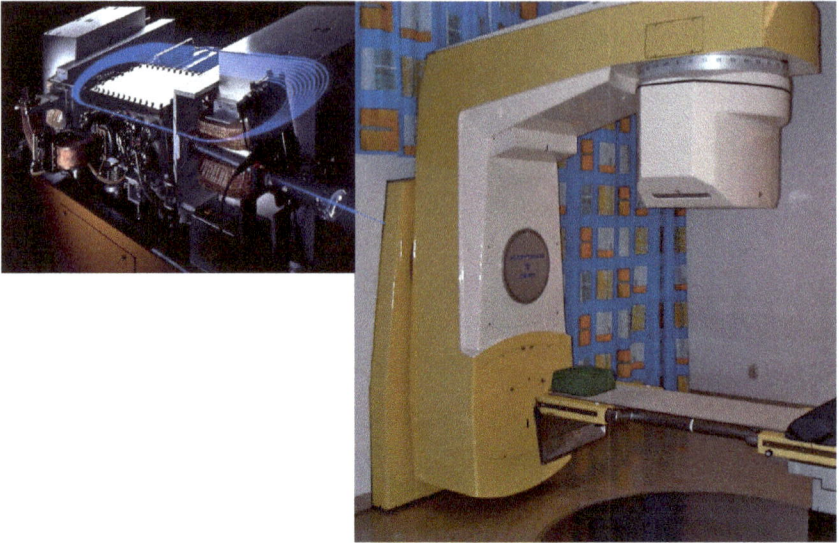

Figure 4.76(b). Photo of the Racetrack Microtron at Karolinska indicating the beam transport to the Gantry with dynamic double focused Multi Leaf collimation and optimized scanning beam dose delivery (cf also Figs. 4.10, 4.21, 4.45 and 4.48).

transport to the treatment head makes it easier to tune the beam; (3) a gantry-mounted racetrack will rotate along with the machine rotation thus the electron source on the target or through the whole transport for that matter will be rotational invariant. Consequently the dose rate and the electron or photon emittance on the isocenter will be independent of rotational angle.

Several engineering challenging design problems have to be solved, namely:

- The gantry must be able to rotate around its rotational axis without the accelerator being affected by vibrations during start or stop and gravitational deformations in the structure. This is a hard problem to solve due to the sensibility of the racetrack structure for mechanical deformations.
- The bending amplitudes are in the 10 micron level for a gantry arm having a mass of several thousand kilos rotating 360°; however, this requires a rigid structure.

4.6.2. *Magnet Alignment and Beam Transport in a Gantry-mounted Racetrack Accelerator*

The racetrack accelerator is an extremely sensitive structure to misalignment for accurate electron transport due to both resonance conditions and position of the beam. A small shift in the distance between the two magnets will cause the electron bunch to be out of phase with the microwave and consequently the beam current decreases or is lost completely. Second, to achieve stability in position of the orbits, it is important that the two main magnets and the linac (linear accelerator, Rosander *et al.* 1982) are correctly aligned. When the main magnets are misaligned, the beam which passes many times between the magnets will drift away in analogy with a beam of light between two nonparallel mirrors. Errors in machining and in mounting of the racetrack are likely to be systematic and will therefore have similar effects. The misalignment of the magnets from the optical linac axis is described by the α, β, and γ angles (Fig. 4.78).

For simplicity, the electrons are assumed to enter and leave the magnet at the same angle to the magnet edge. The angle of the straight part of an orbit will therefore on each revolution tend to increase by 2α and 2β, respectively.

The resulting shifts, Δ, of the beam position after N orbits, if focusing effects of the linac and any other correction magnet are disregarded, are according to

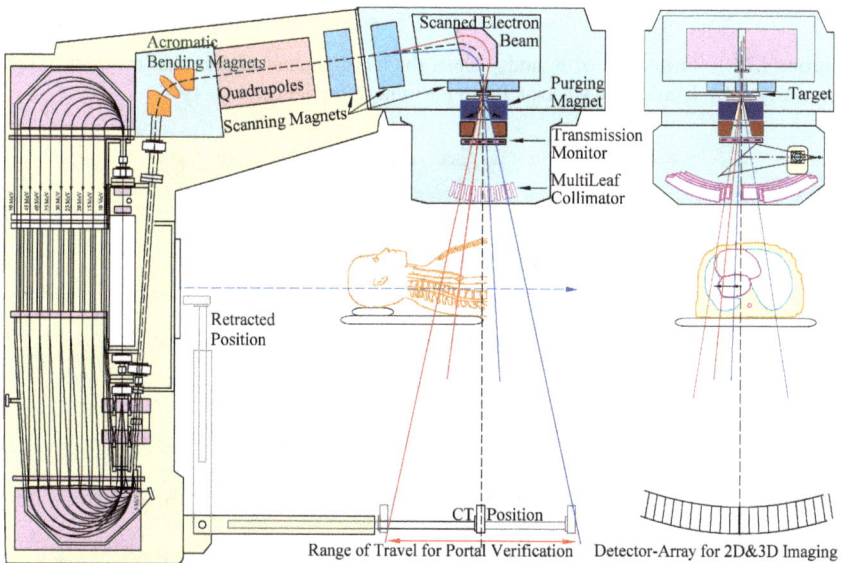

Figure 4.77. A compact gantry mounted version of the MM50 racetrack microtron therapy unit. This new design proposal offers the possibility to generate intensity-modulated large fields with an scanned elementary narrow electron (8 mm) and/or photon (19 mm) beams in combination with dynamic multileaf collimation. For narrow photon beams, the fast purging magnet removes all target-generated leptons and at the same time direct the transmitted primary electrons out of the therapeutic photon beam to a dedicated electron stopper not shown in detail.

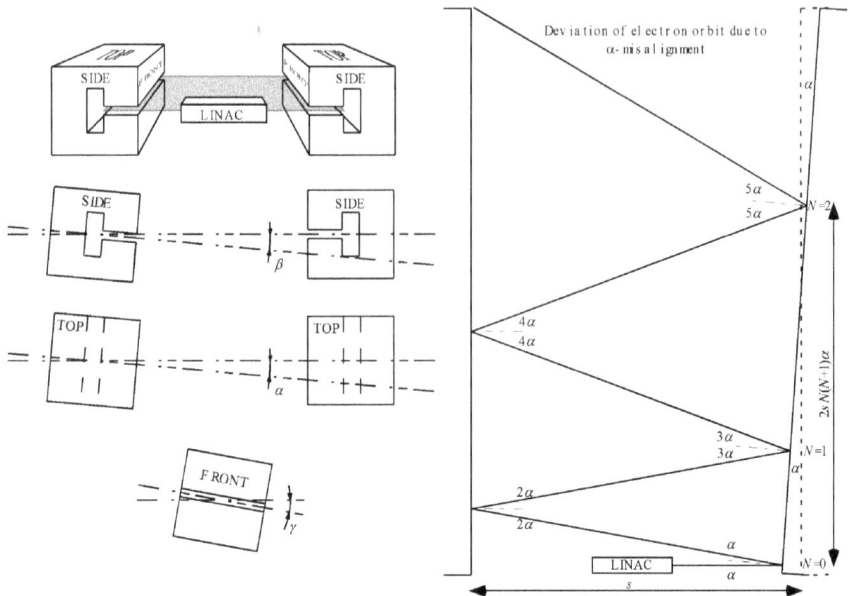

Figure 4.78. The angles α, β, and γ show the misalignment of the main magnets to the linac-axis (left). Example of electron beam position shift (right).

Rosander *et al.*:

$$\Delta_\alpha = 2sN(N+1)\alpha$$

$$\Delta_\beta = \left(\frac{8N+1}{12}\lambda v + 2s\right)N(N+1)\beta \qquad (4.30)$$

$$\Delta_\gamma = \frac{\lambda v}{2\pi}N(N+1)\gamma,$$

where s is the distance between the magnets, λ is the wavelength, v is the mode number, and N is the number of orbits.

This gives for the KTH racetrack, where $s \approx 0.5\,\mathrm{m}$, $\lambda = 0.1\,\mathrm{m}$, $v = 1$, and $N = 15$:

$$\Delta_\alpha = 240\,\mathrm{mm} \times \alpha$$

$$\Delta_\beta = 480\,\mathrm{mm} \times \beta$$

$$\Delta_\gamma = 4\,\mathrm{mm} \times \gamma, \qquad (4.31)$$

MM50 racetrack, where $s \approx 1.12\,\text{m}$, $\lambda = 0.1\,\text{m}$, $\nu = 1$, and $N = 10$:

$$\Delta_\alpha = 246\,\text{mm} \times \alpha$$
$$\Delta_\beta = 320\,\text{mm} \times \beta$$
$$\Delta_\gamma = 1.75\,\text{mm} \times \gamma. \tag{4.32}$$

Assuming that the maximum acceptable spatial deviation of the electron orbit is in the order of 1.5 to 2 mm, computations done at the Department of Electron Physics, Royal Institute of Technology, Stockholm, Sweden, show that the permissible alignment errors are about $|\alpha| < 15\,\mu\text{rad}$ and $|\beta| < 10\,\mu\text{rad}$. The computation has been done without alignment error correction arrangements. With alignment error correction arrangements, that is, small α and β correction magnets in combination with focusing on the beam toward the linac axis when passing through the linac, the alignment error angles are in the order of several hundred μrad. The aim here is to keep them $< 100\,\mu\text{rad}$.

4.6.3. *Shifts in Position of Beam Optics, Quadrupoles*

The effect of shifts in position of the beam optics has not been considered here. As a result of a small shift in the position of a quadrupole, the effects of a misaligned beam can be magnified or reduced by the order of one magnitude. The way the path of the beam is altered depends on the way it enters the quadrupole and the alignment of the quadrupole. A perfectly aligned beam can be misaligned after passing a misaligned quadrupole. A misaligned beam entering a perfectly aligned quadrupole can have a larger or smaller misalignment when it exits. A combination of a misaligned beam and a misaligned quadrupole could magnify or cancel out the misalignment. Measurements on the existing Scanditronix MM50 treatment unit show that the errors vary from day to day depending on current "mood" of the system. The effect of the mechanical displacements of the quadrupoles is a small part of the total error (although small) of the beam transport system.

4.6.4. *Misalignment Contribution from the Earth's Magnetic Field*

Effects on the racetrack from earth's magnetic field have not really been considered here. Since the earth's magnetic field is fairly constant in strength and direction, it is easy to compensate for in a standing racetrack. In a rotating gantry-mounted racetrack, the contribution of the earth magnetic field will have a varying direction in relation to the magnetic fields of the racetrack. This especially affects the transport of the low energetic electrons of the injector beam to the linac. The problem can be solved by covering the electron gun and the transport path to the linac with a mu-metal magnetic shielding.

4.6.5. *Modeling Gantry-mounted Racetrack*

Two different racetracks were modeled mounted in a gantry — the MM50 racetrack manufactured by Scanditronix and the racetrack used for research at the Royal Institute of Technology in Stockholm, Sweden (Rosander *et al.* 1982). They are both of the same weight. However, the KTH racetrack is much shorter than the Scanditronix racetrack due to differences in the design of the first electron orbit. A shorter racetrack such as the KTH racetrack may make the design more favorable for tilting in a non-coplanar fashion.

There are several possible ways of mounting a racetrack into a gantry. A number of different gantry designs were discussed during the process of this work but the gantry chosen has the same type of open design as the treatment units used with the MM50 racetrack today or as for most other isocentric type of machines from, for instance, Varian, Elektra or Siemens, although Elekta has a drum-type gantry. A gantry-mounted racetrack unit model for deformation calculations was made from drawings of existing gantries and their components and measuring components of the MM50 racetrack (Misaghi-Panah 2000). This model should not be regarded as the final design but more as a basic design to make realistic deformation calculations of the whole unit.

The gantry unit is modeled as a gantry arm rotating around a drum axle. The idea of having a drum axle is to use the space for the linac and other equipment. From the gantry arm, a treatment arm is placed perpendicular in one end, called the upper end. The axle is placed with its rotation axis in such a way that it coincides with the isocenter of the treatment beam. The source to isocenter distance was set to be 700 mm (Svensson *et al.* 1998). This distance gives the position of the treatment arm on the gantry arm in relation to the axle and the racetrack components.

The racetrack is placed in the gantry arm with its two main magnets on either side of the rotation axis. The drum axle size is set to be as large as possible between the two main magnets; this will increase the strength of the axle and the whole construction.

In the treatment arm, an achromatic bending magnet bends the beam from the straight line of the extraction pipe of the racetrack into the treatment arm. The angle of the extraction pipe with the optical linac axis is very tight, around 4.5°, in the MM50 design. This angle is too small for the gantry-mounted design. The position of the bending magnet in the upper part of the gantry requires an angle of around 9°. Thus, the lower main magnet has to be widened around 120 mm in order to have the right extraction angle.

In the treatment arm, there are also focusing quadrupoles, a heavy scanning system, a bending magnet, bending the beam toward the isocenter, a purging magnet, and a MLC, all heavy components weighing 30–400 kilos. The geometric data of the essential components, their weights, and their positions in relation to each other are known.

The main magnets of the racetrack and the size of the beam transport system to the treatment head mainly determine the size of the gantry. The dimensions of the gantry arm are mainly determined by the lower main magnet geometry, space is left between the gantry arm walls and the magnets to provide room for support of the magnets. The space between the walls and the magnet is also necessary since tuning of the beam must be performed during installation. The treatment arm is dimensioned to fit the components it holds.

A balance weight to compensate for the weight of the treatment arm is placed behind the lower main magnet. Since the deformations in the components of the treatment arm are less important than those in the main magnets, the mesh density in these parts was kept low, $2 \times 2 \times 2$ cm elements.

4.6.6. *Gantry Deformation*

Classical theory for bending deformation using Timoshenko's beam theory was first used in estimating required material parameters such as material, material thicknesses and dimensions of the gantry. All types of deformations were considered such as self-deformation, shearing, and torsion (see Paul Parsson 2001, for details). For accurate deformation calculation, a finite element method (FEM)-calculation program ABAQUS was used. The FEM calculation was also compared with classical theory for a more simplistic case and was found to be comparable. The misalignment α, β, and γ angles of the magnets from the optical linac axis were calculated for a full rotation of the gantry arm in steps of $15°$ and presented in Fig. 4.79.

The displacement δx, δy, δz of the main magnet from the linac axis along the pole gap was also calculated and never $>50\ \mu$m as shown in Fig. 4.80.

Maintenance hatches influence on deformations. The alignment error angles during maintenance will be influenced by maintenance hatches that will weaken the structure and increase the deformations. Maintenance hatches to access the linac and magnets was included in a subsequent model and the new model was recalculated. The largest deformations in the gantry appear when it is rotated $90°$ or $270°$, as seen in Fig. 4.81. It is, therefore, less likely to be a proper position for the gantry to be in when opening maintenance hatches. A deformation calculation was performed with a $0°$ gantry angle. Since the β and γ misalignment angles are zero with this gantry angle, only α changes when opening the hatches. The changes are small: the change of α of the lower main magnet is very small and α of the upper main magnet increases

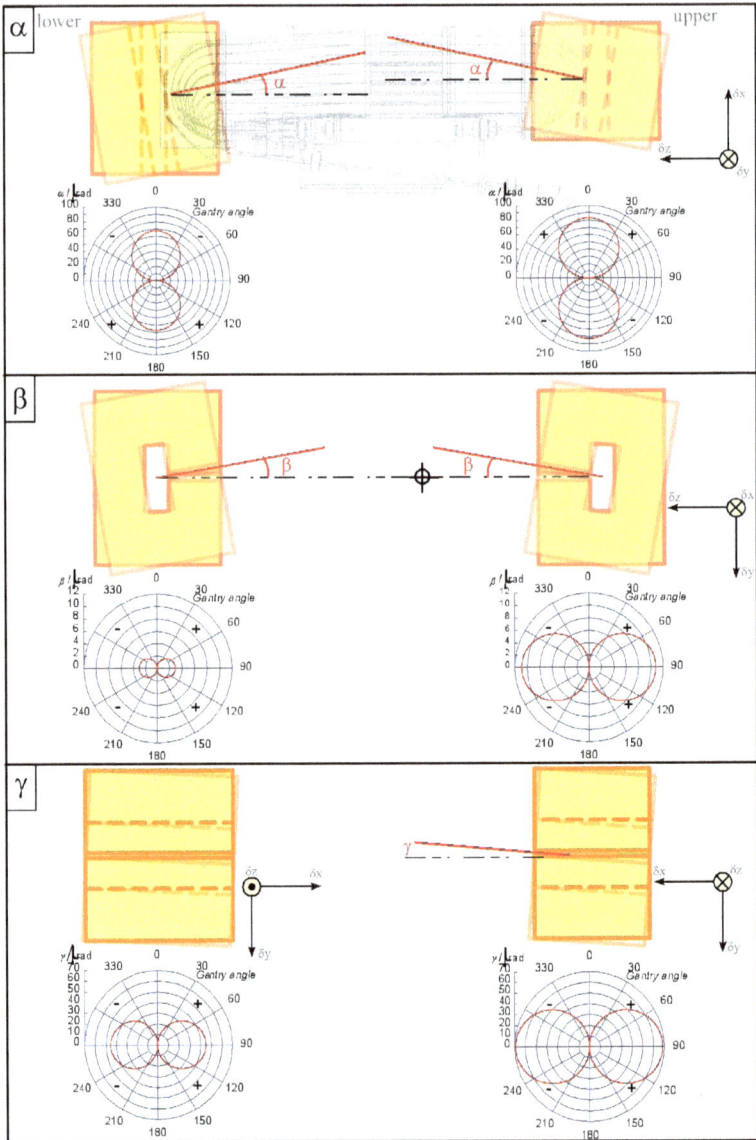

Figure 4.79. Alignment error angles α, β, and γ of main magnets due to deformations for the MM50 racetrack model.

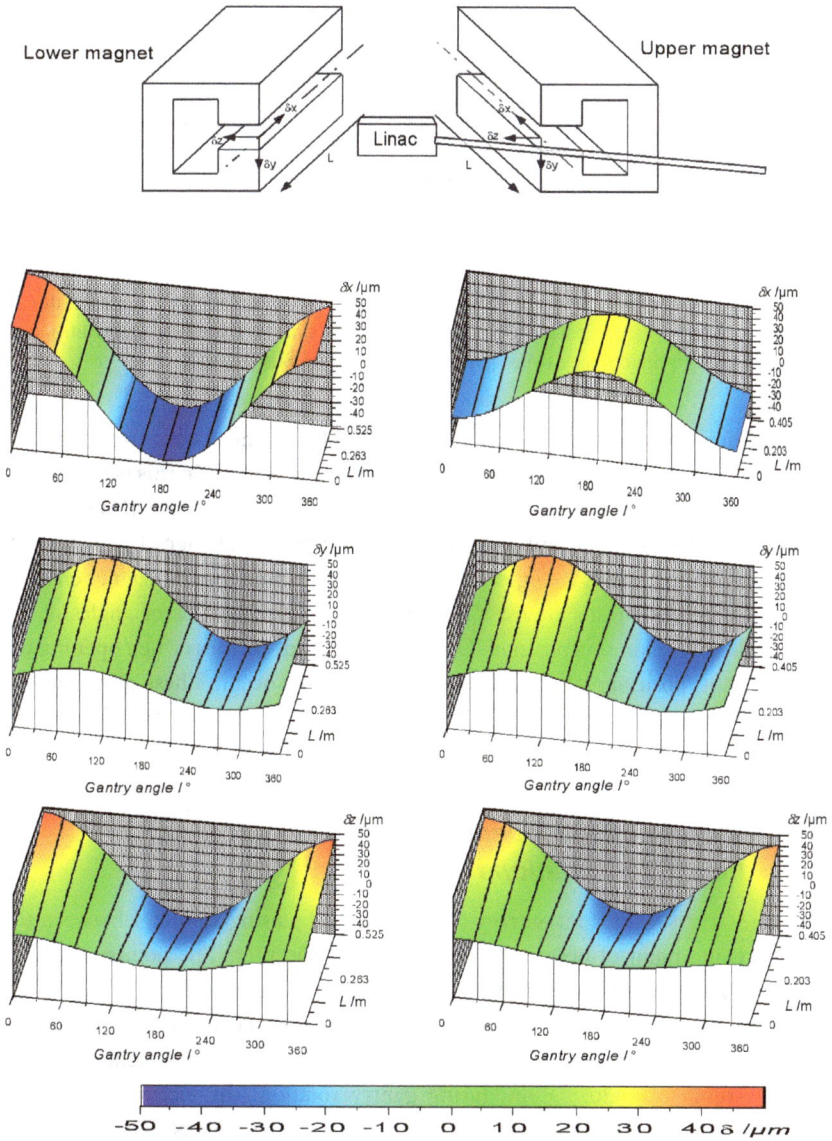

Figure 4.80. Main magnet displacement due to gantry rotation for a gantry-mounted MM50 racetrack.

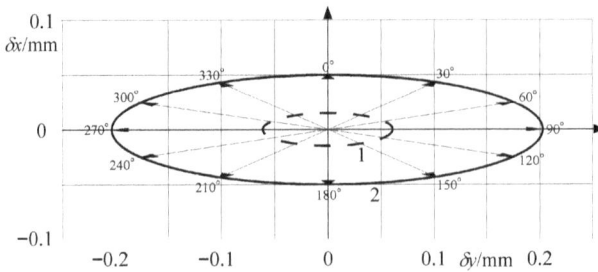

Figure 4.81. Geometric isocenter inaccuracy of electron beam in the xy planes due to deformation versus gantry rotation angle (solid line) is shown. Estimated contribution of geometric isocenter inaccuracy from gantry arm (dotted line) is shown.

by 3%, which shows that this position is quite stable. To check what happens if the hatches are opened in a 90° gantry angle, a calculation was made in this position. In a 90° gantry angle, the α misalignment angle is zero but β and γ are not; β of the lower main magnet changes 55% and β of the upper main magnet changes 114%, while γ of the lower main magnet changes -15% and γ of the upper main magnet changes 74%. As seen, this means almost a doubling of the alignment error angles of the upper main magnet in this position. Owing to the loss of stiffness when opening the hatches in this position, larger unwanted deformations occur. These deformations may be permanent after closing the hatches and can give an asymmetric distribution of the misalignment angles during a 360° rotation of the gantry.

4.6.7. *Inaccuracy in Isocentric Rotation due to Deformations*

The geometric isocenter inaccuracy was calculated by using a pointer between the source (exiting point of beam) and the geometric isocenter. The pointer does not consider errors in beam transport through the system from the linac to the source and it does not undergo deformation and has no material characteristics. It follows the translation and rotation of the node of the treatment arm it is attached to.

As a result of deformation during gantry rotation, the electron beam will perform an elliptic path around the isocenter in the xy planes. The maximum distance from isocenter in the model was 0.2 mm. The contribution of geometric isocenter inaccuracy from the gantry arm alone was estimated to be 30% of the total geometric isocenter inaccuracy. Measurements on the existing Scanditronix MM50 treatment unit show that the isocenter inaccuracy lies within 6 mm radius from isocenter. No data on how the path of the electron beam is performed within the 6 mm radius is available.

4.7. 3D Photon Beam Dose Delivery Monitoring by PET-CT Imaging

4.7.1. In Vivo *Dose Delivery Verification Using Photonuclear Reactions*

The new thin transmission target technique [14] with narrow high-energy bremsstrahlung beams of high intensity in the forward direction is also suitable for 3D *in vivo* verification of dose delivery using PET-CT imaging of the induced $\beta+$ activity [1], [2]. This approach is based on photonuclear reactions in the human tissues to convert normal ^{12}C and ^{16}O to ^{11}C and ^{15}O by knocking out neutrons with the high-energy photons. The generated PET emitters are ideally suited for *in vivo* dose delivery verification, which is an important clinically task that never before has been possible by noninvasive methods. Today, we are using Geant4 with an extensive photonuclear cross-section library to calculate the density of positron emitters in the irradiated medium or patient. Figure 4.82 illustrates the photonuclear PET emitter activity due to the narrow photon pencil beam in Figs. 4.66 and 4.70(a).

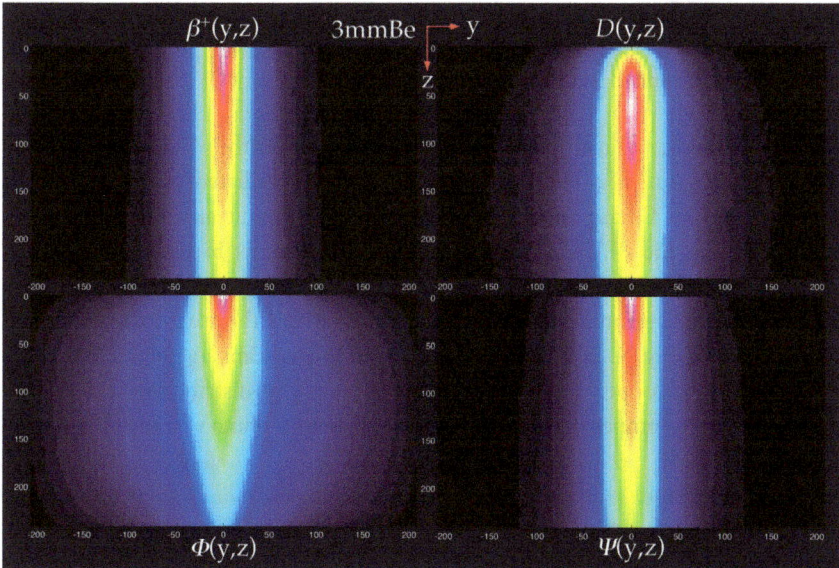

Figure 4.82. This figure illustrates the activity distribution induced by the narrow photon pencil beam in a water phantom. The PET activity distribution ($\beta^+(z)$), the absorbed dose ($D(z)$), the fluence ($\Phi(z)$), and the energy fluence ($\Psi(z)$) in the water phantom are shown. The good agreement between β^+ activity absorbed dose and the energy fluence is particularly clear. The fluence profile is broader because of the background of low-energy photons, which does not contribute much to the energy fluence.

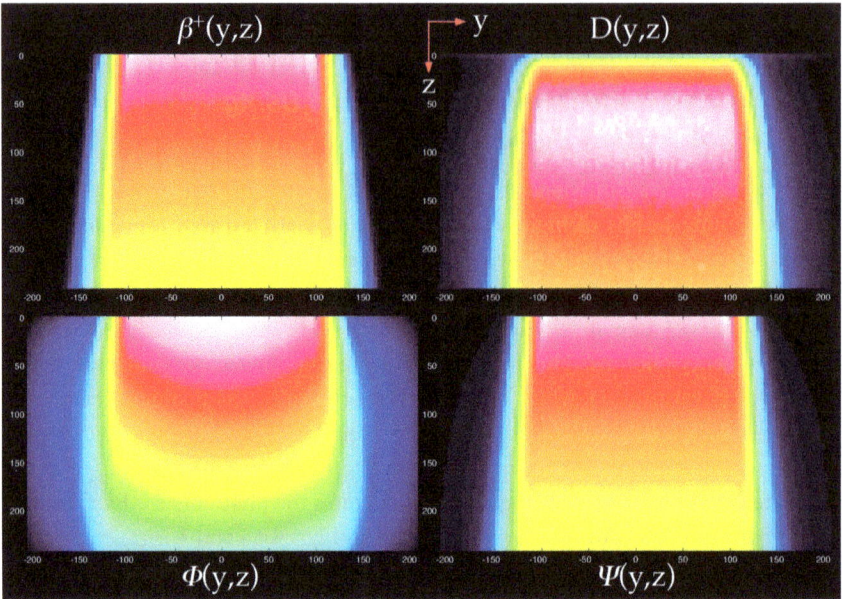

Figure 4.83. The illustration is same as Fig. 4.82 but it uses scanned photon beams so that the absorbed dose is closely uniform. The intensity or energy fluence ($\Psi(z)$) is then slightly over-compensated and so is the PET activity.

Interestingly, this distribution will be spread out over the whole tumor volume during therapy (cf. Fig. 4.83), so all regions being irradiated can be accurately traced by PET-CT imaging in relation to the normal tissue anatomy depicted by the CT part of the PET-CT camera. A broad thick target bremsstrahlung beam would not be so useful here since the mean photon energy will vary considerably over the patient and make it very difficult to interpret the true dose delivery in the patient. In Fig. 4.83, it is seen that if the broad scanned photon beam has a uniform dose (upper right panel), both the activity distribution and energy fluence are over-compensated, whereas the low-energy photon fluence is strongly forward-directed.

4.7.2. *Phantom Studies*

The contour of the PMMA phantom with an overlaid projection of the 3D treatment plan is shown in Fig. 4.84 for a central transaxial plane. In Fig. 4.84(b), the radiation treatment plan for the pig leg is shown. A four-field box technique consisting of four scanned photon beams was used. The calculated isodose levels are shown as dashed lines in the range 10–90% of total dose 10 Gy in steps of 10%. The radiation field at

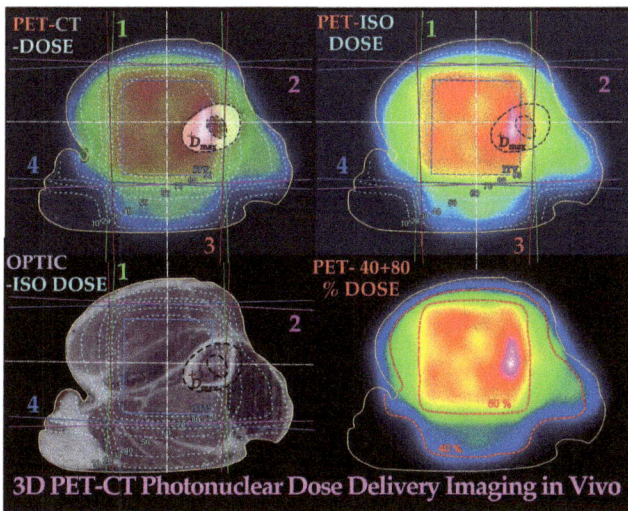

3D PET-CT Photonuclear Dose Delivery Imaging in Vivo

Figure 4.84. 3D PET-CT — ISO dose — optical imaging of a high-energy photon four-field box treatment technique on a frozen pig leg is shown. Totally 10 Gy, for each field a dose of 2.5 Gy, was given using a 50 MV scanned photon beam. (a) Tissue variations within the optical image cut out after treatment. (b) The pig leg was irradiated with a four-field box technique ($10 \times 10 \, cm^2$), marked as numbers from 1 to 4, where 1 denotes the first beam executed and 4 the last. (c) Overlay of PET image plane 12 for frame 1 and CT. (d) Overlay of PET image with activity contours (solid) for plane 12, frame 1 in the range 20–100% of maximum activity signal and isodoses (dashed) in the range 20–100% of total dose.

the PMMA phantom surface was $30 \times 30 \, cm^2$ and $10 \times 10 \, cm^2$ at isocenter on the pig leg. Irradiation time of the single scanned beam on the PMMA phantom was 313 s and the delivery of four scanned beams on the pig leg was completed in a time of 300 s including the extra time needed for gantry rotation. The delivered dose to both the PMMA phantom and the pig leg was 10 Gy at dose maximum (2.5 Gy per field in the pig leg study). The PET acquisition of the PMMA phantom was performed during a total measuring time of 60 min followed by a 15 min 2D transmission scan. The time interval between irradiation on the racetrack microtron and the acquisition of the pig leg was 3 min and 46 s. Positron activity measurements were performed during a total emission and transmission time of (60 + 30) min, split into 12 time frames. The duration of the first three frames were 2 min and the duration of the remaining nine frames 6 min each. During the pig leg study, image fusion was performed using fiducial markers containing radioactivity. Also, by using the transmission rather than emission PET data for image matching, fusion of PET and

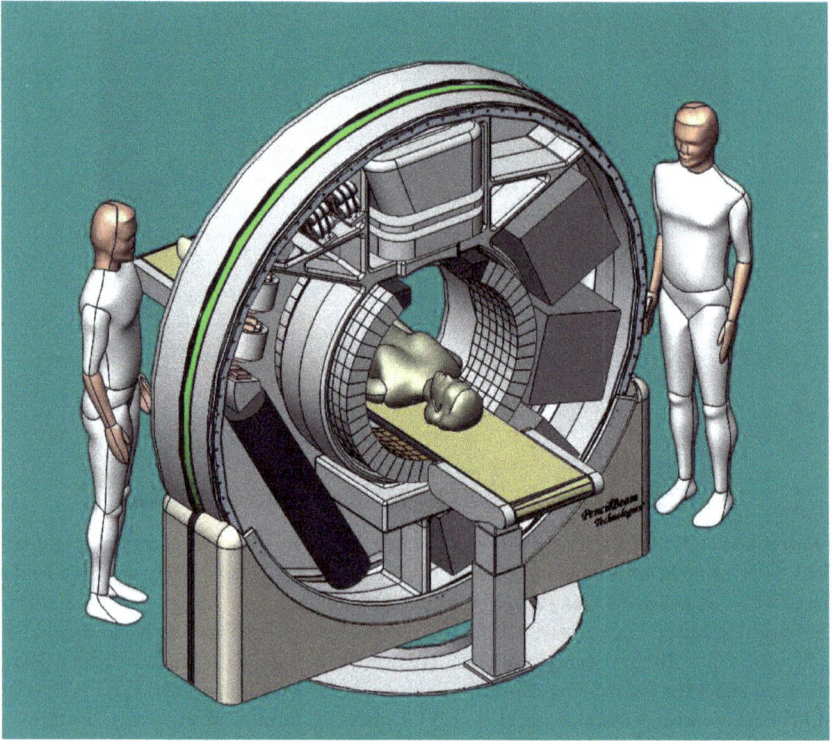

Figure 4.85.   Treatment unit with scanned photon and electron beams for BIOART type (biologically optimized *in vivo* predictive assay based radiation therapy) including a PET-CT camera. Both units include a 3D projection camera for auto setup of the patient with sub-millimeter accuracy and detectors capable of diagnostic and radiotherapeutic CT imaging (this treatment unit is available at TopGrade Healthcare, China).

CT images was easy. In both studies, conventional laser alignment of the object in the treatment room as well as in the PET camera was used.

4.7.3. *Dose Delivery Verification by PET-CT Imaging of Photonuclear Reactions during High-energy Photon Therapy*

Dose delivery monitoring after high-energy photon therapy has been tested using PET during the past decade. The technique is based on the activation of body tissues by high-energy bremsstrahlung photons, preferably with energies >20 MeV, resulting primarily in ^{11}C and ^{15}O but also ^{13}N, all positron-emitting radionuclide produced by photoneutron reactions in the normal parent nuclei ^{12}C, ^{16}O, and ^{14}N.

The induced positron activity distributions were measured in a PET-CT camera about 5–7 min after the end of the treatment. The accelerator used was the racetrack microtron at the Karolinska University Hospital using 50 MV scanned photon beams. Since the measured PET activity changes with time after the irradiation due to the different decay times of the produced radionuclides, the relative concentration of ^{12}C, ^{16}O, and ^{14}N in the irradiated volume could be separated. Most information is obtained from the radionuclides of carbon and oxygen, which are the most abundant elements in soft tissue. The predicted and measured overall positron activity from ^{11}C and ^{15}O agreed almost exactly (-3%), while the predicted low activity originating from nitrogen could be estimated within a factor two.

Based on the results obtained, a high value of a combined radiation therapy (RT-PET-CT) unit is really indicated to fully exploit the high activity signal from oxygen in water-rich tissues immediately after treatment and to avoid patient

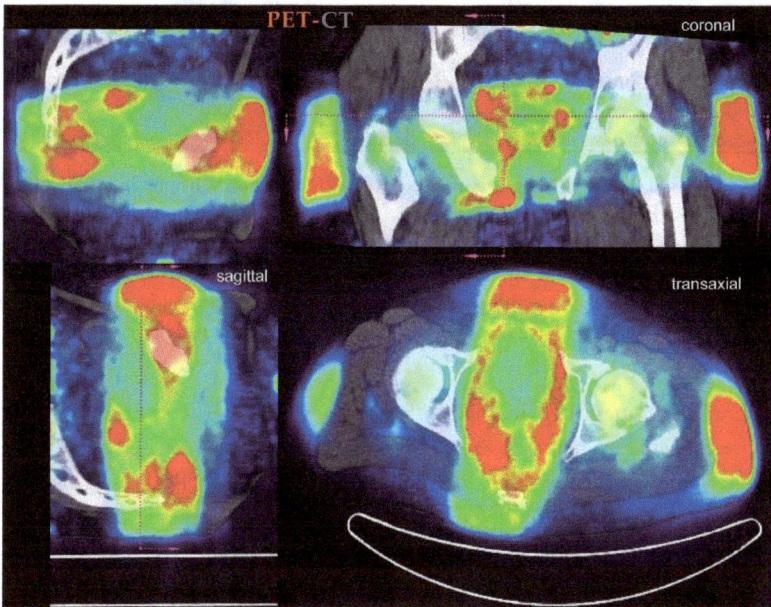

Figure 4.86. *In vivo* dose delivery imaging with 50 MV photon beams in a four-field box technique using the latest generation of PET-CT camera (64 slices CT and ≈4 mm PET resolution, cf. also Chapter 8, Fig. 8.16(b)). Here, the order of delivery can even be seen from the subcutaneous fat signal which is weakest in the posterior field and strongest in the right lateral field delivered last to the patient. Interestingly, the patient were very rapidly positioned on the diagnostic couch so the left side of the body was located outside the couch surface causing a substantial rotation of all the surrounding fatty soft tissues so that it looks like the whole four-field box technique is made with slightly oblique beams.

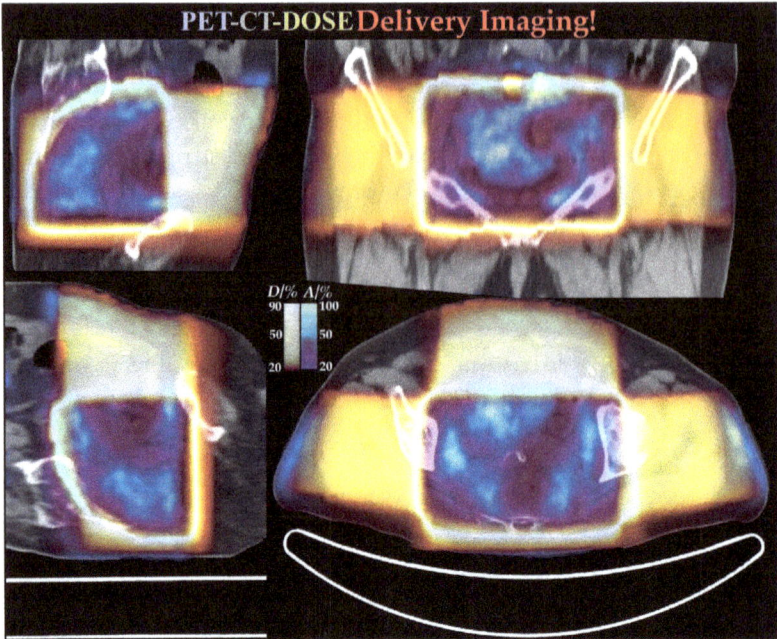

Figure 4.87. *In vivo* dose delivery imaging with 50 MV photon beams in a four-field box technique using the latest generation of PET-CT camera. The yellow to white color scale is from the treatment plan (slightly warped to account for the curved coach on the diagnostic PET-CT) with a whole in the high dose target volume to show the inside tissue activation in blue mainly due to carbon activity in adipose tissue and the intestine in deep purple since most of the oxygen activity have decayed here about 10 min after the end of the treatment.

repositioning as suggested in Fig. 4.85. With a RT-PET-CT unit, a good signal could be collected even at a dose level of a few Gray and the PET acquisition time could be reduced considerably. Dose delivery verification by PET imaging may be feasible on the 5% level provided physical and biological processes such as buildup, decay, and capillary blood flow containing mobile ^{15}O and ^{11}C in the activated tissues, can be taken into account. Finally, very interesting ensamples of clinical dose delivery imaging are shown in Figs. 4.86 and 4.87 really showing the huge potential of clinical dose delivery verification.

Bibliography

Ågren AK, Turesson I, Brahme A (1990) Optimization of uncomplicated control for head and neck tumors. *Int J Rad Oncol Biol Phys* 19:1077–1085.
American Association of Clinical Physics (1983) *Med Phys* 10:741.

American Association of Clinical Physics (1999) *Med Phys* 26:1847.

Anderberg B, Brahme A, Lindbäck S, Lundin L, Svensson H (1985) Unique development of radiation therapy equipment. *Med Teknik* 2:22 (in Swedish).

Andreassen B, Stråät Sj, Holmberg R, Näfstadius P, Brahme A (2011) Fast IMRT with narrow high energy scanned photon beams. *Med Phys* 38:4774–4784.

Andreassen B, Svensson R, Holmberg R, Danared H, Brahme A (2009) Development of an efficient scanning and purging magnet system for IMRT with narrow high energy photon beams. *Nucl Instr and Meth A* 612:201–208.

Andreo P, Brahme A (1986) Stopping power data for high-energy photon beams. *Phys Med Biol* 31:839–858.

Aucouturier J, Huber H et Jaoven J (1970) System de transport du faisceau d'electrons dans le sagittaire. *Rev Tech Thomson-CSF* 2:655.

Bortfeld Th, Bürkelbach J, Boesecke R, Schlegel W (1990) Methods of image reconstruction from projections applied to conformation radiotherapy. *Phys Med Biol* 35:1423–1434.

Bortfeld Th, Kalher DL, Waldron TJ, Boyer AL (1994) X-ray field compensation with multileaf collimators. *Int J Radiat Oncol Biol Phys* 28:723–730.

Brahme A (1972) On the optimal choice of scattering foils for electron therapy, Dept of Electron Phys. Royal Inst. of Technology, TRITA-EPP-72-17.

Brahme A (1975) Investigations on the application of a microtron accelerator for radiation therapy, thesis Dept Radiation Physics, Stockholm University, Sweden.

Brahme A (1977) Electron transport phenomena and absorbed dose distributions in therapeutic electron beams. *14th Int Congr Radiol*, Rio de Janeiro, Brazil.

Brahme A (1979) Scannig system for charged and neutral particle beams. Sv Pat 7904360-0 US4442352.

Brahme A (1980) Neutronkollimator. Sv. Pat. 8000215-7 US4463266.

Brahme A (1985) Multileaf collimator for electrons and photons, U.S. Patent application, US4672212.

Brahme A (1985) Developments of external beam treatment units and the role of high energy electrons and photons. *Teaching Lectures Third Eur Conf On Clin Onc and Cancer Nursing*, Stockholm, p. 46.

Brahme A (1987) Design principle and clinical possibilities with a new generation of radiation therapy equipment. *Acta Oncol* 26:403–412.

Brahme A (1988) Optimization of stationary and moving beam radiation therapy techniques. *Radiother Oncol* 12:129–140.

Brahme A (1995) Treatment optimization using physical and biological objective functions. In: Smith A (ed.), *Radiation Therapy Physics*, 209–246. Berlin: Springer.

Brahme A, Kraepelien T, Montelius A, Nordell B, Reuthal M, Svensson H, Walstam R (1979) Medical applications of a 50 MeV racetrack microtron I, electron, photon and photoneutron therapy. 5th ICMP Jerusalem.

Brahme A, Andreo P (1986) Dosimetry and quality specification of high energy photon beams. *Acta Radiol Oncol* 25:213–223.

Brahme A, Lind BK (1995) Development of treatment techniques for radiotherapy optimization. *Int J Imag Syst Techn* 6:33–42.

Brahme A, Svensson H (1976) Specification of electron beam quality from the central-axis depth absorbed dose distribution. *Med Phys* 3:95–102.

Brahme A, Svensson H (1979) Radiation beam characteristics of a 22MeV microtron. *Acta Rad Oncol* 18:244.

Brahme A, Källman P, Tilikidis A (1995) Development in ion beam therapy planning and treatment optimization. In: Linz U (ed.) *Ion Beams in Tumor Therapy*, 290–299. Weinheim: Chapman & Hall.

Brahme A, Kraepelin T, Svensson H (1980) Electron and photon beams from a 50 MeV Racetrack Microtron. *Acta Rad Oncol* 19:305.

Brahme A, Lind B, Källman P (1990) Inverse radiation therapy planning as a tool for 3D dose optimization. *Phys Med* 6:53–68.

Brahme A, Roos JE, Lax I (1982) Solution of an integral equation encountered in rotation therapy. *Phys Med Biol* 27:1221–1229.

Brahme A, Nyman P, Skatt B (2008) 4D laser camera for accurate patient positioning, collision avoidance, image fusion and adaptive approaches during diagnostic and therapeutic procedures. *Med Phys* 35:1670-1681.

Brahme A and Kempe J (2014) Particle transport theory and absorbed dose. In: Brahme A (ed.), *Comprehensive BioMedical Physics* Vol 8 Ch 2, Major Reference Work Elsevier Oxford.

Breit A (ed.) (1992) *Advanced Radiation Therapy: Tumor Response Monitoring and Treatment Planning*, Springer, Berlin, p. 523.

British Institute of Radiology (1983) *Br J Radiol* (Suppl 17).

Carlsson ÅK, Andreo P, Brahme A (1997) Monte Carlo and analytical calculation of proton pencil beams for computerized treatment plan optimization. *Phys Med Biol* 42:1033–1053.

Carol MP (1992) An Automatic 3D Treatment Planning and Implementation System for Optimized Conformal Therapy by the NOMOS Corporation. *Proc 34th Ann Meeting of the American Society for Therapeutic Radiology and Oncology.*

Carol MP (1995) Beam modulation in conformal radiotherapy. In: Purdy JA and Emami B (eds.), *3D Radiation Treatment Planning and Conformal Therapy*. Madison: Medical Physics Publishing.

Carpender JWJ, Skaggs LS, Lanzl LH, Griem ML (1963) Radiation therapy with high-energy electrons using pencil beam scanning. *Am J Roentgenol* 90:221–230.

Convery DJ, Rosenbloom ME (1992) The generation of intensity modulated fields for conformal radiotherapy by dynamic collimation. *Phys Med Biol* 37:1359–1374.

Cormack AM (1987) A problem in radiation therapy with X-rays. *Int J Radiat Oncol Biol Phys* 13:623–630.

Goward FK, Barnes DE (1946) Experimental 8 MeV synchrotron for electron acceleration. *Nature* 158:413.

Graffman S, Brahme A, Larsson B (1985) Proton radiotherapy with the Uppsala cyclotron. Experience and plans. *EORTC High-LET-therapy Group Meeting Munich Strahlentherapie* 161:764.

Greene D (1986) *Linear Accelerators for Radiation Therapy*. Adam Hilger Ltd.

Grusell E, Montelius A, Brahme A, Rikner G, Russel K (1994) A general solution to charged particle beam flattening using an optimized dual-scattering-foil technique, with application to proton therapy beams. *Phys Med Biol* 39:2201–2216.

Gund K, Paul W (1950) Experiments with a 6 MeV betatron. *Nucleonics* 7(1):36.

Gustafsson A, Lind BK, Brahme A (1994) A generalized pencil beam algorithm for optimization of radiation therapy. *Med Phys* 21:343–356.

Gustafsson A, Lind BK, Svensson R, Brahme A (1995) Simultaneous optimization of dynamic multileaf collimation and scanning patterns or compensation filters using a general pencil beam algorithm. *Med Phys* 22:1141–1156.

Henderson WJ, LeCaine H, Montalbetti R (1948) *Nature* 162:699.

Holmes T, Mackie RT, Simpkin D, Reckwerdt P (1991) A unified approach to the optimization of brachytherapy and external beam dosimetry. *Int J Radiat Oncol Biol Phys* 20:859–873.

Hsieh CL, Uhlmann EM (1956) Experimental evaluation of the physical characteristics of a 45-MeV medical linear electron accelerator. *Med Linear Electron Accelerator* 67: 263–272.

ICRU-35 (1984) Radiation dosimetry: electron beams with energies between 1 and 50 MeV. *International Commission on Radiation Units and Measurements Report* 35, Bethesda, MD, ICRU.

International Atomic Energy Agency, IAEA (1997) *Technical Report series No 381*, Vienna.

Källman P, Lind BK, Brahme A (1992) An algorithm for maximizing the probability of complication free tumor control in radiation therapy. *Phys Med Biol* 37:871–890.

Källman P, Lind BK, Eklöf A, Brahme A (1988) Shaping of arbitrary dose distributions by dynamic multileaf collimators. *Phys Med Biol* 33:1291–1300.

Kapitza SP, Bykov VP, Melechin VN (1960) *Z Eksp Teor Fiz (JETP)* 39:997.

Karlsson M, Nyström H (1993) IAEA-SM-330/54, Book of Abstracts International Symposium on Measurement Assurance in Dosimetry IAEA-SM-330.

Karzmark CJ (1984) Advances in linear accelerator design for radiotherapy. *Med Phys* 11:105–128.

Karzmark CJ, Morton RJ (1998) A Primer on Theory and Operation of Linear Accelerators in Radiation Therapy. Medical Physics Publishing Corporation.

Karzmark CJ, Nunan CS, Tanabe E (1993) *Medical Electron Accelerators*. McGrawHill, New York.

Kerst DW (1941) The acceleration of electrons by magnetic induction. *Phys Rev* 60:47.

Kerst DW (1975) Betatron-Quastler era at the University of Illinois. *Med Phys* 2:297.

Kijewski PK, Chin LM, Bjärngard BE (1978) Wedge-shaped dose distributions by computer-controlled collimator motion. *Med Phys* 5:426.

Larsson B, Liden K, Sarby B (1974) Irradiation of small structures through the intact skull. *Acta Radiol* 13:512–534.

Lax I, Brahme A (1980) On the collimation of high-energy electron beams. *Acta Radiol Oncol* 19:199–207.

Lind B (1990) Properties of an algorithm for solving the inverse problem in radiation therapy. *Inv Probl* 6:415–426.

Lind BK (1991) Radiation therapy planning and optimization studied as inverse problems, thesis, Stockholm University, Sweden.

Lind BK, Brahme A (1985) Generation of Desired Dose Distributions with Scanned Elementary Beams by Deconvolution Methods. *Proc. VII ICMP*, Espoo Finland, p. 953.

Lind BK, Brahme A (1987) Optimization of Radiation Therapy Dose Distributions with Scanned Photon Beams. *Proc Ninth Int Conf on the Use of Computers in Radiation Therapy*, Bruinvis IAD *et al.* (eds), Elsevier, Amsterdam, pp. 235–239.

Lind BK, Källman P, Sundelin P, Brahme A (1993) Optimal radiation beam profiles considering the uncertainties in beam patient alignment. *Acta Oncol* 32:331–332.

Löf J, Lind BK, Brahme A (1995) Optimal radiation beam profiles considering the stochastic process of patient positioning in fractionated radiation therapy. *Inv Probl* 11: 1189–1209.

Löf J, Lind BK, Brahme A (1997) A general code for dynamic and stochastic optimization of radiotherapeutic treatment plans. *Med Biol Engin Comp* 35:920.

Löf J, Lind BK, Brahme A (1998) An adaptive control algorithm for optimization of intensity modulated photon therapy considering uncertainties in beam profile, patient setup, and internal organ motion. *Phys Med Biol* (accepted).

Mackie TR, Holmes T, Swerdloff S, Reckwerdt P, Deasy JO, Yang J, Paliwal B, Kinsella T (1993) Tomotherapy: a new concept for the delivery of dynamic conformal radiotherapy. *Med Phys* 20:1709–1719.

McMillan EM (1945) The syncrotron — a proposed high energy particle accelerator. *Phys Rev* 68:143.

Mejaddem Y, Söderström S (1998) A new system for optimized intensity modulated few photon beam portals using computerized moulding filter techniques. (In progress).

Mejaddem Y, Lax I, Adakkai S (1997) Procedure for accurate fabrication of tissue compensators with high-density material. *Phys Med Biol* 42:415–421.

Metcalfe P, Kron T, Hoban (1997) *The Physics of Radiotherapy from Linear Accelerators*, Medical Physics Publishing.

Misaghi-Panah (2000) Master of Science thesis, Dept Rad Phys Karolinska Institutet.

Mitchell JS, Smith CL, Allen-Williams DJ, Braams R (1953) Experience with the 30 MeV synchrotron as a radiotherapeutic instrument, From the Dept Radiotherapeutics (Director Prof JS Mitchell), University of Cambridge.

Montelius A, Brahme A, Eenmaa J, Lindbäck S, Wootton P (1984) Dose measurements in a p(50)+Be neutron beam using a continuously variable multi-leaf collimator. Strahlentherapie 160:114.

Montelius A, Rikner G (1992) *Radiother Oncol* 24 (Suppl), Abstract No 390.

Montelius A, Jung B, Rikner G, Murman A, Russell K In advanced radiation therapy.

Näfstadius P, Brahme A, Nordell B (1984) Computed assisted dosimetry of scanned electron and photon beams for radiation therapy. *Radiother Oncol* 2:261–269.

Nilsson B (1985) Thesis, Department of Radiation Physics, Stockholm, Sweden.

Nilsson B, Brahme A (1979) Absorbed dose from secondary electrons in high-energy photon beams. *Phys Med Biol* 24:901–912.

Nilsson B, Brahme A (1981) Contamination of high-energy photon beams by scattered photons. *Strahlentherapie* 157:181–186.

Nilsson B, Brahme A (1986) Electron contamination from photon beam collimators. *Radiother Oncol* 5:235–244.

Nilsson B, Sorcini B (1989) *Acta Oncol* 28:537.

Nordell B (1985) *Phys Med Biol* 30:139.

Nordell B, Brahme A (1984) Angular distribution and yield from bremsstrahlung targets. *Phys Med Biol* 29:797–810.

Nordic Association of Clinical Physics (1980) *Acta Radiol Oncol* 19:55.

Paul P (2001) Master of Science thesis, Dept Rad Phys Karolinska Institutet.

Proimos BS (1966) Beam-shapers oriented by gravity in rotational therapy. *Radiology* 87(5):928–932.

Rassow J (1970) Beitrag zur Electronentiefen-Therapie mittels Pendelbestrahlung. *Strahlentherapie* 140:156.

Redpath AT, Vickery B, Wright DH (1976) A new technique for radiotherapy planning using quadratic programming. *Phys Med Biol* 21:781–791.

Rosander S, Sedlacek M, Wernholm O (1982) The 50 MeV racetrack microtron at the Royal Institute of Technology Stockholm. *Nucl Instr Meth* 204:1.

Scharf W (1994) *Biomedical Particle Accelerators*, American institute of Physics, New York.

Schwinger JS (1945) Harvard lectures in LI Schiff, 1946. *Rev Sci Instr* 17:9.

Sedlacek M (1956) Stockholm conference. *G. Borelius and E. Rudberg Ark för Fys* 11:129.

Slepian US Patent 1.645.304.

Sorcinit BB, Andreo P, Bielajew AF, Hyodynmaa S, Brahme A (1995) An improved energy-range relationship for high-energy electron beams based on multiple accurate experimental and Monte Carlo data sets *Phys Med Biol* 40:1135–1159.

Söderström S, Brahme A (1993) Optimization of the dose delivery in few field techniques using radiobiological objective functions. *Med Phys* 20:1201–1210.

Söderström S, Brahme A (1995) Which is the most suitable number of photon beam portal in coplanar radiation therapy. *Int J Radiat Onc Biol Phys* 33(1):151–159.

Söderström S, Eklöf, Brahme A (1999) Aspects on the optimal photon beam energy for radiation therapy. *Acta Oncol* 38:179–187.

Söderström S, Gustafsson A, Brahme A (1993) The clinical value of different treatment objectives and degrees of freedom in radiation therapy optimization. *Rad Oncol* 29:148–163.

Spirou SV, Chui CS (1994) Generation of arbitrary intensity profiles by dynamic jaws or multileaf collimators. *Med Phys* 21:1031–1041.

Staël von Holstein J, Andreo P, Karlsson, M, Nyström, H (1992) *Book of Abstracts*, Sv Läkaresällskapets Riksstämma, p. 258 (in Swedish).

Stein J, Bortfeld T, Dörchel B, Schlegel W (1994) Dynamic X-ray compensation for conformal radiotherapy by means of multileaf collimation. *Rad Oncol* 32:163–173.

Svensson R, Brahme A (1996) Effective source size, yield and beam profile from multilayered bremsstrahlung targets. *Phys Med Biol* 41:1353–1379.

Svensson R, Åsell M, Näfstadius P, Brahme A (1998) Design of target, purging magnet and electron collector for scanned high energy photon beams. *Phys Med Biol* 43:1091–1112.

Svensson R, Källman P, Brahme A (1994) An analytical solution for the dynamic control of multileaf collimators. *Phys Med Biol* 39:37–61.

Svensson R, Larsson S, Gudowska I, Holmberg R, Brahme A (2007) Design of a fast multi-leaf collimator for radiobiologically optimized IMRT with scanned beams of photons, electrons and light ions. *Med Phys* 34:877–888.

Tuve, Breit (1927–1928) *Carnegie Year Book*, Volume 27, p. 209.

Ulsø N, Brahme A (1988) Computer aided irradiation technique optimization. *In US-Scandinavian Symposium on Computer aided Radiation Therapy*, Zink S (ed.), San Antonio, Texas NCI Bethesta, MD 20892.

van Santvoort JPC, Heijmen BJM (1996) Dynamic multileaf collimation without "tounge-and-groove" underdosage effects. *Phys Med Biol* 41:2091–2105.

van Santvoort JPC, Binnecamp D, Heijmen BJM, Levendag PC (1995) Granulate of stainless steel as compensator material. *Radiother Oncol* 49:450–457.

Veksler VI (1944) *Doklady Akad Nauk SSSR* 43:346.

Walz BJ, Perez CA, Feldman A, Demindeck AJ, Powers WE (1973) Individualized compensating filters and dose optimization in pelvic irradiation. *Radiology* 107: 611–614.

Webb S (1994) Optimizing the planning of intensity-modulated radiotherapy. *Phys Med Biol* 39:2229–2246.

Wernholm O, Rosander S, Sedlacek M, Babic H (1968) in *Rosander S, TRITA-EPP-74-16, Royal Inst. Techn.* Stockholm, Sweden.

Wideröe R (1928) *Arch Electrotech* 2:400.

Wiik BH, Wilson PB (1967) Designs of a high energy, high-duty cycle ractrack microtron. *Nucl Instr Meth* 56:197.

Wiik BH, Wilson PB (1970) The racetrack microtron: an electron accelerator for medium energies. In: *Linear Accelerators*, B3.6553, Amsterdam, North Holland. Yorke ED, Ling CC, Rustgi S (1985) *Phys Med Biol* 30:1305.

Fundamentals of Physically and Biologically Based Radiation Therapy Optimization

5

Anders Brahme, Johan Löf, and Bengt K. Lind

5.1. Introduction

Radiation therapy for cancer is today going through a very dynamic development with the introduction of a large number of new treatment principles, new types of treatment units, and new radiobiologically based optimization algorithms for treatment planning. All of these make use of the recent developments in 3D tumor diagnostics, molecular biology of cancer, the fractionation sensitivity of different tissues, and most recently predictive assays of radiation sensitivity. The most efficient but also least developed area of treatment optimization is to use a few ($\sim$3) nonuniform radiation beams directed toward the tumor. Today, patient individual collimation with beam blocks or multileaf collimators protect organs at risk laterally outside the tumor volume. Nonuniform dose delivery also allows protection of normal tissues anterior, posterior, and even inside the target volume by shaping the isodoses tightly around the tumor tissues and thereby also allowing longitudinal protection of normal tissues. Some of the most advanced new algorithms are even treating therapy optimization as an inverse problem where the optimal incident beam shapes are determined directly from the location of gross disease, presumed microscopic tumor spread, and organs at risk. The optimization is then performed such that the probability, P_+, to eradicate all clonogenic tumor cells without severely damaging healthy normal tissues is as high as possible. Already with a few nonuniform beams, the treatment outcome is within a few percentage of what can be achieved with infinitely many coplanar beams in a dynamic mood. In fact, such few field techniques, besides being simpler, more reliable and robust to deliver, will also minimize the damage to normal tissues because of the recently demonstrated sensitivity increase of many healthy normal tissues at low doses (Joiner 1994). With such optimized nonuniform treatments it should be possible to improve the treatment outcome by as much as 20% and more, particularly in patients with a complex local spread of the disease or several organs at risk.

5.2. Fundamentals of Treatment Optimization

5.2.1. *Integral Equations of Radiation Therapy Planning*

The principal problem of radiation therapy planning can be formulated in the form of an integral equation that expresses the resultant dose distribution in the patient for a given incoming radiation field. The most elementary incident radiation beam is a point monodirectional beam $p^m(E, \boldsymbol{\Omega}, r, \rho)$, that describes the energy deposition at r for a given energy E, point ρ, and direction $\boldsymbol{\Omega}$ of incidence and modality or type of uncharged radiation beam such as photons, neutral pions, and neutrons (m). The absorbed dose at a point r in the patient is then given by an integral over the incident energy fluence, differential in energy and angle of such incident beams on points p on the patient surface:

$$D(r) = \oint_S \iiint \sum_m p^m(E, \boldsymbol{\Omega}, r, \rho) \Phi^m_{E,\Omega}(\rho) \mathrm{d}E \mathrm{d}\boldsymbol{\Omega} \mathrm{d}\rho^2, \qquad (5.1a)$$

where the spatial integrals have to be performed over the relevant entrance surface, S, of the patient (cf. Fig. 5.1 and Gustafsson *et al.* 1994, 1995; Brahme *et al.* 1995). The most suitable unit for p^m is for neutral particles the mean specific energy $\bar{z}$ (J kg^{-1}) imparted per unit incident radiant energy R (J), and thus per kilogram. For Ψ^m the most suitable unit is the increment in incident radiant energy dR (J) per unit area dα (m^2), that is, J m^{-2} (ICRU 1980; Lind and Brahme 1992).

For incident charged particles, such as electrons, protons, and heavy ions, it is more natural to use the fluence of the incoming particles rather than the energy fluence as used in Eq. (4.1a). The pencil beam energy deposition equation then takes the form:

$$D(r) = \oint_S \iiint \sum_m \pi^m(E, \boldsymbol{\Omega}, r, \rho) \Phi^m_{E,\Omega}(\rho) \mathrm{d}E \mathrm{d}\boldsymbol{\Omega} \mathrm{d}\rho^2, \qquad (5.1b)$$

where $\Phi^m_{E,\Omega}(\rho)$ is the incident particle fluence differential in energy and angle at a point ρ on the patient surface. The energy deposition kernel $\pi^m(E, \boldsymbol{\Omega}, r, \rho)$ is the mean specific energy imparted at r per charged particle of energy E, direction Ω, and particle type m incident at ρ. For Φ^m, the most suitable unit is per square meter (ICRU 1980) and for π^m it is J kg^{-1}.

For mixed incident radiation fields of neutral and charged particles, Eqs. (5.1a) and (5.1b) may be combined as a sum over all incident radiation modalities. When

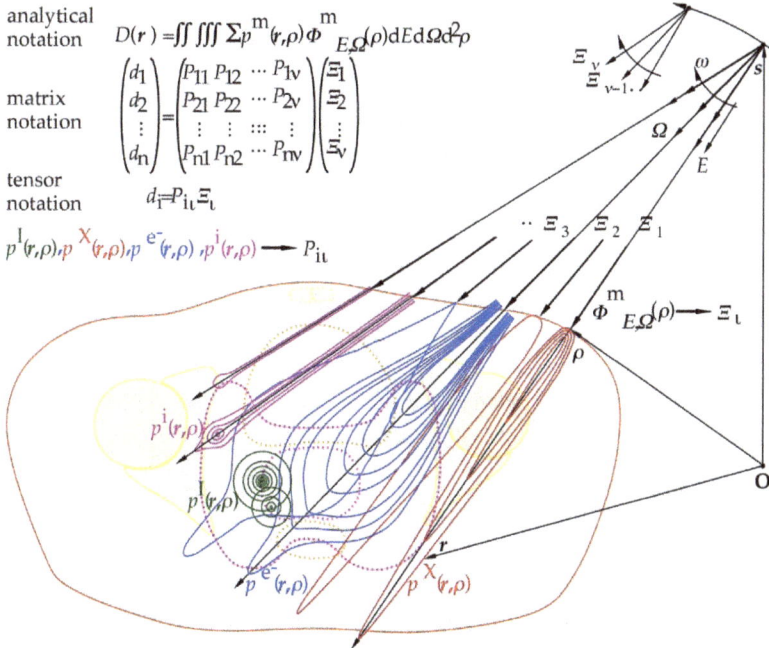

Figure 5.1. Illustration of irradiation geometry used in the optimization of the total dose distribution in the patient delivered by the fluence $\Phi^m_{E,\Omega}$ of electron pencil beams p^{e^-}, and the energy fluence $\Psi^m_{E,\Omega}$ of photon pencil beams p^X. Also illustrated in the figure are ion pencil beams p^i and intracavitary sources p^I. Through the use of accurately calculated pencil beams even taking patient inhomogeneities into account, a very strict optimization is possible considering all major constraints on the dose delivery.

the time dependence is relevant, the temporal variation of D, $\Psi^m_{E,\Omega}$ and $\Phi^m_{E,\Omega}$ also has to be included in this combined equation. The principal unknown quantities to be determined in the optimization of the treatment plan are the incident energy fluence $\Psi^m_{E,\Omega}(p)$ for neutral particles and the incident fluence $\Phi^m_{E,\Omega}(p)$ for charged particles (cf. ICRU 1980; Lind and Brahme 1992; Liu et al. 1993).

For indirectly ionizing particles such as photon or neutron beams, it is more natural to use Eq. (5.1a) in a differential form along the incident rays and then p is replaced by the *point energy deposition kernel*, h, which describes the specific energy distribution by a primary interaction at a point r' in the patient (cf. Mohan et al. 1986; Ahnesjö et al. 1987). The resultant dose distribution then becomes:

$$D(r) = \iiint_V \iiint h(E, \boldsymbol{\Omega}, r, r') f_{E,\Omega}(r') \mathrm{d}E \mathrm{d}\boldsymbol{\Omega} \mathrm{d}r'^3, \qquad (5.2)$$

where the spatial integral has to be evaluated over the whole irradiated volume where the kernel or irradiation density f is larger than zero. The unit for h is the same as for p but the kernel or irradiation density f is expressed in units of the increment in incident radiant energy dR per unit volume dV (cf. Lind and Brahme 1992).

Classical radiation therapy planning is concerned with the evaluation of this type of integral for a given beam kernel p or h and incident energy fluence Ψ or kernel density f, and the resultant dose distribution may be obtained by a direct evaluation of the integrals. This classical approach is, therefore, a forward problem (Fig. 5.2 upper left panel). In treatment optimization, the problem is reversed as again p or h is known, but we need to find the optimal energy fluence Ψ, which generates a desired dose distribution D, or preferably the optimal dose distribution D^* as determined by the treatment objectives. A related well-posed problem is to determine the energy fluence Ψ for a known kernel p and measured dose distribution D (cf. Table 5.1

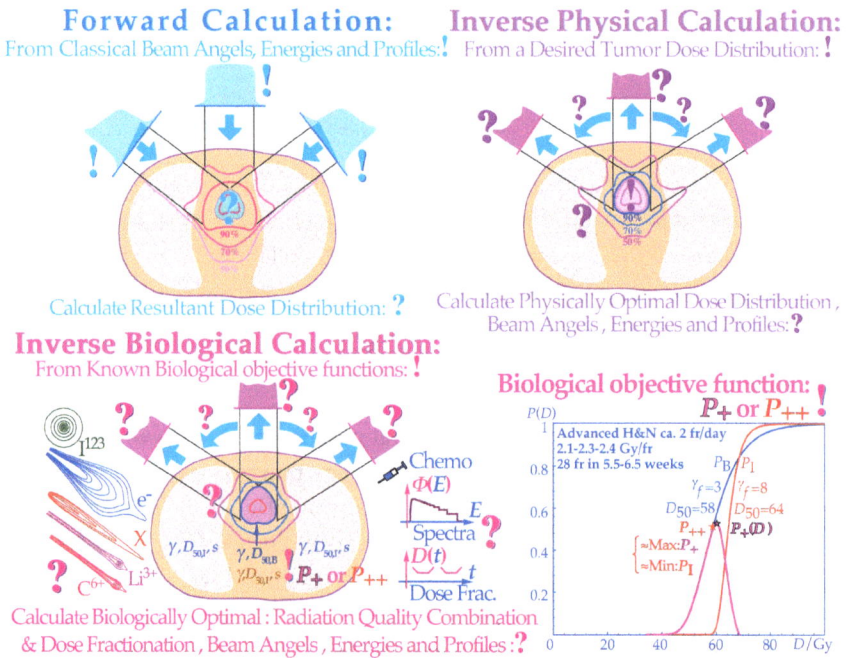

Figure 5.2. Schematic illustration of the difference between conventional forward radiation therapy planning (upper left) and inverse planning. Dose optimization using forward planning is generally a trial and error process, whereas inverse planning directly results in optimal beam profiles and isodose distributions (upper right) with physical or biological objective functions (lower right). The lower left panel indicates the radiobiological data needed for the biological objective function.

Table 5.1. Inverse problems in radiation therapy planning.

Field	Objective	Kernel	Desired quantity	Relation
Brachy-therapy	Generation of a desired dose distribution: $D(\vec{r})$	Point source dose distribution: $d_p(\vec{r})$	Optimal source density distribution: $\varphi(\vec{r})$	$D(\vec{r}) = \iiint d_p(\vec{r} - \vec{\varrho})\varphi(\vec{\varrho})d^3\varrho$
Arc & conformation therapy	Generation of a desired dose distribution: $D(\vec{r})$	Dose distribution in convergent pencil beam point irradiation: $d_c(\vec{r})$	Point irradiation density: $F(\vec{r})$, and incident beam profile	$D(\vec{r}) = \iiint d_c(\vec{r}, \vec{\varrho})F(\vec{\varrho})d^3\varrho$ Fredholm eq. of first kind
External beam therapy	Generation of a desired depth dose curve: $D(z)$	Monocromatic depth dose curve: $d(E, z)$	Optimal incident particle spectrum: Ψ_E	$D(z) = \int d(E,z)\Psi_E\,dE$ Fredholm eq. of first kind
Photon therapy	Determine photon beam spectrum from transmission: $\Psi(z)$	Energy dependence of attenuation coefficient: $\mu(E)$	Incident photon spectrum: $\Psi_E(0)$	$\Psi(z) = \int \Psi_E(0)e^{-\mu(E)z}dE$ Fredholm eq. of first kind
Electron therapy	Determine dose distribution from incident electron fluence: Φ_Θ	Scatter distribution in a thin layer: $\varphi_{\Delta r}(\Theta)$	Spatial dose and fluence distribution: $\Phi_\Theta(\vec{r})$	$\Phi_\Theta(\vec{r} + \Delta\vec{r})$ $= \int \Phi_\Theta(\vec{r})\varphi_{\Delta r}(\Theta - \Theta')d\Theta'$
Beam flattening Beam compensation	Generation of a desired lateral dose distribution: $D(x,y)$	Elementary beam dose distribution: $d(x,y)$	Optimal scanning density distribution: $F(x,y)$	$D(x,y)$ $= \iint d(x - \xi, y - \eta)F(\xi, \eta)d\xi\, d\eta$

(*Continued*)

Table 5.1. (*Continued*)

Field	Objective	Kernel	Desired quantity	Relation
Photon dose planning	Determine dose distribution from incident photon fluence and resultant terma: $T(\vec{r})$	Mean energy imparted point spread function: $h(\vec{r})$	Resultant dose distribution: $D(\vec{r})$	$D(\vec{r}) = \iiint h(\vec{r}-\vec{\varrho})T(\vec{\varrho})\mathrm{d}^3\varrho$
Photon dose delivery	Generate a desired dose profile in a patient: $D(x,y)$	Dose distribution in pencil beam: $d_\ell(x,y)$	Fluence profile in the incident beam: $\Phi(\xi,\eta)$	$D(x,y)$ $= \iint d_\ell(x-\xi,y-\eta)\Phi(\xi,\eta)\mathrm{d}\xi\,\mathrm{d}\eta$
External beam therapy	Generation of a desired lateral dose distribution: $D(\vec{r})$	Collimated slit beam dose distribution: $H(\vec{r})$	Collimator opening density: $F(\vec{r})$	$D(\vec{r}) = \iiint F(\vec{\varrho})H(\vec{r}-\vec{\varrho})\mathrm{d}^3\varrho$
Dosimetry	Determine true dose distribution from a measurement: $D_m(x)$	Detector response function: $S(x)$	True dose distribution: $D(x)$	$D_m(x) = \int S(x-u)D(u)\mathrm{d}u$
Tumor imaging in nuclear medicine	Determine true uptake & increase resolution in measured distribution: $I(\vec{r})$	Point source response: $i_p(\vec{r})$	True uptake: $U(\vec{r})$	$I(\vec{r}) = \iiint i_p(\vec{r}-\vec{\varrho})U(\vec{\varrho})\mathrm{d}^3\varrho$

and Ahnesjö and Trepp 1991). A third situation may also be of interest, namely when the dose distribution D and the irradiation density f or energy fluence Ψ are known and, in principle, it should be possible to determine the associated point energy deposition kernel h or the pencil beam kernel p (cf. Ahnesjö 1984; Chui and Mohan 1988, respectively). These two latter adjoint problems are, in mathematical terminology, often called inverse problems since the unknown is implicitly defined and an inversion of the integral is required to find the solution (Fig. 5.2 upper and lower right panels). In the mathematical literature Eqs. (5.1) to (5.4) are generally called Fredholm integral equations of the first kind (cf. Table 4.1 and Miller 1974).

In a *monodirectional* photon or neutron beam, the equation may for simplicity be integrated over all angles to obtain the fundamental equation of kernel-based radiation therapy planning with indirectly ionizing radiations. In this equation, h is again the point energy deposition kernel but f is often given as the TERMA differential in energy T_E (total energy released per unit mass, to be distinguished from the kinetic energy released per unit mass as specified by the KERMA) and the resultant dose distribution becomes:

$$D(r) = \iint \iint h(E, r, r')\rho(r')T_E(r')\mathrm{d}E \ \mathrm{d}r'^3.$$ (5.3)

This equation is the base for modern forward treatment planning algorithms (cf. Ahnesjö et al. 1987 and Holmes et al. 1991, where the density ρ was included in h for simplicity). By comparison with Eq. (5.2), the irradiation density $f_{E,\Omega}$ can in unit density materials be identified as the TERMA differential both in energy and angle $T_{E,\Omega}$.

If we instead integrate Eq. (5.2) over both energy and angle and assume the incident energy fluence distribution to be spatially invariant in these variables at least over the target volume, the basic equation for inverse kernel-based therapy planning is obtained:

$$D(r) = \iiint h(r - r')f(r')\mathrm{d}r'^3.$$ (5.4)

Here, the kernel h is the mean specific energy distribution per unit incident radiant energy of the elementary radiation field centered at r' and f is the irradiation or kernel density (for more details, see Lind and Brahme 1992, and a related equation for heavy ions formulated by Leemann et al. 1977). This latter convolution-type equation is much easier to handle mathematically than Eqs. (5.1) to (5.3) and it has been extensively used in kernel-based treatment optimization in recent years.

However, the inversion of Eq. (5.4) in radiation therapy is associated with serious mathematical and numerical difficulties. The principal problem is that both $D(r)$ and $f(r)$ are physical quantities (the absorbed dose and the irradiation density, respectively) which by necessity are larger than or equal to zero. The straightforward solution of the integral Eq. (5.4) using Fourier transformation (cf. Brahme 1988) will therefore only work under very special conditions as was shown by Brahme *et al.* (1982).

In fact, for most dose distributions, D, of clinical interest, and for the associated energy deposition kernels, h, there does not even exist an exact solution with $f \geq 0$ which fulfills Eq. (5.4) over the entire volume of the patient. However, over a small volume such as the target volume, it may be possible to find an exact solution, provided their radiation kernel h is appropriately chosen and well confined relative to the desired dose distribution D. These problems make it necessary to solve Eq. (5.4) by approximate methods or by reformulating the principal problem of radiation therapy planning, as will be discussed in more detail in the subsequent sections. For completeness, some of the most important optimization problems of radiation therapy planning are summarized in Table 5.1. Most of them are linked to integral equations like Eqs. (5.1) to (5.4) and they can be handled by the methods discussed in this chapter.

5.2.2. *Well-posed and Ill-posed Optimization Problems*

From the above discussion of the integral equations, it is clear that an exact solution of the inverse problem of radiation therapy planning as formulated by the Eqs. (5.1) to (5.4) is far from trivial and in general is an impossible task. To proceed, there are two alternatives: either the clinical objective to generate a desired dose distribution has to be reformulated, to get a more well-posed formulation, or some numerical method has to be developed in order to solve the inverse problem as accurately as possible.

In either case, it is very useful to transform the relevant integral equation into an algebraic form by discretizing the transport quantities along the coordinates of the free variables. In this discretized form, it is natural to describe the dose distribution by a vector d where its components d_i run over the volume of interest of $D(r)$ in the 3D Cartesian space $\mathbf{R}^3$. With a uniform m-fold sampling along each coordinate, i will run from 1 to $n = m^3$. The convolution integral in Eq. (5.4) may then, for example, be described as a matrix multiplication between the convolution matrix, H, built up by the kernel vector h and the irradiation density vector f. The convolution matrix is a $m^3 \times m^3$, the so-called Toeplitz matrix, where the kernel vector h is shifted one position for each new row in the matrix. With this notation, Eq. (5.4) may be rewritten as:

$$d = Hf.\tag{5.5}$$

For the more complex case of a kernel that changes shape depending on where it is centered (cf. Eq. (5.2), the kernel vector of the convolution matrix will also change from row to row.

Whether we would like to reformulate the problem or find an approximate solution to Eq. (5.4), the best possible solution has to be found using some optimization procedure. To find the best possible solution to the integral equation, the deviations between the desired dose d and the best achievable (Hf) should be minimized according to the most appropriate measure. This measure could either be physical in terms of the deviations in the absorbed dose delivered $(d - Hf)$ or biological when we are trying to quantify the influence of the dose deviations on tumor control and normal tissue reactions. To allow a strict optimization, a scalar measure is required which accounts for all deviations of the components $d_i - (Hf)_i$. Most generally this weighting is mathematically described by a functional F, that is, a mapping from $\mathbf{R}^n$ to $\mathbf{R}^1$. In mathematical terms, we thus have to minimize some functional F of the dose deviations (cf. Eqs. (5.27) to (5.35) below) according to:

$$\begin{cases} \min F(d - Hf) \\ f_i \geq 0 \end{cases}.\tag{5.6}$$

The systematic solution of this problem is discussed in Sec. 5.4 below.

The second and generally preferred alternative is to restate the treatment objectives in order to get a well-posed problem. Instead of trying to find the best way to produce a desirable dose distribution, this goal could be achieved by trying to find the dose distribution which cures the patient with minimal risk for treatment-related morbidity and severe adverse reactions in normal tissues. It is clear that one way to get a well-posed problem is to formulate the clinically relevant treatment objectives as precisely as possible. Owing to the uncertainties in: (a) diagnoses, (b) tumor and target volume specification, (c) identification of organs at risk, and (d) treatment setup, we can at best express the probability to achieve the treatment objectives in the treatments by fractionated radiotherapy. If we, for the moment, call the associated functional P_+, the well-posed problem of radiation therapy may be formulated as

$$\begin{cases} \min F_+^n(d - Hf) \\ f_i \geq 0 \end{cases}.\tag{5.7}$$

In the subsequent sections, we will give examples of the functionals F and F_+^n (cf. Eqs. (5.11)–(5.26)) and discuss algorithms capable of finding the optimal f (cf. Eqs. (5.57), (5.61)–(5.65)). If we replace Hf by the more general integral from Eqs. (5.1) to (5.4), we can see that the optimization problem of radiation therapy can be described as a variational problem where we want to find the incident beam combination which maximizes the functional that specifies the treatment objectives. This kind of formulation will be applied in the special case when the density of clonogenic tumor cells is approximately known, for example, by quantitative diagnostic imaging (cf. Sec. 5.4.5).

5.2.3. *Degrees of Freedom in Radiation Therapy Optimization*

Since the beginning of radiation therapy, a very extensive number of methods, beam qualities, and irradiation techniques have been developed. Throughout this period, photon beams from external radiation sources have dominated the field closely followed by external electron beams and intracavitary and interstitial therapy, with sealed sources. During recent decades, heavy charged particle therapy with neutrons, protons, and heavy ions have also been extensively used. The optimal use of these more exotic beam modalities has been discussed at length in the literature (Raju 1980; Fowler 1981; Brahme 1982; Brahme *et al.* 1991; Källman 1992; Linz 1995 see Chapter 8 for further details) and will therefore not be further discussed here. Similarly, brachytherapy will be left out due to the greatly differing irradiation techniques even if most of the optimization methods that are discussed here apply to both these latter modalities (Brahme 1988; Brahme *et al.* 1990; Holmes *et al.* 1991).

In addition to the choice of radiation modality, the degrees of freedom include dose fractionation schedule, beam energy, beam directions, beam collimation, beam profiles, and the irradiation technique in general as determined by the type of equipment used. Figure 5.3 compares the degrees of freedom in classical treatment optimization as it was developed in the mid-1960s (Hope and Orr 1965; Sterling *et al.* 1965; Bahr *et al.* 1968; Redpath *et al.* 1976; Ebert 1977) with present-day full-blown generalized conformal therapy (Brahme 1988; Bortfeld *et al.* 1990; Brahme *et al.* 1990; Webb 1990; Holmes *et al.* 1991). The former used rectangular wedged fields from a large number, n, of beam directions even though for practical reasons the number was generally <20. Classical conformation therapy as developed by Takahashi (1965) instead used a continuum of generally uniform or blocked beams conforming to the target volumes and the organs at risk as seen from the point of view of the beam source. The ultimate step in the therapy development is to allow full freedom in the shape of the delivered beams with regard to beam

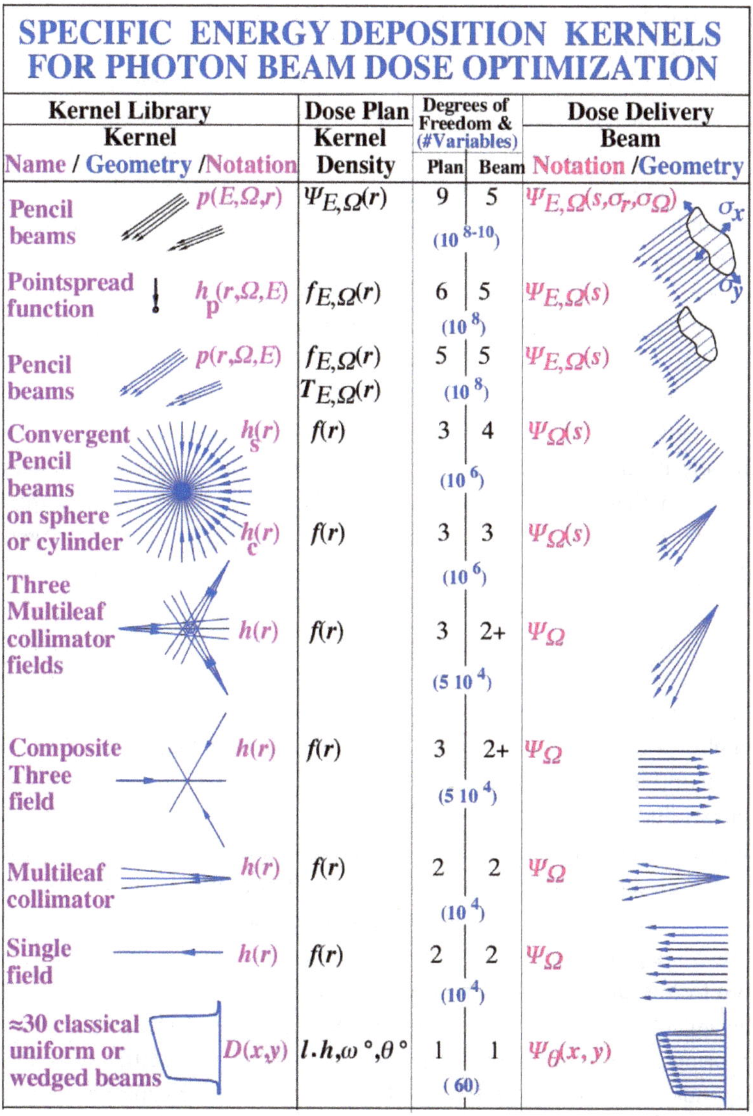

SPECIFIC ENERGY DEPOSITION KERNELS FOR PHOTON BEAM DOSE OPTIMIZATION

Kernel Library — Kernel — Name / Geometry /Notation	Dose Plan — Kernel — Density	Degrees of Freedom & (#Variables) — Plan	Beam	Dose Delivery — Beam — Notation /Geometry
Pencil beams $p(E,\Omega,r)$	$\Psi_{E,\Omega}(r)$	9	5 ($10^{8\text{-}10}$)	$\Psi_{E,\Omega}(s,\sigma_r,\sigma_\Omega)$ σ_x σ_y
Pointspread function $h_p(r,\Omega,E)$	$f_{E,\Omega}(r)$	6	5 (10^8)	$\Psi_{E,\Omega}(s)$
Pencil beams $p(r,\Omega,E)$	$f_{E,\Omega}(r)$ $T_{E,\Omega}(r)$	5	5 (10^8)	$\Psi_{E,\Omega}(s)$
Convergent Pencil beams on sphere or cylinder $h_s(r)$	$f(r)$	3	4 (10^6)	$\Psi_\Omega(s)$
$h_c(r)$	$f(r)$	3	3 (10^6)	$\Psi_\Omega(s)$
Three Multileaf collimator fields $h(r)$	$f(r)$	3	2+ ($5\,10^4$)	Ψ_Ω
Composite Three field $h(r)$	$f(r)$	3	2+ ($5\,10^4$)	Ψ_Ω
Multileaf collimator $h(r)$	$f(r)$	2	2 (10^4)	Ψ_Ω
Single field $h(r)$	$f(r)$	2	2 (10^4)	Ψ_Ω
≈30 classical uniform or wedged beams $D(x,y)$	$l.h,\omega°,\theta°$	1	1 (60)	$\Psi_\theta(x,y)$

Figure 5.3. Overview of the free variables and the degrees of freedom in kernel-based radiation therapy planning. The flat or wedge-shaped beams of conventional treatment optimization are machine-oriented. Most of the other kernels are instead tumor-oriented and have the potential both to improve the accuracy and to optimize the dose delivery.

energy, beam direction, and beam profiles as illustrated by Figs. 5.1 and 5.3 and Eq. (5.1) (Brahme 1987; Lind 1991; Gustafsson *et al.* 1994, 1995). These new degrees of freedom of what might be called generalized conformal therapy will allow a full-blown optimization of the dose delivery also for very complex concave and heterogeneous target volumes. However, such treatments are quite complicated both to plan and to deliver, so a more practical treatment optimization as discussed above requires some treatment parameters or degrees of freedom to be locked to make planning and dose delivery practical and manageable. For example, the beam energy or directions of incidence may be preset to clinically obvious values for the target at hand (Söderström and Brahme 1993; Söderström *et al.* 1993, 1995(a)–(c)).

For many simple target volumes, few field techniques with uniform beams are quite sufficient, whereas for more complex shapes nonuniform dose delivery is generally much more advantageous. In fact, it can be shown that the classical conformation therapy method with uniform beams is almost equal to the fully optimized generalized conformal method only for the special case of homogeneous circular symmetric target volumes. For most other target volumes, nonuniform dose delivery will be clearly advantageous.

The number of degrees of freedom for different irradiation techniques are quantified and illustrated in Fig. 5.3. From this figure, it is evident that the new treatment optimization methods allow several additional degrees of freedom and the number of free variables is several orders of magnitude higher than for classical treatment optimization. In principle, the most general treatment situation is to allow an arbitrary fluence in every direction, from the radiation source to the target volume, for all possible or desirable source positions. Alternatively, this may be expressed such that each point in the target can be irradiated by arbitrary ray intensities from all directions in space depending on the pattern of motion of the radiation source. This means in general five degrees of freedom: either three for the source position and two for the angle of emission or three for a point in the target volume and two for the angle of incidence. On top of this, the energy spectrum and fractionation can in principle vary from point to point. Disregarding these latter variables, this corresponds to about $100^3 \times 100^1 = 10^8$ free variables for coplanar irradiations and about 10^{10} for arbitrary beam directions assuming a spatial and angular grid of calculation with 100 rays along each coordinate. This explains the substantial improvements in the delivered dose distributions with some of the new treatment techniques.

The main methods for nonuniform dose delivery are dynamic multileaf collimation and scanned elementary beams. The dynamic jaw collimation method in principle allows full modulation of the incident beam but at the cost of extended treatment times. Furthermore, it requires that both the upper and the lower jaw pairs are fully asymmetric so that, in principle, a narrow rectangular beam spot could be scanned arbitrarily across the entire radiation field. If very high dose rates were available and the speed of motion of the collimator jaws was very fast, the time required could be reduced but this is not a very realistic method with present accelerator systems. The classical filter and block techniques have the flexibility but they are probably not realistic for more than some three portals per patient. In recent years, several compensator optimization techniques have been developed (Miller 1985; Brix *et al.* 1988; Ulsø and Brahme 1988; Djordjevich *et al.* 1990; Weeks and Sontag 1991; Söderström *et al.* 1993) which are quite useful to handle few field techniques provided suitable beam directions can be identified. In reality, the optimal choice of beam direction is one of the most difficult problems of treatment optimization since it involves a restriction on the phase space of feasible beam combinations.

This cannot be achieved without having tested all possible beam combinations, which in practice is equal to a global optimization. It also accentuates a difficult radiobiological problem, in a way the Scylla and Charybdis of radiation therapy: with a single beam, the small volumes of normal tissues in the entrance region receive a rather high local dose, whereas on the other extreme, with a continuum of arc beams, large volumes receive rather low doses (Popp *et al.* 1975). To allow a strict optimization, realistic radiobiological objective functions capable of distinguishing between these extremes are needed (see Sec. 5.3.2).

5.2.4. *Development of Optimal Treatment Techniques*

Irrespective of the method of dose delivery, a large number of treatment techniques are possible, as illustrated in Fig. 5.4. Most classical treatment techniques are coplanar with all the beams incident in one plane by the rotation of the radiation source mounted in an isocentric rotary gantry. Since there is a significant exponential dose fall-off in most photon beams, despite their high energy, the most common treatment techniques employ symmetric beam configurations to largely eliminate dose gradients over the tumor. Thus, parallel opposed and four-field box techniques are quite common (Fig. 5.4 upper row). However, the symmetrical or nearly symmetrical three-field technique is generally more advantageous since entrance and exit volumes are almost fully separated and the high dose volume in normal tissues surrounding the tumor is significantly reduced. A two-field technique which is more advantageous than the standard parallel opposed configuration is, therefore, to use more oblique nonuniform fields at ~100° to 120° from each other (cf. Söderström *et al.* 1995(a) and (b) and Fig. 3.4 middle row, even if strictly speaking two fields

Classical Coplanar Beam Techniques

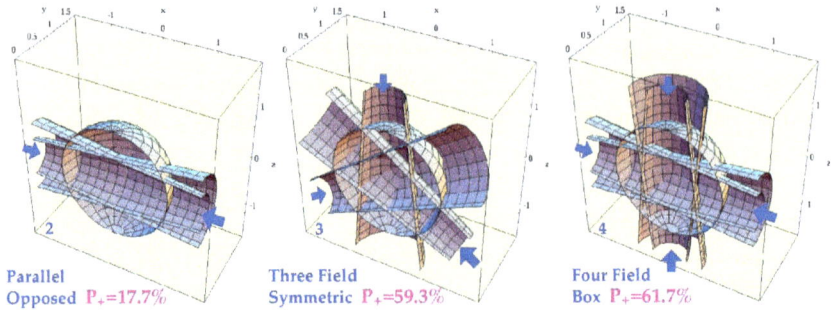

Parallel
Opposed P₊=17.7%

Three Field
Symmetric P₊=59.3%

Four Field
Box P₊=61.7%

Non Coplanar Beam Techniques

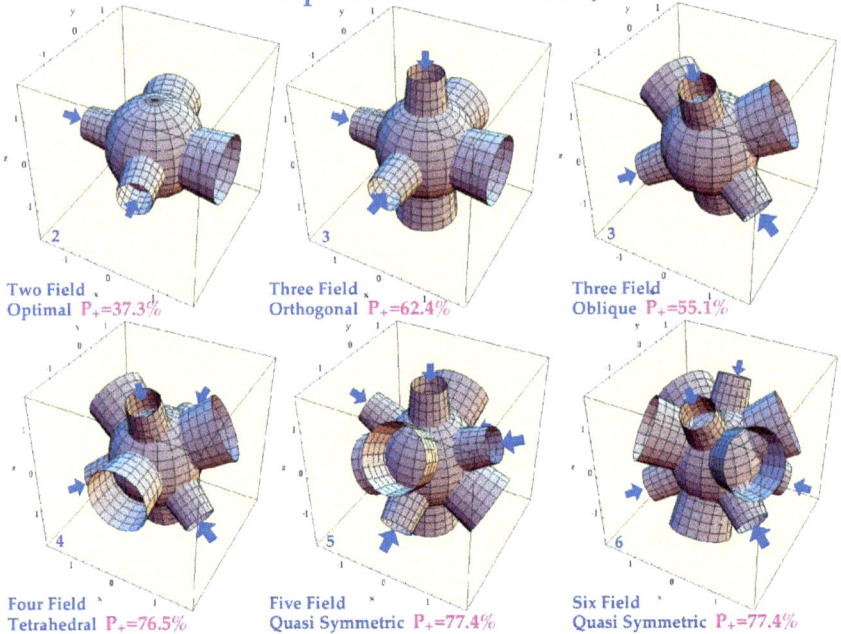

Two Field
Optimal P₊=37.3%

Three Field
Orthogonal P₊=62.4%

Three Field
Oblique P₊=55.1%

Four Field
Tetrahedral P₊=76.5%

Five Field
Quasi Symmetric P₊=77.4%

Six Field
Quasi Symmetric P₊=77.4%

Figure 5.4. Comparison of classical coplanar and more general noncoplanar treatment techniques. The symmetric planar configurations allow, in most cases, a fairly uniform dose to the tumor. The noncoplanar techniques are more advantageous with regard to the dose to normal tissues. But for the few field techniques (≤ 3) to be effective, it is important to use nonuniform dose delivery. The optimal angle between two neighboring fields is often around 100° to 120° to maximize the effect on the tumor and minimize beam interactions in normal tissues (cf. Fig. 6.7d). The resultant complication-free cure P_+ is indicated in each case.

always are coplanar). Similarly with nonuniform dose delivery, it is sometimes better to turn the three fields out of the plane of rotation to better separate the incident and existing beams from each other (first row of noncoplanar techniques in Fig. 5.4).

A much better four-field technique than the standard four-field box is obtained by letting the beams enter through the main tetrahedral directions again reducing the normal tissue dose by non-coinciding entrance and exit regions. There are also five- and six-field techniques which largely keep the symmetry of the noncoplanar three- and four-field techniques. All the treatment techniques in the last row of Fig. 5.4 has the interesting property of generating almost uniform target doses independent of using wedge filters. This is a very valuable property of many semi-optimal few field techniques. No exact symmetric solution is known for five or six fields like the tetrahedral solution for four fields. The close packing problem of circles on a sphere is unsolved for arbitrary numbers of circles (Sailer *et al.* 1993).

The theoretical limits for irradiation with external photon beams for different treatment geometries are compared by their dose distribution kernels in Fig. 5.5. Three external beam kernels are shown and one isotropic internal source, all for an monochromatic energy of 1.25 MeV corresponding to ^{60}Co γ rays. The basic building block for external beam therapy is the point monodirectional energy deposition kernel or pencil beam, represented here by the lowest solid line radial dose profile in Fig. 5.5. If the pencil beam is rotated 2π in the plane, the resultant dose distribution in the plane corresponds to a full arc treatment of a point target (upper solid line). The dose distribution across the plane of rotation is the same as the radial distribution through the pencil beam (lowest solid line). For comparison, the limiting case of a beam without attenuation, is also shown with a fall-off proportional to $1/r$ (upper dotted line).

The ultimate dose kernel is obtained by superposing the contributions from pencil beams from every angle over 4π and the resultant dose distribution corresponds to a true isotropic spherical irradiation (middle solid line). For comparison, the limiting case without attenuation is also shown with a fall-off proportional to $1/r^2$ (lower dotted line). With the same photon energy, but using an internal point source, a dose distribution very much like the isotropic external irradiation is obtained. The main difference can be seen in the small dose buildup near the surface of the focused external beams. It is interesting that for fully isotropic dose delivery it does not matter very much whether internal or external radiation sources are used since the dose distribution is mainly determined by the convergent geometry and not beam attenuation. Thus, by using external convergent beams it is possible to make dose distributions which are almost as good as with internal

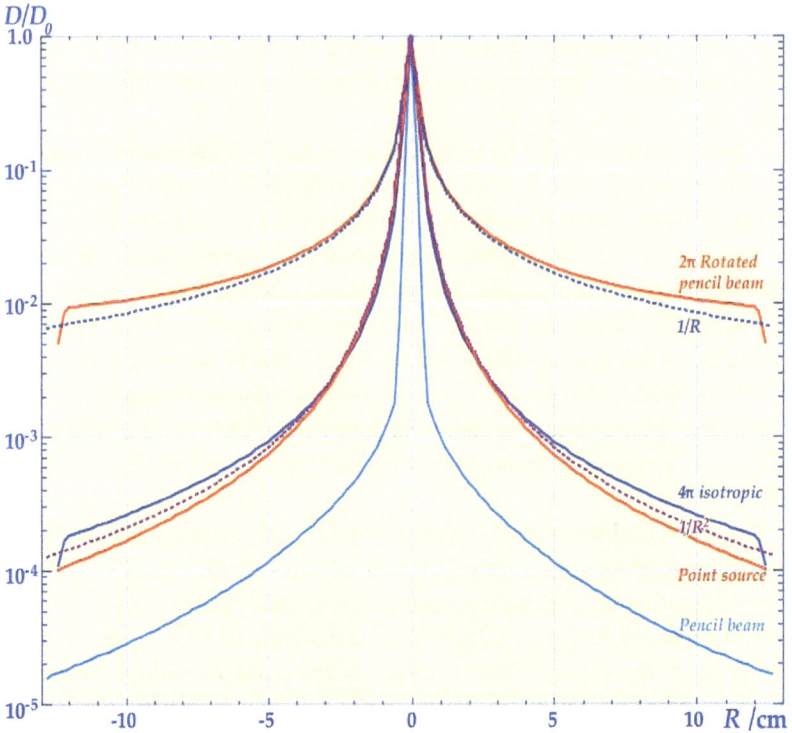

Figure 5.5. The radial dose profiles through four different ^{60}Co beam kernels: point monodirectional pencil beams (lowest solid line), planar rotated pencil beams (upper solid line), 4π isotropically incident pencil beams (middle solid line), and isotropic internally emitting nuclides (middle thin line without build up at the surface). For comparison, the dotted lines correspond to $1/r$ and $1/r^2$ radial dependencies. For the 2π and 4π curves, the deviations from the $1/r$ and $1/r^2$ dependencies are due to photon absorption in the patient.

intracavitary sources. This is basically the way the "γ knife" works with convergent ^{60}Co beams for radiotherapy of brain tumors but the principle may equally well work for a whole body device (cf. Lind and Brahme 1995). This has the considerable dose distributional advantage of a dose fall-off close to $1/r^2$ compared to nonuniform arc therapy with the fan beam devices discussed above, basically only having a $1/r$ dose fall-off as illustrated in Fig. 5.5. Of course, the axial modulation is theoretically better with classical arc therapy as shown by the lowest profile across the plane of rotation.

However, a multiple beam device like the whole-body multi-cobalt device has the advantage of simultaneously delivering all beams with a high dose rate at the focus and considerably lower dose rate everywhere else. These low dose rates imply additional radiobiological protection of healthy tissues around the tumor — an effect which is not obtained by ordinary arc therapy techniques. The whole body multi-cobalt device, therefore, has the interesting advantage of simplifying optimized dose delivery as no rotary parts are needed and the dose delivery kernel is closer to $1/r^2$ than $1/r$.

5.3. Objective Functions for Treatment Optimization

5.3.1. *General*

The primary objective of radiation oncology, and for all medical care for that matter, is to make the quality of life for the patient as high as is reasonably achievable from the time of his or her first contact with the medical system. In principle, some kind of integral measure of the quality of life would be a suitable objective function since a long, healthy, and comfortable life, accounting also for all steps of medical care, is most desirable (cf. Osoba 1993). However, a very long life with severe loss of life quality is not generally desirable (Bush 1982), so some kind of weighted integral measure should, therefore, be most desirable. Obviously, the quality of life of a person has to be a subjective quantity (Campbell *et al.* 1976), but from a medical point of view, it also has a more strict and objective side as a general health index. This index can range from unity, corresponding to perfect health, to zero when the person in medical terms is dead or does not want to live anymore. There are several measures which are aimed at quantifying the condition of the patient, such as the Karnofsky status (Karnofsky *et al.* 1948), but all of them are far too complex and difficult to relate to therapeutic actions to be used directly for therapy optimization. We will, therefore, first discuss some simplifications that need to be introduced to obtain an objective function that is useful for computed radiation therapy optimization.

In practice, two main groups of objective functions have been used, physical and biological. The most commonly employed objective functions are physical and they often describe the properties of the delivered dose distribution in the internal target volumes and affected normal tissues or organs at risk. In fact, the definition of these latter concepts is extremely important for how accurate and precise a treatment can be performed as illustrated in Fig. 5.6. It is fundamental that the target volume and the organs at risk are defined with the narrowest possible safety margins in relation to the anatomical landmarks that are used for setting up the patient for external beam therapy. In principle, the target volume should include all the volumes and margins whose dimensions cannot be affected by changing or developing dose-planning or treatment techniques

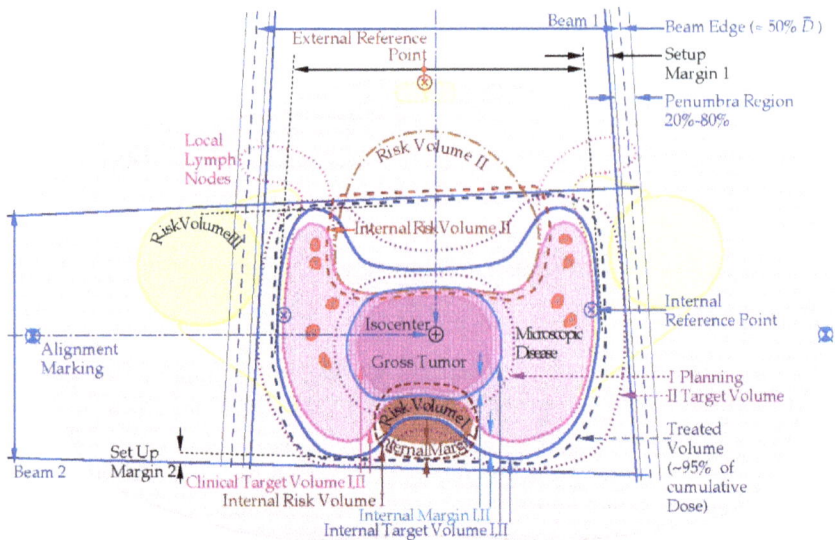

Figure 5.6. Illustration of the internal target volume concept and its relation to the clinical target volume, the target margin, and the treatment setup margin. Different target margins may be needed for the gross tumor and the microscopic disease (clinical target volumes I and II). The external and internal anatomic landmarks are fundamental for defining the target volume in the coordinate system of the patient and for the accurate setup of the therapeutic beams, whereas the internal landmarks are fundamental for simulation and portal imaging.

(including fixation of the patient). Therefore, the internal target volume should be obtained by adding a target margin to the clinical target volume to account for all internal shape changes of organs and their possible motions relative to the anatomical reference points (Aaltonen 1992; NACP 1997). To this target volume, setup margins may later have to be added depending on the treatment technique and the quality of the treatment unit. It is fundamental that the internal target volume is used for dose prescription and reporting as it is the most relevant one for the treatment outcome. The biological objective functions aim at quantifying the probability that the patient will have a desirable treatment outcome. From this point of view, the radiobiological objective functions, therefore, do quantify the quality of life of the patient after therapy, as will be discussed in more detail now.

5.3.2. *Biological Objective Functions*

There are a large number of factors that by necessity make the biological objective functions statistical quantities. First, the vital end point of killing all clonogenic tumor cells to eradicate the tumor makes the beneficial treatment outcome "tumor control" a stochastic quantity. We can only state the probability, P, and standard deviation, σ, for a certain treatment outcome due to the large uncertainty in hitting the very last tumor clonogen.

On top of this uncertainty there are uncertainties both in defining the target volume, so that all tumor clonogens really are included, and dose delivery, so that the target volume is fully irradiated. Furthermore, for a given patient and tumor classification (e.g., according to the TNM system to minimize the influence of the extent of the disease), it is generally not known whether a given tumor is more sensitive or resistant than that of the "mean patient" for which established dose–response relations should be applicable. The same is true for the normal tissues of that patient and both these facts add to the total uncertainty.

In the phase space of clonogenic tumor cells, the "minimax" treatment objective can be formulated: the maximum value of the tumor recurrence probability should be as low as possible throughout the target volume without causing severe damage to normal tissues. This formulation is quite interesting because it solves the inversion problem in a simple way, but it requires information about the density and sensitivity of clonogenic tumor cells, as discussed in Sec. 5.4.5 below.

For the above reasons, it is also often assumed that the probability to severely injuring the patient P_I is statistically independent process from the probability of a beneficial treatment outcome P_B, that is, tumor control. The probability of a truly successful treatment, P_+, can then be expressed since the covariance of P_B and P_I is zero and the product law of statistics can be applied to determine the conditional probability of having tumor control without severe injury:

$$P_+ = P_B - P_{B \cap I} \approx P_B(1 - P_I), \tag{5.8}$$

which is the traditional expression used by most workers (Schultheiss *et al.* 1983; Wolbarst 1984; Cohen 1987; the longer notation NTCP and TCP are sometimes used instead of P_I and P_B, respectively; the compact notation used here was introduced by Cohen (1960), Moore and Mendelsohn (1972) and Andrew (1985)). However, there are a number of factors that can alter this conclusion since it is strongly dependent on the type of dose delivery that is employed and also on whether there might be a true biological correlation between P_B and P_I.

There are several biological mechanisms that can cause such an effect, for example, if the patient happens to have an unusually efficient repair system for double-strand breaks (Schwartz *et al.* 1988; Ågren *et al.* 1990). Since most oncogenes are different from the genes responsible for efficient repair of radiation damage, it is probable that both the tumor and the normal tissue will be more resistant to irradiation. Conversely, if the patient happens to have the ataxia telangiectasia gene, which is one of several known genetic defects quite common in cancer patients (Norman *et al.* 1988; Timme and Moses 1988; Ågren *et al.* 1990), there would be an increased risk for normal tissue injury but also an increased probability to control the tumor. Furthermore, Leibel *et al.* (1991) pointed out that some tumors which show an increased risk for metastatic dissemination might be phenotypically distinct and their presence like the radiation sensitivity may thus be used as a predictive indicator which could influence the treatment decision.

However, there are also other genes, such as the tumor suppressor gene p53, which seem to have a regulatory effect on how DNA damage is handled by the cell (Lane 1992). The mutant p53 genes are one of the most common defects which are found in tumor cells. This may be an important molecular biological explanation why for such tumors radiation therapy has a more severe effect on the tumor cells than on healthy normal tissues with the wild-type p53 which is better capable of handling radiation-induced DNA damage.

Depending on the type of dose delivery and the proportions of genetic variations for the tumor at hand, deviations from the simple expression of Eq. (4.8) may therefore be seen. It is, therefore, advantageous to base the analysis as far as possible on clinical data. In a recent study of head and neck tumors by Ågren *et al.* (1990) a rather uniform dose distribution was used (parallel opposed beams). It was found that P_B and P_I was totally correlated at low doses, and only for large tumors at high doses, the uncorrelated portion δ reached a value of about 20%. The clinically observed P_+ value was then accurately described by

$$P_+ = P_B - P_I + \delta(1 - P_B)P_I, \tag{5.9}$$

for clinically relevant doses as illustrated by the clinical data for large head and neck tumors in Fig. 5.7. A very convenient and radiobiologically coherent parameterization of the dose–response curves (solid curves) is obtained from:

$$P_{B,I} = 2^{-\exp(e\gamma(1-D/D_{50}))}, \tag{5.10}$$

where γ is the normalized dose–response gradient and D_{50} is the dose causing 50% probability of effect (tumor control or severe normal tissue reactions). The above relations are generally based on uniform dose delivery to the various organs and tumors, but can be modified to account also for nonuniform dose delivery (cf. Källman *et al.* 1992b, Löf *et al.* 1995 and Ågren *et al.* 1995). Using Eqs. (5.5), (5.9), and (5.10), the associated functional may now be written as

$$F_+^0(f) = P_+(Hf). \tag{5.11}$$

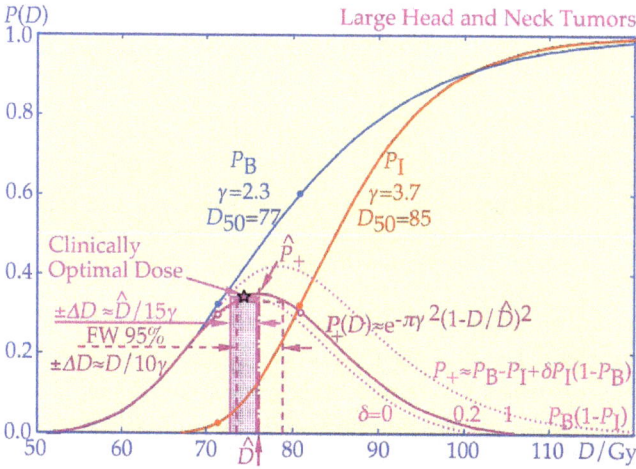

Figure 5.7. Illustration of clinically established dose–response relations for tumor control (P_B) and fatal normal tissue reactions (P_I). The clinically established probabilities for complication-free tumor control (P_+) are indicated by open circles (cf. Eq. (3.9)). The other clinical data points are indicated by solid circles and the curves are fitted using Eq. (5.10).

To predict the probability that a proposed dynamic fully 3D nonuniform dose delivery will result in complication-free tumor control requires in principle three different categories of data. These data should describe (1) the treatment geometry of the patient, that is, the locations and shapes of the relevant tissues during each treatment session, (2) the location of the incident beams and thus the dose distribution delivered to the patient, and (3) the radiobiological parameters that describe the dose–response relationships for the different tumors and normal tissues at every treatment session. The dependency of the treatment outcome on these external and internal treatment parameters may be expressed by:

$$P_+ = P_+(\{D_\nu(r)\}, \{G_\nu(r)\}, \{B_\nu\}), \tag{5.12}$$

where the three categories of information needed for computation of P_+ are represented by three sequences of data that describe the interfractional variations by the fraction number ν. Hence,

$$\{D_\nu(r)\} = D_1(r), D_2(r), \dots, D_n(r) \tag{5.13}$$

denotes the sequence of absorbed dose distributions delivered to the patient during the n fractions. Similarly, the sequence,

$$\{G_\nu(r)\} = G_1(r), G_2(r), \dots, G_n(r) \tag{5.14}$$

describes the variation of the patient geometry throughout the treatment course. The treatment geometry during the vth dose fraction is given by:

$$\{G_v(r)\} = (G_1(r), G_2(r), \ldots, G_{n_0}(r))_v, \tag{5.15}$$

which is a vector-valued function that specifies the cell densities at any point r for the n_0 different tissues or organs that are under consideration. The last sequence,

$$\{B_v\} = B_1, B_2, \ldots, B_n \tag{5.16}$$

is the sequence of vectors that describes the radiobiological properties of the treatment geometry. However, for the purpose of this discussion we assume that the biological parameters are constant from fraction to fraction, that is, $B_v = B$, $(v = 1, 2, \ldots, n)$ where B holds the radiation response of each tissue in some form, for instance:

$$B = (D_{50,1}, \gamma_1, s_1, D_{50,2}, \gamma_2, s_2, \ldots, D_{50,n_0}, \gamma_{n_0}, s_{n_0}). \tag{5.17}$$

Clearly, all of the three categories of treatment data are subject to significant uncertainty. Given a fixed configuration of the beams incident on the patient surface, there will still be a variation in the dose $D_v(r)$ from one treatment fraction to the next due to the difficulty and uncertainty in repositioning the patient on the treatment couch. Likewise, because of internal organ motions and deformations, the treatment geometry $G_v(r)$ will also vary from fraction to fraction as shown in Fig. 5.8. The reason for the uncertainty in the radiobiological parameters is that they are in effect averaged over a population of similar patients which means that the parameters for an individual patient may differ from the mean values. This source of error will, of course, be greatly reduced once predictive assays become a clinical reality.

Before P_+ can be used as an objective function in an optimization procedure, we have to find some way to handle the imperfection of our knowledge about the factors which determine P_+. The classical strategy to account for the uncertainties caused by changes in shape and location of the organs is to add a margin around the tumor tissues to get the internal target volume which is accurately defined in relation to internal and external anatomical reference points (see Sec. 5.3.1). Furthermore, during dose delivery, setup margins are added to the beam portals to safely cover the projection of the strict target volume to account for the uncertainties in the alignment of the anatomical reference points relative to the beams. There is, of course, no established practice yet for dealing with the uncertainties in biological parameters in the context of treatment optimization. However, in analogy to adding a target margin around the target tissues and setup margins to the projection of the

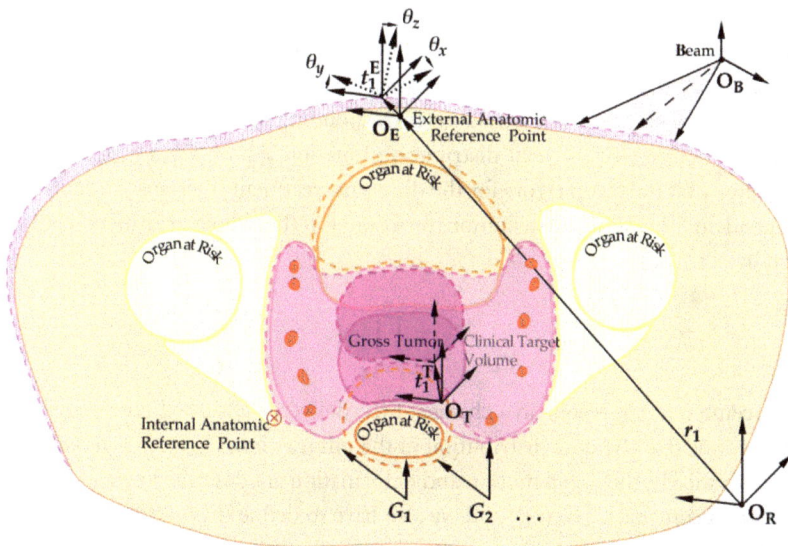

Figure 5.8. Illustration of the coordinate systems and the notations used during treatment optimization to consider stochastic internal organ motions and external setup uncertainties.

strict target volume, it would be reasonable to assume for a given patient that his or her tumor is somewhat more resistant than the average and, at the same time, that his or her healthy normal tissues are more sensitive. Thus, we are considering a worst-case scenario with safety margins on tumor and normal tissue sensitivity just like the geometrical margins used in covering the extremes of the organ and setup movements.

A more advanced technique to find the optimal treatment plan despite the influence of setup errors, organ motions, and uncertainties in radiobiological parameters is to formulate the situation as a *stochastic optimization* problem. The fundamental idea is to include information about the uncertainties in the objective function which will thereby measure the effects of disturbances on the nominal treatment plan and thus on the treatment outcome (Lind *et al.* 1993; Löf *et al.* 1995). In other words, treatment plans that are sensitive to disturbances are punished by this objective function and modified so that the solution to the stochastic optimization problem will be beams that are *robust* with respect to positioning and biological uncertainties.

To formulate the disturbance-sensitive objective function, we have to construct statistical models for the three sets of independent variables that influence the value of P_+. The main source of setup errors is the uncertainty in the repositioning of the patient between fractions which can be described by the translation and rotation of the anatomic coordinate system of the patient O_E in relation to the room fixed system O_R (see Fig. 5.8). Since we can only make a prediction about the *probability* that O_E is at a certain location relative to O_R, we let this position be represented by six stochastic variables (corresponding to the six degrees of freedom for a rigid body) with the *joint* distribution function $\mathscr{F}_S^\nu = \mathscr{F}_S^\nu(T_E, \theta_x, \theta_y, \theta_z)$. If we assume that the setup errors for the different treatment fractions are statistically independent, the setup distribution function for the whole treatment course is given by:

$$\mathscr{F}_S = \prod_{\nu=1}^{n} \mathscr{F}_S^\nu. \tag{5.18}$$

Displacement of O_E results in a change of the delivered absorbed dose distribution which means that the dose distribution in the νth fraction, $D_\nu(r)$, is a *stochastic dose process*. Similarly, the organ motion and deformation are described by a sequence of stochastic geometries, $\{G_\nu(r)\}$, and we also have to define the distribution function $\mathscr{F}_G^\nu$ which assigns a probability to all possible configurations of organ displacements and deformations. Also, assuming that the treatment geometries are independent from fraction to fraction, we have the *geometry distribution function*,

$$\mathscr{F}_G = \prod_{\nu=1}^{n} \mathscr{F}_G^\nu. \tag{5.19}$$

Finally, the uncertainties in the radiobiological parameters are described by letting B have a multivariate normal distribution, that is, $B \in N(\overline{B}, Q)$, where $\overline{B}$ is the vector of expectation values, $EB = \overline{B}$, and Q is the covariance matrix given by $Q_{ij} = \text{Cov}(b_i, b_j)$. The differential of the *biological distribution function* then has the explicit expression:

$$d\mathscr{F}_B = ((2\pi)^{n_0} |Q|)^{-\frac{1}{2}} \exp\left(-\tfrac{1}{2}(B-\overline{B})Q^{-1}(B-\overline{B})^{\mathrm{T}}\right) dB. \tag{5.20}$$

Since the arguments of P_+ are now treated as stochastic quantities, P_+ will itself become a stochastic quantity. For this reason, we consider instead the expectation value of P_+ which gives us the required disturbance-sensitive objective function:

$$EP_+ = \int P_+(\{D_\nu(r)\}, \{G_\nu(r)\}, B) d\mathscr{F}_S d\mathscr{F}_G d\mathscr{F}_B. \tag{5.21}$$

To illustrate the usefulness of this objective function, we study a simple case with setup uncertainties but no organ motion, that is, the patient geometry remains constant throughout the treatment so that $G_\nu = G$, ($\nu = 1, 2, \ldots, n$). If we neglect the small rotations of O_E, the dose distribution for each fraction may then be written as the stochastic dose process:

$$D_v(r) = D(r - T_v), \tag{5.22}$$

where the stochastic displacement vector T_v is described by its probability density $d\mathcal{F}^v_S(t_v) = \varphi_T(t_v)dt_v$. Furthermore, we assume that the uncertainties in biological parameters are accounted for by safety margins, which means that B is a real *nonstochastic* vector. In the first approximation, P_+ is a function of the *total* dose distribution:

$$D^\Sigma(r) = \sum_{v=1}^{n} D(r - t_v), \tag{5.23}$$

which is now a sum of the n individual statistically independent dose fractions and the expectation value of P_+ in this case is given by:

$$F^n_+ = EP_+(\{D_v(r)\}, G(r), B) = \int_{-\infty}^{+\infty} P_+ \left(\sum_{v=1}^{n} D(r - t_v) \right) \prod_{v=1}^{n} \varphi_T(t_v)dt^n. \tag{5.24}$$

This expression, which is a $3n$-fold integral, is quite complex and time–consuming to evaluate but has recently been solved for arbitrary η using a Monte Carlo technique (cf. Sec. 5.4.9 and Löf *et al.* 1995). However, two cases of clinical interest can be handled more easily. For one single dose fraction ($n = 1$), Eq. (5.12) reduces to:

$$F^1_+ = \int_{-\infty}^{+\infty} P_+(D(r - t))\varphi_T(t)dt. \tag{5.25}$$

This equation offers a possibility to find the dose distribution $D(r)$ which in a single irradiation maximizes the expectation value of the probability to cure the patient without fatal complications (cf. Fig. 5.10 below).

For the case with a very large number of small dose fractions, $\varphi_T(t)$ will be very accurately sampled by $T_1, T_2, \ldots, T_n$. In the limit $n \to \infty$, the total dose is given by the sum of a series consisting of infinitesimal dose fractions the limit value of which is a convolution integral with the probability density $\varphi_T(t)$. Since this limit value will be independent of t, it can be brought outside the outer multiple integral and the remaining integral is identical to unity according to:

$$F^\infty_+ = \int_{-\infty}^{+\infty} P_+ \left(\int_{-\infty}^{+\infty} D(r - t)\varphi_T(t_v)dt \right) \prod_{n \to \infty} \varphi_T(t_v)dt^n$$

$$= P_+ \left(\int_{-\infty}^{+\infty} D(r - t)\varphi_T(t)dt \right). \tag{5.26}$$

In the limit with infinitely many treatments, the optimal dose distribution under influence of a stochastic process is well defined. The problem of finding the optimal beam profile $D(r)$ is equal to finding the resultant dose distribution $D*(r)$ which by the influence of the stochastic process maximizes P_+. This simplifies the calculation since the biological objective function does not need to be evaluated inside the integral (see Sec. 5.2.1).

5.3.3. Physical Objective Functions

To allow a strict optimization, the objective function should be a scalar quantity (more precisely a functional) as otherwise two different treatments cannot be compared on the same scale. This property is true for all the radiobiological objective functions presented above. However, to speed up calculations, further simplifications of the objective function are often desirable. This can be achieved by expressing the treatment objectives as simple functions of the dose distribution, but other parameters such as the total treatment time have also been used (Ebert 1977). The success of the dose optimization is directly linked to the ability to define such clinically relevant treatment objectives. The following physical features of importance for tumor control and normal tissue complications are probably the most significant ones and have been used extensively over the years:

(1) Mean energy imparted to target volume and organs at risk ($\overline{\varepsilon}, \overline{D}$).
(2) Dose variance over target volume (σ_D).
(3) Minimum absorbed dose in target volume ($D_{\min}$).
(4) Peak dose to organs at risk ($D_{\max}$).
(5) Conformity of treatment volume to target volume (cf. Fig. 5.6).

When the desired dose distribution, $D(r)$, in the target volume is known, it may also be possible to quantify the difference between it and the best achievable dose distribution, $D^*(r)$. In principle, one could therefore apply the pth norm on this difference to quantify deviations at all points r_i of interest according to:

$$\Delta_p = \left(\sum_i |D(r_i) - D^*(r_i)|^p \right)^{1/p}, \quad p \geq 1 \qquad (5.27)$$

The elliptic norm may also be of interest in this context since it can be regarded as a generalization of Δ_p:

$$\varepsilon_A^2 = d^T A d, \qquad (5.28)$$

where A is a positively definite matrix. For the simple case when $A = I$, the unity matrix, the elliptic norm is precisely equal to the second norm. The second norm is of special clinical interest since it can be shown to be related to the probability of achieving local control for a homogeneous tumor (Brahme 1984, 1992):

$$P_B(D(r)) = P_B(\overline{D}) - \frac{\gamma^2}{2P_B(\overline{D})} \left(\frac{\sigma_D}{\overline{D}} \right) \dots, \tag{5.29}$$

where

$$\overline{D} = \frac{\sum_{i=1}^{n} D(r_i)}{n} \quad \text{and} \quad \sigma_D^2 = \frac{\sum_{i=1}^{n} (D(r_i) - \overline{D})^2}{n}. \tag{5.30}$$

The first two spatial moments of the dose distribution (and thus the second norm) are therefore closely related to the clinical outcome as seen from the expression for the probability to control the tumor. In fact, it can be shown that $\overline{D}$ is also a very suitable quantity for prediction of complication-free tumor control (Brahme 1992) and there are also clinical data supporting this (e.g., Burgers et al. 1985). In addition, the use of $\overline{D}$ minimizes the second norm (Δ_2) which is a very valuable property according to Eq. (5.29). For tissues with a steep threshold-type response, the infinite norm, Δ_∞, may also be of interest as it is a way of quantifying the maximum deviations.

In the above discussion, some arguments are given for the frequent use of the least square deviation as a measure of the deviation from the desired dose distribution in treatment optimization (Δ_2^2) according to Eq. (5.27). The associated functional is given by:

$$F_2(f) = \frac{1}{2} f^{\mathrm{T}} H^{\mathrm{T}} H f - f^{\mathrm{T}} H^{\mathrm{T}} d \left(= \frac{1}{2} (Hf - d)^{\mathrm{T}} (Hf - d) - \frac{1}{2} d^{\mathrm{T}} d \right). \tag{5.31}$$

However, this is only a first step toward treatment optimization since, for example, a narrow hot spot is generally better tolerated than a narrow cold spot, provided the mean energy imparted to the tumor is the same (cf. Brahme 1992). The principal reason for this is that the lower dose inside the cold spot decreases the local overkill much more than an extended cold area, thus making the net recurrence probability with the cold spot significantly larger.

A way out of this clinical dilemma is to not even allow local underdosage but to minimize the local overdosage in all regions where a perfect match is not possible. The appropriate functional for this more clinically relevant objective function has recently been developed by Lind et al. (Lind 1990; Lind and Brahme 1992):

$$F_1(f) = \frac{1}{2} f^{\mathrm{T}} H f - f^{\mathrm{T}} d. \tag{5.32}$$

The last two functionals will be discussed in more detail below (Eqs. (5.58) and (5.59)).

In addition to the above functional, it may also be relevant to look at the entropy (cf. Boltzmann 1927; Shannon 1948) of the deviation from the desired plan using the definition:

$$S(d) = \sum_i (D(r_i) - D^*(r_i)) \ln \frac{D(r_i) - D^*(r_i)}{\overline{D}}. \tag{5.33}$$

This functional is particularly relevant when the microscopic extension of the tumor is uncertain (cf. Fig. 5.4) and it may be desirable to maximize the entropy of the deviation from some prior estimate of the uncertainties in the dose prescription. When such specifications are available they may even be accounted for using the maximum entropy method (Jaynes 1957; Buck and Macaulay 1991). To make comparisons with other measures and objective functions it is sometimes useful to transform $S(d)$ linearly, so it is always less than unity corresponding to maximum information content (Söderström and Brahme 1992) by putting:

$$S(d) = 1 + \frac{S_{\min} - S(d)}{S_{\max}}. \tag{5.34}$$

Another parameter which may also be useful to quantify the clinical value of a beam is its content of spatial frequencies, u. Generally, it is fields with low spatial frequencies that can be most easily produced, so an integral over the absolute value of the low frequency portion of the Fourier transform of the dose distribution, $\hat{D}$, may be a useful quantifier of gross structure in a beam:

$$Z = \int_0^{u_{\max}} |\tilde{D}(u)| du. \tag{5.35}$$

Here, it is natural to set the upper integration limit to correspond to the smallest structure of interest in the target volume, the resolution of the leaf collimator, or the precision of patient setup, depending on the situation, and thus $u_{\max} \approx 1/2\Delta r$ (cf. Söderström and Brahme 1992).

Finally, it should be pointed out that it is sometimes of interest to use the difference in biological effect to quantify the deviations from the desired dose distribution rather than the dose difference as described by Eq. (5.27).

5.4. Mathematical Methods for Treatment Optimization

5.4.1. *Overview*

As discussed above, treatment optimization can follow two routes: either a direct one where a *true inversion* of the integral equation is possible or a more complex one where a global numerical optimization has to be employed depending on the specific treatment objectives. Under simplifying assumptions, an optimization can sometimes be performed by a direct mathematical inversion (see the review by Brahme 1995). We will first look closer at four examples of this technique. In all cases they work when the desired dose distribution is "well behaved." In the first case, it is also assumed that the target volume possesses rotational symmetry and in most cases the risk for damage to normal tissues is disregarded.

Almost all optimization methods use some kind of *iterative algorithm* which successively finds better and better solutions. Traditional "manual" or "visual" optimization belongs to this group and the cumulated experience of the planner helps him to find better and better plans. The iterative algorithms may be either *stochastic* or *systematic* in their way of searching for the optimal solution, whereas a human planar generally exemplifies a mixture of the two.

Stochastic iterative methods. Traditional computer-aided visual treatment optimization is a quasi-stochastic iterative method where better and better treatment plans are found by a trial and error process in which the "adaptive experience" of the planner can help in speeding up the process. There are two slightly more systematic but still basically stochastic methods namely: *inverse Monte Carlo* and *simulated annealing.* The inverse Monte Carlo approach is based on studying the adjoint transport equation in order to find the most suitable solution (Dunn 1981). So far this method has not been applied to radiation therapy planning problems, probably since it is associated with the same existence problems as Eqs. (5.4), (5.36), (5.38), and (5.39).

In recent years, the simulated annealing technique has become quite popular for treatment optimization (Webb 1989; Morrill *et al.* 1990; Mohan *et al.* 1992). The method, which in general is very time-consuming, is based on a stochastic sampling of the total phase space of incident elementary beams (preferably pencil beams). The beam samples are accepted if they increase the objective function, and if not, they might still be accepted with a reduced probability given by the annealing expression, $e^{-\Delta F/kT}$. To start with, the "annealing temperature" T is high and larger deviations are accepted, but as the iteration number increases, it is lowered to obtain a quasi-adiabatic annealing and to buildup the fine structure of the optimal solution (cf. Metropolis *et al.* 1953; Kirkpatrick *et al.* 1983). In principle, the functional F could be any of the objective functions described in Sec. 5.3. Most logical is to use a biological functional (Morrill *et al.* 1990; Mohan *et al.* 1992), but the deviation from a desired dose distribution also may be useful, for example, Δ_2 (Webb 1989). The advantage of the different annealing approaches is that they work well for non-convex problems where systematic approaches may get stuck at local minima. They are therefore not really needed for intensity modulation of fixed beam portals as this is a convex problem but when it comes to the best possible beam portals, they may have a role despite the long computation time.

Systematic iterative methods. Most algorithms that have been developed over the past almost 30 years of treatment optimization belong to the systematic iterative group. In the present chapter, there is not enough room to cover all of them in detail. Some of them have already been mentioned in Sec. 5.2.3 in connection with the degrees of freedom allowed in the optimization. We will briefly discuss them here grouped in temporal order of introduction and according to the principal type of objective function and mathematical technique employed. The objective function could be linear or quadratic or generally nonlinear in the unknown variable and so could the associated constraining conditions.

The earliest techniques used various score functions in an attempt to quantify a good dose plan (Hope and Orr 1965; Hope *et al.* 1967; Van der Laarse and Strackee 1976) mainly aiming at a good dose homogeneity in the tumor and little normal tissue damage. More recently biological scores have also been used (Gremmel *et al.* 1978; Goitein and Niemierko 1988; Niemierko 1993).

The *linear programming* technique has also been very popular over the years with minimal mean energy imparted to normal tissue as the most common objective (Bahr *et al.* 1968; Hodes 1974; Ebert 1977; Mantel *et al.* 1977; Bollman *et al.* 1981; Legras *et al.* 1982; Sonderman and Abrahamson 1985; Langer and Leong 1987; Langer *et al.* 1991; and the review by Rosen *et al.* (1991)).

As soon as the least square deviation from a desired dose distribution was sought, quadratic or nonlinear methods were needed (Redpath *et al.* 1976; McDonald and Rubin 1977; Franke 1980; Starkschall 1984; Bortfeld *et al.* 1990; Djordjevich *et al.* 1990).

A special class of algorithms has recently gained interest. They do not perform a strict optimization but rather a feasibility search for solutions within the limits of the treatment constraints such as minimal and maximal tumor and normal tissue doses. This method has been pioneered by Censor *et al.* (1988a,b) (cf. also Powlis *et al.* 1989; van Santvoort and Huizenga 1991; Starkschall and Eifel 1992).

Owing to their universal applicability, the gradient and pencil beam methods are treated in some detail toward the end of this section. Finally, a discussion of the similarities and differences of radiation therapy optimization and image reconstruction in computed tomography is given.

5.4.2. *Cylindrically Symmetric Target Volumes*

One of the earliest applications of inversion methods in radiation therapy planning was the development of nonlinear wedge techniques for lymph node irradiation in the head and neck region, as shown in Fig. 5.9. Lax and Brahme (1982) developed a strongly nonlinear wedge filter to give a high uniform dose to the nodes and a low dose to the spinal cord using a double arc technique. First, a numerical solution for the wedge profile was found by an iterative procedure but in addition the integral equation for rotationally symmetric arc therapy was also setup (cf. Fig. 5.9):

$$D(r) = \oint d(x)e^{-\mu_p z}d\varphi/\pi. \tag{5.36}$$

By using Laplace transformation on this integral equation, a general inversion formula was derived which allows the calculation of the required beam profile from a given desired dose distribution in the patient under cylindrical symmetry and full 360° rotation:

$$d(x) = \frac{d}{dx}\int_{r_0}^{x} \frac{\cos\left(\mu_p\sqrt{x^2 - r^2}\right)}{\sqrt{x^2 - r^2}}D(r)r\,dr. \tag{5.37}$$

Here, $D(r)$ is the desired radial dose profile in the patient, $d(x)$ is the required dose profile of the incident beam, and μ_p is the practical attenuation coefficient of the beam (Brahme et al. 1982).

Unfortunately, most patients are not rotationally symmetric so this general inversion equation is not generally applicable. However, a straightforward generalization of the theory is possible, namely to arbitrary body shapes, provided the target volume is still cylindrically symmetric. Because the derived dose profiles of the beam Eq. (5.37) pertain to a plane through the center of the target volume, it is just a matter of correcting for the nonuniform beam absorption in each CT slice of the body to obtain the optimal shape of the incident beam at each angle. Obviously this results in angularly varying incident beam profiles in the general case even though the target volume is rotationally symmetric.

However, Eq. (5.37) does not result in physically possible incident dose distributions for arbitrary desired dose profiles, $D(r)$, in the patient since $d(x)$ may take both negative and infinite values. These problems were touched on in Sec. 5.2.2 and are further discussed in Sec. 5.4.5 below and in several recent publications (Brahme et al. 1982; Cormack 1987 and 1995; Cormack and Cormack 1987; Brahme 1988; Cormack and Quinto 1989; Barth 1990; Gustafsson et al. 1994 and 1995; Brahme 1995). A closely related inversion problem, namely how to make an arbitrary picture just by drawing uniform straight lines, was solved by Birkhoff (1940), see also Cormack (1995) and Brahme (1995). If the uniform lines were to be replaced by "pencil" beams and the solution could be generalized for this case, the inverse problem of radiation therapy would be solved.

However, even if this solution is found, it would in general be useless because for most desired dose distributions there is no non-negative solution. Furthermore,

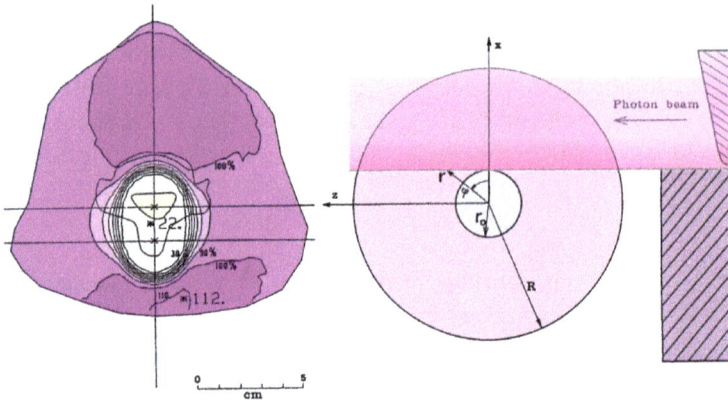

Figure 5.9(a). The irradiation geometry and coordinate system used in the nonlinear wedge technique are shown to the right. The origin of the rectangular and polar coordinate system is located at the isocenter of the therapy machine. The location of the beam block and the nonlinear wedge-shaped filter is also indicated. The resultant dose distribution in a larynx cross-section is shown to the left. The gantry rotation starts with the gantry in a dorsal position, and the treatment is performed during a 180° arc using the ventral rotation center. Thereafter, the treatment table is raised to a distance equal to that between the two rotation centers, and the treatment is continued for the remaining 180° (Lax and Brahme 1982).

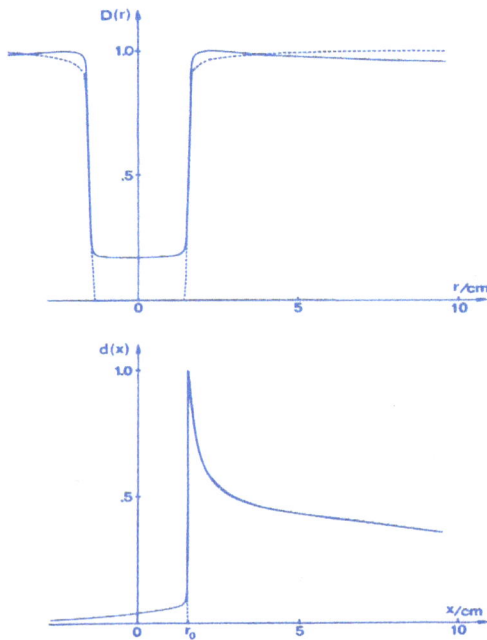

Figure 5.9(b). Both $d(x)$ and $D(r)$ are shown with and without (dashed lines) edge scatter and leakage.

the requirement given by Birkhoff for the existence of a solution in the drawing case is very hard to use since the coefficients of the resultant Fourier series have to be continuous and form a nonnegative continuous function. Such a requirement is of little help when trying to judge whether a desired dose distribution can be produced or not, and it is not useful for finding the best possible solution when no exact solution exists (cf. also Bochner 1948, as discussed below). The recent pencil beam algorithm of Gustafsson *et al.* (1994, 1995) solves this difficult problem by a very efficient numerical algorithm (see Sec. 5.4.7) not only Physically as in Birkhoffs drawing problem but also Biologically by maximizing the complication free cure.

5.4.3. *Direct Fourier Transformation*

Equations (5.3) and (5.4) are strict convolution integrals and can be solved exactly by direct Fourier transformation as the convolutions in Fourier space are given by the product of the Fourier transforms. Therefore, a strict mathematical inversion of Eq. (4.4) results in:

$$f(r) = \mathcal{F}^{-1}\left(\frac{\hat{D}(u)}{\hat{h}(u)}\right), \tag{5.38}$$

where $\hat{D}(u)$ and $\hat{h}(u)$ are the Fourier transforms of the desired dose distribution and the kernel, respectively, and $\mathcal{F}^{-1}$ is the inverse Fourier transform operator (cf. Brahme 1985, 1988; Holmes 1991, see also more recent references in Harder *et al.* 2014). Unfortunately, this method is not generally applicable for arbitrary D and h since f will often take negative values on some interval and may have severe oscillations at the zeros of $\hat{h}$. This latter problem can sometimes be handled by a low-pass filter in Fourier space according to:

$$f_\lambda(r) = \mathcal{F}^{-1}\left(\frac{\hat{D}(u)}{\hat{h}(u)(1 + |\hat{h}(u)|^2)}\right). \tag{5.39}$$

However, in most difficult problems the inverse Fourier transformation method is not very useful due to the negativity problem (Davis 1982; Lind and Brahme 1985). The problem with negative f components is in principle handled by a theorem derived by Bochner (1948), but as in the case of the Birkhoff solution, it is very hard to use it in practice. It was pointed out by Birkhoff (1940) that if erasure or negative functions are allowed, there always exists an exact solution of the drawing problem as in the radiotherapy case given by Eqs. (5.38) and (5.39) (see also Sec. 5.4.5). One of the more interesting applications of Eq. (5.38) in recent years has been worked out by Mackie *et al.* (cf. Holmes *et al.* 1991) where they have replaced $\hat{h}$ by a modified narrower kernel without zeros over the range of interest of $\hat{D}$. The resultant f is only a crude first approximation to the optimal solution but it may serve as a suitable starting point for a more exact optimization similar to the case with Eq. (5.43) below (cf. Holmes *et al.* 1991; Källman *et al.* 1992a).

5.4.4. *Inversion by Taylor Expansion*

In cases where the function being studied is well behaved with multiple continuous derivatives, an approximate inversion by Taylor expansion can sometimes be of value. We will illustrate this on the problem of optimization under uncertainties in beam patient alignment. According to Eq. (5.14), the total dose distribution under continuous irradiation with stochastic beam patient motions may be expressed by:

$$D^{\Sigma}(r) = n \int_{-\infty}^{+\infty} D(r - t)\varphi_T(t)dt. \tag{5.40}$$

By Taylor expansion of D around r and term-wise integration, D^{Σ} is expressed in a series containing the derivatives of D. By assuming φ_T to be rather narrow compared to the fluctuations of D, this series may be inverted, thus giving:

$$nD(r) = D^{\Sigma}(r) - \frac{\sigma^2}{2}\nabla^2 D^{\Sigma}(r) + \frac{\sigma^4}{8}\nabla^4 D^{\Sigma}(r) - \frac{\sigma^6}{48}\nabla^6 D^{\Sigma}(r) + \cdots, \tag{5.41}$$

where σ is the standard deviation of φ_T (cf. Brahme 1981). This method may be used, for example, to correct the desired target dose distribution D^{Σ} for the influence of known uncertainties in beam patient alignment (Lind and Brahme 1992; Löf *et al.* 1995).

5.4.5. *Optimization as a Variational Problem*

The probability of controlling a tumor can be expressed mathematically using Poisson statistics when the initial density, $n_0(r)$, and sensitivity, $D_0(r)$ of clonogenic tumor cells are known. The associated integral equation becomes:

$$P_B = \exp\left(-\iiint_{V_t} n_0(r)e^{-D(r)/D_0(r)}dr^3\right). \tag{5.42}$$

When the risk of causing normal tissue damage is negligible, this equation can be used to derive the optimal dose distribution $D^*(r)$, which eradicates the tumor with a certain probability P_B and, at the same time, gives the lowest possible maximum value of the local recurrence probability in the target volume. This "minimax" problem was treated independently by Fisher (1969) and Brahme and Ågren (1987). The optimal dose distribution under this optimality criterion should obviously have a uniform recurrence density over the whole target volume making the integrand in Eq. (5.42) constant and thus:

$$D^*(r) = D_0(r) \ln \left\{ \frac{n_0(r)V_t}{-\ln P_B} \right\}, \tag{5.43}$$

where V_t is the target volume. The optimal dose distribution in this meaning is thus proportional to the local D_0 value and the logarithm of the tumor cell density. This relation is very useful when $n_0(r)$ can be estimated by 3D diagnostic imaging and D_0 is known by experience or by a predictive assay. But it can also be used with approximate n_0 and D_0 values to find a first estimate of the optimal dose distribution as input to more complex algorithms (cf. Källman et al. 1992a).

This method could be generalized to take normal tissue reactions into account using an equation similar to Eq. (5.42) for the tissue rescuing units of the normal tissues or preferably by using Eq. (5.10) and Eq. (5.8) or (5.9) to express the probability of achieving complication-free tumor control. Similar to the case with Eq. (5.7) above, this results in a variational problem where one wants to maximize P_+ by essentially maximizing the difference between integrals analogous to Eq. (5.42). If we, for simplicity, assume $\delta \approx 0$ and a uniform sensitivity and completely serial organization (cf. Källman et al. 1992a) of surrounding normal tissues, the problem is:

$$\begin{cases} \max P_+ = \exp \left(- \iiint_{V_t} n_0(r) e^{-D(r)^*/D_0(r)} dr^3 \right) \\ \quad + \exp \left(- \iiint_{V_t} v_0(r) \ln\{1 - P_I(D^*(r))\} dr^3 \right) - 1 \\ D(r)^* \geq 0 \end{cases}, \tag{5.44}$$

where v_0 is the relative density of the normal tissues. The condition for Eq. (5.43) to be applicable is thus that the P_I term in Eq. (5.44) is negligible. In clinical practice, the normal tissues are generally dose limiting and the full optimization problem of Eq. (5.44) has to be treated. However, this problem is outside the scope of this chapter.

5.4.6. Gradient Methods

Owing to their universal applicability and general flexibility, the gradient methods will be treated here in more detail. The gradient methods also have the advantage of being much faster than the stochastic methods particularly for convex functionals. Among the successful applications of gradient methods belong the work by Cooper (1978), Lind et al. (1987–1992), Ulsø and Brahme (1988), Bortfeld et al. (1990),

Holmes *et al.* (1991), Söderström and Brahme (1993), Söderström *et al.* (1993, 1995a) Söderström and Brahme (1995b,c), Gustafsson *et al.* 1994, 1995, and Hyödynmaa *et al.* (1996). A natural starting point in the analysis is to make a second order Taylor expansion of the objective function $F(f)$ around some point in the phase space of possible solutions f^k (cf. Sec. 5.2.2):

$$F(f) \cong F(f^k) + \nabla F(f^k)(f - f^k) + \frac{1}{2}(f - f^k)^{\mathrm{T}} \nabla^2 F(f^k)(f - f^k), \quad (5.45)$$

where ∇F is the gradient vector of the objective function:

$$\nabla F^{\mathrm{T}} = \left(\frac{\partial F}{\partial f_1}, \frac{\partial F}{\partial f_2}, \cdots, \frac{\partial F}{\partial f_m} \right), \quad (5.46)$$

and $\nabla^2 F$ is the associated *Hessian* matrix:

$$\nabla^2 F = \left\{ \begin{array}{cccc} \dfrac{\partial^2 F}{\partial f_1^2} & \dfrac{\partial^2 F}{\partial f_2 \partial f_1} & \cdot & \dfrac{\partial^2 F}{\partial f_m \partial f} \\ \dfrac{\partial^2 F}{\partial f_1 \partial f_2} & \cdot & \cdot & \cdot \\ \cdot & \cdot & \cdot & \cdot \\ \dfrac{\partial^2 F}{\partial f_1 \partial f_m} & \cdot & \cdot & \dfrac{\partial^2 F}{\partial f_m^2} \end{array} \right\}. \quad (5.47)$$

From Eq. (5.45) it is clear that the quadratic form in f has its minimum value at f^{k+1} given by:

$$f^{k+1} = f^k - (\nabla^2 F(f^k))^{-1} \nabla F(f^k), \quad (5.48)$$

which is *Newton's* method to find the minimum of the functional F by a systematic iterative search. For most objective functions that are encountered in practice, Newton's method converges rapidly at points near the solution. This is so because sufficiently smooth functions are well approximated by a quadratic form at such points. Unfortunately, the convergence properties usually deteriorate in regions remote from the minimum due to non-quadratic terms in the objective function. Also, for optimization problems with a large number of decision variables it is often too time-consuming to compute the inverse Hessian as required by Newton's method. To deal with these shortcomings, several more practical and efficient methods have been developed. The algorithms that we will consider here all have the common structure that the next iterate is found by minimizing the objective function along some *search direction* v^k, that is,

$$f^{k+1} = f^k + \alpha^k v^k, \tag{5.49}$$

where the search parameter α^k is determined by performing a line search which means to find the α that approximately minimizes the function $\Lambda(\alpha) \equiv F(f^k + \alpha v^k)$. A line search generally involves a sequence of objective function evaluations. The calculation of a search direction, on the other hand, is commonly based on gradient information and a gradient computation often requires about n times the work of one function evaluation. Therefore, the purpose of a line search is to make maximal use of each search direction.

The optimization procedures with the general structure of Eq. (5.49) differ in how the search direction is selected. For example, identifying Eq. (5.48) with Eq. (5.49) gives that the search direction for Newton's method is $v^k = -(\nabla^2 F)^{-1}\nabla F$. Another natural choice for v^k is $v^k = -\nabla F$ since the objective function decreases most rapidly in this particularly direction and the corresponding method is therefore called *steepest descent*. Unfortunately, this method has poor convergence characteristics close to the optimum.

There are two main approaches to improve the convergence of the steepest descent method while avoiding the great computational task of determining the inverse Hessian required by the pure Newton method. These two classes of algorithms are the *quasi Newton* and *conjugate gradient* methods.

The underlying idea for the quasi-Newton method is to use gradient information to construct better and better approximations G^k to the inverse Hessian according to the general iteration formula:

$$G^{k+1} = G^k + \Delta G^k, \tag{5.50}$$

where ΔG^k is a symmetric, low rank correction to the current approximation. This correction is based on information obtained from the Hessian through the basic relation:

$$\nabla F(f^{k+1}) - \nabla F(f^k) \cong \nabla^2 F(f^k)(f^{k+1} - f^k). \tag{5.51}$$

One of the most popular quasi-Newton method is the *Broyden–Fletcher–Goldfarb–Shanno* scheme where the update formula for G^k is given by:

$$G^{k+1} = G^k + \left(1 + \frac{(q^k)^T G^k q^k}{(q^k)^T p^k}\right)\frac{p^k(p^k)^T}{(p^k)^T q^k} - \frac{p^k(q^k)^T G^k + G^k q^k(p^k)^T}{(q^k)^T p^k}, \tag{5.52}$$

where $p^k \equiv f^{k+1} - f^k$ and $q^k \equiv \nabla F(f^{k+1}) - \nabla F(f^k)$ and the search direction becomes $v^k = -G^k \nabla F(f^k)$ in this case.

The second class of algorithms that lie somewhere between the method of steepest descent and Newton's method are the conjugate gradient methods. These algorithms take quite another approach to achieve superlinear convergence without explicit use of the Hessian. In contrast to the quasi-Newton methods, they generate search directions without storing a matrix to approximate the inverse Hessian which is, of course, advantageous in the case of large optimization problems. All conjugate gradient methods are first developed to minimize purely quadratic forms and then modified to account also for general nonlinear functionals. We, therefore, consider the purely quadratic problem:

$$\text{minimize } \frac{1}{2} f^T Q f - b^T f, \tag{5.53}$$

where Q is an $m \times m$ symmetric positive definite matrix. The unique solution to this problem is obtained by differentiating the objective function and setting the gradient equal to zero which yields the linear equation $Qf = b$. A useful notion in this context is that of Q-orthogonality (or Q-conjugacy): two vectors v^1 and v^2 are said to be Q-orthogonal, if it holds that $(v^1)^T Q v^2 = 0$. Assume now that we have somehow constructed a set of m Q-orthogonal vectors $v^0, v^1, \ldots, v^{m-1}$, for example, through a *Gram–Schmidt* procedure. It can easily be shown that these vectors are then linearly independent so that we can expand the solution in terms of them:

$$f^* = \alpha^0 v^0 + \cdots + \alpha^{m-1} v^{m-1}, \tag{5.54}$$

where the α^is are determined by multiplying by Q, taking the scalar product with v^i, and applying Q-orthogonality, that is:

$$\alpha^i = \frac{(v^i)^T b}{(v^i)^T Q v^i}. \tag{5.55}$$

Note that if v^i would have been *orthogonal* instead of Q-orthogonal, the α^is would have been expressed in terms of the unknown solution f^* instead of the known vector. For general nonlinear problems, certain modifications are needed. First, instead of having to predetermine a set of Q-orthogonal directions, we would like to generate those in an iterative manner. One very frequently used conjugate gradient method that determines the direction vectors from successively computed gradient vectors is the *Polak–Ribiere* method, where the next search direction is given by:

$$v^{k+1} = -g^{k+1} + \frac{(g^{k+1} - g^k)^T g^{k+1}}{(g^k)^T g^k} v^k, \tag{5.56}$$

where $g^k \equiv \nabla F(f^k)$ and the α^ks are determined by line search.

To illustrate how the iterative schemes look for different objective functions, we will present them for three of the objective functions in Sec. 5.3. Owing to the fundamental importance of the positivity constraint on the irradiation density f, Lind (1990) showed that the following simplified and modified gradient algorithm is well suited for fast calculations in radiation therapy:

$$f^{k+1} = C(f^k - \alpha^k I \nabla F(f^k)). \tag{5.57}$$

Here, the inverse of the Hessian corresponds to α times the unity matrix, I (cf. also Bortfeld *et al.* 1990) and a positivity operator C has been added to ensure that f never falls below zero during the iterations. It can be shown that the iterative algorithm Eq. (5.57) converges with steepest descent in the quadratic norm Δ_2 since $A^{-1} = I$, in this case (Ortega and Rheinboldt 1970). Equation (5.57) can now be combined with any of the objective functions from Sec. 5.3 above (F_1, F_2, F_+^n) to get a rapidly converging algorithm capable of optimizing different treatment aspects.

(1) With F_1 (Eq. (5.32)), it can be shown that f^∞ generates the dose distribution which is as close to d as possible but never below d. In fact, it can be shown that $Hf^\infty = d$ whenever $f_i \geq 0$ and $Hf^\infty \geq d$ when $f_i = 0$ and according to the *Kuhn-Tucker* theorem (cf. Luenberger 1973) F_1 is minimized by the algorithm (Lind 1990). After insertion of F_1 in Eq. (5.35) we obtain:

$$f^{k+1} = C(f^k - \alpha^k(Hf^k - d)). \tag{5.58}$$

This type of algorithm has earlier been discussed by Schafer *et al.* (1981) in the context of image restoration.

(2) With F_2 (Eq. (5.3)), the algorithm will minimize the deviations between Hf and d in the least square sense and according to Eq. (5.30) the variance of the dose distribution will be minimal and cause a high tumor control according to Eq. (5.29). This algorithm reduces to:

$$f^{k+1} = C(f^k - \alpha^k(H^T Hf^k - H^T d)), \tag{5.59}$$

which is basically the same algorithm as Eq. (5.58) except for the factor H^T (cf. Lind 1991).

(3) Finally, with $-F_+^n$ (Eq. (5.24)) it is possible to maximize the treatment with respect to the expected probability of achieving complication-free tumor control. By using the chain rule, we obtain:

$$\nabla_f(F_+^n(Hf)) = -\frac{\partial Hf}{\partial f}\nabla_{Hf}F_+^n(Hf) = -H^T\nabla_{Hf}F_+^n(Hf). \tag{5.60}$$

For this very important objective function we will only write out the full algorithm for the case of $n = 1$ where Eq. (5.57) takes the form:

$$f^{k+1} = C(f^k - \alpha^k H^T\nabla_{Hf}F_+^1(Hf^k)). \tag{5.61}$$

Similarly for $n = \infty$:

$$f^{k+1} = C(f^k - \alpha^k H^T\boldsymbol{\Phi}\nabla_{Hf}F_+^\infty(Hf^k)), \tag{5.62}$$

where $\boldsymbol{\Phi}$ is the Toeplitz convolution matrix corresponding to φ_T. These last two algorithms differ from the two previous ones in that they are capable of determining the optimal dose distribution and dose level (Hf^∞) directly, whereas Eqs. (5.58) and (5.59) will only generate the desired dose distribution d as truthfully as possible in terms of the treatment objectives (Källman *et al.* 1992b; Lind and Brahme 1992).

There is one remaining problem with all gradient algorithms which is associated with the initial estimate of f^0. If it is chosen outside the convex region of F, convergence is not ensured. A good starting point for Eqs. (5.58) and (5.59) is $f^0 = 0$ or αd which will gain one iteration with Eq. (5.58). For Eqs. (5.61) and (5.62) the initial guess is even more important to ensure convergence and that the algorithm does not get trapped at local maxima. A very good method is to take Hf^0 from Eq. (5.43) and generally approach the optimal solution from above as complications are disregarded in this equation (Källman *et al.* 1992a). All the experience gained with these iterative algorithms is that trapping on local maxima is a very rare event, probably due to the interaction of the large number of variables which may collectively push f away from a local minimum, provided the convex region is not too large compared to the step length of the algorithm. This is particularly true for the pencil beam algorithms, to be discussed next, mainly because of very large number of free variables (see Fig. 5.4 row 2).

5.4.7. *Pencil Beam Methods*

The most general optimization technique would be to take all the degrees of freedom at an advanced radiotherapy department into account as illustrated in Fig. 5.1. By using the finest elemental radiation source, a pencil, or even a point monodirectional beam, different radiation modalities can be combined with a high geometrical resolution and computational accuracy on complex target volumes based on Eqs. (5.1a) and (5.1b). In this way a very large number of beam modalities, beam energies, beam directions, and beam intensity distributions can be combined in a global optimization using pencil beams or more complex composite energy deposition kernels. Compared to other optimization techniques, this method allows direct determination of the incident energy fluence required by the different radiation modalities in the optimization and accurate consideration of all possible degrees of freedom mentioned.

In the numerical formalism, the neutral pencil beams p^m and the charged particle pencil beams π^m are combined in a single pencil beam matrix P, while the corresponding energy fluences $\Psi^m_{E,\Omega}$ and particle fluences $\Phi^m_{E,\Omega}$ can be represented in a single generalized fluence vector denoted Ξ. This generalized fluence could also include the source density of brachytherapy sources or other combined radiation fields such as from a multi-cobalt device or a Piotron with multiple convergent pi-meson beams. The units of the different elements in the P matrix and the Ξ vector will thus depend on whether the particle referred to is charged or neutral and whether brachytherapy or external beam therapy is used. The mapping of Ξ and P on $D(r)$ is illustrated in Fig. 5.1.

If the target volume is irradiated with v_m different beam modalities each with v_E, different energy bins from v_Ω different source positions and the fluence profile for each source position is discretized into v_ω components, then the total number of components in the fluence vector Ξ is $v = v_m v_E v_\Omega v_\omega$. The pencil beam belonging to each fluence component is best discretized on the same grid as the dose distribution vector d. The pencil beam matrix P will therefore consist of $n \times v$ components, where n is the number of components in the vector representation of $D(r)$ as described in Sec. 5.2.2.

The dose distribution vector, the pencil beam matrix, and the fluence vector can, respectively, be denoted d_i, $P_{i\iota}$, and Ξ_ι where the range of i is $1, 2, \ldots, n$ and the range of ι is $1, 2, \ldots, v$. Thus, Latin indices are used for the volume quantities of dimensionality $\mathbf{R}^n$ and Greek indices for the surface quantities of dimensionality $\mathbf{R}^v$.

The pencil beam equations, Eqs. (5.1a) and (5.1b) can now be directly implemented in Eq. (5.57) resulting in the algorithm:

$$\Xi^{k+1} = C(\Xi^k - \alpha^k \nabla_\Xi F(\Xi^k)). \tag{5.63}$$

Unfortunately, F_1 of Eq. (5.32) cannot be used as a physical objective function in the fluence optimization, since Ξ^T is a v-dimensional row vector, while $P\Xi$ and d are n-dimensional column vectors. It is, thus, necessary to use other objective functions such as F_2 and P_+ that are compatible with the dimensions of the relevant vectors and matrices of the new problem (Gustafsson *et al.* 1994, 1995). By analogy with Eq. (5.59) of the previous section, one obtains the following for F_2:

$$\varXi^{k+1} = C(\varXi^k - \alpha^k P^{\mathrm{T}}(P\varXi^k - d)). \tag{5.64}$$

Similarly, for P_+ the iterative scheme can be written as

$$\varXi^{k+1} = C(\varXi^k - \alpha^k P^{\mathrm{T}} \nabla_{P\varXi} P_+(P\varXi^k)). \tag{5.65}$$

As discussed in more detail by Gustafsson *et al.* (1994), both of these equations are similar to Eqs. (5.59) and (5.61) above, but differ in that they can consider a very large number of beam kernels as specified by P. They, therefore, can take almost all important degrees of freedom of modern radiation therapy planning into account.

5.4.8. *Similarities and Differences in Treatment Plan Optimization and Tomographic Imaging*

In connection with the optimization of arc therapy, Lax and Brahme (1982, see Sec. 5.4.2) found that there are fundamental similarities between the optimization method employed and the image reconstruction techniques used in computed tomography. More specifically, they found that the optimal wedge filter for rotation therapy was reminiscent of the spatial filter used in filtered back projection (cf. Fig. 5.9). Furthermore, the associated integral equation has similarities with that of computed tomography (cf. Eq. (5.36) and Brahme *et al.* 1982). This similarity is even more pronounced when comparing filtered back projection with optimization of generalized conformal therapy, where angular dependent nonuniform incident beams are used to realize a uniform target dose (Brahme 1985, 1988). For example, the attenuated Radon transform is linked to both these problems due to the influence of exponential photon absorption in both cases (As shown in Fig 5.10a cf also Eq. (5.36) and Liu *et al.* 1993). These similarities have recently been pursued by several authors for treatment optimization (Cormack 1987, 1995; Cormack and Quinto 1989; Barth 1990; Bortfeld *et al.* 1990; Holmes *et al.* 1991; Brahme 1995). However, there are several serious differences, which make a strict comparison impossible. First, in tomography, we know that there always exists an exact solution from which the projections were collected, at least in the absence of projection noise, whereas for arbitrary desired dose distributions there does not generally exist an exact solution in the form of a set of incident external beam profiles containing non-negative energy fluence. This is so, for the simple reason that in order to treat the tumor to a high dose it is necessary to pass through normal tissues that then always will get damaged, except possibly for superficial tumors on the skin surface. As seen in Fig. 5.10a the best way to treat a small tumor may be by rotating a narrow beam on it from as many directions as possible.

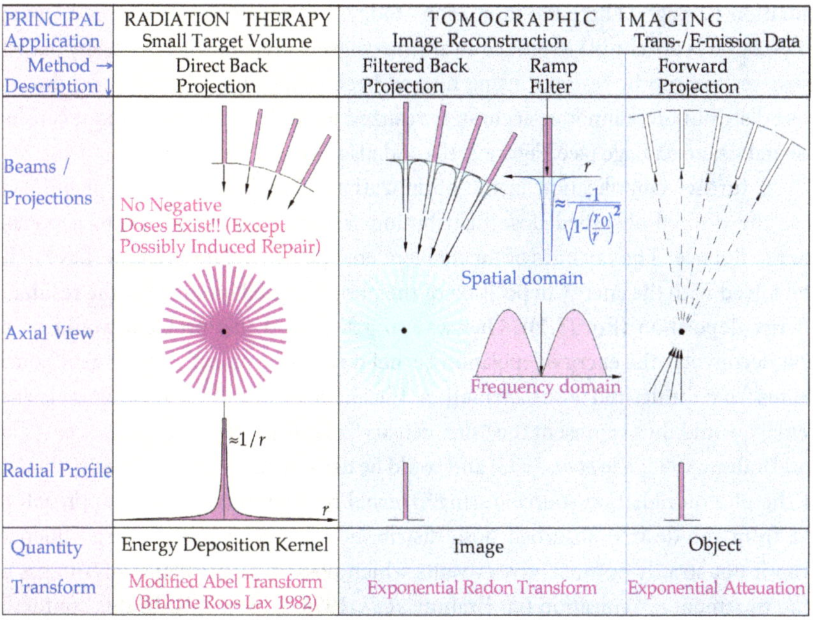

PRINCIPAL Application	RADIATION THERAPY Small Target Volume	TOMOGRAPHIC IMAGING Image Reconstruction		Trans-/E-mission Data
Method → Description ↓	Direct Back Projection	Filtered Back Projection	Ramp Filter	Forward Projection
Beams / Projections	No Negative Doses Exist!! (Except Possibly Induced Repair)		r $\approx \dfrac{-1}{\sqrt{1-\left(\frac{r_0}{r}\right)^2}}$	
Axial View		Spatial domain Frequency domain		
Radial Profile	$\approx 1/r$ r			
Quantity	Energy Deposition Kernel	Image		Object
Transform	Modified Abel Transform (Brahme Roos Lax 1982)	Exponential Radon Transform		Exponential Atteuation

Figure 5.10a. Comparison of the similarities and differences of treatment optimization and image reconstruction. The ramp filter of the latter unfortunately cannot be employed in treatment optimization, because of its negative beam portions. Instead, a convergent planar kernel with essentially a $1/r$ dependence (lower left) may be used in rotation therapy optimizations as also seen in Figs. 5.5 and 5.10b.

An even more fundamental difference is that in tomography the physics of image production comes first and later the theory of image reconstruction is applied to restore the original image from its projections. In radiation therapy, the causality of events is reversed since the desired dose distribution image is known or estimated from clinical knowledge. However, the beam profiles that will give the best conformation to this dose image are unknown. Therefore, in radiation therapy optimization, the mathematical theory has to come first and the physics last, namely when the patient is irradiated by the determined optimal beam shapes. This considerable conceptual difference makes it possible in image restoration to use filter functions in the filtered back projection, which have negative portions as the exponential Radon transform kernel similar to the Abel kernel (cf. Fig. 5.10a left half and Abel 1823 Radon 1917 and Brahme *et al.* 1982) capable of erasing the diffuse tail of each ray line during the back projection. In radiation therapy, this is impossible as a negative energy fluence or dose delivery is physically impossible unless there exists some kind of radiation that could stimulate or induce repair of radiation damage. There are indications today that repair can be induced (Joiner 1994) but it is not simple to induce an arbitrary level of repair. Therefore, any patient cross-section can be restored using filtered back projection, whereas most desired dose distributions cannot be accurately realized in clinical practice without causing normal tissue damage (see Figure 5.10a and also Sec. 5.4.2).

A further complication in the optimization of radiation therapy planning is that the desired absorbed dose distribution is not uniquely related to a certain energy fluence. The ray lines of the incident energy fluence, for example, have to be convolved with the energy deposition of the photon pencil beam to get the resultant energy deposition (Eq. (5.1)). One way to get around this problem would be to first deconvolve the energy deposition kernel due to isotropically convergent pencil beams from the desired dose distribution (Fig. 5.10b right half). The resultant kernel density would then represent the "sink density" of the incident energy fluence (Lind and Brahme 1992; Liu *et al.* 1993) and could be used to determine the energy fluence of the photon radiation source during external rotation therapy. This approach to get from the desired absorbed dose distribution to the required energy fluence, even if not strictly optimal, gives results which possess all the major advantages of true treatment optimization (cf. Brahme *et al.* 1990; Lind 1991). In this context it should also be pointed out that the close $1/r$ dependence of this convergent energy deposition kernel in the plane (cf. Fig. 5.5) is identical to the back projection kernel of the Radon transform (Deans 1983; Natterer 1986).

5.4.9. Treatment Optimization under Consideration of the Stochastic Process of Patient Positioning

Most modern algorithms for radiation therapy optimization make use of nonuniform dose delivery to eradicate the clonogenic tumor cells, while saving the normal tissues. However, these algorithms generally do not take the position uncertainty of the beam relative to the target volume into account. The question posed is: Given that the uncertainties in beam patient alignment is governed by some known stochastic process, how should the incident beams be modified in order to ensure that the desired dose distribution is delivered to the target volume as closely as possible?

An optimal dose distribution which has been computed under the assumption of complete certainty in geometrical location may become far from optimal if somewhat displaced relative to the target volume. This is obvious since such a displacement would probably result in a high absorbed dose being delivered to sensitive organs in the immediate neighborhood of the target tissues at the same time as the target dose is decreased. Taken together, this may result in a loss of local tumor control and increased normal tissue injury.

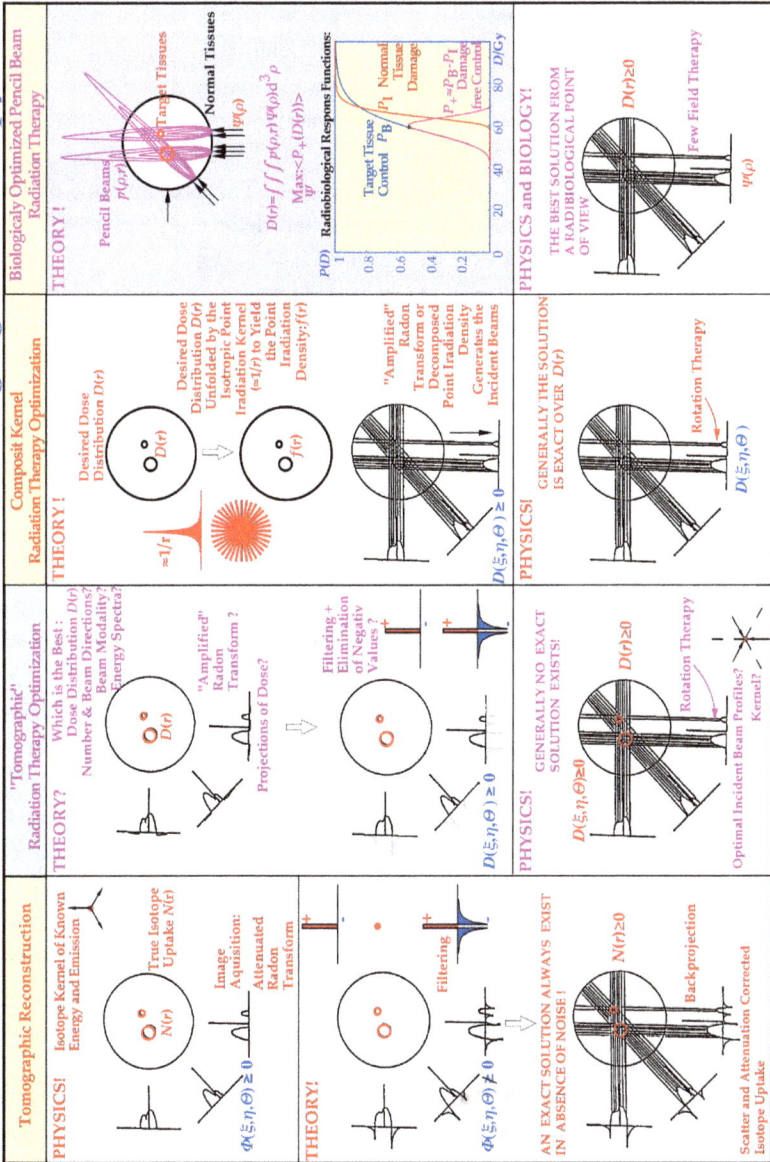

Figure 5.10b. Schematic comparison of the similarities and differences of treatment optimization and image reconstruction. Unfortunately, the ramp filter of the latter cannot be employed in treatment optimization owing to its negative beam portions. Instead, a convergent planar kernel with essentially a $1/r$ dependence (third column) can be used in rotation therapy optimization.

The position of uncertainty stems from two main groups of uncertainties. The first group is due to the internal motions and changes in shape of the different tissues and organs of the human body. These motions and variations in shape may be caused by respiration, heartbeat, peristaltic movements, varying bladder, stomach and intestine contents, etc. These internal dynamic uncertainties have to be accounted for when irradiating the target tissues. This is often done by adding a target margin to the target tissues to obtain a well-defined internal target volume defined in relation to relevant external anatomic landmarks.

The second group of external errors is due to the uncertainty in the relative location of the patient and the anatomical reference points in relation to the therapy beam. This "setup" uncertainty is caused during repositioning of the patient on the treatment coach and the finite mechanical precision in the alignment of a number of devices such as: the initial electron beam from the accelerator, the collimator blocks, the light source and mirror used to simulate the radiation beam, the indication of isocenter, for example, by laser beams and the gantry and treatment couch, but it is also influenced by the accuracy of respective read-out devices and the geometrical resolution of the dose planning system.

By using the expectation value of the complication-free cure as objective function (Eqs. (5.21) and (5.26)), it is possible to develop an algorithm which produces dose profiles that are optimal with respect to a given treatment geometry and robust with respect to a given stochastic process describing the uncertainty in beam patient alignment.

To illustrate the properties of the stochastic optimization, it is here applied to an asymmetric treatment geometry which consists of a one-dimensional target and two organs at risk, named left and right, with low and high radiation sensitivity, respectively, and the tumor grows in the most sensitive organ which still populates 10% of the target volume. It will be illustrated how the optimal dose distribution varies with the number of dose fractions given. This was carried out for a standard deviation σ of the Gaussian distribution governing the beam patient displacements, which was assumed to be 10 mm, a realistic value for the pelvic region. The results of these calculations are shown in Fig. 5.11.

Included in this diagram is the optimal dose when there is no uncertainty at all, that is, $\sigma = 0$. The desired dose is simply a uniform dose over the target region (since the tumor is uniform) and zero doses in the surrounding normal tissues. In this case, the algorithm merely needs to balance the benefit from increasing the dose within the target volume against the resulting injury of the organ in which the tumor grows. The highest achievable P_+ for $\sigma = 0$ is then 0.916 at a total dose level of 63.4 Gy. The other curves are optimal beam profiles for a varying number of dose fractions, n. The first observation we make is that the curves become increasingly

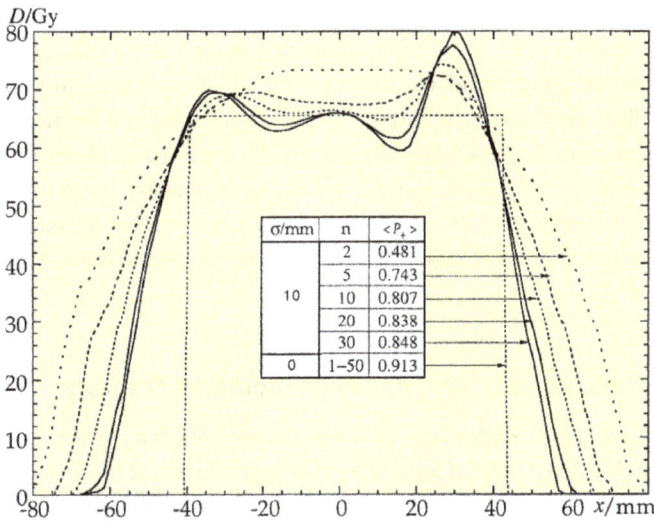

σ/mm	n	$<P_+>$
	2	0.481
	5	0.743
10	10	0.807
	20	0.838
	30	0.848
0	1–50	0.913

Figure 5.11. Illustration of the use of the linear quadratic model for cell survival. Shown are the optimal dose distributions with $n = 2, 5, 10, 20$, and 30 for an asymmetrical case with a sensitive organ at risk to the right and a more resistant one to the left. For comparison, the solution for $n = 30$ when the linear model is used, is also included. The severe oscillations are smoothed out and cold spots are prevented when using this more authentic biological model.

oscillatory with increasing n. This is consistent with the results of Lind *et al.* (1993) who solved the problem for the same treatment geometry but only for the two special cases of one dose fraction and infinitely many dose fractions. For $n = 1$, the present solution is in complete agreement with their result and the tendency of increasing oscillations is confirmed as the previous solution for $n = \infty$ showed oscillations from about 20 to 140 Gy within the target region. The second observation is that the optimal beams for larger n use two complementary methods to mitigate the effects of the uncertainty: the beam width is extended outside the tumor region and overcompensated at the tumor edges. It is also clear that the relative use of these two methods depends on the sensitivity of the neighboring organs: if the organ in question is very sensitive, it is better to overcompensate the beam than extending the dose distribution into the sensitive organ and vice versa. The optimal solution for 30 dose fractions results in the expectation value $EP_+ = 0.854$. This is surprisingly close to the value $EP_+ = 0.916$ obtained when $\sigma = 0$, considering that the uncertainty given by $\sigma = 10$ mm is rather large.

The oscillating behavior seems to be a fundamental characteristic of the optimal beam profiles. The shape of the beam near the target boundaries is of course the most critical in order to achieve robustness with respect to displacements. This fact is also illustrated during the optimization process where it is seen that the algorithm completes the appropriate overcompensation and extension first since this leads to the most rapid increase of EP_+. The two outermost peaks then trigger oscillations toward the interior of the target region. The mechanism for the induction of oscillations is that the parts of the dose distribution just inside the two peaks have to be reduced in order not to cause excessive harm to the healthy tissue there. These local reductions in dose, in turn, require an increase of the dose further into the target region so that local tumor control is not lost there and so on.

5.4.10. *Optimization of Adaptive Radiation Therapy*

Radiobiological optimization in radiation therapy planning often gives rise to nonuniform fluence profiles that tightly conform to the shape of the target volume. The resulting optimal beam profiles may contain quite sharp dose gradients both inside the target and in the transition region between the target and healthy normal tissues. It is, therefore, clear that a biologically optimized dose plan could easily lose its advantages when subject to uncontrolled displacements of the patient relative to the incident beams or if the patient geometry and beam profiles differ considerably during dose delivery from those derived during the treatment optimization process. Hence, for successful delivery of an optimal conformal treatment, it is of paramount importance that the patient be carefully immobilized and that extreme caution is taken when trying to reproduce the treatment setup during the course of treatment. However, even with the most advanced patient fixation methods, there will always remain discrepancies from the planned position, especially with regard to deviations that are due to internal temporal changes in the patient's anatomy. To achieve good treatment results, it is critical that such remaining uncertainties are properly accounted for in the optimization procedure so that the delivered dose distribution is robust with respect to small geometrical changes.

By using portal imaging, an increasing amount of information about the setup errors is obtained during the course of fractionated treatment. It is, therefore, desirable that this information is fed back to the treatment and utilized for setup adjustments in subsequent treatment fractions. For this purpose, extensive efforts have been made to measure and analyze setup uncertainties in patient populations (Hunt *et al.* 1995; Creutzberg *et al.* 1996; Tinger *et al.* 1996; Mavroidis *et al.* 2002) and to develop protocols for inter-treatment correction strategies (Bel *et al.* 1993; El-Gayed *et al.* 1993; Shalev and Gluhchev 1994; Yan *et al.* 1995). Consequently, adaptive feedback algorithms have been developed for determining optimal corrections to both the geometric setup and the beam profiles, given individual patient data obtained by a number of treatment verification methods. The data obtained by such methods as *in vivo* patient dosimetry, portal imaging, and repeated CT, MR, or laser scans are used by these algorithms to optimize the dose delivery during the remainder of the treatment course. The solution of this problem yields the optimal incident fluence Ξ^* to be delivered in every fraction and the corresponding optimal fractional dose distribution d^*, which is then robust with respect to organ motion (see Fig. 5.12). For clarity and speed of computation, the behavior of an adaptive feedback algorithm has been investigated by applying it to a simplified one-dimensional geometry where the tissues are distributed along the x axis (see Fig. 5.13) (Löf *et al.* 1998). The organ motion consists of a tumor that moves in surrounding stationary healthy tissue and whose exact position is unknown at each dose delivery. The displacements of the tumor for the n fractions are modeled by a sequence of stochastic variables, $X_1, X_2, \ldots, X_n$, which are assumed to be independent and normally distributed with zero mean value and known variance σ^2. Here, the only stochastic quantities of the patient geometry are the displacement variables X_ν.

5.4.10.1. *Adaptive interfractional correction of the fluence profiles*

During each treatment session, a fully integrated portal image is assumed to be recorded to detect the mean field edge misalignment and to allow calculation of the realized fluence profiles on the patient surface relative to anatomical landmarks. Since portal imaging generally does not render any changes in organ position or shape compared with the initial simulation or CT scan, the objective throughout the treatment series will be to try to deliver d^* as closely as possible. The purpose of separating the positioning uncertainties into measured and non-measured quantities now becomes clear: the stochastically optimized fluence profiles Ξ^* are shaped to handle the "invisible" internal variations in the patient geometry, whereas the portal imaging information is used to adjust the energy fluence incident onto the patient so that the right dose is delivered despite setup uncertainties. A schematic illustration of the treatment planning, monitoring of treatment, and correction of dose delivery is shown in Fig. 5.12 (Löf *et al.* 1998).

A natural state description in this context is the accumulated fluence profiles on the patient surface from all the beams up to and including the νth fraction, which is denoted by $\Xi^\Sigma(\nu)$. The dynamics of the discrete time system is then given by the

The Adaptive Control Algorithm

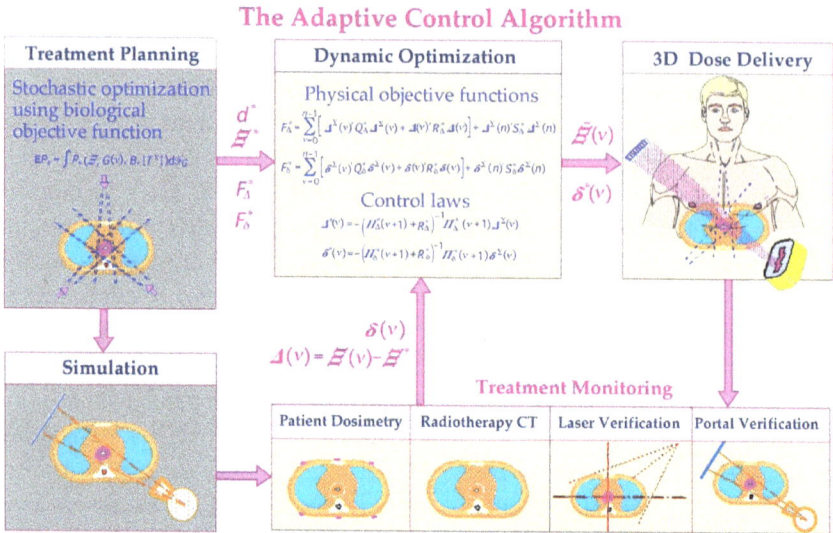

Figure 5.12. Schematic illustration of the adaptive feedback algorithm. Organ motion is accounted for during treatment planning by solving a stochastic optimization problem. The solution of this problem yields the fluence profiles Ξ, and the corresponding dose distribution d^*, which make the treatment relatively insensitive to organ motion. Before the treatment begins, the optimal control functionals, F_Δ^* and F_δ^*, that determine the control laws for interfractional setup adjustments, are also computed (cf. Sec. 5.4.10.3). The aim during the treatment execution is to deliver, as closely as possible, the stochastically optimized fluence distribution despite the influence of setup uncertainties. This is done by monitoring the treatment and modifying the dose delivery based on the information gained. Before each dose delivery the patient's setup position is corrected by $\delta^*(\nu)$ (cf. Sec. 5.4.10.2) and the fluence profiles by $\Delta^*(\nu)$ (cf. Sec. 5.4.10.1). The fluence distribution desired for the next fraction then becomes $\hat{\Xi}(\nu) = \Xi^*(\nu)\Delta^*(\nu)$. The actual fluence distribution $\Xi(\nu)$ delivered and the setup error $\delta(\nu)$ for this fraction are determined by using portal imaging and other monitoring devices and will be used for the computation of subsequent corrections (cf. Fig. 8.29a, for more details on radiation therapy imaging).

linear difference equation (Löf *et al.* 1998):

$$\begin{cases} \Xi^\Sigma(\nu+1) = \Xi^\Sigma(\nu) + \Xi(\nu) & (\nu = 0, 1, \ldots, n-1) \\ \Xi^\Sigma(0) = 0 \end{cases}, \qquad (5.66)$$

where $\Xi(\nu)$ is the energy nuence that has been delivered in the νth fraction. Because of setup variations, it will be impossible to deliver exactly Ξ^* in every treatment session and we would like to compensate for the errors in previous fractions by correcting the subsequent deliveries. Let $\Delta(\nu)$ denote a correction to Ξ^* and let

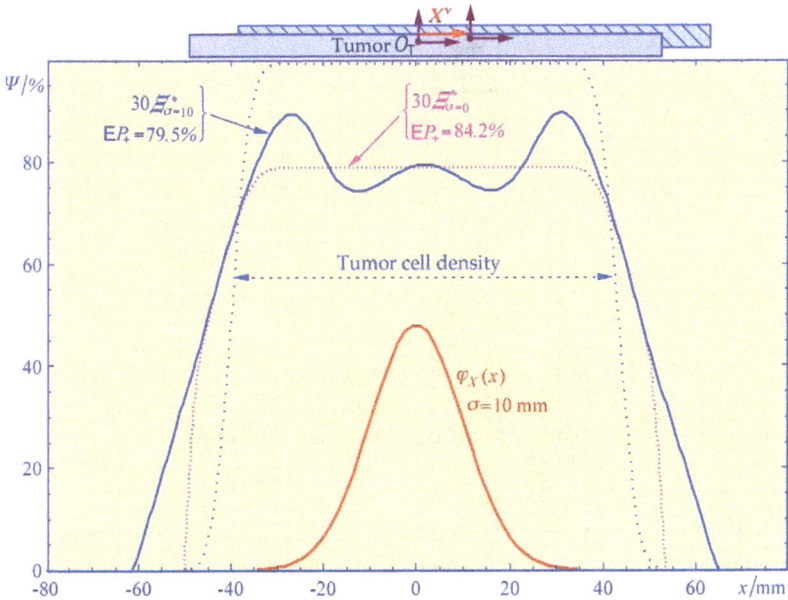

Figure 5.13. Stochastically optimized fluence profile $\Xi^*_{\sigma=10}$ (full curve) accounting for internal organ motion due to the uncertainty in the location of the target volume in the anatomical coordinate system O_A defined by external anatomic reference points. This optimized beam is significantly modified compared to the optimal fluence distribution $\Xi^*_{\sigma=0}$ (dotted curve) for the case without internal organ motion to mitigate the effects of tumor movements. First, the optimal fluence distribution is overcompensated near the tumor edges and second, it is widened relative to the projection of the assumed tumor cell density. Also shown in this figure is the Gaussian probability density $\varphi_X(x)$ for $\sigma = 10$ mm describing the stochastic displacements of the tumor relative to O_A.

$\Delta^{\Sigma}(v)$ represent the deviation from the nominal state according to:

$$\Delta^{\Sigma}(v) = \Xi^{\Sigma}(v) - \Xi^{\Sigma*}(v). \qquad (5.67)$$

The time-dependent vector $\Xi^{\Sigma*}(v)$ is the nominal trajectory through the fluence phase space and is taken as $\Xi^{\Sigma*}(v) = v\Xi^*$, here. However, in the general case, the desired fluence profile might vary from fraction to fraction, as in the shrinking field techniques. Cases like that can be handled by the algorithm in its present form without any modification.

To optimally correct the dose delivery, the functional to be minimized consists of the weighted combination of three terms which expressed the main aspects of the objective, namely to reduce the influence of setup errors. A first term, $\Delta^{\Sigma}(v)' Q_{\Delta} \Delta^{\Sigma}(v)$, is used to prevent too large deviations from the nominal trajectory anywhere through the phase space of fractionated dose delivery, a second term, $\Delta(v)' R_\Delta \Delta(v)$, represents the obvious objective that the corrections should not be larger than necessary, and the last term, $\Delta^\Sigma(n)' S_\Delta \Delta^\Sigma(n)$, should help minimize the difference between the accumulated delivered and the precalculated optimal total incident fluence profile. The relative importance of these objectives is given by the diagonal matrices Q_Δ, R_Δ, and S_Δ, and prior to the start of the treatment an *optimal importance weighting* can be determined for them (see Sec. 5.4.10.3).

Consider now the following dynamic optimization problem:

$$
\begin{cases}
\text{minimize } F_\Delta = \sum_{v=0}^{n-1} [\Delta^\Sigma(v)' Q_\Delta \Delta^\Sigma(v) + \Delta(v)' R_\Delta \Delta(v)] + \Delta^\Sigma(n)' S_\Delta \Delta^\Sigma(n) \\
\text{with respect to strategies of fluence corrections } \{\Delta(0), \Delta(1), \ldots, \Delta(n-1)\}
\end{cases}
$$
$$(5.68)$$

that is, find the sequence of fluence corrections that minimizes F_Δ. From *optimal control theory* (Stengel 1994), we are now able to compute the fluence delivery desired for the next treatment fraction, taking into account the outcome of all previous dose fractions according to:

$$\hat{\Xi}(v) = \Xi^*(v) + \Delta^*(v), \tag{5.69}$$

where the optimal correction is given by

$$\Delta^*(v) = -(\Pi_\Delta(v+1) + R_\Delta)^{-1} \Pi_\Delta(v+1) \Delta^\Sigma(v), \tag{5.70}$$

and the matrix $\Pi_\Delta(v)$ satisfies the *discrete-time Riccati equation*:

$$
\begin{cases}
\Pi_\Delta(v) = Q_\Delta + \Pi_\Delta(v+1) - \Pi_\Delta(v+1)(\Pi_\Delta(v+1) + R_\Delta)^{-1} \Pi_\Delta(v+1) \\
\Pi_\Delta(n) = S_\Delta
\end{cases}.
$$
$$(5.71)$$

5.4.10.2. *Adaptive interfractional correction of the positions of the incoming beams*

Since, in general, errors in delivered nonuniform dose distributions are smaller than those caused by patient displacements, it is more efficient to treat patient displacements separately. To best exploit the information that is available through portal imaging, we define a second state description, which is complementary to the accumulated energy fluence state. The vector $\delta(v)$ contains the discrepancy, in terms of rotation and translation, between the patient's actual and nominal positions in the vth fraction. This vector describes the patient's setup error, that is, the difference between the measured position derived from portal imaging and the planned position, as given by simulation or digitally reconstructed radiographs. The errors could be determined by using *chamfer matching* or other image matching techniques. The position state description is defined as the accumulated position errors up to and including the vth fraction and is denoted by $\Delta^{\Sigma}(v)$. The dynamics of positioning is then described by the difference equation:

$$
\begin{cases}
\delta^{\Sigma}(v+1) = \delta^{\Sigma}(v) + \delta(v) \quad (v = 0, 1, \ldots, n-1) \\
\delta^{\Sigma}(0) = 0
\end{cases}
, \qquad (5.72)
$$

and to obtain the optimal positioning corrections, the following optimization problem is studied:

$$
\begin{cases}
\text{minimize } F_{\delta} = \sum_{v=0}^{n-1} [\delta^{\Sigma}(v)' Q_{\delta} \delta^{\Sigma}(v) + \delta(v)' R_{\delta} \delta(v)] + \delta^{\Sigma}(n)' S_{\delta} \delta^{\Sigma}(n) \\
\text{with respect to strategies of positioning corrections } \{\delta(0), \delta(1), \ldots, \delta(n-1)\}
\end{cases}
.
$$

$$(5.73)$$

Here, an optimal sequence of positioning corrections is sought such that the functional F_{δ} is minimized. The situation is perfectly analogous to the problem described in Eq. (5.68) above and the importance weighting of the three objectives given by the diagonal matrices Q_{δ}, R_{δ}, and S_{δ}, may also in this case be pre-optimized as described in Sec. 5.4.10.3.

The optimal position correction for an upcoming treatment delivery taking the positioning errors in all the previous treatments into consideration is then given by:

$$
\delta^*(v) = -(\Pi_{\delta}(v+1) + R_{\delta})^{-1} \Pi_{\delta}(v+1) \delta^{\Sigma}(v), \qquad (5.74)
$$

where the matrix $\Pi_{\delta}(v)$ again solves the corresponding Riccati equation:

$$
\begin{cases}
\Pi_{\delta}(v) = Q_{\delta} + \Pi_{\delta}(v+1) - \Pi_{\delta}(v+1)(\Pi_{\delta}(v+1) + R_{\delta})^{-1} \Pi_{\delta}(v+1) \\
\Pi_{\delta}(n) = S_{\delta}
\end{cases}
. \qquad (5.75)
$$

Before each treatment fraction, the patient is first rotated and translated according to $\delta^*(v)$, after which the corrected fluence $\hat{\Xi}(v)$ for the treatment fraction is delivered (see Fig. 5.12). This delivery will, of course, also be subjected to positioning disturbances, so the resulting fluence distribution $\Xi(v)$ and positioning error $\delta(v)$ are measured and their phase space trajectories transferred to the next states, $\Xi^{\Sigma}(v+1)$ and $\delta^{\Sigma}(v+1)$. The new states are subsequently used for correction of the following treatment, and the whole process is repeated so that uncorrected errors are accumulated until finally corrected.

5.4.10.3. *Optimal importance weighting of the two control functionals*

By using a statistical model for the setup errors expected for the treatment site and treatment technique in question, we can even optimize our control laws (Eqs. (5.70) and (5.75)), in the sense that they furnish sequences of treatments that maximize the expectation value of P_+. This requires that the geometric probability distribution F_G and the beam patient setup probability distribution F_Ψ are approximately known. For this purpose, consider the two general quadratic control functionals in the two optimization problems described in Eqs. (5.68) and (5.74). The following quite unusual but interesting optimization problem then has to be solved in order to adapt the control to the particular patient geometry to be treated, the anticipated setup error distribution and the stochastically optimized fluence profile for zero setup errors (Löf *et al.* 1998):

$$
\begin{cases}
\underset{F_\Delta, F_\delta}{\text{maximize}} \; \mathrm{E}P_+ = \iint P_+(\{\tilde{\Xi}(v)\}, \{\tilde{g}(v)\}) \mathrm{d}F_\Psi \mathrm{d}F_G \\
\text{where } \{\Xi(v)\} \text{ minimizes } F_\Delta \text{ and } \{\delta(v)\} \text{ minimizes } F_\delta.
\end{cases}
\qquad (5.76)
$$

In other words, we want to find the two functionals F_Δ^* and F_δ^* such that the sequences of fluence distributions $\{\Xi(v)\}$ and positioning corrections $\{\delta(v)\}$ for which these functionals are minimized at the same time maximizes the expectation value of P_+. Hence, the solution to this optimization problem yields the six matrices Q_Δ^*, R_Δ^*, S_Δ^*, Q_δ^*, R_δ^*, and S_δ^* that determine the optimal objectives F_Δ^* and F_δ^*.

5.4.10.4. *Dynamic optimization using the biological objective function*

At centers where diagnostic quality portal imaging or CT is available in the treatment room such that images can be acquired immediately before or after each treatment, it is possible to use biological objective functions to determine an optimal strategy of fluence profiles. If we measure the current patient geometry before the dose is delivered and measure the exit dose by portal imaging, we can quite accurately reconstruct the dose distribution that was actually delivered to the patient (McNutt *et al.* 1996). For simplicity, we assume a biological model in which the logarithm of the survival fraction is only linearly dependent on dose. The appropriate radiotherapy state space description is then the accumulated dose distribution in the patient up to and including the vth fraction:

$$d^\Sigma(v) = (d_1^\Sigma(v), d_2^\Sigma(v), \ldots, d_{n_0}^\Sigma(v)), \tag{5.77}$$

where $d_j^\Sigma(v)$, $(j = 1, 2, \ldots, n_0)$ is the accumulated dose distribution in the jth organ or tumor. If the linear–quadratic response model is used, the state has to be augmented by the accumulated squares of the fractional dose distributions. This extension of the state is straightforward but unnecessarily complicates the notation in the following equation. The dose distribution in the patient after delivery is first computed in the anatomical coordinate system O_A and then transformed to the local organ coordinate systems according to their location and how respective organ has been deformed. The dose in every fraction is given by $d(v) = P(v)\Xi(v)$ and the dynamics becomes in this case:

$$\begin{cases} d^\Sigma(v+1) = d^\Sigma(v) + P(v)\Xi(v) & (v = 0, 1, \ldots, n-1) \\ d^\Sigma(0) = 0 \end{cases}. \tag{5.78}$$

The probability to achieve uncomplicated tumor cure when v fractions have been delivered is given by:

$$P_+(v) = P_+(d^\Sigma(v)) \quad (v = 1, 2, \ldots, n). \tag{5.79}$$

With the selected state descriptor, we see that in order to be able to compute P_+ $(d^\Sigma(v+1))$, all that is needed is the state in the previous fraction $d^\Sigma(v)$, the fluence to be delivered $\Xi(v)$, and the measured patient geometry $g(v)$ (since it affects the pencil beam matrix $P(v)$) independently of how $d^\Sigma(v)$ was reached. This is an important property of the state descriptor and the objective function, called the *Markov property*, and it is required by the method of dynamic optimization. In every treatment fraction, the aim is to give the best possible treatment $\Xi(v)$ based on the current state $d^\Sigma(v)$ and patient geometry $g(v)$. In other words, we want to be able to evaluate by how much the probability of survival increases for a given fluence profile. A natural measure for this purpose is the resultant *incremental* P_+ defined by:

$$\Delta P_+(d^\Sigma(v), \Xi(v), g(v)) = P_+(d^\Sigma(v+1)) - P_+(d^\Sigma(v)). \tag{5.80}$$

Notice that knowledge of $\Xi(v)$ and $g(v)$ allows for computation of $d^\Sigma(v+1)$ and thereby the value of $P_+(d^\Sigma(v+1))$, which means that $\Delta P_+(d^\Sigma(v), \Xi(v), g(v))$ is well defined. The nonlinear dynamic optimization problem with biological objective function can now be formulated as

$$
\begin{cases}
\text{maximize } P_+(d^\Sigma(n)) = \displaystyle\sum_{\nu=0}^{n-1} \Delta P_+(d^\Sigma(\nu), \Xi(\nu), g(\nu)) \\[2mm]
\text{with respect to strategies of energy fluence particles} \\
\quad \{\Xi(0), \Xi(1), \ldots, \Xi(n-1)\}
\end{cases}
, \qquad (5.81)
$$

that is, find the sequence of fluence profiles $\Xi(0), \Xi(1), \ldots, \Xi(n-1)$ that maximizes the sum of incremental P_+ or the total $P_+(d^\Sigma(n))$.

5.4.10.5. *Implementation of adaptive radiotherapy*

For clarity and simplicity, the present algorithm has been applied to simplified one-dimensional treatment geometries where the healthy and tumor-bearing tissues are distributed along the x axis. During treatment planning, a biologically optimized fluence distribution that accounts for organ motion is determined. In a sample case, where we have a purely random setup uncertainty with $\sigma_{\text{setup}}^{\text{rand}} = 10\,\text{mm}$, the fluence dynamics without and with corrections is shown in Figs. 5.14(a) and 5.14(b), respectively (Löf *et al.* 1998). The Gaussian displacements erode the structure of the incident fluence profile as displayed in Fig. 5.14(a) and the resulting distribution significantly deviates from the desired biologically optimized profile. However, in spite of the very low precision in the patient setup, it is possible to achieve the desirable fluence distribution by continuously adjusting the beam through optimal fluence corrections to compensate for the errors in previous fractions. The final accumulated profile agrees very well with the desired profile, as is seen in Fig. 5.14(b). Taken together, the two strategies of positioning and fluence corrections provide a very efficient method to eliminate the effect of setup uncertainties, both random and systematic. It is also evident that the assumed setup probability distribution F_ψ is not important for the robustness of the control laws. This is so, since the setup uncertainties that were assumed in this hypothetical case were much larger than the standard deviations used for the optimization of the control laws but could still be well accounted for by these controls.

We will apply the algorithm to a more realistic case where both random and systematic setup errors are present at the same time. We study the treatment of a patient who was systematically displaced 9 mm to the right from the onset of the treatment, as described by a translation of the local anatomical coordinate system O_A. The standard deviation of the tumor motion was $\sigma_{\text{org}} = 10\,\text{mm}$ and the

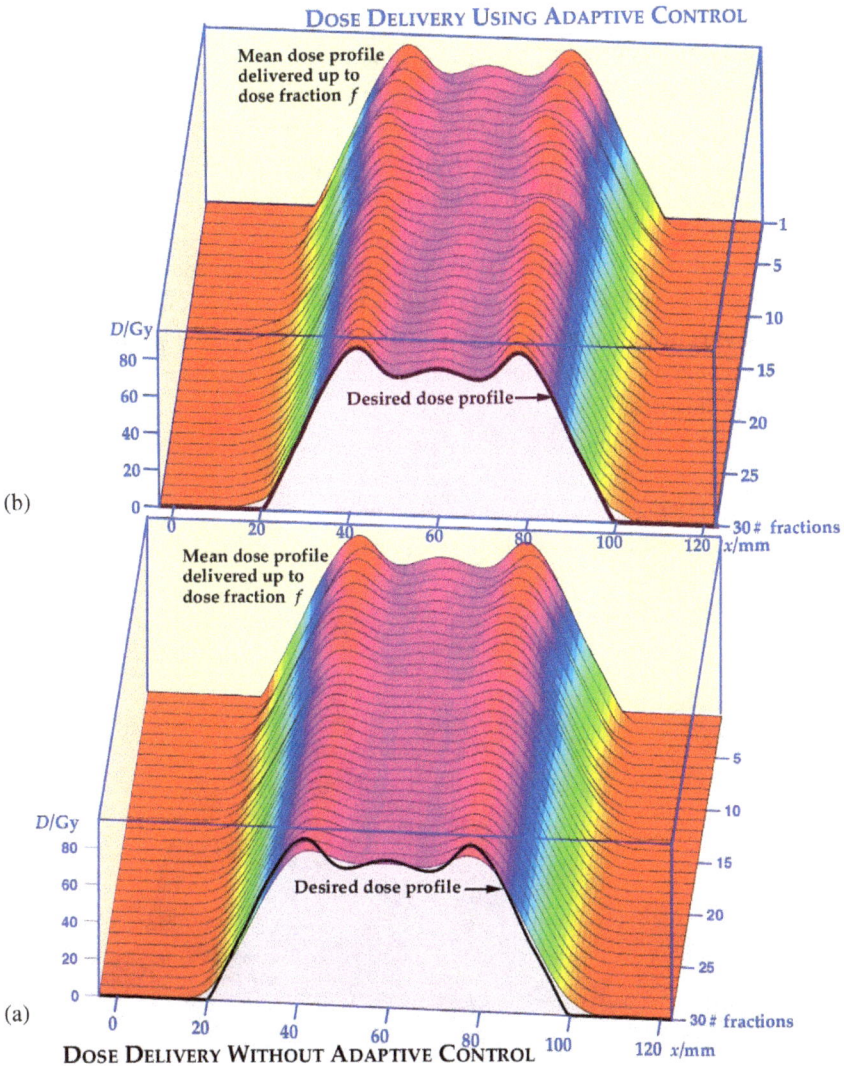

DOSE DELIVERY USING ADAPTIVE CONTROL

Mean dose profile delivered up to dose fraction f

Desired dose profile→

D/Gy

(b)

Mean dose profile delivered up to dose fraction f

Desired dose profile→

D/Gy

(a)

DOSE DELIVERY WITHOUT ADAPTIVE CONTROL

Figure 5.14. (a) The mean accumulated fluence distribution (scaled to $v = 30$) when the setup uncertainties are assumed to contain only a Gaussian random component with standard deviation σ_{setup}^{rand} mm is shown. In this figure, the fluence deliveries are not corrected for the imprecision in patient setup and the accumulated dose profile therefore smooth out due to the random displacements. Consequently, the desired dose profile is not delivered to the patient. (b) The mean accumulated fluence distribution for the same statistical sample is shown as in (a), but in this case the fluence profile is corrected before each dose delivery in order to reduce the effects of previously measured random fluctuations in patient setup. As seen in the figure, the desired fluence profile can readily be achieved by intensity modulating the optimal fluence profiles according to the new adaptive control algorithm.

standard deviation for the random component of the setup error was $\sigma_{\text{setup}}^{\text{rand}} = 2$ mm. The resultant accumulated fluence distribution delivered by the adaptive feedback algorithm is Ξ^{Σ} (30) and it is remarkably close to the nominal fluence distribution $\Xi_{\sigma=10}^{*}$, particularly in view of the rather large systematic setup uncertainty for this patient. The robustness of the algorithm is confirmed by the slight decrease of the expectation value of P_{+}, from 79.5% to 79.3%, due to the setup variations. $\Xi_{\text{noc}}^{\Sigma}$ (30) is the resulting total fluence, if *no control* actions are taken. However, the desired overall shape is retained due to the smallness of the random fluctuations. In a case where the systematic error in position is much larger than the corresponding random component, the main strategy of the algorithm will be to eliminate the systematic component by positioning corrections, while the fluence corrections will play a subordinate roll.

The time evolution of the accumulated fluence distribution $\Xi^{\Sigma}(v)$ during the 30 dose fractions is displayed in Fig. 5.15 (Löf *et al.* 1998). Again, we see that the first few fractions end up in the wrong position but that the algorithm soon detects the systematic error to eliminate it so that the total fluence almost exactly

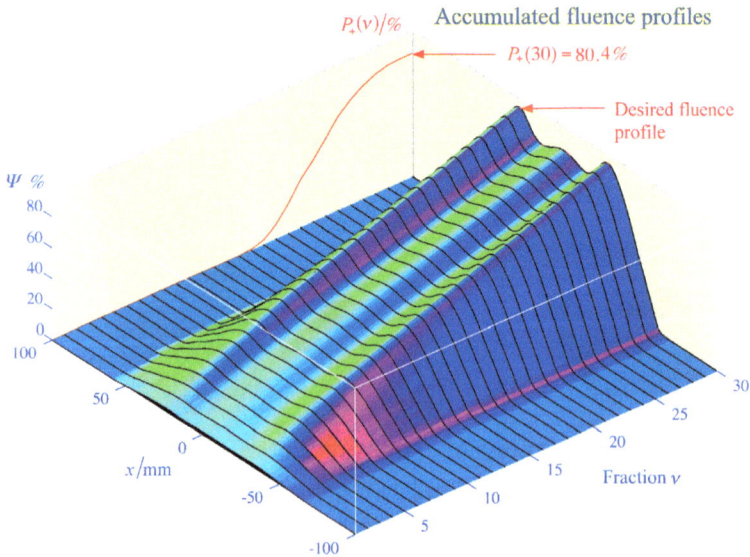

Figure 5.15. The temporal evolution the accumulated fluence $\Xi^{\Sigma}(v)$ for a single patient is illustrated. The desired total fluence profile $\Xi^{\Sigma*}(30)$ is almost exactly achieved despite the setup errors, due to the profile and position corrections generated by the algorithm. The first few dose fractions end up displaced to the right in the figure due to this patient's systematic error. Also shown is the increasing probability of achieving complication-free tumor control as the treatment proceeds.

coincides with the desired fluence profile. As the treatment proceeds, an increasing amount of the dose is delivered to the target volume and when all efforts are made to limit the dose in the surrounding organs at risk, the probability of survival for the patient should correspondingly increase. The probability of uncomplicated tumor control as a function of treatment fraction, $P_+(v)$, is also plotted in Fig. 5.15 and it appears that benefit from the treatment becomes non-zero first after the seventeenth fraction and that the increase of $P_+(v)$ levels out in the last few fractions to reach its maximum value $P_+(30) = 80.4\%$ for this sample patient.

Examining the increase of the biological objective function for several different beams and treatment strategies, it is found that the least successful method in the case that geometrical uncertainties exist is, of course, dose delivery using the same beam shape $\varXi^*_{\sigma=0}$ without corrections. The resulting accumulated fluence is denoted by $\varXi^{\Sigma,noc}_{\sigma=0}(v)$. A better technique is obviously to apply optimal margins to account for organ motion and setup uncertainties. The resulting fluence is called $\varXi^{\Sigma,noc}_{uniform}(v)$. The best strategy is to use the combination of the stochastically optimized beam $\varXi^*_{\sigma=10}$ to account for organ motion and the adaptive control algorithm to correct for the setup uncertainties, giving the resulting accumulated beam sequence $\varXi^\Sigma(v)$.

5.5. Simultaneous Optimization of Beam Orientation, Intensity Modulation, and Dose Delivery Varians in Radiation Therapy Using P_{++} Optimization Strategy

During the past decades, radiotherapy planning and treatment optimization have gone through a very dramatic development. The resultant improvements in treatment outcome have both been due to the development of new treatment principles and treatment objectives such as inverse therapy planning using biological or physical objective functions, and due to the introduction of new treatment techniques using dynamic or segmented multileaf collimation and narrow scanned pencil beams. The gain in treatment outcome is largely obtained through an increased therapeutic window since the mean dose to the tumor can be increased at the same time as the dose distribution in the most sensitive normal tissues is substantially decreased by the use of intensity-modulated beams (Brahme 1999).

Simultaneous optimization of the angle of incidence of the beams and the intensity modulation of the constituent pencil beams using one and the same objective function is often quite difficult to manage since only that part of objective function that describes the intensity modulation is generally a convex function (Brahme 1995). The angular dependence of the objective function is often concave with multiple local extreme values and fairly large fluctuations as a function of the angle of incidence also in multiple beam configurations (cf. Söderström and Brahme 1993, 1999). Using previously developed generalized pencil beam optimization codes, it was possible to scan the whole phase space of complication-free cure as a function of angle of incidence for 1, 2, and 3 beam portal techniques with photons (Söderström *et al.* 1993–1999) and 1 and 2 portal techniques with electrons or electrons combined with photons (Åsell *et al.* 1999). These studies have clearly demonstrated the existence of multiple extreme values often as many as 2–4 with a single beam and 8 with two simultaneous beams. At low-electron energies, the phase space looks very different due to the finite range of the electrons but at high electron energies beyond about 30 MeV and at all photon energies, the shape of the phase spaces is fairly similar. An interesting early observation with very flexible intensity-modulation devices, such as multiple narrow scanned photon, proton, and electron was that the angle of incidence became less and less important as the number of beams increased. However, these early studies were all limited to 2D planning using one single slice representing the average target shape. The new algorithm presented here is more efficient and accurate and capable of handling full 3D patient geometries with intensity and angle of incidence optimization in the same computational time as the older algorithms did 2D geometries with just intensity optimization. In true 3D treatment planning, the optimization of the angle of incidence is a more important degree of freedom than previously found in 2D geometry. This is partly due to the lower accuracy and 2D restriction of previous pencil beam algorithms in exhaustive search calculations but also due to the fact that the optimal beam directions for each slice through the patient may differ. The considerably increased accuracy obtained by closely integrating the optimization code with an advanced clinically proven forward treatment planning code has also been an important factor in the present development of a clinically mature and very flexible optimization code.

In its present form only coplanar optimization with full freedom in selecting the angle of incidence is possible. However, the optimization can be combined with one or several fixed noncoplanar beam directions to find the best coplanar augmentation. In an upcoming release, full noncoplanar optimization will be available. The optimization code is very flexible as it is written in an object-oriented code (C++, Orbit, RaySearch Lab, Stockholm, Sweden; www.raysearchlabs.com/ Default____4.aspx?epslanguage =EN) to facilitate developments and simplify the integration with modern forward treatment planning codes.

5.5.1. *Target Volume and Organs at Risk*

The patient used for the present study of new treatment optimization methods had a complex prostate tumor with local spread to the seminal vesicles. Since the target had

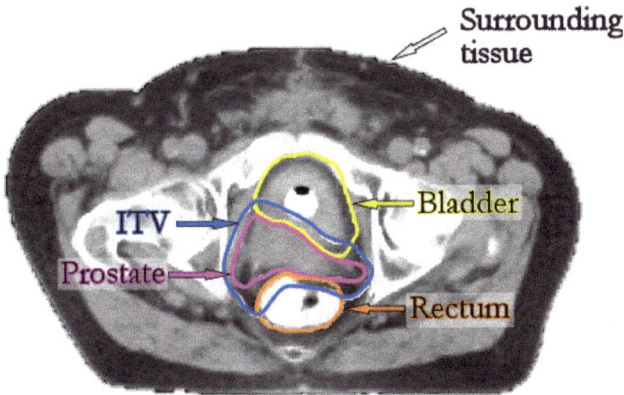

Figure 5.16. A slice through a prostate cancer patient is shown. There are two target volumes: the prostate and the internal target volume. The organs at risk volumes are the rectum, the bladder, and the surrounding tissues.

a quite asymmetrical shape with concave regions, it provided a challenging case for intensity-modulated radiotherapy. The most critical organs at risk were the bladder and the rectum, which both were in the close vicinity of the tumorous tissues. All other normal tissues were treated as one organ at risk called the currounding tissue and its response was assumed to be typical for small bowel (cf. Ågren *et al.* 1995). The gross tumor was called the prostate, and in order to account for organ motion the internal target volume (ITV, Aaltonen 1992) was defined as an expanded volume always enclosing the gross tumor independent of rectal and bladder status. Part of the radiotherapeutic challenge was that the internal target volume also included parts of the organs at risk making radiobiological treatment optimization the ideal method to maximize the outcome.

Sometimes it is desirable to optimize the treatment with respect to a wider range of variation of the sensitivity of the patient making the treatment more robust with regard to the exact radiation sensitivity of the patient. It can be shown that approximately 50% of standard deviation in sensitivity should be subtracted from the 50% response dose for the normal tissues approximately as shown in Table 5.1.

The employed dose–response data for the tumor and the normal tissues averaged over a large patient population and are given in Table 5.2.

5.5.2. *Beam Delivery Technique*

One important property of the Orbit framework is that it makes it possible to optimize the machine parameters directly. A common approach is to optimize the

Table 5.2. Radiobiological parameters for the robust prostate treatment.

Tissue	$\dfrac{D_{50}}{\text{Gy}}$	γ	$\dfrac{\alpha/\beta}{\text{Gy}}$	$\dfrac{\alpha}{\text{Gy}}$	$\dfrac{\beta}{\text{Gy}^2}$	s	End points
Rectum	55	3.1	3	0.095	0.032	0.7	Proctitis, necrosis, stenosis
Bladder	60	3.0	3	0.085	0.028	0.2	Contracture, volume loss
Surrounding	54	3.1	3	0.098	0.033	1.5	
Gross tumor	78	4.5	10	0.136	0.014		Control
ITV	78	4.5	10	0.136	0.014		Control

Table 5.3. Radiobiological parameters for an average prostate treatment.

Tissue	$\dfrac{D_{50}}{\text{Gy}}$	γ	$\dfrac{\alpha/\beta}{\text{Gy}}$	$\dfrac{\alpha}{\text{Gy}}$	$\dfrac{\beta}{\text{Gy}^2}$	s	End points
Rectum	80	2.2	3	0.048	0.016	0.7	Proctitis, necrosis, stenosis
Bladder	80	3.0	3	0.065	0.022	0.2	Contracture volume loss
Surrounding	60	2.0	3	0.058	0.019	1.0	
Gross tumor	78	4.5	10	0.136	0.014		Control
ITV	78	4.5	10	0.136	0.014		Control

fluence profiles of the beams and then convert these profiles to machine settings. This will always lead to suboptimal solutions, since one does not know beforehand if the obtained fluence profiles are possible to realize with the available treatment unit. Provided that there exists a sufficiently accurate model for the beam transport through the treatment head and means to calculate how a change in a machine setting affects the dose distribution in the patient, the optimization problem can be solved directly in terms of the selected degrees of freedom of the treatment unit and the solution is by definition deliverable and truly optimal in terms of the available degrees of freddom, at least when it concerns plain intensity modulation. Another important feature is that all degrees of freedom that are available for optimization within the Orbit framework can be combined with each other in any fashion. This allows for simultaneous optimization multileaf collimator settings, beam weights, beam energy, and beam directions.

5.5.3. *Treatment Optimization Objectives and Criteria*

It is a well-established fact in modern radiation therapy that by modulating the radiation intensity over the beam cross-section, one can achieve dose distributions where the high dose region is well tailored to the shape of the target volume at the same time as normal tissues are effectively spared. Since the surrounding normal tissues are better protected by such dose distributions, the dose delivered to the tumor can be increased without increasing the radiation-induced damage in the normal tissues. The intensity modulation of the therapeutic beams is usually calculated by solving an optimization problem that contains a mathematical formulation of the desired effects of the treatment delivery. The desired effects can be stated directly in terms of certain properties of the delivered dose distribution, for instance, a minimum dose level in the target and maximum dose levels in organs at risk. This is referred to physical optimization, since the problem is formulated in terms of physical properties of the dose distribution. When the desired effect instead is stated in terms of the clinical outcome of the treatment, we have a biological optimization problem. In this case, we do not pose any *a priori* requirements on the resulting dose distribution, but let the patient geometry, relative radiation sensitivities of the involved tissues, tissue architectures, etc. determine the optimal treatment delivery.

To illustrate the clinical value of beam angle optimization, three different sets of clinically relevant optimization criteria have been investigated. The first set is unconstrained and uses the biological objective function P_+. The second set uses the same objective function but has additional constraints on the maximum allowed risk for complication in the organs at risk. In the third set, a new unique approach is taken. First, the overall complication-free cure is maximized, globally. In the second phase, the most severe complications are minimized by allowing a slight reduction of complication-free cure. This procedure thus offers a unique possibility of maximizing the treatment outcome at the same time as complications are minimized which otherwise theoretically is an impossibility.

5.5.3.1. *Maximization of P_+*

The unconstrained problem of maximizing the probability of complication-free tumor cure when varying the set ξ of free variables is noted:

$$\left\{ \underset{\xi}{\text{maximize}}\, P_+(D), \right. \tag{5.82}$$

and is one of the most powerful formulations in radiation therapy optimization. The solution will be the optimal tradeoff between eradication of all tumor clonogens and avoidance of severe normal tissue complications. The obvious advantage with this objective function is that the probabilities of curing the tumor and avoiding fatal side effects are combined in one single scalar function, which quantifies the best possible combination of these two desirable end points can be determined. The only disadvantage is that it is only strictly valid for complications that are directly comparable with failure of tumor cure, milder complications are accepted to a greater extent than fatal complictions and a recurring tumor.

5.5.3.2. Maximization of P_+ with P_I constraints

In some situations, it is desirable to be able to define upper bounds on the complication risks when maximizing the uncomplicated tumor control:

$$\begin{cases} \text{maximize} & P_+(D) \\ \quad\quad \xi \\ \text{subject to} & P_I^j(D) \leq p^j, \quad j \in I_O \end{cases} \tag{5.83}$$

This is useful, for instance, if the patient demands that some particular complication must be avoided at the expense of higher probability to get another complications or even reduced probability of curing the tumor. This feature enhances the capability to tailor the treatment to the the objective or subjective desires of the clinician or the patient. Once having solved this set of optimization criteria, we can directly determine how sensitive the solution is with respect to a change in the predefined tolerance level p^j from the corresponding lagrange mulpliers calculated during the optimization. If the lagrange multiplier λ^{j*}, associated with the constraint $P_I^j \leq p^j$ is relatively large, then relaxing the restriction on complication probability will result in a relatively large improvement in P_+ which may be more desirable if the complication is mildly tolerable.

5.5.3.3. Maximization of P_+ followed by constrained injury relaxation

Assume that we have solved the unconstrained problem and obtained the optimal objective function value $\hat{P}_+$ together with $\hat{P}_B$ and $\hat{P}_I$. An interesting question is whether it is possible, by sacrificing some small amount of uncomplicated tumor cure probability, say ΔP_+, to achieve significant reduction of the normal tissue injury. This is a reasonable proposal since at $\hat{P}_+$ the rate of increase of P_B and P_I are equal and therefore by a slight reduction in dose, P_B and P_I are reduced by about the same amount in absolute terms, whereas P_+ almost remains unchanged. By varying the intensity modulation and angles it may thus be possible to find a more clinically acceptable combination of tumor cure and fatal complictions only sacrificing a minimal amount of complication-free cure ΔP_+. The problem to be solved in this case is then

$$\begin{cases} \underset{\xi}{\text{minimize}} & P_I(D) \\ \text{subject to} & P_+(D) \geq \hat{P}_+(\hat{D}) - \Delta P_+ \end{cases}, \qquad (5.84)$$

that is, minimize the overall normal tissue complication probability under the constraint that the probability for uncomplicated tumor control is greater than or equal to $P_+ - \Delta P_+$. In practice, since P_I is minimized in the last step, the full relaxation in complication-free cure will generally be observed ($P_+(D) \equiv \hat{P}_+(\hat{D}) - \Delta P_+$). Since we then simultaneously almost maximizes tumor cure and minimizes normal tissue injury as much as is possible without severe loss in cure we often call this approach P_{++}.

5.5.4. *Result with Increasing Number of Beam Portals*

Since the number of degrees of freedom increases directly proportional to the number of beams and their cross-sectional area, the clinical outcome of a treatment in the first approximation will improve with the number of beams. However, there is always an associated cost with each beam delivery, in terms of treatment time, quality assurance, and increased complexity of the treatment setup. Therefore, an interesting question is how many beams are sufficient for giving a close to optimal treatment plan, that is, to find the number of beams for which another beam has only an insignificant effect on the quantity that measures the quality of the treatment. In the following subsections, we investigate this question for three different beam delivery techniques where the degree of intensity modulation varies from no modulation at all (i.e., uniform beams) to fully modulated fields where each pencil beam can be individually controlled.

In this section, beam direction optimization is combined with full intensity modulation of the fluence profiles. Figure 5.17 shows how P_+, and the associated complication and tumor control probabilities, vary with the number of beams for these degrees of freedom. For this case, P_+ is, of course, a monotonically increasing function of the number of beams. When there are only a few beams, each additional beam improves the treatment outcome greatly. As the number of beam portals become >6, the curve levels off and the P_+ value only increases 3% when going from six to nine beams. The drastic improvements in P_+ for the first few beams is mainly due to the reduction of the normal tissue complication probabilities for the rectum and the surrounding tissues, as each additional beam allow for better protection of the organs at risk. The dashed red curve shows the P_+ values when

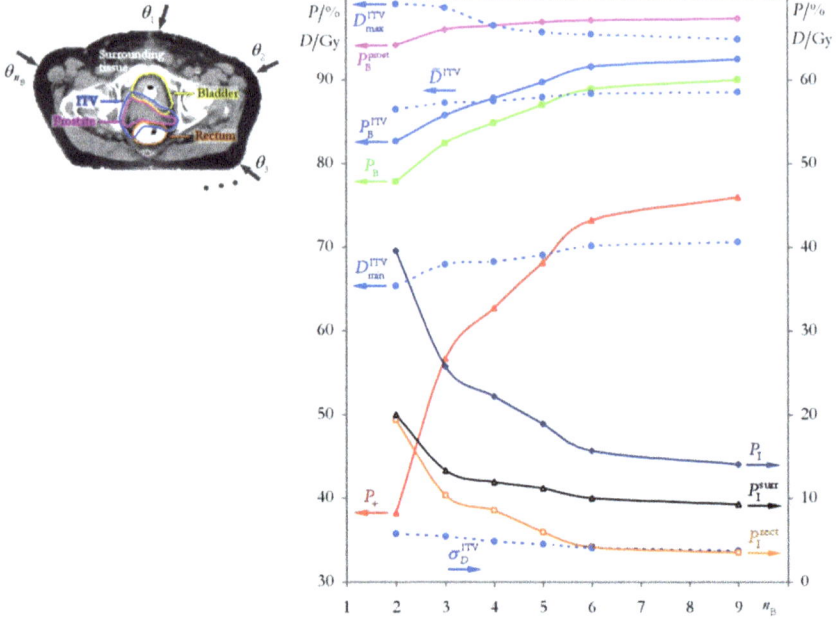

Figure 5.17. This diagram shows the variation of the probabilities for complications, tumor control, and complication free cure with the number of beam portals, when both the beam directions and the intensity profiles are simultaneously optimized. The arrows next to the curves indicate which axis the data points refer to. When the number of beams are below six, the probability for complication free cure increases significantly for each added beam portal, while for higher number of beams the improvement per added beam starts to levels off. It is also seen in the diagram, that the improvements in P_+ are mainly due to better protection of the organs at risk. The dotted curves show the dose statistics for the Internal Target Volume.

the beams have been intensity modulated, but the gantry angles were manually preselected and kept fixed during the optimization. Quite expectedly, there is an overall increase of P_+ with the number of beams as seen in Fig. 5.18. However, for badly chosen directions with parallel opposed beams, P_+ may even decrease locally (see, e.g., the six-field plan in Fig. 5.19). The fixed angles of the lower curve were also used as the initial values for the beam direction algorithm, so for each configuration it is possible to read out from the diagram how much the treatment result can be improved by tuning the gantry angles. In some situations, the initial directions happen to be well chosen, for example, for two and five beams, and the increase of the P_+ value is only marginal. In other cases, such as for four and six beams, the initial directions are poor and the beam directions algorithm finds much

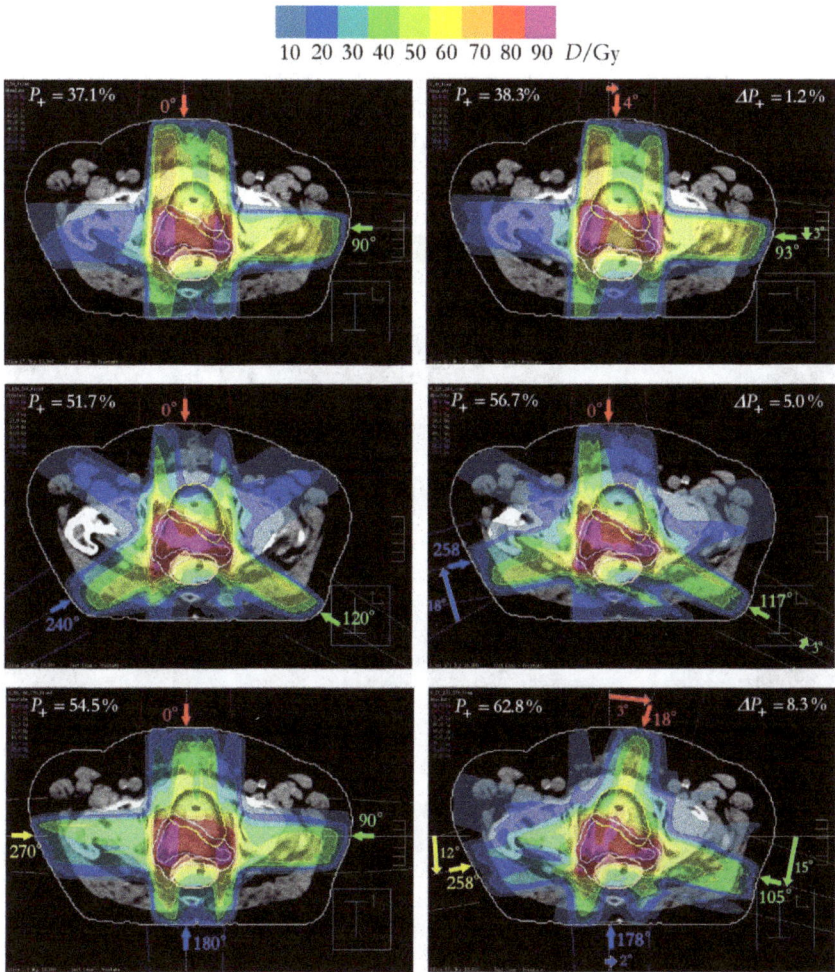

Figure 5.18(a). The treatment plans in the left column were optimized using fixed beam directions and full intensity modulation. For the treatment plans shown in the right column, simultaneous optimization of beam directions and intensity modulation was applied. The gantry angles of the fixed beam plans were used as starting angles for the beam direction algorithm and the changes in directions are indicated as arrows in the figures to the right. The obtained P_+ values are shown for each plan, and also the increased DP_+ when the angles free variables. For the 2-beam case, the selected fixed angles are quite close to optimal and the treatment can only be marginally improved by modifying the angles. For 3 beams, there is a more significant improvement of the dose distributions and P+ increases by 5%. For the classical 4-box technique, there is an even more pronounced improvement when letting the optimization algorithm select the angles. The resulting dose distribution is much

(Continued)

Figure 5.18(a). more conformal with the target volume and P_+ increases by 8%. It is also clear in this case how the algorithm seeks to align the beams along preferred directions for which large volumes of the target tissues can be irradiated without having to pass through sensitive structures (an illustrative example of this is the beam entering at 105°).

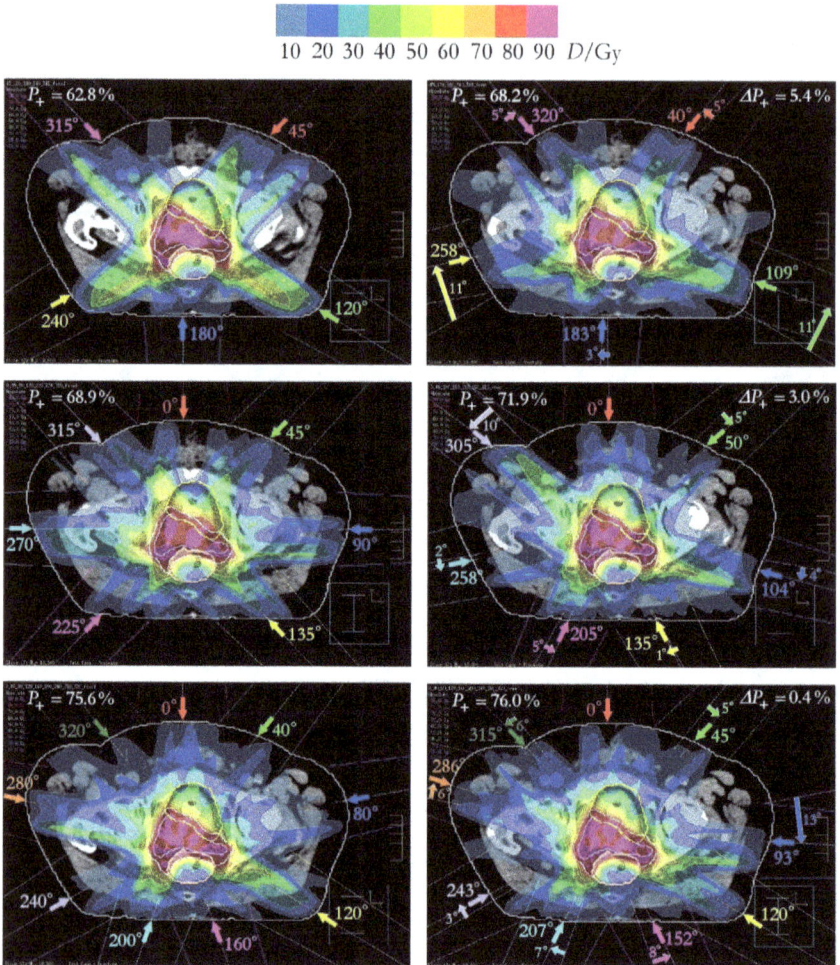

10 20 30 40 50 60 70 80 90 D/Gy

Figure 5.18(b). With increasing number of portals, beyond five beams, there is a reduced benefit with angle of incidence optimization mainly for an odd number of beam portals (cf. Fig. 5.18(a)).

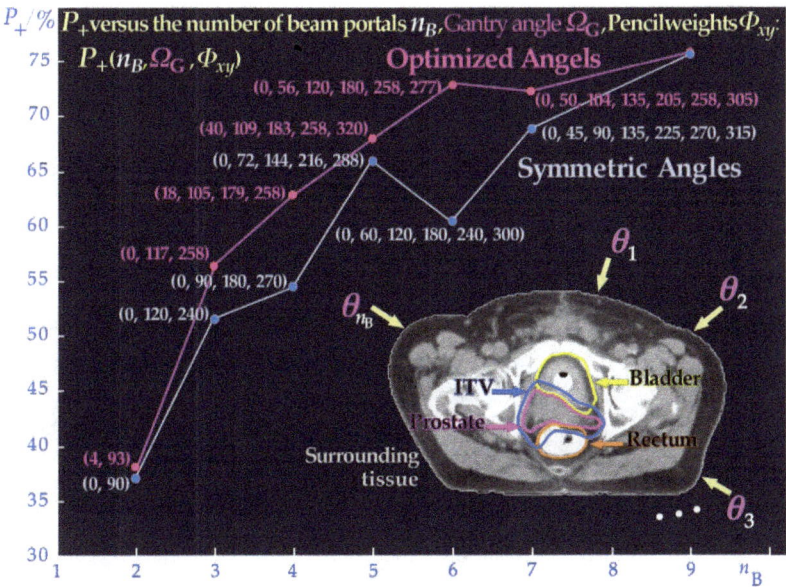

Figure 5.19. Owing to the avoidance of parallel-opposed beam portals, there is a reduced benefit with angle of incidence optimization for an odd number of beams with increasing number of beam portals beyond 5 as also seen in more detail in Fig. 5.18.

better directions. For nine beams, the equidistantly distributed beams give almost as good a result as with optimized directions.

5.5.5. *Maximization of P_+ with P_I Constraints*

The result of clinical optimizations of P_+ with different acceptable levels of rectal injury is shown in Figs. 5.20 and 5.21 for three- and five-field plans, respectively. It is clearly seen in Fig. 5.21 that if a low level of about 4% injury is acceptable for the patient, there is negligible loss in the maximum possible complication-free cure that can be achieved for the same patient.

5.5.6. *The P_{++} Strategy: Maximization of P_+ Followed by Constrained Injury Minimization*

The P_{++} strategy is as close as it is possible to come to a simultaneous maximization of P_+ and a minimization of P_I and it can effectively be achieved by a first optimization of P_+ followed by a constrained injury minimization where the maximum value of P_+ is allowed to be slightly reduced by about a fraction of 1% as shown in Figs. 5.22 and 5.23. It is striking in Fig. 5.23 that a rather small

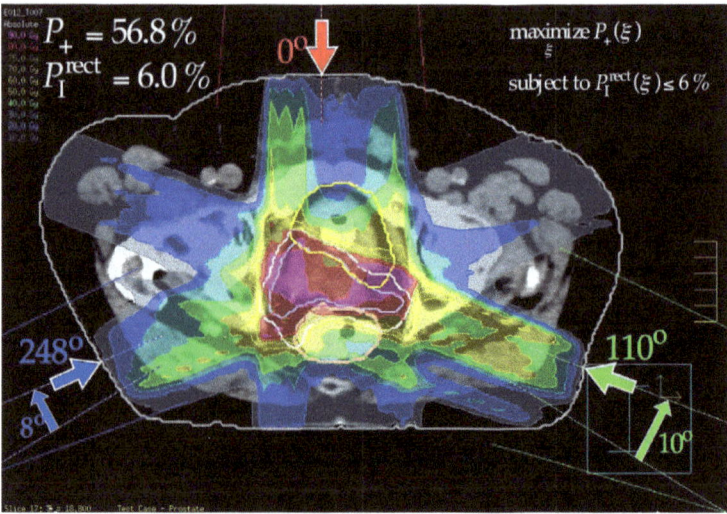

Figure 5.20. A three-field plan where the probability for rectal injury is below 6%.

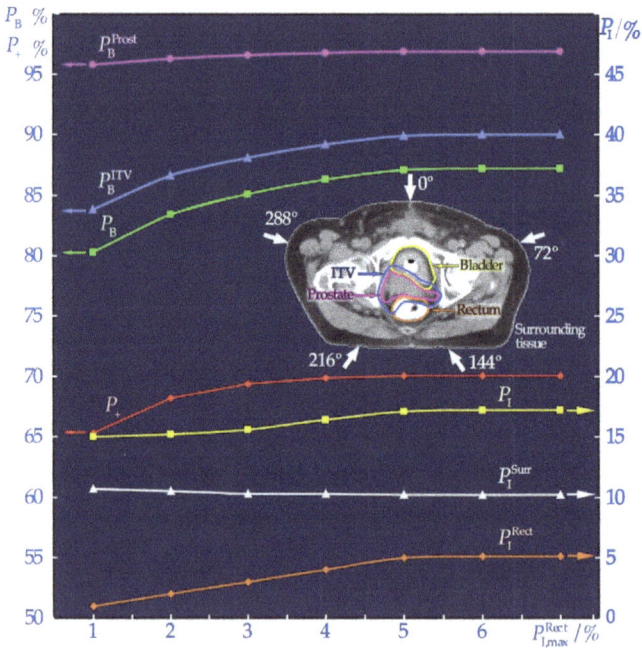

Figure 5.21. The increase in P_+ with increasing acceptable risk for rectal injury is shown.

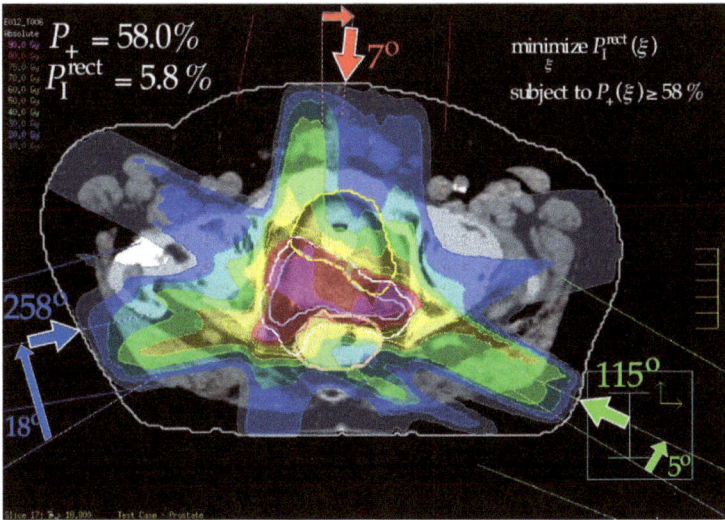

Figure 5.22. Illustratation of the lowest P_I value achievable with a three-field technique, provided P_+ is $\geq 58\%$. A slight improvement is seen compared to Fig. 5.20.

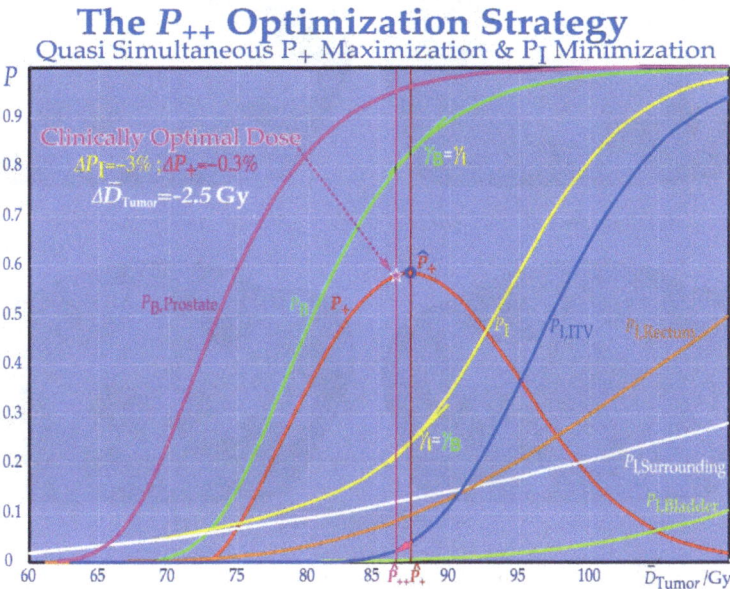

Figure 5.23. Illustration of the most clinically optimal treatment strategy where the complication-free tumor cure is allowed to decrease 0.3% at the same time as the normal tissue injury is reduced by as much as 3% which may be advantageous if the normal tissue injury is of a severe nature.

relaxation in P_+ can generate a substantial reduction in the risk for normal tissue injury. Obviously, this also results in an even larger loss in tumor cure, but since we are very close to the maximum of P_+, there is a very small associated loss in P_+. The somewhat analogous approach in Fig. 5.24 is less effective as a very large loss in P_+ seems to be necessary.

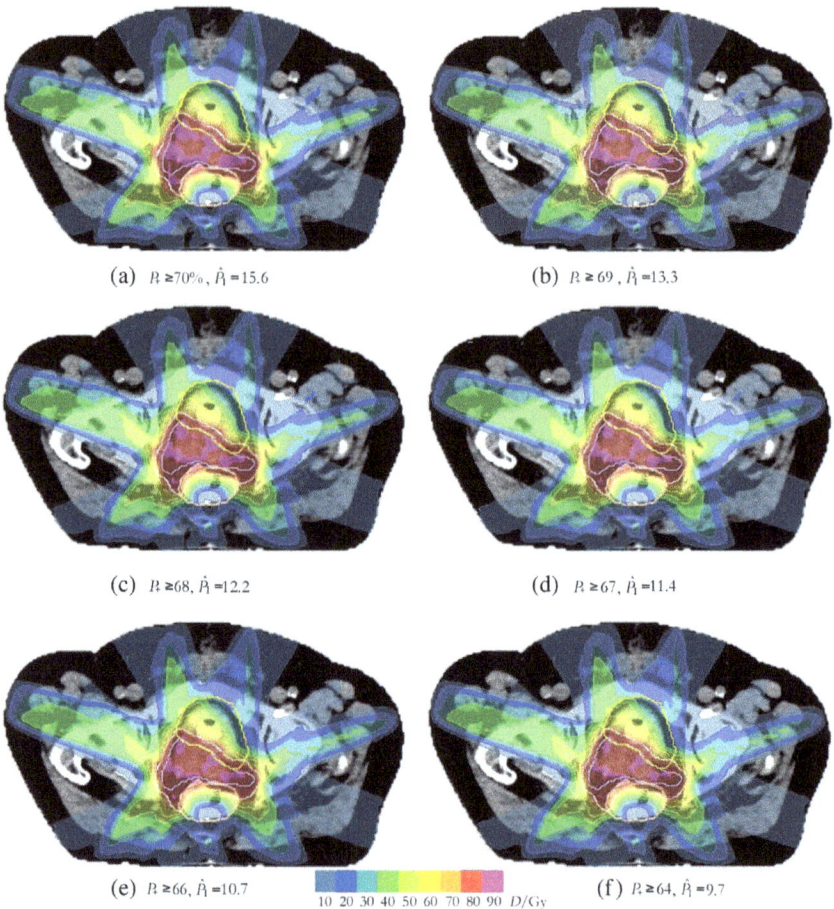

(a) $P_+ \geq 70\%$, $\hat{P}_I = 15.6$ (b) $P_+ \geq 69$, $\hat{P}_I = 13.3$

(c) $P_+ \geq 68$, $\hat{P}_I = 12.2$ (d) $P_+ \geq 67$, $\hat{P}_I = 11.4$

(e) $P_+ \geq 66$, $\hat{P}_I = 10.7$ 10 20 30 40 50 60 70 80 90 D/Gy (f) $P_+ \geq 64$, $\hat{P}_I = 9.7$

Figure 5.24. Illustration of the optimal dose distributions in a transversal slice of the prostate patient when the global normal tissue complication probability is minimized but with an inequality constraint on the probability for complication-free cure.

5.5.7. *Maximization of P_+ with Constrained Relative Standard Deviation in Tumor Dose*

Figures 5.25 and 5.26 clearly show how important it is to allow some flexibility in the dose delivery to the target volume when the treatment is optimized. This allows

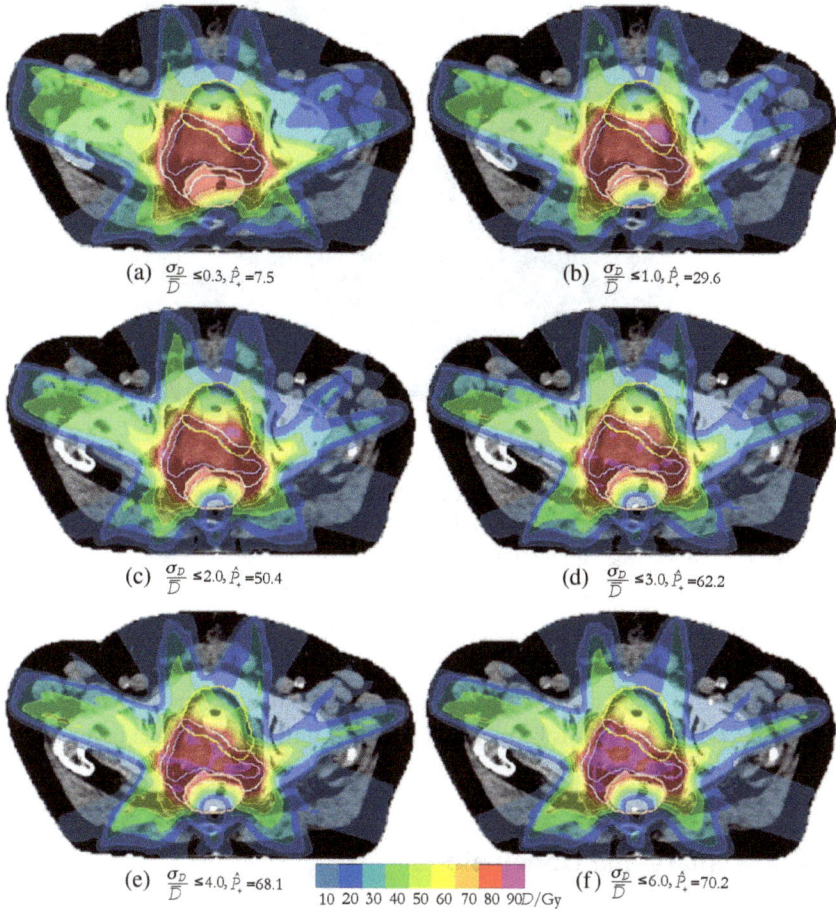

(a) $\frac{\sigma_D}{\bar{D}} \leq 0.3, \hat{P}_+ = 7.5$ (b) $\frac{\sigma_D}{\bar{D}} \leq 1.0, \hat{P}_+ = 29.6$

(c) $\frac{\sigma_D}{\bar{D}} \leq 2.0, \hat{P}_+ = 50.4$ (d) $\frac{\sigma_D}{\bar{D}} \leq 3.0, \hat{P}_+ = 62.2$

(e) $\frac{\sigma_D}{\bar{D}} \leq 4.0, \hat{P}_+ = 68.1$ 10 20 30 40 50 60 70 80 90D/Gy   (f) $\frac{\sigma_D}{\bar{D}} \leq 6.0, \hat{P}_+ = 70.2$

Figure 5.25. Illustration of the optimal dose distributions in a transversal slice of the prostate patient for varying dose homogeneity constraints in the internal target volume. It is clear that too strict requirements on the homogeneity in the ITV, lead to increased high dose regions in the surrounding normal tissues, and hence a decreased probability of complication-free cure.

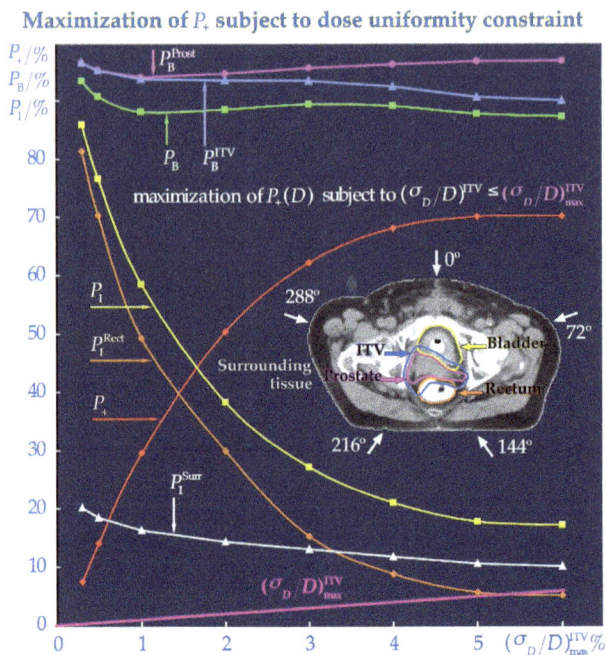

Figure 5.26. The significant increase in complication-free cure as the relative standard deviation in tumor dose is allowed to increase slightly (0.3–6%). The allowance of a relative standard deviation in dose of around 5% (Brahme 1984; Aaltonen 1992) is really the basis for the success of biologically optimized radiation therapy.

the dose to be reduced in the target volume on the side where the most sensitive normal tissues are located and instead to be increased in regions of the tumor where there is minimal risk for severe side effects in surrounding normal tissues. By this technique, the tumor cure is kept high at the same time as the normal tissue injury is as low as possible.

5.6. Summary and Conclusions

In quite general terms the function of nonuniform dose delivery is to protect normal tissues in front of, inside, or beyond the target volume, whereas irregular beam collimation spares organs at risk outside or lateral to the target volume as shown in Fig. 5.27. From the beam's point of view, the function of the beam compensator or the scanned beam is, therefore, to save organs at risk longitudinal to the target volume. The block or multileaf collimation system, on the other hand, saves normal tissues transversal to the target volume. To conclude, it should be pointed out that

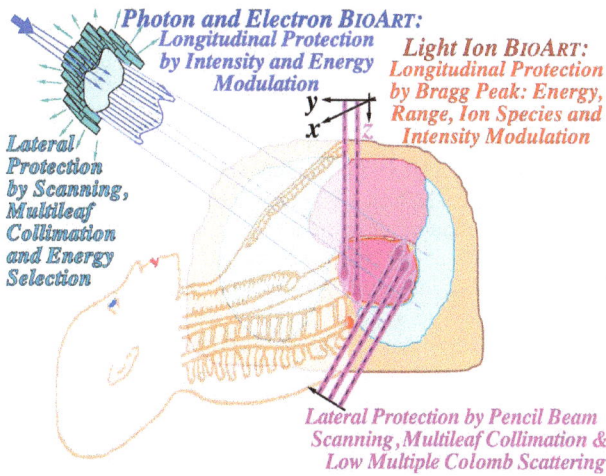

Photon and Electron BIOART:
Longitudinal Protection
by Intensity and Energy
Modulation

Light Ion BIOART:
Longitudinal Protection
by Bragg Peak: Energy,
Range, Ion Species and
Intensity Modulation

Lateral
Protection
by Scanning,
Multileaf
Collimation
and Energy
Selection

Lateral Protection by Pencil Beam
Scanning, Multileaf Collimation &
Low Multiple Colomb Scattering

Figure 5.27. Overview of some of the key degrees of freedom in dose delivery that are available for implementation when using biologically-based treatment optimization methods (BioArt method is described in more detail in Chapter 6 and Fig. 8.26).

nonuniform dose delivery is the single remaining degree of freedom that still has been left largely unused in radiation therapy optimization. Strangely enough, at the same time it also seems to be one of the most powerful degrees of freedom of external beam radiotherapy, at least with regard to its ability to shape the delivered dose distributions to conform to the shape of the target volume (cf. Figs. 5.26 and 5.27). In the future, we will, therefore, most likely see an increased use of nonuniform dose delivery for the optimization of external beam radiation therapy by the new methods discussed here. Furthermore, the more conformal a treatment technique becomes the more precise and accurate the patient setup process should be. In these techniques the dose distribution is so well matched with the radiation sensitivity map of the clinical case that a small misalignment in the setup can very much reduce the optimality and effectiveness of the treatment.

Bibliography

Aaltonen P (1992) An inventory of dose specification in the Nordic centres and a suggestion to a standardized procedure. Radiation Dose in Radiotherapy from Prescription to Delivery. *IAEA*, Leuven, Belgium, Sept 16–20 1991.

Abel NH (1823) *Magazin for Naturvidenskaberne 1*, see also: Abel NH (1881) *Oeuvres Completes I*. Grøndal and Son, Christiania, pp. 2 and 97.

Ågren Cronqvist A-K, Källman P, Turesson I, Brahme A (1995) Volume and heterogeneity dependence of the dose response relationship for head and neck tumours. *Acta Oncol* 34:851–860.

Ågren A-K, Brahme A, Turesson I (1990) Optimization of uncomplicated control for head and neck tumors. *Int J Radiat Oncol Biol Phys* 19:1077–1085.

Ahnesjö A (1984) Application of transform algorithms for calculation of absorbed dose in photon beams. *Proc. 8th Int Conf on the Use of Computers in Radiation Therapy*, IEEE Computer Society, San Diego, CA, pp. 227–230.

Ahnesjö A, Andreo P, Brahme A (1987) Calculation and application of point spread functions for treatment planning with high energy photon beams. *Acta Oncol* 26:49–56.

Ahnesjö A, Trepp A (1991) Acquisition of the effective lateral energy fluence distribution for photon beam dose calculations by convolution models. *Phys Med Biol* 36:973–985.

Andrew JR (1985) Benefit, risk and optimization by ROC analysis in cancer radiotherapy. *Int J Radiat Oncol Biol Phys* 11:1557–1562.

Åsell M, Hyödynmaa S, Söderström S, Brahme A (1999) Optimal electron and combined electron and photon therapy in the phase space of complication free cure. *Phys Med Biol* 44:235–252.

Bahr GK, Kereiakes JG, Horwitz H, Finney R, Galvin J, Goode K (1968) The method of linear programming applied to radiation treatment planning. *Radiology* 91:686–693.

Barth NH (1990) An inverse problem in radiation therapy. *Int J Radiat Oncol Biol Phys* 18:425–431.

Bel A, Van Herk M, Bartolink H, Lebesque JV (1993) A verification procedure to improve patient setup accuracy using portal image. *Radiother Oncol* 29:253–260.

Birkhoff GD (1940) On drawings composed of uniform straight lines. *J de Mathematiques pures et appliqués* 19:221–236.

Bochner S (1948) *Vorlesungen über Fouriersche Integrale*. Chelsea Publication Company, New York.

Bollman R, Schmidt K-P, Tabbert E (1981) Verbesserung der Bestrahlungsplanung in der Hochvolttherapie durch mathematische Optimierung. *Radiobiol Radiother* 22:594–601.

Boltzmann L (1927) *Vorlesungen über Gas-theorie*. Barth, Leipzig.

Bortfeld T, Bürkelbach J, Boesecke R, Schlegel W (1990) Methods of image reconstruction from projections applied to conformation radiotherapy. *Phys Med Biol* 35:1423–1434.

Brahme A (1995) Treatment optimization using physical and biological objective functions. In: Smith A (ed.) Radiation Therapy Physics, 209–246. Berlin: Springer.

Brahme A (1999) Optimized radiation therapy based on radiobiological objectives. *Seminars in Oncology* 9:35–47.

Brahme A (1981) Correction of a measured distribution for the finite extension of the detector. *Strahlenther* 157:258–259.

Brahme A (1982) Physical and biologic aspects on the optimum choice of radiation modality. *Acta Radiol Oncol* 21:469–479.

Brahme A (1984) Dosimetric precision requirements in radiation therapy. *Acta Radiol Oncol* 23:379–391.

Brahme A (1985) Developments of external beam treatment units and the role of high energy electrons and photons, Teaching lecture ECCO 3, Stockholm, p. 46

Brahme A (1987) Design principles and clinical possibilities with a new generation of radiation therapy equipment. *Acta Oncol* 26:403–412.

Brahme A (1988) Optimization of stationary and moving beam radiation therapy techniques. *Radiother Oncol* 12:129–140.

Brahme A (1992) Which parameters of the dose distribution are best related to the radiation response of tumors and normal tissues? Radiation Dose in Radiotherapy from Prescription to Delivery. IAEA, Leuven, Belgium, Sept 16–20, 1991.

Brahme A (1995) Similarities and differences in radiation therapy optimization and tomographic reconstruction. *Int J Imag Syst Technol* 6:6–13.

Brahme A, Ågren A-K (1987) Optimal dose distribution for eradication of heterogeneous tumours. *Acta Oncol* 26:377–385.

Brahme A, Källman P, Lind B (1991) Optimization of the probability of achieving complication free tumor control using a 3D pencil beam scanning technique for protons and heavy ions. *Proc. NIRS Int. Workshop on Heavy Charged Particle Therapy and Related Subjects*, Itano A, Kanai T (eds.), Chiba, Japan, pp. 124–142.

Brahme A, Källman P, Tilikidis A (1995) Developments in ion beam therapy planning and treatment optimization. In: Linz U (ed.) *Ion Beam Therapy*. New York: Chapman & Hall.

Brahme A, Lind B, Kiillman P (1990) Inverse radiation therapy planning as a tool for 3D dose optimization. *Physica Medica* 6:53–63.

Brahme A, Roos J-E, Lax I (1982) Solution of an integral equation encountered in rotation therapy. *Phys Med Biol* 27:1221–1229.

Brix F, Christiansen R, Hancken C, Quirin A (1988) The field integrated dose modification (FIDM): three typical clinical applications of a new irradiation technique. *Radiother Oncol* 12:199–207.

Buck B, Macaulay VA (1991) *Maximum Entropy in Action, A Collection of Expository Essays.* Clarendon Press, Oxford.

Burgers JMV, Awwad HK, Van der Laarse R (1985) Relationship between local cure and dose-time-volume factors in interstitial implants. *Int J Radiat Oncol Biol Phys* 11:715–723.

Bush RS (1982) The compliant oncologist: Franz Buschke lecture. *Int J Radiat Oncol Biol Phys* 8:1019–1027.

Campbell A, Converse PE, Rodgers WL (1976) *The Quality of Life Perceptions, Evaluations, Satisfactions.* Russel Sage Foundation, New York.

Censor Y, Altschuler MD, Powlis WD (1988a) A computational solution of the inverse problem in radiation therapy treatment planning. *App Math Comput* 25:57–87.

Censor Y, Altschuler MD, Powlis WD (1988b) On the use of Cimmino's simultaneous projections method for computing a solution of the inverse problem in radiation therapy treatment planning. *Inv Problems* 4:607–623.

Chui CS, Mohan R (1988) Extraction of pencil beam kernels by the deconvolution method. *Med Phys* 15:138–144.

Chui CS, Yorke E, Hong L (2003) The effects of intra-fraction organ motion on the delivery of intensity-modulated field with a multileaf collimator. *Med Phys* 30:1736–1746.

Cohen L (1960) The statistical prognosis in radiation therapy. *Am J Roentgenol* 84:741–753.

Cohen L (1987) Optimization of dose-time factors for a tumor and multiple associated normal tissues. *Int J Radiat Oncol Biol Phys* 13:251–258.

Cooper REM (1978) A gradient method of optimizing external-beam radiotherapy treatment plans. *Radiology* 128:235–243.

Cormack AM (1987) A problem in rotation therapy with X-rays. *Int J Radiat Oncol Biol Phys* 13:623–630.

Cormack AM (1995) Some early radiotherapy optimization work. *Int J Imag Syst Technol* 6:2–5.

Cormack AM, Cormack RA (1987) A problem in rotation therapy with X-rays: dose distributions with an axis of symmetry. *Int J Rad Oncol Biol Phys* 13:1921–1925.

Cormack AM, Quinto ET (1989) On a problem in radiotherapy: questions of non-negativity. *Int J Imag Syst Technol* 1:120–124.

Creutzberg CL, Althof VGM, de Hooh M, Visser AG, Huizenga H, Wijnmaalen A, Levendag PC (1996) A quality control study of the accuracy of patient positioning in irradiation of pelvic fields. *Int J Radial Oncol Biol Phys* 34:697–708.

Davis AR (1982) On the maximum likelihood regularization of Fredholm convolution equations of the first kind. In: Baker CTH, Miller GS (eds) Treatment of integral equations by numerical methods. New York: Academic Press, pp. 95–105.

Deans RS (1983) *The Radon Transform and Some of Its Applications.* John Wiley, New York.

Djordjevich A, Bonham DJ, Hussein EMA, Andrew JW, Hale ME (1990) Optimal design of radiation compensators. *Med Phys* 17:397–404.

Dunn WL (1981) Inverse Monte Carlo analysis. *J Comput Phys* 41:154–166.

Ebert U (1977) Computation of optimal radiation treatment plans. *J Comput Appl Math* 3:99–104.

El-Gayed AAH, Bel A, Vijlbrief R, Bartelink H, Lebesque JV (1993) Time trend of patient setup deviations during pelvic irradiation using electronic portal imaging. *Radiother Oncol* 26:162–171.

Fisher JJ (1969) Theoretical considerations in the optimization of dose distribution in radiation therapy. *Brit J Radiol* 42:925–930.

Flampouri S, Jiang SB, Sharp GC, Wolfgang J, Patel AA, Choi NC (2006) Estimation of the delivered patient dose in lung IMRT treatment based on deformable registration of 4D-CT data and Monte Carlo simulations. *Phys Med Biol* 51:2763–2779.

Fowler JF (1981) *Nuclear Particles in Cancer Treatment.* Adam Hilger, Bristol (Medical Physics Handbooks 8).

Franke DS (1980) Die Anwendung der mathematischen Optimierung in der Strahlentherapie zur Findung optimaler Bestrahlungstechniken. *Radiobiol Radiother* 21: 668–676.

George R, Keall PJ, Kini VR, Vedam SS, Siebers JV, Wu Q, Lauterbach MH, Arthur DW, Mohan R (2003) Quantifying the effect of intrafraction motion during breast IMRT planning and dose delivery. *Med Phys* 30:552–562.

Goitein M, Niemierko A (1988) Biologically based models for scoring treatment plans. *Proc. Joint US-Scandinavian Symp. Future Directions of Computer-aided Radiation Therapy*, Zink S (ed.), San Antonio, Texas, National Cancer Institute.

Gremmel H, Hebbinghaus D, Wendhausen H (1978) An optimization criterion for dose distributions, minimizing the radiation effect in healthy tissue. *Proc. 6th Int. Conf. Use of Computers in Radiation Therapy*, Rosenow U (ed.), Göttingen, Mylet-Druck, Dransfeld, pp. 199–209.

Gustafsson A, Lind BK, Brahme A (1994) A general pencil beam algorithm for optimization of radiation therapy. *Med Phys* 21:343–356.

Gustafsson A, Lind BK, Svensson R, Brahme A (1995) Simultaneous optimization of dynamic multileaf collimation and scanning patterns or compensation filters using a generalized pencil beam algorithm. *Med Phys* 22:1141–1156.

Harder D, Looe HK, Poppe B (2014) Convolutions and deconvolutions in radiation dosimetry In: Brahme A, ed.: Comprehensive BioMedical Physics Vol 8 Ch 11, Major Reference Work Elsevier Oxford.

Hodes L (1974) Semiautomatic optimization of external beam radiation treatment planning. *Radiology* 110:191–196.

Holmes T, Mackie RT, Simpkin D, Reckwerdt P (1991) A unified approach to the optimization of brachytherapy and external beam dosimetry. *Int J Radiat Oncol Biol Phys* 20: 859–873.

Hope CS, Laurie J, Orr JS, Halnan JS (1967) Optimization of X-ray treatment planning by computer judgment. *Phys Med Biol* 12:531–542.

Hope CS, Orr HB (1965) Computer optimization of 4 MeV treatment planning. *Phys Med Biol* 10:365–373.

Hunt MA, Schultheiss TE, Desobry GE, Hakki M, Hanks GE (1995) An evaluation of setup uncertainties for patients treated to pelvic fields. *Int J Radiat Oncol Biol Phys* 32:227–233.

Hyödynmaa S, Gustafsson A, Brahme A (1996) Optimization of conformal electron beam therapy using energy- and fluence-modulated beams. *Med Phys* In press.

ICRU Report 33 (1980) Radiation quantities and units. International Commission on Radiation Units and Measurements, Bethesda.

Jaynes ET (1957) Information theory and statistical mechanics. *Phys Rev* 106:620–630.

Joiner MC (1994) Induced radioresistance: an overview and historical perspective. *Int J Radiat Biol* 65:79–84.

Källman P (1992) Optimization of radiation therapy planning using physical and biological objective function, thesis, Stockholm University, Sweden.

Källman P, Ågren A, Brahme A (1992b) Tumor and normal tissue responses to fractionated non-uniform dose delivery. *Int J Radiat Biol* 62:249–262.

Källman P, Lind BK, Brahme A (1992a) An algorithm for maximizing the probability of complication free tumor control in radiation therapy. *Phys Med Biol* 37:871–890.

Karnofsky DA, Abelmann WH, Craver LL, Burchenal JH (1948) The use of the nitrogen mustards in the palliative treatment of carcinoma. *Cancer* 1:634–658.

Keall P (2004) 4-dimensional computed tomography imaging and treatment planning. Semin. *Radiat Oncol* 14:81–90.

Keall PJ, Joshi S, Vedam SS, Siebers JV, Kini VR, Mohan R (2005) Four-dimensional radiotherapy planning for DMLC-based respiratory motion tracking. *Med Phys* 32:942–951.

Kirkpatrick B, Gelatt CD, Vecci MP (1983) Optimization by simulated annealing. *Science* 220:671–680.

Lane DP (1992) p53, guardian of the genome. *Nature* 358:15–16.

Langer M, Leong J (1987) Optimization of beam weights under dose-volume restrictions. *Int J Radiat Oncol Biol Phys* 13:1255–1260.

Langer M, Kijewski P, Brown R, Ha C (1991) The effect on minimum tumor dose of restricting target-dose inhomogeneity in optimized three-dimensional treatment of lung cancer. *Radiother Oncol* 21:245–256.

Lax I, Brahme A (1982) Rotation therapy using a novel high-gradient filter. *Radiology* 145:473–478.

Leemann C, Alonso J, Grunder H, Hoyer E, Kalnins G, Rondeau D, Staples J, Voelker F (1977) A 3-dimensional beam scanning system for particle radiation therapy. *IEEE Trans Nucl Bci NS* 24:1052–1054.

Legras J, Legras B, Lambert J-P (1982) Software for linear and non-linear optimization in external radiotherapy. *Comput Programs Biomed* 15:233–242.

Leibel SA, Ling CC, Kutcher GJ, Mohan R, Cordon-Cordo C, Fuks Z (1991) The biological basis for conformal three-dimensional radiation therapy. *Int J Radiat Oncol Biol Phys* 21:805–811.

Lind B, Brahme A (1985) Generation of Desired Dose Distributions with Scanned Elementary Beams by Deconvolution Methods. *Proc VII ICMP Espoo*, Finland, p. 953.

Lind B, Brahme A (1987) Optimization of Radiation Therapy Dose Distributions with Scanned Photon Beams. *Proc 9th Int Conf on the Use of Computers in Radiation Therapy*, Bruinvis IAD, Van der Giessen PH and Van Kleffens HJ (eds), Elsevier, Amsterdam, pp. 235–239.

Lind B, Brahme A (1995) Development of treatment techniques for radiotherapy optimization. *Int J Imag Syst Technol* 6:33–42.

Lind BK (1990) Properties of an algorithm for solving the inverse problem in radiation therapy. *Inverse Problems* 6:415–426.

Lind BK (1991) Radiation therapy planning and optimization studied as inverse problems, thesis, Stockholm University Sweden.

Lind BK, Brahme A (1992) Photon field quantities and units for kernel-based radiation therapy planning and treatment optimization. *Phys Med Biol* 37:891–909.

Lind BK, Källman P, Sundelin B, Brahme A (1993) Optimal radiation beam profiles considering uncertainties in beam patient alignment. *Acta Oncol* 32:331–342.

Linz U. (ed.) (1995) *Ion Beam Therapy*. Chapman & Hall, New York.

Liu S, Lind BK, Brahme A (1993) Two accurate algorithms for calculating the energy fluence profile in inverse radiation therapy planning. *Phys Med Biol* 38:1809–1824.

Löf J, Lind BK, Brahme A (1995) Optimal radiation beam profiles considering the stochastic process of patient positioning. *Inverse Problems* 11:1189–1209.

Löf J, Lind BK, Brahme A (1998) An adaptive control algorithm for optimization of intensity modulated radiotherapy considering uncertainties in beam profiles, patient set-up and internal organ motion. *Phys Med Biol* 43:1605–1628.

Luenberger DG (1973) *Introduction to Linear and Nonlinear Programming*. Addison-Wesley, Menlo Park.

Lujan AE, Larsen EW, Balter JM, Ten Haken RK (1999) A method for incorporating organ motion due to breathing into 3D dose calculations. *Med Phys* 26:715–720.

Mantel J, Perry H, Weinkam JJ (1977) Automatic variation of field size and dose rate in rotation therapy. *Int J Radiat Oncol Biol Phys* 2:697–704.

Mavroidis P, Axelsson S, Hyödynmaa S, Rajala J, Pitkänen MA, Lind BK, Brahme A (2002) Effects of positioning uncertainty and breathing on dose delivery and radiation pneumonitis prediction in breast cancer. *Acta Oncol* 41:471–485.

Mavroidis P, Lind BK, Brahme A (2001) Biologically effective uniform dose ($\overline{\overline{D}}$) for specification, report and comparison of dose response relations and treatment plans. *Phys Med Biol* 46:2607–2630.

Mavroidis P, Stathakis S, Gutierrez A, Esquivel C, Shi C, Papanikolaou N (2008) Expected clinical impact of the differences between planned and delivered dose distributions in helical tomotherapy for treating head and neck cancer using helical megavoltage CT images. *J Appl Clin Med Phys* (Submitted).

McDonald SC, Rubin P (1977) Optimization of external beam radiation therapy. *Int J Radiat Oncol Biol Phys* 2:307–317.

McNutt TR, Mackie TR, Recwerdt P, Paliwal BR (1996) Modeling dose distributions from portal dose images using the convolution/superposition method. *Med Phys* 23:1381–1392.

Metropolis N, Rosenbluth A, Rosenbluth M, Teller H, Teller E (1953) Equation of state calculations by fast computing machines. *J Chem Phys* 21:1087–1092.

Miller DW (1985) Optimization of attenuator shapes for multiple field radiation treatment. In: Paliwal BR, Herbert DE, Orton CG (eds) *Optimization of Cancer Radiotherapy*, AAPM, *Symp Proc No 5, Am. Inst. of Physics*, New York.

Miller GF (1974) Fredholm equations of the first kind. In: Delves LM, Walsh J (eds) *Numerical Solutions of Integral Equations*, 175–188. Oxford University Press, London.

Mohan R, Chui C, Lidofsky L (1986) Differential pencil beam dose computation model for photons. *Med Phys* 13:64–72.

Mohan R, Mageras GS, Baldwin B, Brewster LJ, Kutcher GJ, Leibel S, Burman CM, Ling CC, Fuks Z (1992) Clinically relevant optimization of 3D conformal treatments. *Med Phys* 19:933–944.

Moore DH, Mendelsohn ML (1972) Optimal treatment levels in cancer therapy. *Cancer* 30:95–106.

Morrill SM, Lane RG, Rosen II (1990) Constrained simulated annealing for optimized radiation therapy treatment planning. *Comput Meth Progr Biomed* 33:135–144.

NACP (Nordic Association of Clinical Physicists) (1997) Specification of dose delivery in radiation therapy. Recommendations by the NACP. Aaltonen P, Brahme A, Lax I, Levernes S, Näslund I, Reitan JV, Turesson I. *Acta Oncol Suppl* 10:1–32.

Natterer F (1986) *The Mathematics of Computerized Tomography*. John Wiley, Stuttgart.

Niemierko A (1993) Random search algorithm (RONSC) for optimization of radiation therapy with both physical and biological end points and constraints. *Int J Radiat Oncol Biol Phys* 23:89–108.

Norman A, Kagan AR, Chan SL (1988) The importance of genetics for the optimization of radiation therapy. *Am J Clin Oncol* 11:84–88.

Ortega JM, Rheinboldt WC (1970) *Iterative Solution of Nonlinear Equations in Several Variables*. Academic Press, New York.

Osoba D (1993) *Effect of Cancer on Quality of Life*. CRC Press Inc., Boston.

Popp FA, Bothe B, Goedecke R (1975) Prinzipien zur Optimierung der Bestrahlungsplanung. *Strahlenther* 150:389.

Powlis WD, Altschuler MD, Censor Y, Buhle EL (1989) Semi-automated radiotherapy treatment planning with a mathematical model to satisfy treatment goals. *Int J Radiat Oncol Biol Phys* 16:271–276.

Radon J (1917) Uber die Bestimmung von Funktionen durch ihre Integralwerte langs gewisser Mannigfaltigkeiten. *Ber Sachs Akad Wiss, Leipzig, Math Phys Kl* 69:262–267.

Raju MR (1980) *Heavy Particle Radiotherapy*. Academic Press, New York.

Redpath AT, Vickery B, Wright DH (1976) A new technique for radiotherapy planning using quadratic programming. *Phys Med Biol* 21:781–791.

Rietzel E, Chen GT, Choi NC, Willet CG (2005) Four-dimensional image-based treatment planning: target volume segmentation and dose calculation in the presence of respiratory motion. *Int J Radiat Oncol Biol Phys* 61:1535–1550.

Rosen II, Lane RG, Morrill S, Belli JA (1991) Treatment plan optimization using linear programming. *Phys Med Biol* 18:141–152.

Sailer SL, Rosenman JG, Symon JR, Cullip TJ, Chaney EL (1993) The tetrad and hexad: maximum beam separation as a starting point for non-coplanar 3D treatment planning. *Int J Radiat Oncol Biol Phys* 27:138.

Schafer RW, Mersereau RM, Richards MA (1981) Constrained Interactive Restoration Algorithms. *Proc. IEEE* 69:432–450.

Schultheiss TE, Orton CG, Peck RA (1983) Models in radiotherapy volume effects. *Med Phys* 10:410–415.

Schwartz JL, Rotmensch J, Giovanazzi BS, Cohen MB, Weichselbaum RR (1988) Faster repair of DNA double-strand breaks in radioresistant human tumor cells. *Int J Radiat Oncol Biol Phys* 15:907–912.

Shalev S, Gluhchev G (1994) When and How to Correct a Patient Setup. *Proc. 11th ICCR, the Use of Computers in Radiation Therapy*, A R Hounsell, J M Wilkinson and P C Williams (eds), Christie Hospital NHS Trust, Manchester, pp. 274–275.

Shannon CE (1948) A mathematical theory of communication. *Bell System Tech J* 27:623–659.

Söderström S, Brahme A (1992) Selection of suitable beam orientations in radiation therapy using entropy and Fourier transform measures. *Phys Med Biol* 37:911–924.

Söderström S, Brahme A (1993) Optimization of multiple field techniques using radiobiological objective functions. *Med Phys* 20:1201–1210.

Söderström S, Brahme A (1995b) Which is the most suitable number of photon beam portals in coplanar radiation therapy? *Int J Radiat Oncol Biol Phys* 33:151–159.

Söderström S, Eklöf A, Brahme A (1999) Aspects on the optimal photon beam energy for radiation therapy. *Acta Oncol* 38:179–187.

Söderström S, Gustafsson A, Brahme A (1993) The clinical value of different treatment objectives and degrees of freedom in radiation therapy optimization. *Radiother Oncol* 29:148–163.

Söderström S, Gustafsson A, Brahme A (1995a) Few-field radiation therapy optimization in the phase space of complication-free tumor control. *Int J Imag Syst Technol* 6: 91–103.

Sonderman D, Abrahamson PG (1985) Radiotherapy design using mathematical programming models. *Oper Res* 33:705–725.

Starkschall G (1984) A constrained least-squares optimization method for external beam radiation therapy treatment planning. *Med Phys* 11:659–665.

Starkschall G, Eifel PJ (1992) An interactive beam-weight optimization tool for three-dimensional radiotherapy treatment planning. *Med Phys* 19:155–163.

Stengel RF (1994) *Optimal Control and Estimation*, Dover, New York.

Sterling TD, Perry H, Weinkam JJ (1965) Automation of radiation treatment planning V. Calculation and visualisation of the total treatment volume. *Br J Radiol* 38:906–913.

Su FC, Shi C, Mavroidis P, Goytia V, Crownover R, Papanikolaou N (2008) Assessing four-dimensional radiotherapy planning and respiratory motion-induced dose difference based on biologically effective uniform dose. *Med Phys* (Submitted).

Takahashi S (1965) Conformation radiotherapy, rotation techniques as applied to radiography and radiotherapy. *Acta Radiol* (Suppl. 242).

Timme TL, Moses RE (1988) Review: diseases with DNA damage-processing defects. *Am J Med Sci* 295:40–48.

Tinger A, Michalski JM, Bosch WR, Valicenti RK, Low D, Myerson RJ (1996) *Int J Radiat Oncol Biol Phys* 19:683–690.

Ulsø N, Brahme A (1988) Computer aided Irradiation Technique Optimization. *Proc Joint US-Scandinavian Symp Further Directions of Computer aided Radiation Therapy*, Zink S (ed.), National Cancer Institute, San Antonio, Texas.

Van der Laarse R, Strackee J (1976) Pseudo optimization of radiotherapy treatment planning. *Brit J Radiol* 49:450–457.

Van Santvoort J, Huizenga H (1991) The use of elementary electron beams and inverse planning techniques. *Paper Presented at a Workshop on Developments in Dose-planning and Treatment Optimization*, The Karolinska Institute, Stockholm.

Webb S (1989) Optimization of conformal radiotherapy dose distributions by simulated annealing. *Phys Med Biol* 34:1349–1369.

Webb S (1990) Non-standard CT scanners: their role in radiotherapy. *Int J Radiat Oncol Biol Phys* 19:1589–1607.

Weeks KJ, Sontag MR (1991) 3-D dose-volume compensation using nonlinear least squares regression technique. *Med Phys* 18:474–480.

Wolbarst AB (1984) Optimization of radiation therapy II: the critical-voxel model. *Int J Radiat Oncol Biol Phys* 10:741–745.

Yan D, Wong JW, Gustafson G, Martinez A (1995) A new model for 'accept or reject' strategies in off-line and on-line megavoltage treatment evaluation. *Int J Radiat Oncol Biol Phys* 31:943–952.

Properties and Clinical Potential of Biologically Based Treatment Plan Optimization

6

Anders Brahme, Kerash Moaierifar,
Panayiotis Mavroidis, and Bengt K. Lind

6.1. Introduction

Treatment planning is the process whereby the therapeutic strategy of the radiation oncologist is realized as a set of treatment instructions together with a physical description of the distribution of dose in the patient. The aim of treatment planning can be summarized as follow:

(1) To localize the tumor volume in the patient and to define the target volume for treatment.
(2) To measure the outline of the patient and to define the location of the target volume in relation to external International Commission on Radiation Units and Measurements (ICRU) reference points.
(3) To determine the optimal configuration when irradiating the target volume to a prescribed dose within clinical constraints on normal tissue doses.
(4) To calculate the resultant dose distribution in the patient.
(5) To prepare an unambiguous set of treatment instructions for the radiographers.
(6) Define points and expected doses for dosimetry.

6.2. Single Beam Portals

In single-field treatments, the irradiated volume is well covered by a single beam portal. This technique is quite common with low-energy electron beams for irradiating small superficial targets. In photon therapy, treatments with a single field are generally used only for palliative purpose, a typical example being the irradiation of the spine to relieve pain in patients who have spinal metastases. In general, it is only with low-energy X-rays ($<150\,\text{kV}$) that these kinds of fields have been used

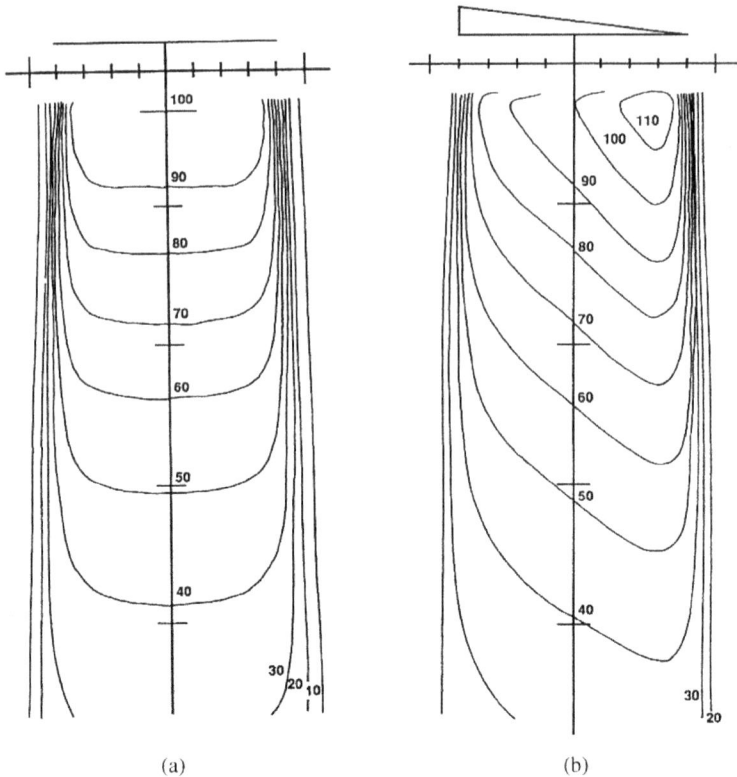

(a)   (b)

Figure 6.1(a).   Isodose distributions for a 6 MV X-ray beam with an 8 × 8 cm field size: (a) open field; (b) field with a 45° wedge.

curatively. The dose distribution in a central plane of single-field irradiation can be described by an isodose chart. A typical dose distribution is shown for 6 MV X-ray beam in Fig. 6.1(a) (Williams and Thwaites 1993).

An *isodose* is a line, which connects points of equal dose. The dose represented by the line is expressed as a percentage of the dose on the central axis at the depth of the dose maximum. By convention, single-field isodose curves are plotted at 10% intervals and are presented on a 1:1 scale.

6.3. Parallel-opposed Beams

Two opposed coaxial beams can be used to irradiate large radiation-sensitive tumor volumes or when the tumor cannot be accurately defined or because the intention is to give a relatively low dose for palliation. Many tumor sites can be irradiated in this way by a parallel-opposed technique. Figure 6.2 shows a typical isodose distribution

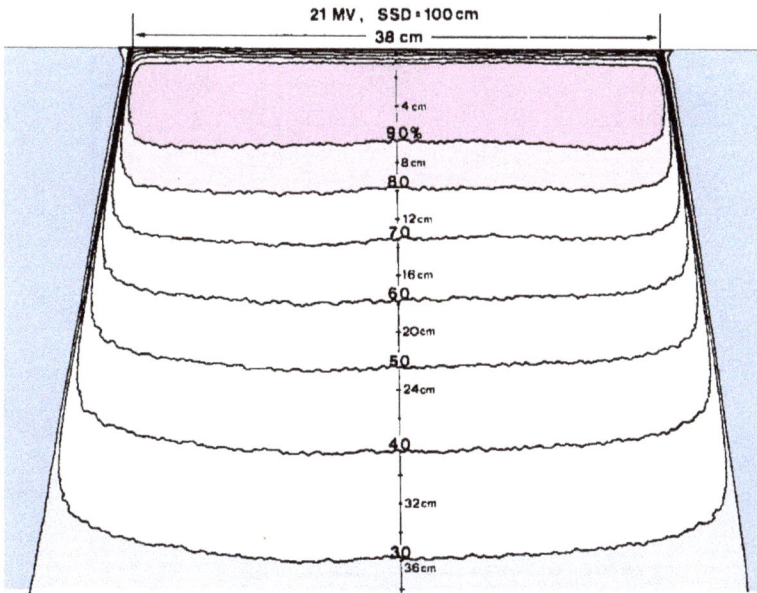

Figure 6.1(b). Isodose distributions for a 21 MV X-ray beam for a beam size of 38 × 38 cm produced with an achromatic flattening filter. The target flattening filter system used makes it possible to produce large uniform beams up to 42 × 37 cm, at **SSD = 100** cm. The uniformity is good at all depths and field sizes (Brahme and Svensson 1979) (for narrow pencil beams see Figs. 3.70(a) and (b) in Chapter 3).

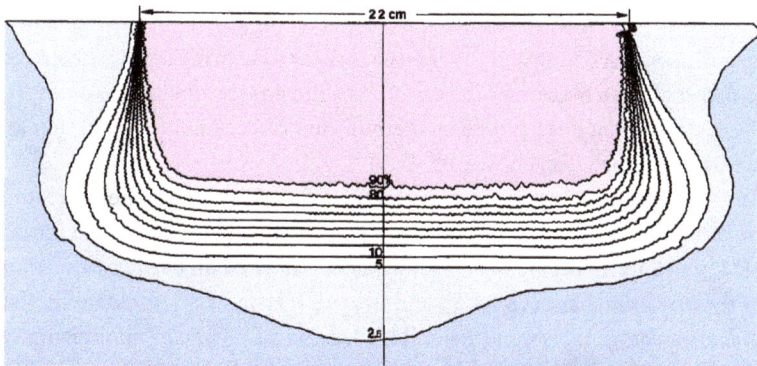

Figure 6.1(c). Typical electron beam isodose distribution (Ep = 21.1 MeV, SSD = 100 cm, size 22 × 22 cm). The measurements where made with a diode in a water phantom. Original curves from the isodose-plotting system are shown. The bump in the 2.5% isodose level in the center of the field is partly due to photons from the scattering foil. The 2.5% level outside the field is caused by photons both from the beam limiters and the scattering foils. The photon background generated in the water phantom amounts to about 1.5% (Brahme and Svensson 1979).

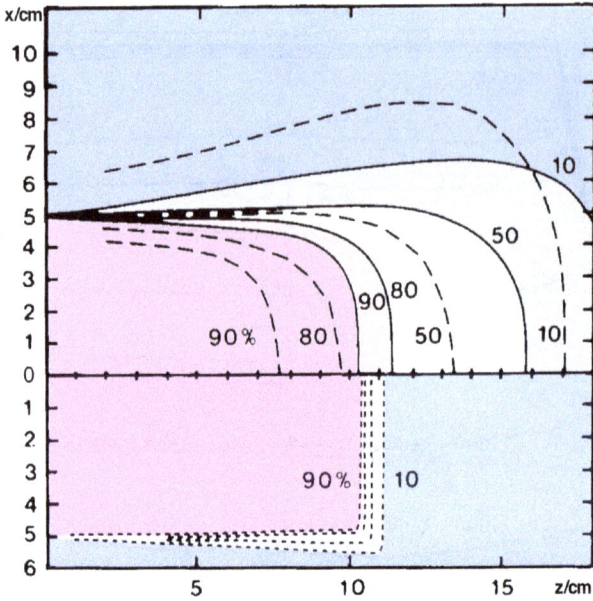

Figure 6.1(d). Comparison of the charged particle isodose distributions in the central plane for a 35 MeV Betatron and 50 MeV scanned Racetrack e⁻ beams with a field size of 10×10 cm compared with a similar therapeutic range proton beam (modified from Koehler and Preston 1972). The 50 MeV scanned electron beam isodose contours were measured by Brahme *et al.* (1980, solid lines, upper panel), the p⁺ 120 MeV cyclotron (dash line, lower panel) and the 35 MeV e⁻ betatron (dash line, upper panel).

with parallel-opposed beams. It can be seen that the dose to the subcutaneous region at the depth of dose maximum is greater than the dose at the central point. It can also be seen that the dose profile at the mid-line is less uniform than that at the surface owing to the effects of scatter.

In many centers, there is a preference for performing these treatments isocentrically. *Isocentric irradiation* involves positioning the patient so that the center of the target volume is coincident with the center of rotation of the treatment unit. When the first beam has been set up and treated, it is necessary to rotate the gantry 180° before treating the second field. This has the advantage of minimizing setup time for the second field and thus the overall treatment time.

6.4. Classical Multiple Beam Treatment Planning

Single fields and to some extent parallel-opposed fields are often used for palliative treatments for which the dose to the target volume is relatively low. Curative

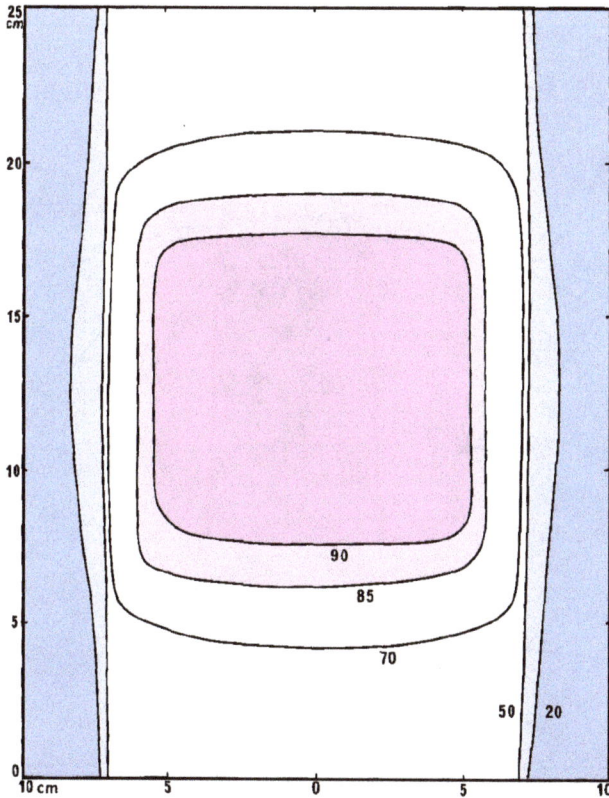

Figure 6.2(a). Isodose contours (percentage) for a parallel-opposed beam irradiation with 50 MeV scanned e$^-$ beams for a patient cross-section of 25 cm and a beam size of 14×14 cm (cf. Figs. 6.1(c)) and 6.7 for the central axis depth dose). Owing to high electron energy, the penumbra is fairly narrow and the influence of inhomogeneity is quite small. (Brahme *et al.* 1980) (for the central axis dose distribution, see Fig. 6.2(c)).

treatments generally require higher doses, and these simple treatment techniques are unsatisfactory because the dose to the tissues overlying the target volume would be excessive and could cause unacceptable early and late radiation effects. In these circumstances, three or more fields are used, although there are some high dose treatments, particularly within the head and neck region, which may only require two fields because the depth of the target volume is relatively small and covers most of the patient cross-section. Optimization of the treatment is required to ensure that a rather uniform dose is delivered to each target volume and that the dose to the surrounding tissues is as low as possible. The field arrangement, that is, the number and orientation of the fields, determines the basic treatment technique.

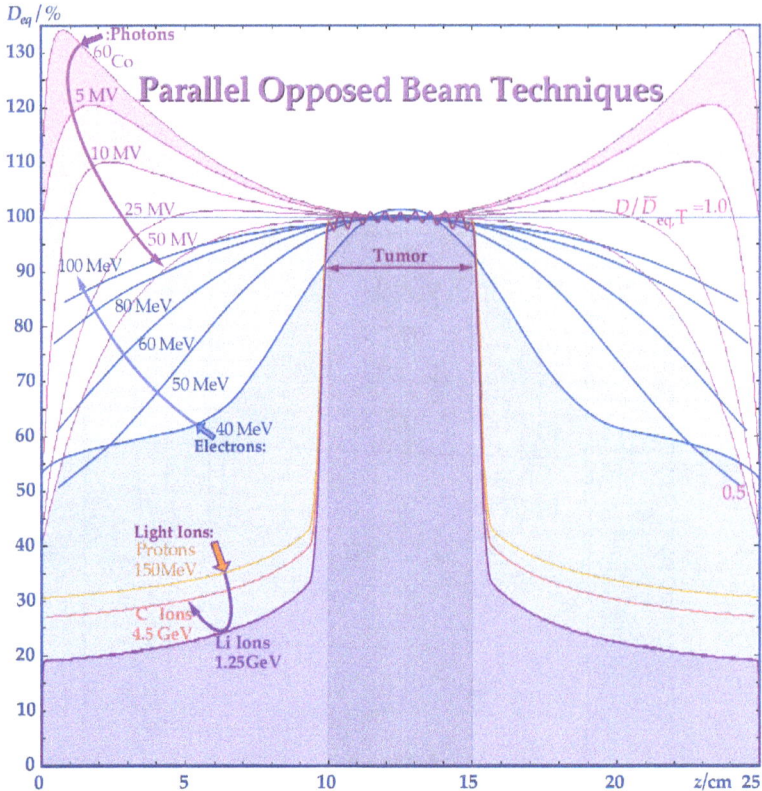

Figure 6.2(b). The central axis depth dose distribution with a parallel-opposed beam irradiation geometry with ^{60}Co-50 MV scanned bremsstrahlung beams, 40–100 MeV e$^-$ beams and p, Li and C ion beams for a patient cross-section of 25 cm and a beam size of 10×10 cm (Brahme *et al.* 1980). It is seen that 50 MV is needed with photons to avoid subcutaneous hot spots, whereas 40+ MeV e$^-$ and ions have always god shallow tissue spearing.

In general, the number of fields required is fewer when biologically optimized intensity modulation is employed but also when high-energy beams are used or the tumor depth is shallow. The orientation of the treatment fields is generally chosen to provide a uniform dose distribution and to avoid irradiation of sensitive structures. However, with radiobiologically optimized dose delivery, all directions are possible, which means that all the resultant dose distributions are examined to find the superior one. Today, it is even possible to optimize the dose delivery using biological objectives to a degree, which is possible to reach with the available treatment technology.

6.4.1. *Four-field Box Treatment*

The internal target volume (ITV) includes not only the tumor and its subclinical spread but also the margins, which have been added to allow for uncertainties in tumor localization inside the patient in relation to external reference points. Figure 6.3 illustrates a four-field box treatment technique. At the edge of the target volume, beams III and IV deliver a dose, which is approximately equal to the dose contributed by those beams to the isocenter due to the exponential quasilinear dose fall-off. The contribution from beams I and II will similarly deliver a uniform dose. However, even if the dose delivery vary in uniform it is not very optimal from a biological point of view since the entrance and exit doses coincides and elevates normal tissue side effects making three-field techniques almost equally good (cf. Fig. 5.4).

For most multiple-field plans, a smaller additional size is required if the fields are positioned uniformly around the patient. However, for the determination of the field length, the size has to be increased to the same extent as would be required for opposed coaxial fields.

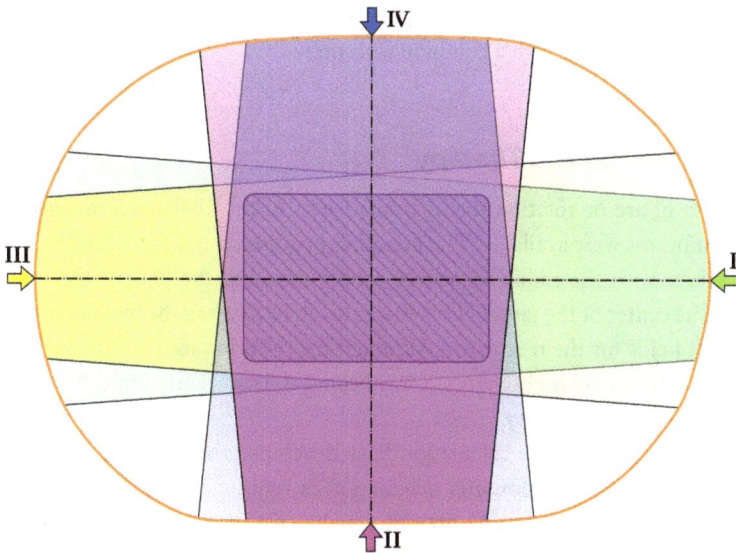

Figure 6.3. Four-field box treatment technique showing the attenuation and location of the geometric edges of the individual beams I–IV. This technique produces easily large uniform dose distributions but it is far from optimal for non rectangular target volumes were organs at risk may receive full therapeutic dose. This was clearly demonstrated in Fig. 5.4 where for example 3 perpendicular beams do better and 3 coplanar beams do almost equally well or better depending on the location of organs at risk.

6.4.2. *Asymmetric Collimators*

Asymmetric collimators are now standard equipment on the majority of currently available accelerators. In addition, all the modern multileaf collimators have this utility. In theory, asymmetric collimators allow the irradiation of any rectangular area within the maximum field size available on the accelerator. Some of the clinical advantages to be gained from the use of asymmetric collimators are as follows:

(1) Field matching with different radiation modalities or qualities: by first delivering the first half of the beam moving one collimator blade to the central axis of the beam, so that it used as an half-beam block, beam divergence at one edge of the beam is eliminated. This requires that all radiation modalities are collimated by the same collimator system (Brahme 1982).
(2) Reduction in treatment volume: in some treatments it is often necessary to ensure that the spinal cord is out of the treatment volume, again after a certain dose has been given. These changes in treatment volume can be achieved simply by changing the position of one or more of the collimator blades without having to change the treatment setup or make any major change to the treatment plan.
(3) Asymmetric collimators are particularly useful in enabling *non-coplanar* techniques to be set up with a common isocenter.

6.4.3. *Classical Arc Therapy*

The origin of arc or rotation therapy data backs to the 1940s before high-energy treatment beams were available. The logical improvement was to use one field aimed at the target with the radiation source rotating around the patient about an axis through the center of the target. This was equivalent to using the maximum possible number of fields for the treatment. Although the need for arc therapy was reduced with the availability of high-energy X-ray beams, it has recently undergone a revival for use in conformal therapy.

Many systems of dose calculation have developed for arc therapy, but now it is widely accepted that computer calculation is required. Any treatment arc can be adequately simulated by a number of equally spaced static fields. Even at mega voltage energies, rotation therapy dose distributions have some merits over dose obtained with fixed-field therapy. For example, skin doses are lower in rotation therapy and high-dose regions are more regular, approximating to cylindrical or ellipsoidal shapes. Then, even the fall-off in dose outside the target region is not as rapid as at a beam edge. At mega voltage energies an adequate tumor dose can usually be achieved with three or four intensity-modulated beams, making the treatment planning and setup considerably simpler than for rotation therapy. For quality assurance and these reasons, fixed-field therapy is preferred.

6.4.4. Conformal Therapy

Conformal treatment techniques are still commonly used in radiotherapy. The principle is to more closely match the shape of the high-dose region to the projection of the target volume, thus minimizing the dose to the surrounding normal tissue and, in particular, the dose to organs at risk. Many treatment sites could benefit from conformal therapy, in particular tumors of irregular shape or distribution, and certain sites (e.g., esophagus), may lend themselves well to clinical assessment of treatment outcome versus the increased dose resulting from the use of conformal therapy. The multileaf collimator is the ideal device for conformal therapy. A modern multileaf collimator consists typically of two opposed sets of 40 leaves, each projecting to a width of 1 cm at the isocenter. The leaves are made of tungsten of sufficient thickness that the primary transmission is <1% and are preferably double focused to account for beam divergence. The leaves can be set over the mid-line of the field so that any beam shape can be set. The position of leaves are set and monitored under computer control.

In general the above method is used to shape static fields. However, treatment machines operating under computer control have made possible the more complex technique of dynamic multileaf collimation. This is achieved by the continuous or multi-segmented variation of some of the treatment parameters during irradiation, producing complex high-dose volume shapes. Dynamic therapy can combine dynamic multileaf collimation and arc therapy where the treatment beam can be switched on or off during the arc.

6.4.5. Three-dimensional Planning

There has been a rapid move toward the development of three-dimensional planning systems to the extent that virtually all-commercial planning systems now offer this capacity. For a planning system to be fully three-dimensional, it must deal with irradiation techniques in three dimensions; it must be able to handle narrow beams and broad beams with a lot of scattered photons and electrons as well as extreme density variation from bones to lungs and air cavities and deal with non-coplanar beams and include rotation of the treatment head and couch. The dose calculation must also be three-dimensional with the use of an appropriate beam model. The greatest effort that has been put into three-dimensional systems is the use of three-dimensional graphic packages to display patient anatomy with beam direction and dose distributions superimposed.

6.5. Treatment Plan Optimization

Treatment plan optimization should be considered as an integral part of the radiation therapy planning process. The optimization algorithm is supplied with the best available data regarding radiation transport physics radiobiological functions for cell kill and clinical response models for normal tissues (Fig. 6.4(a), Söderström 1995, and Fig. 6.4(b), Löf 2000).

The most important free variable in radiation therapy optimization is to use nonuniform dose delivery. It can be obtained by several techniques. The classical approach is to use compensators inserted in essentially uniform flattening filtered therapeutic beams. When the number of individual beams becomes large, separate compensating filters are impractical or even impossible to use in a clinical situation. For such cases, dynamic techniques can be used to generate nonuniform dose distributions. Dynamic techniques include the use of dynamic multileaf collimation and computer-controlled scanned elementary beams of photons, electrons, and light ions.

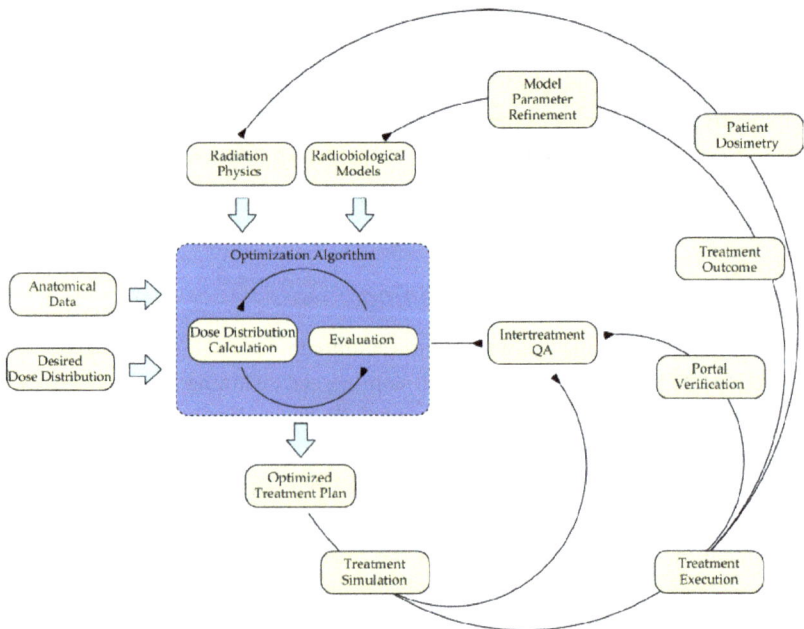

Figure 6.4(a). The radiation therapy process with the treatment plan optimization as an integrated part.

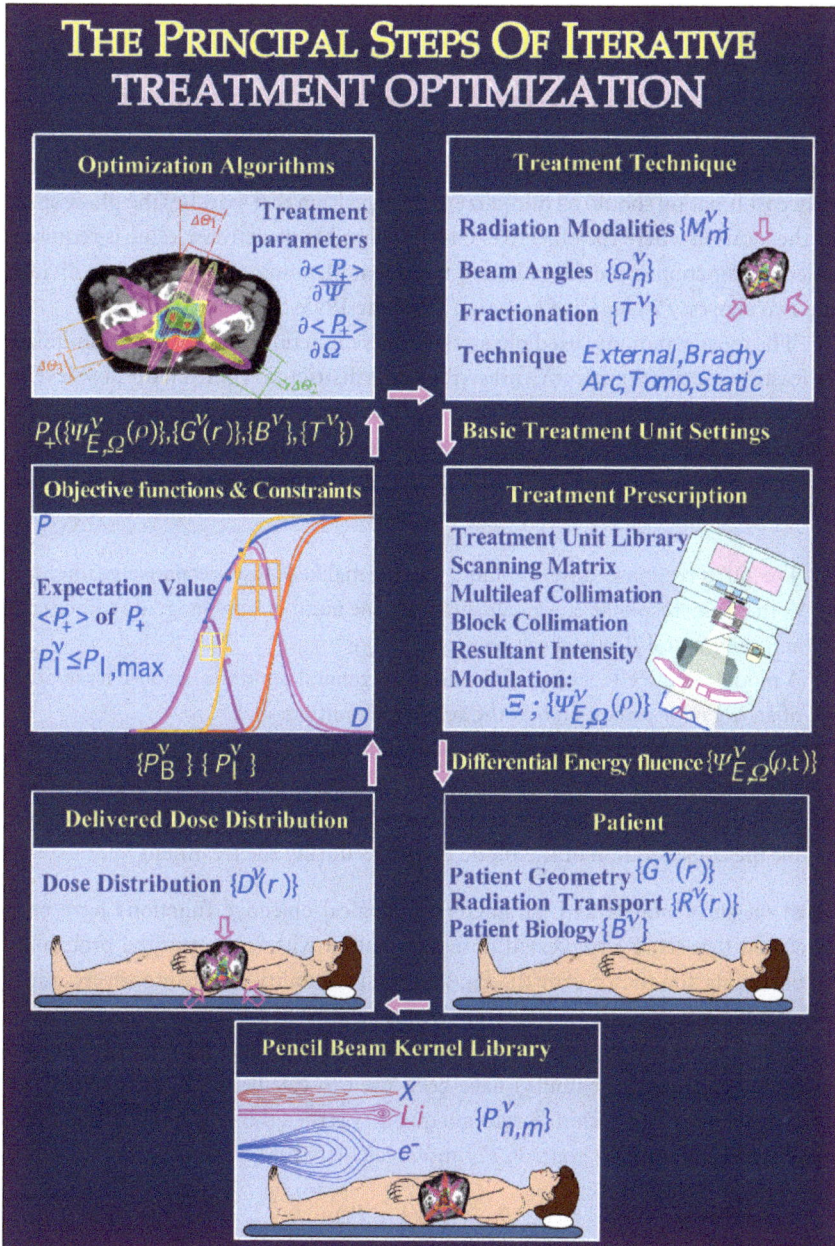

Figure 6.4(b). Illustration of the principal steps of iterative treatment plan optimization.

6.5.1. *Optimization Methods*

The optimization of radiation therapy involves the determination of the optimal values of a large number of variables that influence the outcome of treatment. It requires the identification of an objective or cost function used for evaluation of the calculated dose distributions. In general, an objective function should be maximized and a cost function should be minimized. An algorithm that searches the phase space for the location where the objective/cost function has its extreme value is required. Objective functions used for radiation treatment planning purposes can be divided into two classes: *Physical and biological* (Brahme 1995; Söderström 1995).

The most commonly used physical objective functions is that which minimizes the least square deviation from a prescribed dose distribution or from the dose level at some anatomical regions of special interest. The disadvantages with an optimization of the least square deviation are that the dose tends to become peaked in the central region of the target volume and too low at the edge of the target volume. A biological objective function should include the following:

(1) The 50% response dose, D_{50} and the normalized dose–response gradient, γ of the dose–response curve (alternatively the number of clonogenic tumor cells and their α and β values can be used).
(2) A model for the effect of partial volume or general nonuniform dose irradiation of an organ or tumor such as the relative seriality model.
(3) A model for the description of possible nonuniform sensitivity of the organs or tumors.
(4) A model describing the DNA repairs mechanism and the proliferation of cells, for the consideration of the tissue dynamics during the treatment.

Based on such models and parameters, biological objective functions have been developed that score a suggested dose distribution with respect to the probability of controlling the tumor growth, P_B, and the probability of causing severe injury in normal tissues, P_I. A combination of probability of treatment benefit, P_B (or tumor control probability) and the probability of tumor tissue injury, P_I (or normal tissue complication probability) have been used to calculate a scalar quantity for scoring the dose distribution. A relevant quantity is then the probability of achieving complication-free tumor control, P_+ and is calculated as:

$$P_+ = P_B(1 - P_I), \tag{6.1}$$

if the two probabilities are uncorrelated and:

$$P_+ = P_B - P_I, \tag{6.2}$$

if the responses are totally correlated. If both expressions are combined, for a mixed population, the probability can be expressed as

$$P_+ = P_B - P_I + \delta(1 - P_B)P_I, \tag{6.3}$$

where δ is the fraction of the patient population that has statistically uncorrelated responses P_B and P_I. More recently, the P_+ optimization strategy has been developed.

Biological objective functions have, as shown above, a clear advantage compared to physical objective functions since they are based on accurate biological dose–response data. Biological models, may allow a true biological optimization of a radiation therapy plan. One example is that it has been shown that biological objective functions closely follow the variation of the mean dose of the target volume as it is illustrated in Figs. 6.5(a) and 6.5(b) where the power of intensity modulation is shown for a simplistic target volume.

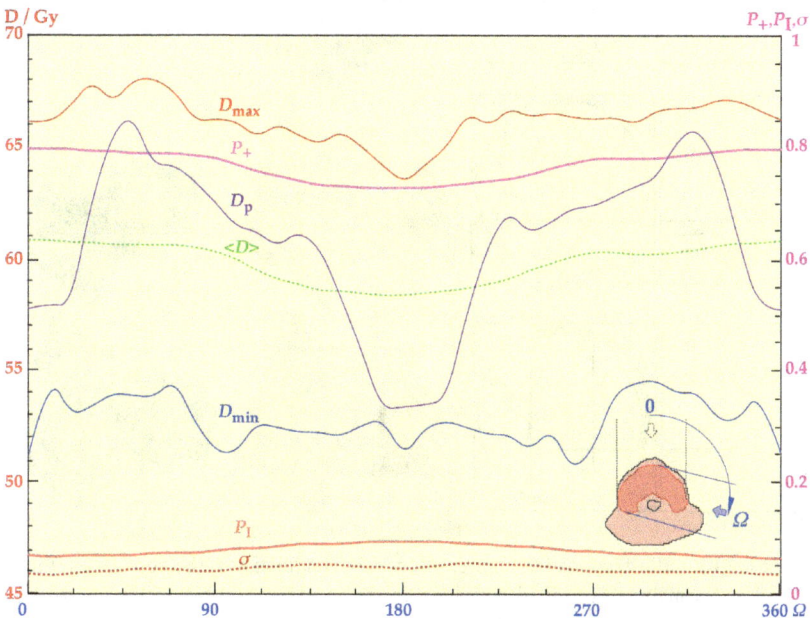

Figure 6.5(a). The relations of different dosimetric parameters with the biological objective function P_+ and the probability of normal tissue injury, P_I, are illustrated. The dose maximum in the target, D_{max}, the mean dose of the target, $<D>$, the minimum dose in the target, D_{min}, the dose at a point in the target, D_p, and the standard deviation of the target dose, σ are also presented.

Figure 6.5(b). Illustration of the power of intensity modulation with a square target inside a cylinder. Normally perpendicular wedged beams or a four-field box technique would be used but here the optimal wedge or planar beam profiles are calculated and the right most figure show how fast the square target can have circular isodoses around it with the strongly nonuniform dose delivery of the upper row. The advantage of full intensity modulation (last panel) is also seen compared to standard conformal (second last panel) or two-, three-, and four-field techniques to the left. Interestingly the nonuniform three-field technique can generate almost a square high dose volume too.

The mean dose of the target volume $<D>$, and the relative standard deviation of the dose distribution (σ_D) in Fig. 6.5(a), may be seen as a compact description of the most important properties of the dose volume histogram (DVH). The main properties of a DVH are very closely described by a normal probability integral, defined by $<D>$ and σ_D. D describes well the 50% effect point, whereas σ_D provides a good characterization of the slope of the DVH around $<D>$.

6.5.2. *Beam Modality*

Even though in this book we are primarily concerned with the optimization of photon beam therapy techniques, it is of interest to point out the possibilities offered by combinations with other modalities such as electron and light ion beams. These modalities can of course be used alone, but also in combination with photon beam therapy. It is preferable to use electrons for the delivery of dose to the gross tumor. Photon beams, with their sharper penumbra, are more suitable for use at the periphery of the target volume. The benefit of a dual modality treatment is illustrated by the phase space diagrams in Fig. 6.6 (Söderström 1995).

Any of the two-field plans with a single modality results in a lower probability of complication-free tumor control than the combined electron and photon beam treatment plans. The main reason for this is that a second treatment modality increases the flexibility of the treatment. Photons provide sharp penumbras and electrons negligible exit dose, so with two modalities both these advantages are available.

It is also noteworthy that there is a very small variation with beam entry direction when a combined electron and photon beam treatment is performed (Fig. 6.6(c)). It is thus of relatively low importance to optimize beam directions if a combined optimized electron and photon beam treatment can be performed. Beams can even be delivered from the same directions. Combined high-energy electron and photon therapy is therefore a very promising technique since the beam portals can also be chosen to maximize other aspects of the treatment such as the efficiency of portal verification or to select beam angles with higher accuracy in dose delivery.

6.5.3. *Beam Cross-section*

Optimization of beam cross-sections aims at restricting the high-dose region to the target volume, while accounting for setup and target margins. The high-dose region should closely wrap around the target volume in three dimensions. This is sometimes referred to as "conformal" radiotherapy. A perfect fit to any target volume can be

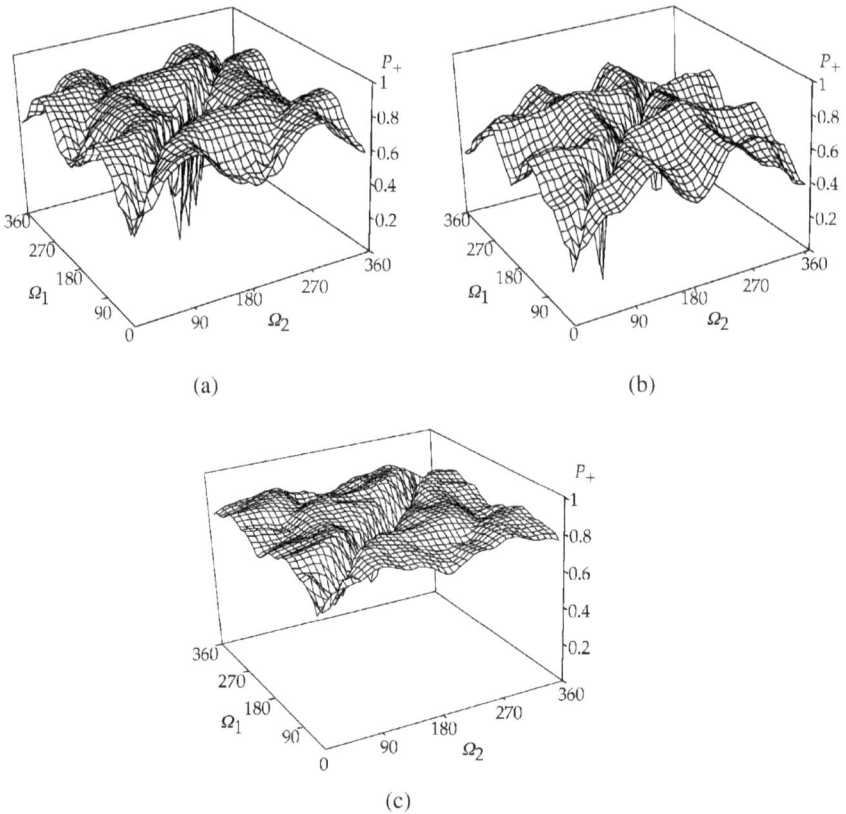

(a)

(b)

(c)

Figure 6.6. (a)–(c) The two-dimensional phase space of P_+ for different treatment modalities are shown (a) 50 MV photons; (b) 50 MeV electrons; (c) 50 MV photons and 50 MeV electrons, simultaneously. It is striking how good results are obtained by combining electrons and photons. This is obtained by combining the advantage of a narrow penumbra of the photon beam with the advantage of a low exit dose of the electron beam. It is also amazing to see that all beam directions are quite useful even if a few local maxima and minima can be seen especially on the diagonal where the two beams coincide and one beam is effectively lost.

obtained with Cerrobend blocks. However, these blocks are quite labor-intensive to use in routine work. For this reason, the multileaf collimator has been explored to define multiple treatment fields geometrically shaped to the projected area of the target volume. The best rotation angle of the multileaf collimator for a given convex target volume is obtained, in most cases, by aligning the direction of the motion of the leaves to the direction in which the target volume has its smallest cross-section (Brahme 1988).

6.5.4. *Beam Intensity Modulation*

A large number of methods for the generation of nonuniform dose delivery are now available. The classical approach is to use compensators. When a large number of individual beams, is used, separate compensating filters are impractical or even impossible to use. In such situations, there are other dynamic techniques that can be used to generate nonuniform dose delivery. These techniques include the use of dual dynamic asymmetric collimator jaw pairs, dynamic multileaf collimation, or computer-controlled scanned elementary bremsstrahlung beams (cf Fig. 3.17 and Söderström 1995). The asymmetric collimator jaw technique is rather a slow process because only a small fraction of the radiation field can be irradiated at many moments.

Dynamic multileaf collimations are much faster than using asymmetric collimator jaws and will only prolong the radiation time by about 30–45%. This is, therefore, one of the most flexible techniques available for nonuniform dose delivery because very steep dose gradients may be generated when the collimator penumbra is sharp. However, the fastest and the most general approach to nonuniform dose delivery is to use computer-controlled scanned beams that, particularly at high electron and bremsstrahlung energies, generate rather narrow beams. Such narrow beams can be used to shape the lateral dose distribution to compensate for inhomogeneities. However, especially in the case of bremsstrahlung beams, with their larger lateral tails, it is not possible to generate such steep dose variations as it is with the dynamic multileaf collimation. To improve the speed of the dose delivery even further, the scanning of electron and bremsstrahlung beams may be combined with dynamic multileaf collimation. Beam intensity modulation is probably the single most important degree of freedom for photon beam radiation therapy. The importance of this parameter is illustrated more clearly in Figs. 6.7(a)–(c) (Söderström 1995; Ferreira *et al.* 2006).

When nonuniform dose delivery is used, a very large increase in the probability of complication-free tumor control is obtained for all beam configurations. In most optimization algorithms, the optimization of beam intensity modulation also includes an optimization of the beam cross-section (cf Fig. 6.7 (d)–(f)).

6.5.5. *Selection of Beam Orientation*

The beam orientation should perfectly be optimized for the exact number of beam portals used. The optimal orientations can be found by search through the entire phase space of P_+. This method of beam selection is computationally very intensive, and when the number of beams exceeds three, strict optimization is not even realistic due to enormous calculation times. Nonuniform dose delivery obviously results in a large improvement in the expected treatment outcome. This can clearly be seen

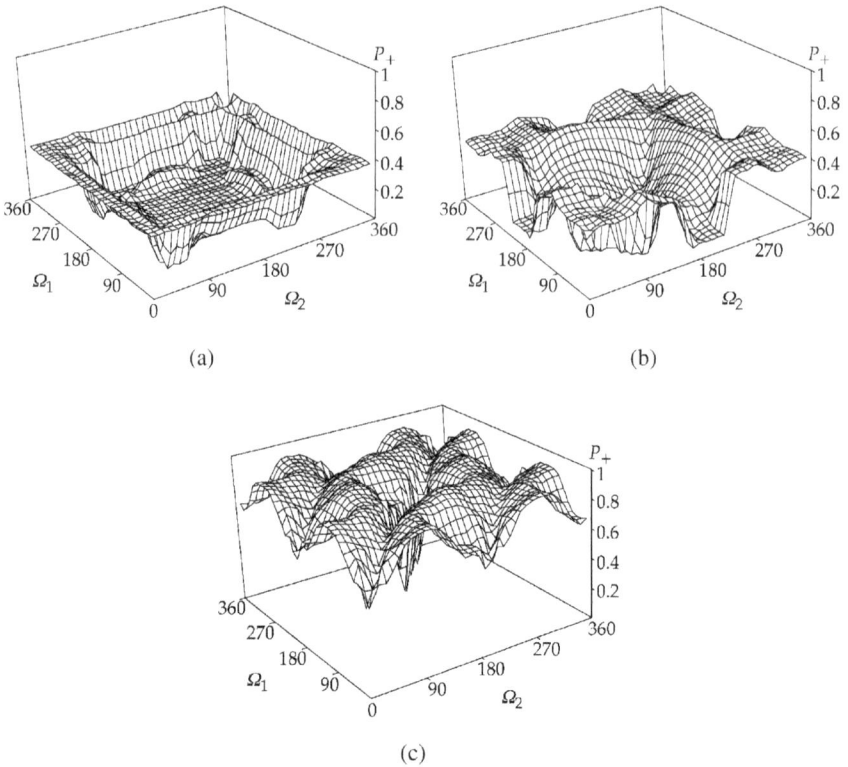

Figure 6.7(a)–(c). Illustration of phase spaces of complication free cure with 2 beam techniques obtained with different degrees of freedom in beam intensity modulation (a) uniform beams; (b) wedged beams; and (c) nonuniform IMRT beams (cf Figs. 6.24–4.27 for 1 and 3 beams).

in Fig. 6.8 (Söderström 1995) where the probability of achieving complication-free tumor has been plotted against the number of nonuniform fields used for a cervix tumor (solid circles).

The relative standard deviation of the delivered dose distribution in the target volume (open circles) is also shown. The obtained P_+ level and the standard deviation of the mean dose to the target when using only uniform fields are also shown (solid and open squares, respectively).

For comparison, two different four-field box techniques with uniform dose delivery are also included: the standard four-field box with lower weight posterior fields compared to the anterior and lateral fields to spare the rectum (open triangle) and a four-field box technique where the two lateral fields are used to boost the gross tumor region (solid triangles). From the curves it is seen that for uniform

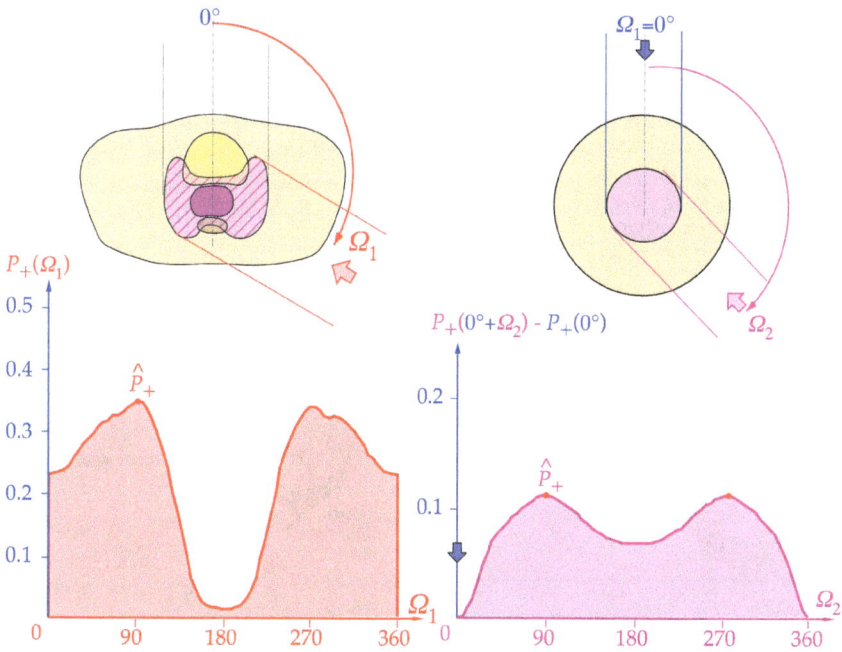

Figure 6.7(d). The variation in complication-free cure with angle for a complex target and organs at risk. In **(e)** the target is rotationally symmetric but a first beam instead makes the most effective direction to be at around 90° to decouple the beam interactions (cf. Fig. 5.4).

beams a three-field configuration, where the beam entry directions, beam weights, and beam widths are subject to optimization, is generally better than a traditional box technique where none of the above parameters are optimized.

The increase in the P_+ level when using nonuniform beams compared to uniform beams is almost independent on how many beams portals are used (cf Fig. 6.8). The relative increase in the P_+ level when using nonuniform beams is ~15% except when only one beam portal is used. A four-field box technique (open triangle) results only in a P_+ level of 20%. A four-field box technique where the two lateral fields are used as boost fields to the gross tumor region (solid triangles) results in a higher P_+ level (52%), since the dose to the gross tumor region may be increased from 55 Gy without boost to 61 Gy with the boost technique.

Figure 6.9 (Söderström 1995) shows the increase in P_+ as a function of the number of beam portals used for three different target volumes: a head and neck target (open circles), a thorax target (solid squares), and a cervix target (solid circles).

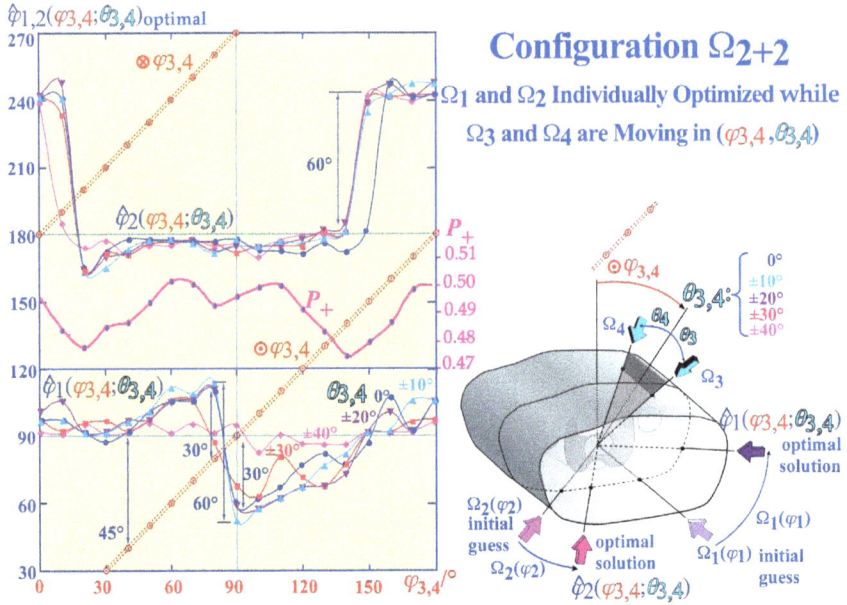

Figure 6.7(f). The variation in the optimal beam direction of φ_1 and P_+ as a function of a continuously increasing set of beams at $\varphi_{3,4}$. Interestingly the shift of φ_1 of about $60°$ when $\varphi_{3,4}$ reaches around $80°$ is obtained to increase the degrees of freedom in intensity modulation indicating that the angular interval between neighboring beams should be at least about $30°$ to $40°$ to minimize the effect of beam coincidence.

The figure shows that independent of target geometry, the P_+ level quickly reaches a value where no more than a few percentage can be gained by adding more nonuniform coplanar beams. It may be noted that the steepest increase in P_+ is obtained for the thorax target. This is due to the small convex shape and the location of the target volume.

Figure 6.10 (Söderström 1995) shows the variation of the minimum, maximum, and mean dose level in the target volume of the head and neck geometry. There is a small but constant increase in the mean dose level from 61.6 to 91.9 Gy with increasing number of beam portals. More apparent is the decrease of the span between minimum and maximum dose from 48 to 72 Gy ($\sigma_D = 9.2\%$) when using only one beam portal, down to 61.6–62.4 Gy ($\sigma_D = 0.1\%$) with 72 beam portals. Both these trends of increasing mean dose to the target volume and decreasing dose variance are followed by an increased probability of achieving complication-free tumor control as shown in Fig. 6.11 (Söderström 1995).

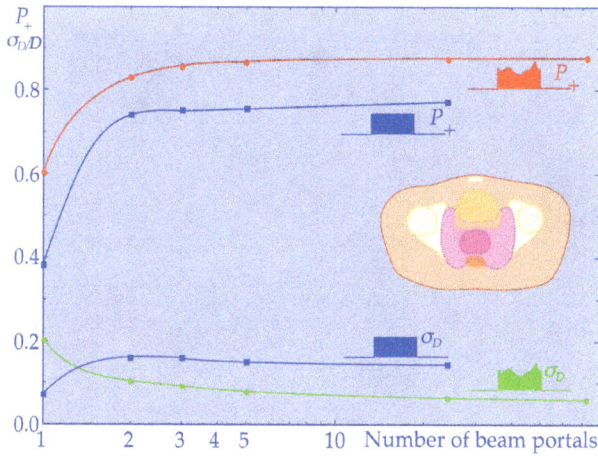

Figure 6.8. Probability of achieving complication-free tumor control plotted against the number of nonuniform fields used for the cervix tumor (red circles). The relative standard deviation of the delivered dose distributions in the target volume is shown in green circles.

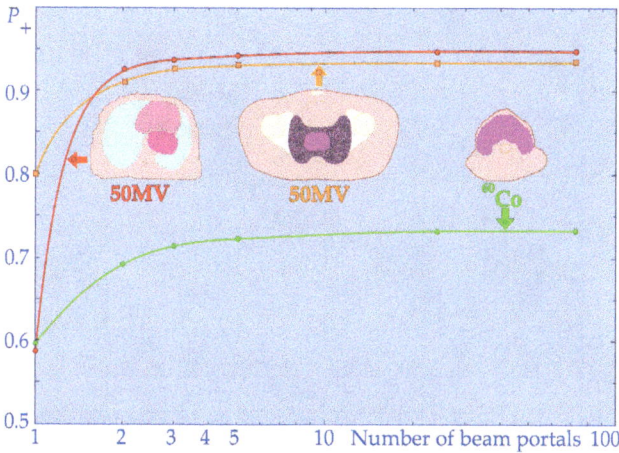

Figure 6.9. Illustration of the increase in P_+ as a function of the number of the beam portals used for three different targets volumes: the head and neck target (green circles), the thorax target (red circles) and the cervix target (brown squares).

The figure illustrates the optimal P_+ level and the standard deviation of the mean dose to the target volume as a function of number of nonuniform beam portals. The close relation between the mean dose and its standard deviation and the resulting P_+ level indicates that these two parameters should be used when

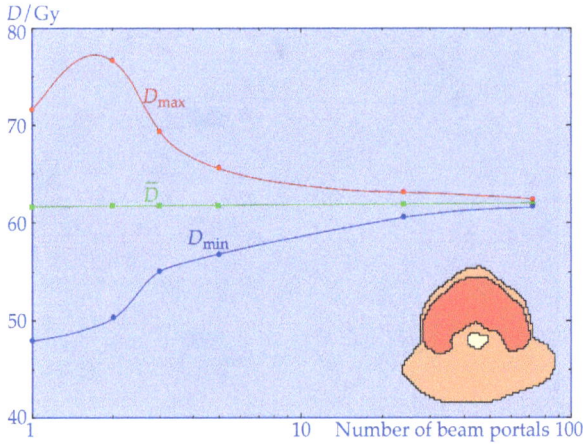

Figure 6.10.　Illustration of the variation of the minimum, maximum, and mean dose level in the target volume of the head and neck geometry.

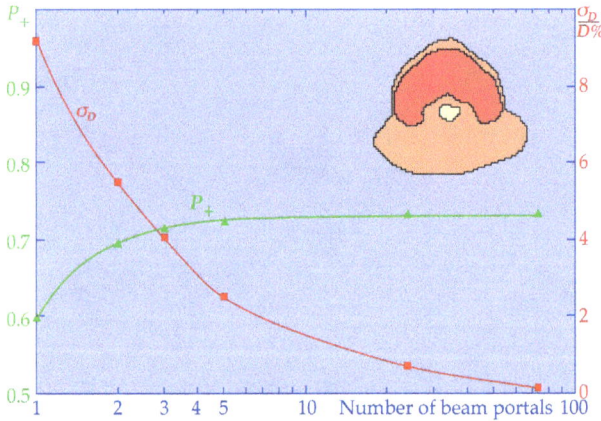

Figure 6.11.　Increasing mean dose to the target volume and decreasing dose variance are followed by an increased probability of achieving complication-free tumor control.

prescribing dose delivery in radiation therapy. In Fig. 6.12 (Söderström 1995) four examples of dose distribution obtained while irradiating the cervix target are shown. The corresponding dose distributional data is shown in Table 6.1 (Söderström 1995).

Using standard four-field box techniques, as shown in Fig. 6.12(a), results in a P_+ level of only 20%. If instead, the two lateral beams are used as boost fields to the gross tumor region, as shown in Fig. 6.12(b), the P_+ level is increased to 52%.

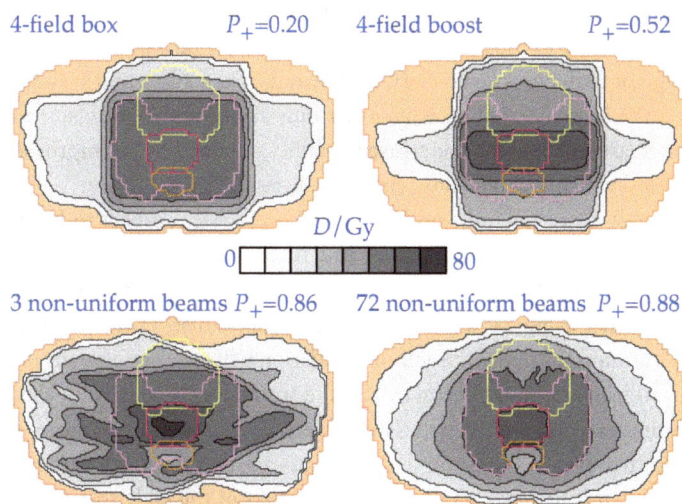

Figure 6.12. Four dose distributions of increasing efficacy for irradiating a cervix target volume are shown.

Table 6.1. The radiobiological data set used in the calculation. The normalized gradient of the dose–response relation, γ, the 50% response dose, D_{50}, and the relative seriality, s.

Organ type	γ	D_{50} [Gy]	s
Head and neck geometry			
Normal tissue stroma	2.76	65.0	1.00
Spinal cord	1.78	60.0	1.00
Tumor	3.00	52.0	—
Thorax geometry			
Normal tissue stroma	2.76	65.0	1.00
Spinal cord	1.78	60.0	1.00
Lung	2.10	24.5	0.006
Heart	3.00	49.2	0.20
Tumor	3.00	52.0	—
Cervix geometry			
Small bowel	1.50	80.0	2.60
Bladder	3.00	80.0	1.30
Rectum	3.00	55.0	0.69
Lymph nodes	3.00	39.0	—
Tumor	4.00	52.0	—

An increased probability of tumor control in the gross tumor region and a reduced probability of injury to the rectum can, thus, be seen (Söderström 1995). However by allowing nonuniform dose delivery, as shown in Fig. 5.12(c), and at the same time optimizing the beam entry directions using only three beams portals, the obtained P_+ may increase to 86%. This increase is mainly obtained by an improved sparing of the rectum, which will now only suffer a 1% risk of severe injury.

Using only three nonuniform beams optimized with respect to beam profiles and beam entry direction is profitable and there is hardly any more to gain by adding more beams. This can be seen in Fig. 6.12(d) which shows the obtained dose distribution when using 72 nonuniform beam profiles distributed in equal angular increments around the patient. Such an irradiation will only increase the probability of complication-free tumor control by 2% and would rarely be the clinician's choice due to the increased complexity in delivery and verification. Even more important is to realize that even if for example continuous arc therapy looks very nice on the treatment plan it is linked with treating a large part of the body to a dose level that potentially may induce secondary cancers. The risk is maximum around 5–10 Gy. It may therefore be the optimal treatment fore elderly people that may never have time to develop a secondary cancer around 20 years down the road, and they often also need minimal acute side effects. For young people with good prognosis and long life expectancy the risk for developing a secondary cancer is higher and should be avoided by a suitable biologically optimized few field techniques that gives a higher dose in the entrance channels eliminating a large part of the risk for a secondary cancer and leaves a larger part of the body largely uniradiated (cf Söderström and Brahme 1996). It should also be pointed out that this section was studied using a fixed patient cross section longitudinally. If the tumor volume was more complex longitudinally, different slices may need different optimal beam directions and in general more beam directions may be desirable for a complex tumor like the prostate case in Section 5.5.6 cf Figs. 5.18 and 5.19.

6.5.6. *Optimal Photon Beam Energy Distribution for Radiation Therapy*

During the past 30 years, optimization of external beam radiation therapy has been mainly focused on different methods to improve the dose delivery by using multiple uniform or wedge-shaped beams. The goal of optimization process has often been to deliver a dose to the target volume as close as possible to the prescribed one or to achieve as small a deviation as possible from the dose constraints for the tumor and the organs at risk. A recent development has been seen for stationary and dynamic multileaf collimation with the goal to deliver truly three-dimensional conformal irradiation by modulation of the photon fluency profiles. Dose optimization with scanned photon and electron beams (Lind and Brahme 1995) is a very flexible and practical technique.

One of the continued developments of radiation therapy equipment to better treat deep-seated tumors has been the development of higher photon beam energies. The most recent high-energy accelerator is the racetrack microtone, which covers a range from 5 to 50 MeV. Higher photon beam energy reduces the skin dose but increases the exit dose as the maximum dose is moved deep below the skin surface due to the increasing energy of the secondary electrons.

6.5.6.1. Treatment geometries

A head and neck target including the locally involved lymph nodes of larynx cancer in addition to gross tumor is used as one test configuration (cf Fig. 6.16 below). In this case, the normal tissue stroma and the spinal cord are the principal organs at the risk. The gross tumor regions and the lymphatic spread are regarded as separate biological structures and are associated with different biological responses. The gross tumor volume is assumed to consist of normal tissue infiltrated to 50% by clonogenic tumor cells and the region of presumed lymphatic spread is assumed to contain a uniform clonogenic tumor cell burden of 10%. When uniform, as well as nonuniform photon beams are used, these targets are treated with a three-field technique using one frontal field and two parallel-opposed fields.

An advanced cervix cancer including locally involved lymph nodes is treated as an example of a deep-seated tumor. In this case, the organs at risk are bladder, rectum, hip joints, small bowel, and the normal tissue stroma. The gross cervix tumor and the involved local lymph nodes are regarded as separate biological structures and therefore associate with different biological responses. The gross cervix tumor is assumed to contain a 50% clonogenic cell infiltration and the local nodes are assumed to have a 10% clonogenic tumor cell infiltration. Independent on beam energy and level of beam modulation this target volume is irradiated by a three-beam technique with one anterior beam and two oblique lateral beams separated by 120° equidistantly around the patient (Söderström 1995).

6.5.6.2. Deep-seated tumors

When treating deep-seated tumors, such as the cervix tumor used here as test case, it might be expected that the extended build-up region at high photon energies

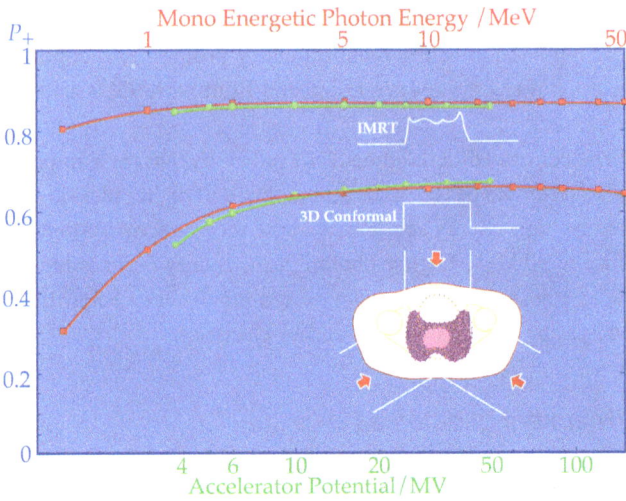

Figure 6.13. The variation of P_+ value with uniform bremsstrahlung photon beams (green) increases steadily with increasing photon energy until 50 MV. However with intensity-modulated monoenergetic beams, an almost constant and much higher P_+ value is achieved.

would result in a better treatment outcome than lower photon energies. This can be seen in Fig. 6.13 (Söderström 1995) where variation of P_+ value with uniform monoenergetic photon beams increases steadily with increasing photon energy.

The increase is primarily explained by the increased depth of the build-up region in high-energy photon beams and thus a better possibility to deliver high doses to the tumor without causing injury. However when the photon energy reaches ∼3 MeV or 10 MV, a gradual saturation sets in with small increase of P_+ level at higher energies. To be precise, a small decrease in the P_+ level may be observed beyond 30 MeV for monoenergetic beams, whereas in uniform bremsstrahlung beams P_+ keeps increasing up to 50 MV. The small decrease of P_+ level at the highest monoenergetic photon beam energies is due to the larger lateral electron scatter and thus the larger penumbra width as shown in Fig. 6.14.

For *very large tumor* masses, the optimal energy will keep increasing, since the dose outside the field edge due to scattered photons is decreasing with energy and so is the importance of penumbra width. An increase of the dose gradient at tissue borders will enable a higher dose in the target volume at the same time as the dose causing normal tissue injury is reduced or constant. To verify this principle, each beam can be subdivided into two energies, one central and one peripheral, in which different monoenergetic beam energies are allowed and the beam energies

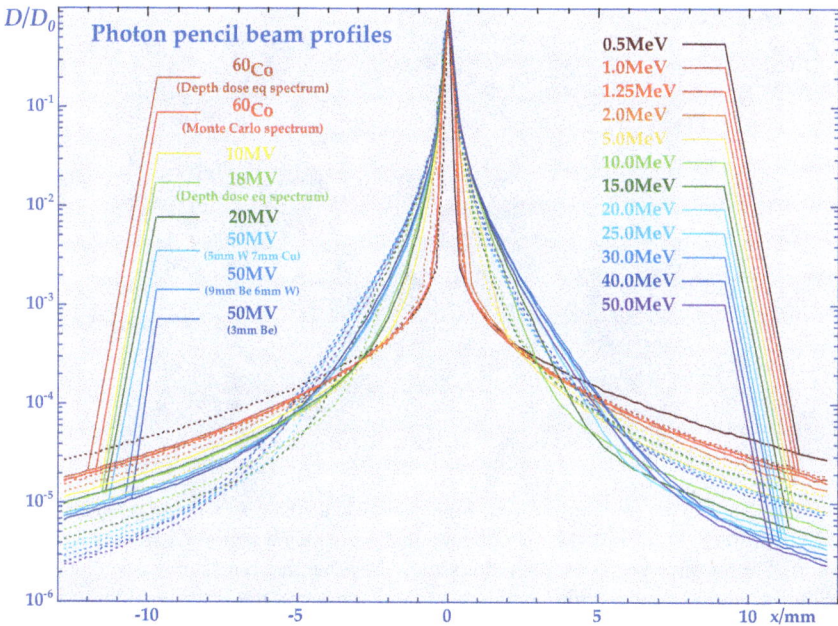

Figure 6.14. The small decrease of P_+ level at the highest monoenergetic photon beam energies is due to the larger lateral electron scatter and thus the larger penumbra width.

are consequently optimized for all regions of the beam simultaneously. The result is illustrated in Fig. 6.15 (Söderström 1995).

Figure 6.15 shows that the peripheral lymphatic regions benefit by the low photon energies around 2 MeV (6 MV) and the central gross tumor region by energies of around 10 MeV (30 MV).

6.5.6.3. Shallow tumors

It is obvious from Fig. 6.16 (Söderström 1995) that low photon energies are most advantageous for this type of treatment. The figure shows the probability of achieving complication-free tumor control, P_+, for treatment of larynx cancer with uniform photon beams (lower curves) and nonuniform photon beams (upper curves) as a function of energy.

Nonuniform dose delivery improves the expected treatment outcome significantly, just as in the case for the deep-seated cervix target. However, both monoenergetic and clinical bremsstrahlung beams give a clear reduction of the P_+ level with increasing beam energy. The P_+ level for monoenergetic beams drops dramatically for energies >10 MeV. This is not the case of the clinical

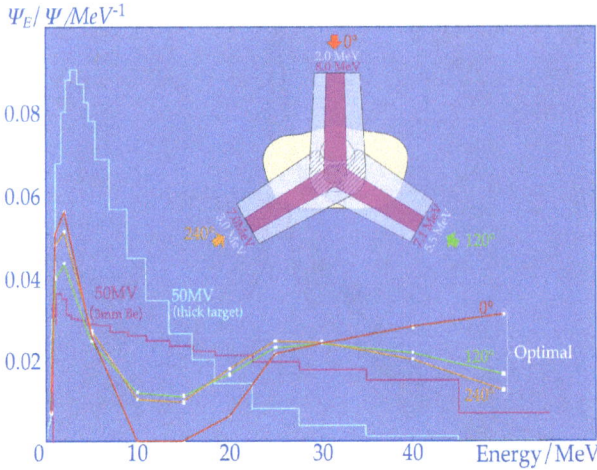

Figure 6.15. Optimal monoenergetic beam energies for the cervix target geometry when each constituent beam is divided in a central and a peripheral region are illustrated in the insert. The lowest energy is used anteriorly since the penumbra is most important here. The optimized energy fluence spectra for each portal of the cervix target (disregarding periferal and central regions) for the 0°, 120° and 240° beam portals. Two energy fluence spectra from clinical beams are also shown for comparison. A 50 MV beam from a 6 mm tungsten target (blue histogram) and 50 MV beam from a 3 mm Be-target (pink histogram) are included for comparison. Interestingly, the low- and high-energy components of the optimal spectra can be delivered by a combination of 50 MV thin and thick target beams or preferably by 50 MV thin target and about 6 MV beams to improve depth dose and penumbra simultaneously! Even better result would be obtained using the high energy spectrum with high dose over the gross tumor and the low energy over the shaded lymphatic spread region.

bremsstrahlung beams due to the presence of large amount of low energy photons in such beams.

6.5.6.4. *Optimal beam entry directions*

As it can be seen in Fig. 6.17 (Söderström 1995) for the cervix target and in Fig. 6.18 (Söderström 1995) for the head and neck target, there are only small variations of P_+ as a function of angle of incidence when varying the beam energy. For the cervix tumor, there is a monotonic increase in P_+ with energy independent of angle of incidence.

This indicates that, to a first approximation, the variation of the optimal beam entry with varying beam energy is small for deep-seated tumors as shown in Fig. 6.17. For shallow target volumes, the optimal angle of incidence varies considerably with energy as seen in Fig. 6.18. It is close to 0° at low photon energies and increases

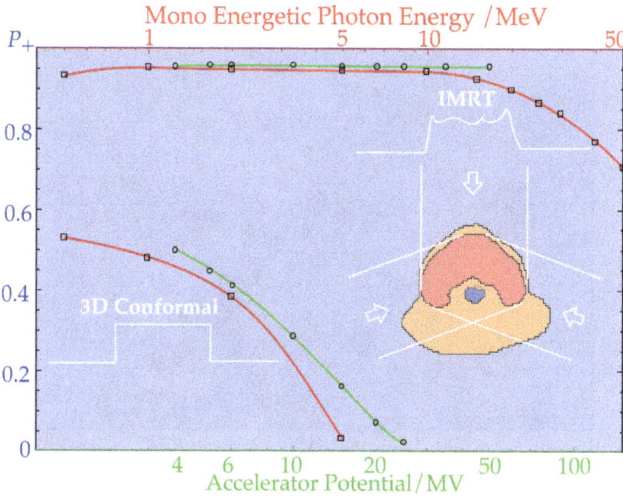

Figure 6.16. Illustration of the probability of achieving complication-free tumor control, P_+, for treatment of larynx cancer with uniform photon beams (lower curves) and nonuniform photon beams (upper curves) as a function of energy.

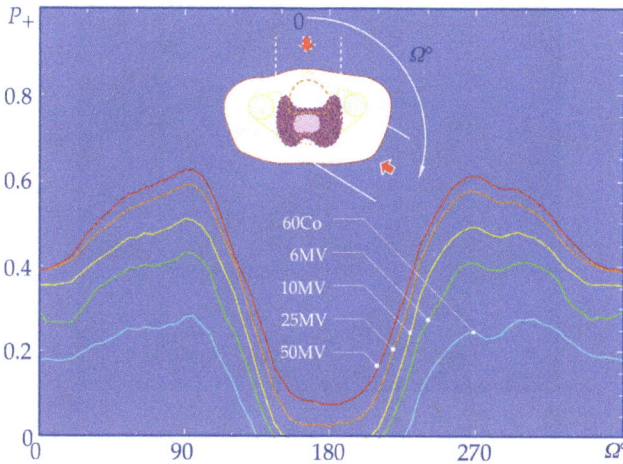

Figure 6.17. For the cervix target, there are always large variations of P_+ as a function of angle of incidence when varying the beam energy.

slowly to about $120°$ at $50\,\mathrm{MV}$ to avoid placing the target volume in the build-up region.

As it can be seen in Fig. 6.19 (Söderström 1995), the optimal photon beam energy for a small deep-seated target volume is as low as $3\,\mathrm{MeV}$. As the volume

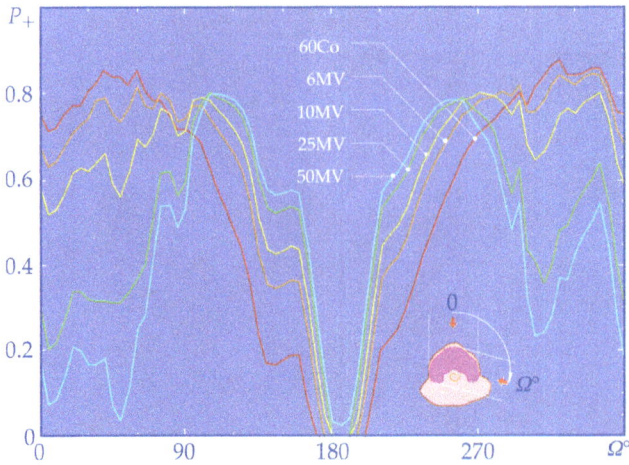

Figure 6.18. For the head and neck target, there are even larger variations of P_+ as a function of angle of incidence when varying the beam energy.

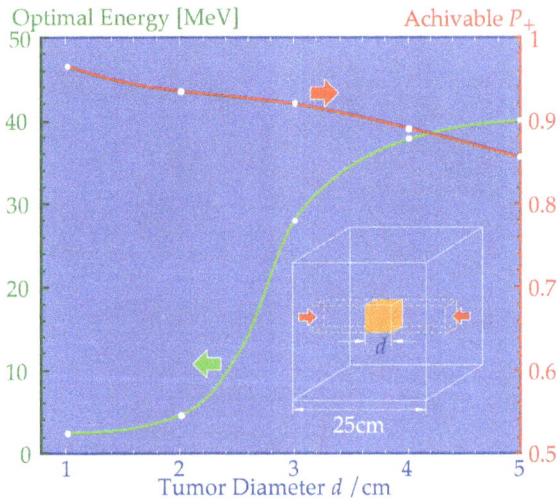

Figure 6.19. The optimal photon beam energy for small deep-seated target volume is as low as 3 MeV but increases rapidly toward 40 MeV from 5 cm and beyond.

of target increases, the optimal energy increases to values >50 MeV for target volumes >125 cm^2 ($d = 5$ cm). The explanation for the low optimal energy for small deep-seated targets is that the *mean cord length*, C of the target volume ($A/V = 4/C$) increases with the tumor radius.

Really small targets have a larger amount of tumor–normal tissue interface relative to their volume and are thus in need of photon beams with a *narrow* penumbra. This is particularly true for targets treated by conformal therapy and not to such a large extent valid for few field treatment techniques. When a small number of fields are used, only a small portion of circumference of a spherical target is tangentially irradiated, where a narrow penumbra is required. If the target, on the other hand, is more cube-like, as much as four-sixth of the circumference may be irradiated by one single field.

A narrow penumbra enables a high dose delivery to the target volume without causing serve injury to the surrounding normal tissue. In addition, the treatment of small tumors produces less normal tissue damage due to the small beam portals and normal tissue volume irradiated. As the volume of target increases, the ratio of tumor to normal tissue interface volume decreases and the need to deliver large doses to the bulky tumors increases. To be able to deliver high doses to large volumes, the dose maximum of photon beam should preferably be reached inside the target volume. Since only one photon beam energy is allowed during the present optimization process, the probability of achieving complication-free control will decrease with increasing target volume. The main reason for this is the increasing risk for normal tissue damage and thus the inability to reach high enough dose in the target volume with a single parallel-opposed pair of photon beams. For multi-portal angle of incident optimization, the more powerful methods of Sec. 5.5 are needed (cf. Fig. 5.18).

6.5.7. *Beam Direction*

A thorough investigation of the merits of different combinations of beam portals are of outmost importance, in order to get a good knowledge of how sensitive the treatment outcome is with respect to the selection of beam entry directions. It is also of interest to investigate how important the optimization of beam orientation really is in few-field radiation therapy.

This investigation is focused in the phase space of P_+. The term *phase space* of P_+ means the value of P_+ obtained as a function of entry direction of used beams (Söderström *et al.* 1995). A method for the selection of beam directions is a direct selection from one-dimensional phase space of P_+. The failure of this method has led to an investigation of the properties and structures of higher order phase spaces.

When looking at the two-dimensional phase spaces, P_+^2, one may observe several lines of symmetry. The principal shape of these symmetry lines presented in the P_+^2 phase space is to a certain extent *independent* of the location of the target volume and the organs at risk (Figs. 6.20(a) and 6.20(b)).

BEAM ORIENTATION DEPENDENT SYMMETRIES

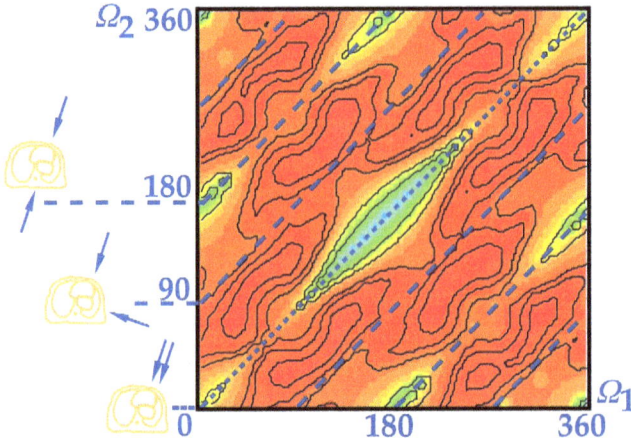

Figure 6.20(a). Symmetry lines in the phase space of P_+ due to a fixed angular increment between Ω_1 and Ω_2.

ATTENUATION & PATIENT GEOMETRY DEPENDENT SYMMETRIES

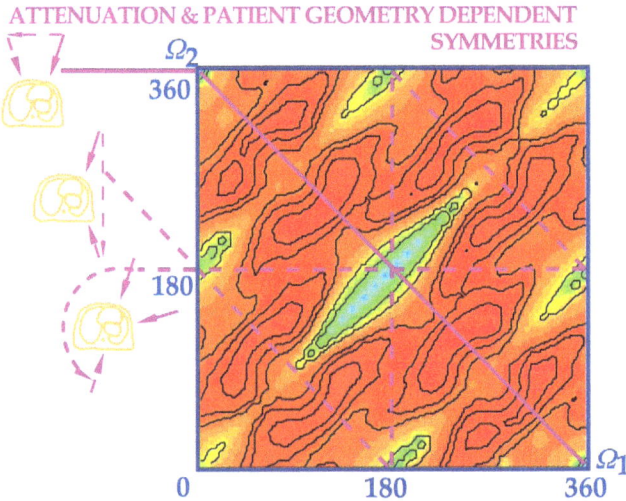

Figure 6.20(b). Symmetry lines in the phase space of P_+ due to attenuation and patient geometry symetries.

Important symmetry lines are not only those that are related to a reduction to a number of beam portals, where $\Omega_1 = \Omega_2$ (solid line in Fig. 6.20(a)), but also those lines representing parallel-opposed beams, where $\Omega_1 = \Omega_2 \pm \pi$ (dashed line in Fig. 6.20(a)). The corresponding lines and plans can also been observed

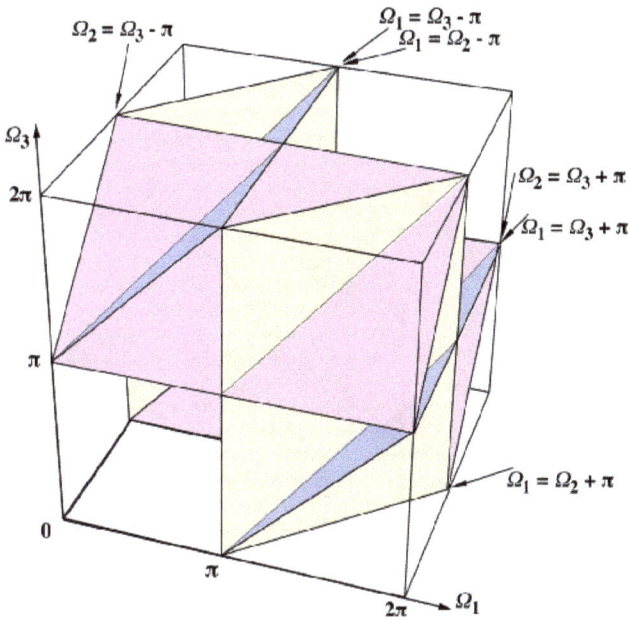

Figure 6.21. Illustration of the six planes in the three-dimensional P_+ phase space representing parallel–opposed beams.

in the three-dimensional P_+^3 phase space as seen in Fig. 6.21 (Söderström *et al.* 1995).

When the two-dimensional phase space P_+^2 is expanded to three-dimensions or P_+^3 by increasing the number of beam portals to three, symmetry lines representing a reduction of the number of beam portals from two to one are expanded to symmetry planes representing a reduction of the number of used beam portals from three to two (Fig. 6.22). Each one of these three planes is actually built up of the complete two-dimensional P_+ phase space. Symmetry lines in the P_+^2 phase space representing parallel-opposed beams become six symmetry planes in the cube of the P_+^3 phase space, as shown in Fig. 6.21. A symmetry line appears at the intersection line of the three planes representing a reduction of the number of used beam portals corresponding to one single incident beam ($\Omega_1 = \Omega_2 = \Omega_3$). Along this line, the lowest possible P_+ level will always appear corresponding to the degenerate of one single field.

The gradient in the P_+ phase space determines how densely the angles of incidence of the beam have to be sampled to find the true optimal beam portals. The maximum gradient of the P_+ phase space surrounding low P_+ areas is increased as

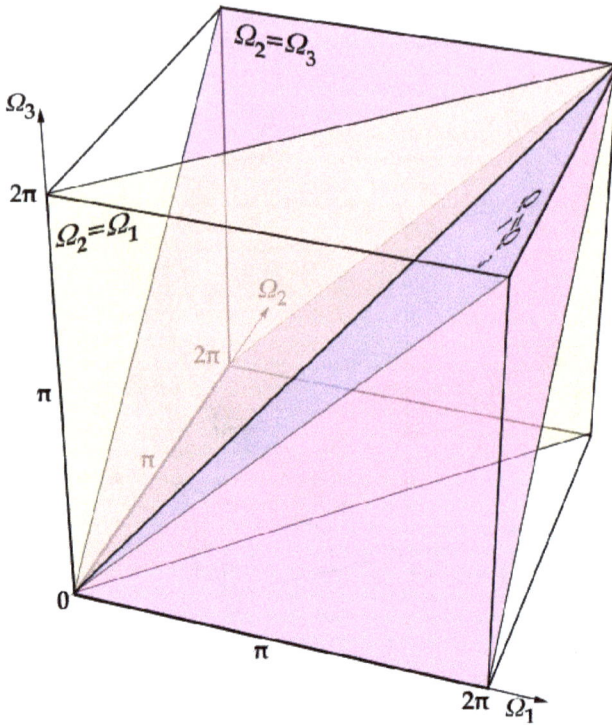

Figure 6.22. Illustration of the three planes in the three-dimensional P_+ phase space representing a reduction of the number of used beam portals from three to two. The line representing a reduction number of used beams portals from three to one is located on the spatial diagonal where the three above-mentioned planes cross each other.

the number of beams is increased. When the number of beams gradually becomes larger, the optimal beam entry directions become gradually less critical, provided that coincident or parallel-opposed beams are avoided. It is also important that optimized nonuniform dose delivery is allowed.

The number of beam portals required to obtain a close to optimal treatment is somewhere between three and five beams. If the number of used beam portals is above this interval, the beam entry directions are relatively unimportant parameters to subject to optimization. Within the intervals (three to five beams) the selection of beam entry directions is relatively important. It is thus of importance to find fast and reliable methods of optimization of beam directions for the number of beam portals within this interval.

6.5.7.1. *Comparison of the merits of uniform, wedged, and nonuniform beams*

When only classical uniform beams are allowed during the optimization, most of these structures of the three-dimensional P_+^3 phase space, resulting from the use of nonuniform beams are lost as seen in the figures at next pages. High P_+ level can only be found when the beam angle combination includes some beam angle close to $\Omega = 0°$. In the one-dimensional phase space for uniform beams, as showed by the dash lines in Fig. 6.23, the maximum P_+ level is located at $\Omega = 0°$ (Söderström *et al.* 1995).

Adding a second beam increases the P_+ level by >2% for the optimum combination of beam directions $\Omega = 75°$ and $\Omega = 345°$ (see Fig. 6.24).

With the optimum three beam combination, $\Omega_1 = 15°$, $\Omega_2 = 270°$, and $\Omega_3 = 345°$, using classical uniform beams, an increase of the P_+ level as little as 0.3% is obtained and no new internal structures in the P_+^3 phase space can be observed as shown in the Fig. 6.25 (Söderström *et al.* 1995).

In Fig. 6.23, the one dimensional P_+^1 due to treatment optimization using *wedged shaped* beams is also shown by dotted lines. It can be seen that it is possible to use

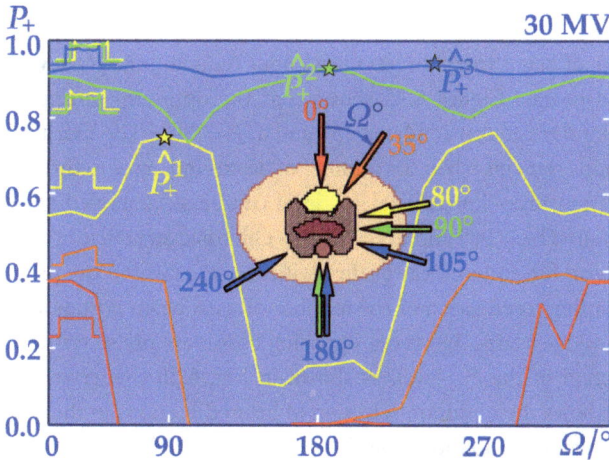

Figure 6.23. The one-dimensional P_+ phase space, $P_+(\Omega)$, for treatment of an advanced cervix cancer with 10 MeV monoenergetic photons using different degrees of freedom in the optimized beam profiles. The red line shows the result when the dose level of the homogeneous beam is the only free variable. The orange line shows the result if there are two free variables for each optimized beam profile, the beam weight, and wedge angle. The yellow green and blue line shows the result if the beam has 64 free variables corresponding to fully non-homogeneous beams with one two and three beam portals respectively (the stars show the optimal angles in each case).

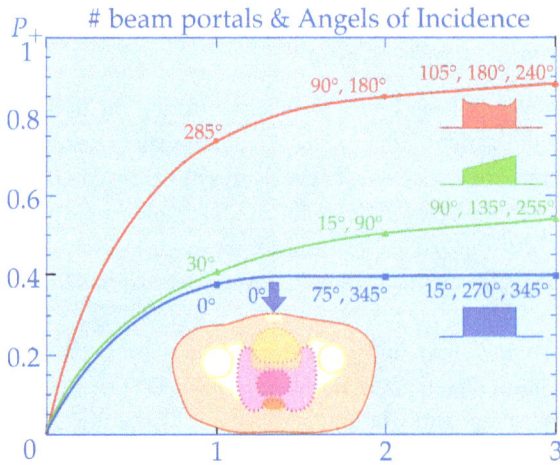

Figure 6.24. The outcome of optimal treatments for the cervix target using increasing number of beam portals when using uniform, wedged, and generally nonuniform beam profiles is shown. In each case, the optimal beam angles of incidence are also given. The results obtained when using homogeneous, wedged, and non-homogeneous beams are indicated by squares, triangles, and circles, respectively.

a wider range of beam entry directions without reducing the P_+ level significantly with wedged beams. Figure 6.25 shows that the optimum settings of the beam entry directions are found at $\Omega_1 = 15°$ and $\Omega_2 = 90°$. The merit of these beam entry directions remains when the number of beams increases to three. The internal structure of the three-dimensional P_+^3 level can be summarized at high levels in beam directions $\Omega_1 = 90°$ and $\Omega_2 = 270°$. The basic structure of P_+ phase space described above for nonuniform beams, with low P_+ level in the planes and along lines representing a reduction of the number of used beam portals are also present when using wedge beams. However, opposing beams are not so critical with respect to the reduction of the P_+ level, as shown in Fig. 6.26 (Söderström *et al.* 1995). Figure 6.26 shows the optimum selection of beam directions in the P_+^3 phase space for wedged beams are found to be $\Omega_1 = 90°$, $\Omega_2 = 135°$, and $\Omega_3 = 255°$ for which a P_+ level of 54% is reached. When allowing *nonuniform* beam profiles in the optimization, all the above-described structures of the P_+^3 phase space are most clearly seen. For the one-dimensional P_+ phase space, the optimum beam direction is around $\Omega = 285°$ as can be seen in Fig. 6.23. When adding a second beam portal during the optimization, the P_+ level increases. The low P_+ level pertaining to $\Omega_1 = \Omega_2$, $\Omega_1 = \Omega_2 + \pi$, and $\Omega_1 = \Omega_2 - \pi$ are also clearly visible in Fig. 6.27 (Söderström *et al.* 1995).

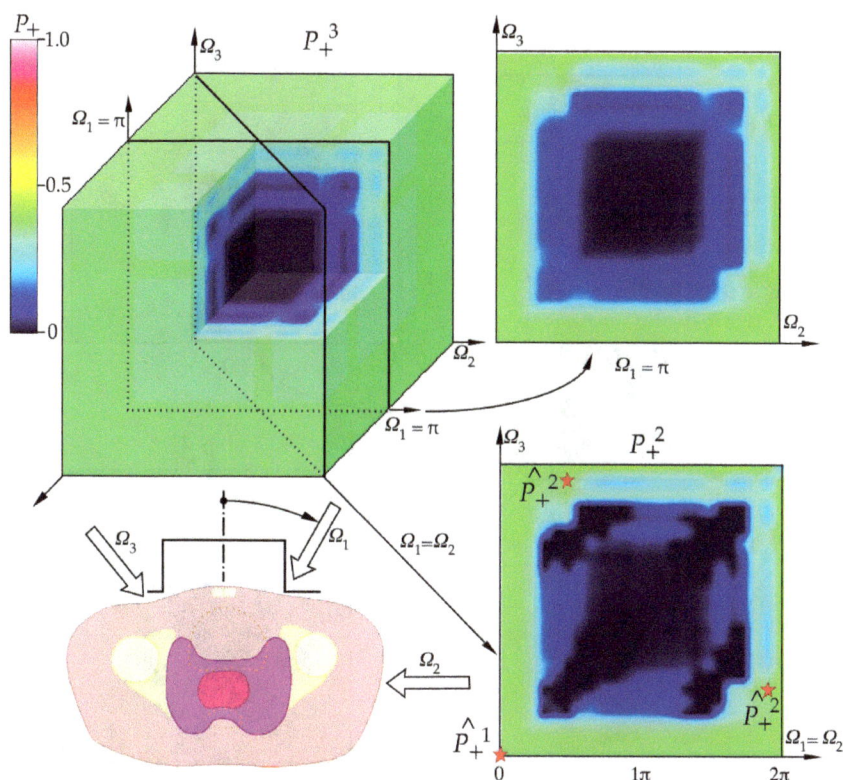

Figure 6.25. The three–dimensional P_+ phase space for the cervix target using only homogeneous beams is shown. The cubical dataset is shown with one corner cut to show the interior structures. One slice of the cube is also extracted to better show the interior structure. To illustrate the properties of the two–dimensional P_+ phase space, the diagonal plane, where $\Omega_1 = \Omega_2$, was also extracted from the cube.

Figure 6.27 shows an optimum three beam dose plan beam with $\Omega_1 = 105°$, $\Omega_2 = 180°$, and $\Omega_3 = 240°$.

6.5.7.2. *Role of large number of beam portals*

It is quite clear from Fig. 6.28 (Söderström 1995) that there is very little to be gained in the probability of achieving tumor control without severe complications when more than approximately three to five beam portals are used. The increase in P_+ when going from 3 to 72 nonuniform beam portals is only between 0.5% to 2.2% for the different patient geometries.

Figure 6.28 (Söderström *et al.* 1995) shows the probability of achieving tumor control without severe complications using increasing number of beams for the

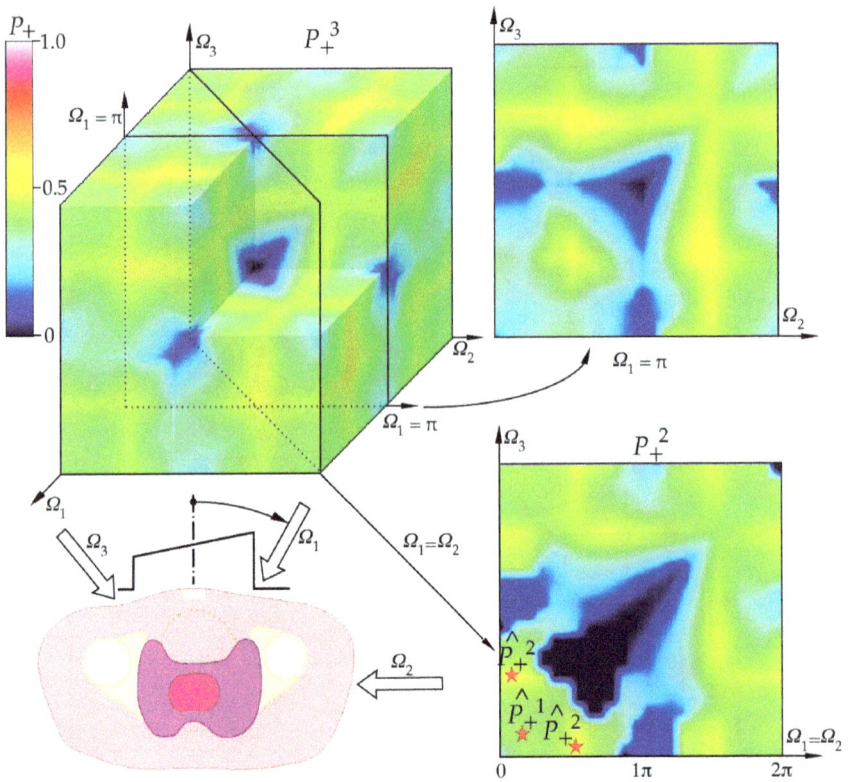

Figure 6.26. The same three-dimensional P_+^3 phase space as in Fig. 6.25 but for optimized wedged beams.

treatment of three different targets. The treatment with one, two, and three beam portals is performed using optimal beam entry directions. Solid squares show the results obtained with the cervix target using 50 MV photons. Solid triangles show the result of the thorax target also using 50 MV photons. Solid circles show the results for the head and neck using ^{60}Co beam. As the figure illustrates, a low number (≤ 3) of nonuniform beams may thus be used without losing P_+ and unnecessarily increasing the total treatment time. The reduction of the number of beam portals that results by using nonuniform fluency profiles may thus decreases the total treatment time.

6.5.7.3. *Clinical factors for optimum treatment*

Table 6.2 (Söderström 1995) shows the increase in the probability to achieve tumor control without causing severe complications in normal tissues when increasing the number of beam portals for the cervix case.

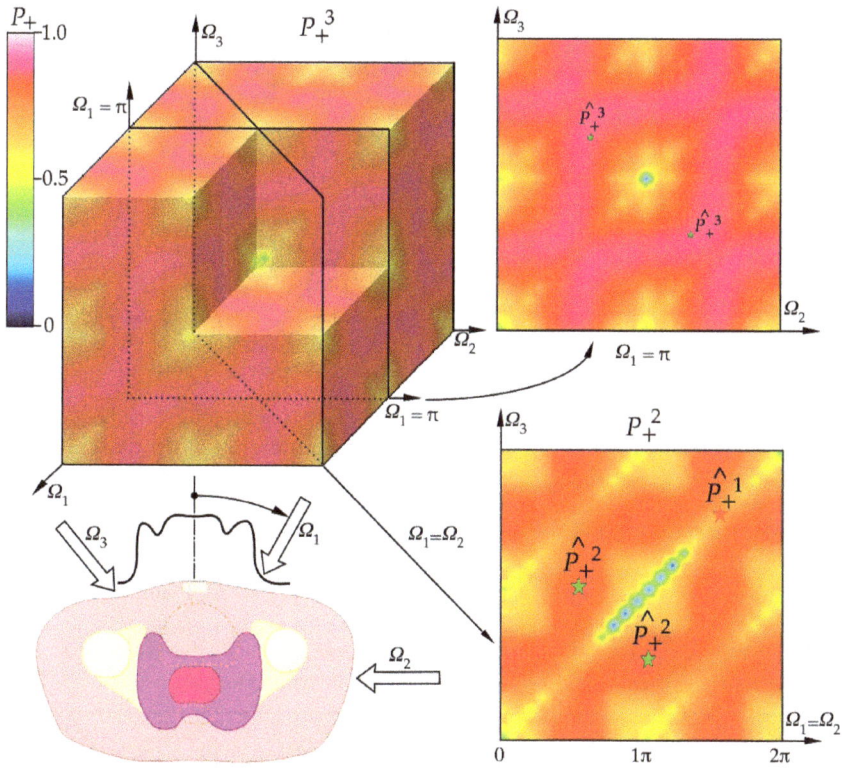

Figure 6.27. The same three-dimensional P_+^3 phase space as in Fig. 6.25 but for optimized non-homogeneous beams. A considerable increase in complication free cure is observed as seen from the color scale.

When going from two to three beam portals, the improvement is mainly to due to an improved probability of controlling in the local lymph nodes. The homogeneity of the dose distributions inside the target are improved as the number portals are increased and the maximum dose may thus be lowered. It should be noted that, when using three beams, the optimum beam entry directions are posterioanterior-oriented despite the fact that the most radiation-sensitive organ at risk, the rectum, is located in the posterior region.

A different situation occurs for the *thorax* target geometry, as shown in Table 6.3 (Söderström 1995). In this case, the heart is the most critical organ obtaining an 11.3% probability injury when one beam is used. When going to two portals, the optimal beam are displaced from the optimal single beam entry direction to two optimal beams, thereby, avoiding injury to the heart increasing the probability of injury to the normal tissue storma by 1.7%. Only minor further improvements

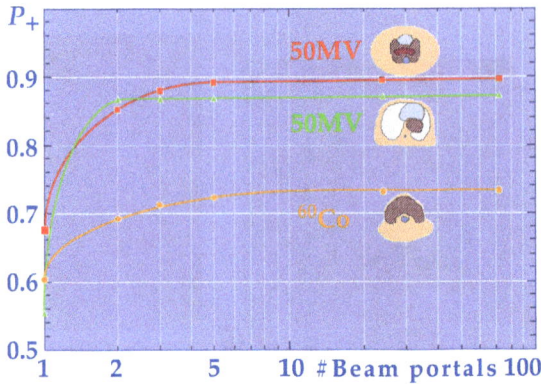

Figure 6.28. Illustration of the probability to achieve tumor control without severe complications using increasing number of beams for the treatment of the three different targets. The treatments with one, two, and three beam portals are performed using optimal beam entry directions. The 24- and 72-beam treatments are performed using beams equidistant in angle. Solid squares show the results obtained with the cervix target using 50 MV photons. Solid triangles show the result for the thorax target also using 50 MV photons. Solid circles show the results for the head and neck target using a ^{60}Co beam.

Table 6.2. The mean, minimum, and maximum dose and standard deviation of dose distribution for the different organs at risk and target volumes for the cervix case.

	$\langle D \rangle$ (Gy)	$D_{\min}$ (Gy)	$D_{\max}$ (Gy)	σ_D (Gy)	$P_0 \cdot P_1$
1 optimized beams, $P_+ = 67.7\%$					
Small bowel	37.5	0.0	89.6	26.0	0.210
Rectum	42.9	40.1	47.9	2.1	0.012
Bladder	53.1	10.7	62.1	10.5	0.002
Lymph nodes	57.3	38.0	84.0	10.6	0.942
Gross tumor	66.3	59.0	73.8	3.8	0.957
2 optimized beam, $P_+ = 85.6\%$					
Small bowel	35.0	0.0	72.5	20.5	0.086
Rectum	40.9	38.0	45.7	2.5	0.006
Bladder	46.7	18.6	58.0	7.4	0.000
Lymph nodes	56.7	39.8	69.3	5.4	0.970
Gross tumor	68.6	60.0	72.5	2.5	0.977
3 optimized beams, $P_+ = 87.9\%$					
Small bowel	33.4	0.1	72.3	18.2	0.082
Rectum	39.1	36.3	43.1	2.1	0.001
Bladder	48.2	43.0	55.5	3.5	0.000
Lymph nodes	57.9	46.2	68.7	4.1	0.983
Gross tumor	68.7	63.5	72.3	2.1	0.979

Table 6.3. The mean, minimum, and maximum dose and standard deviation of dose distribution for the different organs at risk and target volumes for the thorax case.

	$\langle D \rangle$ (Gy)	D_{max} (Gy)	D_{min} (Gy)	σ_p (Gy)	$P_0 \cdot P_1$
1 optimized beams, $P_+ = 55.4\%$					
Normal tissue stroma	18.1	0.0	69.2	25.1	0.061
Spinal cord	0.2	0.1	0.2	0.0	0.000
Lung	15.6	0.0	74.6	25.0	0.021
Heart	32.8	0.2	53.6	19.7	0.112
Target volume	58.6	43.8	67.6	4.3	0.738
2 optimized beam, $P_+ = 86.3\%$					
Normal tissue stroma	19.4	0.0	70.9	25.0	0.078
Spinal cord	8.3	3.0	13.8	6.1	0.000
Lung	14.3	0.0	67.4	15.0	0.002
Heart	28.6	0.6	67.8	13.3	0.002
Target volume	67.7	61.3	70.9	2.0	0.944
3 optimized beams, $P_+ = 85.8\%$					
Normal tissue stroma	19.6	0.0	71.7	24.4	0.079
Spinal cord	8.4	3.0	14.0	6.2	0.000
Lung	14.3	0.1	70.2	14.2	0.009
Heart	37.9	1.2	62.6	12.0	0.005
Target volume	68.2	60.0	71.7	1.9	0.945

can be observed when increasing the number of beam portals to three. All the beams are irradiating the target through the left lung of the patient. The left lung is thus partly satisfied in the treatment to achieve a high probability of tumor control.

The most interesting feature of the *head and neck* irradiation is the movement of the beam entry directions when increasing the number of beam portals (Table 6.4) (Söderström 1995).

The situation here is quite similar to that of the cervix target. In this case, much of the probability of injury to the spinal cord and some of that to the normal tissue stroma may be avoided and at the same time the probability of tumor control can be increase.

The *conclusion* is that the use of optimal nonuniform dose delivery can reduce the number of required beam portals to a low number (2–5) and at the same time as the probability of achieving tumor control without causing severe complications is practically as high as it can be with an infinite number of beams. An important consequence is that a few static fields can be used to deliver an optimal treatment plan without the need of complex rotation techniques. It is, therefore, an important

Table 6.4. The mean, minimum, and maximum dose and standard deviation of dose distribution for the different organs at risk and target volumes for the head and neck case.

	$\langle D \rangle$ (Gy)	$D_{\max}$ (Gy)	$D_{\min}$ (Gy)	σ_D (Gy)	$P_0 \cdot P_1$
1 optimized beams, $P_+ = 60.3\%$					
Normal tissue stroma	48.2	0.0	71.0	15.1	0.158
Spinal cord	42.5	40.4	43.6	0.8	0.047
Target volume	61.4	47.8	71.0	5.4	0.800
2 optimized beam, $P_+ = 59.2\%$					
Normal tissue stroma	45.9	3.8	77.3	15.5	0.152
Spinal cord	35.2	32.3	39.1	2.5	0.005
Target volume	61.8	49.6	77.3	3.4	0.848
3 optimized beams, $P_+ = 71.2\%$					
Normal tissue stroma	44.9	2.5	70.5	16.1	0.146
Spinal cord	33.6	30.5	36.6	2.0	0.002
Target volume	61.9	54.4	70.5	2.6	0.860

task to develop algorithms for optimal selection of beam entry directions for treatments using a limited number of beam portals. Examples showing that the use of sophisticated equipment such as *scanned photon beams* and *computer-controlled compensator production* may not be necessary to obtain a good treatment outcome. It may, thus, be possible to achieve close to optimal treatment results by careful selection of beam portals.

6.5.8. *Dose Calculation and Biological Optimization*

It should be recognized that all studies concerning the number of beams necessary for optimized radiation therapy have been performed in *coplanar* geometry. This will, of course, influence the result. However, the improvement of the dose distribution by the introduction of *non-coplanar* beams is small compared to the improvement obtained by nonuniform dose delivery, which is the principal new feature of optimized radiation therapy that enables a reduction of the number of used beam portals. Nonplanar beams will thus result is a slightly better dose distribution if one or perhaps two additional beam portals are used.

6.5.8.1. *Coplanar configurations*

The results of the P_+ calculations for coplanar multiple beam configurations are shown in Table 6.5 (Weckström 1997).

Table 6.5. The probability of complication-free tumor control, P_+, for different numbers of coplanar beams and for different tumor sizes.

Number of beams	Tumor size (cm)				
	1	3	5	7	9
2	96.4	77.6	57.3	34.5	14.1
3	97.3	82.1	63.4	43.9	24.9
5	98.5	87.3	72.9	55.8	36.6
7	99.0	90.5	77.5	60.1	40.6
9	99.3	91.8	78.7	61.0	40.7
11	99.4	92.3	78.9	61.1	40.7

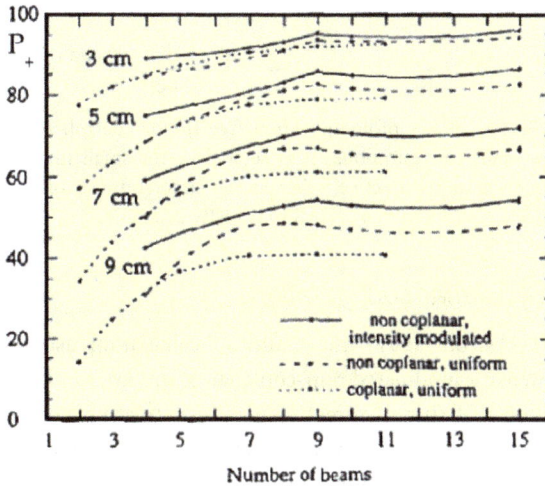

Figure 6.29. Illustration of the probability of complication-free tumor control, P_+, calculated using coplanar beams, non-coplanar beams and non-coplanar intensity-modulated beams.

The values of P_+ are also plotted against the number of fields for each tumor size, as shown in Fig. 6.29 (Weckström 1997).

It is clear from Fig. 6.29 that the probability of achieving a successful treatment outcome is larger in the smaller the target volume. This is because the dose to the normal tissue is reduced with decreasing tumor size for a given target dose level. It can also be seen that P_+ improves slowly when the number of fields are increased, but the increase saturates at a large numbers, both for small and large tumors.

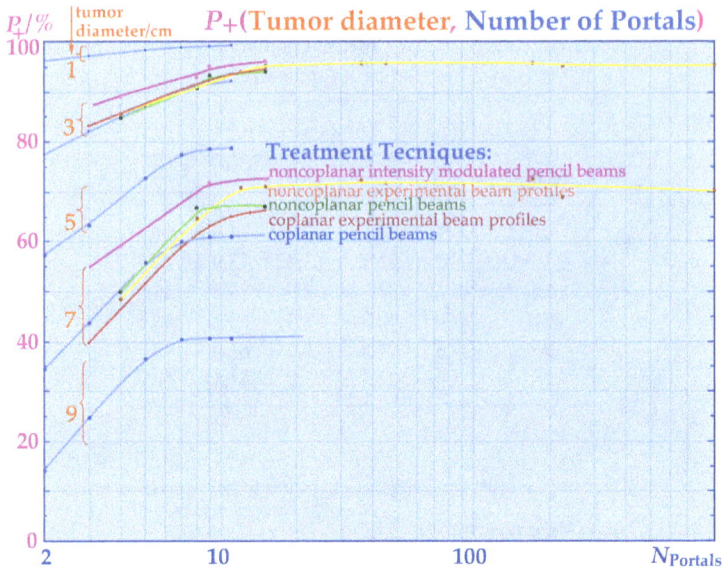

$P_+/\%$ tumor diameter/cm P_+(Tumor diameter, Number of Portals)

Treatment Tecniques:
noncoplanar intensity modulated pencil beams
noncoplanar experimental beam profiles
noncoplanar pencil beams
coplanar experimental beam profiles
coplanar pencil beams

Figure 6.30. The probability of complication-free tumor control, P_+, calculated using different multiple beam configurations. It is seen that, for small tumors ($<\approx 3$ cm), the stereotactic approach is quite effective, whereas for larger tumors, biologically optimized intensity-modulated radiation therapy (IMRT) is really desirable.

6.5.8.2. *Non-coplanar configurations*

In Fig. 6.30 (Weckström 1997) the results of calculation using coplanar, non-coplanar, and intensity-modulated non-coplanar beams are summarized.

It is seen in Fig. 6.30 that a symmetric increase in treatment outcome is obtained going from coplanar to non-coplanar beams and further on by introducing full intensity modulation of the incoming beams.

6.6. Radiobiological Investigation of Non-coplanar Beam Configurations

One of the aims of clinical radiation therapy is to use a small number of beams in order to have a reasonably short treatment time, while simultaneously achieving high treatment outcome and reliable dose delivery. An effective dose delivery may often be obtained with a tetrahedral-like beam configuration (cf Fig. 3.19), due to maximum beam separation. For this reason, two different tetrahedral-like beam configurations with four photon beams of 50 MV from the racetrack microton MM50 have been examined (Ferreirra *et al.* 2003).

The gantry angle, φ, and the non-coplanar angle, θ, define the direction $\Omega(\varphi, \theta)$ of each beam (Figs. 6.31(a) and 6.31(b)). The problem of finding the optimal beam orientations is characterized by a high mathematical complexity, since the optimal direction of one beam depends on the position of all other beams used in the treatment. Thus, all possible beam combinations should ideally be tested. The problem can be studied by utilizing two four-field configurations, denoted here by: Ω_{3+1} and Ω_{2+2}, as shown in Figs. 6.31(a) and 6.31(b), respectively. The first indices represent the number of beams in the transversal plane, whereas the second indicate the number of non-coplanar beams.

In the Ω_{3+1} configuration, the fourth beam, Ω_4, moves along all the possible gantry positions, within a cone of non-coplanar angle from $-40°$ to $+40°$. The use of non-coplanar beam angles beyond this range is mainly prohibited by the extended beam path through the patient, which generally leads a higher dose to the normal tissues. In the Ω_{2+2} configuration, the beams Ω_1 and Ω_2 are in the plane with optimized directions, whereas the beams Ω_3 and Ω_4 are coupled, having the same gantry angle and symmetric non-coplanar angles.

Finally, a configuration similar to configuration Ω_{3+1}, denoted by Ω_{3*+1}, was also studied. In this case, the directions of the three coplanar beams were optimized for each angle of the non-coplanar beam. To compare these beam configurations with coplanardose delivery, treatment plans using three, four, and five coplanar beams optimized in direction and intensity were also examined. These configurations are denoted by Ω_{3+0}, Ω_{4+0}, and Ω_{5+0}, respectively.

Angular optimization is a difficult inverse problem, since the phase space of most relevant objective functions has multiple maxima. Optimization algorithms based on simple gradient methods, such as the one used in this work, may be trapped in local maxima. Consequently, a systematic gradient exhaustive search or a simulated annealing stochastic optimization method, considering the whole phase space, must be used to ensure that the global maximum is found.

6.6.1. Single Non-coplanar Beam, Ω_{3+1}

Figure 6.31(c) shows a systematic overview of the whole phase space of the probability of the complication-free cure, as a function of the angle of incidence of the non-coplanar beam for the configuration Ω_{3+1}. The resultant range of variability for P_+ is slightly $>5\%$, due to a 3% variation in P_B and a 2% difference in P_I (Fig. 6.31(c), P_+) (Ferreirra et al. 2003). The P_+ maxima are located at the gantry angles 90° and 270°, for a coplanar configuration. For these gantry angles, the treatment outcome does not change significantly with increasing non-coplanar angle θ. Local maxima of P_+ can be found, for each non-coplanar angle, at the

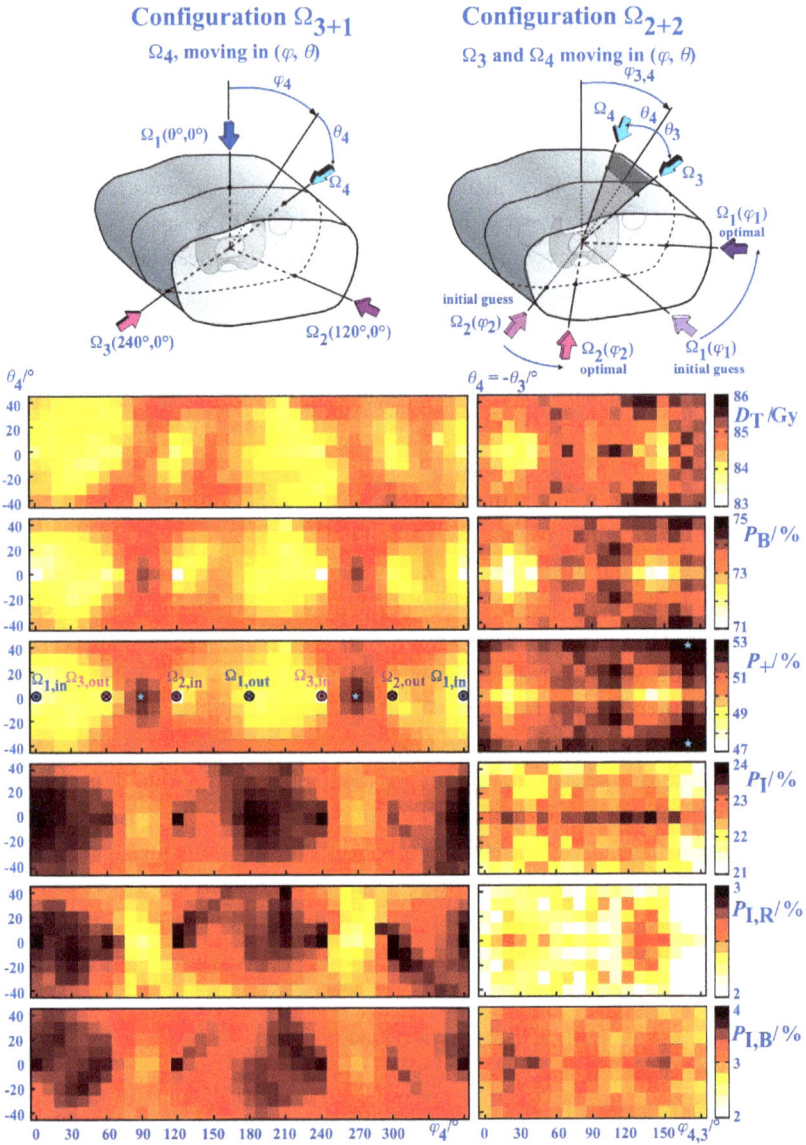

Figure 6.31(a)–(c). Definition of gantry and non-coplanar beam angles φ and θ, respectively. Schematic representation of the studied configurations: (a) Ω_{3+1} and (b) Ω_{2+2}. (c) Phase space of $\overline{D}_{\text{ITV}}$, P_{B}, P_{+}, P_{I}, $P_{\text{I,R}}$ and $P_{\text{I,B}}$, respectively, for the configuration Ω_{3+1} (ITV stands for internal target volume, R for rectum and B for bladder). Ω_{in} and Ω_{out} represents the beam entrance and exit, respectively. (c) Phase space of $\overline{D}_{\text{ITV}}$, P_{B}, P_{+}, P_{I}, $P_{\text{I,R}}$ and $P_{\text{I,B}}$, respectively, for the configuration Ω_{2+2}. On the vertical axes, the two non-coplanar beams travel together obeying $\theta_3 = -\theta_4$. (cf. also Fig. 6.7(f)).

gantry angles 30°, 90°, 150°, 210°, 270°, and 330° (more clearly seen for $\theta = 0°$ in Fig. 6.31(c), P_+). These local maxima are a consequence of maximally separated beams, that is, when there is a 30° separation between the fourth beam and the fixed coplanar beams.

The small bowel has the largest probability of injury among normal tissues, at around 13%, the bladder follows with a probability of around 3.3%, and the rectum of with a probability of 2.6%, as in Fig. 6.31(c), $P_{I,B}$ and $P_{I,R}$; for small bowel, the use of one non-coplanar beam increases the beam path in the small bowel, but the probability of injury is almost constant, independent of the non-coplanar angle. For bladder, for the gantry angles of the local maxima of P_+, the injury for this organ is almost constant with the non-coplanar beam angles (Fig. 6.31(c), $P_{I,B}$). For rectum, this organ is extended in the craniocaudal direction, thus there is almost no difference in the rectum dose or complication probability when the non-coplanar angle is changed.

In conclusion, for this gantry angle, no significant variation in treatment outcome is seen with a non-coplanar orientation (as for the gantry angle $\varphi_4 = 90°$ in Fig. 6.31(c), P_+).

6.6.2. *Two Anti-symmetric Non-coplanar Beams,* Ω_{2+2}

Figure 6.7(f) shows the P_+ phase space for the configuration Ω_{2+2}. The highest P_+ value found is 53%, for the two non-coplanar beams placed at $\Omega_3(170°, +40°)$ and $\Omega_4(170°, -40°)$, and the coplanar beams at $\Omega_1(93°, 0°)$ and $\Omega_2(240°, 0°)$. Comparing the treatment outcome, as measured by P_+ (Fig. 6.7(f)) with the optimal beam positions, it is seen that an equally good treatment can be performed when the preferred positions of around 90° and 180° are being used, either by the coplanar or the non-coplanar beams. This means that these optimal orientations are independent of the angle of the non-coplanar beams. Contrary to what was expected, the lower values in the P_+ phase space, in Fig. 6.7(f), are associated to oblique irradiation directions, such as 20° and 140°, which make them less suitable beam orientations for a treatment with four beams. For small bowel, the major injury, of ~9%, comes from the normal tissue that lies inside the ITV, which cannot be avoided without compromising tumor control. The largest reduction in the injury values for this tissue occurs when going from three to four fields, when a non-coplanar angle different than 0° is used. For bladder and rectum, the gantry angles between 20° and 50° cause the largest bladder injury, whereas the angles between 120° and 150° are the most damaging angles for the rectum (not the angle 180°, as would be expected when using uniform beams). A non-coplanar treatment with these gantry angles increases beam separation, and thus the injury of the individual organs decreases and tumor

Table 6.6. Best results were achieved for the different configurations studied. The treatment configuration Ω_{2+2} was the only non-coplanar configuration, which showed a slightly higher P_+ value compared to the optimal coplanar treatment Ω_{4+0}. The first index represents the number of coplanar beams, while the second is the number of non-coplanar beams. E is the photon beam energy in MV. φ_i and θ_i are the gantry angle and the non-coplanar angle of beam i, with $i = 1, \ldots, 5$, respectively. For the box technique, the treatment outcome was calculated for: uniform (Unif.) and intensity-modulated (IM) beams.

E MV	# Beams	Config.	$P_+\%$	$P_B\%$	$P_1\%$	$\varphi_i\%$	$\theta_1\%$
50	3	Ω_{3+0}	50.1	73.5	23.4	106, 177, 243	0, 0, 0
	4	Box (Unif.)	19.1	55.0	35.9	0, 90, 180, 270	0, 0, 0, 0
		Box (IM)	47.2	70.9	23.7	0, 90, 180, 270	0, 0, 0, 0
		Ω_{3+1}	52.2	74.2	22.0	0, 120, 240, 270	0, 0, 0, 0
		Ω_{4+0}	52.7	74.6	21.9	88, 115, 177, 237	0, 0, 0, 0
		Ω_{3*+1}	52.8	74.6	21.8	96, 120, 238, 357	0, 0, 0, 0
		Ω_{2+2}	53.0	74.7	21.7	93, 240, 170, 170	0, 0, +40, −40
	5	Ω_{5+0}	53.9	74.9	21.0	5, 80, 154, 235, 291	0, 0, 0, 0, 0
10	4	Ω_{4*+0}	52.7	74.3	21.6	88, 116, 237, 356	0, 0, 0, 0
		Ω_{2+2}	51.2	73.2	22.0	5, 92, 180, 180	0, 0, +40, −40

control increases. However, when the preferred directions are used (around 90° and 180°), the choice of a non-coplanar orientation does not significantly change the injury or the tumor control or finally the treatment outcome.

Comparing these results with coplanar beam treatments, fully optimized in intensity and direction, the P_+ values achieved with three, four, and five coplanar beams were 50.1%, 52.7%, and 53.9%, respectively. Their respective optimal directions are given in Table 6.6. It can be observed that none of the dose distributions have hot spots in the organs at risk. The bladder and especially the rectum are very well protected from high doses. The small difference between the P_+ value obtained for the configurations Ω_{3*+1} and Ω_{4+0} is explained by the different optimization process used. For a five coplanar field configuration, the treatment outcome increases by 1% compared to the previous results.

6.7. Physical and Radiobiological Characteristics of Dose-uniformity Regularization Methods

In the past, many studies were published investigating different methods for optimizing three-dimensional dose distributions in external radiotherapy (Brahme and Ågren 1987; Bortfeld 1999; Mavroidis *et al.* 2007). These methods try to improve the therapeutic ratio through dose distributions, which are characterized by a better conformation of dose to the ITV and improved sparing of normal tissues. However, in many cases, the optimization algorithms produced strongly nonuniform dose distributions, sometimes due to technical limitations and unsuitable assumptions, which are very different from the quasi-uniform dose distributions that are typically and easily used in the clinical practice. This problem should be solved by better biologically relevant endpoints and treatment specifications or less accurately by simply regularizing the variation of dose nonuniformity. This is so since the main reason it is possible to improve treatment outcome by IMRT is that the tumor cure is increased by suitable intensity modulation and the normal tissue morbidity is simultaneously reduced. The high dose is moved from regions near the organs at risk to regions in the tumor where it is far from them at the same time as the dose is increased even further since it dose not do much harm there. The beauty of the biological models is that they can quantify the the loss of local cure near organs at risk at the same time as they decide how much the dose can be increased in the tumor core without damaging normal tissues. Good radiation biological models are therefore the very best regularizations method (see Secs. 2.5.5 to 2.5.7 and particularly Fig. 2.18) even though there may be shortcomings in some models when it comes to extreme dose delivery.

Presently, there are a number of such methods that have been proposed in the literature (Webb 2000; Wu *et al.* 2000; Alber and Nüsslin 2002). For example, there have been suggestions to increase the beamlet size used for optimization putting additional constraints on the numerical solutions (Yeboah *et al.* 2002). Another approach to regularize the optimization problem is to utilize the minimal surface membrane analogy or smoothing filter technique (Alber and Nüsslin 2000; Spirou *et al.* 2001). Chvetsov *et al.* (2007) studied the effects of applying a variational regularization technique to equivalent uniform dose (EUD)-based optimization with the L-curve criterion for determining an optimal value of the regularization parameter. This method was proposed because the dose distributions produced by EUD-based IMRT treatment planning optimization typically produce significant hot spots in the target volume (Choi and Deasy 2002; Chvetsov *et al.* 2005). The regularization parameter is found using the L-curve method which is based on the minimization of the residual norm which is a measure of accuracy of the fit and the smoothing norm which is a measure of the smoothness of beam intensity functions (Chvetsov 2005). As a result, the smoothest dose distribution is found for a prescribed value of EUD (Niemierko 1999).

6.7.1. *Tikhonov Regularization*

The variational regularization method known as Tikhonov regularization has been applied to inverse treatment planning for IMRT by three research groups (Tikhonov 1963; Alber *et al.* 2000; Spirou *et al.* 2001; Chvetsov *et al.* 2005). There are two key problems in the variational regularization methods: (1) selection of the stabilizing functional and (2) evaluation of an optimal value of the regularization parameter. The previous published research on inverse planning primarily addressed the selection of the stabilizing functional. The regularization parameter was considered as a fitting coefficient. In this analysis, an L-curve method is applied to evaluate the optimal value of regularization parameter in the radiotherapy optimization problems.

The goal of inverse treatment planning is to find the beamlet weights $w = (w_1, w_2, \ldots, w_M)$ from the following beamlet model (Chvetsov 2005):

$$D(r) = \sum_{m=1}^{M} w_m d_m(r), \qquad (6.4)$$

where $D(r)$ is the desirable distribution of the absorbed dose and $d_m(r)$ is the dose distribution from a beam intensity element ("beamlet"). Equation (6.4) can be solved using a variation regularization method where the beamlet weights are sought as a solution of the following optimization problem (Chvetsov 2005):

$$w_\alpha = \arg\min\left\{\|\sqrt{I}D(r) - D_p(r)\|_2^2 + \alpha\Omega\right\}, \qquad (6.5)$$

where $D_p(r)$ is the prescribed dose distribution, I is the importance factor, and the notation $\|g(x)\|_2$ assumes the following norm $\sqrt{\int |g(x)|^2 dx}$. The stabilizing functional Ω multiplied by a regularization parameter α is added to the least squares objective function to account for the property of ill-posedness of Eq. (6.4). The first order Tikhonov stabilizing functional provides sometimes a good combination of smoothing and approximation for inverse treatment planning with physical objectives. Therefore, the first order stabilizing functional is utilized, which is given by (Chvetsov 2005):

$$\Omega = \|f(\xi)\|_2^2 + \left\|\frac{df(\xi)}{d\xi}\right\|_2^2. \qquad (6.6)$$

In this analysis, the optimal value of the optimization parameter α is of interest, and for this reason the following variables are introduced (Chvetsov 2005):

$$\xi(\alpha) = \|\sqrt{I}D(r) - D_p(r)\|_2, \qquad (6.7)$$

$$\eta(\alpha) = \sqrt{\Omega}. \qquad (6.8)$$

The first variable $\xi(\alpha)$ describes the accuracy of fit of the calculated dose to the prescribed dose distribution. The second variable $\eta(\alpha)$ relates to the stabilizing functional Ω and presents a norm of smoothness of solution. The optimal value of the regularization parameter α is the one which minimizes both variables $\xi(\alpha)$ and $\eta(\alpha)$. The main idea of the L-method is to plot the smoothing norm $\eta(\alpha)$ as a function of the residual norm $\xi(\alpha)$ for all values of the regularization parameter α. The function has an L-shaped form with distinctive vertical and horizontal parts. It follows from the vertical part that the smoothing norm can be practically reduced without changing the residual norm. The optimal value of the regularization parameter can be found in the vicinity of the "corner" of the L-curve where both values of smoothing norm $\eta(\alpha)$ and residual norm $\xi(\alpha)$ achieve their minimum. By substituting in Eqs. (6.5) and (6.7), the dose distribution, $D(r)$ with the EUD, the smoothest solution which meets the EUD-based objective function can be found.

6.7.2. *Clinical Evaluation of a Head and Neck Case*

Different dose-uniformity regularization and treatment plan optimization methods were applied to a sample head and neck IMRT cancer case. The critical structures included the normal tissue stroma, left parotid gland, and spinal cord. In this study, three different dose distributions were obtained by using a dose-based optimization technique, an EUD-based optimization without applying any technique for regularizing the nonuniformity of the dose distribution, and an EUD-based optimization using a variational regularization technique, which controls dose nonuniformity.

The sample patient is irradiated with nine 6 MV beams equally placed between the angles 0° and 360°. Representative slices of the IMRT plans in each case showing structure contours and isodose lines are shown in Fig. 6.32 (Chvetsov *et al.* 2007). The prescription dose to the PTV, all critical organs (spinal cord and parotid gland) and normal tissue was selected to be 100%, 10%, and 0%, respectively. The EUD prescription was defined as 1 for target volumes, 0.1 for critical structures, and 0 for normal tissue. The importance factors have been set as 1 for target volumes, 0.1 for critical structures, and 0.01 for normal tissue stroma. The value of the radiobiological parameter was set to $a = -10$ for targets, $a = 5$ for critical structures, and $a = 10$ for normal tissue.

In this analysis, the effectiveness of the three optimization techniques is compared by evaluating their physical and radiobiological characteristics. Figure 6.33 shows comparisons of the head and neck dose distributions, which were produced using the examined dose nonuniformity regularization methods (Mavroidis *et al.* 2008). In this figure, the treatment plans are compared in terms of DVHs (upper

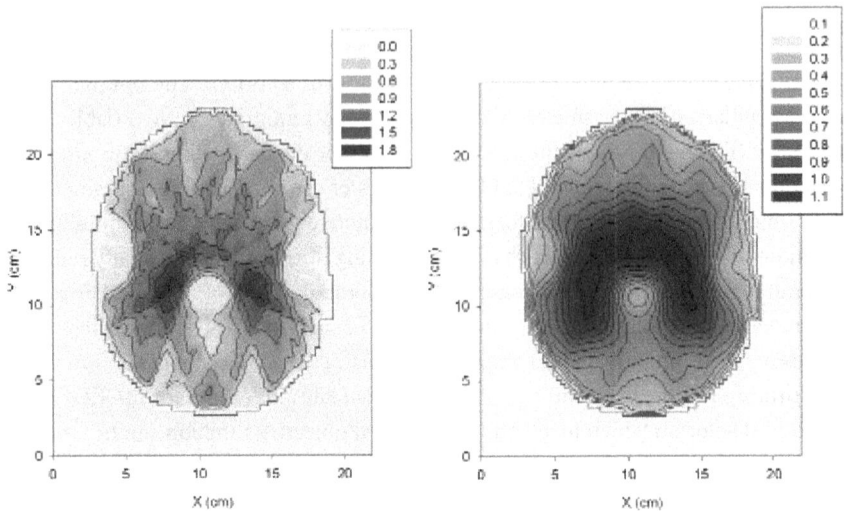

Figure 6.32. Dose distributions in EUD-based optimization for the head and neck case: (a) EUD objective function and (b) EUD objective function with regularization.

diagrams) and dose–response curves (middle diagrams), respectively. In the lower diagrams, the dose–response curves of total control and total complications are presented together with the P_+ curve. In the middle diagrams of Fig. 6.33, the dose–response curves of the ITV and individual organs at risk are normalized to the mean dose in the ITV ($\overline{D}_{\mathrm{ITV}}$). In the lower diagrams of the figures, the dose–response curves have been normalized to the $\overline{\overline{D}}_{\mathrm{B}}$, which forces the response curves of the ITV (P_{B}) of the different treatment plans to coincide. In these diagrams, the same dose distribution is kept at all dose levels and the curves show how tissue responses change with dose prescription. The normalization using $\overline{\overline{D}}_{\mathrm{B}}$ allows the intercomparison of the different modalities on the same basis and gives emphasis to the therapeutic window, which characterizes each treatment plan.

Based on the DVHs, it is clear that the EUD-based optimization without regularization (EUD-based no reg) has larger dose inhomogeneity inside the ITV than the dose-based optimization and the EUD-based optimization with L-curve regularization (EUD-based reg). Between the latter two, dose inhomogeneity is lower for the dose-based optimization. Regarding the organs at risk, the dose-based optimization delivers higher doses than the EUD-based reg to the spinal cord and left parotid. On the other hand, the EUD-based no reg delivers lower mean and maximum doses to all three tissues. For the dose-based optimization and at the optimal dose prescription the P_+ value is 32.9% and the biologically

HEAD & NECK CANCER CASE - Planned:Solid, Delivered:Dashed

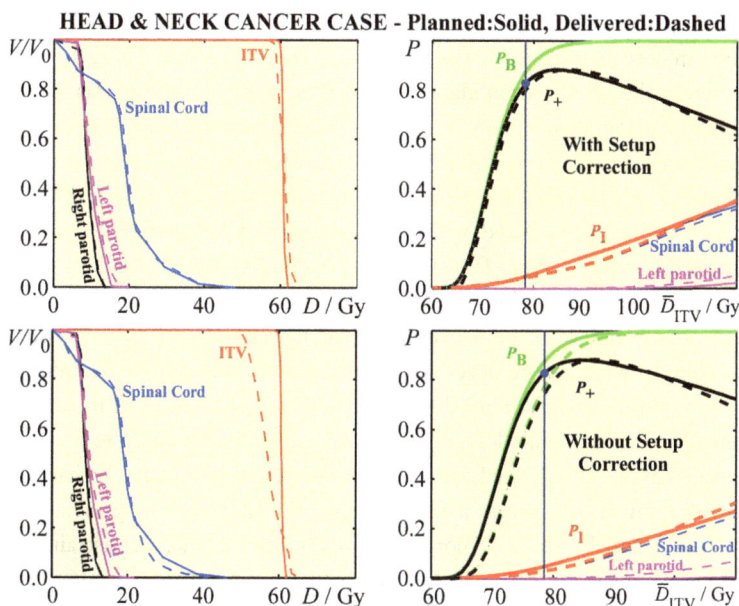

Figure 6.33. Radiobiological evaluation and comparison for the head and neck cancer case. In the left panel, the dose-based optimization method (solid) is compared with the EUD-based reg method (dashed), whereas in the right panel, the EUD-based without regularization method (solid) is compared with the EUD-based reg method (dashed). In the upper diagrams, the DVHs of the ITV and organs at risk are shown for the respective optimization methods. In the middle diagrams, the corresponding dose–response curves are presented. In the lower diagram, the respective $P_+ - \overline{\overline{D}}$ plots are shown, which demonstrate the overall radiobiological evaluation of the different dose distributions.

effective uniform dose to the ITV, $\overline{\overline{D}}_B$ is 72.4 Gy. The total control probability P_B is 79.6% and the total complication probability P_I is 49.0%, which mainly stems from the response probabilities of normal tissue stroma ($P_{NT} = 7.2\%$) and left parotid ($P_{LP} = 44.6\%$). Similarly, for the EUD-based no reg optimization, the P_+ value is 56.4% for a $\overline{\overline{D}}_B$ of 68.8 Gy. The total control probability P_B is 71.9% and the total complication probability P_I is 15.5%, which is almost equal to the response probability of the normal tissue stroma. Finally, for the EUD-based reg optimization, the P_+ value is 67.3% for a $\overline{\overline{D}}_B$ of 75.3 Gy. The total control probability P_B is 87.4% and the total complication probability P_I is 20.1%, which mainly stems from the response probabilities of normal tissue stroma ($P_{NT} = 6.2\%$) and left parotid ($P_{LP} = 14.8\%$).

Regarding the target volumes, the dose nonuniformity in the EUD-based optimization with regularization is larger than in the dose-based optimization.

This finding stems from the fact that in radiobiological terms, there may exist nonuniform dose distributions, which may be as effective as their EUD distribution. In this sense, the optimization algorithm takes advantage of the higher number of degrees-of-freedom provided by the radiobiological measures and finds regularized nonuniform dose distributions that irradiate the target as effectively as the uniform dose distribution and at the same time optimize the dose fall–off toward the organs at risk. Also, the EUD-based with regularization and dose-based optimization algorithms optimize this dose fall–off around the target, differently. This is because the EUD-based optimization takes into account the volume effect of all the involved organs at risk in the proximity of the target and optimizes the dose fall–off accordingly, whereas the dose-based optimization uses partly this information through the dose constraints used to guide it. The radiobiological measures used in evaluating the different dose distributions support this analysis. It has to be mentioned that the EUD-based optimization without regularization is characterized by a higher number of degrees-of-freedom than the EUD-based optimization with regularization and by using a more biologically relevant dose constraint for the normal tissue stroma, this could lead to better results than the latter method. However, the large hot spots produced in the target volume by this method would increase the risk for secondary cancer (Schneider *et al.* 2006). Consequently, by deteriorating physical dose conformation, the EUD-based optimization provides better biological conformation.

In conclusion, this evaluation shows that the EUD-based optimization using L-curve regularization gives better results than the EUD-based optimization without regularization and dose-based optimization. On the other hand, the EUD-based optimization without regularization shows better results than the dose-based optimization in the head and neck case, whereas the opposite has been reported for other cancer sites, which indicates that these two uniformity regularization methods have similar potential of producing treatment plans with small integral doses to the healthy organs and fairly homogeneous doses to the PTV.

6.8. Role of Radiobiological Treatment Planning in Breast Cancer Radiation Therapy

Some of the techniques used in the treatment of early breast cancer with lymphatic spread were selected and biologically optimized using the biological objective function complication-free tumor control probability, P_+. Figure 6.34(a) show, similar to Fig. 6.23 above for the cervix tumor, how the complication-free cure is increased by using more beams on the chest wall. If there are not too severe

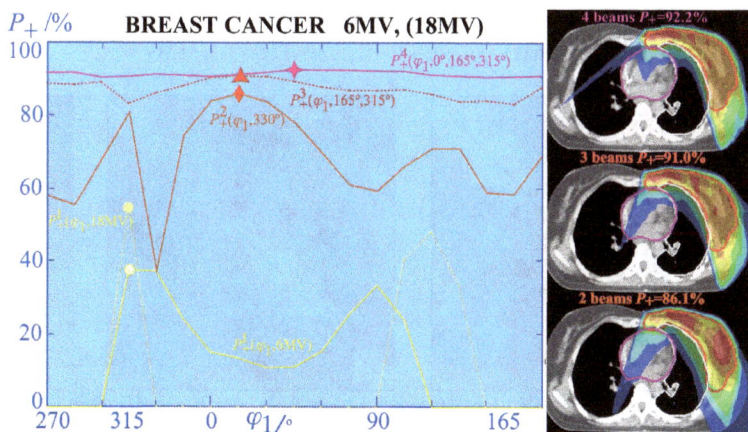

Figure 6.34(a). Dose distributions for two to four beams on a representative patient illustrating how the optimal beam angles and the achievable complication-free cure change with the number of beams. It is seen that the optimal angles depend not only on the applied energy but also on the number of beams so optimal directions with one or two beams may not be so for three or four beams.

changes in geometry along the axis, three to four beams are generally sufficient as seen in Fig. 6.34(a).

Interestingly, there are three major beam combinations that are of main interest as seen in the phase space diagram in Fig. 6.34(b) (cf. Figs. 6.20 and 6.21 above). The set of optimal two beam techniques for breast cancers consist of an anterior pair, a lateral pair, and a tangential pair as shown by the three bumps in the phase space diagram. Obviously, more than two beams may be of interest for many targets as seen in Fig. 6.34(a). However, this may need the use of three- or four-dimensional phase spaces and very long computational times as shown in Figs. 6.25–6.27 for the cervix tumor. A number of more complex beam configurations will therefore be tested and are described below.

Four postoperative patients were used in this comparison, with the same positioning, patient delineation and tissue biology (Ferreirra 2004). In brief, the clinical target volume (CTV) is formed by the remaining breast tissue left by the surgery and the surrounding lymph nodes, that is, axillary, internal mammary chain, infra and supra-clavicular lymph nodes. A margin of about 5 mm was added to the CTV to form the ITV, except at the skin surface. The organs at risk are the heart, the left and right lungs, considered as separate structures, the contralateral breast, the spinal cord, and the remaining surrounding normal tissue. The different techniques are denoted depending on beam direction, where the subscript indicates the region of the target volume mainly covered by that

3 Different Optimal Two Beam Techniques For Breast Cancer

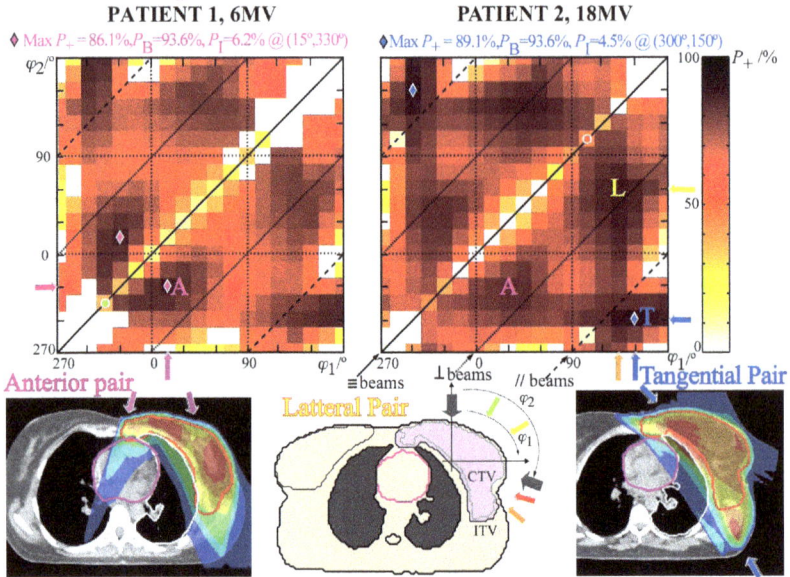

Figure 6.34(b). Dose distributions for two representative patients with phase space diagrams identifying three different optimal two beam techniques for breast cancer with an anterior pair, lateral pair, and tangential pair, A, L and T respectively.

beam (cf. Fig. 6.34(c)) (Ferreira 2004). For example, 2TG+APPAscvax shows that the target volume at the level of the breast was irradiated by two tangential beams (TG). When the entire internal target volume was covered by tangential beams, the subscript "itv" is then added. The target volume from the supraclavicular lymph nodes (scv) to the remaining axillary lymph nodes (ax), not covered by the tangential beams, were irradiated by anteroposterior (AP) and posteroanterior (PA) photon beams.

A technique using only two almost parallel-opposed uniform tangential beams is not adequate for the treatment of the complex target volume produced in a patient with breast cancer with lymphatic spread. As the technique becomes more complex and thus with a larger number of degrees of freedom, an improvement in treatment outcome was seen. Interestingly, the tumor control probability remained almost the same, while the largest benefit was from the reduction of damage to the organs at risk. A significantly lower injury in the left lung and heart was then obtained, while the injury in the surrounding normal tissue was increased and a larger average dose was delivered in the contralateral breast (Fig. 6.35) (Ferreira 2004). With biological optimization, it was not possible to achieve the standard dosimetric

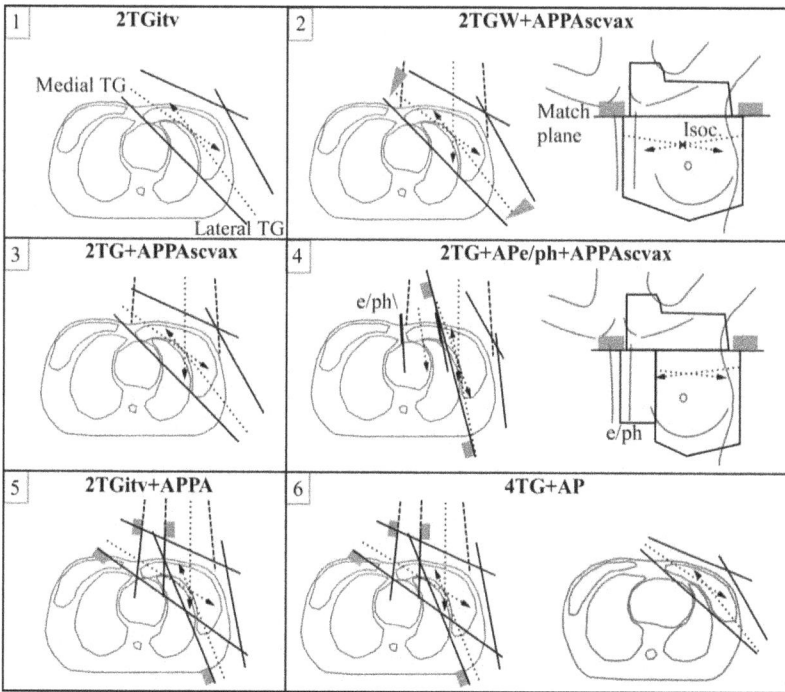

Figure 6.34(c). Schematic representation of the uniform beam techniques investigated. TG stands for tangential beams, AP for anteroposterior and PA for a posteroanterior beam. W indicates that a 15° wedge was used; scv stands for supraclavicular lymph nodes; ax for axillary lymph nodes; and itv for internal target volume; e/ph shows that an abutted electron and photon beams were used.

recommendations (ICRU 50 1994; Aaltonen *et al.* 1997) for the target volume and organs at risk (Fig. 6.35). Indeed the selected group of patients represents difficult cases for uniform beam radiation therapy, due to the large breast, ITV, and large target volume concavity. Furthermore, the main goal of biological optimization is not to achieve a homogeneous dose in the target volume but to obtain the largest treatment outcome for the patient. Therefore, commonly a low dose in the internal margin was obtained, resulting in the low dose minimum in the ITV (Fig. 6.35). Also a large dose maximum in the ITV was generally seen, especially for the multimodality technique due to the hot spot in the junction of the different beams (Fig. 6.36). Even that the reported values of dose maximum and dose minimum may seem quite extreme, the clinical significance of them is not known, since they involve very small volumes. Although, this may be of importance for normal tissues with a serial structural organization, it is less critical for parallel organs. Otherwise, dose

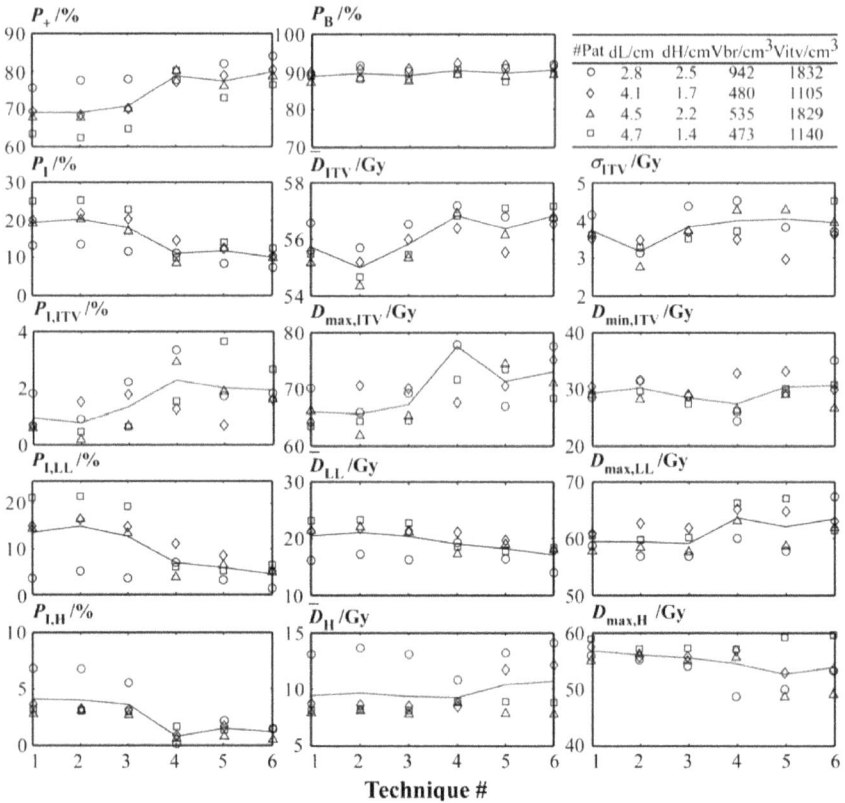

Figure 6.35. Illustration of some response probabilities and the most important dosimetric data in function of the techniques investigated for the four patients used in the study, indicated by different symbols. The solid lines show the respective average values; dL and dH are the maximum distance of left lung and heart inside the target volume curvature, respectively; V_{br} and V_{itv} are the volume of breast and ITV for each patient, respectively.

constraints can be added to the optimization to reduce the maximum dose in the organs at risk and increase the dose minimum in the target volume.

6.8.1. *Early Breast Cancer IMRT*

For the best uniform treatment technique, using five photon beams, an average P_+ value of 80% was obtained. Although, this is a rather complex treatment technique with a large number of uniform beams, presently it is still simpler to deliver than IMRT, requiring a large quality control protocol prior to the delivery. However, a significant improvement in treatment outcome, of around 12%, can

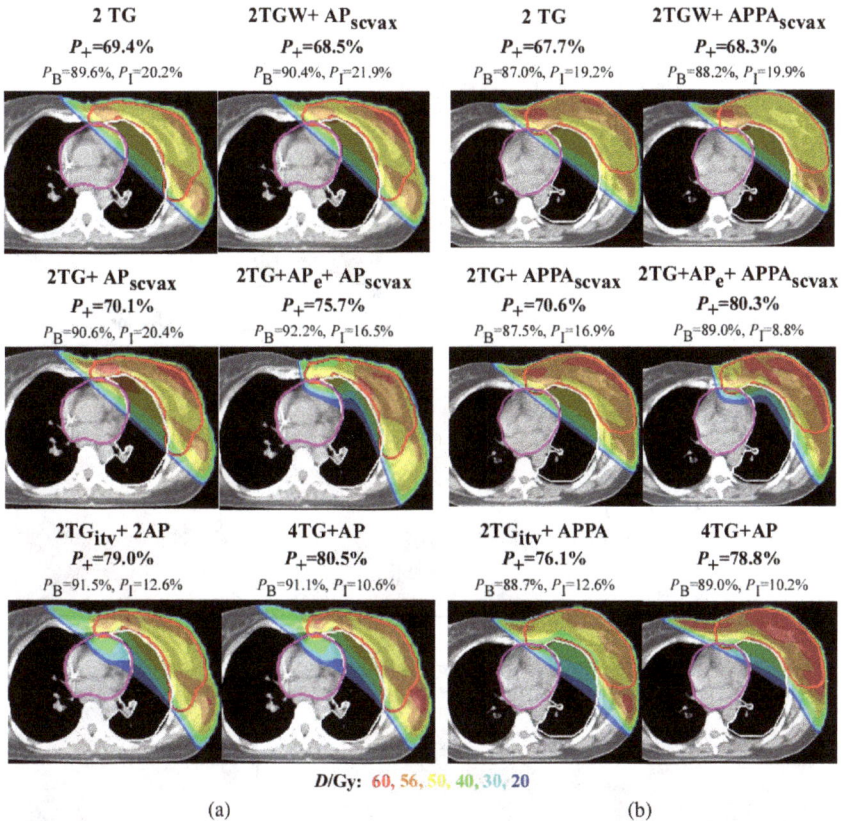

Figure 6.36. Dose distributions for two representative patients for the uniform beam techniques investigated are given. For the patients in the figure and for technique 4(2TG+APe/ph+APPA$_{scvax}$), the anterior photon beam aiming to irradiate the internal mammary chain was removed by the optimization of P_+ and therefore the technique is denoted here only by 2TG+APe+APPA$_{scvax}$. For the patient in (b) this technique had the largest P_+ value.

be obtained by using three or four intensity-modulated photon beams biologically optimized to treat this tumor site. This was due to an increase in tumor control probability and a significant reduction in the probability of injury. For two intensity-modulated photon beam plans and three different beam configurations with almost, the same treatment outcome were found. These use a pair of almost parallel-opposed tangential directions around 300° and 150° (peak T in Fig. 6.37 — 2 BEAMS), one tangential medial beam coupled to an anterior beam (peak A) or a perpendicular beam configuration with an oblique beam around 45° and a lateral tangential beam

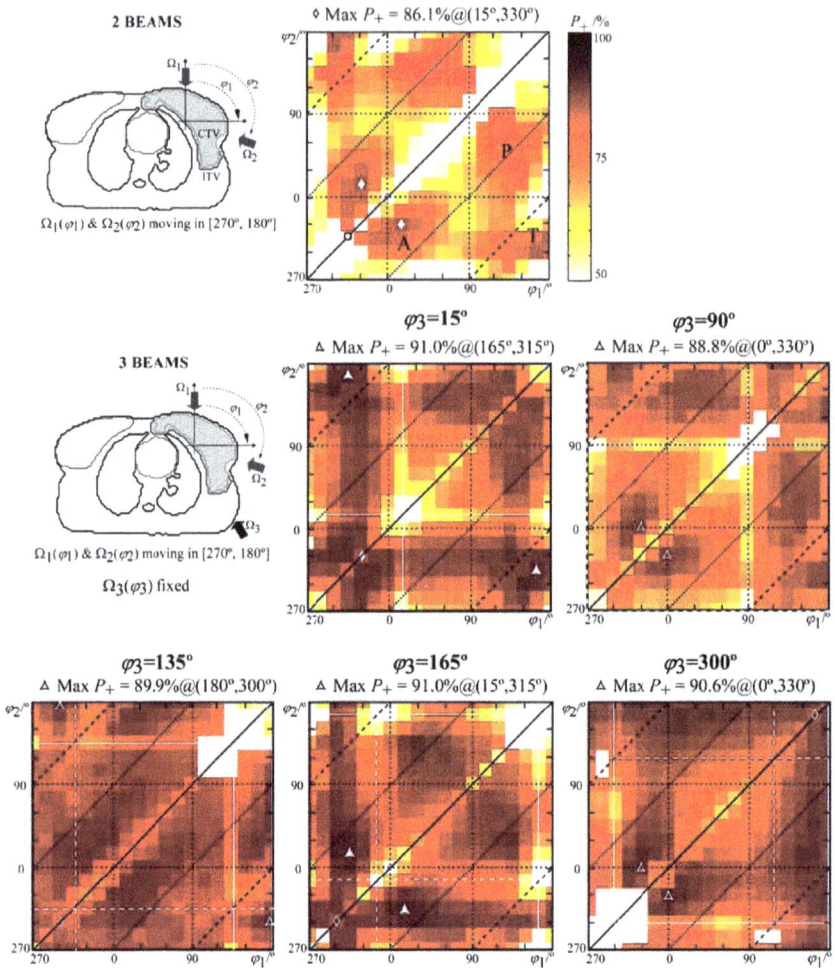

Figure 6.37. The two-beam phase space (above) and several three-beam phase spaces with different gantry directions of beam Ω_3, φ_3 are shown. The solid diagonal lines show plans with one and two beams for the two- and three-beam phase space, respectively, because two beams are coinciding. The diagonal doted lines indicate perpendicular beam configurations and dashed lines indicate parallel-opposed beams. The white solid lines show the direction of the fixed beam Ω_3, while the dashed white lines specify the exit of the same beam. The circle indicates the optimal direction for a single-beam plan. The open diamonds and stars show the maximum P_+ value for each phase space, while the same closed symbols show the maximum P_+ value for a two- and three-beam plan, that is, $\hat{P}_+^2$ and $\hat{P}_+^3$, respectively.

(peak P). Still, standard tangential directions will probably be always preferred, even for IMRT. However, the significant increase in treatment outcome when a third beam is added to the plan, recommends that at least three intensity-modulated beams, biological optimized, placed in the optimal directions, should be used in radiation therapy of breast cancer stage II. For a four intensity-modulated beam plan, only a slight improvement in treatment outcome was obtained and if at least two beams have quasi-parallel opposed tangential directions, angular optimization becomes almost unnecessary. Even so, the third and fourth beams should be placed so as to void parallel-opposed configurations and if possible with large beam separation. To find the optimal directions for a three-beam configuration, an exhaustive search was made fixing one beam in 165°. This approximation was based on the efficiency of this direction for a two-beam plan (Fig. 6.37 — 2 BEAMS) and the particular target geometry favoring tangential orientations (Ferreira 2004). Owing to the efficiency of a three intensity-modulated beam plan in the treatment of this tumor site, further calculations were made to verify if a better plan could be found when using different gantry angles for the fixed beam Ω_3 (Fig. 6.37 — 3BEAMS). Additionally, it is shown that the similarity between the two beam phase space and the three beam phase space when $\varphi_3 = 165°$, stands for other directions of Ω_3. Thus, it is possible to estimate the directions of the local maxima for a three-beam plan from the two-beam phase space, that is, the three-beam phase space local maxima are placed in the same angular intervals of the two-beam phase space local maxima, but shifted by at least 15° to escape to the influence of the fixed beam (Fig. 6.37) (Ferreira 2004). The phase space with $\varphi_3 = 90°$ was simulated to clearly illustrate this feature. Since in the two-beam phase space, no local maxima existed in the neighborhood of 90°, when $\varphi_3 = 90°$, the shape of the resulting three-beam phase space is extremely similar to the two-beam phase space (comparison of the two-beam phase space with the three-beam phase space when $\varphi_3 = 90°$ is shown in Fig. 6.37). Interestingly, now the global maximum was moved to find a perpendicular beam configuration and not to escape to the fixed beam.

Even that a similar treatment outcome was obtained for the local maxima of the different three-beam phase spaces in Fig. 6.37, the optimal directions for a three intensity-modulated beam plan biologically optimized are (15°, 165°, and 315°) with a $\hat{P}_+^3$ value of 91%. Although a 1% lower P_+ value was obtained when $\Omega_3(\varphi_3 = 135°)$, in the directions (135°, 180°, and 300°), this phase space showed some interesting results, since almost the same P_+ values were obtained independently of the orientations of the beams Ω_1 and Ω_2. Thus, when such lateral tangential direction is used, the selection of the beam directions of the additional portals becomes less critical.

6.9. Impact of Different Dose–Response Parameters on Biologically Optimized IMRT in Breast Cancer

Treatment planning systems generate and optimize intensity-modulated beam plans based on user-defined objective functions for the target volume and organs at risk. However, in order to truly maximize the treatment outcome in biologically optimized treatment planning, the objectives should be based on individual patient radiation sensitivity. Unfortunately, the available biological parameters derived from clinical trials reflect an average radiation sensitivity of the examined populations. Deviations from those parameters are thus expected when considering the true individual patient radiation sensitivity.

A breast cancer patient of stage I–II with positive lymph nodes was chosen for analyzing the effect of the variation of individual radiation sensitivity on the optimal dose distribution (Ferreira *et al.* 2008). Thus, deviations from the average biological parameters, describing tumor, heart, and lung response, were introduced covering the range of patient radiation sensitivity. Two treatment configurations of three and seven biologically optimized intensity-modulated beams were employed.

The reference dose distribution, which is the optimal for the average patient biology, showed that in the three intensity-modulated beam plan the complication-free tumor control probability, P_+ was 91% with a probability of injury in the heart and left lung of 0.4% and 1.1%, respectively, and a probability of tumor control of 94.4%, whereas in the seven-beam plan, P_+ was 93.4% with a probability of injury for heart and left lung of 0.4% and 0.7%, respectively, and a tumor control probability of 95.7%. In the three-beam plan, the difference in P_+ between the optimal dose distribution (when the individual patient radiation sensitivity is known) and the reference dose distribution ranges up to 13.9% when varying the radiation sensitivity of the target volume, up to 0.9% when varying the radiation sensitivity of heart and up to 1.3% when varying the radiation sensitivity of lung. Similarly, in the seven-beam plan, the differences in P_+ are up to 11.8%, 1.6%, and 0.9% for the target, heart, and left lung, respectively. When the radiation sensitivity of the most important tissues in breast cancer radiation therapy was simultaneously changed, the maximum gain in outcome was as high as 7.7%.

Our results have shown that the jump from generalized to individualized radiation therapy may increase the therapeutic window for a significant number of patients, provided that these are identified. Even for radiosensitive patients, a simple treatment technique is sufficient to maximize outcome, since no significant benefits were obtained with a more complex technique using seven intensity-modulated beam portals.

There are many variables leading to uncertainties in the derived parameters. The collection of clinical response data requires large randomized trials involving different treatment techniques, which, however, are rarely available. An average interpatient radiation sensitivity is usually estimated for the population. In this way, similar treatment techniques with similar dose distributions result in a range of tissue responses (Tucker *et al.* 1992; Turesson *et al.* 1996). Also, due to the long follow ups required to observe late complications, historically derived parameters from uniform beam treatment techniques based on two-dimensional treatment planning, may not be sufficient. Still, a biologically optimized plan should be based on the individual tissue response to radiation. To assess the individual patient sensitivity before radiation therapy, different forms of predictive assays are being developed. The need of such assays is very high but still much debated, since there are so many factors other than the cellular response to radiation, on which these are based, that may influence patient response to radiation (Bentzen *et al.* 1993; Peters and McKay 2001; Russel and Begg 2002).

Since conventional doses are now determined to avoid severe normal tissue complications for the most radiosensitive patients (Norman *et al.* 1988), the identification of these patients prior to the treatment could reduce their rate of complications, while more radiation-sensitive patients could be prescribed with higher doses (Brahme 2001).

6.9.1. *Range of the Dose–Response Parameters*

To simulate patients of different radiation sensitivity, the values of the model parameters were varied within the range of their confidence intervals as these have been reported in the literature for heart, lung, and ITV. Subsequently, the optimal dose distributions for all patient radiation sensitivities were calculated.

The variation in tumor cell radiation sensitivity was simulated within the range suggested by Okunieff *et al.* (1995) for microscopic disease. Additionally, the $D_{50,\mathrm{ITV}}$ value of 50 Gy was also simulated to explore the possibility to treat the primary tumor in a patient with a breast cancer that for medical reasons cannot be subjected to surgery (Arriagada *et al.* 1993). Thus, $D_{50,\mathrm{ITV}}$ was varied from 30 to 50 Gy, while γ_{ITV} varied from 2 to 4.

Heart is perhaps the most important organ at risk in radiation therapy of breast cancer due to the severity of the clinical endpoint: late cardiac mortality, can be fairly weighed against tumor control when the biological objective function P_+ is used in the optimization of the dose distribution.. The data from Emami *et al.* (1991) and Ågren (1995) suggest dose–response parameters for pericarditis, which describes a rather parallel organ. Thus, the $D_{50,H}$ for heart was varied from 49 to 57 Gy, γ_H from 1 to 3, and s_H from 0.2 to 2. Other data for this endpoint (Eriksson *et al.* 2000; Gagliardi *et al.* 2000) suggest that the relative seriality parameter, s, should be somewhat larger, probably due to an inaccurate modeling of the volume dependence of the heart.

Several researchers have estimated biological parameters for radiation pneumonitis using different biological models (Burman *et al.* 1991; Ågren 1995; Kwa *et al.* 1998; Gagliardi *et al.* 2000; Moiseenko *et al.* 2003; Seppenwoolde *et al.* 2003). However, parameters for symptomatic lung fibrosis, the endpoint of interest, were only estimated by Moiseenko *et al.* (2003) from radiotherapy of thymoma tumors using the Lyman model. It is then suggested that fibrosis is an endpoint with a steeper γ value and a more serial behavior than radiation pneumonitis. Thus, for this organ, variations in $D_{50,L}$ from 26 to 34 Gy, for γ_L from 1 to 3, and for s_L from 0.01 to 0.1 were applied.

There are evidences indicating that patient radiation sensitivity has genetic basis (Andreassen *et al.* 2002). In this case, a patient with a radiosensitive breast tissue might need a lower dose to eradicate all tumor cells, whereas a radiation-resistant patient may expect less severe complications, but at the same time may require higher therapeutic doses to kill all tumor cells. Thus, the optimal dose distribution for patients of increased radiation sensitivity both in the normal and tumor tissues were calculated.

All the dose distributions, which were optimized for patients of different radiobiology, were compared with the reference dose distribution, which is the optimal only for the patient with average radiation sensitivity. The *radiobiological gain*, achieved if individual radiation sensitivity is predicted prior to radiation therapy, was determined by subtracting the response to the optimal dose distribution from the response to the reference dose distribution applied to the same patient. A positive value of ΔP_+ indicates how much is gained by knowing the true patient radiation sensitivity and thus by individualizing biologically optimized radiation therapy. A positive ΔP_B shows an increase in the probability of tumor control when the optimal dose distribution is delivered. A positive ΔP_I indicates that the optimal dose distribution estimates a larger probability of complications than the reference.

In Fig. 6.38, gains in treatment outcome of around 1% were obtained when the radiation sensitivity of the organs at risk was changed individually (Ferreira *et al.* 2008). However, this small improvement in therapy may not be detected clinically. But values as high as 7.7%, as obtained in Fig. 6.38, when the radiation sensitivity of all tissues was simultaneously varied, may become clinically relevant. These may be patients with hypoxic tumor cells or hypersensitivity to radiation like telangiectasia,

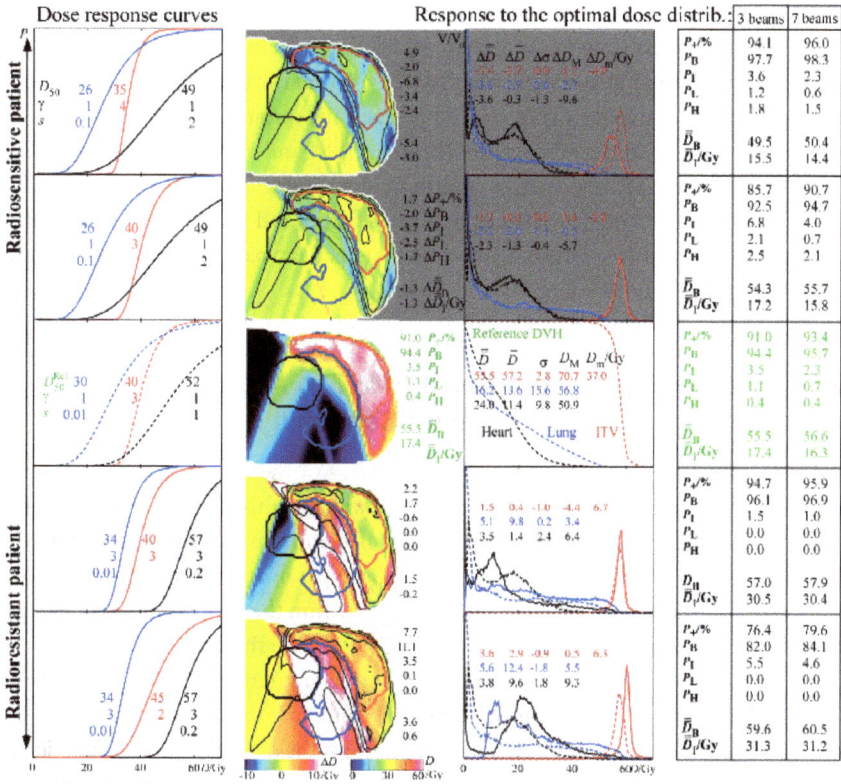

Figure 6.38. The reference dose distribution, shown in the middle, was applied to patients with increasing radiation resistance: ITV, heart, and lung were changed simultaneously following the dose–response relations shown on the left. Thus, the most radiation-sensitive patients are on the top, while the most radiation-resistant are on the bottom. The difference between the optimal and the reference dose distribution is shown in all other cases. In each diagram, the difference between the response to the optimal dose distribution and the reference dose distribution is also shown. Except for the reference patient, where the cumulative DVHs are shown with the dashed lines, the differential DVHs of the optimal dose distributions for patients with different radiation sensitivities are shown by the solid lines.

for example. If the values reported in the literature represent one standard deviation of the radiation sensitivity of all the population, 32% of the patients may significantly benefit from the use of predictive assays to support accurate biological optimization.

There are certainly many uncertainties linked to clinically observed, even established, dose–response relations and associated biological parameters. Even with a well-designed clinical trial, controlling all the parameters involved in the derivation of the dose–response parameters, the statistical process by itself leads to uncertainties in the prediction of individual patient parameters (Lind *et al.* 1999; Bentzen *et al.* 1994). Patient heterogeneities originate the shallow γ values generally reported in the literature. Tumor size or cells heterogeneity within the tumor also reduces the γ value and by consequence varies D_{50}. This raises the question: should these values be corrected for patient heterogeneity (Brahme 1984 and 1994), or should the best derived values be used, when performing individual biological optimization? Finally, the relation between the size of the functional subunits and the voxel size used in the optimization is also fundamental and a user-dependent parameter which is rarely discussed in the literature (Ågren *et al.* 1990).

6.10. Stochastic Optimization of IMRT to Account for Uncertainties in Patient Sensitivity

A biological objective function is defined in terms of the desirable treatment outcome. To calculate the treatment outcome, knowledge of the biological parameters, like the radiation sensitivity and the density of clonogens, is required. In today's clinical practice, the mean values of the response parameters from populations of similar patients are often available. The variance within the population may be large though, and biological data for the individual patient are thereby uncertain.

Uncertainties in the intrinsic biological parameters stem both from the variation in radiation sensitivity between different cells and functional subunits within an organ, as well as between different individuals. With the present objective functions, that is, the probability of achieving complication-free tumor cure, the sources of uncertainty are: (a) the number of clonogens, N_0, (b) the structural organization of the normal tissue described by the relative seriality, s, (c) the intrinsic radiation sensitivity parameters (D_0, α and β), and (d) the delivered dose distribution $D(r)$.

6.10.1. *Methods for Handling Uncertainties*

Much better than various regularizations techniques is to account for physical and biological uncertainties by direct biological methods! To find a dose distribution that is as robust as possible with regard to uncertainties in the radiobiological response parameters, three methods have been compared. These are: (a) stochastic optimization using the expectation value, EP_+, of P_+ as the objective function, (b) using biological "margins" on the radiation sensitivity, that is, assuming the patient to be more sensitive than the mean value for normal tissues and more resistant for the tumor to better account for the variability in radiation sensitivity, and (c) by adjusting the slope, γ, and the location of the dose–response curve, D_{50}, to simulate the dose–response curve of a heterogeneous population. Using the last method,

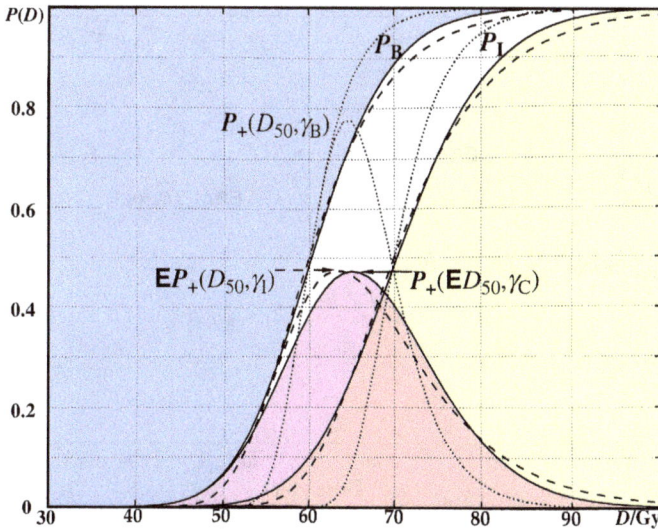

Figure 6.39. Comparison between the expectation value of P/P_+ (dashed curves) and the corresponding fitted Poisson function with the same maximum normalized gradient (full curves) is shown. The probability of benefit and injury are assumed to be totally correlated (Eq. (6.2) is used) in the P_+ calculation. The relative uncertainty of D_{50} is 10%.

the resulting dose–response curve obviously does not have the same shape as that calculated based on EP_+. This is because the expectation value is no longer well represented by a Poisson distribution as shown in Fig. 6.39 (Kåver *et al.* 1999). The dotted curve in Fig. 6.39 shows the response (P_+) of a patient without uncertainties and the dashed curve the expectation value of the response (EP_+), that is, the mean response for a patient population with a 10% relative uncertainty in D_{50}. The full curve in Fig. 6.39 is a Poisson distribution with a slope adjusted so the maximum normalized gradient equals that of EP_+. It can be seen that although the slope and the D_{50} values are approximately equal, the shapes of the curves are different. Using this adjusted Poisson function will generally predict a lower dose at optimal P_+ than the curve describing the expectation value of P_+. In all three cases, the P_B and P_I curves are also shown.

If the probability of tumor control EP_B (or normal tissue injury EP_I), is given by clinical data, the γ of the individual patient can be calculated assuming the individual γ to be constant by unfolding the effect of the varying radiation sensitivity. This results in a higher individual, γ, depending on the level of uncertainty in the parameters describing the biological response, γ_B, as seen in Fig. 6.39. The relation between the individual γ value and the clinically observed γ value, γ_c is shown in

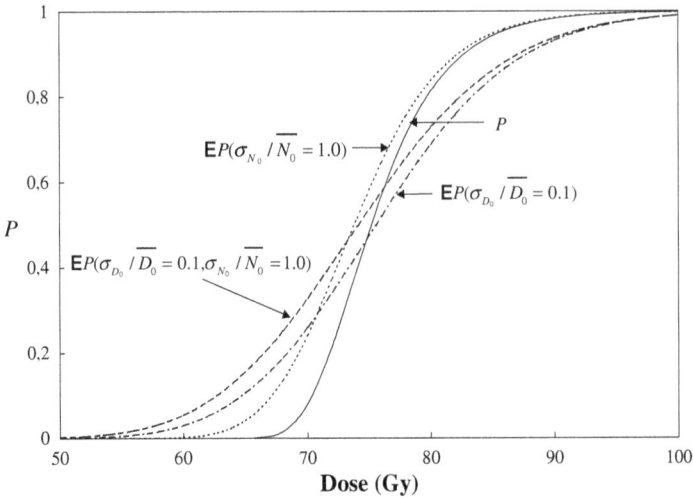

Figure 6.40. The dose–response curve for P and the expectation value of P given that the relative uncertainty in D_0 and N_0 are 0.1 and 1.0, respectively. (The relative uncertainty is the standard deviation divided by the expectation value.)

Fig. 6.40, where D_{50} has been assumed to be normally distributed (Kåver *et al.* 1999). The values are somewhat approximate since the clinical γ used in the Poisson expression does not correspond exactly to the curve obtained by $EP(D)$. However, it gives a good qualitative guidance on the change of the shape of the dose–response curve. The presence of systematic changes in the γ values is not dependent on the specific assumptions made in this case regarding the type of distribution and the level of interpatient variations. We would have to make corrections on γ for any given distribution of D_{50} (or corresponding measures of radiation sensitivity).

The second method using biological response margins on certain biological parameters is analogous to adding geometrical margins to the treated volume to compensate for setup errors and uncertainties in the patient geometry. The optimization is thus executed assuming that the tumor is more resistant than average and that the normal tissues are more sensitive so that the resultant treatment plan is useful over a wider range of variation in radiation sensitivity.

6.10.2. *Uncertainties in Treatment Parameters*

The number of clonogens (or functional subunits (Withers *et al.* 1988)), N_0, and the radioresistance, D_0 (or the radiation sensitivity parameters, α and β), are assumed to be independent. This is not strictly true for large bulky tumors, where D_0 tends to increase at the same time as N_0 increases, but is a reasonable assumption provided that the volume effect is accounted for separately. Furthermore, it is assumed that N_0 has a lognormal distribution:

$$\ln N_0 \in N(E(\ln N_0), \sigma_{\ln N_0}). \tag{6.9}$$

Solving this equation gives the following relation:

$$E(N_0) = \exp\left[E(\ln N_0) + \sigma_{\ln N_0}^2/2\right]. \tag{6.10}$$

Calculating the similar expression for $E(N_0)$ gives us the variance, and the standard deviations can thereby be connected by:

$$\sigma_{\ln N_0} = \sqrt{\ln\left\{[\sigma_{N_0}/E(N_0)]^2 + 1\right\}}. \tag{6.11}$$

The differences between the probability of a response and its expectation value over a range of given uncertainties in D_0 or N_0 are shown in Fig. 6.40. As can be seen in this figure, the clinical dose–response curve is changed due to interpatient uncertainties. Uncertainties in D_0 lead to a decrease in the response for high doses and an increase at low doses, while uncertainties in N_0 only lead to an increased response at low doses comparable to the effect of dosimetric uncertainties in dose delivery (Brahme 1984). It may be noted that with a reasonable dose plan, that is, a high probability of tumor response and a low probability of normal tissue injury, the relative uncertainty in the number of clonogens of the tumor is less important than the relative uncertainty in the number of functional subunits of the normal issue. This is due to the logarithmic influence of the latter. Furthermore, the shift due to uncertainty in D_0 is higher for higher values of N_0 due to the steeper dose–response relation. Uncertainties in these parameters have a much smaller influence on the result than uncertainties in D_0 and N_0. The value of s was varied to simulate different types of organ.

6.10.3. *Clinical Cases*

A pencil beam algorithm has been used in the optimizations to find the fluence profiles, and thereby the dose distribution, that maximizes the expectation value of P_+. To find a solution that is robust with regard to uncertainties in biological parameters, the expectation value of the probability of achieving uncomplicated tumor control has been used as the objective function. To perform a more unrestrained variation of the biological data and other treatment parameters to clarify the effects of uncertainties in biological data, two simple single-beam cases have been used. To see the effects on the result and the optimized dose plan, a more realistic clinical case has also been used (Fig. 6.41) (Kåver *et al.* 1999). The first simple case is a tumor and a separated organ at risk receiving equal homogeneous

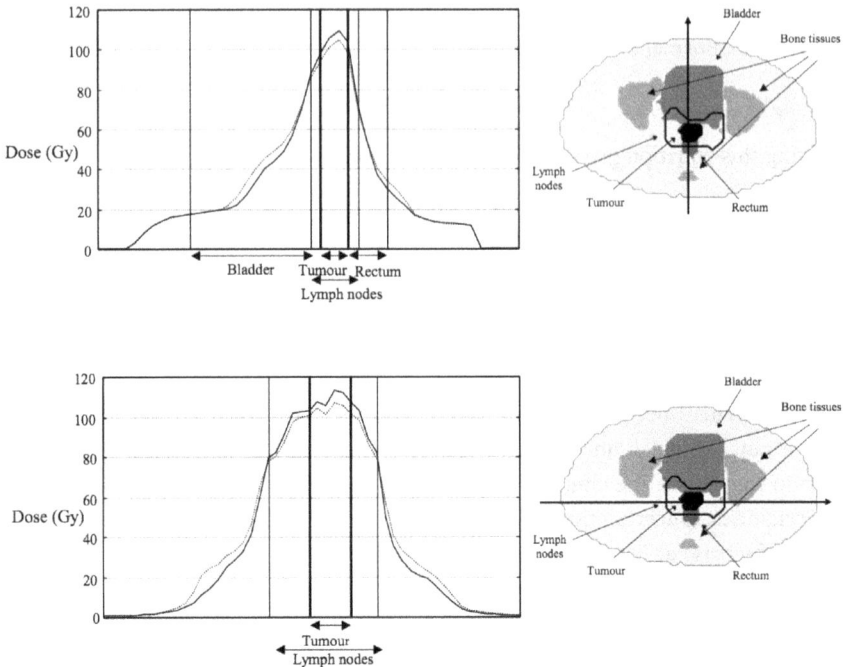

Figure 6.41.   The dose distribution along lines in the same plane is shown. The full and the dotted curves represent doses obtained with and without stochastic optimization, respectively.

dose distributions. The second simple case is a box-shaped target volume (with or without an inhomogeneity in form of a step) growing invasively in the surrounding stroma (the organ at risk). The relative density of the tumor is 30% and the relative seriality of the normal tissue is set to 0.85. The target was irradiated from one direction with a 20 MeV intensity-modulated electron beam.

The more realistic case studied is the cervix case, the geometry of which is shown in Fig. 6.41. The level of uncertainty has been varied in analogy with the superficial target, that is, the relative standard deviation of D_0 has been varied from 0 to 0.1. The corresponding maximum value for the relative standard deviation of N_0 is 1. The optimization is based on the pencil beam algorithm described in Gustafsson *et al.* (1984), with the beams as well as the patients represented by a voxel model (typically a $64 \times 64 \times 64$ matrix of 5 mm cubic voxels).

To compare the effect of uncertainties, two kinds of objective function have been used in the optimization process. The first of these, which may be referred to as *stochastic optimization*, reflects taking uncertainties into account by regarding P_+ as a stochastic variable, using the expectation value of P_+ as objective function,

$(EP_+)_S$ (where s stands for stochastic). The effect of ignoring the uncertainties has been simulated using the deterministic P_+ as the objective function. This has been done with a γ value adjusted to the lowered slope of the dose–response curve due to uncertainties. The resulting expectation value is denoted by $(EP_+)_{ns}$ (where ns stands for non-stochastic). The difference between these values of EP_+, which will be referred to as the *gain using stochastic optimization* (or simply the *gain*), $AEP_+ = (EP_+)_S - (EP_+)_{ns}$ is a measure of how much the effect of uncertainties influences the treatment outcome. The *relative gain* is a measure of the increase in response calculated as a percentage and is defined as:

$$\Delta EP_{+,\mathrm{rel}} = \frac{(EP_+)_s - (EP_+)_{ns}}{(EP_+)_{ns}} \qquad (6.12)$$

6.10.4. *Dependence on Biological Parameters*

From Fig. 6.39, one would expect the gain to increase with increasing γ, as the importance of the linear part of the dose–response is reduced, that is, the part where the EP and the best fitted Poisson curve differ the least. This tendency can be seen in Fig. 6.42, showing the result from simulations, where the gain is plotted against the

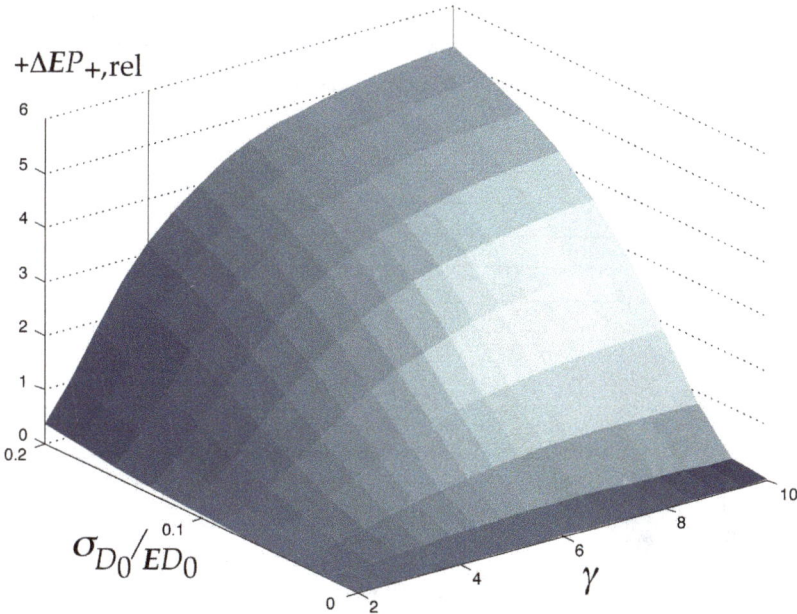

Figure 6.42. Illustration of the relative gain in complication-free cure using stochastic optimization plotted as a function of the γ value and the relative uncertainty in D_0.

γ value and the level of uncertainty in D_0 (Kåver *et al.* 1999). The gain also increases with increasing level of uncertainty. This can be explained by the fact the expectation value of the response probability differs more from the Poisson distribution with increasing level of uncertainty.

The gain increases with higher values of uncertainty. This is not always true for low values of uncertainty in N_0. This could be understood by the curve simulating uncertainty in N_0 in Fig. 6.40 where the low-dose region of the dose–response curve is shifted. This results in a higher mean sensitivity for the normal tissue and thereby a higher P_1. This effect counteracts the effect of uncertainty but is not equally important for higher values of uncertainty and for more complex cases, such as the cervix case. However, the relative gain increases with higher level of uncertainty as seen in Fig. 6.43 (Kåver *et al.* 1999). The relative gain is as high as 8% in the cervix case.

6.10.5. *Dose Profiles*

The first observation to be made, using stochastic optimization, is that the algorithm tries to increase the mean dose to the tumor and decrease the mean dose to organs at risk more and more for an increasing uncertainty in D_0. This can be explained partly by the shift in uncertainty shown in Fig. 6.40, and partly by the fact that the expectation value of response is lower than the corresponding Poisson response at the high end of the dose–response curve. This increase in dose can also be seen

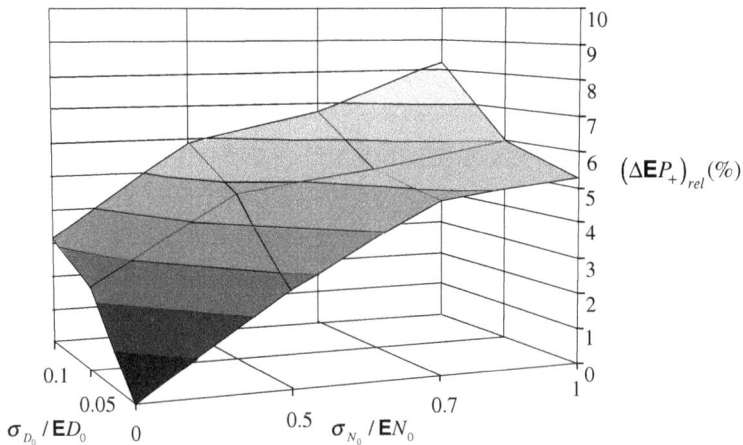

Figure 6.43. Illustration of the relative gain using stochastic optimization in the cervix case for different levels of uncertainty in D_0 and N_0.

with increasing uncertainty in N_0. This is not generally true though, as the dose–response curve is shifted as seen in Fig. 6.40. The field is narrowed down using stochastic optimization and the dose is increased in the target.

In the cervix case, the mean dose to the tumor increases while the mean dose to organs at risk decreases, independent of the level of uncertainty employed. This is expected as the algorithm tries to improve the plan to account for the most sensitive cases in a population. The change is not proportional to the level of uncertainty as in simpler cases. This could be understood by the fact that many counteracting effects interact in a more complex case. Even though the mean dose to the rectum is lower using stochastic optimization, the maximum dose is higher due to higher dose in the adjacent tumor.

In the prostate case, the mean dose to the tumor increases, although some parts of the tumor receive a lower dose in order to lower the dose to the organs at risk. The mean dose to the rectum and the bladder decreases using stochastic optimization. Figure 6.44 shows a cross-section of the dose distribution using non-stochastic optimization and Fig. 6.45 shows the dose difference between dose plans obtained by stochastic and non-stochastic optimization (Kåver *et al.* 1999). Although the dose difference is small (<5 Gy in the cervix case and <2 Gy in the prostate case), the dose plan is not only quantitatively but also qualitatively different. Figure 6.45 shows that a more robust solution is obtained by increasing the weight of the right beam and decreasing the weight of the frontal beam.

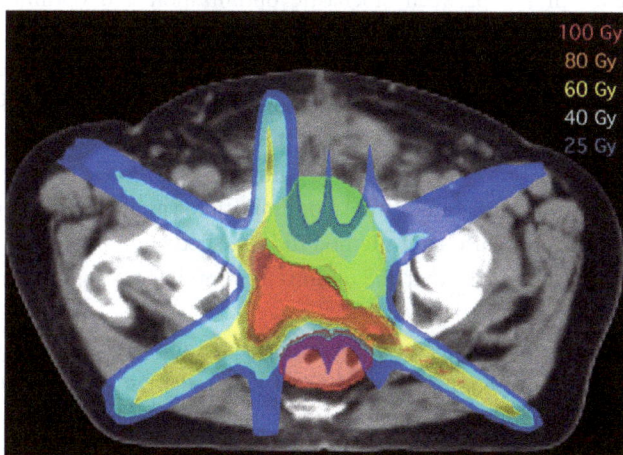

Figure 6.44. An optimized dose plan at a cross-section in the center of the tumor for a prostate case with three coplanar fixed-angle intensity-modulated fields is shown. Organs at risk are the rectum (bright red) and the bladder (green).

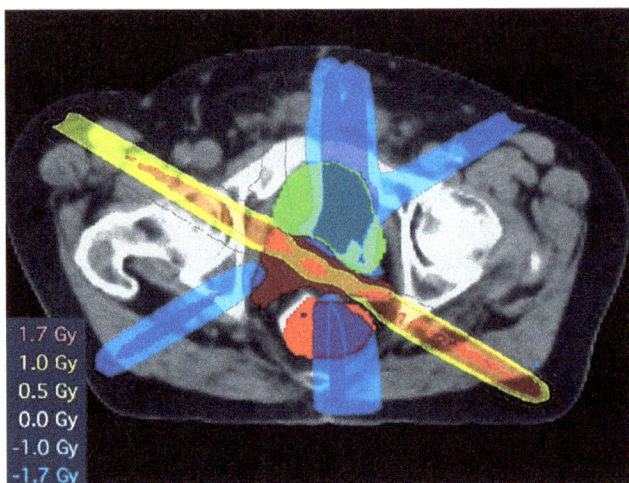

Figure 6.45. The difference between dose plans from stochastic optimizations and non-stochastic optimization of the case in Fig. 6.41 assuming 10% relative uncertainty in N_0 and 100% relative uncertainty in N_0 is shown.

6.10.6. *Margins*

For the more difficult cervix case, optimizations with margins were not as efficient. The difference in P_+ between stochastic optimization and optimization with margins did not show a clear minimum for any fixed margin on D_{50}, and the margin optimization resulted in an, at best, 5% lower P_+. This could be because more organs are influenced by the dose distribution. The margins added or subtracted were of equal size for all organs and tumors. It is reasonable to believe that this is not the optimal combination of margins. Differentiating the size of the margins gives a drastic magnification of the problem. An optimization with the margins treated like variables using a finite difference method did not succeed in improving the result either. The result indicated that the optimization problem is highly complex and non-convex.

Most dose planning and dose optimization is done ignoring clinical uncertainties of the optimization process such as varying radiation sensitivity, dosimetric uncertainties, and beam-patient alignment uncertainties. Stochastic optimization is a general tool to account for such uncertainties. By taking the uncertainty in biological response parameters into account during the optimization process, a more robust solution with tolerance for individual variations in sensitivity is obtained, compared to using the most probable dose–response relation.

The method of using biological margins on radiation sensitivity often gives a result that is close to that obtained using stochastic optimization, even though this requires a careful choice of the response margin size. Although stochastic optimization is more time-consuming, it will always be preferable to optimizations with margins. Furthermore, stochastic optimization assures us that we have found the optimal robust nonuniform dose delivery.

The absolute gain using stochastic optimization is a few percent and the relative improvement compared with non-stochastic optimization is 5–10%. The extent of this gain varies with the interpatient variability as well as with the difficulty and complexity of the clinical case studied. Even when the difference is small, there is a strong desire to make the treatment plan more robust with regard to individual variations in patient sensitivity.

Bibliography

Aaltonen P, Brahme A, Lax I, *et al.* (1997) Specification of dose delivery in radiation therapy, Recommendations by the NACP. *Acta Oncol* 36 (Suppl. 10):1–32.

Ågren AK (1995) Quantification of the response of heterogeneous tumors and organized normal tissues to fractionated radiotherapy, PhD thesis, Stockholm University, Sweden.

Ågren-Cronqvist AK, Brahme A, Turesson I (1990) Optimization of uncomplicated control for head and neck tumors. *Int J Radiat Oncol Biol Phys* 19:1077–1085.

Alber M, Birkner M, Nüsslin F (2002) Tools for the analysis of dose optimization: II. Sensitivity analysis. *Phys Med Biol* 47:N265–N270.

Alber M, Nüsslin F (2000) Intensity modulated photon beams subject to a minimal surface smoothing constraint. *Phys Med Biol* 45:N49–N52.

Andreassen CN, Alsner J, Overgaard J (2002) Does variability in normal tissue reactions after radiotherapy have a genetic basis-where and how to look for it? *Radiother Oncol* 64:131–140.

Arriagada R, *et al.* (1993) Radiotherapy alone in breast cancer. Analysis of tumour and lymph node radiation doses and treatment-related complications. The experience of the Gustave–Roussy Institute and the Princess Margaret Hospital. *Radiother Oncol* 27:1–6.

Bel A, van Herk M, Bartelink H, Lebesque JV (1993) A verification procedure to improve patient set-up accuracy using portal images. *Radiother Oncol* 29:253–260.

Bentzen SM (1994) Radiobiological considerations in the design of clinical trials. *Radiother Oncol* 32:1–11.

Bentzen SM, Overgaard M, Overgaard J (1993) Clinical correlations between late normal tissue endpoints after radiotherapy: implications for predictive assays of radiosensitivity. *Eur J Cancer* 29A:1373–1376.

Bortfeld T (1999) Optimized planning using physical objectives and constrains. *Sem Rad Oncol* 9:20–34.

Boswell S, Tomé W, Jeraj R, Jaradat H, Mackie TR (2006) Automatic registration of megavoltage to kilovoltage CT images in helical tomotherapy: an evaluation of the setup verification process for the special case of a rigid head phantom. *Med Phys* 33: 4395–4404.

Brahme A (1982) Physical and biological aspects on the optimum choice of radiation modality, Department of Medical Radiation Physics, Karolinska Institutet, Stockholm, p. 475.

Brahme A (1984) Dosimetric precision requirements in radiation therapy. *Acta Radiol Oncol* 23:379–391.

Brahme A (1988) Optimal setting of Multileaf Collimators in stationary beam radiation therapy. *Strahlentherapie* 164:343–350.

Brahme A (1994) Which Parameters of the Dose Distribution are Best Related to the Radiation Response of Tumors and Normal Tissues? *Proc. Int. Semin Europe, the Middle East and Africa Organized by the IAEA (Leuven)*, pp. 37–58.

Brahme A (1995) Treatment optimization using physical and biological objective functions. In: Smith A (ed.) *Radiation Therapy Physics*, 209—246. Berlin: Springer.

Brahme A (2001) Individualizing cancer treatment: biological optimization models in treatment planning and delivery. *Int J Radiat Oncol Biol Phys* 49:327–337.

Brahme A, Ågren AK (1987) Optimal dose distribution for eradication of heterogeneous tumors. *Acta Oncol* 26:377–385.

Brahme A, Kraepelien T, Svensson H (1980) Electron and photon beam characteristics from a 50 MeV Racetrack Microtron. *Acta Radiol* 19:305–319.

Brahme A, Svensson H (1979) Radiation beam characteristics of a 22 MeV microtron. *Acta Radiol Oncol Radiat Phys Biol* 18:244–272.

Burman C, *et al.* (1991) Fitting of normal tissue tolerance data to an analytical function. *Int J Radiat Biol Phys* 21:123–135.

Choi B, Deasy JO (2002) The generalized equivalent uniform dose function as a basis for intensity-modulated treatment planning. *Phys Med Biol* 51:2353–2365.

Chvetsov AV (2005) L-curve analysis of radiotherapy optimization problems. *Med Phys* 32:2598–2608.

Chvetsov AV, Calvetti D, Sohn JW, Kinsella TJ (2005) Regularization of inverse treatment planning for intensity modulated radiation therapy. *Med Phys* 32:501–514.

Chvetsov AV, Dempsey JF, Palta JR (2007) Optimization of equivalent uniform dose using the L-curve criterion. *Phys Med Biol* 52:5973–5984.

Creutzberg CL, Althof VGM, Huizenga H, Visser AG, Levendag PC (1993) Quality assurance using portal imaging: the accuracy of patient positioning in irradiation of breast cancer. *Int J Radiat Oncol Biol Phys* 25:529–539.

Emami B, Lyman J, Brown A, Coia L, Goitein M, Munzenrider JE, Shank B, Solin LJ, Wesson AM (1991) Tolerance of normal tissue to therapeutic irradiation. *Int J Radiat Oncol Biol Phys* 21:109–122.

Eriksson F, Gagliardi G, Liendberg A, Lax I, Lee C, Levitt S, Lind B, Rutqvist LE (2000) Long-term cardiac mortality following radiation therapy for Hodgkin's disease: analysis with the relative seriality model. *Radiother Oncol* 55:153–162.

Ferreira BC (2004) Biological optimization of angle of incidence and intensity modulation in breast and cervix cancer radiation therapy, PhD thesis, Stockholm University, Sweden.

Ferreira BC, Mavroidis P, Adamus-Gorka M, Svensson R, Lind BK (2008) The impact of different dose–response parameters on biologically optimized IMRT in breast cancer. *Phys Med Biol* 53:2733–2752.

Ferreira BC, Svensson R, Lind B, Johansson J, Brahme A (2006) Effective beam directions using radiobiologically optimized IMRT of node positive breast cancer. *Phys Med* 22:3–15.

Ferreira BC, Svensson R, Löf J, Brahme A (2003) The clinical value of non-coplanar photon beams in biologically optimized intensity modulated dose delivery on deep-seated tumors. *Acta Oncol* 42:852–864.

Gagliardi G, *et al.* (2000) Radiation pneumonitis after breast cancer irradiation: analysis of the complication probability using the relative seriality model. *Int J Radiat Oncol Biol Phys* 46:373–381.

Gustafsson A, Lind BK, Brahme A (1984) A generalized pencil beam algorithm for optimization of radiation therapy. *Med Phys* 21:343–356.

International Commission on Radiation Measurements, ICRU Report 50 (1994) *Prescribing, Recording, and Reporting Photon Beam Therapy, 50*, ICRU Publications, Bethesda, MD, USA.

Kåver G, Lind BK, Löf J, Liander A, Brahme A (1999) Stochastic optimization of intensity modulated radiotherapy to account for uncertainties in patient sensitivity. *Phys Med Biol* 44:2955–2969.

Kwa SL, *et al.* (1998) Radiation pneumonitis as a function of mean lung dose: an analysis of pooled data of 540 patients. *Int J Radiat Oncol Biol Phys* 42:1–9.

Lind BK (1991) Radiation therapy planning and optimization studied as inverse problems, thesis, Stockholm University, Sweden.

Lind BK, Brahme A (1995) Development of treatment techniques for radiotherapy optimization. *Int J Imag Syst Technol* 6:33–42.

Lind BK, Källman P, Sundelin B, Brahme A (1993) Optimal radiation beam profiles considering uncertainties in beam patient alignment. *Acta Oncol* 32:331–342.

Lind BK, Mavroidis P, Hyodynmaa S, Kappas C (1999) Optimization of the dose level for a given treatment plan to maximize the complication free tumour cure. *Acta Oncol* 38:787–798.

Lin SH, Sugar E, Teslow T, McNutt T, Saleh H, Song DY (2008) Comparison of daily couch shifts using MVCT (Tomotherapy) and B-mode ultrasound (BAT system) during prostate radiotherapy. *Technol Cancer Res Treat* 7:279–286.

Löf J (2000) Development of a general framework for optimization of radiation therapy, PhD thesis, Stockholm University, Sweden.

Mackie TR, Balog J, Ruchala KJ, Shepard D, Aldridge JS, Fitchard EE, Reckwerdt P, Olivera GH, McNutt T, Metha M (1999) TomoTherapy. *Sem Radiat Oncol* 9:108–117.

Mavroidis P, Axelsson S, Hyödynmaa S, Rajala J, Pitkänen MA, Lind BK, Brahme A (2002) Effects of positioning uncertainty and breathing on dose delivery and radiation pneumonitis prediction in breast cancer. *Acta Oncol* 41:471–485.

Mavroidis P, Ferreira BC, Papanikotaou N, Svensson R, Kappas C, Lind BK, Brahme A (2006) Assessing the difference between planned and delivered intensity-modulated radiotherapy dose distributions based on radiobiological measures. *Clin Oncol* 18:529–538.

Mavroidis P, Ferreira BC, Shi C, Lind BK, Papanikolaou N (2007) Treatment plan comparison between helical Tomotherapy and MLC-based IMRT using radiobiological measures. *Phys Med Biol* 52:3817–3836.

Mavroidis P, Komisopoulos G, Lind BK, Papanikolaou N (2008) Interpretation of the dosimetric results of three uniformity regularization methods in terms of expected treatment outcome. *Med Phys* 35:5009–5018.

Mavroidis P, Lind BK, Brahme A (2001) Biologically effective uniform dose ($\overline{\overline{D}}$) for specification, report and comparison of dose response relations and treatment plans. *Phys Med Biol* 46:2607–2630.

Mavroidis P, Lind BK, Brahme A (2002) $\overline{\overline{D}}$, an effective uniform dose linked to the probability of response. *Phys Med Biol* 47:L3–L9.

Mavroidis P, Lind BK, Van Dijk J, Koedooder K, De Neve W, De Wagter C, Planskoy B, Rosenwald JC, Proimos B, Kappas C, Danciu C, Benassi M, Chierego G, Brahme A (2000) Comparison of conformal radiation therapy techniques within the dynamic radiotherapy project "DYNARAD". *Phys Med Biol* 45:2459–2481.

Meeks SL, Harmon Jr JF, Langen KM, Willoughby TR, Wagner TH, Kupelian PA (2005) Performance characterization of megavoltage computed tomography imaging on a helical tomotherapy unit. *Med Phys* 32:2673–2681.

Mitine C, Dutreix A, van der Schueren E (1991) Tangential breast irradiation: influence of technique of set-up on transfer errors and reproducibility. *Radiother Oncol* 22:308–210.

Moiseenko V, Craig T, Bezjak A, Van Dyk J (2003) Dose-volume analysis of lung complications in the radiation treatment of malignant thymoma: a retrospective review. *Radiother Oncol* 67:265–274.

Niemierko A (1999) A generalized concept of equivalent uniform dose (EUD) (abstract) *Med Phys* 28:1100.

Norman A, Kagan AR, Chan SL (1988) The importance of genetics for the optimization of radiation therapy. A hypothesis. *Am J Clin Oncol* 11:84–88.

Okunieff P, Morgan D, Niemierko A, Suit HD (1995) Radiation dose-response of human tumors. *Int J Radiat Oncol Biol Phys* 32:1227–1237.

Orton NP, Jaradat HA, Tomé W (2006) Clinical assessment of three-dimensional ultrasound prostate localization for external beam radiotherapy. *Med Phys* 33:4710–4717.

Peters L, McKay M (2001) Predictive assays: will they ever have a role in the clinic? *Int J Radiat Oncol Biol Phys* 49:501–504.

Pouliot J, Lirette A (1996) Verification and correction of setup deviations in tangential breast irradiation using EPID: gain versus workload. *Med Phys* 23:1393–1398.

Russell NS, Begg AC (2002) Predictive assays for normal tissue damage. *Radiother Oncol* 64:125–129.

Schneider U, Lomax A, Pemler P, Besserer J, Ross D, Lombriser N, Kaser-Hotz B (2006) The impact of IMRT and proton radiotherapy on secondary cancer incidence. *Strahlenther Onkol* 182:647–652.

Seppenwoolde Y, *et al.* (2003) Comparing different NTCP models that predict the incidence of radiation pneumonitis. Normal tissue complication probability. *Int J Radiat Oncol Biol Phys* 55:724–735.

Söderström S (1995) Radiobiologically based optimization of external beam radiotherapy techniques using a small number of fields, PhD thesis, Karolinska Institutet, Stockholm, Sweden.

Söderström S, Eklöf A, Brahme A (1995) Few field radiation therapy optimization in the phase space of complication free tumor control. In: Purdy JA and Emami B (eds), *3D Radiation Treatment Planning and Conformal Therapy*, 57–74. Medical Physics Publishing.

Söderström S, Brahme A (1996) Small is beautiful — and often enough. *Int J Radiat Oncol Biol Phys* 34:757–759.

Söderström S, Eklöf A, Brahme A (1999) Aspects on the optimal photon beam energy for radiation therapy. *Acta Oncol* 38:179–187.

Spirou SV, Fournier-Bidoz N, Yang J, Chui CS, Ling CC (2001) Smoothing intensity modulated beam profiles to improve the efficiency of delivery. *Med Phys* 28: 2105–2112.

Tikhonov AN (1963) Regularization of incorrectly posed problems. *Dokl Akad Nauk SSSR* 4:1624–1627.

Tucker SL, Turesson I, Thames HD (1992) Evidence for individual differences in the radiosensitivity of human skin. *Eur J Cancer* 28A:1783–1791.

Turesson I, Nyman J, Holmberg E, Oden A (1996) Prognostic factors for acute and late skin reactions in radiotherapy patients. *Int J Radiat Oncol Biol Phys* 36:1065–1075.

Webb S (2000) *Intensity-Modulated Radiation Therapy.* IOP Publishing, Bristol.

Weckström K (1997) Investigation of multibeam treatment units considering physical characteristics and radiobiological implications, Department of Medical Radiation Physics, Karolinska Institutet, Stockholm, Sweden.

Welsh JS, Lock M, Harari PM, Tomé W, Fowler J, Mackie TR, Ritter M, Kapatoes J, Forrest L, Chappell R, Paliwal B, Mehta MP (2006) Clinical implementation of adaptive helical tomotherapy: a unique approach to image-guided intensity modulated radiotherapy. *Technol Cancer Res Treat* 5:465–480.

Welsh JS, Patel RR, Ritter MA, Harari PM, Mackie TR, Mehta MP (2002) Helical tomotherapy: an innovative technology and approach to radiation therapy. *Technol Cancer Res Treat* 1:311–316.

Williams JR and Thwaites DI (1993). Radiotherapy physics in practice, Department of Medical Physics and Medical Engineering, University of Edinburg, Edinburg, UK.

Withers HR, Taylor JMG, Maciejewski B (1988) Treatment volume and tissue tolerance. *Int J Radiat Oncol Biol Phys* 14:751–759.

Wu Q, Mohan R Niemierko A (2000) IMRT Optimization based on the Generalized Equivalent Uniform Dose (EUD). *Proc. 13th Int. Conf. on the Use of Computers in Radiation Therapy*, Schlegel and Bortfeld T (eds.), May 2000, Heidelberg: Springer, pp. 17–19.

Yeboah C, Sandison GA, Chvetsov AV (2002) Intensity and energy modulated radiotherapy with proton beams: Variables affecting optimal prostate plan. *Med Phys* 29:176–189.

BioArt: Biologically Optimized 3D In Vivo Predictive Assay-Based Radiation Therapy

7

Anders Brahme

7.1. Introduction

Background. The fast development of intensity, energy, and radiation quality-modulated radiation therapy (IMRT and QMRT) during the past two decades with photon, electron and light ion beams has resulted in a considerable improvement of radiation therapy, particularly when combined with radiobiologically based treatment optimization techniques. This development and the recent development of advanced tumor diagnostics based on PET-CT imaging of tumor clonogen density and hypoxia open the field for new powerful radiobiologically based treatment optimization methods. The ultimate step is to use the unique radiobiological and dose distributional advantages of light ion beams for truly optimized bioeffect planning where the integral 3D dose delivery and tumor cell survival can be monitored early on in the treatment by PET-CT imaging and be corrected by biologically adaptive therapy optimization methods as already described in Chapters 2, 3 and 5 and in particular in Section 5.4.10.

Purpose. The main purpose of this chapter is to illustrate some principal areas of development of therapy optimization considering the whole therapy chain from tumor diagnostics and patient fixation through therapy planning and treatment optimization to the repeated treatment setups and dose delivery on a patient that hopefully has a shrinking tumor and may lose weight. Finally, it is the integral dose delivery and the biological effect distribution that matters so the shaping of the optimal incident beams is a truly complex inverse problem, which is hard to solve by such crude methods as a prescribed point dose to a planning target volume often used today.

Methods. The above introduction indicates that biologically optimized *in vivo* predictive assay based adaptive radiation therapy (BioArt) is really the ultimate way to perform high precision radiation therapy using checkpoints on the integral dose delivery and the tumor response, and based on this information, to perform compensating corrections of the dose delivery. By using biologically optimized scanned high energy photon or ion beams it is possible to measure *in vivo* the 3D dose delivery using the same PET-CT camera that was used for diagnosing the tumor spread. This method thus opens up the door for truly 4D BioArt where the measured dose delivery to the true target tissues can be used to fine adjust the incoming beams so that possible errors in the integral therapy process are practically eliminated toward the end of the treatment. Interestingly enough, all major error sources can in principle be corrected for in this way such as organ motions, treatment planning errors, patient setup, and tumor responsiveness deviations, as well as dose delivery problems for example due to gantry, multileaf, or scanning beam errors. Since it is possible to quantify the surviving tumor clonogen density after the first week or so of therapy, this information can be used to account for uncertainties in the biological responsiveness of the tumor and really cover all clinical uncertainties at the same time as more accurate dose–response data can be derived from the treatment. With the photonuclear or nuclear–nuclear reactions, the response of the PET-CT camera is truly related to the delivered integral dose with correct temporal averaging not least with ^{11}C treatments. Thus, if only small errors are seen, it is sufficient to adjust the last few treatment fractions. When using PET-CT tumor responsiveness monitoring, it is even possible to account for the uncertainty in known historical biological response data for the disease of the patient and perform a truly optimized radiation treatment since the tumor responsiveness is generally the most unknown quantity of the treatment.

Conclusions. Examples of radiobiologically optimized dose delivery are presented and the above-mentioned new treatment approach is illustrated for clinically relevant targets. The unique properties of light ion therapy in this context are also presented. Using the recently available biologically based treatment optimization algorithms, it is possible to improve the treatment outcome by as much as 10–40% for advanced tumors . The adaptive BioArt process based both on 3D tumor cell survival and dose delivery monitoring has the potential of accurate tumor eradication, not least with 4D geometric Bragg peak scanning and intensity-modulated mixed ion beam dose delivery. There is no doubt that the future of radiation therapy is very promising and gradually more and more patients may not even need advanced surgery but instead could be cured by photon and electron IMRT and ultimately biologically optimized light ion therapy, where the high linear energy transfer and high relative biological effectiveness Bragg peak is solely placed in the gross tumor volume.

7.2. Tumor Imaging

Cancer imaging is at the dawn of a third revolution of accurate tumor diagnostics. During the 1970s and early 1980s, computed tomography (CT) with diagnostic X-rays made a revolution in accurate delineation of normal tissue anatomy as well as gross tumor growth. In the mid-1980s and 1990s, magnetic resonance imaging and spectroscopy (MRI and MRSI) allowed an even more accurate differential diagnostics of soft tissue malignancies with the possibility to distinguish between tumorous tissues, edema, and normal tissues. During both these revolutions, positron emission tomography (PET) was a very interesting research modality capable of pinpointing a number of important functional properties of the tumor such as tumor growth, vascularization, and hypoxia (Van de Wiele *et al.* 2003; Aronson *et al.* 2002). The integration of positron emission and X-ray CT in one PET-CT unit is bringing a third diagnostic revolution to tumor imaging. By combining these two imaging modalities an unprecedented accuracy in the delineation of the tumor on a background of normal tissue anatomy is achieved by more accurate image fusion (Townsend 2002). Fusion of the PET and CT data sets has been performed for at least 20 years, but it was not until the datasets were recorded on the same diagnostic couch, within minutes without repositioning the patient, that the quality of the fused images became really outstanding. Part of the success is also due to the use of fluorodeoxyglucose (FDG), the long half-life of which (110 min) allows a more systematic accumulation in the actively dividing tumor cells, as seen in the left panels of Figs. 7.1, 7.9, 7.14 and 7.15(b). Obviously, this compound is not truly tumor-specific, as all regions with an increased metabolic rate will show an elevated uptake. A certain degree of specificity does exist as the FDG molecule is trapped at a higher rate in many tumors than in normal tissues. However, in the new era of molecular imaging (Gambhir 2002; Sullivan and Hoffman 2001), FDG will soon be followed by more specific tumor markers, allowing an even more accurate imaging of the tumor clonogen density. Methionine and other amino acids are already available as tracers and, although they may be better, but they may not be specific enough, since they will be incorporated in all tissues that are being renewed. For some tumors, more specific markers do exist, such as ^{11}C-Choline, FHBC or fluorodihydrotestosterone for imaging androgen receptors in prostate cancer. The first P_{53}PET images in mice with experimental tumors have recently been published (Doubrovin 2001) but what would really be desirable is, for example, the imaging of regions with mutant P_{53}, which is a characteristic of more than half of all human tumors (cf. Liu and Gelmann 2002).

During the past two decades, a large number of improvements of radiation therapy has taken place resulting in considerable improvements in treatment outcome. As just discussed, diagnostic tumor imaging developed from CT and MRI

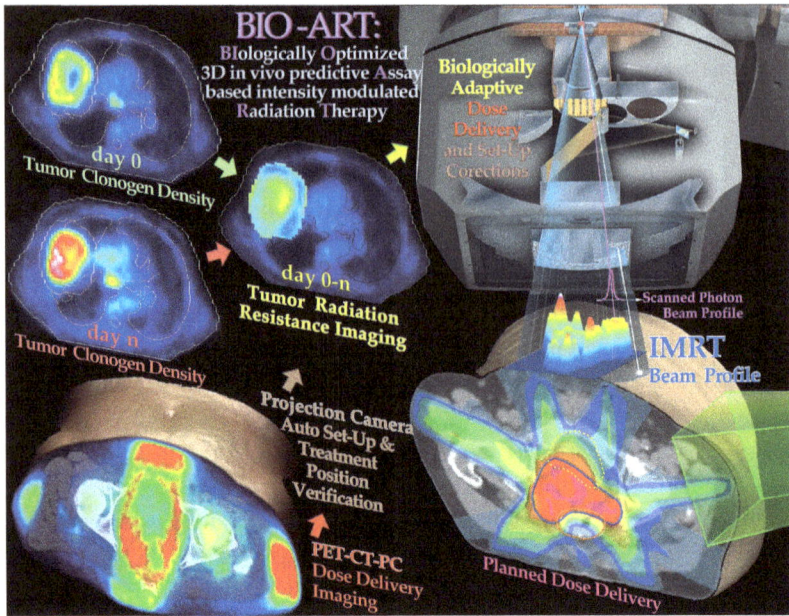

Figure 7.1. The BioArt process where PET-CT imaging is used not only to obtain tumor cell information for biologically optimized intensity-modulated dose delivery but also to monitor the 3D dose delivery *in vivo* inside the patient immediately after a treatment, for example, by photonuclear or ion nuclear reactions (cf. also Fig. 7.10 below) or even primary PET emitter beams of ^{11}C (cf. Fig. 8.38 (d)). In this way the optimal required dose delivery for a truly curative treatment can be calculated after about 10 days of therapy.

to PET-CT and MRSI will allow a more specific tumor diagnostics on a background of normal tissue anatomy. Equally important for the treatment outcome has been the development of radiation therapy from parallel opposed uniform rectangular beams to stationary and dynamic multileaf collimation from optimal directions using intensity modulated scanned pencil beams not only of electrons and photons but, most recently using, also of light ions from protons to carbon. In parallel to these developments, there has been an increasing desire to confine the dose delivery as accurately as possible to a dynamic target volume both by imaging and more dynamic alignment techniques such as breath hold and synchronized treatment techniques and ultimately adaptive treatment delivery where the final treatment fractions are modified to take early misalignments into account (cf. Löf *et al.* 1998).

The ultimate adaptive approach would be to also try to quantify the tumor response *in vivo* before the final dose delivery is decided after about 10 days of therapy using the biologically optimized 3D *in vivo* predictive assay-based radiation therapy (BioArt) approach as schematically shown in Fig. 7.1, making the final treatment

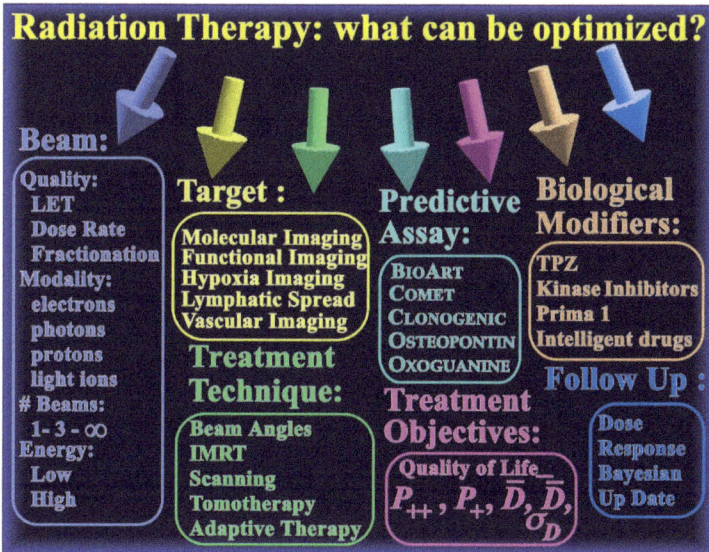

Radiation Therapy: what can be optimized?

Beam:

Quality:
LET
Dose Rate
Fractionation
Modality:
electrons
photons
protons
light ions
Beams:
1 - 3 - ∞
Energy:
Low
High

Target :

Molecular Imaging
Functional Imaging
Hypoxia Imaging
Lymphatic Spread
Vascular Imaging

Treatment Technique:

Beam Angles
IMRT
Scanning
Tomotherapy
Adaptive Therapy

Predictive Assay:

BioArt
Comet
Clonogenic
Osteopontin
Oxoguanine

Treatment Objectives:

Quality of Life__
$P_{++}, P_{+}, \bar{D}, \bar{\bar{D}},$
σ_D

Biological Modifiers:

TPZ
Kinase Inhibitors
Prima 1
Intelligent drugs

Follow Up :

Dose
Response
Bayesian
Up Date

Figure 7.2. Overview of some of the principal factors that may be subject to optimization during radiation therapy. Many of these factors can with advantage be handled by inverse treatment planning approaches as shown more clearly in Fig. 7.3.

biologically optimized using 3D *in vivo* predictive assay-based radiation therapy. In the present overview, this approach and optimization of the fourth dimension of radiation, namely the time–dose fractionation will be discussed in some detail.

But as can be seen from Figs. 7.2 and 7.3, there are a great number of other quantities that can be optimized in radiation therapy some of which will be discussed in connection to the BioArt and fractionation approaches and others have been described in many recent reviews during the past 20 years (Brahme 1992; Brahme 1994; Brahme 1995a–c; Lind and Brahme 1995; Brahme 1999a,b; Brahme 2000; Brahme 2001a,b; Brahme 2003b; Brahme 2004, 2009).

7.3. Optimization of Fractionation Schedules

The classical dose–time fractionation schedule of 30 fractions of 2 Gy in 5 weeks was developed during the latter half of the 20th century as a suitable approach mainly with parallel opposed and four-field box techniques using rectangular beams. With such a conventional dose delivery, the dose to normal tissues is generally of the same order of magnitude as that in the tumor. The mean tumor dose of 2 Gy/fraction is therefore largely determined by the tolerance of surrounding healthy normal tissues to ensure that all sublethal normal tissue damage is fully repaired in the 24 h commonly available before the next treatment. Today it is well known that higher

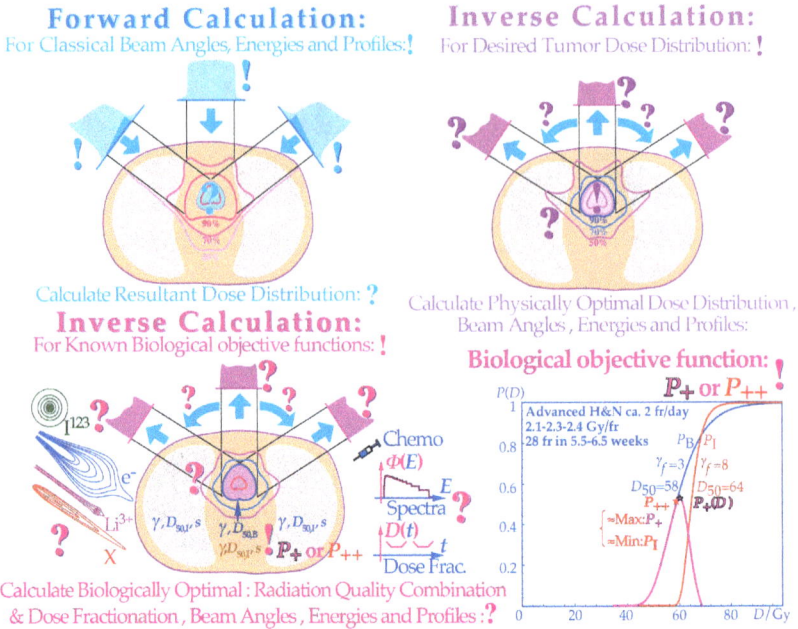

Figure 7.3. The development from traditional forward planning to inverse planning with physical dose objectives and more importantly biological objectives allowing the optimal selection of radiation modalities, intensity, and energy distributions and the time–dose fractionation pattern constitute a major revolution in radiation therapy. The optimization is solely based on known dose–response relations for the tumor and the involved normal tissues as shown in the lower right panel. Once clinically established dose–response contribution to the cell kill can also be taken into account.

doses per fraction are likely to induce severe normal tissue damage in late responding organs as first explained by Withers *et al.* (1988) due to the slow cell turnover and extensive shoulder region of the cell survival curve of such tissues. Today we also know that it is not unlikely that several normal tissues are also linked to low dose hypersensitivity as first discussed by Singh *et al.* (1994) using a computer-controlled microscopic survival assay. Data from a lung epithelial cell line are presented in Fig. 7.4 showing that both low and high doses per fraction may be associated with increased normal tissue damage per unit dose to the tumor. In fact, Fig. 7.4 shows that doses in the range from 1.5 to 2.5 Gy/fraction may be the most tolerated range for this cell line largely in agreement with the standard fractionation schedule of 30 times 2 Gy. The cell survival plotted in Fig. 7.4 shows that at 0.3 Gy/fraction, $D_{0,\text{eff}} = 1.3$ Gy and raises to 2.3 and 3.1 Gy at 1.1 and 2 Gy/fraction, respectively, just to decrease to 2.7 and 2.4 Gy at 3 and 4 Gy/fraction, respectively. Since a large

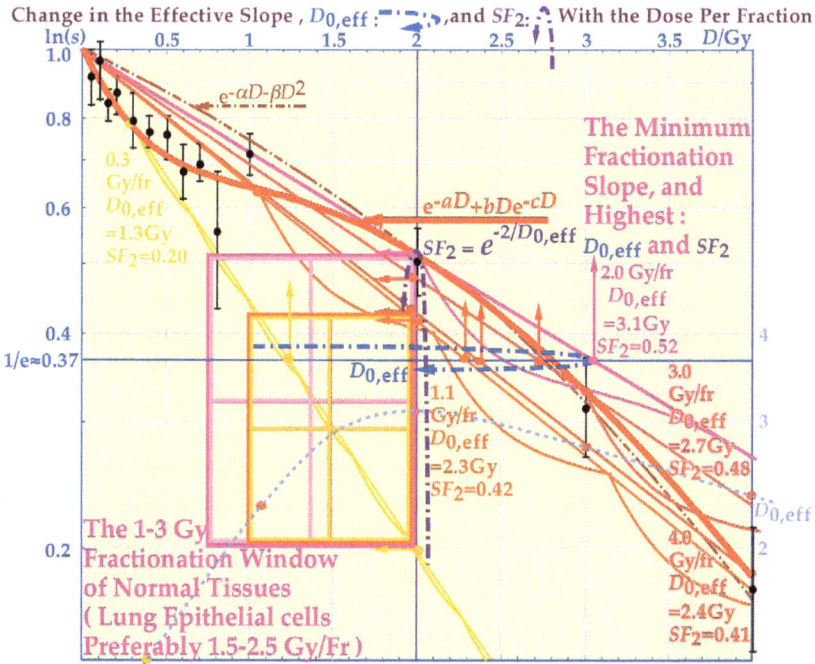

Change in the Effective Slope , $D_{0,eff}$: ▪ ◂ ▸ , and SF_2 : ✦ With the Dose Per Fraction

$e^{-\alpha D - \beta D^2}$

0.3 Gy/fr $D_{0,eff}$ $=1.3$Gy $SF_2=0.20$

$e^{-aD+bDe^{-cD}}$

$SF_2 = e^{-2/D_{0,eff}}$

The Minimum Fractionation Slope, and Highest : $D_{0,eff}$ and SF_2

2.0 Gy/fr $D_{0,eff}$ $=3.1$Gy $SF_2=0.52$

$D_{0,eff}$

1.1 Gy/fr $D_{0,eff}$ $=2.3$Gy $SF_2=0.42$

3.0 Gy/fr $D_{0,eff}$ $=2.7$Gy $SF_2=0.48$

$D_{0,eff}$

4.0 Gy/fr $D_{0,eff}$ $=2.4$Gy $SF_2=0.41$

The 1-3 Gy Fractionation Window of Normal Tissues (Lung Epithelial cells Preferably 1.5-2.5 Gy/Fr)

Figure 7.4. The cell survival of many normal tissues such as the lung epithelial cells in this figure possess a low dose hypersensitivity which makes low doses per fraction induce more tissue damage than the standard 2 Gy. Similarly high doses per fraction induce more damage too — so there is an optimal fractionation window in normal tissues at around 1.5–2.5 Gy/fraction (gray dashed line).

$D_{0,eff}$ corresponds to a higher radio resistance a dose of 2 Gy/fraction for the cell line in Fig. 7.4 produces the least harm to the tissue at a given desired dose level to a nearby tumor. It is therefore reasonable to assume that there is a *fractionation window* in most normal tissues that causes minimal damage at doses per daily fractions in the range 1.5–2.5 Gy, because generally very many small dose fractions or a few very large doses produce more acute or late damage to normal tissues than the standard 2 Gy in 30 fractions. One may think that small deviations in the $D_{0,eff}$ may not be too important, but a small difference repeated 30 times gives a strong exponential effect. A 10% reduction in $D_{0,eff}$ results in an ∼20-fold reduction in cell survival when delivered in 30 fractions.

However, when the dose delivery in the tumor and the normal tissues are no longer similar such as with very many narrow beams on a small tumor (sometimes referred to as stereotactic radiation therapy) or in more general terms with intensity-modulated radiation therapy (IMRT) not only the total dose and the dose per fraction but also the dose rate will vary considerably between the tumor and normal tissues. This converts the historical double trouble to a double or even *triple advantage* since the dose to the tumor can be increased and/or the dose to normal tissues can be decreased. The most natural approach would be to take out this clinical advantage by a dose escalation in the tumor keeping the dose per fraction to normal tissues constant at around 1.5 to 2.5 Gy. IMRT would allow very high tumor doses; hence, many more of the hypoxic tumors would become curable. In other cases, where the dose to the normal tissue stroma in the tumor bed may not allow this, the number of fractions could instead be reduced so part of the clinical advantage is taken out as a lower effect to normal tissues both in the tumor bed and elsewhere by fewer treatment fractions, at the same time as the tumor get a more effective treatment by the local high doses per fraction.

Figure 7.5 illustrates very interesting ways to improve the fractionation schedule of the standard 2 Gy/fraction over 6 weeks by modifying the dose delivery either on Fridays and Mondays when there is a whole weekend for repair of sublethal damage and when not much is left on the Monday. The figure shows that preferably higher doses per fraction should be given both Fridays and Mondays and even in

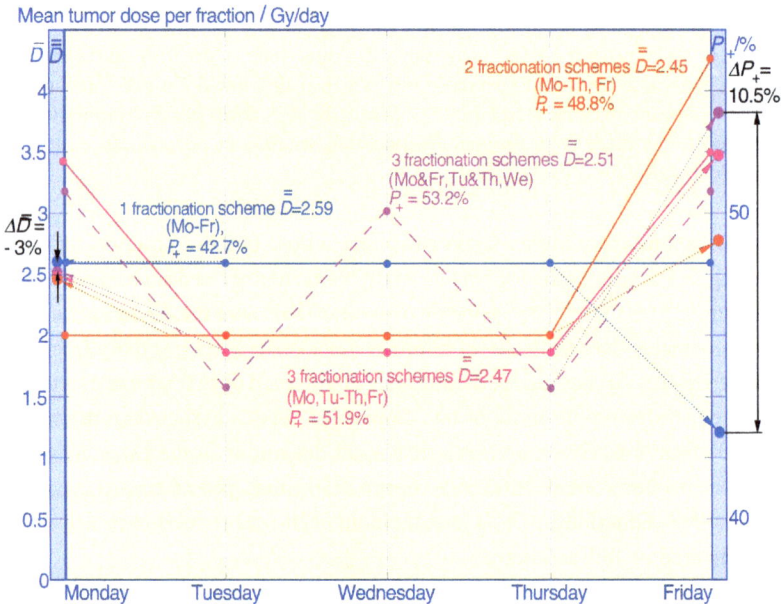

Figure 7.5. Classical fractionation uses a fixed dose around 2 Gy in 30 fractions (blue straight line). By giving higher dose per fraction on Fridays or Fridays and Mondays, an almost 10% increase in complication-free cure is seen (red and purple curves). A slight further improvement results if the dose on Wednesdays is also increased (dashed curve).

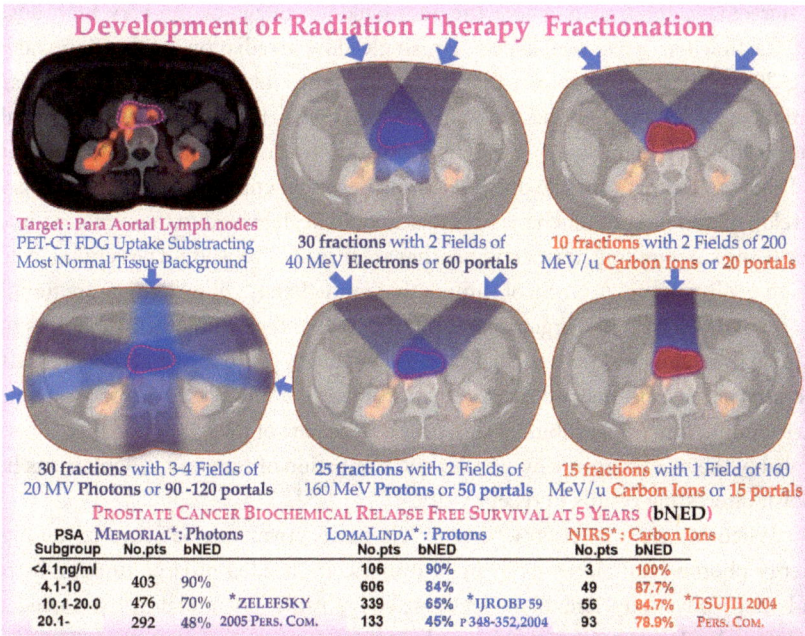

Figure 7.6. Dose distributions using different types of radiation are shown. In the example, it is assumed that the target is included in the red broken line (the target volume — the para aortal lymph nodes — has here been detected by PET-CT FDG uptake). The lower left panel shows a treatment using the classical "cross fire technique" with photon beams; a rather large volume of normal tissues will be irradiated outside the target volume. The upper middle and right panels are irradiations carried out using electrons and protons with a finite range; the irradiated normal tissue volume is then further reduced. The lower right panels show different possibilities using carbon ions. Higher biological effective dose can be achieved to hypoxic or radiation-resistant tumors (indicated in red), with rather limited effective dose to normal tissues. A reduction of the number of fractions — compared with the use of photons, electrons, and protons — is an advantage and improves local tumour control for at least some tumours. The lower table in the figure shows the improvement in biochemical relapse-free control going from protons to photon IMRT and carbon ions.

the middle of the week (Wednesdays) to account for the longer repair time and subsequent lower level of sublethal damage resulting in an improvement of the complication-free tumor cure of $>10\%$ at the same time as the mean tumor dose is reduced.

With well-optimized electron and photon IMRT, as few fractions as 20 may be sufficient for many tumor sites (cf. upper half of Fig. 7.6) and even substantially fewer fractions can be used for small tumor stages where the incoming beams can be quite narrow so they do not need to overlap much in normal tissues. Unfortunately, this reduction of the number of beam portals has not yet been used much except for in stereotactic treatments where the additional advantage of the very high doses per fraction has been established. By single dose irradiation in the dose range of $>20\,\text{Gy}$, it has been shown that many different and even quite radioresistant tumors are effectively eradicated (Shiau *et al.* 1997; Herfarth *et al.* 2001) mainly due to the very high dose per fraction and due to the associated high dose rate and short treatment time. So one interesting development of radiation therapy is to develop methods where such massive doses could be delivered to the tumor without sacrificing surrounding normal tissues that as far as possible should be kept intact for a high quality of life. A group of tumors that immediately come to mind are tumors in organs of very parallel organization (cf. Fig. 2.8) of their functional subunits such as lung, liver, and kidney. In such organs, a small tissue compartment associated with the internal margin may be sacrificed without severe loss of organ function since surrounding functional subunits take over a large part of their functionality. This is probably the main reason why stereotactic irradiation of tumors in such organs has shown interesting results in recent years.

When higher dose protractions in the tumor, than can be achieved by high energy photons and electrons in small tumors, are needed in large tumors, better and more physically and biologically selective radiation modalities are needed. Fortunately, due to a very ambitious light ion program in Japan, such radiation modalities have already been developed for clinical use. The recent attempts to better treat prostate cancer by photon IMRT at Memorial Sloan Kettering Cancer Center in New York, by protons at Loma Linda outside Los Angeles, and by carbon ions at Chiba just east of Tokyo are compared in Fig. 7.6. It is seen that for the more severe form of the disease with PSA levels $>20\,\text{ng/ml}$ before treatment, the probability of biochemical relapse-free cure 5 years after treatment with these three treatment techniques and radiation modalities are 48%, 45%, and 79% respectively. This result clearly indicates that the light ions heavier than protons have an important role to play in the treatment not least of radiation-resistant and hypoxic tumors, and they also have significant advantages when it comes to reducing the dose to normal tissue and therefore reducing the number of fractions as seen in the upper panels of Fig. 7.6.

The importance of large dose fractions in the use of high ionization density ions was already demonstrated by Denekamp *et al.* (1976), where they showed in murine zenografts that doses as high as $10\,\text{Gy/fraction}$ of neutrons had the highest relative biological effectiveness (RBE) in the tumor relative to the value in skin and other normal tissues. Similar results were more recently demonstrated by Koike *et al.* (2002) Ando *et al.* (2005) for the spread out Bragg peak of carbon ions (cf. Chapter 8 Figs. 8.8 (h)–(i)). The whole phenomenon is most likely linked to the large shoulder of the cell survival curve for photons especially for late responding normal tissues

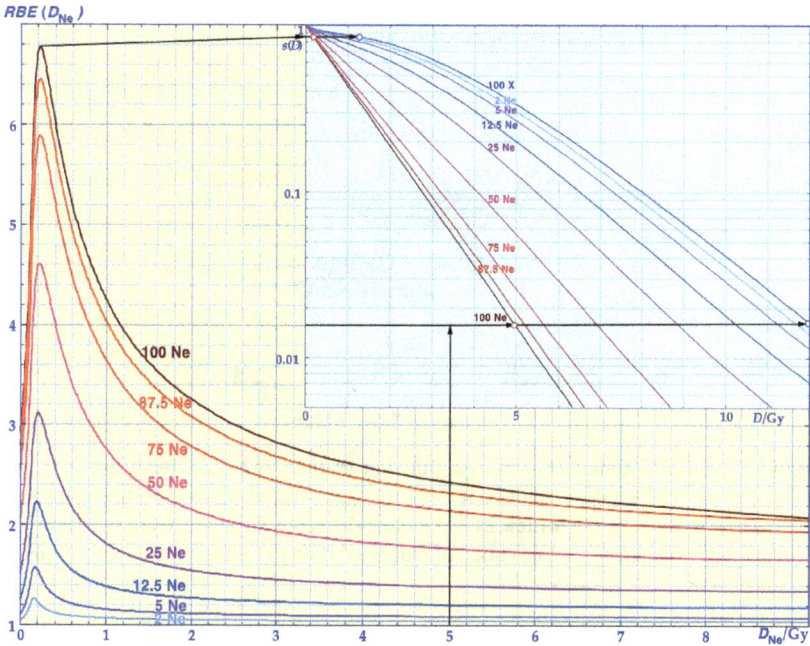

Figure 7.7. The cell survival curves for different combinations of X-rays and neon ions (pale blue insert) are shown. It is seen that low doses per fraction result in a very high RBE (~6), whereas high doses beyond 5 Gy result in an RBE of ≤2.5. A high dose per fraction will therefore be better tolerated by normal tissues, whereas the tumor is less protected since its cell survival has a less curvy shape (based on experimental data from Ngo *et al.* (1981) cf Fig. 8.8e).

with slow cell turnover and many G_0 and G_1 cells. These cells have long repair times before they go into circulation and high survival with photons but are largely killed by light ions of a high linear energy transfer (LET). Therefore, the effective RBE in normal tissues is very much lower at high doses and low survival levels than at low doses as shown in Fig. 7.7. However, the tumor cells which generally are dividing more rapidly and thus are linked to a smaller shoulder already in their photon survival curves so the RBE of the tumor will generally be high at both low and high doses giving a strong clinical advantage of delivering high doses per fraction to the tumor.

It is, therefore, interesting that many of the new clinical advantages of the light ions are obtained with as few as 1–4 high dose treatment fractions during 1 week, such as with lung and hepatocellular cancers, where for the non–small-cell lung tumors today, it get close to 98% local control 5 years after the treatment as seen in Fig. 7.8. The treatment schedule was first started as 18 fractions during 6 weeks but

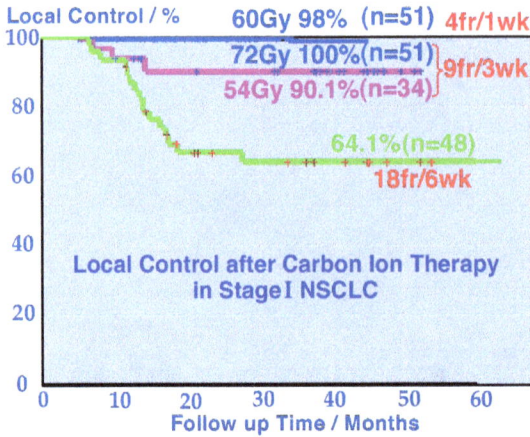

Figure 7.8. Local tumor control as a function of time and dose escalation in stage 1 non-small-cell lung cancer is shown. It is clearly seen that these severely hypoxic tumors are effectively controlled by as few as 4 fractions delivered during 1 week — a significant improvement compared to conventional radiation therapy. Doses are given in photon equivalent Gray.

for reasons of treatment efficiency it was shortened to 9 fractions in 3 weeks with an improvement in local control from about 65% to 90%. This significant improvement triggered a further reduction of the treatment to 4 fractions in 1 week with 98% local control. Despite this successful result today, a single fraction of 46 Gy is being successfully tested. Similar but not quite as pronounced improvements are seen with stereotactic high photon dose irradiations probably due to the larger problem with these often very hypoxic tumors using low LET radiations.

7.4. BioArt Approach

PET-CT is probably the ultimate tool for accurate tumor imaging and 3D *in vivo* predictive assay of radiation sensitivity. By imaging the tumor twice during the early course of therapy, it should be possible to quantify both the tumor cell density and the responsiveness to therapy through the rate of loss of functional tumor cells. This new information is ideal for use together with biologically based therapy optimization and makes it possible accurately to quantitate the dose–response relation, at least for the bulk of the tumor cells. Since the tumor responsiveness is available after about one-and-a-half weeks of therapy, the information is also ideal for use with adaptive therapy where all forms of deviations from the original treatment plan can be accurately monitored and corrected, for since they generally influence the still functional, but mainly doomed tumor cell compartment. Thus, practically all therapy uncertainties such as: (1) the geometric misalignment of the therapeutic beam with the tumor, (2) deviations of the delivered dose distribution from the planned delivery whether due to (3) an erroneous treatment planning algorithm or (4) treatment equipment uncertainties and even (5) deviations in the anticipated responsiveness of the tumor of the patient based on historical response data, can be taken into account. Fortunately, when a larger tumor cell compartment than expected is seen, an increased dose during the remainder of the treatment should always be delivered independent on whichever combination of the above deviations was the true reason. With high-energy photon and ion therapy, it is also possible to image the integral dose delivery *in vivo* during or immediately after a treatment using PET-CT imaging as seen in Fig. 7.1 lower left insert. The high-energy photons >20 MeV produce positron emitters through photonuclear reactions in tissue which are proportional to the photon fluence and thus approximately also to the absorbed dose. Light ion beams, the ultimate radiation modality with regard to physical and biological selectivity, instead produce PET emitters through direct nuclear interactions in tissue, but can also be used as radioactive beams consisting of intrinsic PET emitters such as ^8B, ^{11}C, ^{13}N, and ^{15}O. These radioactive beams allow more accurate imaging of the Bragg peak distribution and thus indirectly the absorbed dose. The most universal feedback for adaptive radiation therapy would then be to use the measured image of mean dose delivery during the early part of the treatment while revising the treatment plan based on the initially planned dose distribution and the observed radiation responsiveness of the tumor as seen after the first 10 days of therapy (cf. Fig. 7.16). By this so-called BıoArt approach, radiation therapy optimization may become an almost exact science, where the patient's true individual radiation response, considering hypoxia and general radiation resistance, as well as possible dose delivery and planning errors, is taken into account.

For the analyses below, we will for simplicity assume that good markers of tumor clonogen density will be available and that we can correct for the nonuniform tumor vasculature so that we can account for the reduced uptake of tracers in regions of low vascular density. Through FDG, we already have a useful surrogate tumor marker, and the vasculature could be accessible by known tracers such as ammonia ($^{13}NH_3$) or water ($H_2{}^{15}O$). We can thus image the tumor spread just before treatment to accurately delineate the clinical target volume based on visualized tumor uptake along well-known pathways of microscopic lymphatic invasion (Grigsby *et al.* 2001; Kiricuta *et al.* 2001 and Fig. 7.9). During radiation therapy about 50% of the tumor clonogens are lethally damaged by each treatment fraction. Therefore, after the first week of therapy, there would be only about 2^{-5} or about 3% of clonogens left. After the first weekend break in the treatment, there is a unique possibility to image the still functional tumor clonogen distribution. This opens up the possibility to

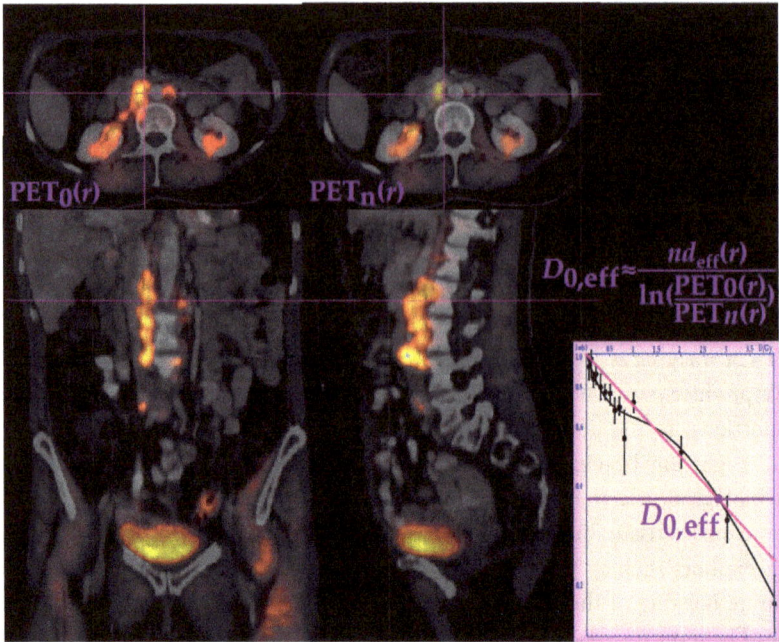

Figure 7.9. PET-CT imaging of the paraaortic lymph nodes of a cervix cancer patient. Accurate tumor imaging on the background of normal tissue anatomy is essential for successful therapy. By monitoring tumor responsiveness during the first weeks of therapy, BioArt therapy is possible. Accurate tumor imaging is a necessity for precision light ion therapy.

quantitate how many of the tumor cells locally remain functional after the first few treatment fractions. PET-CT would then give us the possibility to make a real 3D *in vivo* predictive assay of tumor responsiveness and predict what modifications to the already delivered dose distribution are needed for a truly curative outcome. At each point in time, it is essential to distinguish the finally surviving tumor cells from the temporarily functional but doomed tumor cells that are imaged by PET, even though many of them are lethally damaged and will disappear in a week or two, as will be discussed in more detail below.

Furthermore, using high-energy photons (>20 MV) or light ions, it is possible to use the PET-CT device to image in 3D the mean dose delivery in the patient superimposed on normal tissue anatomy. With high-energy photons, this is achieved by photonuclear reactions mainly in normal oxygen and carbon atoms producing ^{15}O and ^{11}C — both PET emitters. The resultant image is largely proportional to the photon fluence and thus to the absorbed dose as seen in the lower left panel

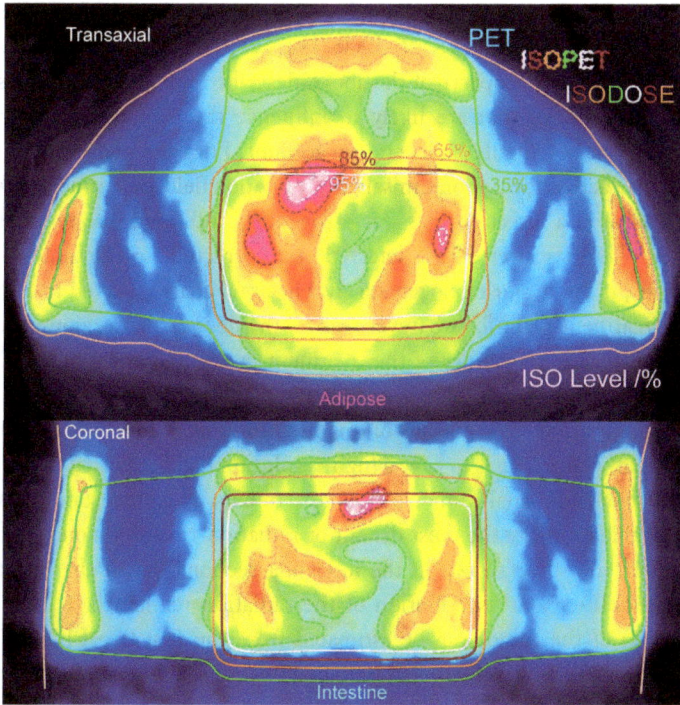

Figure 7.10. Comparison of the PET and isodose distributions for a prostate patient receiving a dose of around 5 Gy/fraction corresponding to the yellow color on the PET color wash. The lower left panel is the dose plan on a pelvic CT scan with the corresponding PET-CT-based treatment plan and PET activity is seen here and also in the upper row inferior and superior of the more central lower slice.

of Figs. 7.1 and 7.10 (Janek 2002, 2013). During ion therapy, the direct nuclear reactions in tissue will similarly generate PET emitters but it is also possible to use a primary radioactive beam of positron-emitting light ions such as ^{11}C (cf. Fig. 8.38(d)). In the latter case, the image information will be more specifically proportional to the ion beam stopping distribution and it will thus largely image the Bragg peak density, which is a key factor of interest in treatment planning and can be accurately used in calculating the dose delivery (cf. Brahme *et al.* 1988; Brahme *et al.* 1991; Brahme *et al.* 1995). Obviously, 3D PET-CT dose delivery monitoring has its greatest value when using intensity-modulated beams that generally avoids organs at risk and produces a largely uniform dose in the tumor, as shown in the biologically optimized plan in the lower right panel of Fig. 7.1 (Löf *et al.* 1998). This advantage will be an important factor in the future development of the clinical advantages of light ion and IMRT photon and electron therapy, particularly when narrow scanned beams are used for the dose delivery (Brahme *et al.* 1988; Brahme *et al.* 1991; Brahme *et al.* 1995; Kraft 2000; Tsujii *et al.* 2002; Brahme 2002).

A very interesting property of the above-described imaging techniques is that all major uncertainties that are likely to influence the tumor response and the dose delivery can potentially be taken into account. For example, would a varying radioresistance across the tumor, a misaligned radiation field, or a tumor growth that is less extensive in a direction that was erroneously indicated due to annihilation in flight or photon scatter all be detected and corrected for (Brahme 2002 and 2003a). Thus, if the tumor clonogen survival is larger than expected in a region of the initial treatment plan, for whatever reason, the dose to that region needs to be increased during the remainder of the treatment period. Unfortunately, when the reverse is true, that is, when there is a too low signal in a region, it is more difficult to modify the treatment for reasons of quantitation uncertainty; hence, one of the most important properties is to have a high-resolution high-sensitivity PET system capable of visualizing small tumor cell compartments. It is uncertain whether we will ever be able to image the most distant or radiation-resistant clonogenic tumor cell. However, it will be shown in the following sections that if the sensitivity is sufficient, the 3D spatial distribution of the radiation resistance (Kiricuta 2001; Nilsson 2002) of the tumor can be estimated from the first two PET-CT tumor images and this can be used to accurately quantify the dose–response relation for tumor cure. Given this patient individual dataset and the increasing knowledge about the dose responsiveness of surrounding normal tissues, a true, radiobiologically based treatment optimization should be possible using this BIOART approach.

7.5. Radiation Biology of Functional Tumor Cells

7.5.1. *Cell Survival*

The microenvironment of an arbitrary tumor can be quite complex, with microscopic variations of tumor clonogenicity, nutrient, and oxygenation pattern as well as cellular density, DNA content, and genetic makeup. The surviving clonogen distribution after irradiation by a dose distribution function $D(r)$ can generally be written as

$$n(r) = \int n_{D_e}(r)S(D(r))\mathrm{d}D_e, \tag{7.1}$$

where $n_{D_e}(r)$ is the spatially dependent clonogen density differential in D_e and D_e is the effective radiation resistance. In the simplest form of response modeling, the cell survival for fractionated irradiation is $D_e = D_{0,\text{eff}}$ that is the effective D_0 value (cf. Nilsson 2002; Brahme 1984) causing a cell survival reduction of $1/e$ or

$\approx 37\%$. The clonogen density distribution function $n_{D_e}(r)$ differential in D_e gives the local density of tumor clonogens of different effective radiation resistance D_e. For a uniform dose or constant dose per fraction D and effective radiation resistance D_e, a purely exponential cell survival is obtained

$$S(D) = e^{-D/D_e}. \tag{7.2}$$

A more accurate cell survival expression also at low and high doses is given by:

$$S(D) = e^{-\bar{a}D} + \bar{b}De^{-\bar{c}D}, \tag{7.3}$$

as derived by Lind *et al.* (2003), where

$$\bar{a} = \frac{\sum a_i D_i}{\sum D_i}; \quad \bar{b} = \frac{\sum b_i D_i}{\sum D_i} \quad \text{and} \quad \bar{c} = \frac{\sum c_i D_i}{\sum D_i} \quad \text{and} \quad D = \sum D_i,$$

and D_i is the dose fraction of radiation quality i with cell survival parameters a_i, b_i, and c_i as seen in Fig. 7.11.

From this expression and Eq. (7.2), D_e can be calculated at an arbitrary dose level or dose per fraction D according to

$$D_e = -\frac{D}{\ln S(D)} = 1 \Big/ \left(\bar{c} - \frac{\ln(\bar{b}D + e^{(\bar{c}-\bar{a})D})}{D} \right). \tag{7.4}$$

Thus, as derived by Lind *et al.* (2003) $D_e = 1/(\bar{a}-\bar{b})$ at low doses, whereas generally $D_e = 1/\bar{c}$ at high doses and there is a continuous transition between these values at intermediate doses. The mean density distribution of surviving clonogens may thus be rewritten as

$$n(r) = \int n_{D_e}(r)e^{-D(r)/D_e} dD_e. \tag{7.5}$$

However, at least for milder types of lethal damage, it may take several cell divisions before a damaged cell becomes necrotic or in general nonfunctional. The functional fraction at time t after a first irradiation at time t_1 to dose D may, therefore, be approximated as

$$f_1(t, D) = S(D) + (1 - S(D))e^{-(t-t_1)/\tau}, \tag{7.6}$$

where τ is the time constant of loss of functionality for lethally damaged cells after an irradiation assuming the fast and slow repair times of sublethally damaged cells is much shorter than τ. The time τ is of the order of a few doubling times of the cells. Thus, a period of about 2 weeks after the irradiation $f_1(t, D) \approx S(D)$,

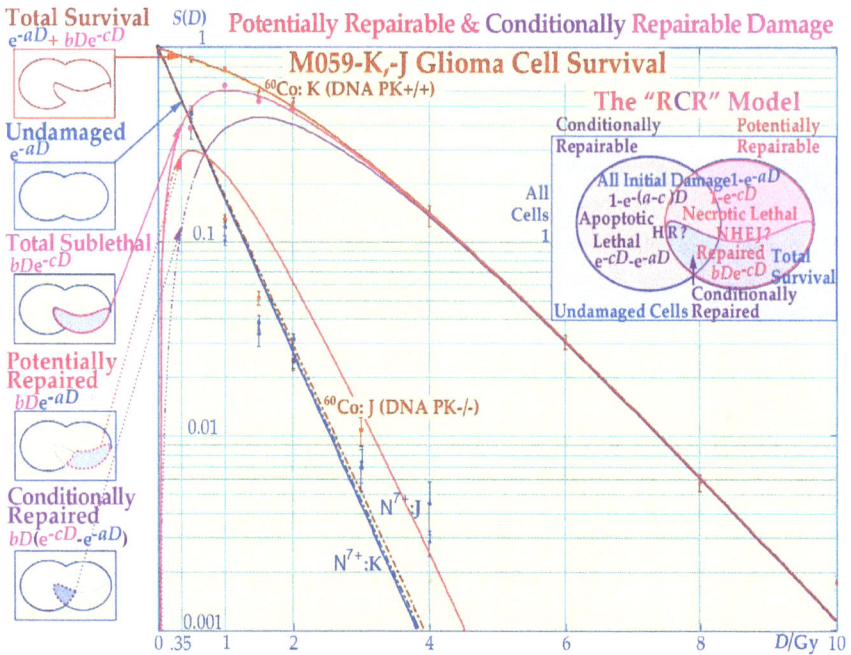

Figure 7.11. The new model for cell survival separates potentially repairable from conditionally repairable damage, which needs the former to be effective. The left margin describes the different subcompartments of cells in a probability diagram form. With arrows to the associated form of the cell survival curve for two different cell lines, one mutated and one normal with regard to the nonhomologous end joining gene DNA-PK. Both cell lines were irradiated by ^{60}Co and nitrogen ions. It is seen that both radiation modalities produce straight repair-free cell survival curves for the mutant cell line (M059J, $(DNAPK^{-/-})$), whereas a substantial repair takes place for ^{60}Co in the clonogenic survival of the repair-efficient human glioma cell lines M059K $(DNAPK^{+/+})$.

since all doomed or lethally committed cells have been lost and complete repair and misrepair of sublethally damaged cells have taken place. Thus, when the second dose fraction is delivered, some of the lethally damaged cells from the first fraction are still functional, but will eventually be lost a few time constants t later as shown in Fig. 7.12.

However, some of the undamaged cells or fully repaired cells from the first treatment $(S(D))$ may be lethally hit by the second irradiation as $t = t_2$, so the functional fraction will now be

$$f_2(t, D) = S^2(D) + S(D)(1 - S(D))e^{-(t-t_2)/\tau} + (1 - S(D))e^{-(t-t_1)/\tau} \qquad (7.7)$$

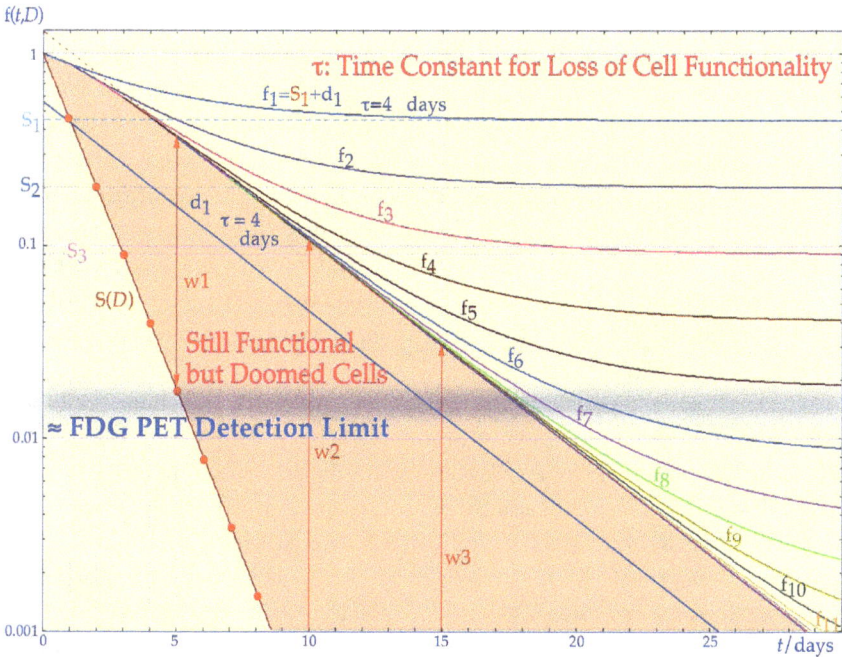

Figure 7.12. The decrease in the finally surviving and the doomed but still functional cell compartments (shaded) during daily fractionation with 2 Gy is illustrated. Also shown is the decay of the lethally hit compartment assuming a time constant of loss of $\tau = 4$ days. The decrease in the functional compartment if the irradiation is interrupted day $n = 1 = 11$. It is seen that the doomed but still functional compartment is more than one order of magnitude larger than the finally surviving fraction per week of therapy (w1, w2, etc.). All cells above $S(D)$ are doomed cells that are eventually going to die. It is seen that the fully functional and doomed cells are below the PET detection limit after 1 and 2 weeks, respectively.

Obviously, some of the cells that are still functional but doomed by a lethal hit by the first dose fraction may also be hit by the second fraction and thus, in reality, are probably somewhat more rapidly lost from the functional compartment. This process is disregarded here, so only the undamaged and fully repaired compartment is suffering anew loss of functionality, which is shifted in time from the first treatment. By generalization, the above formula can be used to calculate the ratio of the functional to the truly surviving cell compartments after n fractions of a dose distribution $D = D(r)$:

$$\frac{f_n(t, D)}{S^n(D)} = 1 + \frac{(1 - S(D))}{S^n(D)} e^{-t/\tau} \sum_{i=1}^{n} S(D)^{i-1} e^{t_{i-1}/\tau}. \qquad (7.8)$$

With daily fractionation, the sum becomes a geometric series since $t_{i-1} = i - 1$ and Eq. (7.9) can be reduced to the following closed expression:

$$\frac{f_n(t, D)}{S^n(D)} = 1 + \frac{(1 - S(D))(S^{-n}(D) - e^{n/\tau})}{1 - S(D)e^{-1/\tau}} e^{-t/\tau}. \tag{7.9}$$

As shown in Fig. 7.12, the functional but doomed tumor cell compartment rapidly becomes many times larger than the finally surviving cells and decreases by a rate indicated by τ quite early after the start of the treatment. This is because the first cell survival term in Eq. (7.9) is considerably smaller than the term assigned to dying but still functional cells, since they decrease more slowly than the long-term cell survival given by $S^n(D)$.

After 1 week of therapy, the dying but still partly functional cells are predominant by about one order of magnitude. This is clearly seen in Fig. 7.12, where the loss rate of functional cells quite rapidly dominates over the ordinary long-term cell survival. This could also be understood by comparing the time constants τ of the order of a several days (typically 2 weeks are needed to determine the true long-term clonogen survival), whereas a survival of about 50% per dose fraction of 2 Gy corresponds to a τ value of about 1.3 days during daily fractionation. Not even the hypoxic tumor fraction will significantly influence the situation unless it is very large and also totally dominates the early cell kill (cf. Nilsson et $al.$ 2002). Since it is possible to image the hypoxic tumor cell compartment separately, for example, by fluoromisonidazole, the decay products of which are trapped in hypoxic tumor regions, we will generalize Eq. (7.9) to account also for hypoxic tumors. First, real hypoxic regions may require an almost threefold dose for eradication, so their survival function $S_h(D)$ will be quite different from that of tumor cells with more normal oxygenation $S(D)$. Furthermore, the total density of tumor cells $n(r)$ will contain a hypoxic fraction $n_h(r)$ with different cell survival so that the number of functional cells after n fractions will be

$$n_{f_n}(t, D) = (n(r) - n_h(r))f_n(t, D) + n_h(r)f_n^h(t, D), \tag{7.10}$$

which after consideration of the difference in cell survival becomes:

$$n_{f_n}(t, D) = (n(r) - n_h(r)) \left(S^n(D) + \frac{(1 - S(D))(1 - S^n(D)e^{n/\tau})e^{-t/\tau}}{1 - S(D)e^{-t/\tau}} \right)$$

$$+ n_h(r) \left(S_h(D)^n + \frac{(1 - S_h(D))(1 - S_h^n(D)e^{n/\tau})e^{-t/\tau}}{1 - S_h(D)e^{-t/\tau}} \right). \tag{7.11}$$

For simplicity, we assume that the rate of loss of functionality of a hypoxic tumor cell is similar to that of a well-oxygenated cell (which may not be the case at least in the early part of the treatment). On this assumption, it is interesting to observe that the time constant of loss of the surviving hypoxic cells and the doomed cells that are still functional but slowly dying because of lethal damage or misrepair is almost comparable with a loss rate of about 20–25% per day.

Therefore, a similar time constant of several days applies to the last three terms in Eq. (7.11). For a fractionation schedule of 2 Gy per day, a large long-lasting compartment of doomed tumor cells are also formed for totally hypoxic tumors with high τ values. This is largely due to their higher radiation resistance, so there are more undamaged cells surviving for longer, continuously feeding a quasi-functional but dying population of approximately the same loss rate as the undamaged cells. Since fewer cells are killed per fraction and they are lost at about the same rate ($\tau \approx 3$), there is a smaller difference between the quasi-functional cells and the true surviving cells in fully hypoxic tumors. However, in most tumors, the hypoxic compartment is fairly small, perhaps $\leq 20\%$, so in general it will only have a small influence on the early imaging and on the quasi-functional cells — especially when their reduced metabolic activity is taken into consideration (as seen in Fig. 7.13; cf. Nilsson *et al.* 2002).

In Fig. 7.13, it is also clearly seen that the hypoxic compartment becomes dominant after a few weeks of therapy unless serious reoxygenation take place. With the same loss rate of functionality, a more continuous shift from the well oxygenated to the hypoxic tumor is seen. However, very small hypoxic percentages have a stronger influence than small oxic compartments, as seen for a constant $\tau = 3$ days in Fig. 7.13.

7.5.2. *PET-CT Imaging*

The glucose analog FDG is assumed to be actively taken up by functionally cycling tumor cells to support their metabolic activity. Unfortunately, the tumor vasculature is not well developed and better vascularized tumor volumes are likely to have a higher uptake than poorly vascularized tumor regions. Furthermore, there are diffusion processes in the tissues and the tumor that may supply FDG also to regions with poor vasculature. In addition, the tissue vasculature is influenced by the irradiation. Fortunately, this is a rather slow process as is known, for example, from radiation therapy of arteriovenous malformations and normal tissues with a rich vasculature. The PET image will, therefore, be due to multiple processes in the patient that may have to be taken into account in a more detailed analysis of the images.

In functional imaging, the recorded information at quasi-equilibrium is generally a product of two processes: (1) the exposure of response elements to the imaging substance and (2) the density of response elements in the tissue. The

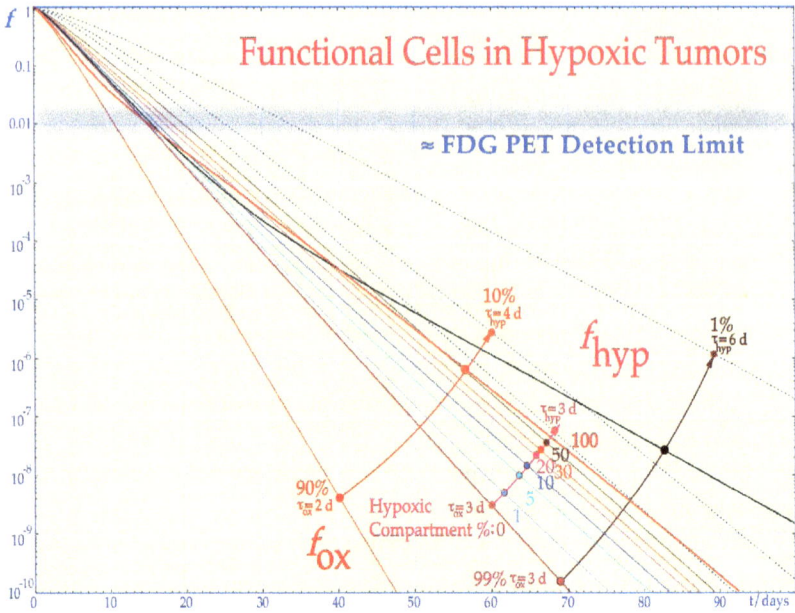

Figure 7.13. Change in the functional cell compartment of partly hypoxic tumors with $\tau = 3$ days and hypoxic fractions of 0, 1, 5, 10, 20, 30, 50 and 100%, respectively, are shown. The change in curve shape with different time constants for oxic and hypoxic cells $\tau = 2$ and 4 days with 10% hypoxia and $\tau = 3$ and 3 days with 1% hypoxic cells is also shown (hockey stick curve).

density, $\rho(r)$, of the imaging substance is generally determined by the vascular density and the diffusion of the active component in the tissues surrounding the vessels. The vascular density is best described by the fluence density, $\Phi(r)$, of vessels in the tissue (Nilsson *et al.* 2002). This is defined as the path length, dl, of vessels per unit volume, dV, in the tissue according to:

$$\Phi(r) = \left(\frac{dl}{dV}\right)_r \equiv \frac{dN}{da}, \qquad (7.12)$$

where the last equality indicates that it is identical to the number of vessels, dN, crossing a sphere of cross-sectional area, da. Small molecules such as glucose or FDG diffuse rather freely in tissue and result in a uniform 3D Gaussian density distribution except where active transport or consumption mechanisms such as lymphatic flow may change it. The density of the imaging substance, $\rho(r)$, may therefore be approximated by a convolution of the vascular fluence density and a quasi-Gaussian substance diffusion kernel $\delta(r)$ according to:

$$\rho(r) = \Phi(r) \otimes \delta(r). \qquad (7.13)$$

It is now possible to approximate the density of trapped positron emitters owing to the metabolic activity in the functional tumor cells. If the PET image accurately quantitates the density of positron emitters in the tissue such that scatter and attenuation of the annihilation photons are accurately taken into account, the imaged emission density $PET_0(r)$ before therapy is given by:

$$PET_0(r) \propto \rho(r) \cdot f_0 = \rho(r) \cdot f_0(0, 0, r), \tag{7.14}$$

where f_0 is the density of functional tumor cells before the first treatment. Similarly, the PET image after n treatments is also well approximated by:

$$PET_n(r) \propto \rho(r) \cdot f_n(t, D, r), \tag{7.15}$$

since at least over the first few weeks $\rho(r)$ is not appreciably changed by the radiation response. If the delivered dose to normal tissues were to be as high as the dose to the tumor, or even higher, it would be possible to obtain an additional uptake due to radiation-induced metabolic activity in such high-dose normal tissue regions. With well-designed intensity-modulated dose delivery and using a suitable conformal radiation modality, this effect should generally not be significant, at least not during the first 2 weeks of therapy (Erdi et al. 2000; Brahme 2002). Furthermore, it has recently been shown that some critical normal tissues such as skin have a suppressive response to radiation for as many as 3 weeks of therapy (Tureson et al. 2001), so in time there should be a diagnostic window of this order of magnitude where the therapeutic response of the tumor should be available for predictive assay. In fact, according to Fig. 7.12, the density of functional tumor cells may be of the order of 30–1% between the first and third week of therapy and therefore accessible to diagnostic evaluation. Beyond 3 weeks, the density of functional tumor cells rapidly becomes too low and the accelerated proliferation in normal tissues and tumor and the microphage invasion is too high for accurate imaging functional tumor cells with FDG. Interestingly enough, two early imaging sessions may be sufficient to quantify the therapeutic responsiveness of the surviving tumor cells, since by assuming the proportionality constants of Eqs. (7.15) and (7.16) to be identical, we obtain:

$$\frac{f_n(t, D, r)}{f(0, 0, r)_0} = \frac{PET_n(r)}{PET_0(r)} = P_n(r) \tag{7.16}$$

It should thus be possible to use PET-CT images of the tumor during the first 10 days of therapy to image the spatial variation of the tumor responsiveness over the most critical part of the target volume. This is where the density of tumor cells is the highest and where the hypoxic compartment is most significant. Since $f_0(0, 0, r) \equiv n_0(r)$ that is the initial density of tumor cells and $f_n(t, D, r) = n_0(r)f_n(t, D(r))$,

Eq. (7.16) may be rewritten using Eq. (7.9) as

$$P_n(r) = S^n(D(r)) + \frac{(1 - S(D(r)))(1 - S^n(D(r))e^{n/\tau})}{1 - S(D(r))e^{-1/\tau}} e^{-t/\tau} \qquad (7.17)$$

This is an equation of $n + 1$ order in S, provided τ is known and at $t = n$, that is, at the time of PET imaging it reduces to:

$$(1 - e^{-1/\tau})S^{n+1}(D(r)) + P_n(r)S(D(r))e^{-n/\tau} + e^{-n/\tau} - P_n(r) = 0. \qquad (7.18)$$

We thus need to find the roots S_k of a polynomial $Q_{n+1}(S)$ of the type:

$$Q_{n+1}(S) \equiv S^{n+1} + aS + b = 0. \qquad (7.19)$$

Equations (7.18) and (7.19) cannot be found analytically for $n > 3$, but can be solved numerically using Newton's method, such as through the recursion relation:

$$S_{i+1}(r) = \frac{(1 - e^{-1/\tau})S_i^{n+1} + P_n(r)e^{-n/\tau}S_i + e^{-n/\tau} - P_n(r)}{(n + 1)(1 - e^{-1/\tau})S_i^n + P_n(r)e^{-n/\tau}}. \qquad (7.20)$$

Obviously, a number of alternative root-finding algorithms could also be used here, such as the Jenkins–Traub algorithm. Ideally, one could start with the expected survival distribution based on the delivered dose distribution and expected biological response parameters, as discussed in Sec. 7.1. When doing so, the change in S as a function of the iterations will be a measure of how closely the biological modeling agrees with the observed clinical response and the difference between the converged values will be a measure of the total error. Quite rapid convergence is also obtained by starting with a constant value of $S \approx 0.5$ over the whole volume where reliable PET data are available.

7.5.3. *Simplified Analytical Solution*

To illustrate some of the principal phenomena and major merits of the present approach, we will treat the case with fast loss of functionality ($\tau \ll 1$). For this case, Eq. (7.17) reduces to

$$S^n(D(r)) = P_n(r). \qquad (7.21)$$

By combination with Eq. (7.2), this allows a direct determination of the spatial variation of effective radiation resistance D_e according to:

$$D_e(r, n) = \frac{\overline{nD(r)}}{\ln(P_n(r))}, \qquad (7.22)$$

and thus

$$S(D(r)) = e^{-\frac{D(r)}{\overline{nD(r)}} \ln\left(\frac{\text{PET}_n(r)}{\text{PET}_0(r)}\right)}, \qquad (7.23)$$

where $\overline{nD(r)}$ is really the mean dose distribution delivered during the n first treatments taking into consideration uncertainties such as in patient setup, organ motions, and dose delivery. Ideally, it should be monitored by a direct *in vivo* dose delivery measurement using PET-CT, as discussed in more detail in Sec. 7.4.2, but conventional portal imaging with pencil beam energy deposition kernel corrections may also be a useful but more labor-intensive approach.

7.6. Biologically Optimized 3D *In Vivo* Predictive Assay-based Adaptive Radiation Therapy

7.6.1. *Tumor Imaging*

From the above analysis, it is clear that PET-CT imaging (as shown in Figs. 7.9 and 7.10) combined with appropriate analytical and radiobiological models is a method potentially capable of giving at least four different kinds of information about the tumor:

(1) Geometric information about the location of the tumor on the background of normal tissue anatomy.
(2) Information about the initial density of tumor cells.
(3) Spatial radioresistance distribution of the tumor cells (cf. Fig. 7.15).
(4) Spatial variation of the rate of loss of tumor cell functionality.

Of this information, (1) to (3) are of key importance for radiation therapy optimization, whereas the rate of loss of functionality (4) is needed to more accurately analyze the tumor responsiveness to obtain (3) and (4). In addition to the above information, gained by direct tumor imaging by FDG, or preferably by more specific tumor tracers, a number of more specific methods are available for tumor characterization:

(5) The vasculature and diffusion of small molecules in tumor and normal tissue can be studied, for example, using $H_2^{15}O$, $^{11}CH_4$ and NH_3 (cf. Fig. 7.14).
(6) The degree and extent of hypoxic regions in the tumor can be imaged using fluoromisonidazole. This information is invaluable for more accurate treatment optimization for choice of dose per fraction and radiation modality using higher effective doses of electrons or photons or even LET beams such as neutrons or the heavier of the light ions in the case of severe hypoxia (cf. Fig. 7.13).

(7) A further possibility with PET-CT imaging in radiation therapy is to visualize the integral dose delivery and tumor vasculature in the patient during or after the treatment (cf. Figs. 7.1 and 7.10).

Often (6) and (7) are intertwined due to the high interstitial pressure inside the gross tumor (Sjöblom *et al.* 2004) as seen in Figs. 7.14 and 7.15 below where both the blood flow and the metabolism are significantly reduced in the tumor core. The high pressure reduces the blood flow so the diagnostic compounds do not even reach the tumor in sufficient amounts (cf. Fig. 8.23 below). Here it may be useful to first give a platelet-derived growth factor (PDGF) antagonist to open up the vasculature before the treatment is started (Sjoblom *et al.* 2004). Otherwise, it may take as long as 1 week before the vascular circulation in the tumor core is started to facilitate reoxygenation (cf. Fig. 7.15).

Finally, the PET-CT camera is the ideal device for advanced virtual CT simulation, allowing unprecedented accuracy in the simulation of the planned dose delivery and producing ideal fused images of the expected portal verification views with the projected tumor cell density from the PET images superimposed on

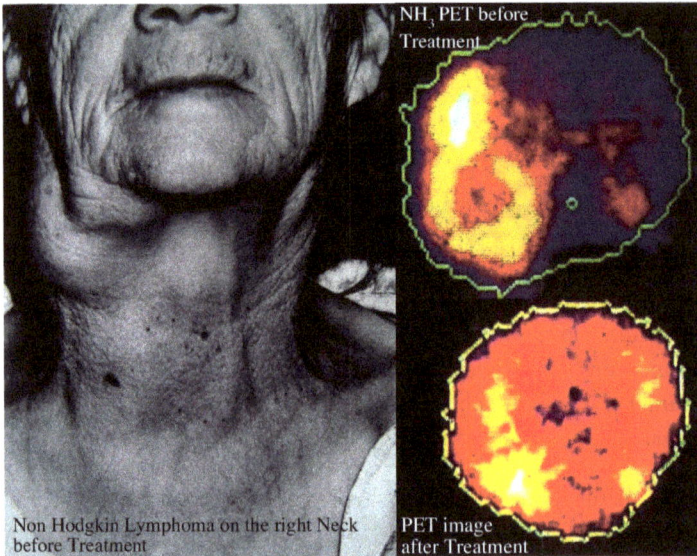

Figure 7.14(a). A woman with a large non-Hodgkin lymphoma on the right neck. The upper ammonia PET image of the tumors before treatment shows a more active tumor just behind the primary tumor. The PET image after treatment indicates that the smaller posterior node is probably cured whereas the strongly hypoxic node will probably reoccur since the tumor cells were protected by the high internal pressure and therefore some hypoxic tumor cells from the large node seem to survive the treatment (de Schryver *et al.* 1987).

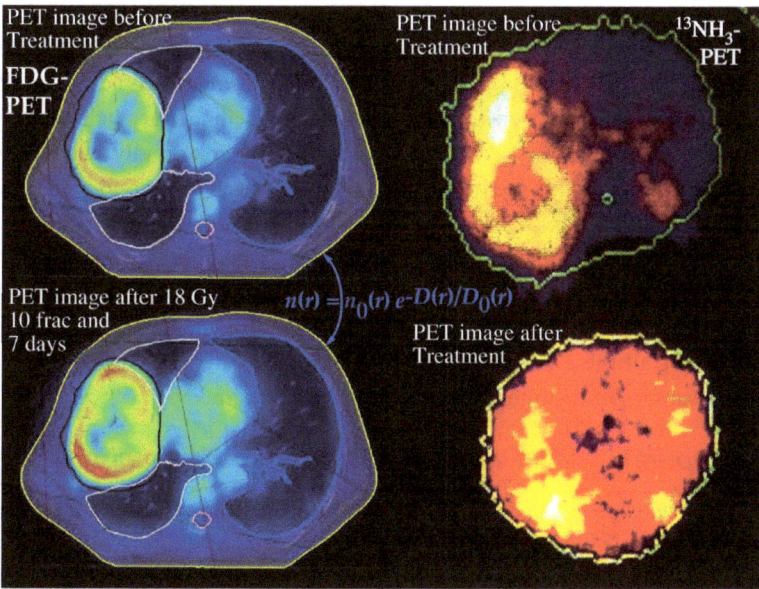

Figure 7.14(b). A lung tumor with similar low vascularization in the core of the tumor as the head and neck tumor in Fig. 7.14(a). It is seen that FDG uptake is increased after a target dose of 18 Gy due to a decreased tumor cell burden (cf. Chapter 8 Fig. 8.23(a))!

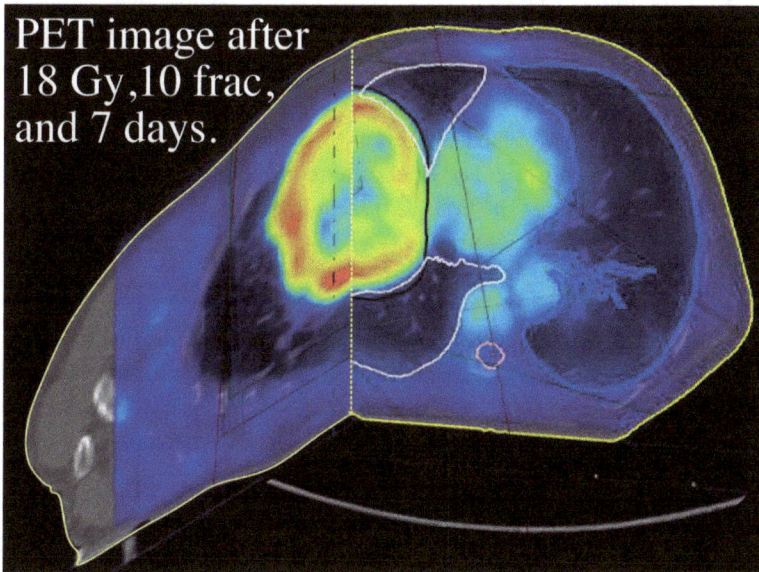

Figure 7.15. A big lung cancer with very similar properties as the larger neck node in Figs. 14(a) and 14(b).

the CT background. This information is of prime importance when comparing the alignment of intensity-modulated dose delivery with a highly heterogeneous tumor and/or a tumor of complex shape with extensive microscopic spread. PET will probably remain the most sensitive detection method for small tumor cell masses even though MRI and more recently MRSI are rapidly allowing new improved tumor-imaging methods. In the future, we will therefore need very effective interfaces among the three main types of radiation therapy activities and different equipment, namely those for diagnostic imaging, therapy planning, and dose delivery, in order to ensure optimal communication and interaction between them. A dedicated 3D image handling and display workstation will therefore be needed where all these activities can be monitored, controlled, and reviewed for optimal assurance of treatment quality during radiation therapy.

7.6.2. *Optimal Dose Distribution for Tumor Eradication*

It is possible to express the optimal dose distribution required to eradicate a tumor to any desired probability P_B with uniform recurrence probability once the density and radioresistance of the clonogens are known (Brahme and Ågren 1987). This makes use of the first three data sets of the above since:

$$\hat{D}(r) = -D_e(r) \ln\left(\frac{n_0(r)V_t}{-\ln(P_B)}\right), \tag{7.24}$$

where V_t is the tumor volume as recorded by $PET_0(r)$ subtracting the metabolic rate of the normal tissue's background. Based on the initial PET-CT image, we have information about the initial tumor cell density provided we know the density of the imaging substance $\rho(r)$. This distribution could be imaged separately if we could image the fluence density of blood vessels in the body as given by Eq. (7.14). This is possible, for example, using, $H_2^{15}O$, so after convolution with a quasi-spherical diffusion kernel, we can rewrite Eq. (7.24) according to:

$$\hat{D}(r) = D_e(r) \ln\left\{\frac{-PET_0(r)V}{\rho(r)\ln(P_B)}\right\}, \tag{7.25}$$

where V now is an unknown correction factor from the proportionality in Eq. (7.14). By inserting Eq. (7.22) for $D_e(r)$, we can express the optimal dose distribution based on three PET-CT images PET_0 and PET_n from above and PET_v (r) for the vascular image convolved with the diffusion kernel $\rho(r)$ (cf. Kraft 2000) according to:

$$\hat{D}(r) = \frac{\overline{nD(r)}}{\ln\left(\frac{PET_n(r)}{PET_0(r)}\right)} \ln\left\{\frac{-PET_0(r)V}{PET_v(r)\ln(P_B)}\right\}. \tag{7.26}$$

This interesting equation gives us the dose distribution that should ideally give a uniform recurrence probability inside the tumor of $1 - P_B$ of say 1%, 5% and thus a tumor cure of 99–95%. However, it does not consider the sensitivity of normal tissue surrounding the tumor — so it is well suited for treatment optimization using physical objective functions which should aim at producing $\hat{D}(r)$ as closely as possible with lowest possible dose outside the tumor (cf. Brahme 1995a). Obviously, this optimization problem is best solved using light ions between helium and carbon since they deliver the least effect in the normal tissues for a given effect in the tumor volume (Brahme *et al.* 2001b; Brahme *et al.* 2001a). However, rather than looking at radiation therapy as a biologically inspired, inverse problem where a desirable physical dose distribution is produced, we know today that the best treatment results are obtained by considering it as the true biological optimization problem (Brahme 1999b; Brahme 2000; Brahme 2001a).

7.6.3. *Dose–Response Relation of the Tumor*

What we really want to know is how to find the dose distribution that maximizes the probability to cure the patient without unacceptable or severe normal tissue side effects (Brahme 2000; Brahme 2001a). To this end, we need to quantitate the probability of eradicating the tumor for a given dose delivery while at the same time we need to know the risk for severe normal tissue damage as discussed in detail elsewhere (Ågren *et al.* 1990; Ågren *et al.* 1995; Brahme 1999b and c). Fortunately, the PET-CT data discussed here can be accurately used to quantify the tumor cure probability. According to binomial and Poisson statistics, the probability of eradicating all clonogenic tumor cells is closely equal to $e^{-\overline{N}}$, where $\overline{N}$ is the mean number of surviving clonogens, assuming $\overline{N}_0$ is large (Brahme 1995a; Brahme 1999b; Lind *et al.* 2001). Based on the PET-CT data, we can thus express the tumor cure probability according to:

$$P_B(D(r)) = e^{-\int n_0(r)e^{-D(r)/D_e(r)d^3r}} \qquad (7.27)$$

Interestingly enough, this expression includes exactly the same biological data as those used in Eq. (7.24), so they are already available from the same PET images that we have just discussed. In fact, Eq. (7.27) can be directly rewritten in terms of these data:

$$P_B(D(r)) = e^{-\iiint_{V_t} \frac{PET_0(r)V}{PET_v(r)V_t} e^{-\frac{D(r)}{nD(r)} \ln\left(\frac{PET_n(r)}{PET_0(r)}\right)} d^3r} . \qquad (7.28)$$

The interesting property of this expression is that it includes the patient's individual initial tumor spread and vasculature data as well as *in vivo* predictive assay data on tumor radioresistance to make the tumor response prediction and treatment optimization as reliable as possible. In true treatment optimization, we now have to combine the probability of curing the patient P_B with the probability of avoiding severe injury or adverse reactions P_I. If they were statistically independent processes, the probability to achieve a complication-free cure would be simply $P_+ = P_B(1 - P_I)$. However, in most clinical studies adverse reactions in normal tissues are often linked to local control and a more accurate expression for P_+ is given by

$$P_+ = P_B - P_I + \delta(1 - P_B)P_I, \qquad (7.29)$$

where δ is of the order of 0.2 or lower for many tumors (Ågren *et al.* 1990; Brahme 1995a). It has been pointed out that Eq. (7.29) may make too crude a compromise between the additional cure and injury as the dose is increased, because some forms of severe treatment-related morbidity may be considered less desirable than a recurring tumor. However, the statistically independent expression is even worse, as it corresponds to $\delta = 1$ and the last uncured injury term in Eq. (7.29) $((1 - P_B)P_I)$ is then five times larger. Recently, an even better way to solve this clinical optimization problem has been developed. This so-called P_{++} optimization strategy first maximizes P_+ by intensity modulation, selection of optimal beam directions, and radiation modalities (Löf *et al.* 1998; Brahme 2002, see Chapters 5.5.3.3 and 5.5.6). Then, the achieved P_+ is relaxed by a fraction of 1% and this $P_{+,min}$ value is used as a constraint on P_+ not allowing lower values while instead the injury P_I is minimized. This strategy almost simultaneously optimizes P_+ and minimizes P_I by selecting slightly modified angles of incidence and intensity modulations. A reduction of P_+ of say 0.5% may result in a decrease in complication probability of as much as 5%. This approach is, of course, highly desirable in countries where the risk of legal action is likely even after slight overtreatment.

7.6.4. *Mean Dose Delivery Monitoring*

Equations (7.22), (7.26) and (7.28) indicate a new way to consider uncertainties in dose delivery, which until now have only been approximately accessible by patient *in vivo* dosimetry and more indirectly by portal imaging. However, for high-energy photon and ion beams, it is possible to image the dose delivery *in vivo* using PET-CT. High-energy photons of ≥ 20 MeV have sufficient energy to knock out neutrons from carbon, nitrogen, and oxygen nuclei in the irradiated tissues of the patient. The remaining ^{11}C, ^{13}N, and ^{15}O nuclei are all PET emitters that can be imaged in the patient during or immediately after treatment to visualize the mean dose delivery (Nordell 1983; Janek 2002). Since this photonuclear activation depends on the amount of C, N, and O atoms in various tissues and the half-life of the associated PET nuclides are quite different (20, 10, and 2 min, respectively), the induced tissue activity will not be strictly proportional to the absorbed dose even though both the activation and the absorbed dose are proportional to the photon fluence. To be more precise, the activation is mainly produced in the energy range of the gigantic photonuclear resonances between 20 and 30 MeV, whereas all photon energies contribute to the absorbed dose according to:

$$\text{PET} = k \int_{E_t}^{\infty} \sum_{Z=6}^{8} \frac{\sigma_{\gamma,n}(E)\Psi_E}{t_{1/2} \cdot E} \, dE, \tag{7.30}$$

$$D_{\text{eq}} = \int_{0}^{\infty} \frac{\mu_{en}(E)}{\rho} \Psi_E \, dE, \tag{7.31}$$

where E_t is the threshold of the photonuclear reactions just below 20 MeV. This latter fact also makes the dose distribution due to photons in the energy range 20–30 MeV slightly different from that including all energies. Furthermore, the absorbed dose is really delivered by the secondary electrons set in motion by the photons, so dose build-up phenomena are not accurately imaged by photonuclear PET activation, even though this might seem to be the case, for example, in the build-up region owing to the finite resolution of most PET cameras. Taking all these factors into account, it is possible by appropriate selection of the time of imaging after therapy to obtain fairly accurate mean dose delivery pictures (see Figs. 7.1 and 7.10), except possibly in the build-up region when using high-resolution PET cameras. It is interesting to notice that the photonuclear reactions *in vivo* can also be used dynamically to image tumor and tissue vasculature since the rapid loss of [15]O after treatment in well-vascularized regions will result in an increased loss of activity beyond the normal 2 min half-life of [15]O (cf. Nüssbaum *et al.* 1983).

This type of imaging would benefit from real-time PET between accelerator pulses since the accelerator duty cycle is only 0.1%. However, the γ-ray bursts during the 5 μs accelerator pulses may totally saturate the sensitive electronics of a normal PET camera.

In ion beams, a much wider range of nuclear reactions are possible making true dose delivery imaging quite complex. However, when using a therapeutic beam of a positron emitting radionuclei, such as ^{11}C, ^{13}N, or ^{15}O, it will be possible to image primarily the Bragg peaks of the ions that are of key importance when using 3D biologically optimized Bragg peak scanning (cf. Brahme 1988; Brahme *et al.* 1995). Furthermore, for radioactive beams, every ion will contribute to the imaging, thus making the process less dependent on the wide spectrum of possible nuclear reactions between projectile and tissue (cf. Kraft 2000; Sobolevski *et al.* 2002; Brahme 2003a). The possibility to do real-time PET imaging with ions improves the treatment accuracy and allows an accurate quality assurance of the dose delivery, particularly when the ion beams stopping powers of the tissues are uncertain (Kraft 2000).

7.6.5. *BioArt Planning*

The availability of PET-CT tumor imaging with FDG or more tumor-specific tracers opens up the possibility for truly biologically optimized therapy. However, at the onset of treatment, tumor responsiveness data are not generally available, so the initial treatment plan will be mainly based on the measured density distribution of tumor cells and historic data on the radioresistance of the given tumor type and stage. Clearly, as our knowledge about altered genes and their influence on the radiation responsiveness of the tumor and the normal tissues increases, it should be possible to use a genetic screening assay to more accurately estimate the radiation response rather than just using the tumor type and stage. This is already considerably more information than is available in classical radiotherapy and it will of course be useful data for biologically optimized therapy where only the clonogen density is estimated from the tumor type and growth pattern. Fortunately, normal tissue data can be taken from historic data more safely than can tumor data, even though deviations may occur depending on the genetic predisposition of the patient. In this context, the use of stochastically optimized therapy is a useful tool when trying to make the treatment as robust as possible taking into consideration a wider range of variability in tumor and normal tissue sensitivities (Kåver *et al.* 1999). A more reliable treatment plan can thus be made preferably using biologically optimized intensity-modulated therapy planning and the patient is treated for about 3 treatment fractions before a repeated PET-CT image is recorded to obtain tumor responsiveness data early on in the treatment. About three more treatment fractions later, the last PET-CT dataset for the predictive assay of the tumor responsiveness is recorded to gain a better feeling for the influence of the rate of loss of functionality of the doomed tumor cells recorded on the images. These last two image sets should be recorded preferably after a relaxed weekend in order to reduce the effects of physical activity and radiation damage on normal tissue as imaged by FDG. Beyond the first 2 weeks of therapy, the functional tumor cell compartment is rapidly reduced and normal tissue response effects are accelerated (Tureson *et al.* 2001); hence, meaningful imaging is no longer feasible, unless a major part of the tumor has been missed during the treatment.

After the first 2 weeks of therapy, we should thus be able to have patient-specific data on the responsiveness of the tumor. This data set should then be used for a totally revised treatment plan taking not only the new biological response data into account but also the new information recorded on the tumor cell distribution after 2 weeks of therapy. Ideally, this should be the final treatment plan where necessary corrections due to factors such as possible beam tumor misalignments, deviations in tumor sensitivity from historic data, and the local tumor responsiveness due to hypoxia should be taken into account. Preferably, if mean dose delivery monitoring was performed, this information, too, should be used for comparison with the initially planned delivery, and corrections should be made for the average motion of internal organs. In this way, the new adaptive dose delivery starting in week 3 will be much more reliable, with a high probability of achieving complication-free cure and correcting for possible errors in the execution of the initial treatment plan during the first 2 weeks of treatment. The new high contrast tumor imaging racers based on fluorinated thymidine analogs seem to be a good starting point for this development (Shields *et al.* 1998; Krohn *et al.* 2001).

In Figs. 7.14 to 7.16 a first application of the new BioArt procedure is illustrated showing a common situation for larger bulky tumors. These tumors

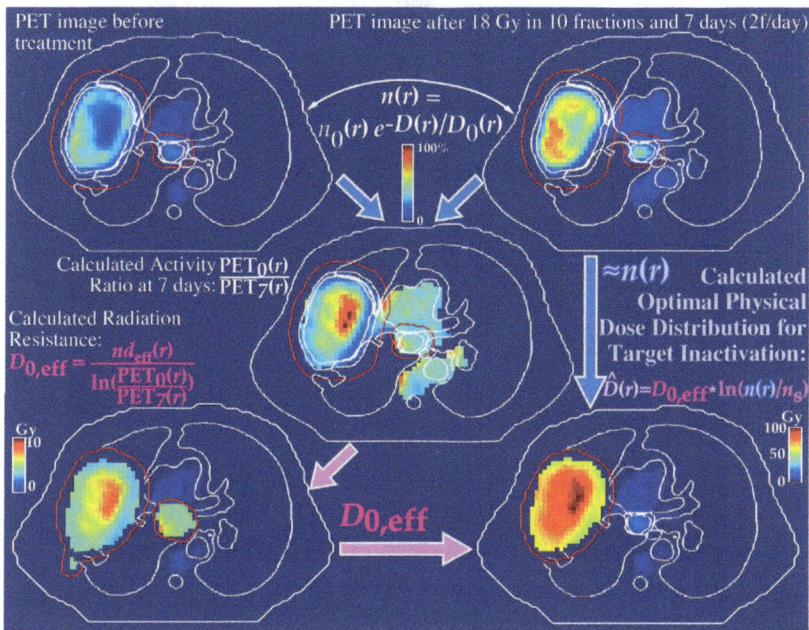

Figure 7.16. A lung cancer with very similar properties as the larger neck node in Fig. 7.14 with a large hypoxic tumor probably due to too high interstitial tumor pressure is shown (cf. Fig. 8.23). Based on the FDG PET images before therapy and after the first week of treatment, it was possible to calculate the optimal dose distribution (cf Eq. 5.43) and the $D_{0,\text{eff}}$ variation across the tumor in the lower panel. It is clearly seen that the strongly hypoxic tumor core needs substantially higher doses for eradiation if no pharmaceutical product such as PDGF antagonists is used to release the internal tumor pressure (Sjöblom *et al.* 2004).

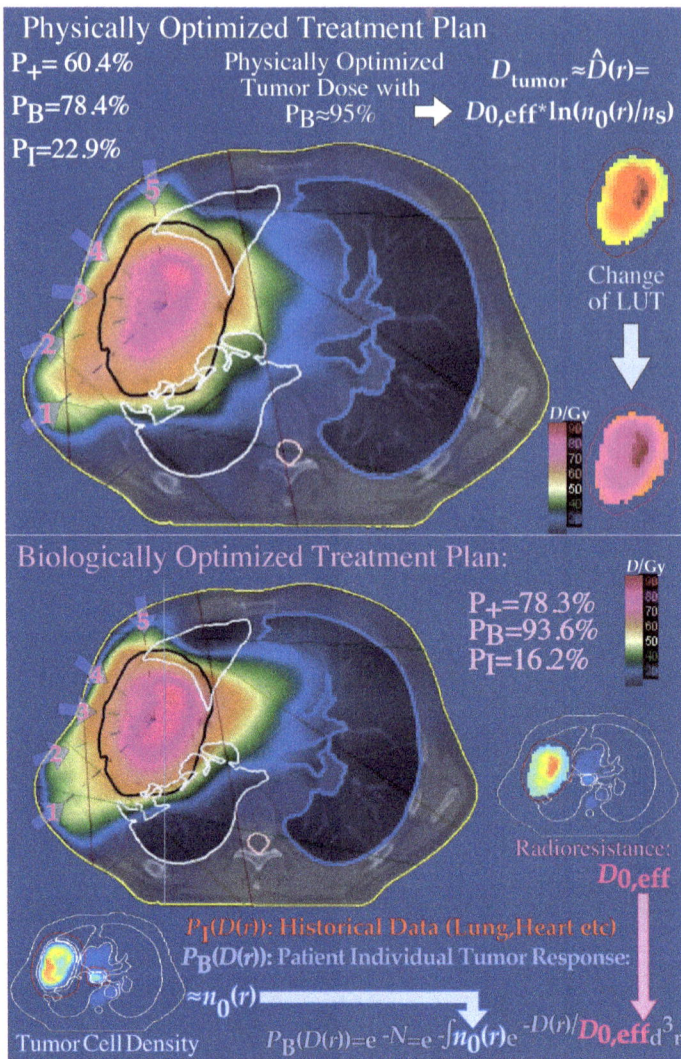

Figure 7.17. (a) The lung tumor in Fig. 7.14–15 is treated by physically and (b) biologically optimized IMRT, where the latter results in ≈ 20% better complication-free cure.

are commonly associated with large interstitial tumor pressures resulting in a high degree of internal hypoxia and significantly reduced vascularization except at the tumor periphery as seen in both Figs. 7.14 and 7.15. Figure 7.14 is from a study by de Schryver (1987) illustrating the response of two non-Hodgkin lymphomas on the neck, one large hypoxic with minimal uptake in the core but with a better

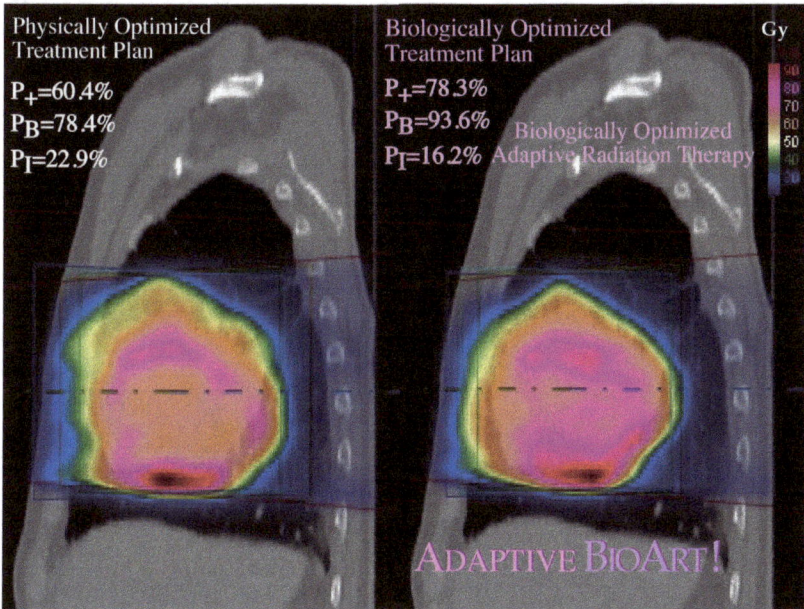

Figure 7.18. A lateral cut through the lung tumor in Fig. 7.17 showing that the part of the tumor close to the diaphragm needs higher dose both by physically and biologically optimized IMRT probably because the lower part of the tumor have been partly moving out of the therapeutic beam.

vascularized periphery, whereas the smaller more active one has a more active central uptake of $^{13}NH_3$ ammonia. After a rather conventional treatment the smaller more active tumor seems largely controlled, whereas the large hypoxic tumor was debulked but most likely not controlled with its high uptake after therapy due to the high initial hypoxic fraction.

The large lung tumor in Fig. 7.14(b) and 7.15 is very similar in appearance to the large neck tumor in Fig. 7.14(a) and was treated in Maastricht in connection with the 6th Framework program BioCare (de Ruysscher *et al.* 2005). For this tumor, the FDG uptake both before and after 1 week of therapy was available as shown in the figure allowing calculation of the spatial variation of the radioresistance of the tumor as illustrated in the lower panel according to Eqs. (7.16) and (7.22). This clearly shows that these bulky tumors should be given a nonuniform dose delivery to allow a more uniform eradication also of the more hypoxic tumor cell compartment. As seen from the FDG uptake after the first week, the radiation induced-cell kill releases the pressure in the tumor so that more oxygen nutrients and FDG can reach the core and make the rest of the treatment more effective.

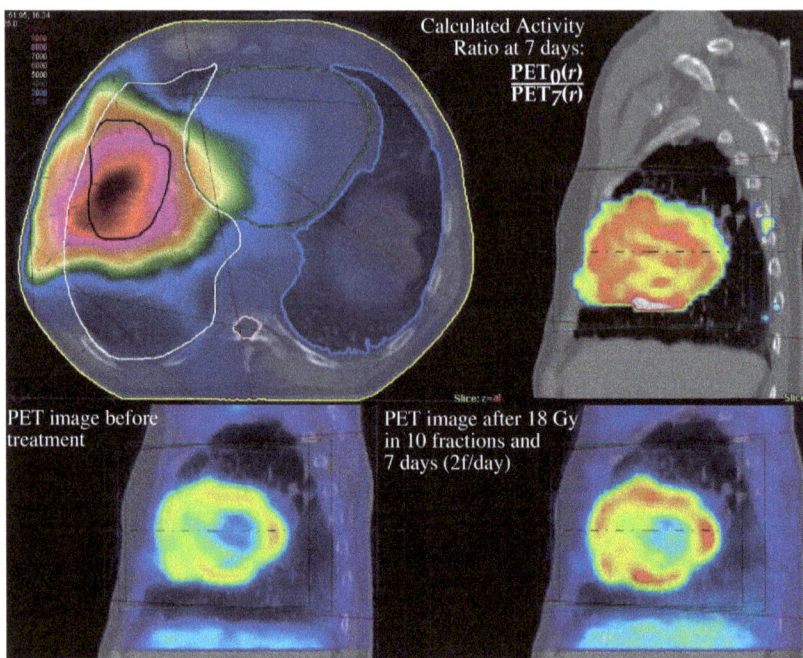

Figure 7.19. A lateral cut through the lung tumor in Fig. 7.18 showing that the part of the tumor close to the diaphragm needs considerably higher dose.

An alternative approach to IMRT would be to use a PDGF antagonist as mentioned above to release the pressure in the tumor before the radiation treatment is started. This would considerably reduce the total dose needed for tumor cure and would significantly reduce the normal tissue side effects and of course simultaneously increase the complication-free cure substantially. In the future, we, therefore, need to more accurately quantify the combined effect of PDGF and radiation on large tumors where radiation alone may otherwise require light ions to maximize tumor cure, minimize the effect of hypoxia, and minimize the normal tissue morbidity.

7.7. Conclusions

There is no doubt that with the introduction of PET-CT imaging and the use of FDG and more tumor-specific tracers, there is a considerable potential to improve radiation therapy. This is particularly true using biologically optimized radiation therapy planning where the PET images are really the missing link for accurate tumor characterization and delineation, tumor responsiveness determination, and treatment response monitoring. The potential to make a real *in vivo* predictive assay of tumor responsiveness is probably the ultimate step in accurate radiation therapy. Even if this new imaging technique is a quantum leap in gross tumor imaging, it will still be very difficult to image the most resistant tumor clonogens since there are probably very few of them and they are probably located in poorly vascularized areas and are therefore unlikely to show up clearly on the images. Hypoxia imaging will therefore still be a very useful complement, particularly for very resistant tumors, even though the gross tumor response during the first weeks of therapy basically takes the initial level of hypoxia into account but the sensitivity is reduced due to the poor vasculature. The possibility to image both the integral dose delivery and the tumor response *in vivo* gives us the opportunity to perform real biologically adaptive therapy where practically all types of clinical error sources in principle could be picked up as long as they influence the dose delivery or the tumor response. To derive maximum therapeutic benefit, it is essential to introduce these new methods with biologically optimized therapy planning, the ultimate tool for combining biomedical, molecular and clinical knowledge with advanced radiation physics and biomedical computing. In fact, the ultimate treatment unit should combine advanced intensity-modulated dose delivery with PET-CT imaging in one integral device as shown in Figs. 4.85 and 8.39b to improve and simplify imaging in direct connection with the treatment, and even during treatment between accelerator pulses. This latter type of real-time PET-CT imaging is particularly useful, with radiation modalities such as high-energy photons and light ions, which allow *in vivo* dose delivery monitoring. If the repeated tumor imaging is done just before the next radiation therapy session, it will then even be possible to image the tumor on the background of the just delivered dose distribution allowing the ultimate form of tumor dose delivery verification.

References

Ågren A, Brahme A, Turesson I (1990) Optimization of uncomplicated control for head and neck tumors. *Int J Radiat Oncol Biol Phys* 19:1077–1085.

Ågren CA, Källman P, Turesson I, Brahme A (1995) Volume and heterogeneity dependence of the dose-response relationship for head and neck tumours. *Acta Oncol* 34:851–860.

Ando K, Koike S, Uzawa A, *et al.* (2005) Biological gain of carbon-ion radiotherapy for the earlyresponse of tumor growth delay and against early response of skin reaction in mice. *J Radiat Res* 46:51–57.

Brahme A (1984) Dosimetric precision requirements in radiation therapy. *Acta Radiol Oncol* 23:379–391.

Brahme A (1992) Biological and physical dose optimization in radiation therapy. In: Fortner JG and Rhoads JE (eds) *Accomplishments in Cancer Research*, 265–300. New York: Lippincott.

Brahme A (1994) Optimization of radiation therapy for cancer patients. *FORUM, Trends in Exp Clin Med* 4:569–584.

Brahme A (1995a) Treatment optimization using physical and biological objective functions. In: Smith A (ed.) *Radiation Therapy Physics*, 209–246. Berlin: Springer.

Brahme A (1995b) Guest editorial: Optimization of the three-dimensional dose delivery and tomotherapy. *Int J Imag Syst Technol* 6:1.

Brahme A (1995c) Similarities and differences in radiation therapy optimization and tomographic reconstruction. *Int J Imag Syst Technol* 6:6–13.

Brahme A (1999a) Optimized radiation therapy based on radiobiological objectives. *Sem Oncol* 9:35–47.

Brahme A (1999b) Biologically based treatment planning. In: Mustakallio Centennial Symposium Helsinki. *Acta Oncol* (Suppl. 13):61–68.

Brahme A (1999c) Optimized radiation therapy based on radiobiological objectives. *Sem Oncol* 9:35–47.

Brahme A (2000) Development of radiation therapy optimization. *Acta Oncol* 39:579–595.

Brahme A (2001a) Individualizing cancer treatment: biological optimization models in treatment planning and delivery. *Int J Radiat Oncol Biol Phys* 49:327–337.

Brahme A (2002) Development of radiobiologically based therapy optimization and target definition. *Proc Second Int Symp on Target Volume Definition in Radiation Oncology. Inst Radiat Oncol*, Kiricuta IC (ed.), St Vincenz-Hospital, Limburg, Germany.

Brahme A (2003a) Biologically optimized 3-dimensional in vivo predictive assay based radiation therapy using positron emission tomography-computerized tomography imaging. *Acta Oncol* 42:123–136.

Brahme A (2003b) Recent advances in light ion radiation therapy. *Int J Rad Onc Biol Phys* 58:603–616.

Brahme A (2004) Development of Radiobiologically Based Therapy Optimization and Target Definition. *The Fourth Int. Symp. on The Lymphatic System — New Developments in Oncology and IMRT*, Kiricuta IC (ed.), Limburg, Germany, May 13–15, p. 155.

Brahme A (2009) Potential developments of light ion therapy: the ultimate conformal treatment modality. First NIRS International Open Laboratory Work- shop, November 17, 2008, Chiba, Japan, Tsujii H (ed.), *Radiol Sci* 52:8–31 (http://www.nirs.go.jp/info/report/rs-sci/pdf/200902.pdf).

Brahme A, Ågren A (1987) On the optimal dose distribution for eradication of heterogeneous tumors. *Acta Oncol* 26:377–385.

Brahme A, Källman P, Lind B (1988) Optimization of proton and heavy ion therapy using an adaptive inversion algorithm. *Radiother Oncol* 15:189–197.

Brahme A, Källman P, Lind B (1991) Optimization of the Probability of Achieving Complication-free Tumor Control using a 3D Pencil Beam Scanning Technique for Protons and Heavy Ions. *Proc Tokyo, World Congress Med Phys Biomed Engineer*, Kyoto, Japan, July 7–12.

Brahme A, Källman P, Tilikidis A (1995) Developments in ion beam therapy planning and treatment optimization. In: Linz U (ed.) *Ion Beams in Tumor Therapy*, 290–299. Weinheim: Chapman & Hall.

Brahme A, Lewensohn R, Ringborg U, Amaldi U, Gerardi F, Rossi S (2001a) Design of a centre for biologically optimised light ion therapy in Stockholm. *Nucl Instr Meth Phys Res* B184:569–588.

Brahme A, Nilsson J, Belkic D (2001b) Biologically optimized radiation therapy. *Acta Oncol* 40:725–734.

De Ruysscher D, Wanders S, Minken A, *et al.* (2005) Effects of radiotherapy planning with a dedicated combined PET-CT-simulator of patients with non-small cell lung cancer on dose limiting normal tissues and radiation dose-escalation: A planning study. *Radiother Oncol* 1–6.

De Schryver A, Schelstraete K (1987) Toepassingen der Positronenemissietomografie (PET) in de Oncologie. *Verhandelingen van de Koninklijke Academie voor Geneeskunde van Belgie* 6:413–431.

Denekamp J, Harris SR, Morris C, Field SB (1976) The response of a transplantable tumor to fractionated irradiation. II. Fast neutrons. *Radiat Res* 68:93–103.

Doubrovin M, Ponomarev V, Beresten T, *et al.* (2001) Imaging transcriptional regulation of p53 dependent genes with positron emission tomography in vivo. *PNAS* 98: 9300–9305.

Erdi YE, Macapinlac H, Rosenzweig KE, *et al.* (2000) Use of PET to monitor the response of lung cancer to radiation treatment. *Eur J Nucl Med* 7:861–866.

Forssell Aronson E, Kjellén E, Mattson S, Hellström M (2002) Medical imaging for improved tumour characterization, delineation and treatment verification. *Acta Oncol* 41:604–614.

Gambhir SS (2002) Molecular imaging of cancer with positron emission tomography. *Nat Rev Cancer* 2:683–693.

Grigsby PW, Siegel BA, Dehdashti F (2001) Lymph node staging by positron emission tomography in patients with carcinoma of the cervix. *J Clin Oncol* 19:3745–3749.

Herfarth KK, Debus J, Lohr F, *et al.* (2001) Stereotactic single-dose radiation therapy of liver tumors: Results of a phase I/II trial. *J Clin Oncol* 19:164–170.

Janek S (2002) 3-Dimensional patient dose delivery verification based on PET-CT imaging of photonuclear reactions in 50 MV scanned photon beams, MSc thesis, Dept Med Rad Phys, Karolinska Institutet and Royal Institute of Technology.

Janek Strååt S, Björn Andreassen, Cathrine Jonsson, Marilyn E Noz, Gerald Q Maguire Jr, Peder Näfstadius, Ingemar Näslund, Frederic Schoenahl and Anders Brahme (2013) Clinical application of in vivo treatment delivery verification based on PET/CT imaging of positron activity induced at high energy photon therapy. *Phys Med Biol* 58:5541–5553.

Kåver G, Lind BK, Löf J, Liander A, Brahme A (1999) Stochastic optimization of intensity modulated radiotherapy to account for uncertainties in patient sensitivity. *Phys Med Biol* 44:2955–2969.

Kiricuta IC (ed.) (2001) *Proc First Int Symp on Target Volume Definition in Radiation Oncology*, Inst of Radiat Oncol, St Vincenz-Hospital, Limburg, Germany.

Koike S, Ando K, Uzawa A, *et al.* (2002) Significance of fractionated irradiation for the biological therapeutic gain of carbon ions. *Radiat Prot Dosim* 99:405–408.

Kraft G (2000) Tumor therapy with heavy charged particles. *Prog Part Nucl Phys* 45: S473–544.

Krohn KA, Mankoff DA, Eary JF (2001) Imaging cellular proliferation as a measure of response to therapy. *J Clin Pharmacol* (Suppl 2001):96S–103S.

Lind B and Brahme A (1995) Development of treatment techniques for radiotherapy optimization. *Int J Imag Syst Technol* 6:33–42.

Lind BK, Nilsson J, Löf J, Brahme A (2001) Generalization of the normalized dose-response gradient to non-uniform dose delivery. *Acta Oncol* 40:719–724.

Lind BK, Persson LM, Edgren MR, Hedlöf I, Brahme A (2003) Repairable conditionally repairable damage model based on dual Poisson processes. *Radiat Res* 159.

Liu MC, Gelmann EP (2002) P53 gene mutations: case study of a clinical marker for solid tumors. *Sem Oncol* 29:246–257.

Löf J, Liander A, Kåver G, Lind BK, Brahme A (1998) ORBIT — a general object oriented code for radiotherapy optimization. *Radiother Oncol* 48(Suppl 1):S69.

Ngo FQH, Blakely EA, Tobias CA (1981) Sequential exposures of mammalian cells to low and high-LET radiations. *Radiat Res* 87:59–78.

Nilsson J, Lind BK, Brahme A (2002) Radiation response of hypoxic and generally heterogeneous tissues. *Int J Radiat Biol* 78:389–405.

Nordell B (1983) Production of photoneutron beams and radionuclides by photonuclear reactions using a 50 MeV racetrack microtron, thesis, Stockholm University.

Nüssbaum GH, Purdy JA, Granada CO, Emami B, Sapareto SA (1983) Use of the Clinac-35 for tissue activation in non-invasive measurements of capillary blood-flow. *Med Phys* 10:4.

Shiau C-Y, Sneed PK, Hui-Kuo GS, *et al.* (1997) Radiosurgery for brain metastases: relationship of dose and pattern of enhancement to local control. *Int J Radiat Oncol Biol Phys* 37:375–383.

Shields AF, Grierson JR, Dohmen BM, *et al.* (1998) Imaging proliferation in vivo with [F-18] FLT and positron emission tomography. *Nat Med* 11:1334–1336.

Singh B, Arrand JE, Joiner MC (1994) Hypersensitive response of normal human lung epithelial cells at low radiation doses. *Int J Radiat Biol* 65:457–464.

Sjoblom T, Yakymovych I, Heldin CH, Ostman A, Souchelnytskyi S (2004) Smad2 suppresses the growth of Mv 1 Lu cells subcutaneously inoculated in mice. *Eur J Cancer* 40:267–274.

Sobolevski N, Gudowska I, Andreo P, Belkic D, Brahme A (2002) Interaction of Ion Beams with Tissue-like Media; Simulations with the SHIELD-HIT Monte-Carlo Transport Code. *Proc Shielding Aspects of Accelerators*. Stanford Linear Accelerator Center, California.

Sullivan DC, Hoffman JM (2001) In vivo imaging of gene expression. *Sem Radiat Oncol* 11:37–46.

Townsend DW (2002) Imaging function and anatomy with a combined PET-CT scanner. Paper Presented at Queens Medical Center, Honolulu, March 28. See also Townsend DW, Cherry SR (2001) Combining anatomy and function: the path to true image fusion. *Eur Radiol* 11:1968–1974.

Tsujii H, Morita S, Miyamoto T, *et al.* (2002) Experiences of carbon ion radiotherapy at NIRS. In: Kogelnik HD, Lukas P, Sedlmayer F (eds.) *Progress in Radio-oncology VII*, 393–405. Bologna, Italy: Monduzzi Editore.

Tureson I, Bernefors R, Book M, *et al.* (2001) Normal tissue response to low doses of radiotherapy assessed by molecular markers. *Acta Oncol* 40:941–951.

Van de Wiele C, Lahorte C, Oyen W, *et al.* (2003) Nuclear medicine imaging to predict response to radiotherapy: a review. *Int J Radiat Oncol Biol Phys* 55:5–15.

Withers HR, Taylor JMG, Maciajewski B (1988) Treatment volume and tissue tolerance. *Int J Radiat Oncol Biol Phys* 14:751–759.

Physical, Biological, and Clinical Background for the Development of Biologically Optimized Light Ion Therapy

8

Anders Brahme and Hans Svensson

8.1. Introduction

Biologically optimized intensity-modulated photons, electrons, and light ions represent the ultimate development of radiation therapy where the absorbed dose and biological effect to normal tissues can be adjusted to be as low as possible from a radiation physical point of view, at the same time, as the therapeutic effect on radiation-resistant tumor cells is as high as possible from a radiation biological point of view. With light ions, the border region between the clinical target volume and surrounding healthy normal tissues can be set as narrow as physically possible, the required number of treatment fractions can be substantially reduced, and the curative gain factor for hypoxic tumor cells can often be more than doubled compared to photons, electrons, and protons (cf. Figs. 8.1–8.5, 8.13, and 8.14). Taking all this information into account, the cost effectiveness of light ions per patient cured is often up to two times higher or similar to that of advanced conventional radiation therapy, and about three times higher than that for proton therapy. The only problem with the light ions is the large capital cost requiring an initial investment in the order of €100–150 million. Beside the increased therapeutic efficiency and short treatment time often with as few as 1–12 fractions in a couple of weeks, the major clinical advantages of light ion therapy are an increased therapeutic outcome in terms of improved local tumor control and quality of life, and a substantially increased patient survival as well as a significantly reduced risk for adverse normal tissue reactions.

As our knowledge about the molecular biology of cancer is rapidly improving, we continuously need better tools for diagnostic molecular imaging to match the improved treatment accuracy and therapeutic efficiency with light ion therapy.

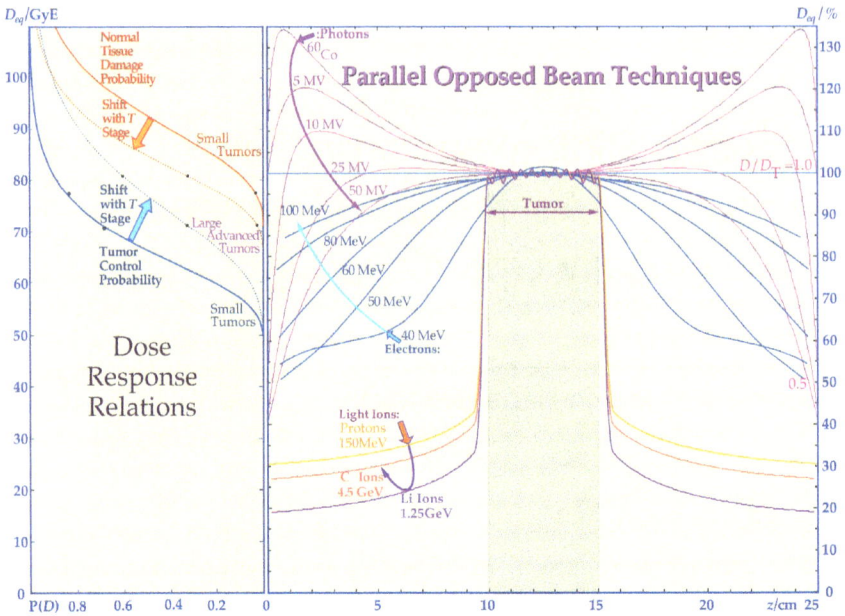

Figure 8.1. Comparison of the biologically effective dose distributions when irradiating a deep-seated tumor using parallel opposed photon, electron, and light ion beams is shown. This is probably the best geometry for a serious comparison of the therapeutic properties and quality of different radiation modalities. It is clearly seen that the normal tissues surrounding the tumor are considerably less damaged with the lightest ions around lithium (cf. also Figs. 8.2, 8.4, and 8.9). For hypoxic radiation-resistant tumors, the clinical advantage is even larger for the light ions beyond helium. For well-oxygenated tumors, the difference is less significant and protons and electrons can be used with rather small differences in clinical response since both generally deliver doses that are below the threshold for significant normal tissue damage as shown by the schematic dose–response curves in the left panel. In both panels, a vertical effective dose scale is used in Gray-equivalent and percentage, respectively. In addition to the longitudinal dose distribution shown here, the lateral penumbra (Fig. 8.4(e)), the biological effectiveness, and the degree of hypoxia or radiation resistance (Figs. 8.3–8.5) should be considered when selecting the optimal treatment energy and modality. It is seen that quite high photon or electron beam energies are generally needed for optimal results with deep-seated tumors as seen more clearly in Fig. 8.2.

Malignant tumors are our major life-threatening disease at least up to the age of about 65 years and as many as 50% the young generation today may be diagnosed with cancer sometime during their life. A comprehensive cancer center and a center of excellence for advanced radiation therapy should, therefore, be focused on two unique developments that will considerably improve our ability to cure cancer patients and maximize their quality of life.

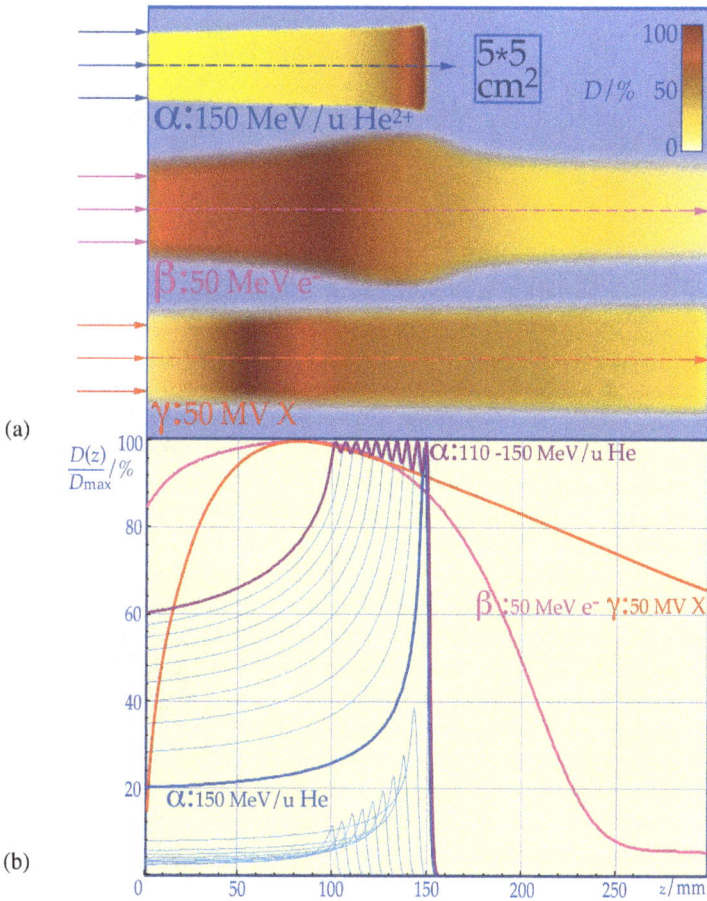

Figure 8.2(a). Planar- and **(b)** central axis depth-dose distributions for helium ion (α), electron (β), and photon (γ) beams are shown. It is seen how much sharper the penumbra is and that very little dose is deposited beyond the tumor depth with light ions (α, upper panel, upper beam). Contrary to photons and electrons, a monoenergetic ion beam can only be used for a very small tumor. With the required energy modulation for a 5 cm thick tumor between 10 and 15 cm depth, the central axis dose distribution for ions is more similar to those for very high-energy electrons and photons (β, lower panel e-and γ, X-ray beams).

First, the sensitivity, resolution, and field of view of modern positron emission tomography-computerized tomography (PET-CT) and magnetic resonance spectroscopic imaging (MRSI) cameras should be improved as far as possible so that they really become a unique, sensitive, and fast tool in the early detection and screening of tumor spread (cf. Fig. 8.26(b)). Furthermore, they should also be used during initial treatment phase to evaluate the radiation resistance of the tumor *in vivo* by

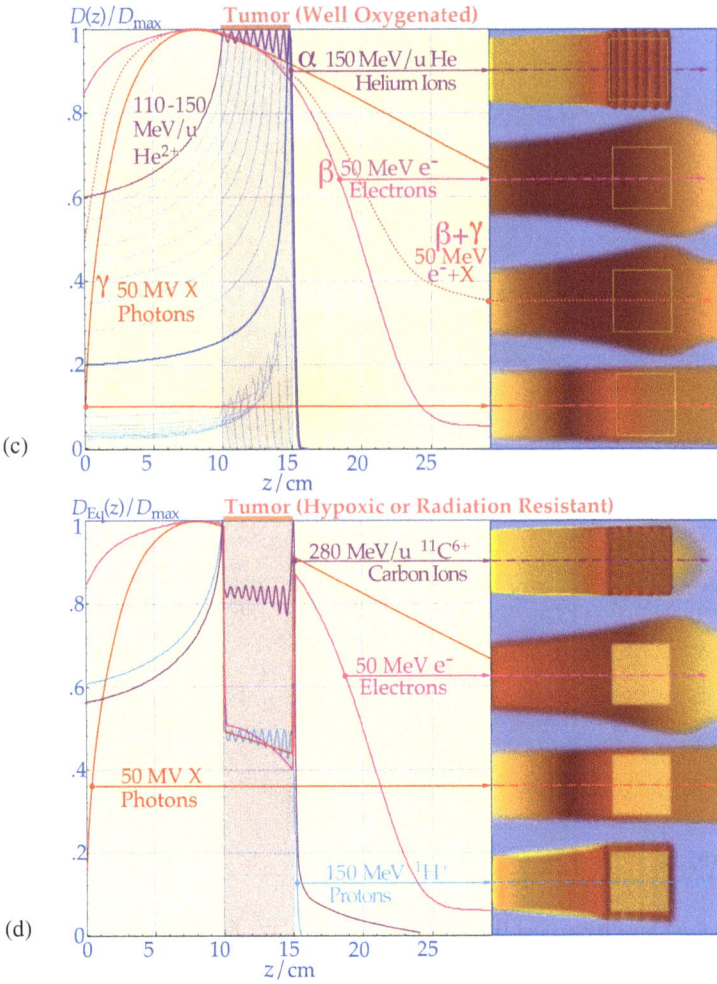

Figure 8.2(c). With a 5 cm thick tumor located between 10 and 15 cm depth, the energy of the proton, helium and carbon ions have to be modulated over a 110–150 MeV/u energy range, whereas electrons and photons can be used directly as seen in (a) and (b). With range modulation, the total dose distribution to the tumor is better defined with less dose to surrounding normal tissues, even though the skin spearing is best with the photon beam and very good also for e⁻ + X-rays. **(d)** If the tumor is of low oxygen content (hypoxic), the effective dose with photons electrons and protons will be reduced between 2 to 3 times due to reduced toxicity since less oxygen radicals are being formed. This effect is much lower with carbon ions where the direct cell kill dominates due to the much higher ionization density (cf. Fig. 8.4, below) and there is less dependence on oxygen radical-mediated cell kill and thus tumor oxygenation status. As seen in (d) the higher OGF of carbon ions makes them significantly more effective (cf. Fig. 8.13) in hypoxic and otherwise radiation-resistant tumors generally requiring fewer beam portals (generally 2 are sufficient) rather than 3–5 with low *LET* beams (cf. also Fig. 8.26(a)).

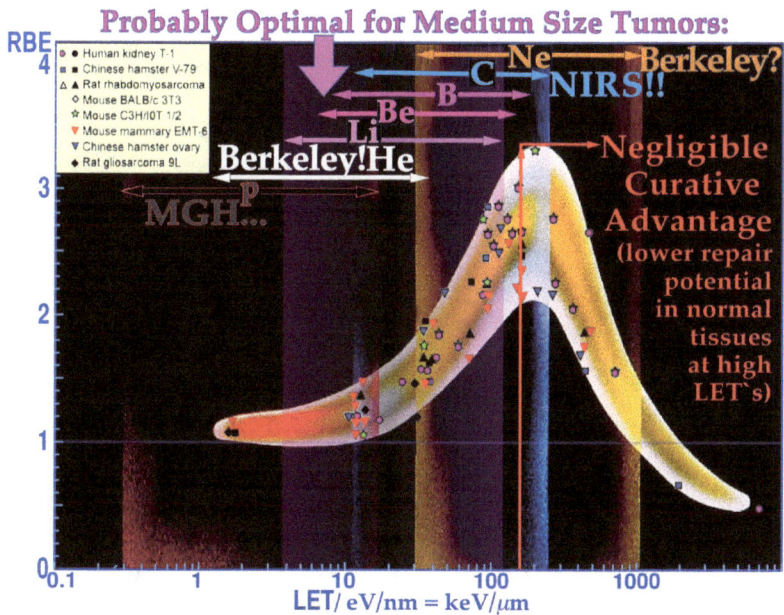

Figure 8.3(a). Comparison of the RBE and *LET* ranges available with protons, lithium, carbon, and neon ions is shown. It is seen that the range from lithium to carbon ions is most interesting, especially for hypoxic tumors (see also Eqs. (8.8) and (8.13) and Figs. 8.7(f) and 8.8(j)), modified from Blakely and Chang (2009). In general, *LET*s far beyond the RBE peak should be avoided to minimize normal tissue damage in the entrance, plateau and fragmentation tail regions (cf. Fig. 8.4).

repeated PET-CT imaging during the first week to 10 days of therapy (Brahme 2003) and then do biologically adaptive corrections during the last part of the treatment. This was one of the key goals of the Sixth Framework Program, BIOCARE, coordinated by Karolinska Institutet, and discussed in Chapter 7. MRSI, even if the sensitivity is lower, should be used for screening purposes since the radiation dose to normal tissues is really minimal and of low energy (RF fields). A third potential diagnostic development that should be mentioned here is stereoscopic phase-contrast (SP) X-ray imaging, since it has a very interesting potential in tumor diagnostics due to improved resolution and contrast at significantly lower doses compared to diagnostic CT. It also has the potential to allow advanced molecular imaging of tumor properties and treatment responses (Zhou and Brahme 2008, 2010). With any of these diagnostic approaches, the initial image of tumor spread should be used as the base for treatment response monitoring during the early phase of therapy and allow accurate *in vivo* predictive assay of radiation responsiveness

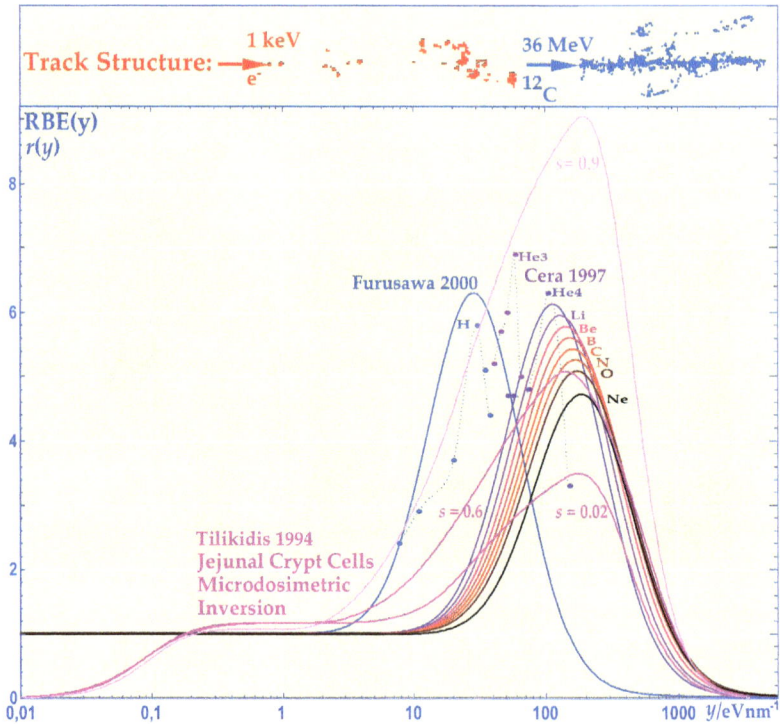

Figure 8.3(b). Variation of the RBE and track structure as a function of the ionization density or LET (in eV/nm) of light ion beams is shown. The highest biological effectiveness is seen for ionization densities between 25 and 200 eV/nm ($=$ keV/μm), which are achieved by track-end electrons and Bragg peak carbon ions as illustrated above. The broad peaks are based on microdosimetric inversion of jejunal crypt cell survival and are based on a wide set of neutron data (Tilikidis and Brahme 1994). Analytical ion-specific response data from Furusawa (2000) are also included based on his analytical formula which has wider spread than the track segment data of Cera *et al.* (1997) as expected (cf. Fig. 8.4(h)). It is seen that the microdosimetric data cover partly the hydrogen influence since it is largely based on neutron beam irradiations that generate many secondary protons. The proton data pertain to low-energy protons with negligible range straggling contrary to high-energy protons (cf. Fig. 8.4(h)). The microdosimetric inversion data of Tilikidis cover a wider range of particle species resulting in a broader RBE peak more similar to Fig. 8.3(a).

and consequently biologically based adaptive therapy optimization (Brahme 2003, 2009).

Second, to maximize the therapeutic response of the tumor and minimize eventual adverse normal tissue reactions, biologically optimized photons, electrons, and light ions are the ultimate therapeutic modalities delivering high densities of

DNA lesions in genomically unstable tumor cells and largely only induce a low density of repairable lesions in normal tissues (cf. Chapter 1 and Brahme 2004). To maximize the therapeutic outcome, it is important to be able to accurately quantify the therapeutic properties of the beams in terms of linear energy transfer (LET), relative biological effectiveness (RBE), oxygen enhancement ratio (OER), oxygen gain factor (OGF), apoptotic fraction (A_{Fr}), and the dose–response relation (DRR) for the tumor and affected normal tissues (cf. Figs. 8.1–8.5, 8.13–8.15, and 8.22–8.24).

Since both the diagnostic and therapeutic methods have millimeter resolution, a comprehensive cancer center with light ion therapy will represent a quantum leap in our ability to accurately treat malignant tumors. It is, therefore, very important that such centers are realized as soon as possible to make full use of the clinical advantages and fast developments of light ion therapy and molecular tumor imaging as well as of molecular genomics and proteomics of cancer and to make these methods clinically available for the benefit of our cancer patients. It is one of the few areas where a substantial investment in new diagnostic and therapeutic methods is cost effective and rapidly brings improved treatment results and quality of life for our patients and the health and cancer care systems.

8.2. Present Use of Radiation Therapy for Malignant Tumors

8.2.1. *Cancer Statistics*

Cancer registration is well developed in Sweden and we will use data from the present population of 9 million inhabitants to illustrate the cancer problem. The Swedish Cancer Registry has been active since 1958. Since that time, a continuous increase of cancer patients has been registered, at an annual incidence increase of 1–2%. At present, about 55,000 new malignant tumors are diagnosed every year, approximately equally distributed between genders. Time trend analyses indicate continuous increase with at least 1% per year and probably more since new efficient treatment methods such as intensity-modulated radiation therapy (IMRT) and light ion therapy will significantly increase the clinical indications. Cancer survival trends over the past decades have been positive with a significantly prolonged survival, which is an effect of both improved early diagnosis and more effective treatments. During the 1990s, a small but statistically significant decrease in mortality has been observed. A combination of increased incidence and prolonged survival has influenced the prevalence figures. At present, Sweden has more than 300,000 living cancer patients and the prevalence is increasing by about 3% per year. According to a recent investigation, 47% of the Swedish cancer patients receive radiation therapy, which is low compared to other industrialized countries. The increasing number of patients, particularly at an old age, as well as the increasing indications to treat with radiation creates an increasing need for efficient radiation therapy facilities with IMRT and light ions, presently at a rate of several percentage per year.

8.2.2. *Radiation Therapy Treatment Strategy*

The principle of radiation therapy is to deliver as high dose as possible to surely eradicate the target (i.e., the tumor plus an internal margin taking into consideration the diagnostic accuracy and positional uncertainties of internal organ motions) and as low dose to healthy tissues as possible to avoid severe side effects. A compromise between these two mutually conflicting goals must generally be found. Improved imaging techniques make it possible to determine the location of the tumor volume more accurately. The best treatment techniques must be found which can increase the dose to the tumor and restrict radiation effects in surrounding normal tissues as far as possible. Furthermore, the tumor cells are often resistant to conventional photon, electron, and proton radiations. Light ion beams are, therefore, the ultimate treatment modality since they are more effective than other types of radiation to concentrate the absorbed dose and the therapeutic effect to the irradiated tumor. More importantly, the biological effect to eradicate hypoxic and otherwise radiation-resistant tumor cells is considerably increased when the narrow high dose and densely ionizing Bragg peak of light ions heavier than protons is deposited in the tumor with a reduced risk of damaging surrounding normal tissues as can be seen from Figs. 8.1 to 8.4. Similar to biologically optimized IMRT with photons and electrons, the best form of therapy optimization with light ions has as treatment objective to maximize the quality of life or the probability to cure the patient from his tumor without inducing severe normal tissue morbidity (P_+ strategy, cf. Sec. 5.5.3.1) or even simultaneously maximizing the cure and minimizing the probability of severe injury (P_{++} optimization strategy, cf Secs. 5.5.3.2 and 5.5.6). With severely hypoxic tumors, a single beam portal is never sufficient and 2–3 Li-C ion beam portals or 3 and more photon beams will generally be needed for curative treatment as seen by comparing Figs. 8.1 and 8.2.

8.3. Therapeutic Properties of Light Ion Beams

Compared to low LET photon, electron, and proton beams, the clinical properties of light ion beams for radiation therapy are much more versatile and complex as discussed in more detail in the following sections. A large part of the detailed specific information is presented in graphical form in the Figures and Figure captions for simplicity and clarity.

8.3.1. *High Dose to the Tumor and Low dose to Healthy Normal Tissues*

Light ions are charged nuclear particles with rest masses several thousand times higher than that of an electron. A much higher accelerating potential is, therefore, needed for ions than for the electrons presently dominating in therapy accelerators of today. The six lightest ions from protons to carbon ions are of major clinical interest even though ions up to neon and argon have been used at Berkeley. Light ions have dose distributional properties, which are very advantageous for radiation therapy as seen in Figs. 8.1 and 8.2. When they pass through superficial tissues much less energy is absorbed locally than at the end of the range. The peak initial energy of the ions is, therefore, selected so that the distal part of the tumor is located at the end of the primary particle range. Figures 8.1, 8.2, and 8.4 show the dose, energy deposition density, and biologically effective dose distribution for conventional photon and electron beams compared with the distribution using different light ion beams. With increasing atomic number, the penumbra gets narrower and the longitudinal range straggling is also lower, so more of the energy is deposited in the tumor by light ions of increasing atomic weight (cf. Figs. 8.2, 8.3, 8.4(e)–(h)). The high energy deposition at the end of the ion range is due to a velocity resonance causing a high energy transfer to the tumor cells when the speed of the ion is almost equal to the speed of the electrons in the tumor tissue and there is a very high probability that energy is transferred from the multiply charged ions to the electrons as they travel longer distances together toward the end of the particle range (Bragg peak, cf. Figs. 8.2, 8.3, and 8.4). When the atomic weight gets too high, the amount of particle fragments increases, so the dose beyond the Bragg peak gets too high as seen in Figs. 8.4(a) and (b). Therefore, particles heavier than carbon should be used very carefully with sensitive normal tissue structures located in front of and beyond the tumor.

When the relative depth–dose curves in Fig. 8.4(b) are converted to relative biological effect curves in 8.4(c) based on the microdosimetric RBE data in Fig. 8.3(b), the advantages with the lightest ions beyond protons but below carbon are clearly visible. A reduction of the relative plateau level biological effect of about 50% is seen for lithium and beryllium. In Fig. 8.4(d), the fast initial reduction in lateral penumbra and longitudinal range straggling when going from protons to helium and more slowly going from helium to lithium and carbon is illustrated. Since the penumbra of helium ions (α-particles) is already half of that for protons, helium ions are really the particle of choice in the quasi-low *LET* region. The influence of these effects on the Bragg peak of pencil beams is shown in Figs. 8.4(f)–(g), demonstrating that for each dose addition to a small part of the tumor volume

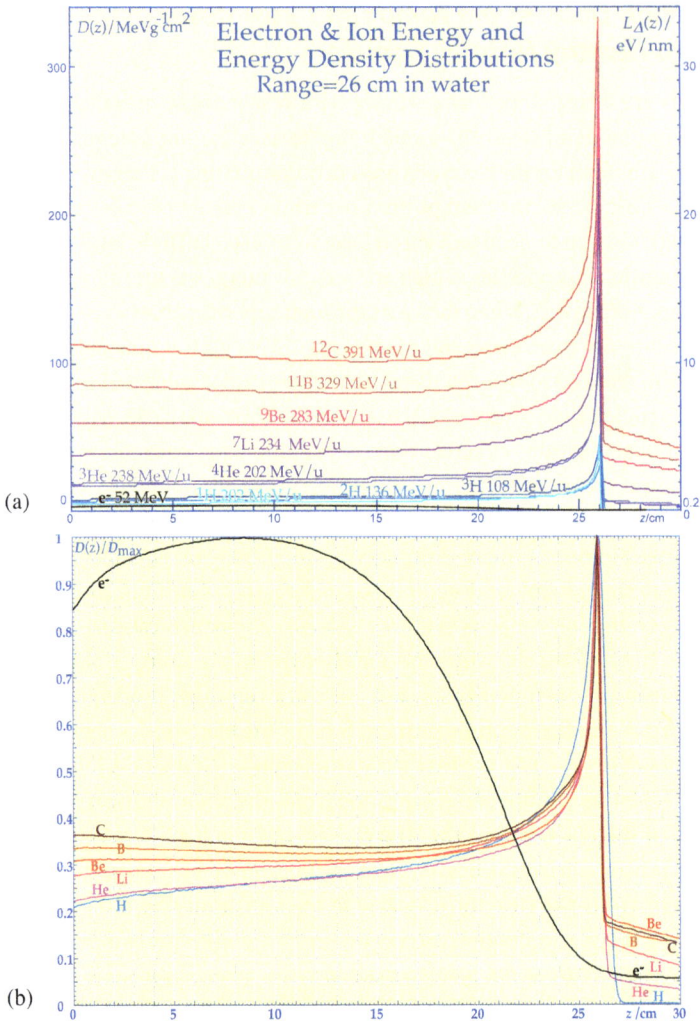

Figure 8.4(a). Comparison of the energy deposition density distributions obtained in water when irradiating with broad plane parallel–charged particle beams of electrons and light ions is shown. It is seen that with increasing atomic number, the *LET* in the entrance or plateau region as well as in the deep Bragg peak and beyond increases. The optimal therapeutic effect ratio between tumor and normal tissue is generally obtained between helium and carbon ions (cf. Figs. 8.4(c) and 8.9(d). **(b)** The dose distributions in (a) is here normalized to the absorbed dose at dose maximum showing the high dose distribution similarity between light ions from protons to carbon.

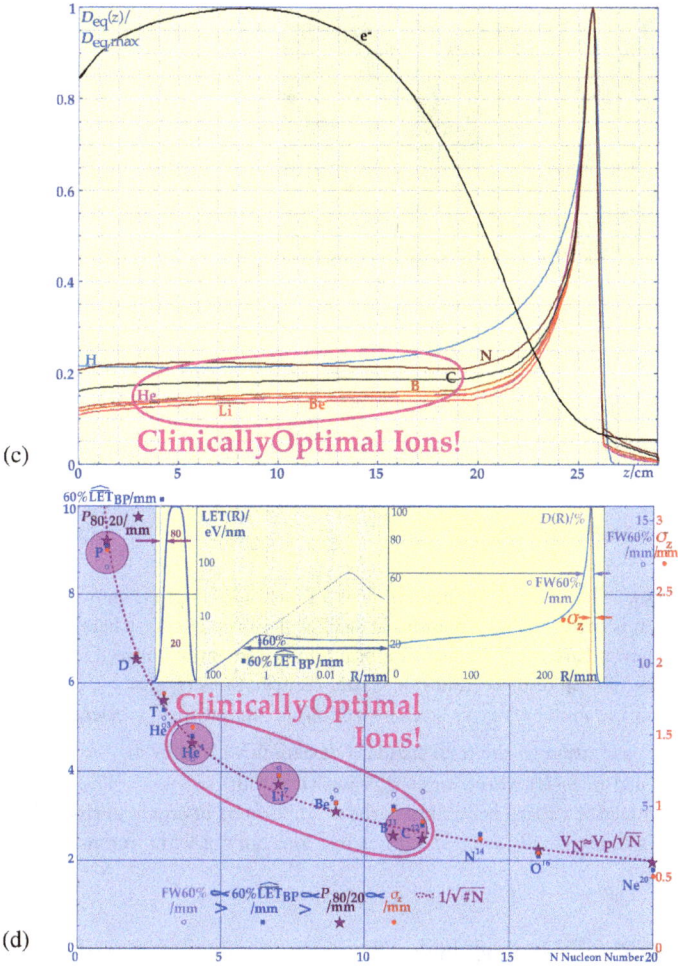

Figure 8.4(c). The biological advantage of lithium to boron ions are illustrated more clearly here by combining the energy deposition in Figs. 8.4(a) and (b) with the biological effectiveness in Fig. 8.3 and normalizing to the obtained biologically effective dose at the Bragg peak. For a given dose in the tumor and Bragg peak, the lowest biological effect in the plateau region with normal tissues is obtained by lithium and beryllium ions (cf. also Fig. 8.9). **(d)** Variation of the penumbra width and longitudinal range straggling for light ions of increasing nucleon number. A sharp reduction to 50% of the wide half value width of protons is seen for helium and almost one-third for lithium and even less for beryllium and beyond. Interestingly, the lateral scattering and penumbra decrease in almost the same fashion as the longitudinal straggling measured as the 60% width of the high LET region of the Bragg peak rather accurately proportional to the inverse square root of the nucleon number. The 60% width is approximately the half width of these latter quantities as seen in the insert and Fig. 8.4(h) (cf. Kempe *et al.* 2007).

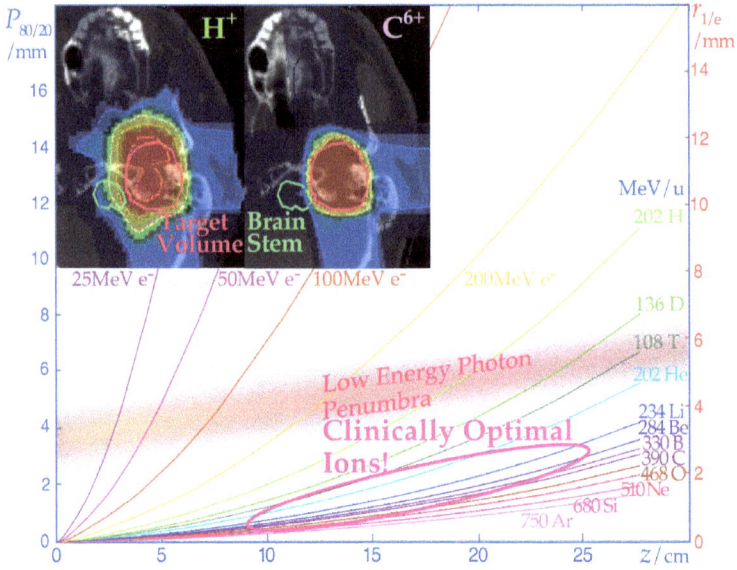

Figure 8.4(e). The increase in penumbra as a function of depth is shown. With electrons and protons it is better than for photons at shallow depths (<5 cm), whereas light ions from helium and beyond are needed to get significant improvements compared to photons at large tumor depths. On top of these, multiple scatter contributions ($r_{1/e} = \sigma_r$, cf. Fig. 8.4(f), (g)) the part from the initial effective source size (σ_0) of the intrinsic accelerator beam should be added in quadrature so the total standard deviation is given by $\sigma_{tot} = \sqrt{\sigma_r^2 + \sigma_0^2}$. This part is included in 8.4(f) where $\sigma_0 = 2,5$ mm ($1/e$ width = 5 mm). The insert shows the clinical advantage of a sharp penumbra in the neighborhood of organs at risk. The brainstem in this case is almost totally avoided with carbon ions but not with protons (courtesy Jürgen Debus).

require that about four times higher dose has to be given to normal tissues in front of the tumor with protons as compared to light ions from lithium to carbon. This is a severe dose delivery drawback for small radiation-resistant tumors and perhaps less of a dosimetric problem for large tumors where the broad beam Bragg peak dose level is more easily reestablished. However, large tumors often have extensive hypoxic regions so protons are not generally the radiation modality of choice, and lithium to carbon ions are more often indicated for larger hypoxic tumors. The clear improvement in tumor coverage and normal tissue avoidance is seen in Figs. 8.4(d)–(g), where also the reduction in penumbra width is seen beyond helium ions. One may ask why there is such a large difference in biological effect between protons and other light ions (cf. Figs. 8.2, 8.3, and 8.4) even though their normalized dose distributions are fairly similar (cf. Fig. 8.4(b)). This is partly due to the same

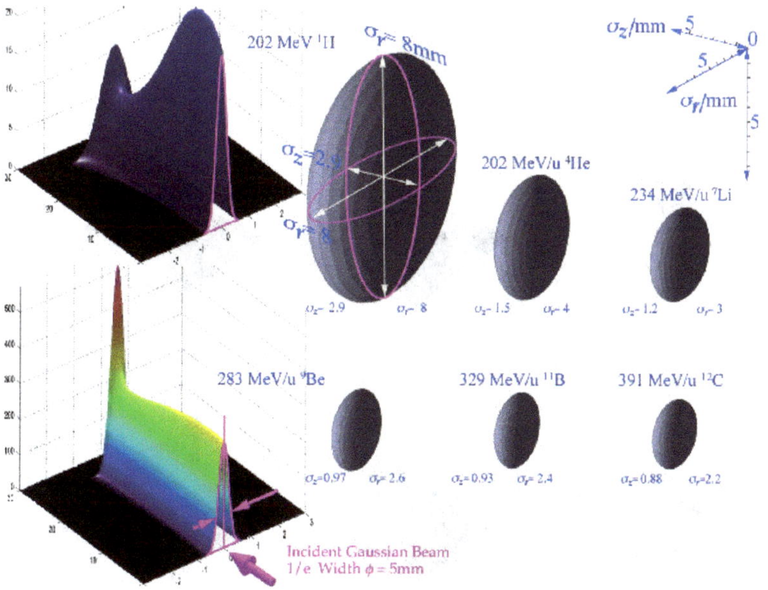

Figure 8.4(f). Illustration of how the Bragg peak is almost gone in a narrow proton beam at 26 cm but not so in a lithium — carbonion beam ($1/e$ width 5 mm). Also shown are the longitudinal and radial modulation of the high dose and LET Bragg peak distribution due to range straggling (σ_z) and multiple Coulomb scattering (σ_r). Beyond helium, the change is more gradual and slow. The large range straggling of protons tend to erase the small high LET component of each individual proton as seen in Fig. 8.4(h). The ellipsoidal volumes of range straggling and multiple scatter correspond to the $1/e$-region over which the true Bragg peak is distributed in a narrow point monodirectional pencil beam (cf. Fig. 8.4(e) at 26 cm depth). Figure 8.4(g) show the more clinically relevant effect in Gaussian pencil beams.

phenomenon that reduces their Bragg peaks in narrow beams (Figs. 8.4(f)–(g)) and increases their range straggling (Fig. 8.4(d)) as explained in Fig. 8.4(h)). The dotted curves show the variation of the stopping power along a light ion path as it slows down. It is seen that it is only during the last 45 microns or so (or 3 to 5 cell diameters) that the LET is >20 eV/nm and a true high LET effect is obtained with protons (cf. also Fig. 8.3). However, in a high-energy proton beam, the range straggling is about 1% of the range and is therefore generally about 2–3 mm (cf. Fig. 8.4(h)) and thus the range straggling almost completely dilutes and erases the high LET regions of the individual proton tracks. The mean LET of the remaining Bragg peak in Fig. 8.4(h) is only about 4 eV/nm for protons but as high as 20 eV/nm for helium, 45 eV/nm for lithium, and about 150 eV/nm for carbon ions as seen in the figure. To compare the dose delivery ability of the light ion beams with those

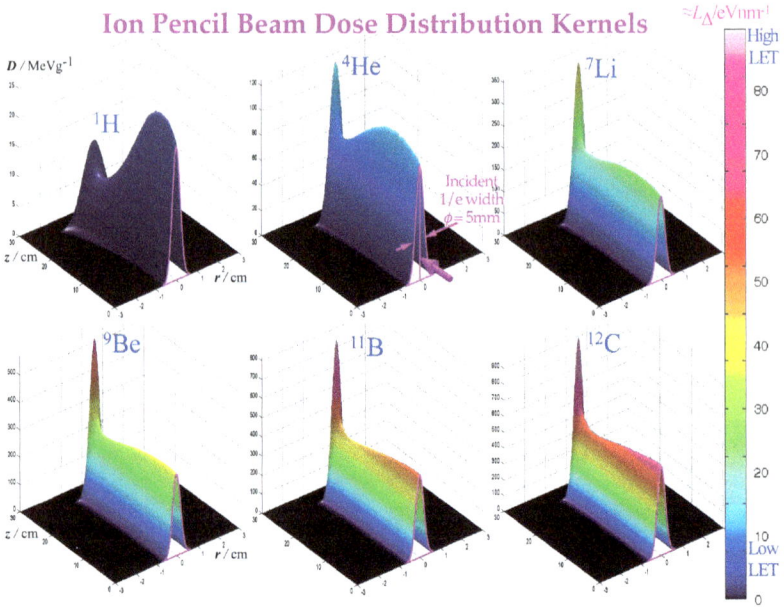

Figure 8.4(g). Illustration of the clinical value of different 5 mm $1/e$ width light ion pencil beams for biologically optimized therapy planning. With protons, the dose to normal tissues in front of the tumor is twice the tumor dose due to significant multiple scatter (cf. Figs. 8.4(e) and (f)), whereas it is only a small fraction of the tumor dose for the light ions from lithium and above. The color scale illustrates to some extent the ionization density and thus the additional increased biological effect in the tumor, which comes as an equally important advantage on top of the physical dose distributional advantage shown in the figure. From carbon and higher, the increasing LET in the plateau type entrance region have to be considered when maximizing the complication-free cure.

of low LET photons and electrons, some of the key dose distribution kernels are compared in Fig. 8.4(i). It is seen that even the most systematic use of narrow photon beams for IMRT in coplanar 2π or stereotactic 4π geometries cannot produce as accurately confined dose distributions as narrow light ion beams beyond helium. Furthermore, their dose distributions are not only sharper and more well confined so are their biological effectiveness as seen from the Bragg peaks in Figs. 8.4(a)–(c), (h), and (j), their unique biological effects in Fig. 8.5, and their apoptotic induction in Fig. 8.9 below.

Similar to high-energy photon beams light ion beams are always associated with a significant neutron production as illustrated in Fig. 8.4(j). Owing to nuclear reactions, the cross-section σ_N and number of neutrons increases with increasing atomic number of the projectile as the number of fragmentation reactions and

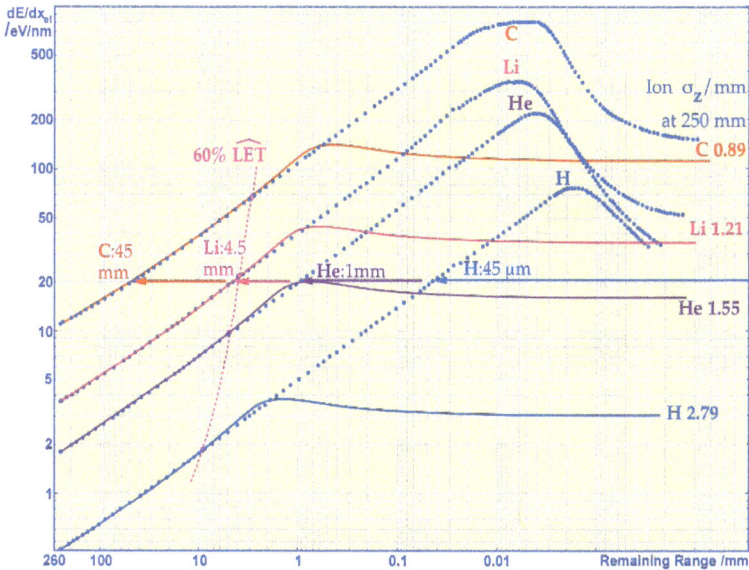

Figure 8.4(h). Description of how the high Bragg peak RBE of a single low-energy proton (cf. Figs. 8.3, 8.4(d)–(f), and 8.6(d)) over some 45 μm is completely lost by range straggling over almost 3 mm in a high-energy beam of many particles. The dotted curves correspond to the mean value along a single ion track, whereas the solid lines account for range straggling in a beam with many ions. The smaller range straggling and higher peak *LET* and wider high *LET* width of heavier light ions make their Bragg peaks have true high *LET* properties beyond helium. Interestingly, the peak mean *LET* is about 1 mm upstream of the practical range independent of ion, whereas the region with an *LET* > 20 eV/nm increases rapidly with ion charge from 45 μm with protons to 45 mm with carbon ions, while the region > 60% of the peak average *LET* slowly decreases with ion charge.

consequently the number of neutrons increases with nuclear size. However, the number of ions needed to eradicate a hypoxic tumor simultaneously decreases rapidly since the *LET* and RBE increases with atomic charge and mass (cf. also Fig. 8.10). Figure 8.4(j) shows, in fact, that the fluence reduction increases both the dose and the effective dose per unit neutron energy fluence generated and thus steadily decreases the neutron production with increasing atomic number of the ion at tumor lethal doses. From the point of view of neutron and light fragment production the ideal ion is generally somewhere between lithium and carbon as seen in Fig. 8.4(j).

It is also seen that both the mean absorbed dose (D_{Bp}) and dose equivalent ($D_{Eq,Bp}$) at the Bragg peak per unit energy fluence of neutrons generated (Ψ_N) increase rapidly with the atomic number, indicating that the absolute neutron production reduces steadily with atomic number. The dose of fragments is

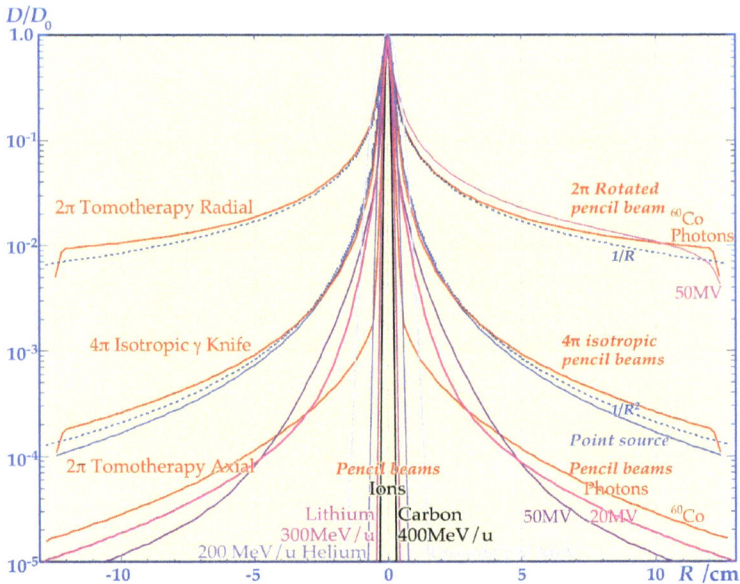

Figure 8.4(i). Comparison of photon and ion beam radial dose profiles for elementary point monodirectional pencil beams as well as for 2π and 4π steradian rotated convergent photon pencil beams are shown. It is seen that ions from about lithium and heavier (cf. Fig. 8.4(d)) always allow a much more accurate dose delivery than photons and protons with a penumbra which is reduced to between half and one-third of present values for photons (cf. also Figs. 8.4(e)–(g)). The longitudinal range straggling is shown in Figs. 8.4(f)–(h) and the fragmentation tail is shown in Figs. 8.4(a)–(c) and result in a longitudinal deterioration of the beam quality, setting in mainly from beryllium and higher as also seen in Fig. 8.4(j). To get ion pencil beam kernels that are sharper than ^{60}Co, photons ions heavier than helium are needed as deep therapy photons have generally sharper penumbra than proton beams as seen in the figure. The protons kernels were derived by Brahme *et al.* (1989) and the photon kernels by Eklöf *et al.* (1999).

unfortunately also increasing. Therefore helium and lithium ions are indicated to be most optimal for pediatric tumor where the dose and biological effect to normal tissues surrounding the tumor should be minimized.

8.3.2. *Small Cell Cycle Variation and Generally No Dose Rate Effect on Cell Survival*

With low LET radiation, the effect on the cell is very much dependent on the phase of the cell cycle at which the cells are irradiated, as shown in Figs. 8.5(a) and (b). Many tumor cells are in the radiation resistant S-phase during the irradiation and are therefore difficult to eradicate by low LET electrons, photons, and protons. High

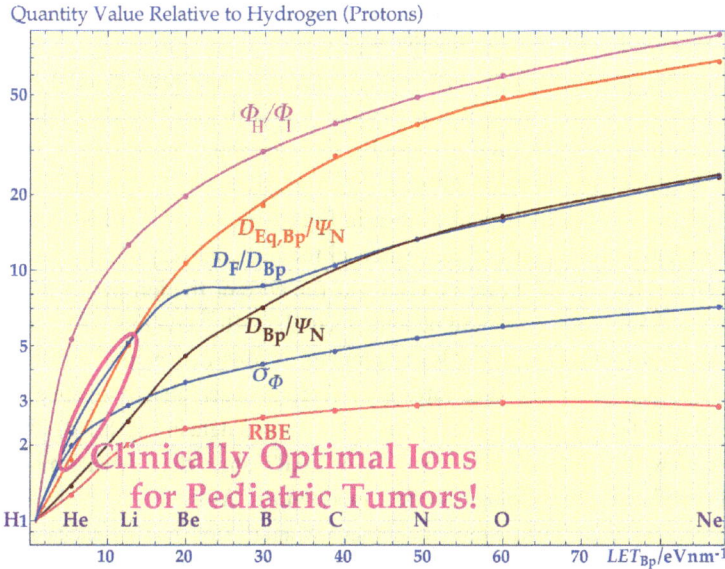

Figure 8.4(j). Variation of the RBE, the Bragg peak, and fragment tail doses, D_{Bp} and D_F, per unit neutron energy fluence, Ψ_N, and fluence, Φ, at tumor lethal doses of light ions as a function of the atomic number (elemental symbols), Z, or more specifically their mean Bragg peak LET values is shown. The values are given in relation to protons (hydrogen ions) and the curves show how much higher the fluence of protons is compared to other light ions up to neon, how much higher the therapeutically effective dose in the Bragg peak is per generated unit of neutron energy fluence, the ratios of tail fragment dose compared to the Bragg peak dose, the Bragg peak absorbed dose per neutron energy fluence, the increasing variance in particle fluence due to a lower fluence of ions, and the mean ion Bragg peak RBE. Interestingly, at the same degree of cell kill, protons produce the highest neutron energy fluence partly because their mass is close to that of the neutron. From beryllium and above, the fragmentation is rather high so helium and lithium are probably the most ideal ion species for the treatment of pediatric malignancies. So even if the cross-section per ion to produce neutrons goes up slowly with atomic number about 5, 15, and 40 times higher fluence of protons is needed compared to helium, lithium, and carbon ions making the total neutron production the highest with curative doses of proton therapy.

LET radiation, such as at the end of the range of light ions, gives almost the same cell survival independent on the cell cycle phase. The light ions are therefore more efficient for eradicating tumor cells since they can no longer hide, for example, in the radiation-resistant late S-phase and early G_2 phase or in the mid G_1- and S-phases and avoid efficient cell kill. This is a very important property since rapidly growing tumor cells have a significant portion of their cells in these resistant phases (cf. Figs. 8.5(a) and 8.25) where non-homologous end joining (NHEJ) and especially sister chromatid exchange are effective high fidelity repair processes.

Quite similar to the cell cycle dependence, the dose rate effect at low LET protects cells exposed at low dose rates as seen in Fig. 8.5(c). This dose rate effect is due to a more efficient removal of sublethal damage when the damage is afflicted at a lower rate so there is less risk for damage interactions. With high LET radiation, this effect is almost entirely gone as seen in Figs. 8.5(c) and (d). This is quite clear from a radiobiological point of view since all the dose delivery of light ions occurs at an extremely high dose rate especially at the core of the ion path where about 10^6 Gy are delivered in a fraction of a picosecond ($\gg 10^{20}$ Gy/min). At a low mean dose rate in an ion beam just a lower number of ions deliver the dose but all of them at the same high local dose rate. Therefore, no observable dose rate effects are expected with light ions as seen from the experimental data in Fig. 8.5(d). From this point of view, light ions are ideally suited for IMRT since the intensity modulation is well translated into a biological effect modulation. At extremely high doses and dose rates, there may be dose rate phenomena when the ionization density is so high that individual ions interact. This is generally not the case during fractionated light ion radiation therapy.

8.3.3. High RBE in the Tumor and Low RBE in the Normal Tissues

Particle radiations such as light ions transfer energy to tissue through low energy secondary electrons of varying density that locally cause a very high ionization density in the irradiated tissues. For light ions, the ionization density increases about four to fivefold at the end of the range (cf. Figs. 8.3(a) and (b), 8.4(a)–(c), 8.5(f), 8.7(a), and 8.8(l)). Photons, electrons, and protons produce a rather low and uniform ionization density at all depths and are generally considered to be low LET radiations even though each proton posses a narrow Bragg peak with elevated LET (cf. Figs. 8.4(a) and (h)). High LET radiations increases the RBE (cf. Figs. 8.5(e), (n), and (p)) particularly at the end of the range at the so-called Bragg peak (cf. Figs. 8.4(h), 8.6, and 8.7). The RBE is around 2 to 5 for light ions at the Bragg peak with an ionization density of around 50–150 eV/nm and this densely ionizing region should only be located in the tumor during therapy to avoid damage to normal tissues. This means that for a given effect in the tumor, it requires the dose to be higher by a factor of about 3 with low LET radiations such as photons, electrons, or protons to get a similar therapeutic effect. With lightest ions beyond protons, the most favorable therapeutic situation occurs for radiation-resistant tumors against the conventional low LET radiations, since the RBE is then high at the tumor depth,

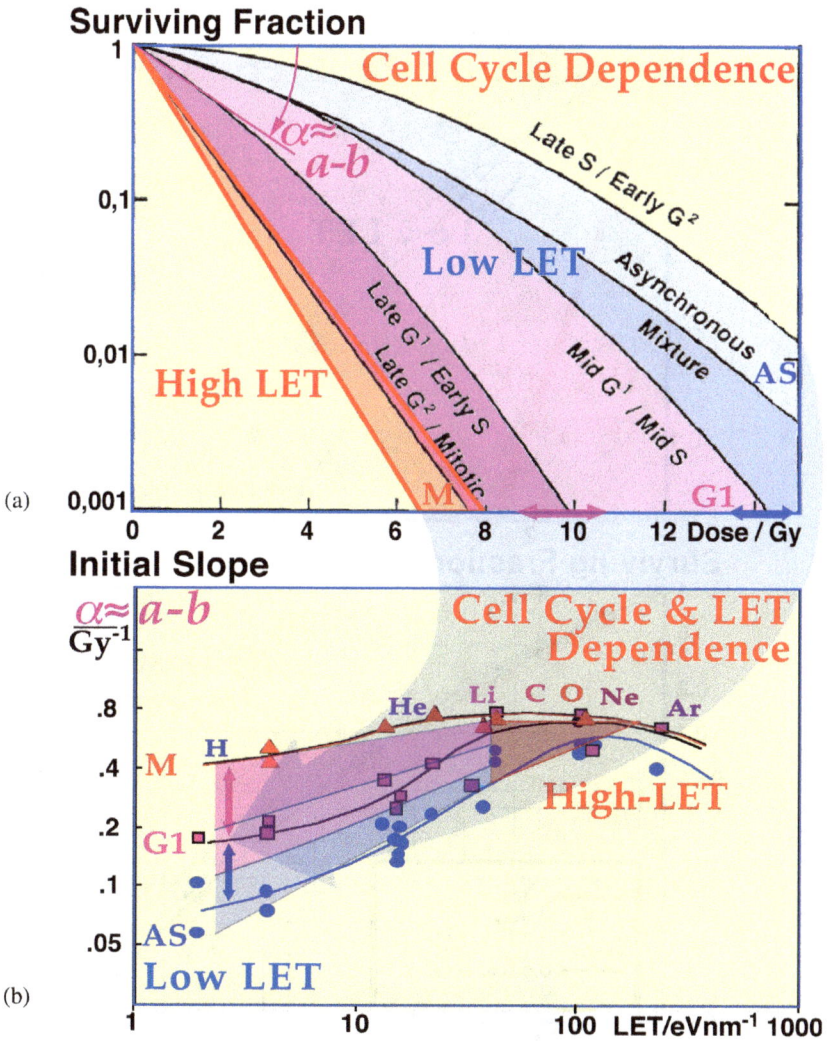

Figure 8.5(a). Illustration of the cell survival that varies quite a lot with cell cycle phase with sparsely ionizing photons, electrons, and protons. The synthesis phase and Gap1 are very radiation-resistant, where as the late Gap2 and the mitotic phases are generally very sensitive. Instead, all phases are almost equally sensitive with densely ionizing light ions and tumor cells cannot escape cell kill as they do with photons and protons in, for example, the S phase, which is commonly occupied by rapidly dividing tumor cells. **(b)** Illustration of the variation of the initial linear cell survival curve slope α with LET illustrating the LET dependence in Fig. 8.5(a) more clearly. The large variability in α (cf. Eqs. (8.3), (8.20), and (8.30), below) at low LET is reduced to negligible amounts as the LET reaches 50–200 eV/nm for lithium ions and beyond (modified from Hall and Giaccia (2006)).

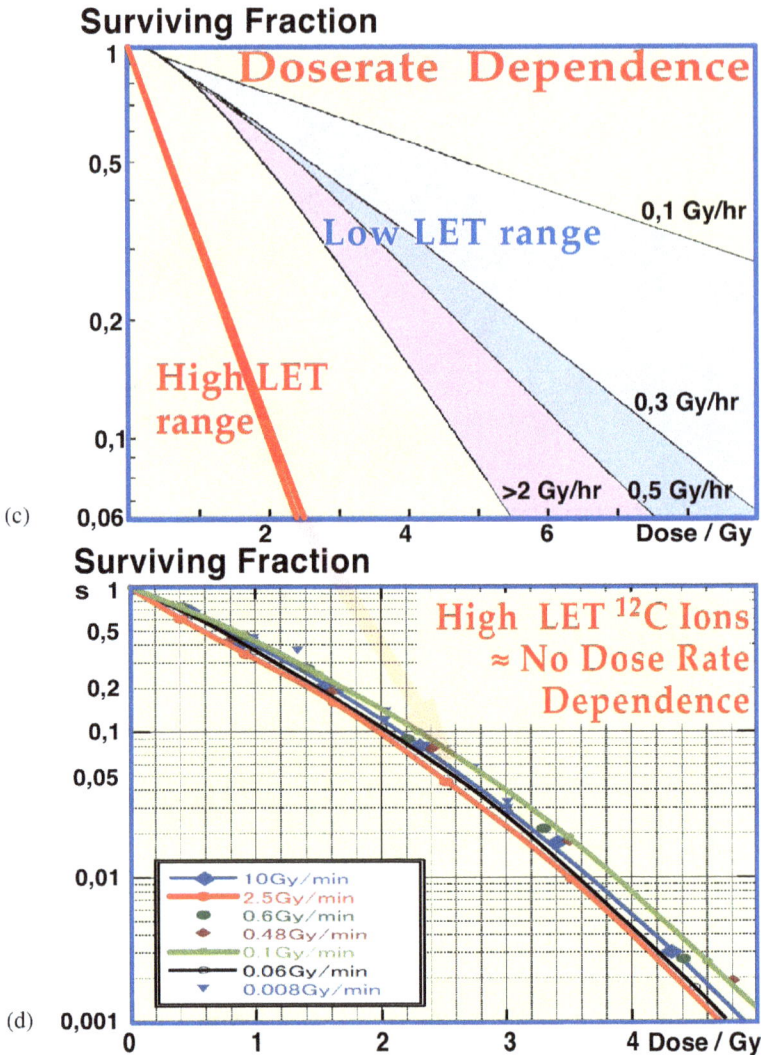

Figure 8.5(c). Since most of the lethal cell kill with densely ionizing ions is along single ion tracks, there is practically no dose rate dependence on the cell survival, contrary to what is generally seen for sparsely ionizing photons and electrons. **(d)** For carbon ions, as expected from Fig. 8.5(c), there is no significant dose rate dependence between 0.01 and 10 Gy/min (modified from Hall and National Institute of Radiological Sciences (NIRS), respectively).

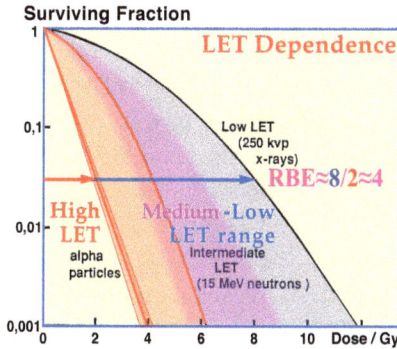

Figure 8.5(e). The RBE is measured as the dose ratio at a given survival level and may reach values between 3 and 5 around the Bragg peak of light ions but is only around 1.5 in the plateau region. At *LET* values >70–120 eV, the RBE decreases as seen in Figs. 8.3, 8.5(f) and 8.8(l).

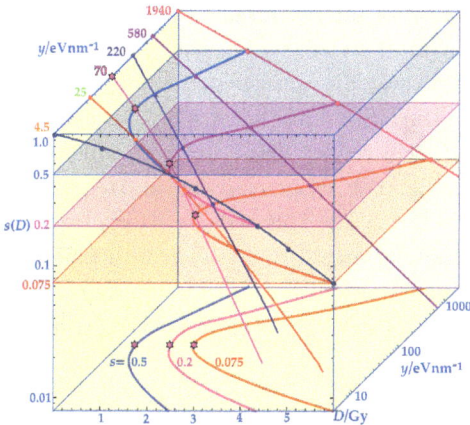

Figure 8.5(f). This 3D plot shows how the shape of cell survival curves changes with increasing *LET* (the oblique-axis pointing into the paper) starting from curvy with X-rays to more and more straight and steep (cf. Figs. 8.5(e) and (g), and 8.8(a)) toward 70–100 eV/nm. Beyond this *LET* peak, the ions loses effectiveness toward the very high *LET* values due to saturation of the inactivation cross-section (Fig. 8.9(b)), increasing radical–radical recombination and over kill phenomena. The peak effectiveness at 70 eV/nm is clearly seen at each survival level as projected at the bottom *y–D* plane (cf. also Fig. 8.3 and 8.8(l), data from Todd 1964).

and as low as possible in the plateau region, when passing through normal tissues as clearly seen particularly for lithium and beryllium ions in Figs. 8.1, 8.4(a) and (c), 8.8(j), and 8.9(c). From a therapeutic point of view, a high LET and RBE is not a strong therapeutic advantage on its own as was seen for the neutrons since an increased therapeutic efficiency in the tumor was counteracted by an increased damage to normal tissues too, unless combined with low LET photons in mixed treatment schedules (cf. also Fig. 8.33(b)). To make a beam as useful as possible from a therapeutic point of view (Brahme *et al.* 1980, 1982, 1983), a high and even LET and dose is desired in the target volume at the same time as these quantities should be as low as possible in the normal tissues. These goals are best fulfilled by lithium and beryllium ions, at least for small tumors (cf. Fig. 8.9), whereas for larger hypoxic tumors boron and carbon and sometimes even oxygen ions may be suitable.

Figures 8.8(a)–(o) show how well the new RCR (potentially repairable and conditionally repairable damage) model (see Eq. (8.20), below and Lind *et al.* 2003) describes the interaction of different types of DNA damage. The model is based on the interaction of the two major DNA damage and repair pathways of mammalian cells: the most severe damage resulting in dual double strand breaks (DSB) (cf. Fig. 8.7) and multiple damaged sites require homologous recombination (HR), whereas most plain DSBs can be handled by NHEJ. The potentially repairable lesions are thus easily repaired by NHEJ, however, with low fidelity at high dose rates, whereas the conditionally repairable lesions also require the high fidelity repair of HR (Brahme and Lind 2008). This is the case for a large part of the high LET damage causing multiple damaged sites such as dual DSBs (cf. Figs. 8.7(a)–(c)), which have a high risk of misrepair requiring HR to step in during the G_2M phase to really make sure that the misrepair is eliminated and corrected by HR.

8.3.4. *Low Dependence on the Tumor Oxygenation Status*

The biological effect of low-to-medium LET radiation is dependent on the local oxygen concentration in the cell since oxygen radicals then mediate a substantial part of the cell kill. The OER might be defined as the increased radiation resistance in absence of oxygen in the cell. The OER is generally between 2 and 3 for low LET photons, electrons, and protons (cf. Figs. 8.5(g)–(i)). Some of the tumor cells are often living under hypoxic conditions and at least some of these cells are therefore more resistant to conventional low LET radiations.

Often 2.5–3 times more dose of a low LET is, therefore, needed for eradicating hypoxic tumor cells as compared to well-oxygenated tumor cells. The OER is lower

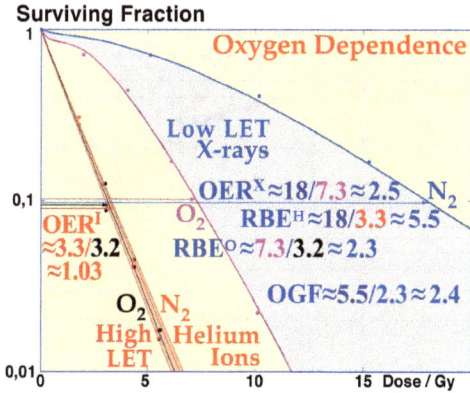

Figure 8.5(g). The cell survival curves of high (Bragg peak helium ions) and low *LET* (X-rays) radiations under well-oxygenated (O_2) and hypoxic conditions (N_2) are shown. The reduced dependence on the oxygenation status of the tumor when using Bragg peak light ions versus sparsely ionizing electrons, photons, and protons are clearly seen (modified from Raju (1980)). The OER is almost unity for Bragg peak helium ions, whereas X-rays need to deliver about 2.6 times higher doses for the same cell survival and tumor response under hypoxic conditions at the 10% survival level. In a therapeutic helium ion beam, the OER is higher due to range straggling and dilution of the high *LET* effect (cf. Fig. 8.4(h)).

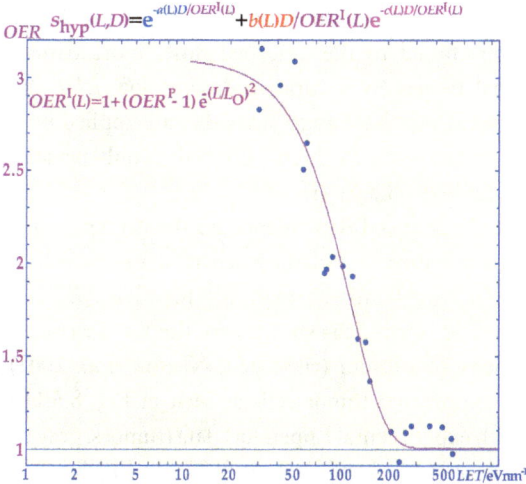

Figure 8.5(h). The *LET* dependence of the OER for carbon ions of different *LET* is shown. The analytical Eqs. (8.17)–(8.19) are based on the cross-section and biological effectiveness in Figs. 8.3(a) and (b) and 8.8(j) and (l) and describe the experimental data (dots) of Furusawa *et al.* (2000) very well.

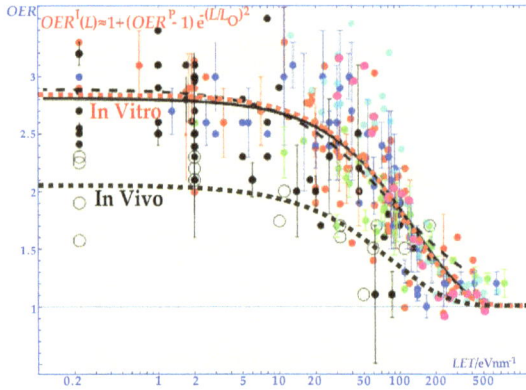

Figure 8.5(i). The *LET* dependence of the OER for different cell lines *in vitro* (circles) and *in vivo* is shown. *In vivo* a spectrum of oxygen tensions is always available as shown in Figs. 8.5(j) and (k) lowering the effective OER at low *LET*s. Pink dots: Furusawa *et al.* (2000), red dots: V79, blue: T1, green: R1, cyan: HSG (fine dashed lines according to Eq. (8.19), modified from Wenzel and Wilkens (2011)).

for high *LET* radiation, which favor the use of light ions with hypoxic tumors (OER ≈ 1.5–1.7). Because the number of DSBs and thus the genetic damage is approximately proportional to the absorbed dose, more damage is inflicted to surrounding normal tissues by a curative dose of low *LET* than by high *LET* radiation. Narrow local light ion Bragg peaks that are applied in the tumor volume further improves this fact since the high LET is then only present there and not in the normal tissues (cf. Figs. 8.5(g)–(i)).

In Fig. 8.5(g), the reduced dependence on the presence of oxygen and more generally on the dominance of radiation-sensitive tumor cells is clearly shown for Bragg peak helium ions and schematically for high *LET* ions in general (cf. Figs. 8.5(h)–(i)). In Fig. 8.5(j), it is shown how the low vascular density and high vascular heterogeneity of a tumor (right half, Nilsson *et al.* 2002) produce a high number of poorly oxygenated tumor cells as seen in Fig. 8.5(k) (solid curves) in good agreement with experimental Eppendorf data (tumors: gray histogram, normal tissues: open histograms; Lind and Brahme 2007). In Fig. 8.5(l), it is seen how the oxygenation curves in Fig. 8.5(k) are converted to cell survival curves of decreasing steepness as the hypoxic fraction increases. These curves are further converted to dose–response curves in Fig. 8.5(m), requiring higher and higher doses as the hypoxic fraction increases. In Fig. 8.5(n), it is furthermore shown how a small hypoxic fraction in a tumor significantly reduces the slope of the DRR with low *LET*

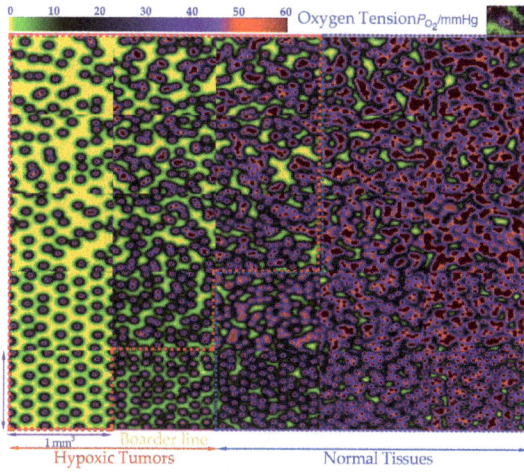

Figure 8.5(j). Illustration as to how a reduced density and increased randomness in the geometric distribution blood vessels reduce the oxygenation particularly at large distances from the vessels and increase the number of cells with significant hypoxia ($P_{O2} < 5\%$, Nilsson *et al.* 2002). The color look up table (upper left scale) was adopted from that of hypoxic tissue markers (upper right insert, courtesy van der Kogel).

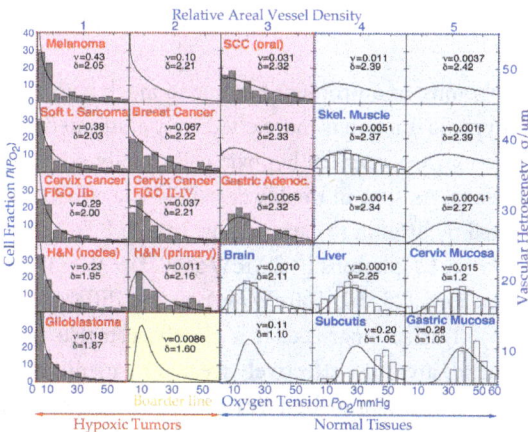

Figure 8.5(k). The vascular model of oxygen diffusion as seen in Fig. 8.5(j), (and given here by the solid line cellular oxygenation curves) agrees very well with many clinically observed Eppendorf data sets for tumors (pink panels with gray histograms in left half) and normal tissues (blue panels with open histograms in right half). It is clearly seen that most tumors have a significant low oxygenation and radiation-resistant cell fraction as opposed to most normal tissues that are well oxygenated and radiation-sensitive as seen in Fig. 8.5(l).

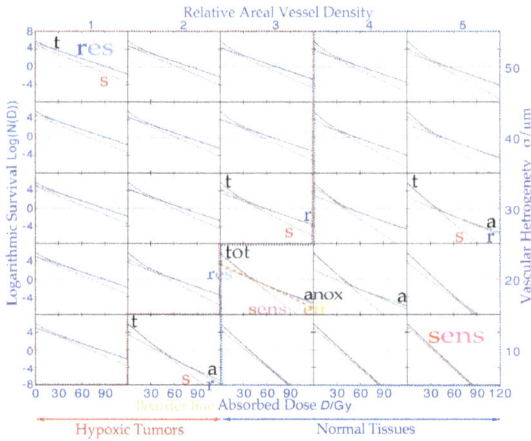

Figure 8.5(l). The logarithmic cell survival as a function of the delivered low LET photon dose decreases more steeply with increased vascular density and decreased vascular heterogeneity as described by the model of oxygen diffusion in Figs. 8.5(j) and (k). Hypoxic tumors have a very shallow response, whereas well-oxygenated normal tissues have a steep response and are quite sensitive. The sum of the effective, sensitive, and a resistant cell lines describe the total response of the complex cell distribution in Fig. 8.5(k) very well. For low LET photons, the resistant response will generally dominate at high curative doses (cf. Fig. 8.5(n)). For high LET ions, the sensitive response will instead practically dominate throughout as shown by the associated dose–response curves in Figs. 8.5(m) and (n).

beams (electrons, photons, and protons) but not so much with lithium and carbon ions since a small hypoxic fraction is more effectively eradicated by the medium-to-high LET dose fraction, whereas the hypoxic cells totally dominates the response for low LET radiations. As seen in the lower panels, the effect of the hypoxic cells are largely eliminated by lithium to carbon ions. In Fig. 8.5(o), the variation in the effective DRR for low LET photons (≈ 2 MeV/gcm^{-2} = 0.2 eV/nm) and medium-to-high LET carbon ions (25, 50, 100 eV/nm) on the different forms of hypoxic tissue in Figs. 8.5(j)–(l) is summarized. It is clearly seen that the hypoxic tumors in the upper left corner are most effectively treated by medium LET carbon ions, whereas well-oxygenated tumors are best treated (requiring the lowest effective dose) by low LET photons, electrons, or protons. This again demonstrates the advantage of using medium LET ions in the range 20–40 eV/nm at least for hypoxic tumors. In Fig. 8.5(p) finally, the cell survival for a range of different human tumors are shown indicating the varying effect apoptosis and senescence in different tissues, and in Fig. 8.5(q) the variation in RBE and OER for different ions is summarized

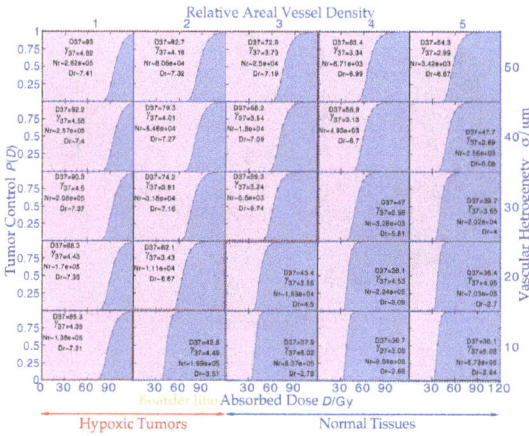

Figure 8.5(m). The cell survival data in Fig. 8.5(l) correspond to the DRRs in this figure when irradiated with low-*LET* photons. It is seen that most tumors require quite high doses for cure almost into the 100 Gy regions, whereas some normal tissues are damaged already at the 40 Gy level. The effect of high-*LET* beams is shown in Figs. 8.5(n) and (o).

indicating the most useful clinical properties at intermediate *LET*s between lithium and boron.

8.3.5. *Simultaneously Low OER and High Dose, LET, and RBE in the Tumor*

From the above discussion of Figs. 8.3, 8.4, 8.5 and Figs. 8.12–8.14 below, it is clear that the light ions from lithium to carbon possess very interesting therapeutic properties as they combine in the Bragg peak all the properties one would like in a hypoxic tumor at the same time as they mainly have a low *LET* character in the surrounding normal tissues as summarized in Fig. 8.5(q). For tumors in organs where an intact internal normal tissue stroma is important for survival, the lightest ions with low to slightly elevated *LET* are most advantageous (cf. Figs. 8.4, 8.9, and 8.14). A very high *LET* may also be less advantageous for young patients when there is a considerable life expectancy after therapy, because there may be a marginally increased risk of a secondary malignancy. The high *LET* Bragg peak should then be used only in the gross tumor which should be exposed to a high dose level from each beam so that even if a secondary tumor is induced there is a high probability that it will also be sterilized by the same high local dose level that is used in sterilizing the primary tumor.

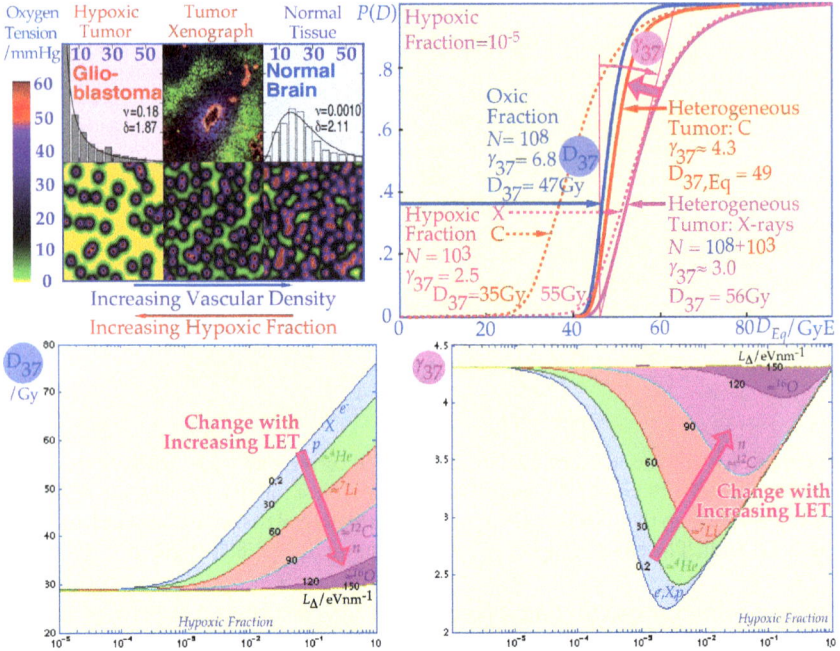

Figure 8.5(n). With increasing hypoxia, the D_{37} (and D_{50}) the low LET dose causing 37% (and 50%) tumor cure increases rapidly. However, the normalized slope of the DRR (γ_{37}) first decreases at low LET values as the hypoxic fraction increases due to the increasing tumor heterogeneity and a dominating small hypoxic compartment and then increases as most of the tumor clonogens become hypoxic. The upper left panel show that the degree of hypoxia in a glioblastoma tumor is much larger than in the normal brain which contributes to the radiation resistance of these tumors and make light ions the modality of choice at least for the primary tumor (cf. Figs. 8.5(j) and (k)). The upper right panel shows how a small hypoxic compartment can totally dominate a low LET treatment, whereas with high LET carbon ions, they are no problem and only marginally increase the effective dose needed for cure. The loss in γ_{37} and increase in D_{37} and D_{50} are substantially smaller with high LET lithium or carbon ions as shown in the lower pair of panels. The D_{37} and D_{50} value rises steadily and may be increased 2–3 times, whereas the γ_{37} value at first is reduced due to the increased heterogeneity and then reverse and increase as the tumor becomes uniform and almost totally hypoxic. Many clinical tumors are close to the lowest γ value (≈ 2) and carbon ions bring this low point back up to $\gamma \approx 3.5$ region with a steeper and more effective dose response (cf. Figs. 8.23(d)–(f)).

To understand the unique clinical properties of light ions, their radiation biological effects are of key importance. With conventional low ionization density or low LET radiations, the microscopic dose distribution on the subcellular scale is fairly uniform except for the effects of single low energy electron track ends or

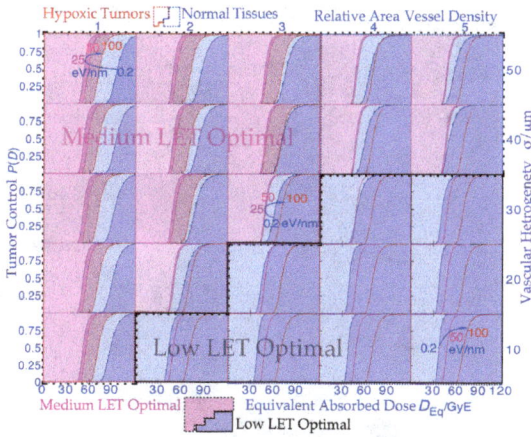

Figure 8.5(o). Overview of how the effective DRR is influenced by the *LET* of the beam from 0.2 eV/nm photons via 25, 50 to 100 eV/nm carbon ions on each tissue hypoxia type taken from Figs. 8.5(j)–(l). Interestingly, the medium *LET* from 20 to 50 eV/nm are most efficient for hypoxic tumors, whereas X-rays are optimal for well-oxygenated non-radiation-resistant tumors.

δ-rays as shown in Figs. 8.6, especially 8.6(c), and 8.7. With ions, the central ion path functions as a source of low energy δ electrons, and at the Bragg peak they are produced at a very high density so that multiple δ electrons contribute to the very high local energy deposition density. In Fig. 8.6(b) the mean radial dose profile across a proton and oxygen ion path is shown, indicating the somewhat larger radial inactivation radius and cross-section with oxygen as compared with protons (cf. also Figs. 8.3 and 8.8(k)). Accurate treatment planning with light ions, therefore, needs to consider the degree of hypoxia in the tumor as discussed in more detail in the last mentioned two references (Nilsson *et al.* 2002; Lind and Brahme 2007). In addition to the increased therapeutic effect on hypoxic tumors, medium *LET* (25–50 eV/nm) ions have a significantly increased probability to induce programmed cell death or apoptosis as seen in Figs. 8.5(p) and 8.9(a).

Beside senescence, this is the mildest form of tumor cell inactivation with negligible inflammatory responses. Lithium and beryllium ions are most interesting in this respect since the apoptotic response is almost only induced at the Bragg peak. Such beams are therefore ideal to, for example, eliminate small oligometastases without much damage to surrounding normal tissues.

A first impression of the radiobiological and clinical consequences of the radial profiles in Fig. 8.6(b) can be obtained by comparing experimental *RBE* values with the radial profiles obtained. This is done in the lower right panel of

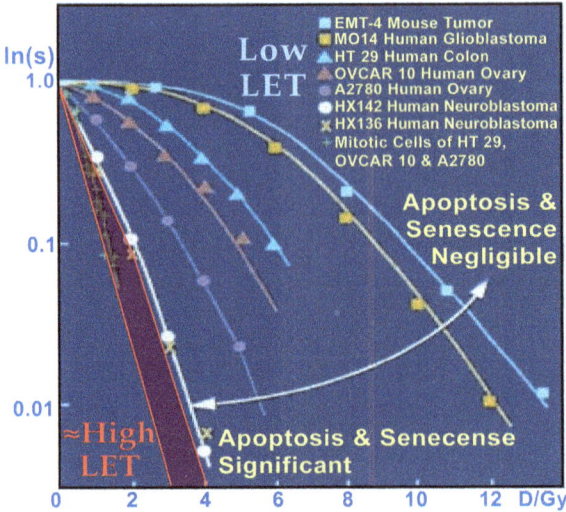

Figure 8.5(p).   The wide range of cell survival in human tumors for low *LET* radiations and the commonly observed associated smaller range of variability for high *LET* Bragg peak ion beams are shown (modified from Hall and Giaccia (2006)). Apoptosis and senescence are key non-repairable inactivation mechanisms of the high *LET* beams that cause almost straight cell survival curves.

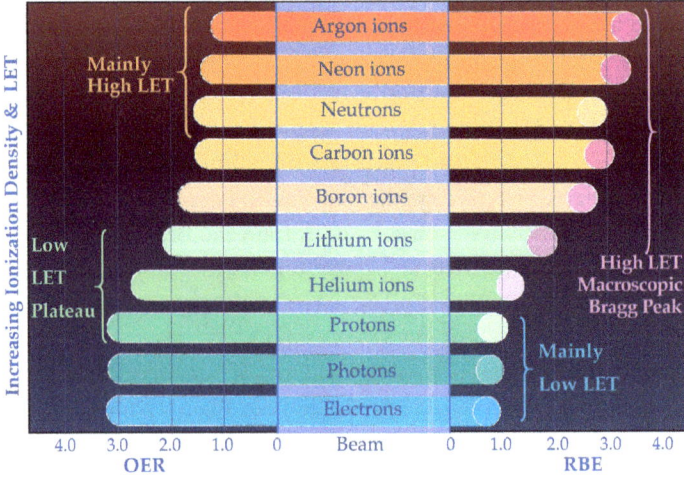

Figure 8.5(q).   Overview of how the RBE and OER change with particle species. For the ions the mean values for an extended Bragg peak are given. For clinical use, a large difference in OER and RBE at the extended Bragg peak and the plateau are most important. For small tumors, this difference is highest for lithium and beryllium ions (modified from Raju (1980)).

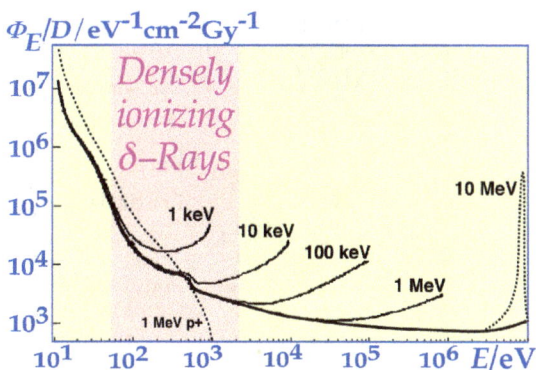

Figure 8.6(a). Illustration of the similarity of the electron slowing down spectrum per unit dose of low *LET* radiations with arbitrary initial photon or electron energies >100 keV. Since the high *LET* track end part of the spectrum between 50 eV and 1 keV is independent of the initial energy in this range, the absorbed dose describes the biological effect in the low-*LET* to high-energy range very well. Low-energy photons and electrons have a higher biological effect (cf. Fig. 8.7(a)) and marginally so has the high-energy proton beams ($RBE \approx 1,1$). In low-energy proton beams, the range straggling is small and the high RBE can be seen in *in vitro* cell cultures (cf. Fig. 8.3(b)). At higher *LET*s the δ electron density increases even more along the track mainly due to the short mean free path of the ions and the radial convergence of the δ electrons on the ion track as seen in Fig. 8.6(b).

Fig. 8.6(c), where experimental RBE data for low-energy protons and α particles are plotted as a function of their respective LET values. Interestingly, they both reach almost the same maximum RBE of around 6, but at quite different LET — between approximately 25 and 30 eV/nm for protons and 100–120 eV/nm for helium. This large difference in LET can easily be understood since the electron spectra (Fig. 8.6(a)) and ion radial dose profiles (Fig. 8.6(b)) of energies around 1 MeV per nucleon are almost identical to the values used in the upper right panel. The hydrogen ions in Fig. 8.6(c) (lower left panel) should therefore have almost the same relative radial dose distribution and electron spectra (cf. Fig. 8.6(a) and upper right panel in Fig. 8.6(c)); the only difference is that at a given dose, the fluence of ions is four times higher with protons than with α particles due to a four times higher LET for the latter. Therefore, the only difference in the microscopic energy deposition between hydrogen and helium ions is that with helium a single ion does the same damage as four protons would do on the same path, whereas the four protons are almost always well separate in the case of hydrogen, as illustrated in Fig. 8.6(c) (upper left panels). There is also a slight difference in the secondary electron spectra per unit dose as seen in the upper right panel in Fig. 8.6(c), corresponding to about 30% more electrons with helium in the sub-kiloelectron-volt range. This may partly

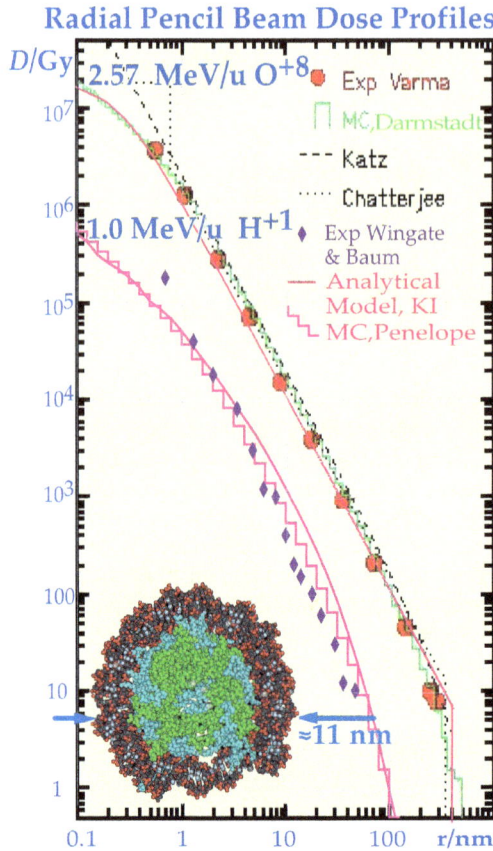

Radial Pencil Beam Dose Profiles

Figure 8.6(b). Comparison of experimental and calculated radial dose profiles through point monodirectional proton and oxygen beams near the Bragg peak is shown. It is seen that the central axis dose is lower and the radial width narrower for the protons. This is also the reason why protons have a lower effective cellular cross-section, whereas oxygen ions have reached the saturation value where the cross-section is practically equal to the molecular size of the DNA in the cell nucleus. Interestingly, the high dose region is of a size comparable to the beads on a string nucleosome fiber of 10 nm and the more condensed 30 nm fiber. For protons a comparison is made between the accurate distorted wave calculated profile (solid line: analytical and histogram: Monte Carlo and experimental data: rhombic (Wiklund *et al.* 2008)). The similar high *RBE* in the Bragg peak region of track segments of all ions is explained by this similarity in individual ion beam profiles. However, for protons, the effect is largely erased in clinical beams with substantial range straggling (cf. Fig. 8.4(h)).

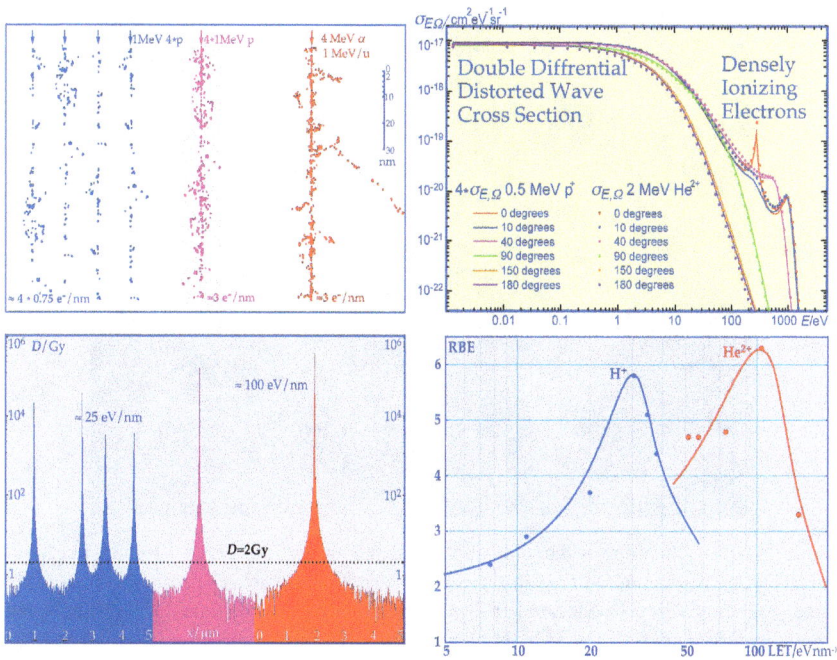

Figure 8.6(c). Comparison of proton and helium ion tracks near the Bragg peak (1 MeV p$^+$ 4 MeV/u He^{2+}, upper left panel, of Paretzke (1987)) is shown. Owing to the higher charge of helium ions (+2) compared to protons (+1), the stopping power is almost fourfold higher at the same velocity, since the electron slowing down spectra are almost identical for one helium ion and four protons travelling on the same path (middle track, upper and lower left panels, Wiklund *et al.* 2008). Interestingly this explains why helium ions have their *RBE* peak at fourfold higher *LET* value, 100–120 eV/nm instead of 25–30 eV/nm for protons (lower right panel, cf. Cera 1997). The upper right panel show that the double differential cross-section based on accurate distorted wave cross-sections for secondary electron production is slightly higher for helium ions than for protons and so is the *RBE* when the "four protons" interact on the same path (higher probability for local severe DNA damage). These facts explain the slightly higher peak value for the helium ions in lower right panel.

explain the slightly higher peak in RBE value with αparticles than with protons, as seen in the lower right panel of Fig. 8.6(c). In addition, there is also a difference in cell killing due to the higher local energy deposition density since it increases the probability of generating locally concentrated multiply damaged sites. Such sites are known to be more lethal than simple, easily repaired DSBs that are repaired though the processes of NHEJ and HR. The most common events are the dual DSBs, which are hard for the cell to repair correctly since four free DNA ends are produced within a 5–15 nm region. Several correlated track experiments (Bird 1979; Kellerer 1980;

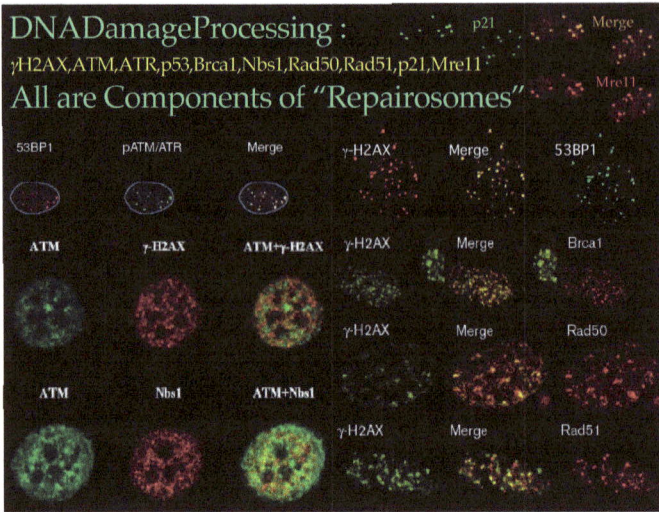

Figure 8.6(d). Illustration of photon-induced repair enzyme foci or repairosomes randomly induced at a rate of about 40/Gy in the cell nucleus. The foci are repairosomes since all the repair gene products are collocated at these sites of the repair machinery. Interestingly, less than one of these are lethal at 2 Gy (cf. Figs. 8.7(e)–(f)).

Zaider 1983) have shown increased cell killing with unusually close light-ion tracks. Thus, taken together, the latter two phenomena indicate a somewhat higher RBE for helium ions at the RBE maximum.

The severe damage on supercoiled DNA first wound two times around nucleosomes in the cell nucleus is shown in close-up in Fig. 8.7(b). Most toxic are the ≈ 700 eV electrons that deposit a dose in the neighborhood of the track of around 10^6 Gy as seen in Figs. 8.7(a)–(c). In front of the Bragg peak, the density of electrons is lower and even more so in the high energy entrance region requiring many more ions to deliver a dose of around 2 Gy (lower right panel of Fig. 8.10(c)). This phenomenon explains why the medium *LET* beams are most efficient in inducing apoptosis as seen in Fig. 8.9(a) and thereby eradicating hypoxic tumor cells. With a low *LET*, too few severe direct cell kill events are obtained and at high *LET* too few ions are available at a given dose even though they produce very severe damage. Figure 8.7(c) shows that the most probable DNA fragment length at high doses is around 78 base pairs, corresponding to a single turn around the nucleosome. This can easily be obtained by δ electron track ends that randomly hit the DNA at any point on the periphery of the nucleosome and then often produce dual DSBs. This

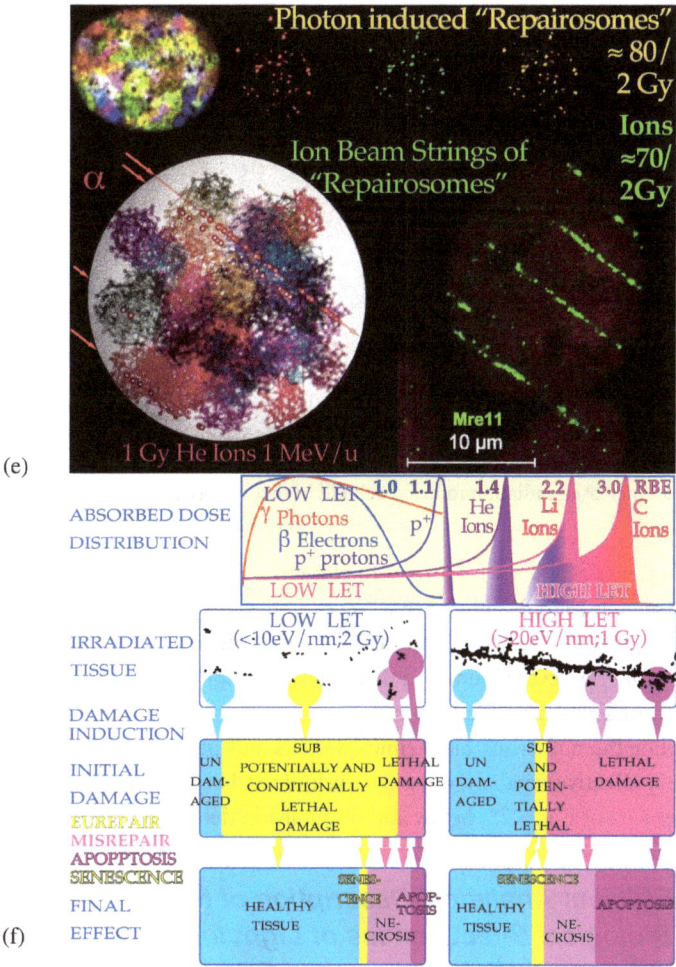

Figure 8.6(e). Fluorescence *in situ* hybridization (FISH) photographs of the different chromosomes in a cell nucleus (upper left corner, courtesy T Heiden). A similar color scheme is used for helium ion track simulation (lower left, Chatterjee *et al.*, Berkeley). Upper right panel shows that Bragg peak ions with slightly fewer foci are induced, but they are perfectly aligned along the ion path (lower right panel, Darmstadt). Interestingly about 75 DSBs are produced both at low and high *LET*s. **(f)** Comparison of the type of damage and repair induced by low (right column) and high *LET* beams (left column) is shown. It is seen that much more of the sublethal damage is repaired at low *LET* and that the apoptotic cell kill is higher with a high *LET* (cf. also Figs. 8.5(h) and 8.9) due to higher probability of dense ionization clusters along the ion tracks (second row of panels). The high *LET* portion and RBE rapidly increase after protons (essentially low *LET*) to helium, lithium, and carbon ions.

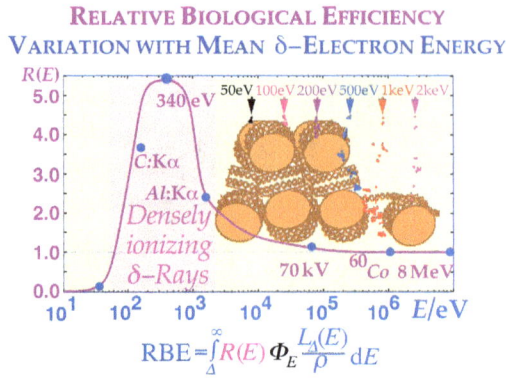

Figure 8.7(a). The cell inactivation induced by low energy δ electrons at increasing geometrical resolution from Figs. 8.7(a)–(e) is shown. It is seen that electron energies between 200 eV and about 1 keV can produce the largest concentration of clustered damage to DNA, which results in an increased radiation induced cell kill in that energy range (Brahme *et al.* 2001).

will very often make DNA fragments of close to a single nucleosomal DNA turn in length as seen in Figs. 8.7(b) and (c).

In Fig. 8.7(d), the wide panorama of DNA damage types are summarized indicating that the severe blunt end dual nucleosomal DSBs and multiply damaged sites are most severe and the high proportion of them make Bragg peak light ions so toxic. For the numerous biochemical lesions and plain single strand break and DSBs, very effective repair systems exist as indicated in the lower rows of Fig. 8.7(d) and upper part of Fig. 8.25, below and discussed in detail below Eq. (8.6).

8.3.6. *Efficient Analytical Description of the Biological Effectiveness of Light Ions at Low and High Doses and LETs*

The cell survival and biological effect is generally dependent on a very large number of factors from the properties of the radiation quality via the physical–chemical interaction of the beam with DNA and other subcellular structures such as the cell membrane to the cellular response determined by multiple genetically controlled pathways for DNA damage surveillance, cell cycle control, and damage interaction and repair. This makes it dependent on many secondary factors such as the heterogeneities of the tissue and its functional organization as well as the time course of irradiation and environmental factors such as the oxygenation, temperature, and nutritional status and their possible variation before, during, and after the irradiation and also on the distribution of cells in different phases of the cell cycle etc. (cf.

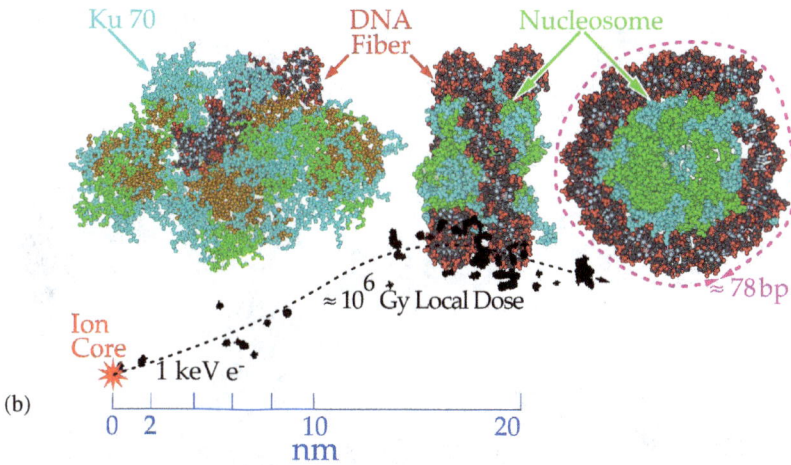

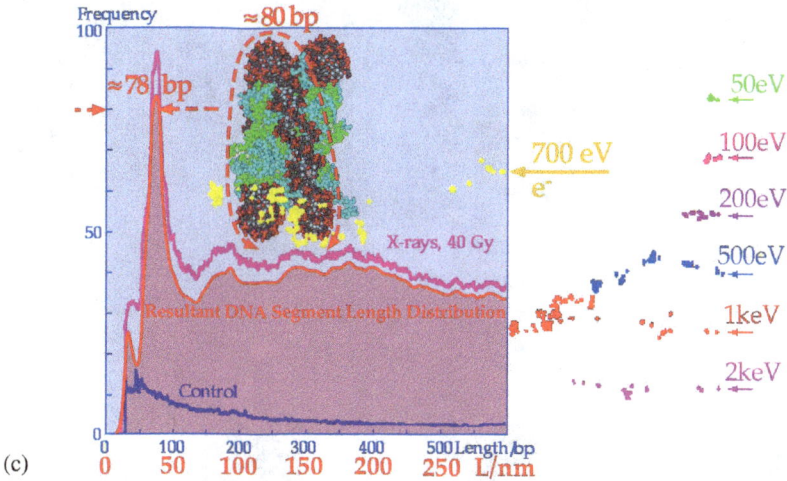

Figure 8.7(b). Molecular close-up view showing that most of the lethal cell damage of densely ionizing ions is induced by low energy δ electrons in the 200 eV to 1 keV energy range that can generate severe difficult-to-repair DNA damage in the cell nucleus such as dual DSBs at the periphery of the nucleosome (cf. Brahme *et al.* 1997). **(c)** Shows that at high doses of low *LET*, the most common DNA segment length corresponds to a single turn of DNA around a nucleosome as expected from the dual DSBs at the periphery of a nucleosome. Interestingly, the 78 base pair fragments are about twice as common as all other fragment sizes and they should be expected to be even more common with high *LET* beams having substantially more secondary δ electrons in the low kiloelectron-volt to sub-kiloelectron-volt energy range with very high probability of inducing dual DSBs. The low RBE of low *LET* radiations in the late S-phase may be due to a more open chromatin with fewer nucleosomes and thus less dual DSBs as well as a more efficient repair using fully developed sister chromatid exchange.

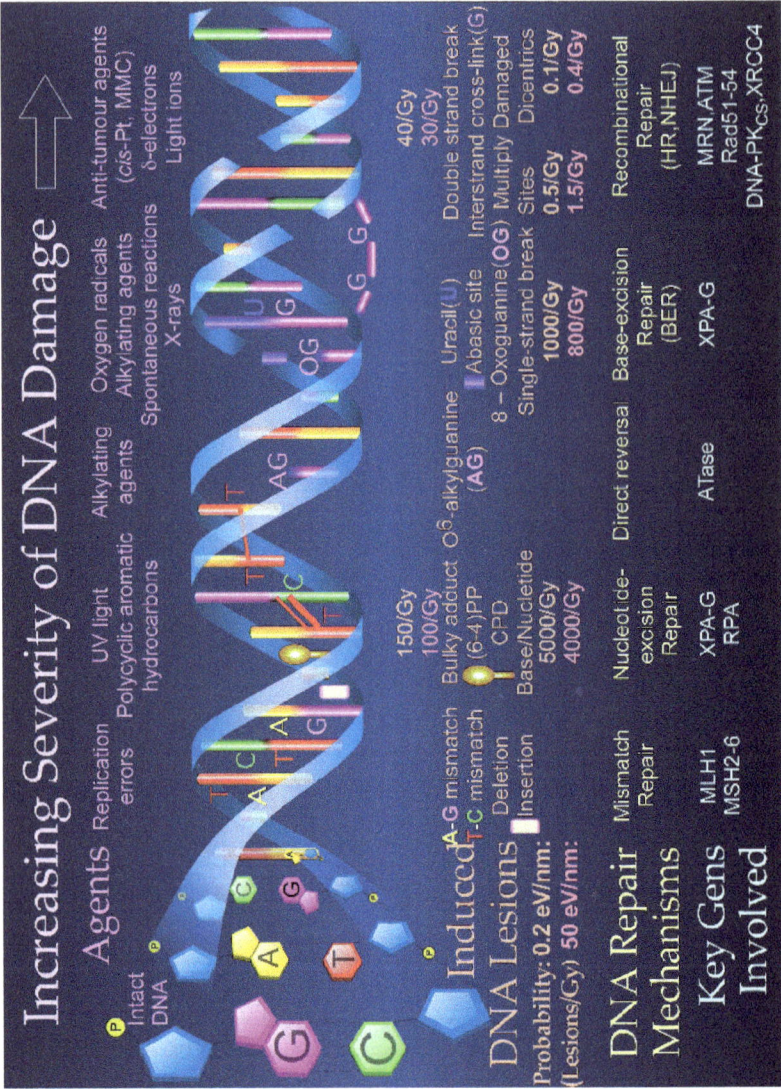

Figure 8.7(d). The spectrum of DNA damage after exposure to different agents (modified from Hoeijmakers (2001)) is shown. Most damage types are efficiently repaired by the cell except more complex blunt DNA end and close dual DSB type damage (cf. Fig. 8.7(b)) induced mainly by high *LET* radiations.

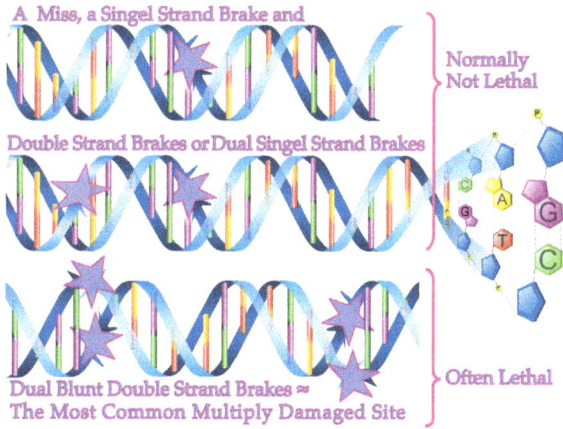

Figure 8.7(e). Illustration of the difference between generally nonlethal single strand and sticky DSBs and often lethal dual DSBs particularly when the damage is inflicted by low energy δ electrons that mainly produce blunt DNA ends without sufficient information where the strand ends really belonged.

Fig. 8.14, below). Owing to this enormous complexity, any accurate study must, by necessity, make simplifying assumptions to make the analysis manageable and easily understandable. In this chapter we assume for simplicity a homogeneous cell line of asynchronously dividing cells being irradiated by a fixed LET and dose, which is commonly the case in cell culture or homogenous tissue experiments.

The linear quadratic (LQ) cell survival model is rather accurate in the classical fractionation region from 1.5 to 4 Gy per fraction where the curvature of the survival curve is rather well described by the LQ expression $S = e^{-\alpha D - \beta D^2}$. However, as the range of applications widens, the classical LQ cell survival model becomes less and less accurate even if Kellerer and Rossi derived it based on single and dual event cell kill. But they did not consider the cellular repair processes in detail, which we now know, are essential to consider getting the shouldered survival curve shape right. This is particularly true at the very high doses employed with single fractions in the dose range from about 5 Gy and above where the traditional β-term saturates and the high dose logarithmic cell survival becomes essentially linear in dose (Scholz *et al.* 1997; Guerrero *et al.* 2004; Park *et al.* 2008). This applies both to low and high *LET* beams (Matsufuji *et al.* 2008). Probably the main problem with the LQ model is that the quadratic term in dose is used to describe the cellular repair due to its curvature. However, it is a negative exponential term so if the cell line has large repair capacity the term is large but reducing the survival due to the negative sign which is totally counter intuitive and instead the linear term has to be reduced to get the survival right. Therefore, the link to understanding the single and dual event cell kill is totally lost since the dominating influence of repair is not really considered in the model. Also in the low dose region, the LQ model breaks down as seen in Fig. 8.8. This is the case in many normal tissues with intact cell cycle regulation, where low dose hypersensitivity is a major concern. This is because side effects in normal tissue at low doses per fraction (<2 Gy) can be much more severe than expected based on the simple LQ model (cf. Marpels *et al.* 2004). Even if a rather complex modification of the LQ model was introduced, recently, there is an urgent need for a cell survival model, which describe the response very well at low, medium, and high doses. The Repair–Miss–Repair model of Tobias (1985) did that to some extent and the Lethal–Potentially Lethal model of Curtis (1986, 1988, 1992) did it even better, but their mathematical formulations are quite complex. In simplified form it may be approximated by: $S = (1 + aD)^b e^{-cD}$ which have some similarity with the LQ expression above but even more so with the Repairable Conditionally Repairable (RCR) model discussed in further detail below cf Eqs. (8.7) and (8.20). More recently, several researchers tried to improve the survival model often leading to rather complex equations as those of the first three references above. The RCR survival model (Lind *et al.* 2003) has the advantage that the mathematics is quite simple and experimental and clinical data are surprisingly accurately described. Furthermore, it has the advantage of being able to separate the two types of damage that normally are repaired by the NHEJ and HR processes that are the key repair pathways of mammalian cells. Therefore, the RCR model has an interesting connection to modern molecular cell biology (cf. Lind and Brahme 2010).

8.3.6.1. *Cell survival curve*

The cell inactivation is in the first approximation generally exponential since for a given fluence the cell kill is proportional both to the number of cells exposed and the fluence density according to:

$$-dN = N\sigma_i d\Phi, \tag{8.1}$$

where σ_i is the coefficient of proportionality namely the inactivation cross-section. This differential equation is directly integratable with the simple solution:

$$N(\Phi) = N_0 e^{-\sigma_i \Phi}, \tag{8.2}$$

so the surviving fraction or survival curve (cf. Figs. 8.5 and 8.8) is in the first approximation given by:

$$S = \frac{N}{N_0} = e^{-\sigma_i \Phi} = e^{-D/D_0} \approx e^{-\alpha D}, \tag{8.3}$$

where $D_0 \approx 1/\alpha$ are the traditional dose proportionality constants of exponential cell survival. In fact, the number of hits can be described more precisely as a binomial (cf, Fig, 8.7(h)) or Poisson process where the probability for exactly ν hits or DSBs is given by:

$$P(\nu) = (\sigma_i \Phi)^\nu e^{-\sigma_i \Phi}/\nu!, \qquad (8.4)$$

and thus the probability for no hit ($\nu = 0$) and consequently for survival is given exactly by Eq. (8.3), whereas the probability for precisely one hit ($\nu = 1$) is similarly given by

$$P(1) = \sigma_i \Phi e^{-\sigma_i \Phi}, \qquad (8.5)$$

and the probability for precisely two hits ($\nu = 2$) similarly is given by

$$P(2) = (\sigma_i \Phi)^2 e^{-\sigma_i \Phi}/2. \qquad (8.6)$$

The interesting experimental fact from Figs. 8.6(d) and (e) is now that at the common low LET therapeutic dose of 2 Gy, there are about 75 repairosomes or DSBs induced in each cell nucleus, whereas the tumor cell survival for effectively cured tumors at that dose is about 50% ($SF_2 \approx 0.5$, cf. Fig. 8.5(c)). However, according to Eq. (8.3), a survival of 50% correspond to a mean lethal hit number of $-\ln(0.5) = 0.693 \approx 0.7$ which surprisingly is <1 lethal hit per cell and thus out of the 75 DSBs <1% of the DSBs are lethal. If we for simplicity assume that all DSBs are correctly repaired which is not far from the truth since >99% of them are correctly repaired, the survival would then according to Eqs. (8.3) and (8.5) simply be a dual Poison process:

$$S = P(0) + P(1) = e^{-\sigma_i \Phi} + \sigma_i \Phi e^{-\sigma_i \Phi} = (1 + \sigma_i \Phi)e^{-\sigma_i \Phi} = (1 + D/D_0)e^{-D/D_0}. \qquad (8.7)$$

So in this way, we have a survival model that also takes into account the important property of most cell lines to very effectively repair DSBs as seen in Figs. 8.6(e) and 8.7(f), (g) and (h). In Figs. 8.7(g) and (h), it is clearly seen how the repair of most DSBs result in a shouldered cell survival curve.

For a given fluence, Φ_E, differential in energy E and restricted mass stopping power $L_\Delta(E)$, the absorbed dose D is in the first approximation given by

$$D = \int_\Delta^\infty \Phi_E \frac{L_\Delta(E)}{\rho} dE = \int_\Delta^\infty \Phi_E \frac{L_\Delta(E)}{\rho} dE \cdot \frac{\Phi}{\int \Phi_E dE} \equiv \frac{\overline{L_\Delta}}{\rho} \cdot \Phi, \qquad (8.8)$$

where $\overline{L_\Delta}$ by definition is equal to the fluence mean restricted stopping power.

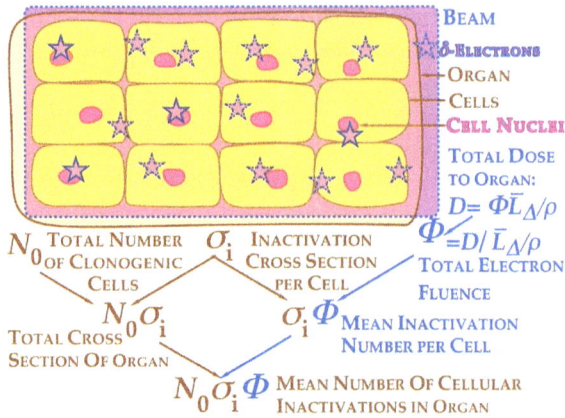

Figure 8.7(f). Illustration of the simple model used in calculating the response of an organ consisting of a large number of individual cells (N) each with an inactivation cross-section (σ_0 per cell), when exposed to a radiation beam as specified by its fluence (Φ/particles cm^{-2}).

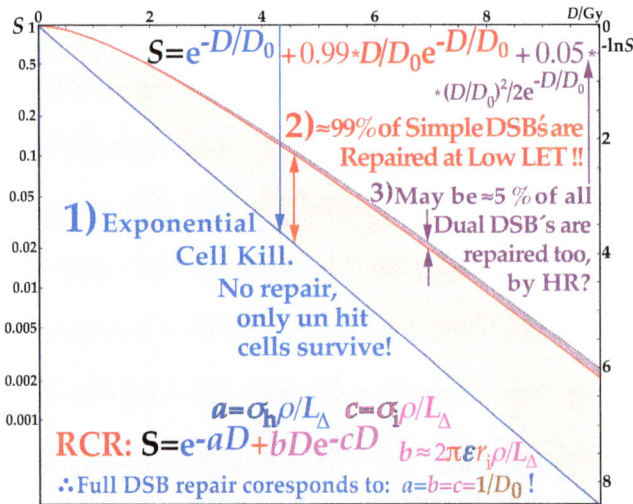

Figure 8.7(g). Interestingly, the cell survival is well described by the simple assumption that all missed cells survive (Straight line) and all SSB and almost all single DSB damaged cells are repaired (Shouldered red solid line and the Equation) even though a small fraction of the cells may suffer bystander effects, and others unsuccessful DSB repair or successful repair of multiply-damaged sites (Dotted line). These secondary processes can be more accurately accounted for by the RCR model, where all the parameters can deviate from $1/D0$ depending on the repair properties of the cell line and its genetic make-up.

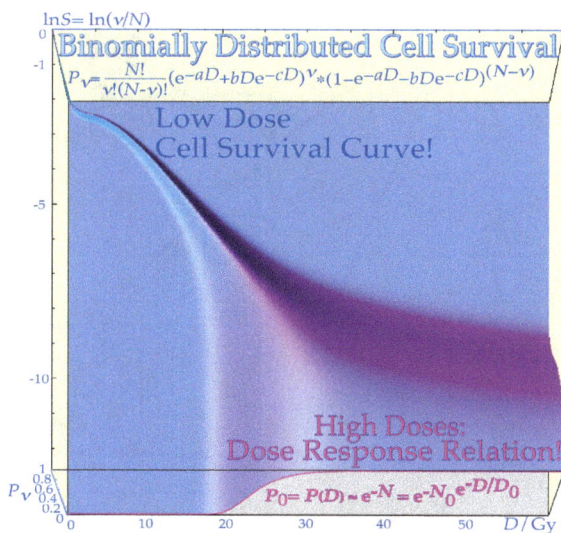

Figure 8.7(h). The binomial and in the first approximation, the Poisson distribution describe the cell kill and the cell survival very well. The probability, P, at which a certain fraction of the cells survive, S, at a given dose D is shown as a 3D surface. At zero dose, 100% of the cells survive (the peak at the origin), whereas at high doses and low survival probabilities, almost all the cells are eradicated leading to tumor control. Outside these areas the probability mass is spread out over a wider range of survival levels and the resultant probability is lower. At low doses, the shouldered cell survival curve dominates, whereas at high doses the DRR is the most useful and sufficient concept to describe the radiation effects. The tumor cure and the inactivation cross-section and the RBE are all described well by the Poisson expression (see also Eqs. (8.3), (8.11), (8.16) and Figs. 8.3(a) and (b)). Interestingly the effective repair of DSBs causes most of the curved shoulder of the cell survival curve (cf. Fig. 8.7(g)).

8.3.6.2. *LET dependence of the inactivation cross-section*

Based on Eq. (8.8), it is possible to determine relation between the parameters σ_i, D_0, and α of Eq. (8.3) according to:

$$\sigma_i = \frac{\overline{L_\Delta}}{\rho D_0} \approx \alpha \frac{\overline{L_\Delta}}{\rho}, \tag{8.9}$$

so the inactivation cross-section for clonogenic survival can be determined from the logarithmic slope $1/D_0$ of the cell survival curve at high doses. It is basically made up of at least five components according to:

$$\sigma_i = \sigma_A + \sigma_S + \sigma_M + \sigma_{Au} + \sigma_N, \tag{8.10}$$

namely the cross-section for inducing apoptosis (programmed cell death), senescence (permanent cell cycle arrest), mitotic catastrophe (a common cause

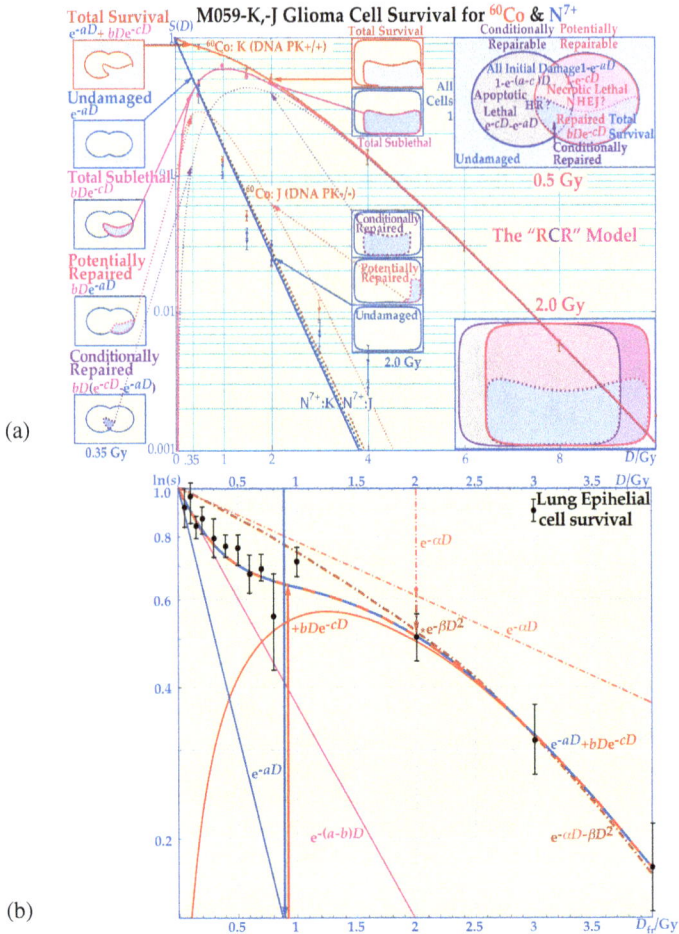

Figure 8.8(a). The survival of glioma cells after irradiation with increasing doses of ^{60}Co and nitrogen ions (N^{7+}) is shown. The two cell lines M059 J & K of which J is repair-deficient (DNA-PK is lost) respond identically to N^{7+}, whereas ^{60}Co damage is efficiently repaired by the K cells. The responses are well described by the new potentially RCR damage model. This also clearly show that cell lines that differ substantially for low LET radiations (^{60}Co) may even have almost identical cell survival for high LET beams (N^{7+}). The small inserts indicate the fractions of the irradiated cells that belong to each subgroup. Interestingly, the conditionally repaired cells are rather few at low doses ($\sim$0.35 and 0.5 Gy), whereas they dominate the survival at 2 Gy. **(b)** Description of the cell survival of low LET by experimental data (Singh *et al.* 1994) and exponential (linear Eq. (8.3)), LQ and RCR models. The LQ model describes the increasing loss of cells with dose by a quadratic term which is valid only at rather low doses, whereas the RCR model is based on damage repair and accounts for the initially increasing cellular repair bDe^{-cD} beyond the survival of unhit cells as the dose increases ($a \approx 2.5\,\mathrm{Gy}^{-1}$; $b \approx 1.3\,\mathrm{Gy}^{-1}$; and $c \approx 0.8\,\mathrm{Gy}^{-1}$).

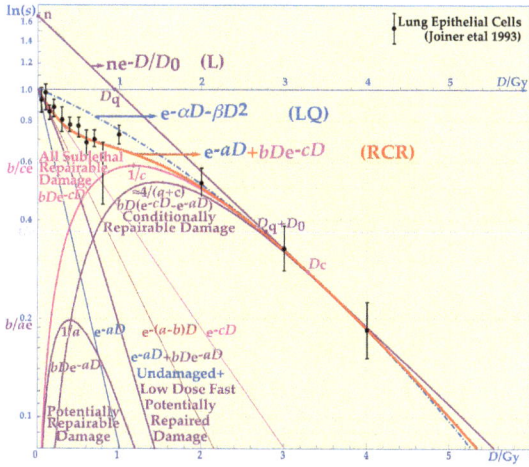

Figure 8.8(c). More detailed comparison of the linear (L) linear-quadratic (LQ) and the RCR models for a normal tissue with clear low-dose hypersensitivity. Interestingly, the RCR model describes all the low-dose hypersensitivity, the shoulder around 2 Gy, and the quasi-exponential high-dose behavior very well. The low-dose region is well described by a single exponential approximation of the RCR model according to Eq. (8.30).

of death in fractionated radiation therapy of genetically unstable tumor cells), autophagy (the cell digest itself), and necrosis.

The cell survival data behind Fig. 8.8(j) illustrate how the inactivation cross-section, σ_i, defined by Eq. (8.9) saturates approximately at the physical size of the cell nucleus (σ_n) as the ionization density reaches the 150–200 eV/nm region and is so high that a nuclear passage generally leads to cellular inactivation.

To be more accurate, this cross-section depends not only on the quasi-circular projected cross-section of the cell nucleus (πr_n^2, cf. Fig. 8.8k) and the width of the radial ion beam dose profile (r_i) at the level of the inactivation dose, generally located in the range above several Gray (cf. Fig. 8.6(b)) but also on the radial *LET* distribution of the ion and the mean cord length when crossing the nucleus.

According to the above discussion (cf. Eqs. (8.3)–(8.7)), the inactivation cross-section in the high *LET* region in the first approximation can be expected to have a dual Poisson process type *LET* dependence since missed cells and single DSB cells generally survive and in the first approximation $\sigma_i(L)$ is of the form:

$$\sigma_i(L) = \sigma_\infty(1 - e^{-\lambda L}(1 + \lambda L)) \approx \sigma_\infty\left(1 - \left(1 - \lambda L + \frac{\lambda^2}{2}L^2\right)(1 + \lambda L)\right)$$

$$\approx \sigma_\infty \frac{(\lambda L)^2}{2} + \cdots \approx \sigma_\infty\left(1 - e^{\frac{-(\lambda L)^2}{2}}\right), \tag{8.11a}$$

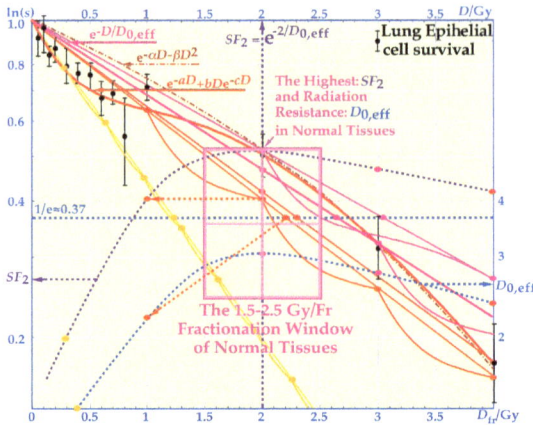

Figure 8.8(d). The low–dose fast potentially repairable damage may cause a small low–dose hypersensitivity plateau on the cell survival curve (as seen less clearly in Fig. 8.8(a) and more clearly in Fig. 8.8(b)), whereas at high doses, the conditionally repaired damage dominate the survival curve shape. In normal tissues, there is therefore a fractionation window around 2 Gy ($\approx$1.5–2.5 Gy) per fraction where the least detrimental response is obtained ($SF_2 \approx 0.52$ and $D_{0,\mathrm{eff}} \approx 3.1$ Gy) for a given dose level in the surrounding tumor volume. The lower dashed curve show how SF_2 and $D_{0,\mathrm{eff}}$ varies with the dose per fraction on the horizontal axis having clear maxima near 2 Gy. This is probably the main reason why in classical radiation therapy, where the tumor and normal tissue dose is often rather similar, the dose per fraction should preferably be around 2 Gy. For light ions with largely a low LET in normal tissues, such as lithium to carbon ions, the normal tissue dose should preferably also be in this range unless there is a substantial high LET dose spill over to critical normal tissues surrounding the tumor region. With good dose delivery, the tumor dose could simultaneously be two to three times higher and in addition the biological effectiveness is similarly increased. This makes the total increase in therapeutic effect in the tumor from about three up to more than six times particularly for smaller tumors.

where the first expression is based on the assumption that cell kill is described well by a dual or higher order Poisson process mainly induced by dual or higher order DSBs but also higher order events such as multiply damaged sites (cf. Figs. (8.6) and (8.7), Brahme *et al.* 1997, 2011; Virsik *et al.* 1981).

As seen in Fig. 8.8(j), the second part may be a suitable low and high dose approximation and was previously used in the theory of dual radiation action (Kellerer *et al.* 1974). It is well-known today that about 70–80 DNA foci or repairosomes are generally induced at DSBs by 2 Gy of low or high LET radiation (cf. Figs. 8.6(e) and (f)). However, less than one of them may be lethal since the survival fraction at 2 Gy is generally around $SF_2 = 0.5$ which is higher than the value with one lethal event, on average giving $SF_2 = e^{-1} \approx 0.37$. Most of the

DSBs are thus repaired as discussed above and in general a close pair of local DSBs is at least needed for lethality as shown in Figs. 8.7(b)–(e). A close pair of DSBs may result in a number of lethal repair problems such as losing a short section of DNA or getting chromosomal crosslinks. Equation (8.11a) is therefore based on Poisson statistics as discussed above and it is assumed that cells that have none ($e^{-\sigma_i \Phi} = e^{-\alpha D} = e^{-\alpha L \Phi / \rho} = e^{-\lambda L}$ (cf. Eqs. (8.9), (8.20)–(8.22) below)) or just a single DSB ($e^{-\lambda L} \lambda L = \sigma_i \Phi e^{-\sigma_i \Phi}$) generally survives and two or more hits are commonly needed for lethality thus including the dual DSBs and the more general multiply damaged sites as commonly recognized lethal events. The dual DSBs induced by δ electrons are in fact the most common lethal event as shown in more detail in Figs. 8.7(b) and (c) and by Brahme et $al.$ (1997). The dual DSBs on the periphery of a nucleosome probably requires full nucleosomal dismantling before the initial repair gene products Ku70 and Ku86 can bind to the free DNA ends since they are almost as large as the nucleosome itself. In this process the precise DNA geometry before the damage was inflicted, may be lost leading to lethality either if it happens in a gene of key importance for cell survival, or if dicentric chromosomes are generated. According to the above discussion, $\sim 0.9\%$ of the DSBs are lethal and thus the inactivation cross-section would be slightly more accurately given by:

$$\sigma_i(L) = \sigma_\infty (1 - e^{-\lambda L}(1 + 0.991\lambda L)), \qquad (8.11b)$$

which is only a small modification of the dual Poisson expression in Eqs. (8.7) and (8.11a), but actually adds a small linear term in λL ($0.009\lambda L$) to $\sigma_i(L)$ in Eq. (8.11a) (cf. also Eq. (8.11c) below and Eq. (8.16)).

Thus, $\sim <1\%$ of the DSBs are lethal. Interestingly, we recently estimated the number of sub-kiloelectron-volt δ electron track ends in the cell nucleus at 2 Gy of low LET electrons and photons and found it to be around 1.5. At low LET a large portion of the lethal events may thus be caused by such low energy track ends, which we know, can make complex DSBs and there are always a constant fraction of them per unit dose (Berger and Seltzer 1969; ICRU 1984). This is so since the slowing down spectrum at low kiloelectronvolt energies is almost the same per unit dose independent of initial electron or photon energy provided the initial energy is above a few hundred kiloelectronvolt. This thus explains the constant RBE ≈ 1.0 for such beams and that the absorbed dose is a good measure of biological effect with low LET radiations.

The constant σ_∞ in Eq. (8.11) is the maximum cross-section reached at LET values around 100–200 eV/nm as seen in Figs. 8.6(b) and 8.8(k) and it is approximately given by:

$$\sigma_\infty \approx \pi (r_n + r_i)^2. \qquad (8.12)$$

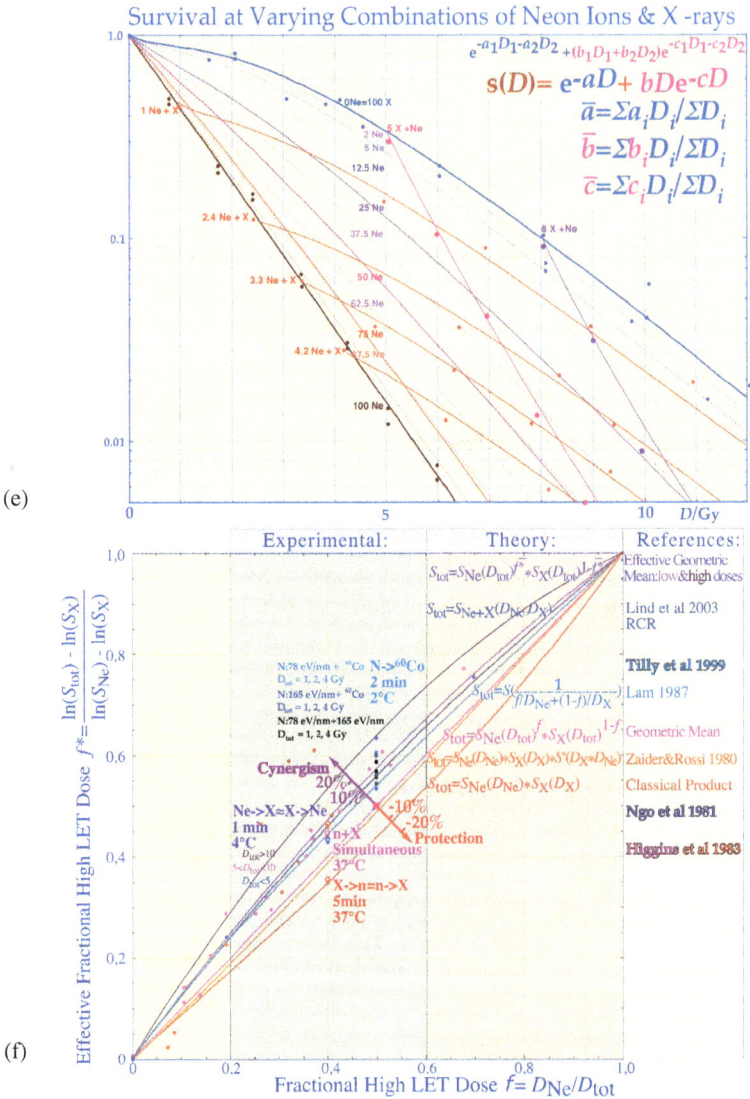

(e)

(f)

Figure 8.8(e). The cell survival after varying combinations of neon ions and X-rays is illustrated, where solid lines represent RCR model and dots represent experimental data (Ngo *et al.* 1981). The formula included in the figure show how the dose weighted mean values of *a*, *b*, and *c* of the different radiations, *i*, combine to give the total effect (cf. Lind *et al.* 2003). **(f)** Variation of the effective high *LET* dose fraction (vertical scale) with the real delivered high *LET* dose fraction (horizontal scale) is shown. The convex upper curve illustrate the data in 8.8(c) (dots) and indicate a clear synergistic action between low and high

(Continued)

Figure 8.8(f). *LET* when delivered almost simultaneously, whereas the lower concave curve show reduced effectiveness when the high and low *LET* fractions are delivered at different time intervals, several minutes apart, allowing significant repair of sublethal damage between the first and second radiation type. This plot clearly shows that a short treatment interval is desirable with scanned light ion beams to get almost a 30% increase in the synergistic effect between the low and high *LET* dose fractions such as between the peak and plateau part of the dose delivery. This is automatically obtained by fast dynamic range shifters or ridge filters but not necessarily by a single swept pencil beam that is gradually scanning the tumor volume starting from its distal edge and thus may cause a 20–30% reduction in therapeutic effect due to fast repair processes in the plateau region. A scanning system that irradiates the tumor many times during a few minutes treatment session or preferably a fast longitudinal scan may be the way to relieve this problem. The study of Ngo *et al.* (1981) and Higgins *et al.* (1983) indicate that the order of high and low *LET* fractions does not matter much. Interestingly, the synergistic effect is highest at high doses per fraction as seen from the uppermost convex curve and associated high dose data points in the figure (cf. Fig. 8.8(i)).

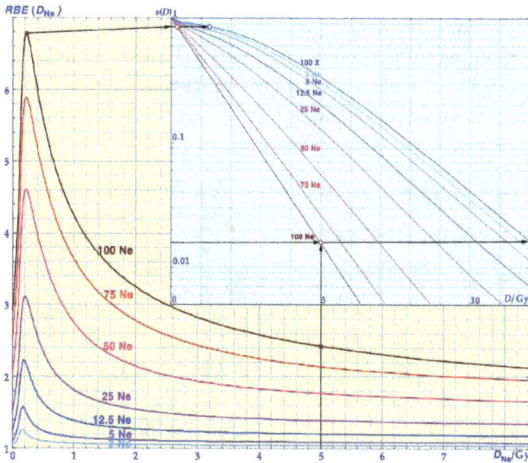

Figure 8.8(g). Illustration of how the RBE decreases with the applied increased dose level since the curvature of the low *LET* reference cell survival curve is high at low doses (cf. Figs. 8.8(a)–(c)). When the cell line shows low dose hypersensitivity (cf. Figs. 8.8(b) and (c)) with the reference X-rays, the high RBE disappears at the lowest doses as described very well by the RCR model.

These expressions are further used in connection with the induction of apoptotic cell kill below.

The low *LET* approximation of this expression (Eq. (8.11a): $\sigma_i \approx \frac{\sigma_\infty(\lambda L)^2}{2}$) accounts for the reduced cross-section and risk for cellular inactivation at low *LET*

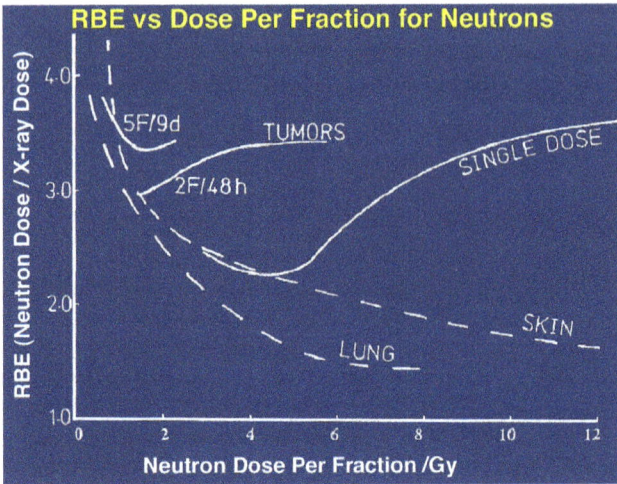

Figure 8.8(h). Illustration of experimental RBE variation for neutron beams in normal tissues (skin and lung) and tumors (Denekamp *et al.* 1979). For the tumors, the RBE stays high at high doses probably due to their genomic instability and lower ability to repair even low *LET* damage resulting in less pronounced survival curve shoulder at low *LET*. Since neutrons and carbon ions are biologically rather similar in the tumors (see Figs. 8.5(e), 8.10, 8.14, and 8.22), a similar situation can be expected also for carbon ions (cf. 8.8(i)).

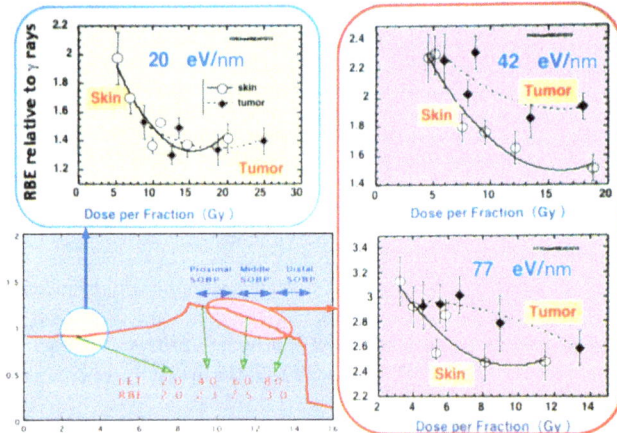

Figure 8.8(i). A detailed study of the dose dependence of the *RBE* for carbon ions in the plateau and spread out Bragg peak (SOBP) regions of the depth dose curve. It is seen that there is no difference in the plateau region but a clear *RBE* advantage in the SOBP region. Thus, provided the normal tissue dose is within the therapeutic window between 1.5 and 2.5 Gy (see Fig. 8.8(d)), fractional tumor doses around 5–8 Gy should be very effective.

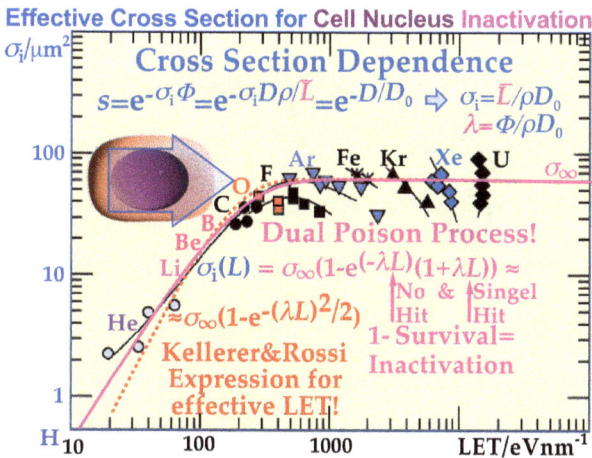

Figure 8.8(j). The cell kill as expressed by the cell inactivation cross-section (cf. Eq. (8.9)) saturates at high *LET* beyond carbon and oxygen ions (modified from Kraft *et al.* (1987)). At lower ionization densities and *LET*s, the ion track is not dense enough to inactivate the cell with high degree of probability after passage through the nucleus. Since the inactivation cross-section is quasi-constant >200 eV/nm, the fluence density of ions determines the survival in that region. As the cross-section saturates, the peak biological effectiveness appears and at higher *LET*s, the biological effectiveness decreases because of a quasi-constant cross-section and an increasing probability of radical–radical recombination as secondary electrons are generated more and more closely together and the so-called overkill effect sets in. The solid and dashed curves are taken from Eq. (8.11a) and its approximation at high *LET*s, respectively, and describe the average response of the multiple experimental data sets very well.

values since the cell may have many crossing charged particles without a severe energy deposition and cell inactivation (e.g., for electrons, protons, or helium ions).

8.3.6.3. *Relative biological effectiveness*

Equation (8.11a) also gives us a very interesting expression for the *LET* dependence of the *RBE* as it is defined as the ratio between the low reference (L_0) to high *LET* isoeffect doses and it is thus assuming the simple exponential survival according to Eqs. (8.3), (8.9), and (8.11) given by:

$$RBE(L) = \frac{D(L_0)}{D(L)} = \frac{D_0(L_0)}{D_0(L)} = \frac{\sigma_i(L)/L}{\sigma_i(L_0)/L_0} = \frac{(1 - e^{-\lambda L}(1 + \lambda L))/\lambda L}{(1 - e^{-\lambda L_0}(1 + \lambda L_0))/\lambda L_0},$$

(8.13)

and this expression describes the shape of the *RBE* peak as shown quite well in Figs. 8.8 and 8.9. The *RBE* peak is located at $L_{max} = 1.79/\lambda$ and the full width

$e^{-\sigma_i \Phi} = e^{-D/D_0} \approx e^{-\alpha D}$

$\sigma_i = L_\Delta / \rho D_0 \approx \alpha L_\Delta / \rho$

$\alpha \approx a - b = \sigma_i \rho / L_\Delta$

$a \propto \sigma_h \quad b \propto \varepsilon \quad c \propto \frac{\sigma_i + \sigma_h}{2}$

For ions: $c \approx$
$a - b/2 =$
$\approx a - b \frac{(\sigma_h + \sigma_i)}{2} \rho / L_\Delta$
$e^{-\sigma_h \Phi} = e^{-aD}$
$a = \sigma_h \rho / L_\Delta$
$a \gtrless 1/D_0$
$\approx \alpha$

Ion Dose Kernel   **Mean Minimal Inactivating Overlap**   **Cell Nucleus**

Ion Track   **Mean Inactivating Radius:** $\rho_i = \rho_h - \varepsilon$

$\rho_h \quad \rho_i \quad \varepsilon \quad r_n$

$r_i = r_n + \rho_i$

$\sigma_i = \pi r_i^2$

$r_h = \rho_h + r_n = r_i + \varepsilon$

$\sigma_h = \pi r_h^2 = \sigma_i + 2\pi r_i \varepsilon + \pi \varepsilon^2 > \sigma_i$

$\sigma_i \approx \alpha L_\Delta / \rho$

$\approx (a-b) L_\Delta / \rho = \sigma_h - b L_\Delta / \rho$

$b = (\sigma_h - \sigma_i)\rho / L_\Delta = \pi \varepsilon \rho (2r_i + \varepsilon)/L_\Delta \quad b \approx 2\pi r_i \varepsilon \rho / L_\Delta$

Mean Hit Radius = ρ_h

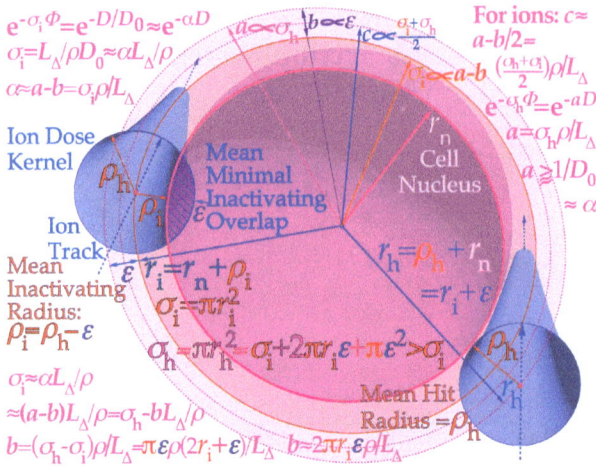

Figure 8.8(k). Illustration of the relation between the inactivation cross-section σ_i and the cross-section for a nuclear hit, σ_h, and their dependence on the quasi-spherical nuclear size and the approximately cylindrical mean radial dose profile of the ion track. The cross-section saturates at high LETs since the radial dose profile and the minimal inactivation overlap becomes quasi-constant in the high LET Bragg peak region as seen in Fig. 8.6(b). The maximum radial excursion of secondary electrons (cf. Fig. 8.6(b)) is denoted ρ_h, whereas the effective ion inactivation radius is ρ_i and is significantly smaller. In Fig. 8.6(b) ρ_h is at a dose level <1 Gy, whereas ρ_i normally is at a dose level significantly >10 Gy making σ_h significantly larger than σ_i and a larger than c in Eqs. (8.10)–(8.12), (8.21)–(8.23), and (8.25).

at half maximum, FWHM $\approx 3.5/\lambda = 1.96 L_{max} \approx 2 L_{max}$. Interestingly, $RBE\ (L)$ in a logarithmic diagram, that is as a function of the logarithm of the LET, is approximately shaped as a Gaussian function as seen in Figs. 8.3 and 8.9(f). Equation (8.11) should strictly be used for ions so L_0 is the LET of very high-energy ions that are generally generating high-energy δ electrons with a quite low LET.

In the low LET region, the RBE is more accurately derived based on the LET, the mean ionization energy $\overline{W}$, the mean cord length, $\bar{l}$, and the effective cross-section, σ, of the cell nucleus to get ionized by a passing particle fluence. The average number of ionizations produced when crossing the cell nucleus is then given by:

$$n = \frac{\overline{l L}}{W(L)} \Phi(L)\sigma(L)$$

$$= \frac{\sigma(L)\rho \bar{l}}{W(L)} D(L) = \frac{\overline{l L_0}}{W_0} \Phi_0 \sigma_0 \approx \frac{\rho \bar{l} \sigma_0}{W_0} D_0 \approx \frac{\rho V}{W_0} D_0 = \frac{m}{W_0} D_0, \quad (8.14)$$

where the middle expression holds for the reference radiation of $LET L_0$ such as ^{60}Co and the last parts identifies the relation to the absorbed dose and cell nuclear volume and mass. For low LET range in the first approximation, the survival is approximately the same at the same number of induced dispersed ionizations and DSBs in the cell nucleus. The RBE is then given by the ratio of the applied absorbed doses associated to the fluencies Φ according to:

$$RBE(L) = \frac{D_0}{D(L)} = \frac{\Phi_0 \overline{L_0}}{\Phi(L)\overline{L}} = \frac{\overline{W_0}\sigma(L)}{\overline{W(L)}\sigma_0},\tag{8.15}$$

where Eqs. (8.8) and (8.14) was applied in the last step in Eq. (8.15). Since $\overline{W}$ varies rather slowly with the LET, the RBE is slowly varying with LET in the low LET region between 0.2 and 2 eV/nm. The RBE increases quasi-linearly in the region of increasing LET beyond a few eV/nm according to Eq. (8.11) as seen by comparison with Eq. (8.11). According to Eq. (8.15) this is not quite true at the lowest LET values (0.2 eV/nm $= 2$ MeV/gcm^{-2}) where the RBE is almost constant. To describe the RBE dependence over the whole LET range from the lowest (0.2 eV/nm, relativistic electrons) to the highest values toward 10^4 eV/nm and higher for heavy ions, it is natural to add a small exponential term to more accurately account for the low almost constant RBE value independent on LET at low quasi-random ionization densities rather than the densely ionizing track ends making dual DBS particularly common around the Bragg peak as described by the last dual Poisson term:

$$RBE(L) = \frac{\overline{W_0}\sigma(L)}{\overline{W(L)}\sigma_0}e^{-L/L_l} + k(1 - e^{-L/L_h}(1 + L/L_h))L_h/L,\tag{8.16}$$

where k is given implicitly by Eq. (8.13), and the characteristic high and low LET values are $L_h = 1/\lambda$ and $L_1 \approx 2L_h\overline{W_0}\sigma(L)/k\overline{W(L)}\sigma_0 \leq L_h$ to have a smooth RBE transition to high LETs somewhat depending on ion species and cell line as seen in Figs. 8.3(a) and 8.8(l). The function in Eq. (8.3) thus describes the RBE dependence on the LET very well over a wide LET range by combining the different low and high LET contributions by δ electrons.

The first term is thus due to high-energy low LET δ rays and the last essentially due to sub-kiloelectronvolt δ electrons with dual DSB induction capability as described in detail in Fig. 8.6. Based on Eqs. (8.11b) and (8.16) it is possible to express the cross-section more accurately taking also the high-energy δ electrons into account according to:

$$\sigma_i(L) = \frac{\overline{W_0}\sigma(L)L}{\overline{W(L)}\sigma_0 L_h}e^{-L/L_l} + k(1 - e^{-L/L_h}(1 + 0.991L/L_h)),\tag{8.11c}$$

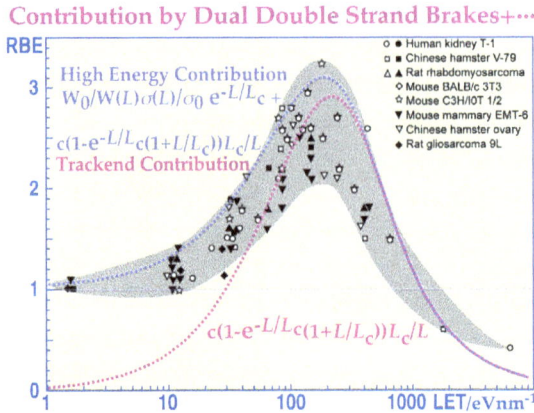

Figure 8.8(l). As the cross-section in Figs. 8.8(j) and (k) saturates, the peak biological effectiveness or *RBE* appears since the cross-section cannot increase with the *LET* any more, and at higher *LET*s the biological effectiveness decreases because of a quasi-constant cross-section an increasing probability of radical–radical recombination as secondary electrons are generated more and more closely together and the "overkill" effect sets in. The dashed curves are taken from Eqs. (8.11) and (8.16) and describe the average response of the multiple experimental data sets very well (modified from Blakely and Chang (2009)).

which makes the expression more valid also for helium, protons, and low *LET* photons and electrons.

8.3.6.4. *Oxygen enhancement ratio*

Interestingly, based on this expression for the *LET* dependence, it is possible to estimate also the *LET* dependence of the *OER*, as shown in Figs. 8.5(g) and (h). If we know the *RBE* at a given survival level for well-oxygenated and hypoxic conditions: $RBE_O = D_O^P/D_O^I$ and $RBE_H = D_H^P/D_H^I$ and furthermore also the generally quite high low *LET OER* for photons defined by $OER^P = D_H^P/D_O^P$ for the same cell line (cf. Fig. 8.5(h)), it is straight forward to calculate the associated ion *OER* according to:

$$OER^I = D_H^I/D_O^I = OER^P RBE_O/RBE_H. \qquad (8.17)$$

Here, the *RBE* ratio can be recognized as the $1/OGF$ where $OGF = RBE_H/RBE_O$ describing the reduction of dose possible going from photons to light ions. So from the known maximum $OER^P \approx 3$ for low *LET* photons and electrons, it is possible through Eq. (8.17) to calculate the ion OER^I using Eq. (8.16) for the oxic and hypoxic cells (index O and H, respectively) according to:

$$OER^I = OER^P \frac{\frac{\overline{W_0\sigma(L)}}{W(L)\sigma_0}e^{-L/L_{OI}} + k_O(1 - e^{-L/L_O}(1 + 0.991L/L_O))L_O/L}{\frac{\overline{W_0\sigma(L)}}{W(L)\sigma_0}e^{-L/L_{HI}} + k_H(1 - e^{-L/L_H}(1 + 0.991L/L_H))L_H/L}.$$

(8.18)

By power expansion similar to Eq. (8.11a) and assuming the OER is a simple dose-modifying factor (cf. Fig. 8.5(h)), this expression can be approximated very well by the simple expression:

$$OER^I = 1 + (OER^P - 1)e^{-(L/L_O)^2},$$

(8.19)

which describes the experimental data of Furusawa et al. (2000) in Fig. 8.5(h) very accurately (cf. also Antonovic et al. 2013).

8.3.6.5. Repairable and conditionally repairable damage model

The simple cell survival expressions in Eq. (8.3) do not really take the effect of the complex cellular repair systems into account. During the 1960s and 1970s, this was taken into account using the extrapolation number, n, and defining a quasi-threshold dose, D_q, at which the extrapolated survival was unity as shown in Fig. 8.8(c). To describe the cell survival more accurately at low doses the linear quadratic survival model was later introduced as also shown in the figure. Unfortunately, at the very lowest and highest doses there are still deviations as seen by comparison with the experimental data points in Figs. 8.8(b) and (c). Interestingly, the effect of the repair system could be taken into account more accurately using the recently developed RCR cell survival model which includes a clear cut repair term that can be seen as a generalization of Eq. (8.7) above (Figs. 8.8(a)–(e); Chapters 2 and 3; and Lind et al. 2003; Lind and Brahme 2010) according to:

$$s = e^{-aD} + bDe^{-cD},$$

(8.20)

where the second term includes all cells that are able to correctly repair their damage and the first term only includes those cells that survives due to absence of a hit, so:

$$e^{-aD} \approx e^{-\sigma_h \Phi},$$

(8.21)

and in analogy with Eq. (8.9) we have:

$$a \approx \frac{\sigma_h \rho}{L_\Delta}.$$

(8.22)

Here the cross-section for a hit, σ_h, is larger than σ_i according to

$$\sigma_h \approx \pi(r_n + r_h)^2, \tag{8.23}$$

where r_h is outside the sublethal ion hit radius that generally is well below the 1 Gy dose level (cf. Figs. 8.6(b) and 8.8(k)). Here, a is only weakly dependent on the LET at least in the low LET region, since high-energy electrons, photons, or protons can cross a cell nucleus with a small but finite probability for a hit and the fairly long range of the generated secondary electrons may cause a hit when the primary particle do not even cross the cell nucleus.

Interestingly, in analogy with Eq. (8.11) the second term in Eq. (8.20) indicate that many of singly hit damage sites (proportional $\lambda L e^{-\lambda L}$ or $\sigma_i \Phi e^{-\sigma_i \Phi}$) are really correctly repaired by the effective cellular repair processes NHEJ and HR and survive the irradiation intact as seen in Figs. 8.8(b) and (c). The traditional term "sublethal damage" thus includes many of the conditionally repairable damage events.

In fact, the RBE is close to unity as long as the fluence of low energy δ electrons (cf. Fig. 8.7) per unit dose is quasi-constant for these particles (cf. ICRU 1984; Tilly et al. 2002). Since σ_i and r_i are smaller than σ_h and r_h, a must be larger than c as is generally shown by experimental data (cf. Figs. 8.8(a)–(e) and Lind et al. 2003). According to the behavior of Eq. (8.20) at low doses $s \approx 1 - (a - b)D$ (cf. Eq. (8.30) below) and thus $\sigma_i \approx (a - b)\overline{L_\Delta}/\rho$ at low doses by comparison with Eq. (8.3). Interestingly, according to Eq. (8.22), it is then possible to express b as $b = (\sigma_h - \sigma_i)\rho/\overline{L_\Delta}$ which based on the notations in Fig. 8.8(k) can be reduced to $b \approx 2\pi r_i \varepsilon \rho/\overline{L_\Delta}$ and thus b is related to the minimal inactivation overlap ε where the mildest inactivations take place.

However, in order to make the survival less than unity, a does not only have to be larger than c but also:

$$1 \geq e^{-aD} + bDe^{-cD} \rightarrow bD \leq e^{cD} - e^{(c-a)D} \rightarrow$$

$$0 \leq (a - b)D + \frac{c^2 - (c - a)^2}{2}D^2 + \frac{c^3 - (c - a)^3}{6}D^3 \cdots. \tag{8.24}$$

A sufficient condition to fulfill this property is thus $a \geq b$ and $\forall n : c^n - (c - a)^n \geq 0$. This last expression is always fulfilled for odd values of n, since $a \geq c$ and it is then equal to $c^n + (a - c)^n \geq 0$. For even n, $|c| \geq |a - c|$ which is fulfilled independent of the value of n if:

$$a \geq c \geq a/2. \tag{8.25}$$

These expressions show that even if Eq. (8.20) seems to have three independent parameters, that is one more than the LQ relation, they obey several conditions (cf. Lind et al. 2003, for further details). In fact from Eq. (8.7) in the first approximation

$a = b = c = 1/D_0$. However, depending on the repair capacity of the cells not all DSB are repaired and some Dual DSB may be repaired by HR so such differences will modify a, b and c from their approximat $1/D_0$ values! Interestingly, Eq. (8.20) is very well suited to describe the cell kill both for photons and light ions and various combinations thereof as shown in Figs. 8.8(e) and (f) (cf. Lind *et al.* 2003). The experimental data for the interaction of X-rays with neon ions and neutrons of Ngo *et al.* (1981) agrees very well with this new analytical formula (cf. Eqs. (8.26) and (8.29)). The experimental data in Fig. 8.8(e) expressed as the effective high *LET* dose fraction (see vertical axis in Fig. 8.8(f) and Lind *et al.* 2003) are replotted in Fig. 8.8(e) as a function of the real high *LET* dose fraction indicating a strong synergistic effect between the low and high *LET* dose fractions especially when delivered simultaneously.

When two radiation modalities are combined on a patient, such as low *LET* photons, electrons, and protons and high *LET* light ions, it is essential that the sublethal damage of all modalities get a possibility to interact with each other as shown in Figs. 8.8(e) and (f). In radiation therapy planning, it is very important to be able to calculate this interaction accurately. This is the case not least with light ions where the high energy plateau part of the dose distribution often is of rather low *LET*, whereas the Bragg peak region has a high *LET* and often is combined with the plateau part specially when using a spread out Bragg peaks (SOBP, cf. Fig. 8.13). As derived in the initial RCR publication (Lind *et al.* 2003) when there is a long time ($\sim$24 h) between irradiations, the survival level is the product of the individual low and high *LET* survival levels (cf. the lowest curve in Fig. 8.8(f)) according to:

$$s_{A+B} = s_A \cdot s_B, \tag{8.26}$$

whereas if the irradiations occur simultaneously, the interaction is instead given by adding the different "dose effects" directly according to:

$$s_{A+B} = e^{-a_A D_A - a_B D_B} + (b_A D_A + b_B D_B)e^{-c_A D_A - c_B D_B}. \tag{8.27}$$

This expression really shows how the RCR parameters of different ions and *LET*s interact. Here, the probability of no hit (first term) is similar to Eq. (8.20) according to the simple exponential hit theory (Eqs. (8.1)–(8.3)). However, the last conditional sublethal repair term is more complex and describes the interaction of damage events that can be significant with high doses of low *LET* as seen in Figs. 8.8(e) and (f). Moreover, when there is a time interval between the irradiations, the sublethal repair has to be considered. Interestingly, Eq. (8.27) can be generalized to an arbitrary number of radiation modalities n such as the situation during SOBP irradiations:

$$s_n = e^{-\overline{a_n} D_{tot}} + \overline{b_n} D_{tot} e^{-\overline{c_n} D_{tot}}, \tag{8.28}$$

where

$$\overline{a_n} = \sum_{i=1}^{n} a_i D_i \Big/ D_{tot},$$

$$\overline{b_n} = \sum_{i=1}^{n} b_i D_i \Big/ D_{tot},$$

$$\overline{c_n} = \sum_{i=1}^{n} c_i D_i \Big/ D_{tot},$$

$$D_{tot} = \sum_{i=1}^{n} D_i. \tag{8.29}$$

These equations have been used not only in Figs. 8.8(e) and (f) but also in Fig. 8.12(c) to calculate a combined optimal scanning pattern with two different light ions (Li and C) to simultaneously obtain uniform cell survival and absorbed dose.

This is the best way to irradiate extended uniform tumor masses, since the regular SOBP method, which is very useful for low LET protons, results in a rather heterogeneous distribution of the radiation quality with helium and heavier ions. This is because a very low, low-LET dose, and very high, high-LET dose, are combined in the distal — and vice versa in the anterior — target volume resulting in a very high microscopic heterogeneity in the target volume and potential clinical problems, at least for quasi-uniform tumors, as discussed in more detail in connection with Figs. 8.10, 8.11, and 8.13 below.

8.3.6.6. *LET dependence of a, b, and c*

Based on Eqs. (8.20)–(8.23) and the discussion about b above, the survival with full repair is therefore well approximated at low doses by:

$$s \equiv e^{-(a-b)D}(e^{-bD} + bDe^{-(c-a+b)D}) \approx e^{-(a-b)D}\left(1 + \left(a - c - \frac{b}{2}\right)bD^2 \ldots\right), \tag{8.30}$$

as seen in Figs. 8.8(b) and (c) and indicating a relation between $a-b$ and $(-a+c+\frac{b}{2})b$ with α and β of the LQ model, whereas at high doses the relevant expression is:

$$s \equiv (e^{-(a-c)D} + bD)e^{-cD} = bD\left(1 + \frac{e^{-(a-c)D}}{bD}\right)e^{-cD} \approx bDe^{-cD}, \tag{8.31}$$

because the exponential term inside the parenthesis rapidly disappears at high doses since $a \geq c$ and thus

$$\frac{d \ln s}{dD} \approx -c + 1/D. \tag{8.32}$$

At very high doses according to Eqs. (8.3) and (8.9), we then have:

$$c \approx 1/D_{0,\infty} = \sigma_i \rho / \overline{L_\Delta}, \tag{8.33}$$

and thus according to Eq. (8.11a)

$$c(L) = \sigma_\infty \rho (1 - e^{-\lambda L}(1 + \lambda L)) / \overline{L_\Delta} = c_0(1 - e^{-L/L_h}(1 + L/L_h))L_h/L, \tag{8.34}$$

where $c_0 \approx \sigma_\infty \rho \lambda$, and it is assumed that the energy spread of the quasi-mono energetic incident ions is so small (often <1%) that $L \approx \overline{L_\Delta}$. The shape of this equation is seen in Figs. 8.3, 8.8(n)–(o) and 8.9(c)–(f) and describes both the RBE (cf. Eqs. (8.13) (8.15), and (8.16)) and apoptosis (cf. Eq. (8.39) below) curves quite well.

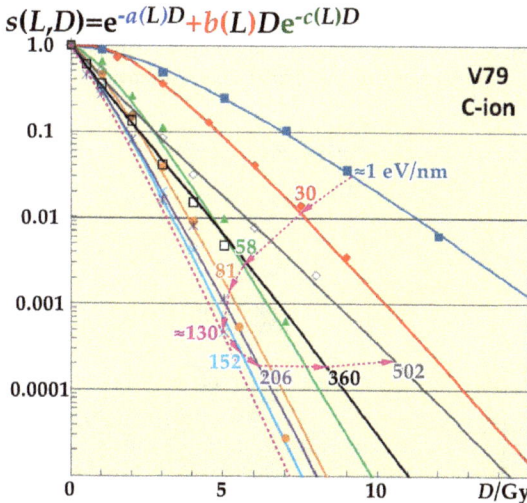

Figure 8.8(m). Illustration of the LET dependence of the cell survival curve going from X-rays (~ 1 eV/nm) to 500 eV/nm carbon ions. The solid line curves were determined by fitting the LET dependence of the RCR model to the experimental data points for V79 hamster cells (Furusawa *et al.* 2000). The disappearance of the shoulder of the cell survival curve going to LET values beyond 150 eV/nm is quite clear. The associated LET dependence of the parameters a, b, and c are shown in Fig. 8.8(n). The correlation coefficient of the fit is 0.99 (modified from Wedenberg *et al.* (2010)).

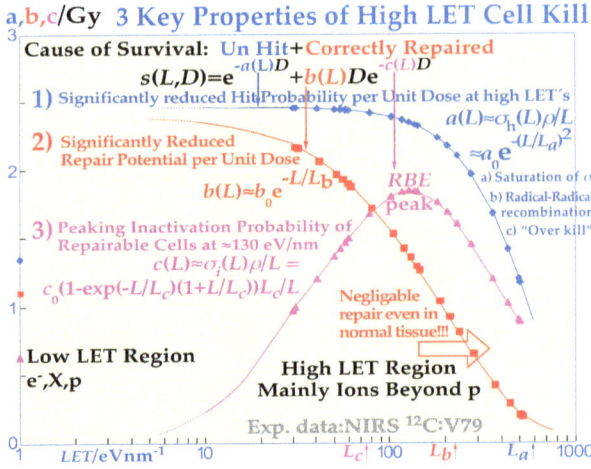

a,b,c/Gy 3 Key Properties of High LET Cell Kill

Cause of Survival: Un Hit+Correctly Repaired

$$s(L,D)=e^{-a(L)D}+b(L)De^{-c(L)D}$$

1) Significantly reduced Hit Probability per Unit Dose at high LET's

$$a(L)\approx\sigma_h(L)\rho/L$$
$$\approx a_0 e^{-(L/L_a)^2}$$

2) Significantly Reduced Repair Potential per Unit Dose

$$b(L)\approx b_0 e^{-L/L_b}$$

RBE peak

a) Saturation of σ_i
b) Radical-Radical recombination
c) "Over kill"

3) Peaking Inactivation Probability of Repairable Cells at ≈130 eV/nm

$$c(L)\approx\sigma_i(L)\rho/L =$$
$$c_0(1-exp(-L/L_c)(1+L/L_c))L_c/L$$

Negligable repair even in normal tissue!!!

Low LET Region
e⁻,X,p

High LET Region
Mainly Ions Beyond p

Exp. data:NIRS ^{12}C:V79

LET/eVnm⁻¹   10 L_c 100 L_b L_a 1000

Figure 8.8(n). The *LET* dependence of the RCR parameters a, b, and c for the cell line and survival curves in Fig. 8.8(m) is illustrated. It is interesting to note that the curves describe the three key properties of light ions: (1) the total hit probability $a(L)$ is almost constant at low *LET*s and decreases as the cross-section saturates (cf. Fig. 8.8(j)), whereas, (2) the repair potential $b(L)$ is exponentially decreasing with *LET*, and (3) $c(L)$ express the *LET* dependence of the RBE peak near 130 eV/nm. The *LET* dependent parameters a, b, and c reproduce the survival curves connecting the experimental data points in Fig. 8.8(m) almost perfectly (modified from Wedenberg *et al.* (2010) and Antonovic *et al.* (2013)).

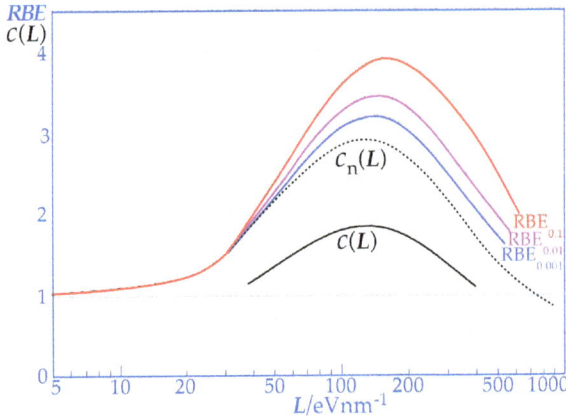

Figure 8.8(o). Illustration of the similarities between the *LET* dependence of the high dose low survival RBE and the parameter $c(L)$ (Eq. (8.35)) normalized at 30 eV/nm (cf. Eq. (8.10)) but also the probability of inducing apoptotic cell kill (Fig. 8.9). The figure is based on data in Figs. 8.9(a)–(c).

A more accurate RBE expression is obtained by adding a small exponential term as in Eq. (8.16) above. This may be particularly useful when the LET range of validity is extended to the lowest LET values. A more general term than in Eq. (8.34) may be useful to make the description $c(L)$ more accurate also at low LETs according to:

$$c(L) = c_0(1 - e^{-L/L_h}(1 + L/L_h))L_h/L + c_1 e^{-L/L_c}, \qquad (8.35)$$

where $L_c \approx 2c_1 L_h/c_0$ to make $c(L)$ quasi-constant at low L values.

To make a larger than c (cf. Eq. (8.25)) and $c(L)$ quasi-constant at low LET values (since the hit cross-section Eqs. (8.22) and (8.23) and RBE are approximately constant then, as seen in the discussion between Eqs. (8.15) and (8.16)) and slowly decreasing at very high LET values where the inactivation cross-section is saturated (Fig. 8.3) and radical–radical recombination and overkill sets in, it is natural to add a small exponential term to Eq. (8.35) to eliminate the linear increase in $c(L)$ at very low L values (cf. Eqs. (8.11), (8.34), and (8.35)) to also make $a(L)$ quasi-constant at low L values according to:

$$a(L) \approx \sigma_\infty \rho(\kappa e^{-\frac{\lambda L}{2\kappa}} + (1 - e^{-\lambda L}(1 + \lambda L))/L)$$

$$= a_0 e^{-L/L_a} + c_0(1 - e^{-L/L_h}(1 + L/L_h))L_h/L, \qquad (8.36)$$

where κ, L_a and a_0 can be selected to fit the low LET value and $L_a \approx 2kL_h = 2a_0 L_h/c_0$ and $a_0 \approx k\sigma_\infty \rho$ so that the last half of Eq. (8.36) is identical to Eq. (8.34) and the additional term makes $a(L)$ quasi-constant up to LET values toward L_a (cf. Fig. 8.8(n)). This expression was recently demonstrated for the complete LET set by Furusawa et al. (2000) (cf. Antonovic et al. 2012).

At low LETs, the expression may be simplified further by power expansion similar to Eq. (8.11) according to:

$$a(L) = a_0 e^{-(L/L_d)^2}, \qquad (8.37)$$

where

$$\frac{a_0}{L_d^2} \approx \frac{a_0}{2L_a^2} - \frac{c_0}{3L_h^2},$$

as recently demonstrated for a reduced LET set (Wedenberg et al. 2010). To fulfill Eqs. (8.24) and (8.25) the value of κ and a_0 should also not be too high so that the range of validity of the L dependence is not restricted too much (cf. Eq. (8.25)).

The remaining LET-dependent parameter $b(L)$ describes the amount of repairable damage at increasing LET values. With increasing LET, the shape of the cell survival curve gets more straight with a decreasing shoulder and curvature.

This means that either $a - c \approx b/2$ (cf. Eq. (8.30), which then results in no second-order term in D) or that b is very small. A further condition, which is related to this, is that the low dose logarithmic slope ($a - b$, cf. Eq. (8.30)) should be similar to that at high doses (c, cf. Eq. (8.32)), which are corresponding to $a - c \approx b$. Obviously, if all these conditions are fulfilled simultaneously, an almost straight cell survival curve should be expected as generally seen at high LET values.

This is obtained if based on Eq. (8.30):

$$b(L) \approx 2(a(L) - c(L)) = b_0 e^{-L/L_b} + b_1 e^{-L/L_a}, \qquad (8.38)$$

where $b_0 \approx 2a_0$ and $b_1 \approx 2c_1$. This equation really shows how the repairable damage is exponentially decreasing with the LET. Interestingly, this also means that the low LET slope ($a - b$) and high LET (c, cf. Eqs. (8.30) and (8.31)) slopes are getting more and more close to each other. This is so since $a - b \approx c$ implies that $a - c \approx b$, which is rapidly approached when b gets smaller at high LETs and $a - c \approx b/2$ as seen from Eq. (8.30) above. Furthermore, this means that there is negligible low dose hypersensitivity (requiring $b/2 \geq a - c$ cf. Eq. (8.30)) and that there is a synergistic interaction between low and high LET damages (cf. Figs. 8.8(e) and (f) and Lind et al. 2003).

The Eqs. (8.35), (8.36), and (8.38) have been used in Figs. 8.8(l) and (m) to compare with experimental data for V79 cells irradiated by a wide range carbon ion LETs. A very good fit is seen over the whole LET range showing that the LET dependence of the parameters a, b, and c are in good agreement with the above-derived expressions. Interestingly, the c parameter is in rather close agreement with the LET dependence of the RBE as seen in Figs. 8.8(l), (n), and (o) and discussed in more detail by Wedenberg et al. (2010) and Antonovic et al. (2012).

As seen from Fig. 8.8(m), the cell survival curves get more linear as the LET increases. This is, according to Fig. 8.8(n), due to the fact that fewer DNA lesions can be faithfully repaired due to increasing severity of the damage and therefore a quasi-exponential decrease of the b-term describing the repairable compartment. This is because less damage is repairable which is the case when the apoptotic, senescence, and mitotic catastrophe pathways become important (cf. Figs. 8.5(o) and (p) and Eq. (8.10) above). Interestingly, this is the case with senescence, which leads to a permanent cell cycle arrest and no further cell divisions and therefore no clonogenic survival even though the cell is alive and potentially operational for some of its functions. This may therefore be the best way of eradicating tumor cells without getting instantaneous massive cell kill. With increasing LET, these types of cell inactivation will dominate more and more and make the cell survival quasi-exponential and linear in a logarithmic diagram as clearly seen in Figs. 8.5(e) and (f), 8.8(e) and 8.8(l), and 8.9(b). It is, therefore, interesting to study the number of cells that are lost by an apoptotic or senescent response and due to their reduced inflammatory effects, making apoptosis and senescence the ideal pathways for tumor eradication.

8.3.6.7. *LET dependence of apoptosis induction*

Assuming that the *LET* dependence of apoptosis is similar to that of inactivation since both are induced by particles with sufficiently high *LET* producing severe DNA damage (cf. Nakamura 2004; Vreede and Brahme 2009), the apoptotic cross-section can then be used together with Eq. (8.11) to formulate an expression for the *LET* dependence of the A_{Fr} according to:

$$\sigma_A(L) = \sigma_{A\infty}(1 - e^{-\lambda_A L}(1 + \lambda_A L)) \approx \sigma_{A\infty}\left(1 - e^{-\frac{(\lambda_A L)^2}{2}}\right), \qquad (8.39)$$

where $\sigma_A(L)$ is the apoptotic part of the inactivation cross-section dependent on the *LET*, $\sigma_{A\infty}$ is the maximum achievable apoptotic cross-section, and $1/\lambda_A = L_A$ is related to the *LET* where maximum apoptosis is induced (cf. Eq. (8.34)). The assumption that two severe events on a single site are needed to trigger the cell to go into apoptosis, corresponds well to the inactivation cross-sections for the dual-Poisson expression as seen in Fig. 8.9(e) (cf. Vreede and Brahme 2009).

To model the dose dependence of apoptosis, an equation analogous to Eq. (8.3) can be used. The saturation of A_{Fr} can then be approximated by:

$$A_{Fr}(D) = A_\infty\left(1 - e^{-\frac{D}{D_A}}\right) = A_\infty(1 - e^{-\sigma_A \Phi}), \qquad (8.40)$$

where $A_{Fr}(D)$ is the induced apoptotic cell fraction at the dose D and D_A is a *LET*-dependent constant related to the cross-section σ_A similar to Eq. (8.9). An example of an experimental fluence dependence is shown in Figs. 8.9(a)–(c). A_∞ and D_A were determined though a weighted least square fitting of the data from Meijer *et al.* (2005). The value of the constants can be seen in Table 8.1.

Table 8.1. Fluence and dose–response parameters.

Quantity	Unit *LET*	40 eV/nm	80 eV/nm	160 eV/nm
A_∞	%	50	40	19
D_A	Gy	0.50	0.79	0.43
σ_A	μm^2	1.3	1.6	3.4
$A_\infty \sigma_A$	$\% \mu m^2$	65	64	64.6

At higher *LET*, fewer particles contributes to the same dose and therefore the lower *LET* values have a steeper initial slope as a function of dose than the higher *LET* ions. However, when looking at the A_{Fr} versus fluence, all *LET*s have the same number of passing particles and there seems to be very small difference in the initial slope between the different *LET*s as seen in Fig. 8.9(b). It is also seen that the rapid saturation of the A_{Fr} is due to a fast reduction of the cell survival with increasing fluence. The A_{Fr} as a function of the doses and *LET* in analogy with Eq. (8.20) is given by:

$$A_{Fr} = A_\infty(1 - \exp(-D\sigma_{A_\infty}\rho(1 - e^{-\lambda_A L}(1 + \lambda_A L))/(\lambda_A L))). \qquad (8.41)$$

Interestingly the Eqs. (8.11) and (8.21) that were derived for inactivation of cells also seems to describe their apoptotic response quite well as shown in Figs. 8.9(e) and (f). However, the apoptotic response does set in at a lower *LET* as a certain energy deposition and *LET* value need to be reached before apoptosis is more readily induced. Below that value, the apoptosis is low and above that value less apoptosis is induced at a given dose level since fewer high *LET* events take place. Therefore, as seen in Figs. 8.9(e) and (f), the highest apoptotic yield is obtained around 30–50 eV/nm somewhat dependent on the ions species.

8.3.7. *High Apoptotic Cell Kill in the Tumor but not in Normal Tissue*

With intermediate *LET* ions, the fluence with sufficiently high *LET* and lethality, to maximize the apoptotic cell kill, is highest at a given dose level as seen in Figs. 8.9(c) and (d). This is advantageous from a clinical point of view since apoptotic tumor cell kill does not cause as large an inflammatory response in normal tissue. Similar to the *RBE* in Fig. 8.3, the apoptosis has a peak at around 30–80 eV/nm that is at lower *LET* values than that of the *RBE* peak at 100–200 eV/nm as seen in Fig. 8.9(f). This is because the number of high *LET* events that are needed to induce apoptosis at a given dose level decreases rapidly with increasing *LET*, as explained in Figs. 8.9(d) and (e). This effect is due to the decreasing number of apoptotic events per unit dose at high *LET* values. In Fig. 8.9(g), a very interesting property of the medium *LET* light ions like lithium and beryllium is illustrated. These ions induce a local apoptotic response (programmed cell death) only in the Bragg peak region where the *LET* is high. Therefore they eliminate tumor cells with natures preferred method characterized by highest possible local efficiency and minimal side effects. At the same time, the DNA damage in the plateau region and the fragmentation tail beyond the Bragg peak triggers cell cycle arrest and DNA repair as indicated by the upper left insert (cf. also Fig. 8.25). Thus, normal

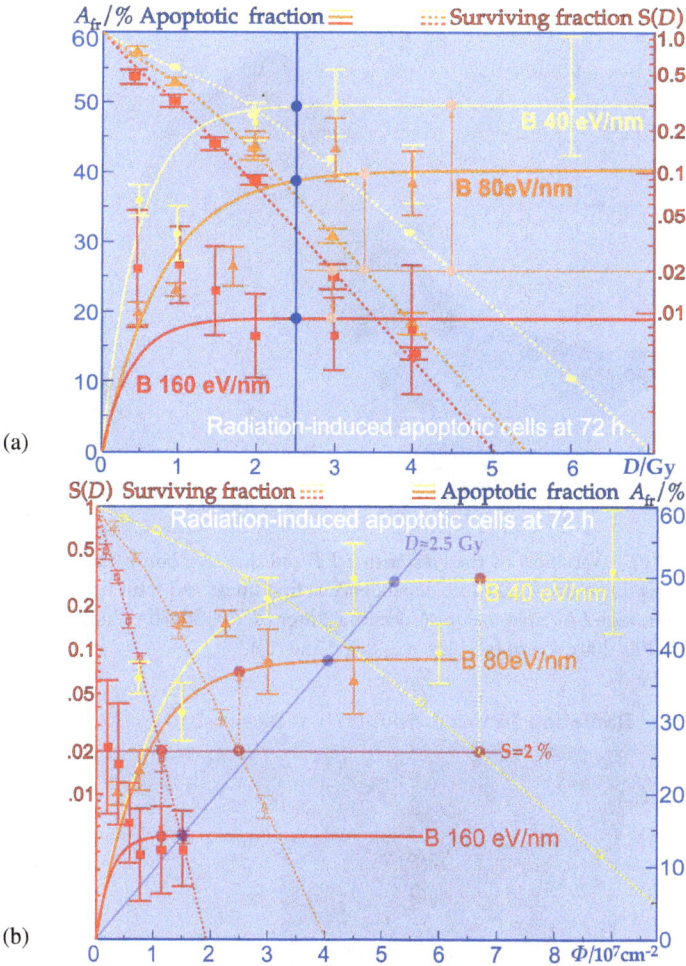

Figure 8.9(a). The saturation of the induced A_{Fr} with dose, in cells irradiated with boron ions, with three different *LET*s 72 h after irradiation is shown. The data is fitted to Eq. (8.19) with a weighted least square method. It can be seen that the saturation value is reached at approximately 2.5 Gy. **(b)** The A_{Fr} plotted against ion fluence for boron ions at three different *LET*s is shown. A_∞ was determined though a weighted least square fitting of the data from Meijer *et al.* (2005). Of special interest is that the initial slope is almost the same for all *LET*s. The saturation at high doses sets in when there are very few remaining viable cells that may contribute (cf. the associated dashed survival curves) and/or when the anti-apoptotic signaling is strong. Interestingly, the slope of the increase is almost constant independent of the *LET* when plotted as a function of the fluence since the number of severely hit cells is then almost the same independent of *LET*. For comparison, the associated cell survival curves are also included as dashed lines.

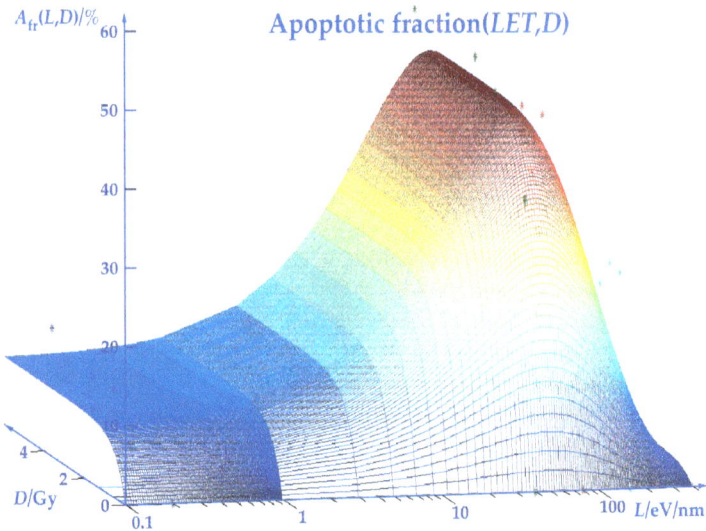

Figure 8.9(c). Variation of the A_{Fr} with *LET* and dose for boron ions (cf. Eqs. (8.40) and (8.41)) is illustrated. The dose dependence is first linear and saturates at doses beyond 3–10 Gy. The high *LET* apoptotic peak occurs at lower *LET* ($\approx$30–80 eV/nm) than the normal *RBE* peak (100–200 eV/nm) (cf. Figs. 8.3, 8.9(a) and (b)).

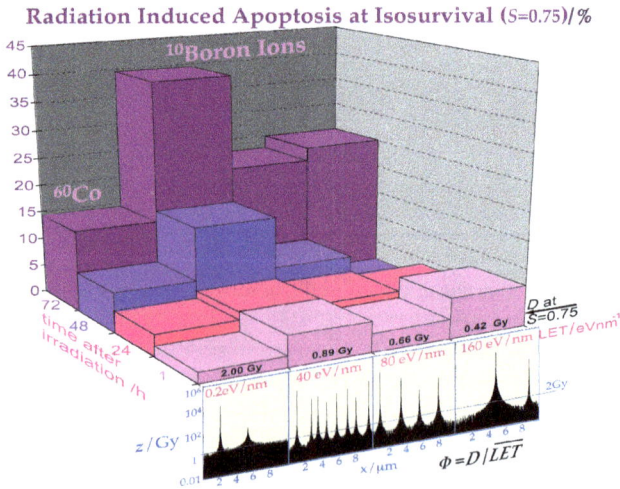

Figure 8.9(d). The fraction of cells that are lost by an apoptotic response or the so-called programmed cell death is maximal at intermediate *LET*. This is because the highest number of lethal track ends or δ electrons per unit absorbed dose are generated at intermediate *LET* values ($\approx$25–50 eV/nm) as indicated by the lower scale (cf. also Fig. 8.6(c)).

tissue will generally recover between treatment fractions, whereas the tumor cells are directly eradicated by apoptosis or slowly during fractionated radiation therapy by accumulated DNA damage eventually leading to mitotic catastrophe. In tumors with a mutant P_{53} pathway, the A_{Fr} is lower (cf. Fig. 8.9(e)). However, with light ions, there still is a significant A_{Fr} due to other apoptotic pathways normally not active with low LET radiations as seen in Fig. 8.25. In P_{53} proficient tumor cells, the lithium and beryllium ions will induce local apoptosis mainly at the Bragg peaks as seen in Fig. 8.9(g). Lighter ions do not produce much, whereas carbon and heavier ions produce a significant level of apoptosis both in front of and behind the Bragg peak (Fig. 8.9(h)). The lithium–boron ions thus possess a unique, geometrical precision in their apoptotic potential as also shown in Figs. 8.9(i) and (j) based on Eq. (8.27). Owing to the saturation effect of the apoptosis at high doses (cf. Figs. 8.9(a)–(c)), this advantage is highest at low doses as seen in Figs. 8.9(d) and (e) where the peak to plateau ratio may reach 15 at low doses of lithium ions (0.1 Gy).

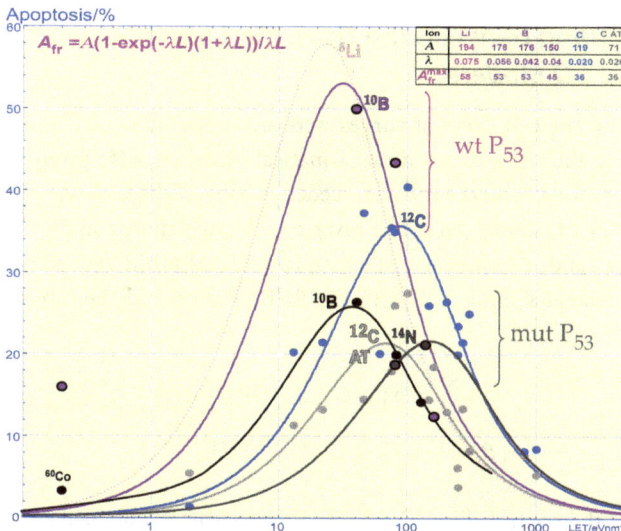

Figure 8.9(e). The variation in the induction of apoptotic cell kill as a function of LET and the status of the P_{53} pathway of the cells are shown. When some part of the P_{53} pathway is mutant, the A_{Fr} is reduced to about half its value for normal wt P_{53} and non-AT cell lines that are intact on the ATM gene upstream of P_{53}. Interestingly, the A_{Fr} peaks at lower LET values due to the higher flux density of ions and apoptotic events per unit-absorbed dose as seen in Figs. 8.9(d) and 8.6(c) than the RBE.

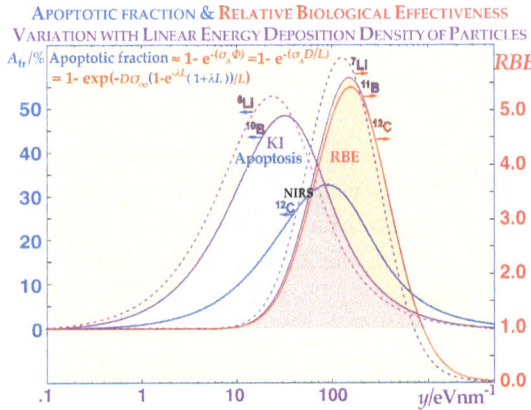

Figure 8.9(f). Comparison of the apoptotically induced programmed cell death with the clonogenic survival measured by the *RBE* as a function of increasing *LET* values is shown. Owing to the higher A_{Fr} at median *LET* with more ion tracks per unit dose, the A_{Fr} peaks at lower *LET* than the *RBE* (cf. Figs. 8.3 and 8.8(l); compare also with the *LET* dependence of the microscopic cross-section for cell inactivation in Fig. 8.8 (j)).

8.3.8. *Relation between the RCR, the Linear, and LQ Models*

The shapes of the cell survival curves were described in some detail in Figs. 8.7 and 8.8 above and the associated mathematical functions were given. To facilitate their use we will also give the relation needed to convert the RCR model parameters to the linear or LQ model data. The inverse problem is more complex but has been discussed by Lind *et al.* (2003). At therapeutic doses in the few Gray region, the LQ model coincides quite well with the RCR relation and the α and β are then given by:

$$\alpha = c - 2\ln\left(bD_c + e^{-(a-c)D_c}\right)/D_c, \qquad (8.42)$$

$$\beta = \ln\left(bD_c + e^{-(a-c)D_c}\right)/D_c^2, \qquad (8.43)$$

where D_c is the central dose where exact agreement is obtained by direct fit as seen in Fig. 8.8(b) ($D_c \approx 2.5\,\text{Gy}$ here). However, at high doses, the LQ model substantially deviates from the quasi-exponential RCR model accounting more accurately for repair (see Eqs. (8.14), (8.21), and (8.22)). This is due to a too high influence of the β term at high doses. Clearly, the LQ model is not generally suitable to describe the response neither at very low nor at very high doses as seen in Figs. 8.8(a)–(h). Interestingly, Eq. (8.32) at high doses shows that $c \approx 1/D_0 \approx \alpha$

since the logarithmic term is rapidly reduced as D_c becomes larger and larger (cf. Eq. (8.27)).

Based on the property that the survival and its first derivative at the central absorbed dose D_c should be equal, n and D_0 of the linear model in Fig. 8.8(c) can similarly be obtained as follows:

$$D_0 = (bD_c + e^{-(a-c)D_c})/c\left(bD_c - \frac{b}{c} + ae^{-(a-c)D_c}\right), \tag{8.44}$$

$$n = (e^{-aD_c} + bD_c e^{-cD_c})e^{c\frac{bD_c - \frac{b}{c} + ae^{-(a-c)D_c}}{b + e^{-(a-c)D_c}/D_c}}, \tag{8.45}$$

indicating that n increases with D_c and D_0 approaches $1/c$. From these two values, D_q can be derived from $D_q = -D_0 \ln \frac{1}{n}$, indicating that D_q increases slowly with increasing central dose D_c. For several applications, the linear model is sufficiently accurate such as for fractionated radiation therapy (cf. Brahme 1984).

8.3.9. *Fewer Microscopic Cold Spots and Higher Microscopic Uniformity of the Energy Deposition at Intermediate LETs*

The standard deviation of the microscopic energy deposition density as a function of the absorbed dose and object size is presented in Figs. 8.9–8.11. Interestingly, around lithium and beryllium, the highest portion of the dose at the Bragg peak is delivered at intermediate energy deposition densities (cf. also Figs. 8.8, 8.13, and 8.14). Figures 8.10(a) and (b) most importantly demonstrate the effect of increasing microscopic standard deviation (σ_μ) on the DRR as the *LET* increases beyond the lightest ions. The shallower normalized dose–response gradient at high *LETs* in Fig. 8.10(b) (Brahme 1984; Lind *et al.* 2001) is caused by the microscopic heterogeneity which increases the risk that some tumor cells may be missed due to microscopic cold spots even at high normally curative doses as also clearly seen in Fig. 8.10(c) especially clear in the upper right panel at 400 eV/nm. The increased randomness may also cause increased cell kill along the tracks at low doses due to random high dose events (cf. Fig. 8.10(b); Lindborg and Brahme 1990; Tilikidis and Brahme 1994).

8.3.10. *Selection of Optimal Treatment Technique and Particle Species*

Based on the different dose distributional and biological properties of light ions discussed above, it is clear that their optimal usage require some careful considerations as briefly discussed in Figs. 8.12–8.14. For example, should the

(g)

(h)

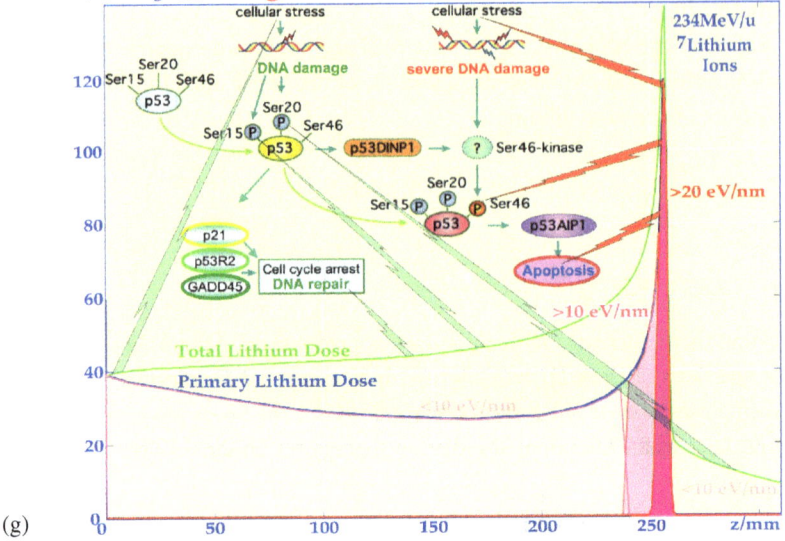

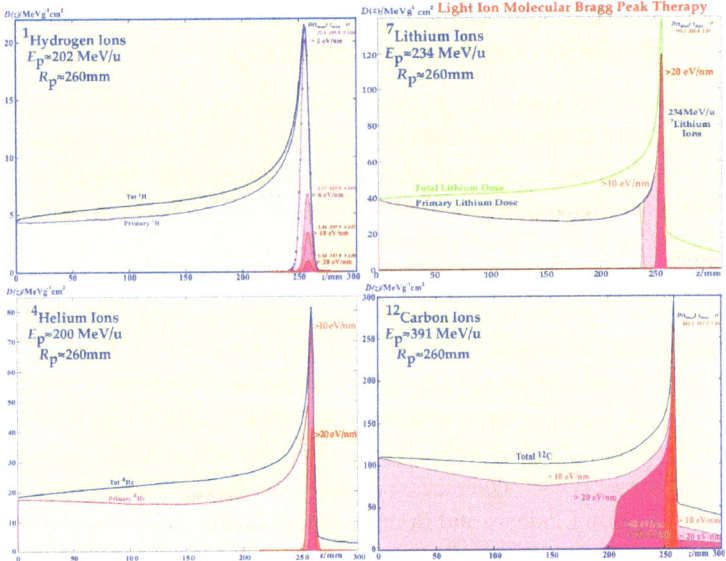

Figure 8.9(g). The DNA damage in the cell nucleus is recognized by proteins that signal their response by phosphorylation of the P_{53} gene product. Low ionization density damage generally phosphorylates the serine 15 and 20 sites on P_{53} which leads to a cell cycle block and initiation of DNA repair by P_{21} and GADD45. With lithium ions this mechanism dominates in normal tissues in front of and behind the tumor. At the Bragg peak, the ionization density is high which leads to more severe unrepairable DNA

← ──

(Continued)

Figure 8.9(g). damage which commonly phosphorylates the serine 46 site on P_{53} and thereby triggers an apoptotic response, eliminating tumor cells with severe DNA damage (Nakamura 2004). With lithium ions, a high LET is only present in the Bragg peak where tumor cells are effectively eliminated by apoptosis, whereas in all normal tissues, the low ionization density triggers a cell cycle block and DNA damage repair. **(h)** With protons and helium ions a too low ionization density is reached at the Bragg peak, whereas carbon ions have a high LET almost 5 cm in front of the Bragg peak and also in the fragmentation tail behind it, so a too little or too wide apoptotic region is obtained with these ions as seen in Fig. 8.9(c). The high LET component increases rapidly from only a few percentage for protons and about 50% for helium, and 80% for lithium. For carbon, the high LET region extends several centimeters in front of the Bragg peak and also behind it. Beryllium and boron ions are located between lithium and carbon and are of interest for intermediate tumor sites.

Bragg peak dose delivery mainly be located in the gross tumor so that negligible high LET dose falls on sensitive normal tissues outside the tumor.

Furthermore, it is better from a microdosimetric point of view to generate a rather uniform microscopic energy deposition density on the cellular scale in the tumor as shown in Figs. 8.10 and 8.11 and discussed in Fig. 8.7 starting with a slightly lower LET at the distal tumor edge as shown in Figs. 8.13(a)–(d). This could basically be done in two different ways by starting with a suitable Bragg peak at the distal tumor edge and gradually increasing the atomic weight and Bragg peak LET of the ion used as in Fig. 8.13(b) or by just mixing two different ions species so that the mean LET stays approximately constant as in Figs. 8.13(c) and (d). This latter approach requires that most of the Bragg peak dose is of very high LET, such as carbon ions, in the anterior part of the tumor, whereas the distal part mainly requires a lower LET such as helium or lithium ions. In fact, mixing lithium and carbon ions may be a very good way to achieve close to ideal microscopic energy deposition density distribution for medium-to-large size tumors where small oligo–metastasis are best treated by lithium ions alone.

8.3.11. *Selection of Optimal Radiation Quality and LET*

Both the biological and dose distributional properties of different radiation modalities from low-energy photons through high-energy electrons and photons to neutrons and light to heavier ions are summarized in Figs. 8.14 and 8.15 indicating that the intermediate LET light ions are most advantageous in most respects. The quite complex Fig. 8.14(a) summarizes the LET (upper horizontal scale) or high LET dose faction (lower scale) dependence of a number of biological parameters showing that most of the high LET advantages are obtained already at around

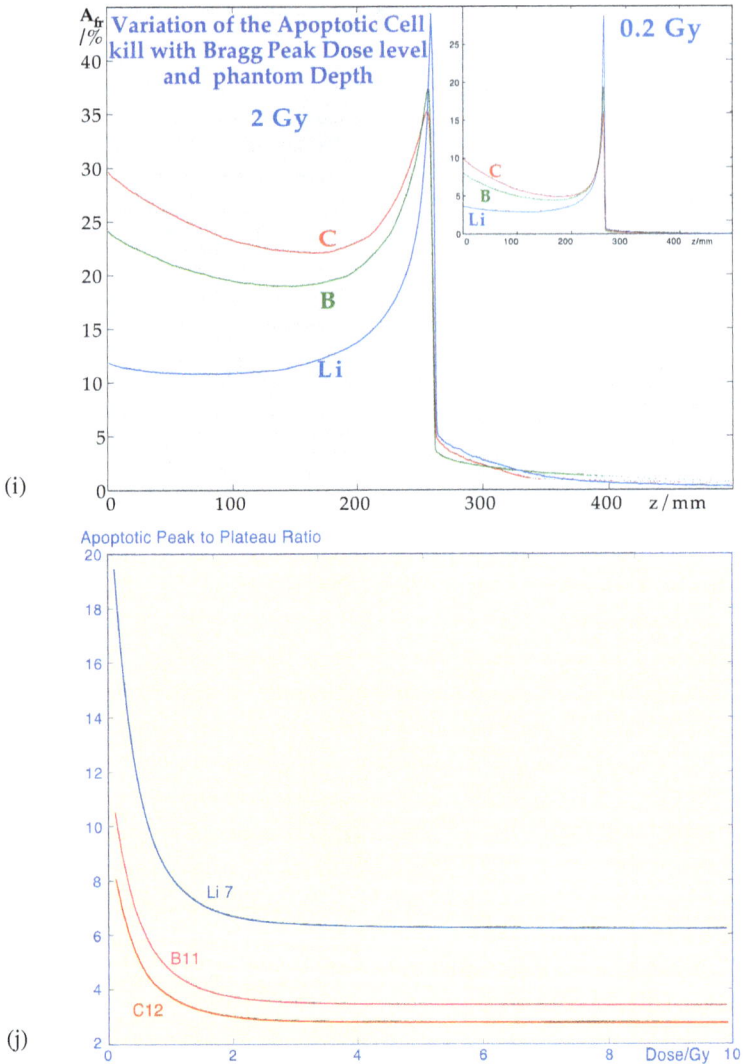

Figure 8.9(i). Based on the response in Fig. 8.9(c) with the initial exponential rise with dose (cf. Fig. 8.9(a) and (b)), the distribution of the apoptotic cell kill varies quite a lot across the dose distribution. It is seen that lithium ions are expected to have a significantly higher Bragg peak apoptosis at the same time as the plateau level apoptosis is as low as possible due to the low plateau LET in lithium beams. At lower dose levels (0.2 Gy inserted), this phenomenon is even stronger though the peak level of apoptosis is also lower. **(j)** The change in the apoptotic peak to plateau ratio, for the light ions in Fig. 8.9(e) with the Bragg peak fractional absorbed dose level is shown. Unfortunately, the low dose per fraction that produces maximum A_{Fr} requires more work with fractionated dose delivery in the clinic.

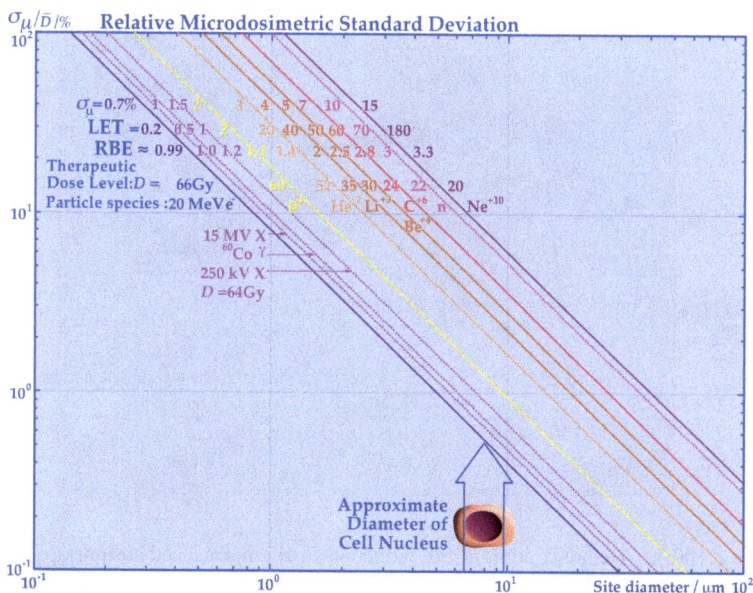

Figure 8.10(a). The increase of the microscopic standard deviation of the mean energy imparted, at the local absorbed dose level required for tumor cure, with decreasing object size and increasing LET or RBE of the ions starting from high energy electrons and photons through protons, helium, lithium, carbon, neutrons, and neon ions is shown. With the highest LET beams like neutrons and neon ions, the microscopic heterogeneity σ_μ is so high that microscopic cold spots may leave some tumor clonogens unhit at otherwise normally curative dose levels. This probably explains the problem with neutron therapy where the increase in dose beyond the low dose RBE to cure the tumor resulted in a severely increased level of normal tissue damage.

30–50 eV/nm or with as little as about one-third of the dose in the form of very high LET neon ions.

This is the case not only with the OGF, the OER, RBE, and A_{Fr} but also with the physical quantities like σ_μ, the microscopic standard deviation in absorbed dose delivery (cf. Figs. 8.10(a) and (b)), the D_{50} dose causing 50% probability of tumor cure, and the maximum clinically observed normalized steepness of the DRR γ_{Clin} is still high. The clinically most useful LET range is thus in this intermediate region and not at very high or low LET values. Within an extended Bragg peak region, the LET is reduced, as by necessity some parts of the target volume will receive low LET plateau ion dose. With a rather uniform tumor with regard to cell density and sensitivity, it is thus desirable to produce a uniform LET distribution with a mean LET value around 40 eV/nm. Interestingly, this intermediate LET region also maximizes the apoptotic cell kill, so tumor cells are more effectively eliminated

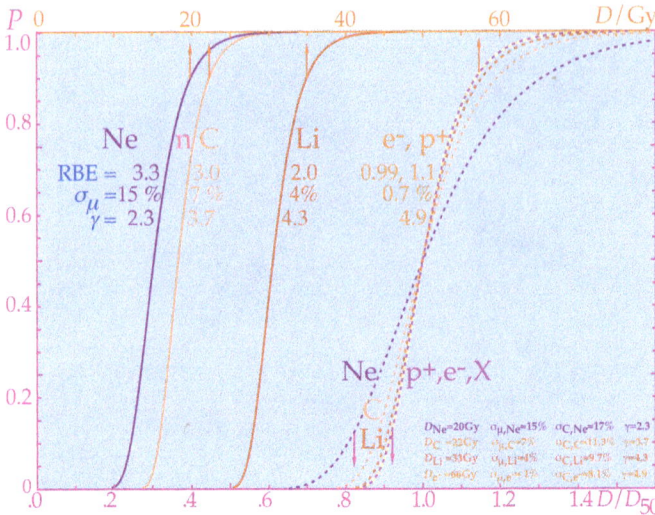

Figure 8.10(b). Demonstration of the reduction in the normalized steepness, γ, of the DRR as the microscopic standard deviation is increased. A steeper DRR generally increases the therapeutic window of radiation therapy since the absorbed dose distribution and the associated therapeutic effect over the therapeutic window can be modulated with greater efficiency with a steep DRR for tumor cure and normal tissue damage.

without too much inflammatory response in normal tissues as shown by the dotted (^{10}B) and shaded (^{12}C) curves in Fig. 8.14(a). The lower LET at the Bragg peak with boron ions as compared to carbon ions indicate the importance of the fluence density of ions with sufficiently high LET for induction of this type of process (see also Figs. 8.9(c)–(f)).

In Fig. 8.14(b), it is furthermore seen that these ions are also among the most cost-efficient ones in clinical use, since the energy needed is lower and the number of treatment fractions are quite low, commonly about one-half to one-third of the number with low LET photons electrons and protons. Optimal particle species are found between helium and carbon for most tumor sites. Figure 8.14(a) is based on data from Figs. 8.1, 8.3 and 8.5, and 8.4, respectively indicating the penumbra width, the biological advantage in a hypoxic- or radiation-resistant tumor, as well as the tumor to normal tissue dose ratio on the three independent, perpendicular axis. The associate cost of an installation and the associated treatment for a single patient are also included. Obviously, treatment modalities with a good tumor to normal tissue dose ratio are most advantageous (deep part of the figure) and so are those with a low penumbra (to the right in the figure) and fairly high biological effectiveness up in the tumor region (high up in the figure)

Microscopic Dose Distribution of X-rays & C^{+6} Ions

$\bar{D} \approx 2$ Gy

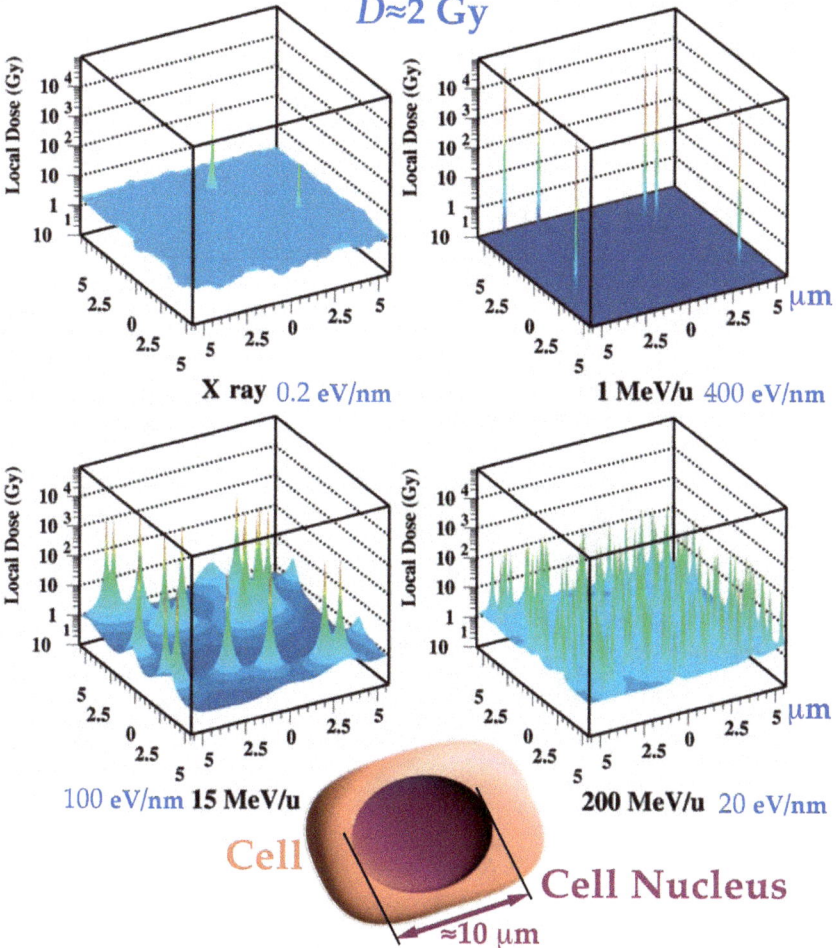

Figure 8.10(c). Illustration of the degree of heterogeneity in the microscopic dose distribution of ^{60}Co and carbon ions (C^{+6}) at a dose delivery of 2 Gy (modified from Schultz (2001)). The dose peaks for low *LET* radiations like ^{60}Co is due to one or two single track-end electrons whereas light ions in the entrance region have very many tracks, fewer toward the Bragg peak and very few very local high doses at the Bragg peak (400 eV/nm). Obviously, the number of tracks in a real beam decreases slowly with depth in an exponential fashion due to nuclear interactions that remove primary carbon ions from the beam (cf. Fig. 8.9(g)). Even though the number of carbon ions in the beam at the Bragg peak is only about one-fifth of the incoming ions their dose is about fourfold to fivefold. Therefore, some 30 carbon ions per 100 μm^2 are needed in the entrance region to get an entrance dose of about 0.5 Gy and a Bragg peak dose of 2 Gy instead of the $\approx$120 ions shown in the lower right panel to get 2 Gy.

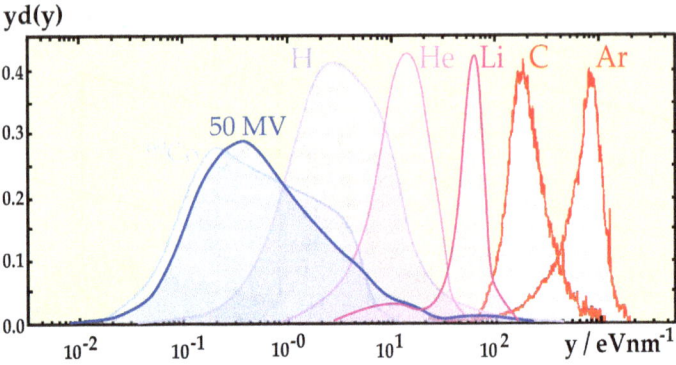

Figure 8.11(a). Microscopic energy deposition spectra of low energy and high energy photons, protons, helium, lithium, carbon, and argon ions. It is the *LET* range around 30–50 eV/nm corresponding to the lithium and beryllium peaks that is of highest interest for apoptotic cell kill as seen in Figs. 8.9(c)–(e).

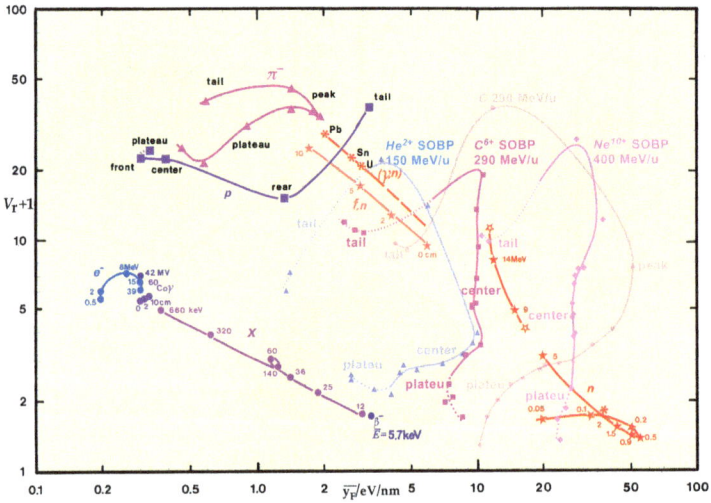

Figure 8.11(b). Dependence of the variance of the energy deposition ($V_r + 1 = \overline{y_D}/\overline{y_F}$, where $\overline{y_D}$ is the dosimetric mean y value) as a function of the frequency mean lineal energy ($\overline{y_F}$) and different locations on the depth dose curve. Interestingly the SOBP is of low variance except near the distal edge where range straggling and very high *LET* Bragg peak $\overline{y_F}$-values combine to make V_r high. For the SOBP, the frequency mean lineal energy or stopping power is fairly constant both for carbon and neon ions but the plateau in front of it and tail has lower y_F-values. The unmodulated carbon beam has a much higher peak $\overline{y_F}$ value at its Bragg peak. Neutron, pi-meson, and proton beams all have a rather high variance in energy deposition. This has the advantage that the region of lowest energy deposition is not so low (few microscopic cold spots) but instead the mean *LET* is low and so is the biological effectiveness (cf. Brahme (1982) updated with recent He, C, and Ne data from NIRS).

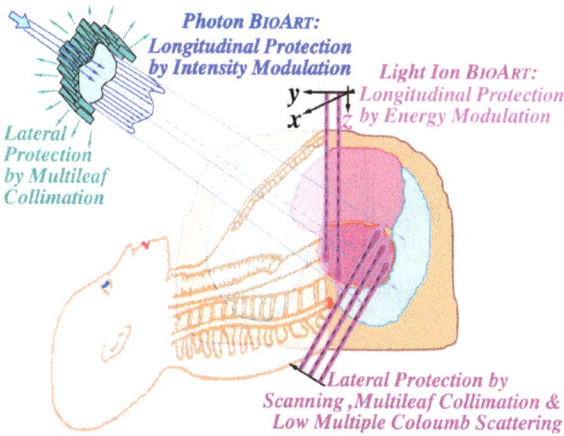

Figure 8.12. Illustration of the difference and similarity between biological optimization with photon and light ion beams. The light ions have advantages both with regard to low multiple scattering penumbras and sharp practical range (cf. Figs. 8.35(b) and (c)).

so hypoxic tumors are effectively eradicated. Too high biological effectiveness results in few ions and a high microscopic randomness in energy deposition with resulting lower apoptotic yield and shallower DRRs. It is clear that the group of ions from helium to carbon is most interesting for radiation therapy both from a clinical-dose distributional, biological, and economical cost effectiveness point of view.

The different clinical factors that are affected by the ionization density finally are summarized in Fig. 8.15. Owing to the almost straight cell survival curve with medium-to-high LET and the low dose to organs at risk, make dose fractionation much more flexible and efficient. In general 1–12 fractions are sufficient with light ions and the treatment outcome is often better with a low number of fractions (cf. Fig. 8.23 below).

Both the biological and dose distributional properties of different radiation modalities from low-energy photons through high-energy electrons and photons to neutrons and light ions are summarized in Figs. 8.14 and 8.15, indicating that the intermediate LET light ions are most advantageous in most respects. The quite complex Fig. 8.14(a) summarizes the LET (upper horizontal scale) or high LET dose fraction (lower scale) dependence of a number of biological parameters showing that most of the high LET advantages are obtained already at around 30–50 eV/nm or with as little as about one-third of the dose in the form of high LET neon ions. With Bragg peak carbon ions this reduction could not be as great as the LET is lower than for neon. With Bragg peak, carbon ions probably about half the

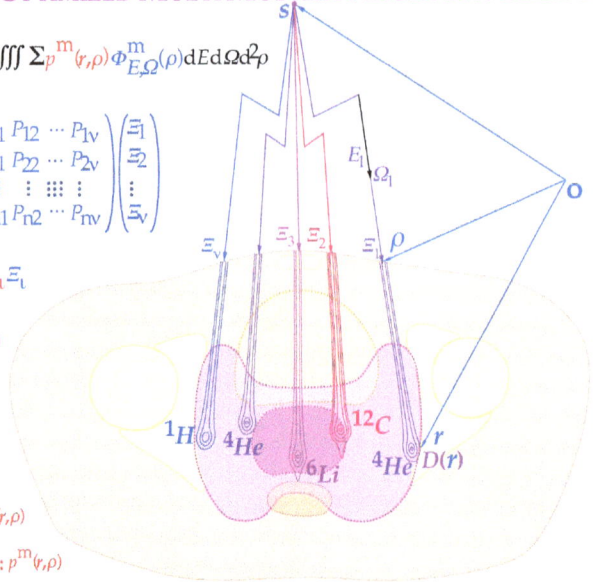

Figure 8.13(a). Different light ion dose distribution kernels are advantageous in different regions of the target volume (here cervix cancer with locally involved lymph nodes). Helium and protons may be most useful in the periphery of the clinical target volume due to their lower *LET* in a region where the tumor cell density is relatively low. Lithium ions may be best in the distal gross tumor region where rectum is downstream of the target volume and may otherwise receive the more toxic fragmentation tail of carbon ions. Finally the anterior gross tumor should preferably receive Bragg peak carbon ions so a uniform medium *LET* ($\approx 40\,\text{eV/nm}$) could be reached throughout the gross tumor.

dose could be low *LET*. The 30–50 eV/nm region is close to ideal not only for the *OGF*, the *OER*, *RBE*, and A_{Fr} but also with the physical quantities like σ_μ, the microscopic standard deviation in absorbed dose delivery (cf. Figs. 8.10(a) and (b)). Furthermore, the D_{50} dose causing 50% probability of tumor cure is still low and the maximum clinically observed normalized steepness of the DRR γ_{Clin} is still high. The clinically most useful *LET* range is thus in this intermediate *LET* region and not at very high or low *LET* values. Within the SOBP region, the *LET* is reduced, as by necessity some parts of the target volume will receive low *LET* plateau ion dose. With a rather uniform tumor with regard to cell density and sensitivity it is thus desirable to produce a more uniform *LET* distribution with a mean *LET* value around 30–40 eV/nm everywhere as could be achieved using mixed lithium and carbon beams (cf. Fig. 8.13 and particularly Fig. 8.13(c)). Interestingly, this

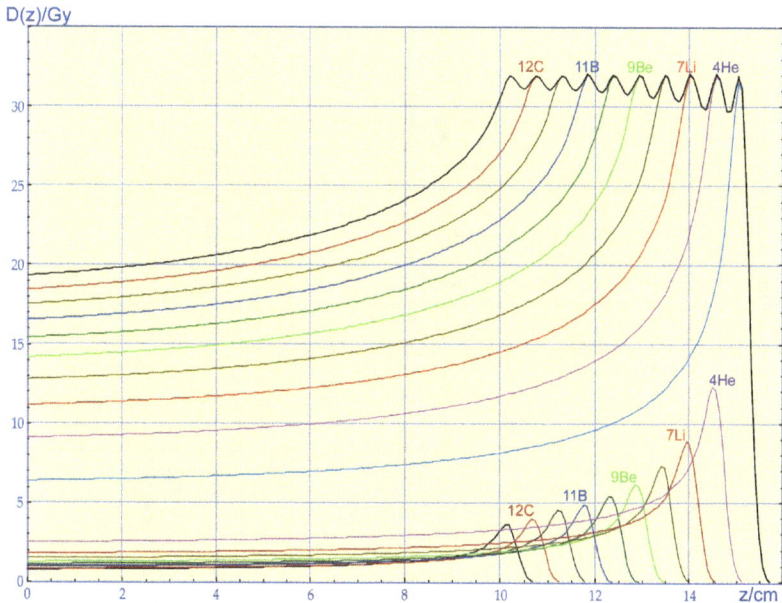

Figure 8.13(b). To get a more uniform biological effect distribution in the target region (10–15 cm depth), it is better to start with a lower distal LET such as lithium or helium ions and increase the Bragg peak LET toward the anterior part of the tumor using carbon ions. In this way, a more uniform LET distribution with a smaller variation in the microscopic standard deviation in the energy deposition is obtained (cf. Figs. 8.8 and 8.10(a) and (b)).

intermediate LET region also maximizes the apoptotic cell kill so that tumor cells are more effectively eliminated without too much inflammatory response in normal tissues as shown by the dotted (^{10}B) and shaded (^{12}C) curves in Fig. 8.14(a). The lower LET at the Bragg peak with boron ions as compared to carbon ions indicate the importance of the fluence density of ions with sufficiently high LET for inducing an apoptotic response (see also Fig. 8.8) and should be advantageous at least in medium size tumors.

A more classical multi-parameter overview of different particle species is given in Fig. 8.14(b) where it is seen that the medium LET ions are also among the most cost-efficient ones in clinical use, since the energy needed is lower and the number of treatment fractions are quite low, commonly about one-half to one-third of the number with low LET photons electrons and protons. Optimal particle species are found between helium and carbon for most tumor sites. The three independent perpendicular axis in Fig. 8.14 are plotted: the penumbra width (x-axis), the biological advantage for hypoxic- or radiation-resistant tumor (vertical axis), and the tumor to normal tissue dose ratio (z-axis) in SOBP beam. The associated costs

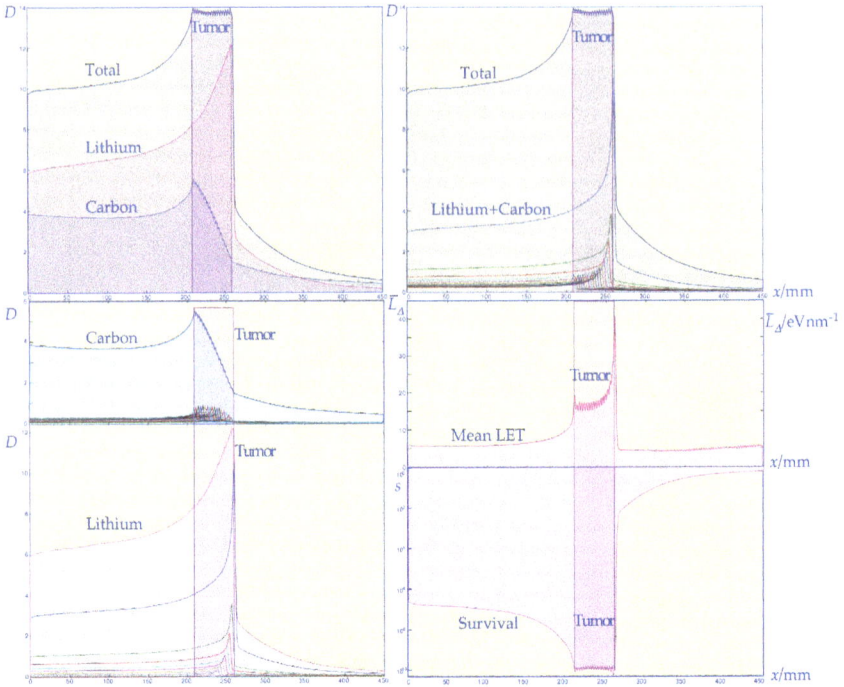

Figure 8.13(c). The generation of a quasi-uniform absorbed dose and cell kill distribution between about 21 and 26 cm of depth by combining lithium and carbon ions in suitable ratios to make the cell kill and survival quasi-uniform is shown. The small local variations in absorbed dose are due to a somewhat too large longitudinal range modulation (~3 mm) used to clearly illustrate the applied mechanism combining lithium and carbon ion Bragg peaks at each depth. The different panels show the total absorbed dose and the carbon dose and the lithium dose in the upper row, whereas the cell survival and mean *LET* distribution is shown below. Interestingly, by combining lithium and carbon ions, a uniform biological effect, survival, and absorbed dose can be obtained for a uniform tumor. The survival was calculated for simultaneous irradiation based on Eqs. (8.22), (8.28), (8.31), and (8.32), whereas the mean *LET* is more difficult to interpret as it is based both on lithium and carbon ions but the *LET* variation is low and it peaks just downstream from the distal part of the tumor where the dose is low.

of a whole installation and for the treatment of a single patient are also indicated. Obviously, treatment modalities with a good tumor to normal tissue dose ratio are most advantageous (deep part of the figure) and so are those with a low penumbra (to the right in the figure) and fairly high biological effectiveness in the tumor region (high up in the figure) so hypoxic or generally radiation-resistant tumors

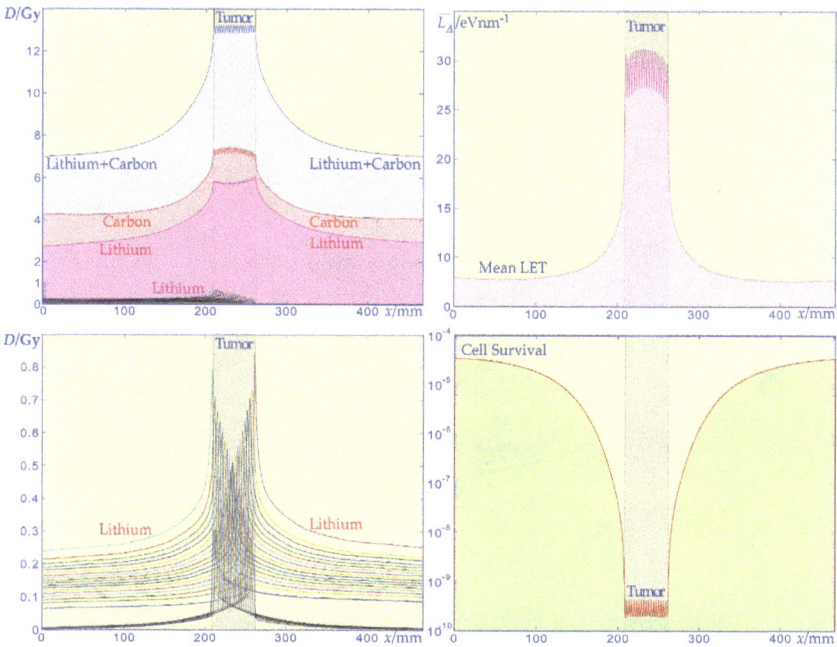

Figure 8.13(d). A common technique to reduce the quality variation is to use parallel-opposed beams. This is even more efficient than using two different ion species as seen on the *RBE* variation which is now almost negligible and the peaks have disappeared because of the dose from the posterior beam. It is also seen that the mean *RBE* is increased even if about half the dose is delivered by lithium ions.

are effectively eradicated. Too high a biological effectiveness results in too few ions, a lower apoptotic cell kill, and a high microscopic randomness in energy deposition with a resulting shallower DRR. It is clear that the group of ions from helium to carbon is most interesting for radiation therapy both from a clinical-dose distributional, biological, and economical–cost effectiveness point of view. The low normal tissue *RBE* at high doses, almost straight cell survival curve in the tumor with medium–to–high *LET* and the low dose to organs at risk, make dose fractionation much more flexible and high doses per fraction very efficient. In general 1–12 fractions are sufficient with light ions and the clinical treatment outcome is often better with a low number of high dose fractions. Finally, Fig. 8.15 summarizes some of the most important differences between low and high *LET* beams. Interestingly, the treatment outcome is often best with a few of fractions (2–12 cf. Fig. 8.23(c) below).

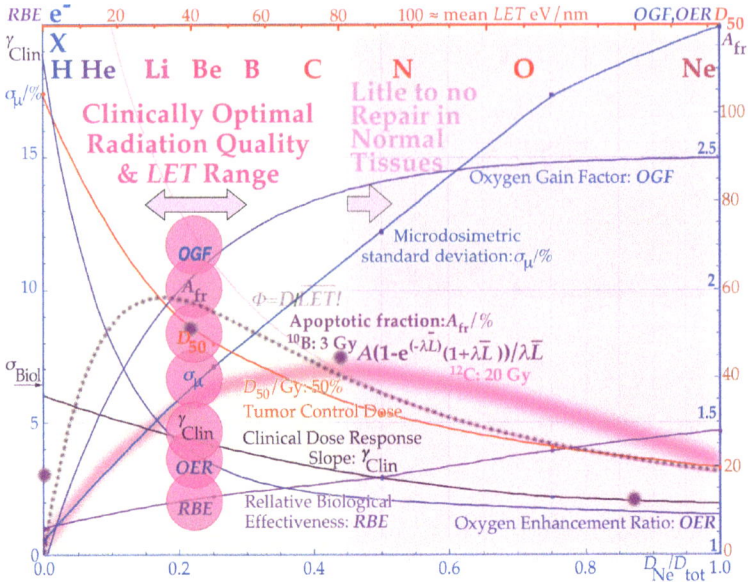

Figure 8.14(a). Selection of optimal radiation quality range of the *OER*, *RBE*, A_{Fr} (cf. Fig. 8) and the dose causing 50% tumor cure as a function of the ionization density (*LET*, upper scale) or densely ionizing dose fraction (lower scale) is shown. Lithium to carbon ions are the most interesting and useful light ions for radiation therapy. It is seen that like the *RBE*, the peak A_{Fr} occurs at increasing *LET* as the atomic number of the ions increases. Also cell lines with a mutant P_{53} pathway have a lower level of induced apoptosis. About 50% the A_{Fr} is induced by P_{53}-independent pathways (cf. Fig. 8.9(e) and Curtis 1976).

8.3.12. *Verification of the Position of the High Dose Volume using PET or PET-CT Imaging*

The high-dose high *LET* volume should as far as possible be restricted to the tumor (cf. Figs. 8.1, 8.4, 8.19, 8.31, 8.33, 8.35, and 8.40). The side effects are critically dependent on the dose and volume of irradiated normal tissues. Large margins must often be applied, as there are uncertainties in the patient setup, in the diagnostic procedure, and in the dose delivery. Modern imaging techniques make it possible to improve the accuracy in gross and clinical target volume delimitation. The irradiated volume and dose distribution could, for example, be directly monitored using real-time PET in light ion and high-energy photon facilities, or immediately after treatment using PET-CT imaging (cf. Figs. 8.16, 8.17, and 8.38(c) and (d)). This means that the real volume that has been irradiated could be determined and adaptive corrections in the positioning of the beam set-up could be performed in proceeding treatments based on PET-CT imaging. It may even be possible to

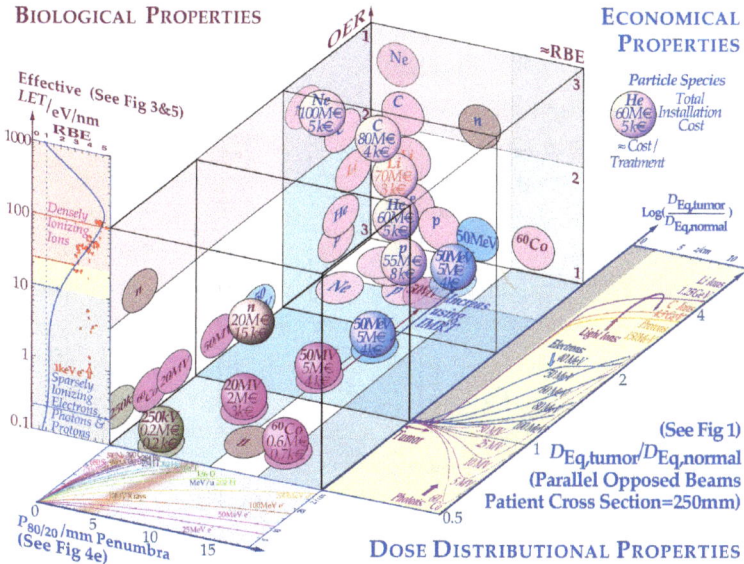

Figure 8.14(b). Comparison of the radiobiological effectiveness (effective *LET*, *RBE*, and *OER*) and the lateral and longitudinal dose distributional properties (penumbra) and tumor to superficial tissue dose ratio for uniform parallel-opposed beam pelvic irradiation using different radiation beam modalities is shown. The higher the spherical indicator the more effective the beam is for eradication of hypoxic and generally radiation-resistant tumors. The approximate costs per typical installation and per patient treated are also indicated. The increase in the tumor to normal tissue dose ratio using electron and photon IMRT are indicated. Similar improvements using biologically optimized intensity-modulated- or radiation quality-modulated-light ion radiation therapy should be possible (Brahme 2009).

monitor the radiation responsiveness of the tumor (Figs. 8.26 and 8.27) to fine adjust the desired dose delivery (BIOART, Brahme 2003, 2009) depending on the response of the individual tumor volume.

Interestingly the excentric gantry in Figs. 8.36–8.38 is designed such that it can use ^{11}C beams produced in the energy-modulating decelerator for carbon therapy. This makes Bragg peak dose delivery imaging about 50 times more sensitive, since the ^{11}C peak is now almost 50 times higher than with a ^{12}C beam as seen in Fig. 8.38(c). Since this imaging technique is available both with high-energy photons and light ions, fast PET-CT imaging should be available in the treatment to allow fast docking of the patient to both the therapeutic and diagnostic modalities as shown in Fig. 8.36 where light ions are planned to be combined with high-energy photons for *in vivo* dose delivery imaging based on PET-CT in the treatment area.

PARAMETER	QUANTITY	LOW LET	HIGH LET	Fig #
TRACK STRUCTURE	$z = d\varepsilon/dm$			8.6 d, 8.7 b-c 8.8 j-l
ELECTRON SLOWING DOWN SPECTRA PER UNIT DOSE	$\Phi_E = ds/dV$ $D = d\bar{\varepsilon}/dm$			8.6 a, 8.7a
MICRODOSIMETRIC SPECTRA	$y \cdot d(y)$			8.11 a,b
ELECTRON MULTIPLICITY	P_n			8.6 d
RELATIVE BIOLOGICAL EFFICIENCY	Cell Survival $RBE_S = \dfrac{D_L}{D_H}$			8.5 e,f
OXYGEN ENHANCEMENT RATIO	$OER_S = \dfrac{D_{N_2}}{D_{O_2}}$			8.5 g-j
CELL CYCLE DEPENDANCE	$\alpha(\tau)$			8.5 a,b
DOSE RATE DEPENDANCE	$\dot{D}$			8.5 c,d
DOSE FRACTION DEPENDANCE	α/β			8.5 m-o
DOSE RESPONSE RELLATION	Therapeutic $RBE_P = \dfrac{D_L}{D_H}$ $< RBE_S$			8.22, 8.23

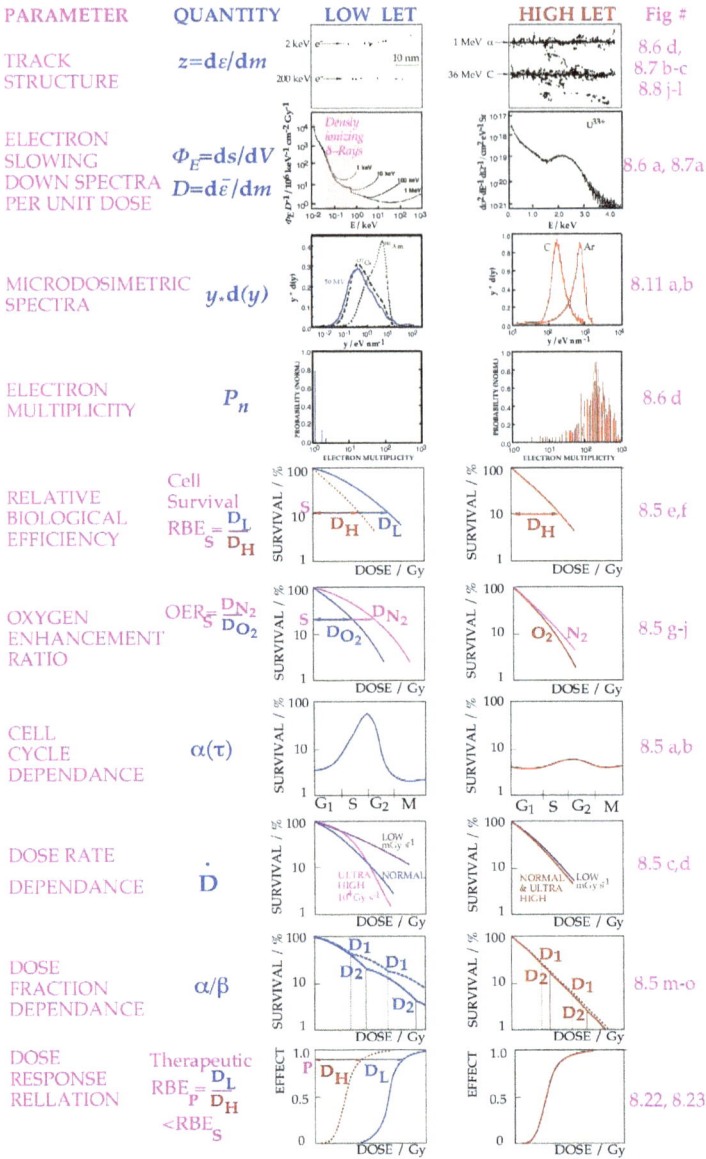

Figure 8.15(below). Illustration of the difference in track structure, electron slowing down spectra, microdosimetric spectra, and electron multiplicity distributions per event as well as cell survival curve shape between low and high *LET* beams. In all cases, the larger shoulder of low *LET* and almost straight exponential fall-off for high *LET* beams is seen explaining the different biological effects as a function of dose, dose rate, *RBE*, *OER*, and time dose fractionation. It is particularly important to understand the difference between the low dose cell survivals *RBE* (row 5) and that due to the high dose therapeutically fractionated DRR (lowest row of panels, cf. Figs. 8.10 and 8.23(d)).

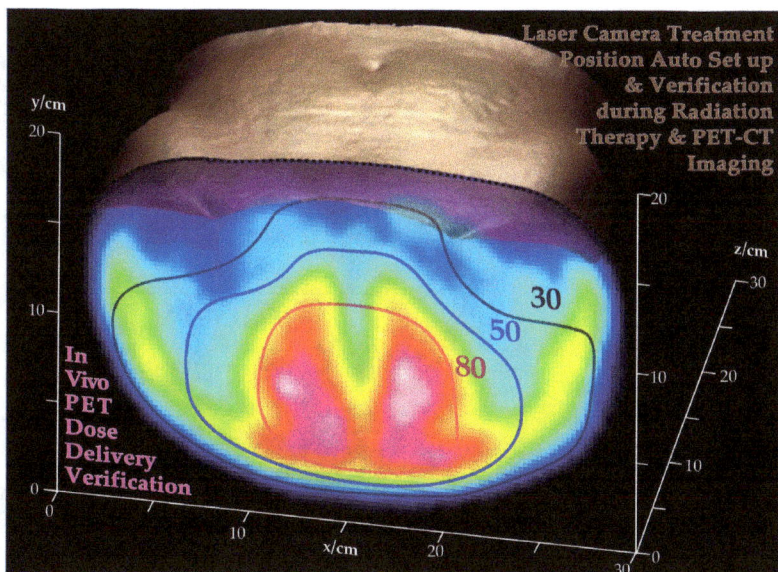

Figure 8.16(a). Dose delivery imaging using high-energy photons which through photo nuclear reactions knockout neutrons from ^{12}C and ^{16}O to produce positron emitting ^{11}C and ^{15}O that are here imaged with a low resolution PET camera ($\approx$8 mm). With high-energy photons (here 50 MV), the photo nuclear ^{11}C and ^{15}O activity is close to the delivered dose distribution and the cold spot in rectum with this four-field box treatment of a preoperative cancer rectum is normally not considered during therapy planning. In the present case, it is caused by bowel gas and causes an unacceptable 10–15% dose reduction on the rectal wall. For this reason high-energy photons and PET-CT imaging should be available in a light ion center (cf. Figs. 8.17 and 8.36). The laser camera (LC) was used in the treatment room for auto setup and treatment position verification (cf. Fig. 8.28).

8.3.13. *Summary of the Clinical Value of Different Light Ions Species*

The clinical value of different ions depends both on their dose-distributional and their radiobiological properties as summarized below and in Figs. 8.13, 8.14, and 8.15.

Dose distribution properties. With light ions considerably improved, physical dose distributions are obtained compared to photons, electrons, and protons

- about threefold reduction in longitudinal range straggling between protons and lithium ions (Fig. 8.4(d)).
- about threefold reduction in lateral scatter (Fig. 8.4(e)).

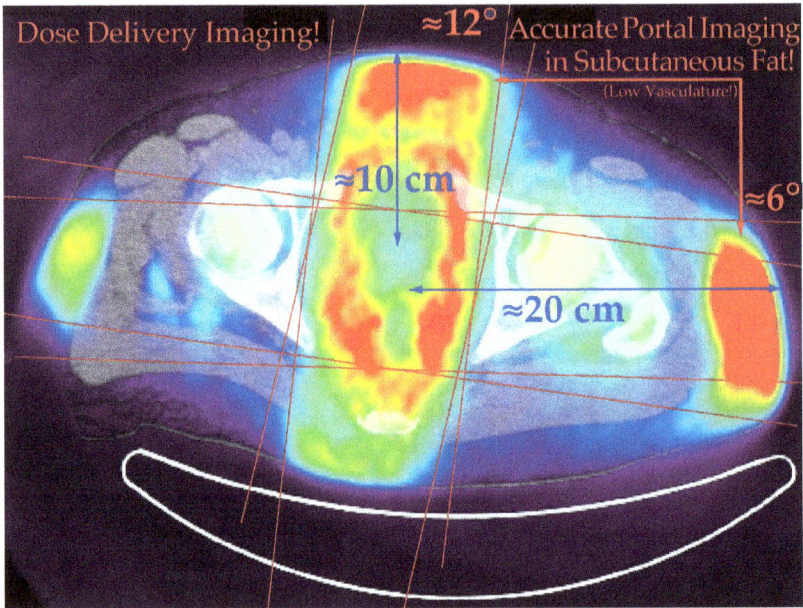

Figure 8.16(b). *In vivo* dose delivery imaging with 50 MV photon beams in a four-field box technique using the latest generation of PET-CT camera (64 slices CT and ≈4 mm PET resolution, cf. Fig. 8.23(a)). Here, the individual radiation fields are more clearly seen than in Fig. 8.16(a) and their order of delivery can even be seen from the subcutaneous fat signal which is weakest in the posterior field and strongest in the right lateral field delivered last to the patient. Interestingly, the patient were very rapidly positioned on the diagnostic couch, to not lose too many counts, and the left side of the body happened to be partly located outside the couch top causing a substantial rotation of all the surrounding subcutaneous fatty soft tissues so that it looks like the whole four-field box technique is made with severely tilted beams. Owing to the lower carbon content in the bladder and prostate, these organs are not contributing much to the PET activity. Instead, the surrounding fatty regions that enclose these organs are clearly seen. Similar rotations of the subcutaneous fatty tissues are commonplace in radiation therapy but fortunately they are generally of a much smaller extent. The clinical consequence of the rotation is an undesirable broadening of the effective penumbra width mainly in surrounding normal tissues and in the tumor region. It is interesting to note that the subcutaneous fat, present in all patients, can work as a portal imaging device and record the induced ^{11}C activity at least for 10 min after the treatment to allow accurate PET-CT imaging of the mean dose delivery profiles of the entrance portals of the beams as seen here in a transaxial cut (cf. Chapter 3, Fig. 3.84, for further images). This portal activation imaging principle induced here by high-energy photons would work also for light ions (cf. Fig. 8.38(d)) both for primary ^{11}C and ^{12}C beams.

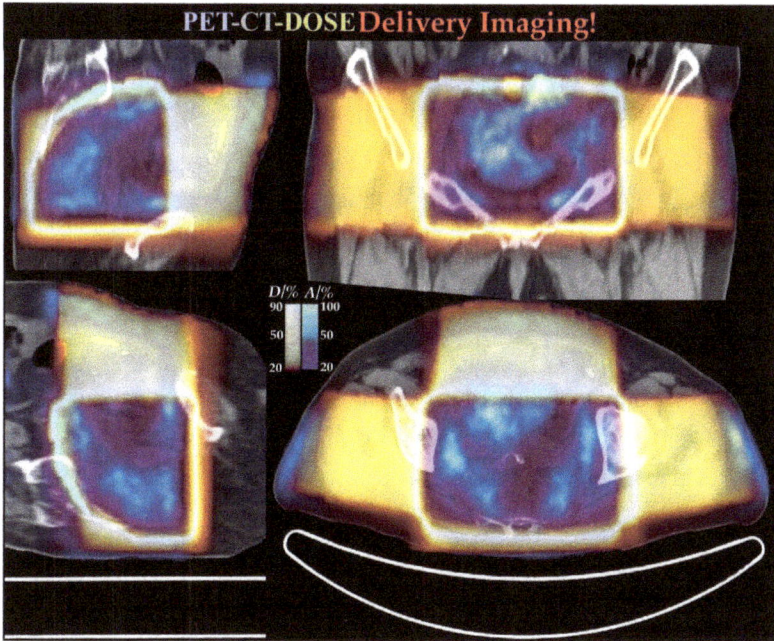

Figure 8.16(c). *In vivo* dose delivery imaging with 50 MV photon beams in a four-field box technique using the latest generation of PET-CT camera. The yellow to white color scale is from the treatment plan (slightly warped to account for the curved coach on the diagnostic PET-CT) with an opening in the high-dose target volume to show the inside tissue activation in blue mainly due to carbon activity in adipose tissue and the intestine in deep purple since most of the oxygen activities have decayed here about 10 min after the end of the treatment.

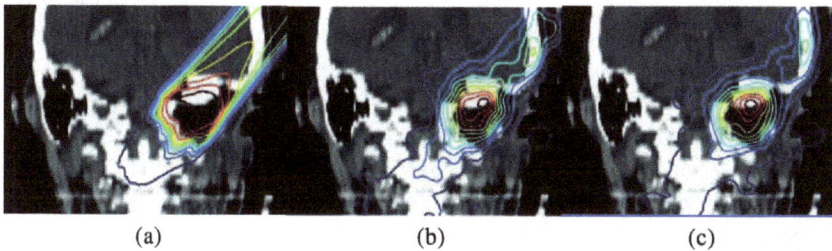

Figure 8.17. PET-imaging during carbon ion radiation therapy. **(a)** Dose distribution of the treatment plan; **(b)** reconstructed PET activity based on (a). **(c)** Real-time PET-CT image recorded during therapy (courtesy T Haberer). It is possible to predict the delivered activity distribution by changing the pencil beam dose delivery kernel to one that describes the $\beta+$ activity distribution produced by the pencil beam (Brahme *et al.* 2006 and Figure 4.82).

- This results in the possibility to have a threefold reduction in penumbra and dose outside well-defined tumor volumes (Figs. 8.4(d)–(g)).
- A lower effective dose and biological effect is deposited in front of and distal to a small tumor with the lightest ions (Figs. 8.1, 8.4, and 8.9(g) and (h)).
- For 5 mm half-width pencil beams, ideally suited for biologically optimized energy, and intensity-modulated light ion therapy, the tumor dose addition is approximately twice the entrance dose for the light ions, whereas it is only half the entrance tissue dose for protons (cf. Fig. 8.4(g)).
- To increase the tumor dose in a small tumor volume, about four times higher proton dose thus needs to be delivered to anterior dose limiting tissues by protons compared to light ions from lithium to carbon (Figs. 8.4(f) and (g).
- Lower biologically effective doses are delivered to normal tissues in parallel-opposed beam techniques (cf. Figs. 8.1 and 8.14(b)) above:

  8 times lower compared to photons;
  2–3 times lower compared to electrons; and
  1.5 times lower compared to protons.
- PET-CT imaging allows for more accurate 3D *in vivo* dose delivery verification applies also to high-energy photons and a more accurate determination of the radiation responsiveness and tumor cell density data (applies to all radiation modalities).

Biological response properties. Light ions have considerably improved biological selectivity due to:

- Co-localization of high dose, high LET, low OER, and high RBE (cf. Figs. 8.4(a)–(c) and 8.5(q)).
- Co-localization of high dose, high RBE, and low OER is obtained in the Bragg peak for ions from lithium to carbon and above (cf. Figs. 8.13 and 8.14).
- This results in a more than doubled therapeutic effect or OGF in hypoxic and radio-resistant tumors (Fig. 8.14(a)).
- Many radiation-resistant tumors can be locally controlled with reduced normal tissue morbidity and increased cure due to the simultaneously improved biological and physical selectivity (Figs. 8.4, 8.12, and 8.14).
- Negligible variation in sensitivity with cell cycle phase is found (Figs. 8.5(a) and (b) and 8.15).
- A considerably steeper DRR is seen in hypoxic and/or radiation-resistant tumors and consequently often an increased therapeutic window between tumor cure and normal tissue complications can be obtained, as seen in Figs. 8.22 and 8.23.
- Considerably reduced variation in radiation sensitivity between oxic and hypoxic and otherwise radiation-resistant tumor cell lines is seen (Fig. 8.5).

Light ions between lithium and carbon deliver the ideal LET spectrum for eradication of hypoxic and low LET radiation resistant tumor cells and for preferential induction of apoptosis in the tumor (cf. Fig. 8.9, 8.13 and 8.14).

Maximum apoptotic cell kill probability per unit dose at a given cell survival level is obtained for intermediate LET ions (Figs. 8.3, 8.10, and 8.13).

The low fractionation sensitivity of light ions reduces the need for a high number of radiation therapy sessions typically 1–12 are used instead of 25–35 for low LET beams (Figs. 8.15 and 8.23).

- Negligible dose rate effect makes the biologically optimized light ion IMRT the preferred treatment for many advanced tumors (Figs. 8.5(c) and (d)).
- More generally, light ions are ideally combined to make radiation quality-modulated radiation therapy (QMRT, Fig. 8.33(b)), for example, combining helium or lithium and carbon (cf. Fig. 8.13). In fact the spread out Bragg peak (SOBP) technique is not well suited for high RBE light ions due to the large variation in microscopic energy deposition.

With lower doses in intensity-modulated dose delivery, light ions will deliver a more linear therapeutic response, whereas low LET beams suffer from the double or triple trouble effect, since when the dose is low, so is the dose rate and the dose per fraction giving a more strongly nonlinear response with photons, electrons, and protons. Expressed differently: the ions have always the same high dose rate along their tracks, largely independent of the mean dose level and thus instead triple advantage (cf. Figs. 8.5, 8.8, and 8.15).

Clinical properties. As a consequence of the significant physical and radiobiological advantages, several clinical advantages can be identified:

- The sharp penumbra and small range straggling makes local high dose and effect treatments possible with minimal side effects in normal tissues (cf. Figs. 8.1, 8.4(e)–(i), and 8.26(a)).
- The whole treatment can often be delivered in a week or two so there is not time to develop severe side effects during the treatment, largely due to advantageous dose distribution and clinical advantages of high fractional tumor doses (cf. Figs. 8.8 and 8.23(c)).
- A higher apoptotic cell kill and senescence in the tumor will reduce the inflammatory response in normal tissues (cf. Fig. 8.9).
- A higher induction of senescence (permanent cell cycle arrest) may be an even milder way to sterilize tumor growth without the need to immediately kill all

tumor cells and avoid acute side effects when large amounts of dying cells are disposed.

- Narrow Bragg peak beams can be used to effectively treat oligo-metastasis with minimal side effects in surrounding normal tissues not least with the uniquely apoptotic Bragg peaks of lithium and beryllium ions (cf. Fig. 8.9).
- With 3D pencil beam scanning, there is increased flexibility in shaping the tumor dose distribution from almost any direction and thus less need for a rotary gantry (cf. Fig. 8.34(c)).
- Accurate verification of the dose delivery is possible with PET-CT imaging, not least using the lightest PET emitter ions ^{11}C and ^{8}B (cf. Figs. 8.38(c) and (d)).

In general, the clinical value of different ions depends on the target size and the location and sensitivity of surrounding normal tissues. When sensitive structures are located distal to the tumor, the lightest ions are preferred due to their negligible fragmentation tail, whereas lateral organs at risk require the heavier light ions such as carbon due to their small penumbra as demonstrated in Figs. 8.4(d), (e) and 8.13(a). Most of the properties of light ions are advantageous for radiation therapy when compared to classical photons and electrons. Among the few negative aspects are the high installation cost and the need for heavy equipment due to the large magnets needed to bend and scan the ions beams. This can partly be compensated for by using a single excentric gantry surrounded by 3–6 treatment rooms as demonstrated in Figs. 8.36, 8.37, and 8.38 below.

8.4. Clinical Indications

8.4.1. *Radiation-resistant and Hypoxic Tumors*

Candidates for light ion therapy are, in particular, tumors that are hypoxic or in general radiation resistant to conventional low *LET* radiations. About 90–95% of the world's clinical experience on the use of carbon ions for radiation therapy comes from the NIRS in Chiba, Japan with almost 5000 patients treated till 2010. Clinical examples with special encouraging results from Japan are summarized in Fig. 8.18 and include: bone and soft tissue sarcomas, Fig. 8.22, non-small-cell lung cancers, Fig. 8.23, hepatocellular carcinoma, prostate cancers, Fig. 8.24, malignant melanomas, cf. Figs. 8.19 and 8.20, head and neck cancer, and adenocarcinomas of the pancreas. Other suggested candidates are: locally advanced tumors with low metastatic potential, pediatric cancers, cordomas, Fig. 8.33, low grade gliomas, anaplastic cancers of the thyroid, locally recurrent tumors in previously irradiated areas, low grade sarcomas, and salivary gland tumors.

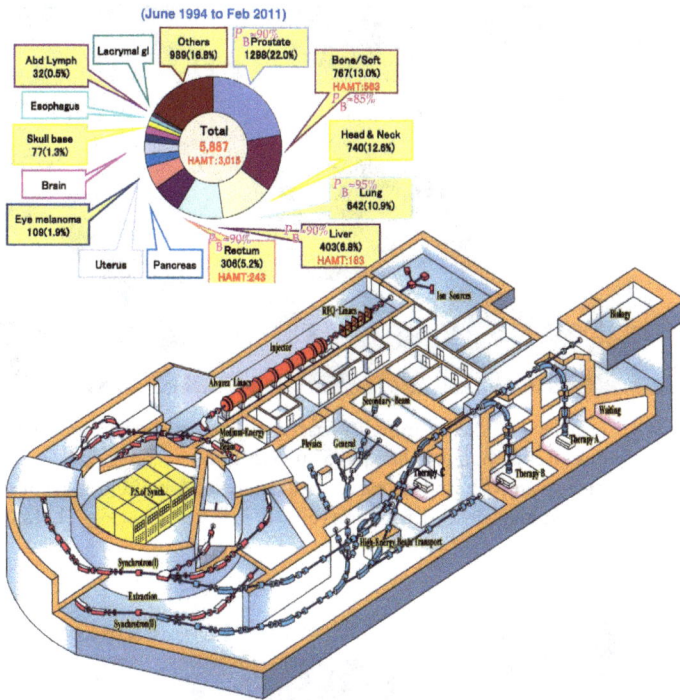

Figure 8.18. Illustration of cross-section through the HIMAC accelerator at NIRS Chiba, Japan where about 95% of the world's experience on light ion therapy has been collected. The upper left panel gives the patient distribution registered for carbon ion therapy at NIRS in Japan early 2011. The recent HAMT treatments are "highly advanced medical treatments" refined based on more than 10 years experience on the clinical outcome of advanced carbon ion therapy.

8.4.2. *Targets Located Close to Organs at Risk*

Skull base, spinal, and para-spinal tumors, some head and neck ocular and brain tumors benefit substantially by light ions due to their outstanding geometrical selectivity and the risk for significant hypoxic tumor cell compartments in these tumors. Figure 8.23(e) illustrates the improvement going from photon therapy by electron accelerators and the γ-knife when treating arteriovenous malformations a substantial improvement in obliteration probability is seen with the helium ions used at Berkeley especially at high doses probably due to the increased induction of apoptosis in these beams (cf. Figs. 8.9(h)–(j)). In Fig. 8.23(f), the improvement of the treatment of cordomas going from photons and protons to helium ions and finally to carbon ions is quite significant, and there are indications that it is important how the dose delivery is performed as discussed below (cf. also Figs. 8.38(j) and (l)).

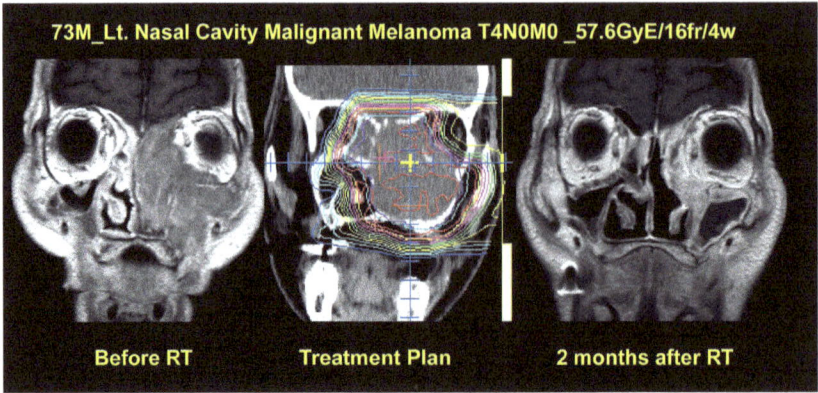

Figure 8.19. Coronary CT slices through a nasal cavity of a malignant melanoma patient before and 2 months after carbon ion therapy (57.6 GyE in 16 fractions during 4 weeks) are shown. The treatment plan with isodose curves is also shown. A rapid local recovery is seen by the very well-defined high-dose distribution (cf. also Fig. 8.20, courtesy H. Tsujii).

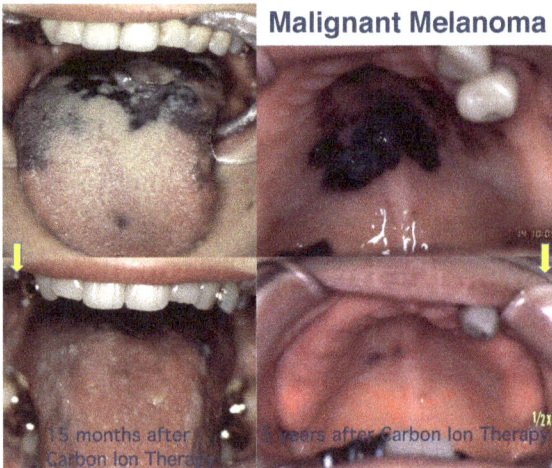

Figure 8.20. Clinical results from Chiba with radiation-resistant malignant melanoma tumors with long-lasting local control, 15 months to the left and 5 years to the right are shown.

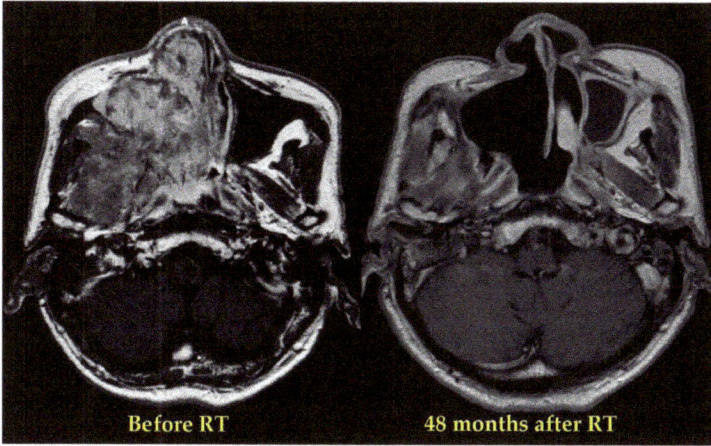

Figure 8.21. Transversal CT scans before and 4 years after the end of therapy are shown. In both cases, a significant improvement practically unobtainable with conventional low *LET* radiation therapy is seen. A very significant long-lasting recovery is seen for this tumor, which is hard to treat by photons and electrons.

8.4.3. *Tumors of Complex Local Spread*

Tumors of complex local spread such as mesotheliomas, cervix cancer, and brain tumors may benefit substantially from the improved geometrical selectivity of the lightest ions. This is seen in Figs. 8.23(d) and (e) where the latter figure show the significant improvement in treating cordomas with photon IMRT, protons, and most importantly with carbon ions compared to conformal photons. A steeper dose response is seen for carbon ions compared to photons and protons similar to the curves for lung cancer in Figs. 8.23(c) and (d), probably due to a significant influence of hypoxic tumors cells. Interestingly, there is a significant difference between the results at GSI and NIRS possibly due to the biological effect difference between fast lateral but rather slow longitudinal pencil beam scanning and instantaneous longitudinal range shifting using ridge filters to get a more uniform synergistic biological effect by simultaneous low and high *LET* radiation interaction (cf. Figs. 8.8(e) and (f)). The slow longitudinal scanning may cause a reduced biological synergism between the low and high *LET* dose fractions as demonstrated in Figs. 8.8(e) and (f), when the time span between the high and low *LET* dose fractions is long compared to the very fast initial repair phase of NHEJ (cf. text above Eq. (1)). Part of the low *LET* sublethal damage will then be repaired before the high *LET* Bragg peak arrives at the end of the treatment reducing the efficiency by some 10–20% as seen in Figs. 8.8(f), 8.23(e), and 8.38(j) and (l)).

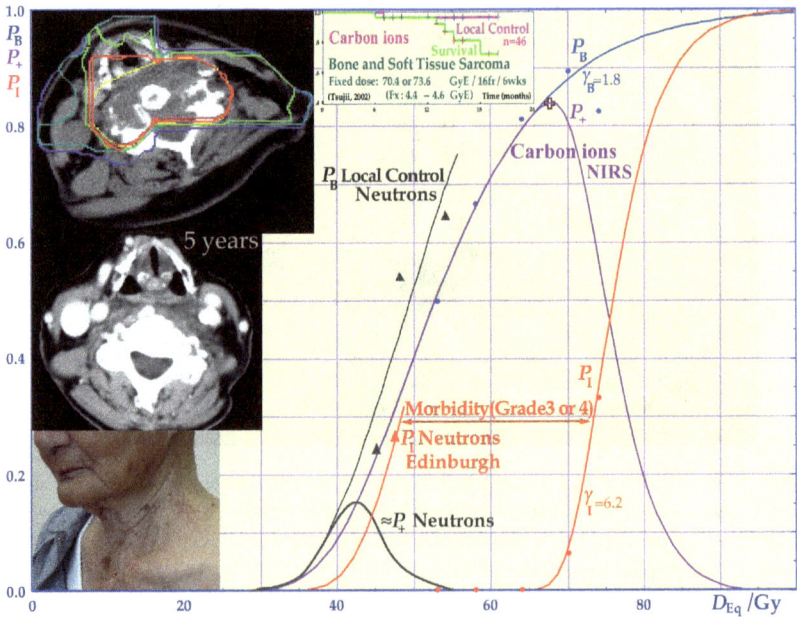

Figure 8.22. Bone and soft tissue sarcoma are very effectively treated by carbon ions similar to neutrons (P_B). However, the normal tissue morbidity is much lower with ions since the *LET* in normal tissue is much lower for carbon ions resulting in a significantly increased complication-free cure ($\approx 85\%$ for carbon ions, and about 15% for neutrons). The very nice cosmetic results and healing of the bony structures is seen in the left inserts.

8.4.4. *Estimation of the Potential Number of Patients*

Figures from the French, Etoile, and German, HICAT, groups applied to Sweden ($\sim$9 million people) would give between 2000 and 5000 patients/year ideally suitable for light ion therapy out of some 50,000 new cancers per year. A small proportion of these tumors may be treated by IMRT photons, electrons, or protons, for example, microscopically invasive tumors in sensitive normal tissues. When the tumor cells are expected to be extra radiation-resistant, it may be advantageous to use the next lightest ions: helium and lithium.

Quite generally one can estimate that in the future about one-third of all radiation therapy patients with small well-oxygenated tumors can be cured well by conventional conformal therapy. One-fourth to one-third of all the radiation therapy patients having well-oxygenated medium-to-medium-large tumors will need IMRT, with photons. The remaining 40% or so large-to-medium sized hypoxic tumor patients will benefit from light ion therapy which provides efficient cure and minimal normal tissue damage and obtain a high quality of life after the treatment. This is

(a)

(b)

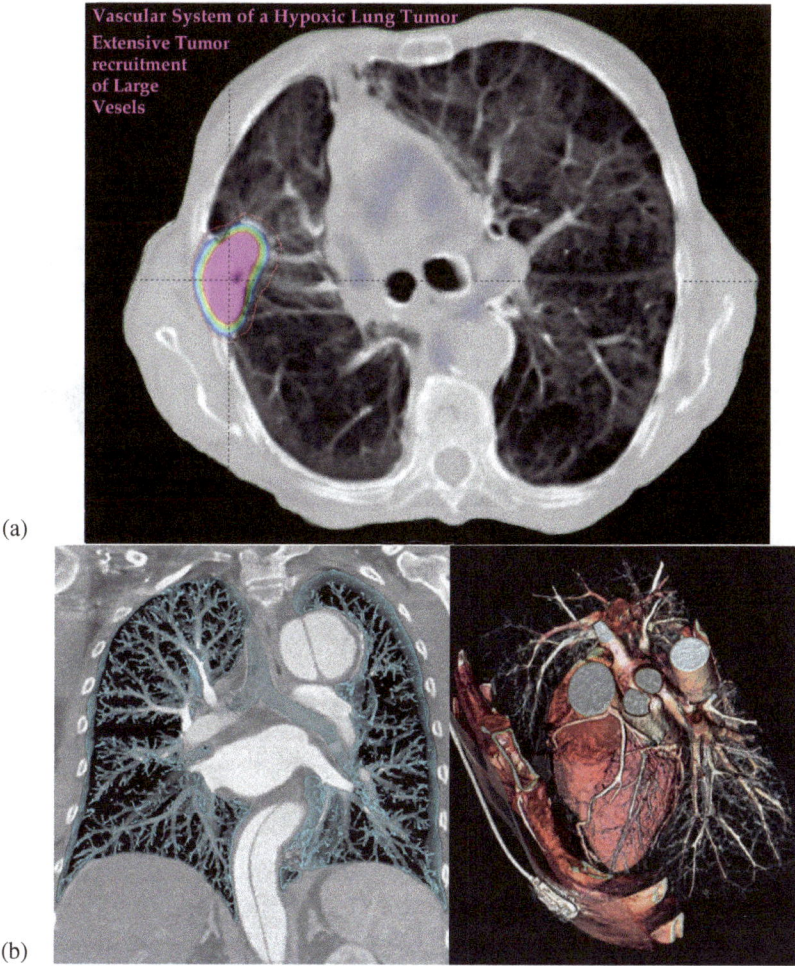

Figure 8.23(a). Illustration of how a large lung tumor can be severely hypoxic since it may be supported peripherally by large high-pressure vessels with hypoxic venous blood directly from the heart. The high pressure tend to close the blood flow from the core of the tumor making it even more hypoxic, so a medium-high *LET* is the ideal therapeutic modality as seen in Figs. 8.23(b) and (c). The lung vasculature and tumor could be imaged very accurately here since the PET-CT camera was the very first of a new generation with a 64 slices CT unit, at Karolinska, which is capable of practically freezing a large part of the lung vasculature. **(b)** Normal lung vasculature, as also seen in the right half of Fig. 8.23(a), where the vessels (direct arteries from the heart with venous blood) get finer and finer toward the chest wall. As suspected from the right half, the short thick arteries from the heart to the left will give a very high vascular pressure to the periphery of the tumor in Fig. 8.23(a) making it difficult for the FDG and other tracers to reach the tumor core (cf. also Fig. 8.27; Chapter 6 and Brahme 2009).

(c)

(d)

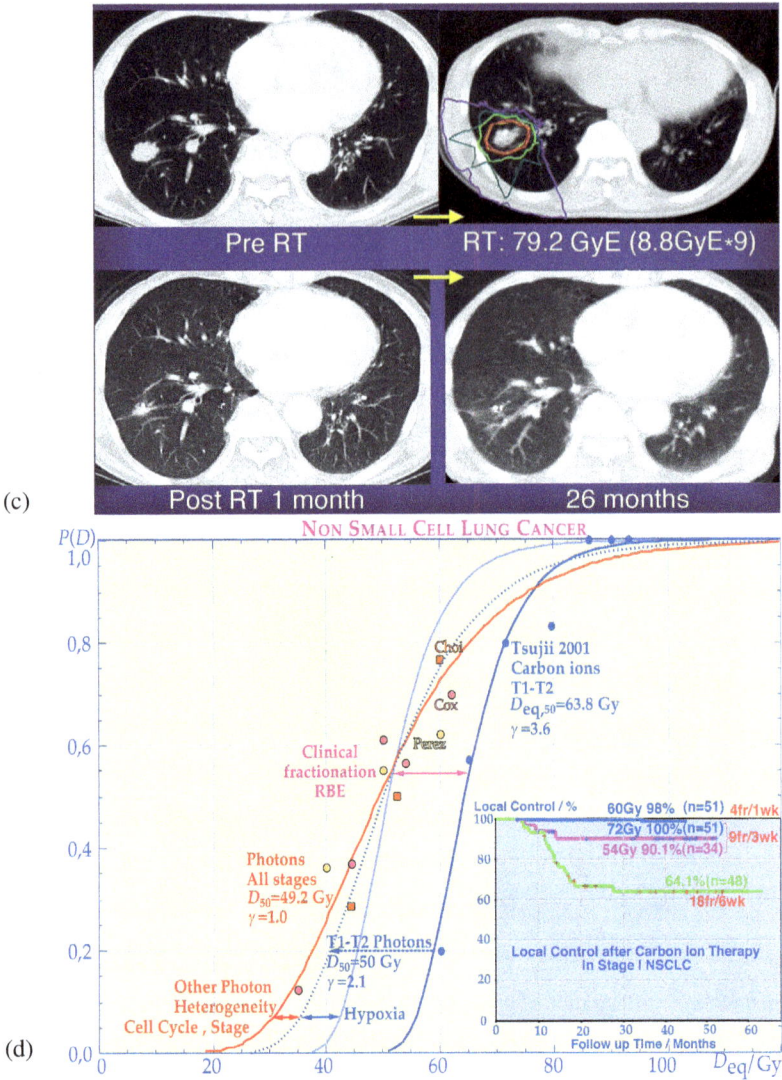

Figure 8.23(c). Clinical results from Chiba with an advantageous steep DRR for lung tumors using carbon 12 ions. Dose delivery and post-therapy normal tissue response as detected by CT scanning. **(d)** Illustration of the DRR of carbon ions compared to photons. The local tumor control as a function of time and dose escalation in Stage 1 non-small-cell lung cancer is shown in the lower right panel. It is clearly seen that these severely hypoxic tumors are effectively controlled by as few as 4 fractions delivered during 1 week. Today only a single four-field treatment delivering about 46 GyE carbon ion dose is used for these lung tumors. There is significant improvement in efficiency, cost effectiveness, and cure compared to conventional radiation therapy. The increased steepness improves

(*Continued*)

Figure 8.23(d). the therapeutic window and is due to a more efficient kill of hypoxic tumor cells. The change in normalized dose–response slope as a function of hypoxic fraction and ion species is shown in Figs. 8.5(m) and (n) above. Most of the loss in dose–response slope for photons is due to hypoxia. However, the variation in sensitivity over the cell cycle (cf. Fig. 8.5a) and with tumor stage affects the lower dose–response slope for photons.

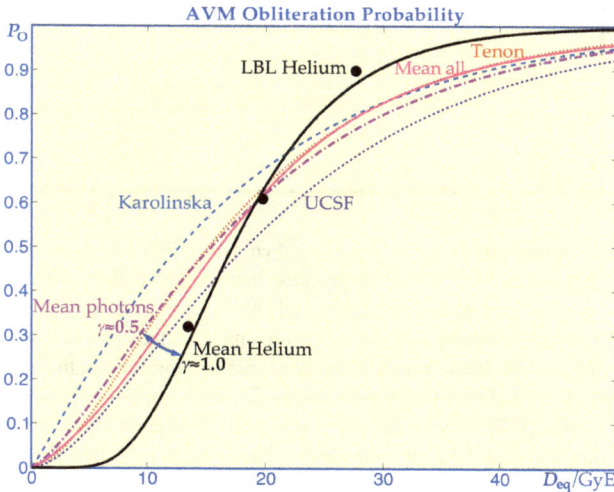

Figure 8.23(e). A similar increase in steepness of the tumor DRR to that in Fig. 8.23(c) is shown, but here the early helium ion treatments at Berkeley (UCSF) for the treatment of arterial-venous malformation (AVM) is seen. A clear-cut improvement in dose response is seen partly due to the increased apoptotic cell kill (cf. Fig. 8.9) with intermediate *LET* Bragg peaks in small-to-medium sized target volumes. Light ions are probably the modality of choice also for these non-malignant, but life-threatening malformations.

clearly seen in Fig. 8.24(b) where the percentages of the patients with a recurrent tumor and treatment complications are clearly reduced when switched from photon and proton IMRT to light ions.

8.5. Important Areas of Research and Development with Light Ions

8.5.1. *Molecular Genetics to Individualize Cancer Treatment*

Today, our knowledge about the molecular genomics of cancer is increasing at an unprecedented rate. We know that in all cancers at least one of the growth-stimulating genes or one of the genes of the DNA surveillance and repair

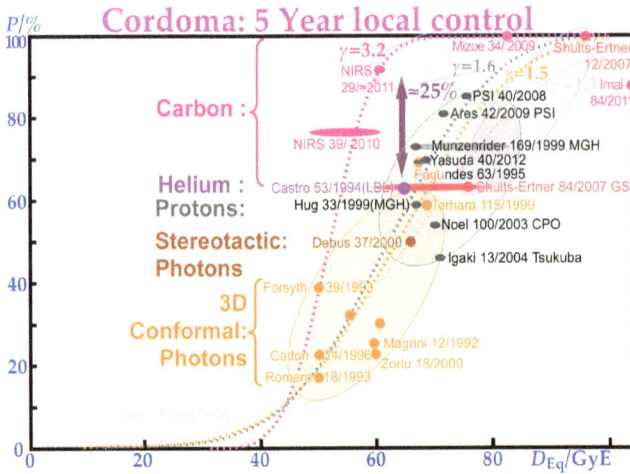

Figure 8.23(f). Comparison of the 5-year local control DRRs for treatment of cordomas by conformal photons, IMRT photons, protons, helium and carbon ions is shown. The right shallow curve with photons (conformal and IMRT) and protons pertains to low *LET* radiations, whereas the left curve pertains to carbon ions. The higher γ value and local control with carbon ions may be due to tumor hypoxia similar to lung tumors in Fig. 8.23(b). The increased tumor control of about 20% with a ridge filtered beam (NIRS) compared to lateral Bragg peak scanning (GSI) may be due to the increased synergistic effect by simultaneously delivering the low and high *LET* components (cf. Figs. 8.8(e) and 8.38(j) and (l)).

pathways are mutated (cf. Fig. 8.25). A number of new drugs have been developed, such as kinase inhibitors and molecules that can recover the apoptotic potential that is generally lost in tumor cells. Many of the other new methods such as gene therapy will benefit greatly from the development of radiation therapy for controlled debulking of gross tumor volumes prior to the application of molecular therapy to allow truly effective adjuvant therapies. Presently, we quantify the synergistic effect of the restitution of the mutant P_{53} protein in tumor cells by Prima 1 (see Fig. 8.25) and radiation therapy and have observed significant increases in cell kill and senescence by photons and expect further increased response by light ions. A number of genetic mutations may significantly alter the responsiveness of tumor cells and increase their genomic instability (cf. Figs. 8.14, 8.25, and 8.26), such as DNA-PK, ATM, and P_{53} (Brahme 1999). Based on microarray techniques, these genetic defects can be located and the ideal ion species should in the future be selected and used in biologically optimized treatment schedules. Light ion therapy is still at an early developmental stage, even though the first treatments were done more than 50 years ago, so the development of advanced adjuvant treatments is a very important potential future development to minimize adverse reactions in the patients and maximize their the quality of life.

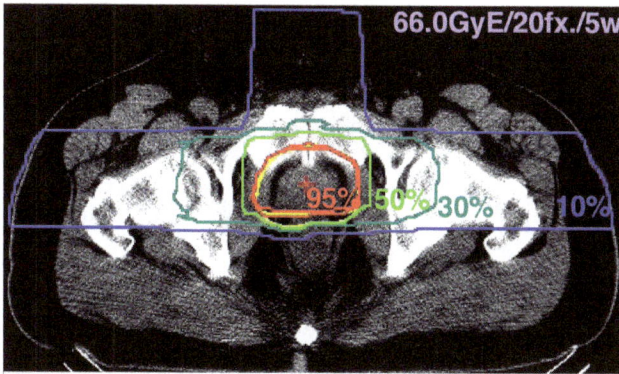

Figure 8.24(a). Improvement in biochemical relapse-free control of prostate cancer by switching from conformal and IMRT photon therapy to protons and carbon ions is shown. For the largest, more complex, tumors (PSA >20) more than a doubling is seen with photon IMRT (and protons) and more than a fourfold increase with carbon ions compared to 3D conformal treatments used in the early 1990s (21− >89%). Interestingly, by using biologically optimized radiation therapy and the BIOART procedure, almost as good result may be in reach (cf. Fig. 8.33 below).

8.5.2. *Improved Diagnostic Imaging by PET-CT and Real-time PET*

In the early days of the development of inverse radiation therapy planning (IRTP) and IMRT, the goal was to shape suitable physical dose distributions by dynamic scanning of pencil beams or by dynamic multileaf collimation (Brahme *et al.* 1979–1989). However, it was very soon realized that true optimization of radiation therapy required that the right dose delivery should produce the best possible clinical combination of tumor cure and adverse normal tissue side effects. Thus, a radiation biological optimization of the treatment outcome is really the key problem of radiation therapy optimization. This could be done in terms of the effective quality of life for the patient during and after therapy or more specifically maximizing

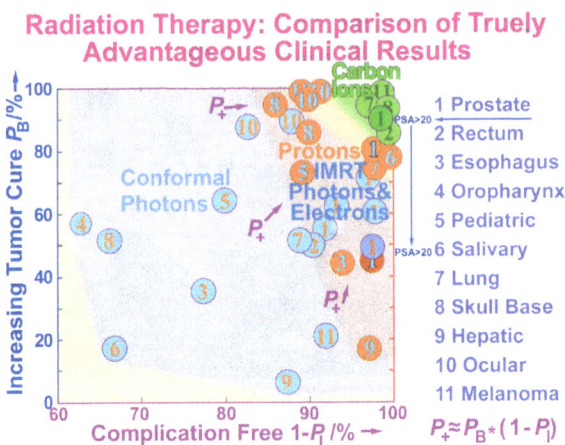

Figure 8.24(b). A more detailed view of how the complications are reduced and the recurrences are diminishing as we switch from conventional conformal photon therapy to IMRT photons and electrons and carbon ions is shown. Interestingly, IMRT photons and protons again do almost equally well, with carbon ion therapy stickling out as the almost perfect modality having only a few percentage complications and recurrences in most cases. Clearly, even if the data are mainly non-randomized, they give a similar indication of the merits of light ions as in Fig. 8.24(a) where the low PSA data are also included in more detail. In theory, with statistically independent tumor and normal tissue responses, the complication-free cure is given by: $P_+ = P_B * (1 - P_I)$. However, with commonly clinically correlated responses, one would expect $P_+ \approx P_B - P_I$ (Brahme 1999). Both expressions should preferably be located in the upper right corner of the diagram (modified from Mayo Clinic).

tumor cure and minimizing treatment-related side effects. For many years, this was effectively achieved by maximizing the probability of achieving complication-free tumor cure (P_+) but can today be best achieved by first maximizing P_+ and thereafter minimizing the probability of treatment-related injury requiring P_+ to stay high during this process as seen in Fig. 8.26(d). These treatment objectives are much more advanced than what is commonly mentioned as biologically optimized ion therapy where just a uniform cell kill is produced in the target volume largely disregarding normal tissue side effects. In fact, most of the time the unavoidable radiation effects in healthy normal tissues surrounding the tumor limit the outcome and it is important to take this into account in a true biologically optimized treatment approach since we never know the true radiation sensitivity of the tumor. The ultimate development of radiation therapy is to try to measure its radiation response during the initial phase of therapy and then try to biologically adapt the treatment to the observed tumor response. For this purpose, we are developing a new approach, which we call BIOART "the art of life" which uses biologically optimized 3D *in vivo*

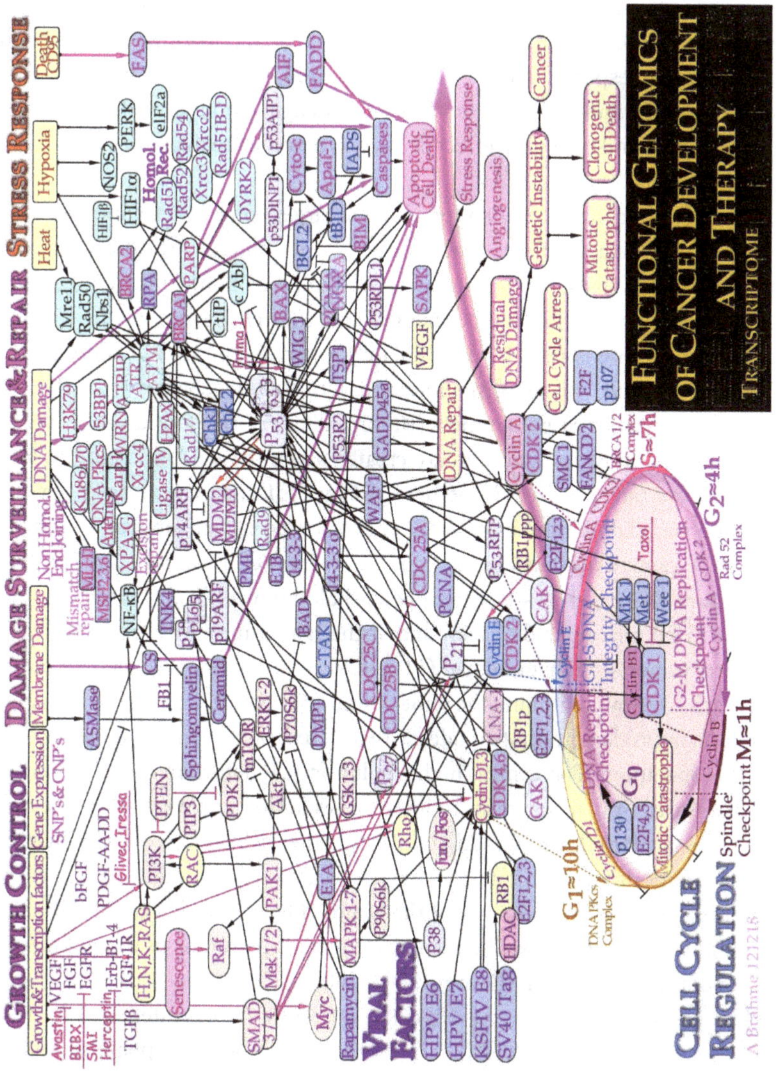

Figure 8.25. Illustration of some of the major genetic pathways involved in tumor development and tumor therapy using forced local DNA damage induction by low and high *LET* radiations and molecular therapies (e.g., Taxol, Herceptin, Gleevec, and Prima). All tumors are genetically instable since at least one of the growth control or DNA damage surveillance and repair pathways are always mutated. This is the bases for all radiation therapy, where induced DNA damage hits the Achilles' heel of the tumor cell. Tumor cells are very often mutant in the P_{53} pathway and may then evade an apoptotic response by photon therapy, whereas light ions may still induce one, for example, through the ceramide, PARP, or death receptor pathways (cf. Fig. 8.9(e)).

predictive assay-based radiation therapy as illustrated in Fig. 8.26(c) and discussed in more detail in Chapter 7. This is a truly biologically adaptive approach where the early tumor response is picked up by repeated 3D PET-CT imaging, patient position imaging, and PET-CT dose delivery imaging, and after the first week of therapy the dose and biological effect delivery is modified to maximize the treatment outcome (Fig. 8.26(c)).

Advanced imaging is needed for accurate target volume determination. Based on a PET-CT image of the para-aortal lymph nodes, Fig. 8.26(a) (cf. also Figs. 8.26(c)–(e)) show the increased effectiveness of light ion therapy, because the number of beam portals can be about 10 times lower than for low LET photons, electrons, and protons (cf. also Fig. 8.23(c)). New diagnostic modalities, such as PET-CT and MRSI, will also give a measure on the tumor cell density and hopefully in addition the tumor responsiveness (Brahme 2003, 2009). The advantage of some light ions is that positron emitters are created along the ion track and not least at the end of the range and could be detected by PET imaging. It has been shown that the delivered dose from high-energy photons and carbon ions can be registered with a PET camera used inside the light ion facility (cf. Figs. 8.16, 8.17, 8.36, and v38). Figure 8.26(b) shows how an open whole body PET-CT camera could be designed to get a large field of view and at the same time a high-resolution CT in the central tumor section and equal CT and PET resolution elsewhere. This latter effect is achieved by combining the same detector for both CT and PET imaging, for example, using avalanche photo diodes with different operation mood during the CT and PET operation. The central high resolution CT detection could even use the phase contrast X-ray mechanism to increase resolution and contrast in the central gross tumor region. This method could be further refined to compare diagnostic and therapeutic information at different times during the treatment schedule to determine the radiation resistance of the tumor. The treatments could therefore be highly individualized, taking the responsiveness of the tumor in three dimensions into account using BIOART (Brahme 2003, 2009).

The scientific basis for the BIOART approach is presented in Fig. 8.26(c) illustrating how repeated PET-CT imaging during the early phase of radiation therapy can be used to estimate the tumor radiation responsiveness and mean dose delivery during the treatment. This information can then be used for biologically based inverse treatment planning as shown in Fig. 8.26(d) and derive the optimal dose delivery with available radiation modalities (Fig. 8.26(e)). Figure 8.26(e) really shows how different radiation modalities such as photons, electrons, and light ions can be combined to give a high optimal therapeutic tumor effect at the same time as normal tissue side effects are kept as low as possible. The BIOART technique is more clearly illustrated in Figs. 8.27(a)–(e) for photon therapy of an advanced lung

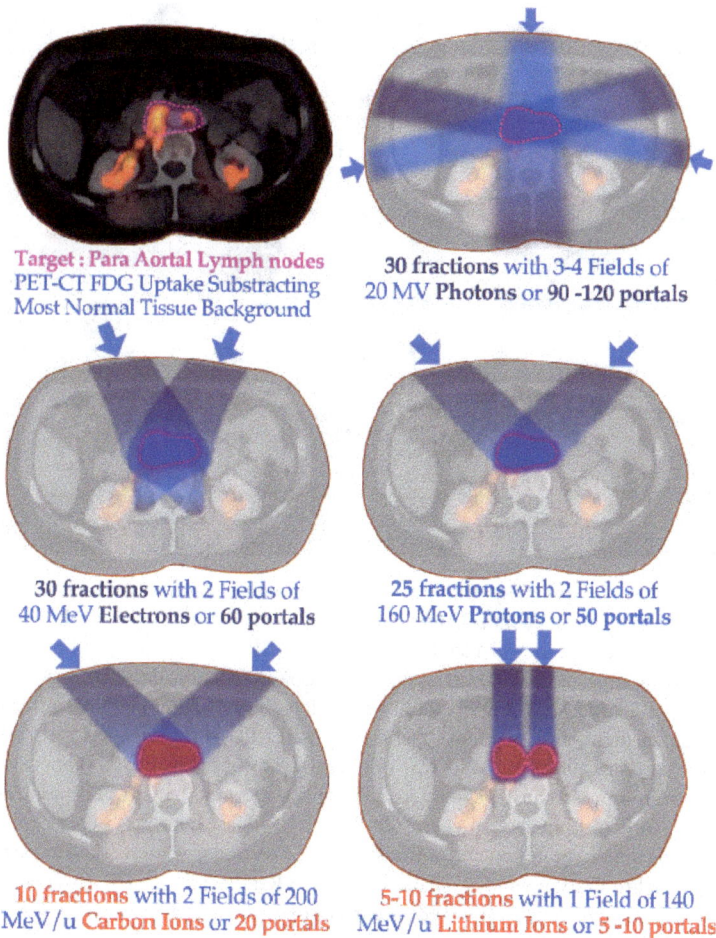

Target : Para Aortic Lymph nodes
PET-CT FDG Uptake Substracting
Most Normal Tissue Background

30 fractions with 3-4 Fields of
20 MV **Photons** or 90 -120 portals

30 fractions with 2 Fields of
40 MeV **Electrons** or 60 portals

25 fractions with 2 Fields of
160 MeV **Protons** or 50 portals

10 fractions with 2 Fields of 200
MeV/u **Carbon Ions** or 20 portals

5-10 fractions with 1 Field of 140
MeV/u **Lithium Ions** or 5 -10 portals

Figure 8.26(a). Illustration of how the efficiency of radiation therapy increases when switched from photons with ≈100 portals to single-charged particles such as electrons and protons with ≈50 portals and negligible exit dose (middle row) and light multiply charged ions such as lithium and carbon ions, where 5–20 portals often suffice. With lithium and carbon ions, the advantage of increased apoptotic cell kill and induction of senescence in the tumor region is illustrated in red (cf. Figs. 8.9(c)–(j)).

cancer, but could have worked equally well or even better with light ions such as ^{11}C. Therefore, Figs. 8.26(c)–(e) really show how different radiation modalities such as photons, electrons, and light ions can be combined to give a high therapeutic tumor effect and at the same time as normal tissue side effects are kept as low as physically and biologically possible.

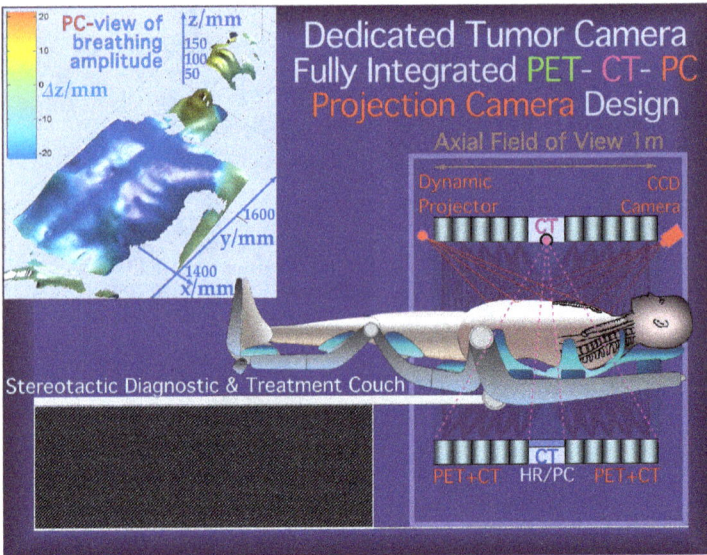

Figure 8.26(b). Cross-section through a dedicated high-resolution open PET-CT tumor camera where the central opening is introduced to allow a high resolution in the central CT region. The PET detectors will also detect peripheral CT photons but at a lower resolution sufficient for optimal PET reconstruction using CT attenuation data. Ultimately a PET resolution in the order of 1–2 mm should be possible by correcting positron diffusion and annihilation in flight. It is planned to integrate a 4D PC (Projection Camera) with the system for accurate auto setup and real-time correction of the patient's motions such as breathing to improve reconstruction accuracy with a dynamic patient. The central high-resolution X-ray imaging section could ideally use PS-Phase contrast Stereoscopic X-ray imaging to maximize resolution and contrast in the gross tumor volume (the auto setup and breathing synchronization unit is available through C-Rad Positioning AB, Sweden; www.c-rad.info/content.php?categoryID=14&parentID=0).

8.5.3. *Accurate Patient Fixation and Registration of Organ and Tumor Movements*

High-precision dose delivery using pencil beam scanning require accurate patient setup and control of tumor and organ movements to achieve accurate dose delivery to the target volume, (see Figs. 8.27, 8.28, and 8.30–8.33). A laser scanner and Projection Camera (PC, cf. Fig. 8.26b) has been developed for use with conventional as well as light ion therapy (Brahme *et al.* 2008). When gated therapy is used to irradiate a lung tumor during a specific phase of the breathing cycle the PC can be used for gating the treatment and largely eliminate movement uncertainties. Breath-hold monitoring or real-time imaging and treatment synchronization of the

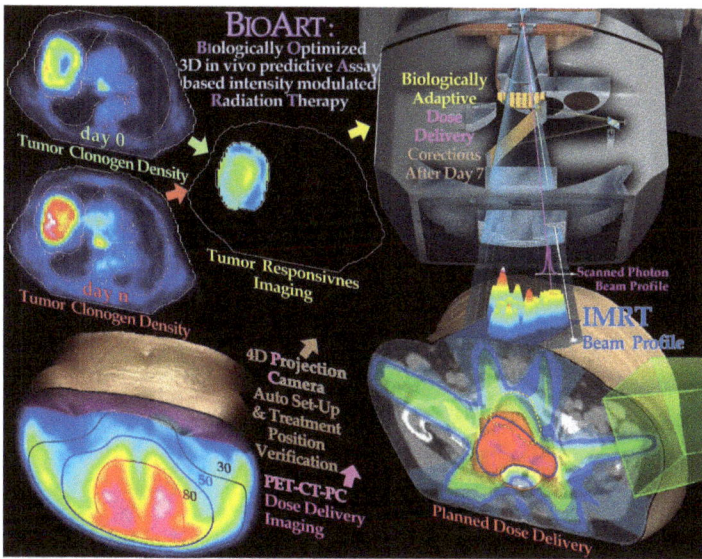

Figure 8.26(c). Illustration of how PET-CT information can be used to verify and allow the BIOART-approach (cf. Figs. 8.26(d) and (e) and 8.27). By measuring the tumor cell kill early on in the treatment 3D *in vivo* predictive assay information about radiation responsiveness is obtained for accurate prediction of the optimal dose delivery based on known hypoxic status (cf. Fig. 8.5(k)). Both with light ions and scanned high energy photon beams, 3D *in vivo* PET-CT dose delivery monitoring is possible to further optimize the treatment based on observed mean dose delivery which may differ from the planned dose delivery without considering organ and patient motions (cf. Figs. 8.16, 8.17, and 8.26(b)). Both these data sets when used together will allow a high degree of therapy optimization where practically all major sources of treatment error can be picked up as long as they influence tumor cell survival and can thus be corrected for using biologically optimized adaptive treatments techniques, "BIOART," more clearly demonstrated in Figs. 8.26e and 8.27 and Chapter 7.

motion of tumors and organs at risk is the final goal. This is perfectly possible with scanned pencil beams where the accelerator pulse can be delivered when chest wall or breathing motions are in same phase interval as derived by 4D biologically optimized treatment planning.

8.5.4. Radiation Physics: Macro and Microscopic Dose Delivery, Nano and Microdosimetry, LET, and Secondary Electron Spectra

The interaction of different light ions with tissues will have to be studied more accurately both theoretically, clinically and experimentally. Track segment data are of great importance for the determination of the biologically effective dose to different

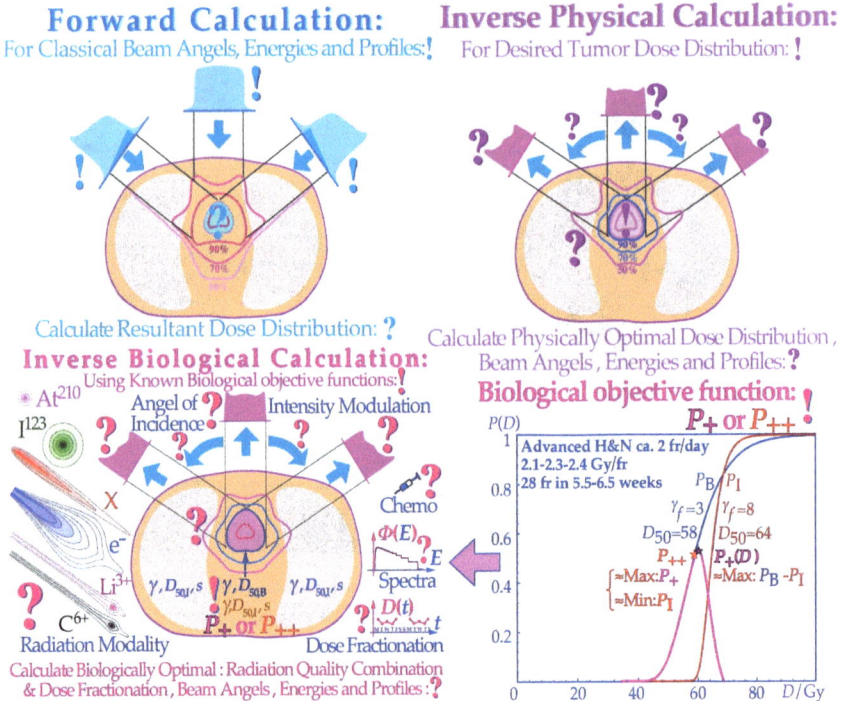

Figure 8.26(d). Illustration of the development from classical forward therapy planning to inverse therapy planning using physical dose (upper right panel) and biological (tumor and normal tissue response cf. Fig. 8.23) treatment optimization objectives. The ultimate step in therapy development is to do biologically optimized light ion therapy where the optimal weights of different light ion species are selected to maximize the complication-free cure (cf. lower right panel and Figs. 8.12 and 8.23).

tumors and tissues (cf. Figs. 8.3, 8.7–8.11, 8.15, and 8.23) (cf. also Nikjoo *et al.* 2009), even though some of the principal clinical consequences microdosimetry are known and understood (cf. Tilikidis and Brahme 1994 and Figs. 8.10(a) and (b)).

8.5.5. *Radiation Biology: OER, RBE, Apoptosis, Senescence, and DRRs*

Radiation physical and biological data will be used as input to theoretically calculate OER and RBE values (cf. Figs. 8.3, 8.5, and 8.8). Calculated data can be compared with experimental cell survival data and with results from clinical trials (cf. Figs. 8.22 and 8.23). An important area of development is how to convert the large amount of photon DRRs to such relations for light ions, considering organ radiation resistance, the distribution of hypoxia, and the microscopic heterogeneity in dose delivery (cf.

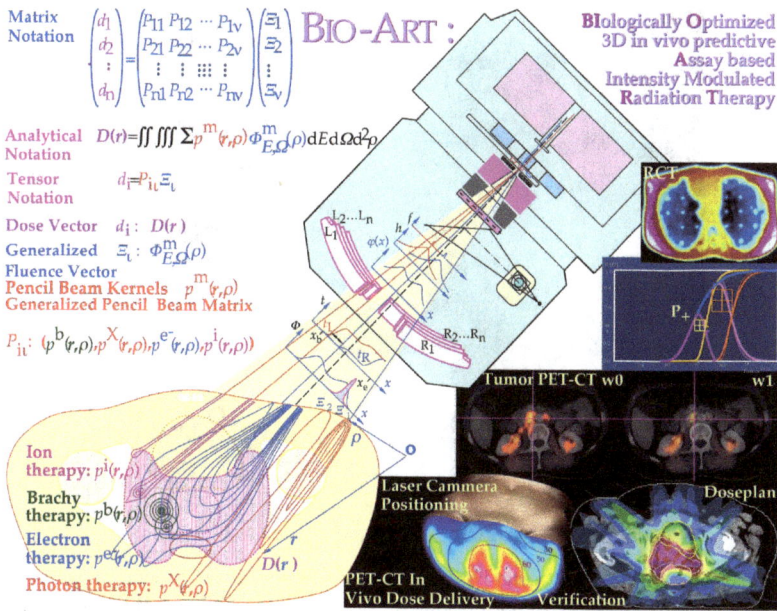

Figure 8.26(e). More detailed description of how the dose delivery in the lower panel of Figs. 8.26(c) and 8.26(d) can be calculated by analytical, matrix, or tensor type equations. From the dose distribution the probability of tumor cure (cf. Fig. 8.23) and normal tissue damage can be calculated and the iterative procedure in Fig. 8.31 can be applied to maximize the probability that the patient is cured without severe normal tissue reactions. The lower right panels, from top to bottom, illustrate how the process can be made more accurate by cone beam radio therapeutic computed tomography, accurate dose response data, 3D *in vivo* predictive assay of radiation responsiveness and *in vivo* IMRT dose delivery imaging and sub-millimeter resolution laser or projection camera auto setup.

Nilsson *et al.* 2002; Lind *et al.* 2007; and Figs. 8.5(i) and (j), and 8.10). A review of the different physical, biological, and clinical advantages of different beams is given in Table 8.2. As important as the classical *OER*, *OGF*, and *RBE* are the induction of milder types of cellular inactivation such as apoptosis and senescence (cf. Fig. 8.9). These end points need to be studied much more carefully in the future to really minimize possible adverse normal tissue side effects.

8.5.6. *Physical and Biological Treatment Optimization*

Computer programs for physical and biological optimization of the dose delivery have been developed at the Department of Medical Radiation Physics at Karolinska Institutet (cf. Chapters 5 and 6 and Figs. 8.3, 8.8, 8.13, 8.14, 8.27, 8.29, 8.30, and 8.33), and they are widely use today. Such programs are also needed for the assessment of optimal treatments with light ions based on the accurate description of the

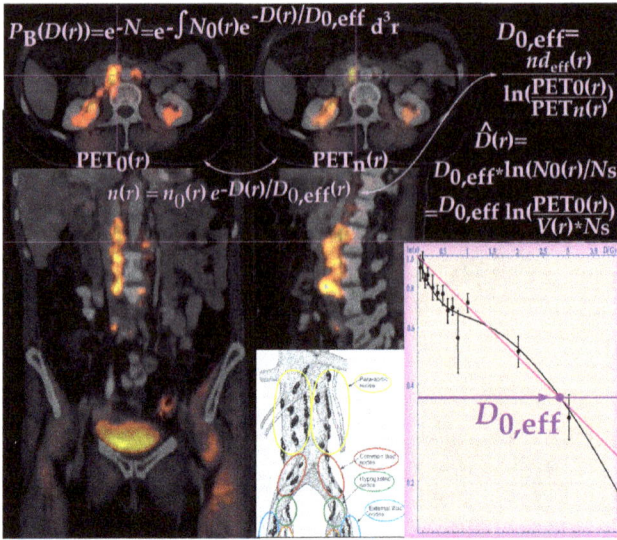

Figure 8.27(a). The para-aortal lymph nodes of a cervix cancer patient as visualized by PET-CT imaging. Accurate tumor imaging on the background of normal tissue anatomy is essential for successful therapy in close agreement with the lower central textbook insert. By monitoring tumor responsiveness during the first weeks of therapy, the BIOART approach (cf. Figs. 8.26(c) and 8.33) is possible. The response data can be used to quantify the individual radiation responsiveness of the tumor in terms of $D_{0,\text{eff}}$ (cf. also Eq. (8.3) above). Such accurate tumor imaging is almost a necessity for advanced biologically optimized light ion therapy.

tumor response (cf. Fig. 8.8(n) and Brahme 2011). The calculations should be based on physical and biological data and the results should be compared with the present clinical experience from Chiba and Darmstadt. Most important are the tumor DRRs that characterize the probability to achieve tumor cure at high doses as seen in Figs. 8.22, 8.23(b), (d), and (e) but also those of the normal tissues as seen in Fig. 8.22. It is particularly important to integrate methods such as the BIOART approach into the clinical light ion procedures as shown in Fig. 8.27 for photon therapy.

8.6. Development of Comprehensive Cancer Centers

Beside a strong multidisciplinary cancer care and advanced equipment for tumor diagnostics and therapy, a comprehensive cancer center needs advanced research departments for medical radiation biology, medical radiation physics, tumor pathology, and molecular oncology, working in close connection with clinical oncology. A comprehensive cancer center should also include laboratories for experimental

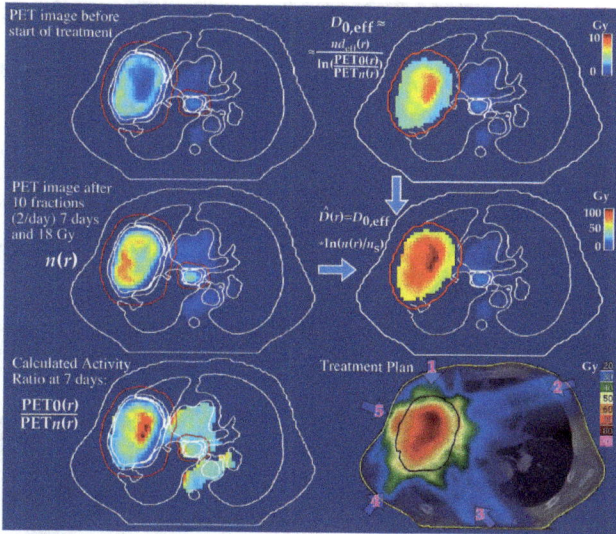

Figure 8.27(b). Dose responsiveness imaging applied on a large lung cancer is illustrated. By taking the ratio of the tumor FDG uptake before therapy and after a week of treatment and 18 Gy of tumor dose, in this case, it is possible to quantify the change in tumor uptake and also the effective radiation resistance, D_0 or $D_{0,\text{eff}}$. From $D_{0,\text{eff}}$ and the tumor cell density, it is possible to estimate the optimal dose level required for tumor eradication (middle right panel, cf. Ågren and Brahme 1986) without considering the risk for normal tissue damage which generally is low for light ion and photon IMRT treatments (cf. Brahme 2009, 2010).

cancer research and basic tumor biology research. The infrastructure for clinical research facilitates and the development of translational research programs require a well-developed clinical trial unit, population-based registries for outcome research, resources for quality of life studies, a well-functioning program for biobanking and technical platforms for genomics and proteomics. With strong research resources in medical radiation physics, the total environment is well optimized for research with the aim to develop comprehensive radiation therapy. The addition of facilities for advanced light ion therapy to more conventional accelerators for 3D conformal, IMRT, and tomotherapy treatments as well as advanced brachytherapy techniques, hyperthermia, photodynamic therapy, and BNCT may be useful to create a complete set of resources for an advanced cancer treatment facility. Coordination of imaging facilities (MRSI, PET-CT, and PS (phase contrast stereoscopic imaging), cf. Fig. 8.26(b)) for an advanced tumor diagnostic center is also essential as seen in Fig. 8.29(a). PET-CT is probably the ultimate tool for accurate tumor imaging and 3–4-dimensional *in vivo* predictive assay of radiation sensitivity. There is a need for large axial field of view (≈ 1 m) advanced PET-CT-based dedicated tumor cameras as

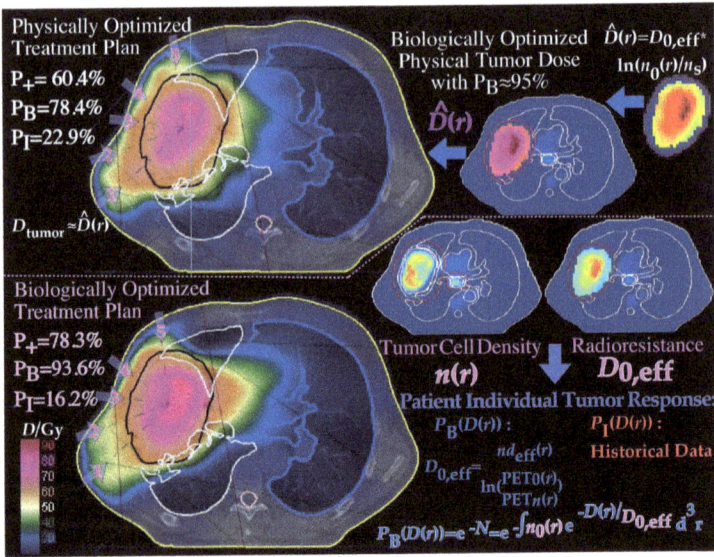

Figure 8.27(c). The optimal dose distribution in Fig. 8.27(b) can be used as objective for inverse physical therapy optimization (upper half figure). Even more accurately, the tumor cell density, $n(r)$, and estimated $D_{0,\text{eff}}(r)$ can be used for biologically effective dose delivery optimization using the clinically observed tumor response and historically observed dose–response data for normal tissue side effects that are systematically much more similar between patients than the effective tumor response that can vary substantially (Brahme 2009). Interestingly, the complication-free cure increases from 60% to almost 80% by introducing the biological optimization method based on 3D *in vivo* predictive assay.

seen in Fig. 8.26(b), allowing accurate sub-millimeter phase contrast X-ray imaging in the central region and ultra-high sensitivity PET imaging in the entire volume of possible tumor infiltration. The integrated PC makes it possible to project all the PET data to any phase of the breathing cycle to maximize tumor sensitivity in dynamic regions. By imaging the tumor twice during the early course of therapy, it is also possible to quantify both the tumor responsiveness to therapy and the rate of loss of functional tumor cells. Development of molecular radiobiology and genomics will be used in establishing methods to predict treatment outcome. A large patient population and collaboration with other centers makes a large number of patients available for routine treatment and research. Within a comprehensive cancer center of this type, a centre of excellence for clinical radiation therapy will create fruitful environment for research and development to which a center of excellence on advanced research for radiation therapy should also be integral part. The rapidly increasing body of genetic information that is now becoming available, not least through projects like the Human Genome Project and the proteomic database

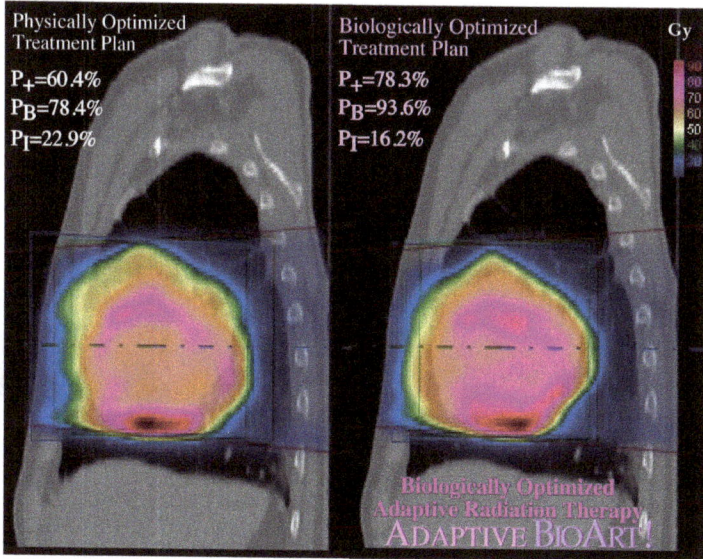

Figure 8.27(d). Interestingly, the derived final dose delivery in Fig. 8.27(c) for both physical and biological optimization results in dose distributions where the increased tumor cell survival in the lower part of the tumor near the diaphragm is compensated by increased dose delivery during the last few weeks as seen in this perpendicular slice through the tumor in Fig. 8.27(c). Most likely the tumor border was partly outside the beam due to diaphragm motions causing increased tumor cell survival at the lower tumor boundary. This is thus a clear-cut example how the BIOART approach can be used for biologically optimized adaptive radiation therapy and for correcting unexpected effects of patient motion, tumor radiation resistance or hypoxia (Brahme 2003, 2009).

will rapidly increase our knowledge about genes that affect radiation sensitivity and will improve our understanding of the clinical radiation responsiveness as seen in Figs. 8.26 and 8.29(b). Different organs of the human body have often significantly differing responses to partial or heterogeneous irradiation, depending on the functional organization of the different tissues and the part of the genome that is actively transcribing (see Chapter 2). In addition, different radiation types have widely varying biological effects depending on their ionization density as shown above in Fig. 8.5 and Table 8.2.

8.7. Development of Advanced Biologically Optimized Radiation Therapy

The first steps in the development of IMRT were taken during 1970s by developing a scanning beam system for electron and photon therapy with 5 to 50 MeV peak

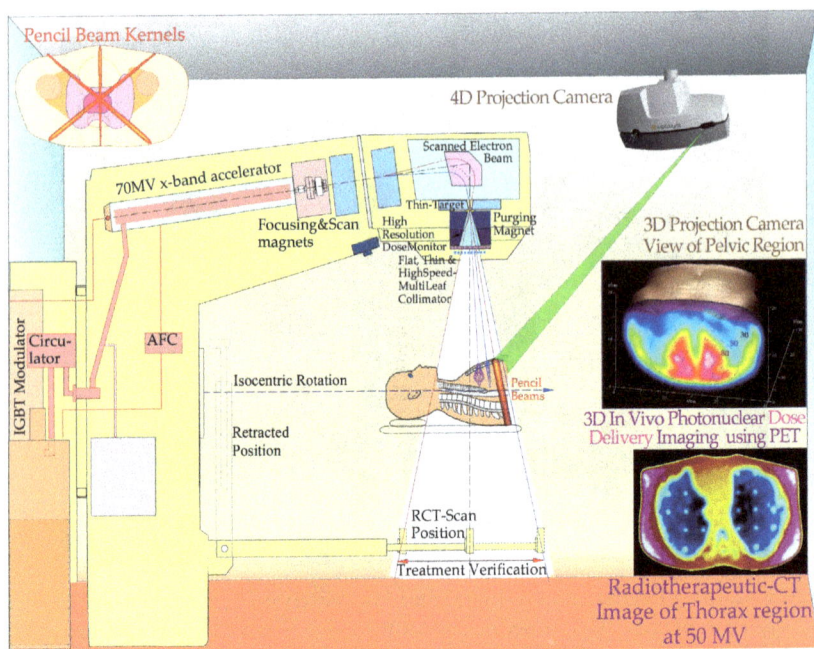

Figure 8.28. Illustration of 3–4D optical patient imaging by the laser- or projection-camera imaging on a newly developed treatment unit using narrow high-energy scanned photon beams allowing PET-CT imaging for *in vivo* patient dose delivery monitoring as shown by the inserted pelvic dose distribution (cf. room α in Fig. 8.36, these auto setup units are available through C-Rad Positioning AB, Sweden: www.c-rad.info/content.php? categoryID=14&parentID=0). The laser- or projection-camera will allow sub-millimeter accurate auto setup of the patient before treatment and *in vivo* patient dose delivery monitoring both for light ion and high-energy photon therapy (Brahme *et al.* 2008, this treatment unit is available through TopGrade Healthcare, China: www.topgradehc.com/ html/fangshezhiliao%20-%20en.html).

beam energies (Brahme *et al.* 1979–1992). With the arrival of the first computers, we realized that rather than just making uniform scanning patterns with triangle wave generators, we could modulate arbitrary dose distributions by computer control. We, therefore, developed a number of algorithms to optimize the dose distribution and even the biological effect on the tumor and reduce it as much as possible in the normal tissues as illustrated in Figs. 8.30(a) and (b). In the first approximation, biological optimization is a one-dimensional problem where one wants to give a high dose to the tumor such as a prostate cancer and negligible dose to surrounding normal tissues such as rectum to the right. To the left a more resistant tissue was assumed with small risk of complications. The resultant optimal one-dimensional scan profile given in

Table 8.2. Comparisons of clinical parameters of radiation therapy modalities.

Parameter Beam	Classical photons 4–10 MV	IMRT 5–25 MeV	High energy 50 MeV	Protons	Light ions
Cost-effectiveness gain factors					
Total Treatment Time/w	6–7	5–6	4–6	4–6	1–3
Number of fractions	30	25	20–25	25	1–15
Number of beam portals/fraction	4	4	3	3	1–3
Number of treatment rooms/accelerator	1	1	2	3	4–8
Serviceable lifetime/years	15	15	20–25	30	30–50
Cost/treatment k€	3	7–10	7	25	7–10
Cost/cure k€	10	15	10	35	10
Cost/M€ per installation with four treatment rooms	4	5	12	75	80
Quality of life gain factors					
5-year biochemical relapse free control, prostate	20	50	50–70	45–70	90
Complication-free cure: $P+$/% for H&N	45	55–60	60–65	55–65	65–75
OGF	1	1	1	1.05	1.2–2.5
Cell cycle gain factor	1	1	1	1	1.2–2
Apoptotic–senescence gain factor	1	1	1	1	1.2–2
Dose–response slope, γ, for hypoxic tumors	1–2	1–2	1–2	1–2	3–5
Relative microscopic standard deviation, %	1	1	1	1.5	3–7
Dose delivery gain factors					
D_{eq}, 5 cm/D tumor/ %	40	30–40	30	35–40	25–30
D_{eq} tumor/D_{eq} normal (parallel-opposed broad beams)	0.9	0.9–1	1.5	2.5	3–4
D_{eq} tumor/D_{eq} normal(5 mm pencil beam)	0.6	0.7	0.8	0.5	1.7–3.5
Penumbra $P_{80/20}$/mm at 20 cm depth	6	4–8	8	12–15	3–5
In vivo dose delivery monitoring	—	—	+++	+	+−+++

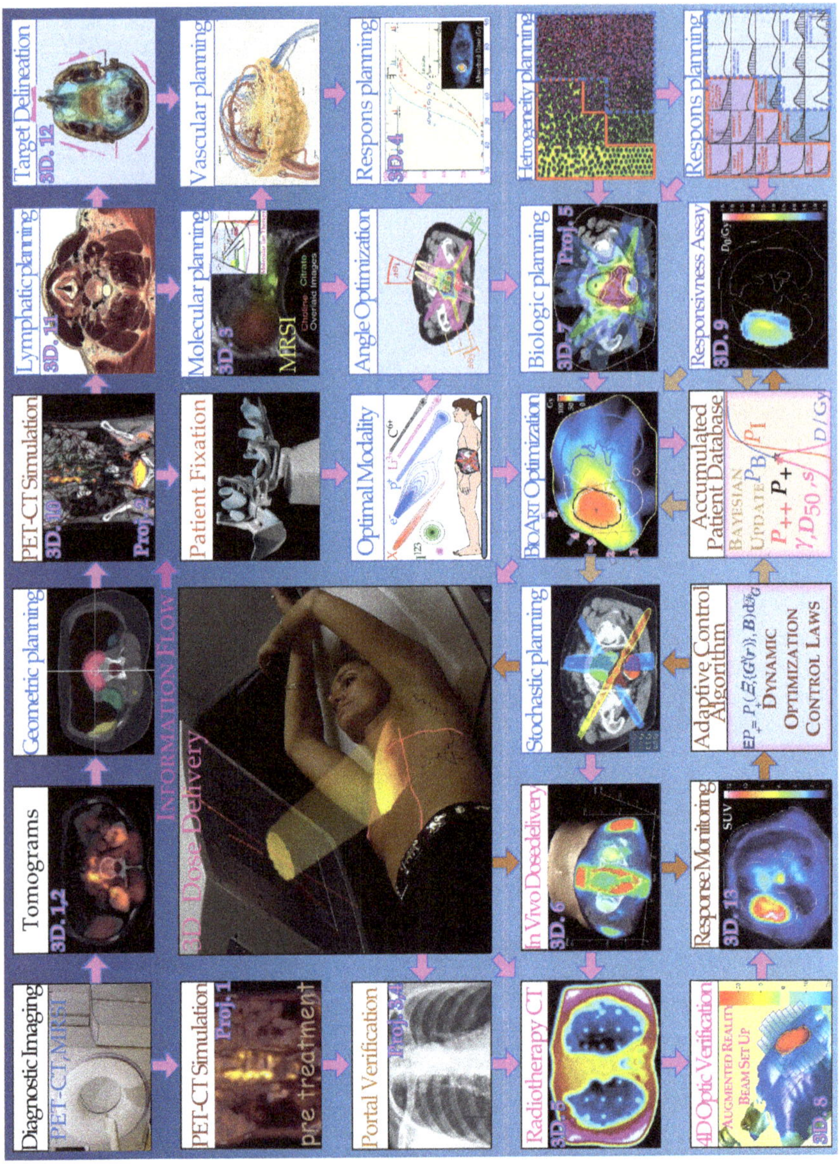

Figure 8.29(a). Overview of the multifaceted array of tumor diagnostics, therapy planning, and treatment verification approaches available for biologically optimized radiation therapy today. Beside the standard tomographic and projective techniques such as ordinary and phase contrast X-rays, diagnostic and therapeutic CT, MRI, MRSI, PET, and PET-CT, multiple molecular approaches are available for imaging hypoxic tumors and vasculature. In addition diagnostic and therapeutic X-rays may be used for treatment setup verification

(Continued)

Figure 8.29(a). and so may PET-CT be used for *in vivo* dose delivery verification and tumor responsiveness monitoring. For advanced tumors as many as possible of these advanced diagnostic techniques should be used to ensure an optimal treatment considering the multifaceted clinical background. Equally important is to have an accurate protocol for treatment follow up to ensure that the clinical response information gained during each treatment is accurately fed back to historical dose response data (lower row).

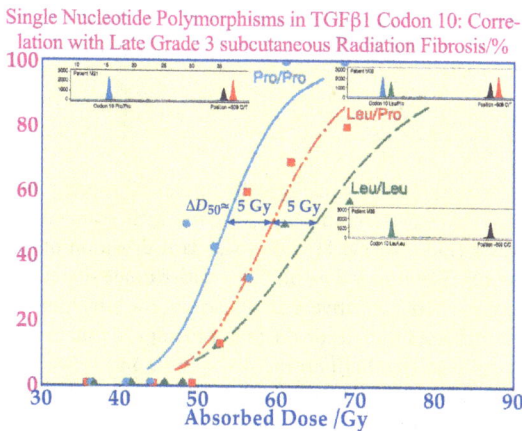

Figure 8.29(b). Interestingly, a single nucleotide polymorphism of a gene may cause a large change in radiation sensitivity as seen here for the probability induction of Grade 3 subcutaneous fibrosis with different genotypes on codon 10 of the TGF β 1 gene (cf. Fig. 8.25; modified from Andreassen *et al.* (2003)).

Fig. 8.30(a) was calculated to maximize the probability to cure the tumor without inducing severe damage to normal tissues. Interestingly, this goal was achieved by a scanning density (lower 4 peaked curve), which generated a quasi-uniform dose throughout the tumor except near the right sensitive edge where the dose inside the tumor first had a hot peak away from the tumor edge and a narrow colder region near the organ at risk. In this way, reducing the cell survival near the tumor edge, but allowing a very narrow cold spot near the risk organ border, to minimize complications, ensured the tumor cure. This clearly illustrates the power of local biologically optimized dose delivery to maximize tumor cure and minimize severe normal tissue complications. The scanning pattern in this case applies almost equally well to electrons, protons, or light ions, possibly the cold region should be even colder with a high *LET* beam since more rectal damage is expected then. In fact, this simple example shows another interesting development where a high *LET* beam should be

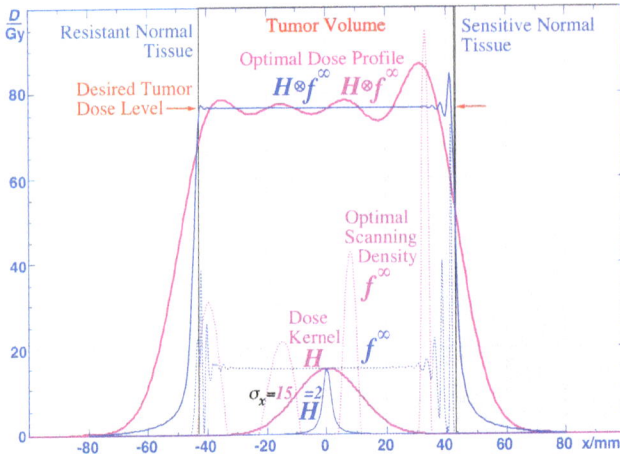

Figure 8.30(a). Illustration of how biologically optimized therapy planning and inverse therapy planning use biological treatment objectives with scanned intensity-modulated Gaussian pencil beams (Dose kernel H with a standard deviation of 2 mm (e.g. 20 cm C) or 15 mm (e.g. 35 cm p)). The tumor dose on the sensitive tissue side (to the right) is lowered to avoid adverse reactions in the sensitive normal tissue at the same time as the dose close the resulting cold spot is increased well inside the tumor to improve tumor cure. On the resistant side, the opposite is true as seen both on the dose profile and converged scanning density profile (f^∞) which peaks very close to the tumor border thus maximizing the complication-free cure, that is, the probability to eradicate the tumor without severe normal tissue damage. The algorithm for deriving the optimal scanning pattern in this oversimplified one- or two-dimensional case was developed in the early-1980s (Lind and Brahme 1985).

replaced by a low *LET* beam to minimize spillover to organs at risk especially at the tumor edge where the degree of hypoxia is generally lower and maximum *LET* is not generally needed. Such a case is illustrated in Fig. 8.30(b) where a tumor with a hypoxic core was treated by neutrons and photons to optimize the total dose delivery. The biologically optimized delivery was based on radiobiological cell survival data from Louvain la Neuf for neutrons (≈ carbon ions) and photons (≈ protons), so the case is almost applicable to a carbon ion and proton plan even though the penumbra of carbon ions is much sharper and would result in sharper dose borders then in Fig. 8.30(a). Interestingly, the high *LET* radiation component is only used in the hypoxic tumor volume, whereas a low *LET* photon or proton dose would be optimal in the well-oxygenated microscopically invasive tumor periphery. A low *LET* only treatment would result in much higher peak effective doses in the tumor core and thus a higher probability of normal tissue damage in the incoming beams. Of course a one-dimensional approximation is not sufficient to treat a large tumor and the full three-dimensionality need to be considered in such cases. From a conception

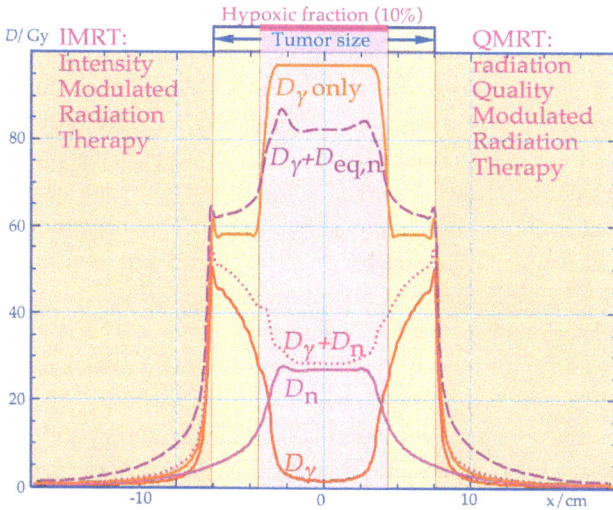

Figure 8.30(b). By combining a low and a high ionization density radiation, it is possible to optimize the local radiation quality by intensity modulating each *LET* component separately. In this case the neutron or carbon ion dose D_n is high in the hypoxic tumor, whereas the well-oxygenated part is best treated by lower ionization density photons. This is a generalization of IMRT to QMRT where by the local radiation quality and dose is optimized to eradicate the tumor as effectively as possible without causing adverse normal tissue reactions. This approach is particularly important with hypoxic tumors that today can be imaged quite well, for example, using PET-CT imaging (cf. Figs. 8.12(a) and 8.27(a)–(d)) with ^{18}F-Misonidazol. Interestingly, the technique in Fig. 8.30(a) can be combined with that in Fig. 8.30(b) to make truly biologically optimized treatment plans (Brahme *et al.* 1989).

point of view Fig. 8.30(b) illustrates a new step in the development of IRTP where the radiation quality instead of just the intensity is modulated to optimize radiation therapy (QMRT instead of IMRT).

The fast development of energy and IMRT during the last two decades using photon and electron beams has resulted in a considerable improvement of radiation therapy, particularly when combined with radiobiologically based treatment optimization techniques as seen in Figs. 8.24, 8.31(a) and (b). This has made intensity-modulated electron and photon beams more powerful than conventional uniform beam proton therapy (cf. e.g. Fig. 8.24(a)). To be able to cure even the most advanced hypoxic and radiation-resistant tumors of complex local spread, intensity-modulated light ion beams are really the ultimate tool and in clinical practice even two to three times more cost effective than just proton therapy. This development and the recent development of advanced tumor diagnostics based on PET-CT imaging (cf. Figs. 8.26–8.29) of the tumor clonogen density opens the field for new powerful radiobiologically based treatment optimization methods. The ultimate step is to use the unique radiobiological and dose distributional advantages of light ion beams for truly optimized bio-effect planning where the integral 3D dose delivery (cf. Figs. 8.27, 8.29, and 8.32) and tumor cell survival can be monitored by PET-CT imaging and corrected by adaptive therapy optimization methods. Beside the "classical" approach using low ionization density photons, electrons, and hydrogen ions (protons, but also possibly deuterons and tritium nuclei) the intermediate lithium, beryllium or, boron ions, and high ionization density carbon ions induce the least detrimental biological effect to normal tissues for a given biological effect in a small volume of the tumor. They may therefore be key particles for curative therapy in the future. In a more cost-efficient approach, referred patients will first be given a high dose high precision "boost" treatment with carbon ions during 1 week preceding the final treatment with conventional radiations in the referring hospital. The rationale behind these approaches is to reduce the high ionization density dose to the normal tissue stroma inside the tumor and to ensure a more microscopically uniform dose delivery with a strong early effect on radiation resistant and hypoxic tumors. BIOART (cf. Figs. 8.26(c) and (e)) is really the ultimate way to perform high-precision radiation therapy using checkpoints of the integral dose delivery and the tumor responsiveness, and based on this information, performing compensating corrections of the dose delivery during the last part of the treatment so-called adaptive therapy. By using biologically optimized scanned high-energy photon or ion beams, it is possible to measure *in vivo* the 3D dose delivery using the same PET-CT camera that was used for diagnosing the tumor spread. This method, thus, opens up the door for truly 3D biologically optimized adaptive radiation therapy where the measured dose delivery to the true target tissue and their therapeutic response, can be used to fine adjust the incoming beams so that possible errors in the integral therapy process are eliminated toward the end of the treatment. Interestingly enough, practically all major error sources can be corrected for in this way such as organ motions, treatment planning errors, patient setup errors, and dose delivery problems due to gantry, multileaf or scanning beam errors. As long as it is possible to quantify the surviving tumor clonogens after the first week or two of therapy, this information can be used to also account for uncertainties in biological response data and really cover all key clinical uncertainties at the same time as more accurate dose–response data can be derived during follow up (Fig. 8.29(a)). The response of the PET-CT camera is related to the truly delivered integral dose (cf. Figs. 8.27, 8.29, 8.31, and 8.32) with correct temporal averaging, thus if only small errors are seen, it is sufficient to adjust the last few treatment fractions. When using PET-CT tumor response monitoring, it is even possible to account for the uncertainty in biological

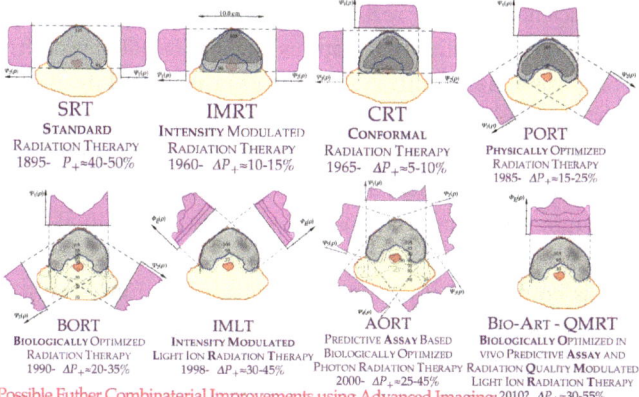

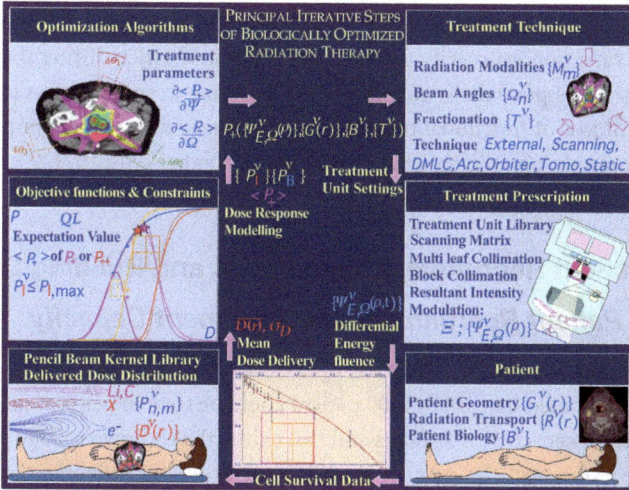

(a)

(b)

Figure 8.31(a). Eight steps in the development of external beam radiation therapy. The bottom row shows a number of new imaging techniques some of which, such as PET-CT imaging, may significantly improve the outcome of radiation therapy particularly when used in biologically optimized approaches. **(b)** Illustration of the numerical techniques needed in iterative procedures for biologically optimized therapy planning and inverse therapy planning using physical (dose) and biological (quality of life or tumor and normal tissue response objectives, cf. Fig. 8.23) treatment objectives (soft ware available at RaySearch Labs, Sweden: www.raysearchlabs.com/Default____4.aspx?epslanguage=EN).

response of the patient and it may even be possible to convert radiation therapy to an exact science and use real-time *in vivo* predictive assay to perform truly biologically optimized radiation therapy using the BIOART approach (Brahme 2003, 2009). Thus, using the recently available biologically based treatment optimization algorithms, it is possible to improve the treatment outcome for advanced tumors by as much 10–40% (Brahme 2005). The adaptive radiotherapy process based both on 3D tumor cell survival and dose delivery monitoring has the potential of percentage accuracy in tumor response and dose delivery, not least with 3D geometric Bragg peak scanning and intensity-modulated ion beam dose delivery.

There is no doubt that the future of radiation therapy is very promising and gradually more and more patients may not even need advanced surgery but instead could be cured by photon and electron IMRT, and for radiation-resistant and severely hypoxic tumors using biologically optimized light ion therapy, where the high *LET*–high RBE Bragg peak is solely placed in the gross tumor volume. PET-CT is still a few orders of magnitude more sensitive than other diagnostic modalities and probably the ultimate tool for accurate tumor imaging and 3D *in vivo* predictive assay of radiation sensitivity. By imaging the tumor twice during the early course of therapy, it is possible to quantify both the tumor responsiveness to therapy and the rate of loss of functional tumor cells. MRSI may also become important in this respect even though the molecular sensitivity is higher with PET.

8.8. Patient Recruitment, Equipment, and Building Design

8.8.1. *Patient Recruitment and Treatment Capacity*

The facilities planned for Heidelberg, Pavia, and Lyon have three treatment rooms and it is estimated that 1000–1200 patient could be treated per year in each place. In the Austrian center with 5 rooms and two shifts, the estimate is 2300 patients/year. In these estimates, the new information on a drastic reduction of the number of treatment fractions needed per patient, as shown in the trials in Japan, was not considered. At least a twofold increase in the treatment capacity and a cost-effectiveness superior to that of advanced electron and photon therapy may be expected. With conventional radiotherapy equipment in Sweden, in an average, 338 patients are treated per year using one treatment unit and one shift (8 am to 4 pm). The main difference, when only curative intended treatments are considered for conventional radiation and ion treatments, is that for the former many more fractions must be delivered. In the recent SBU report (Ringborg *et al.* 2003), it was shown that the most common schedule was 25 fractions. In Japan, the very successful carbon ion treatments of lung tumors was very successful with only 4 fractions during 1 week but today they are only using one single treatment fraction. Further, for some

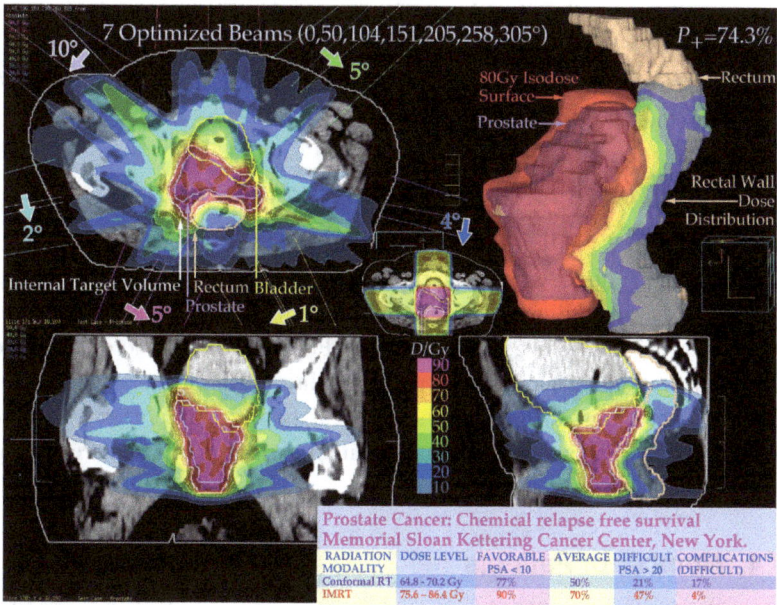

Figure 8.32. Illustration of the advanced 3D dose delivery possible by low *LET* photon beams using biologically based intensity and angle of incidence radiation therapy optimization (BIO-IMRT). The red to pink region is the high dose tumor volume accurately enclosing a prostate tumor substantially enlarged by tumor growth.

patients only a boost will be given with light ions. However, it must also be taken into account that with conventional therapy, 46% of the treatments are given with palliative aim, and much less fractions are needed, generally between 5 and 15. In conclusion, taken all these facts into consideration it seems possible to treat 500 to 800 patients per room/year with light ions.

8.8.2. *Conventional Radiotherapy and Biologically Optimized IMRT*

Most small non-hypoxic tumors are well treated by conventional uniform photon and electron beam dose delivery as described in Fig. 8.31(a) and Table 8.2. More advanced tumor stages and tumors of complex local spread (Fig. 8.27) may require BIO-IMRT (Figs. 8.29–8.32). The most advanced severely hypoxic tumors benefit substantially by biologically optimized PET-CT imaging (Figs. 8.27 and 8.30) using 3D *in vivo* predictive assay-based radiation therapy (BIOART, Figs. 8.26 and 8.27). The latest biological optimization algorithms are a prerequisite for these developments (cf. Figs. 8.31 and 8.32) not least when multiple treatment modalities are desirable. The important ability to optimize the complication-free cure by QMRT using light ion therapy is illustrated in Fig. 8.30 by first scanning the therapeutic beam so that the high LET component in normal tissue is least detrimental and finding the optimal high and low LET dose combinations (cf. Figs. 8.10, 8.26(c) and (d), and 8.30(b)).

8.8.3. *Connection to Conventional Radiotherapy*

As an example the Light Ion Center at Karolinska University Hospital (Fig. 8.36) is planned to be in direct connection to the present radiotherapy center with eight accelerators covering the energy range from 2 to 50 MeV for photons and electrons. Interestingly the high-energy photons allow PET-CT based photonuclear dose delivery verification *in vivo* similar to the possibilities with the nuclear interactions of light ions (cf. Figs. 8.16, 8.17, and 8.28). It is of key importance to have a good connection between conventional radiation therapy infrastructure and light ion therapy for careful patient selection.

8.8.4. *Number of Patients*

After running a new facility for about 3 years, it seems realistic to treat about 1500–2000 patients/year with four treatment rooms and one eccentric gantry (Figs. 8.37–8.38). At 5 years more than double this value may be possible. Further increases in patient numbers treated are possible by using an initial boost therapy where hypoxic tumors are massively eradicated in the early phase of the treatment with light ions and gradually low LET radiations are used to eliminate eventual remaining microscopic tumor cells as discussed in more detail in a recent publication (Brahme *et al.* 2001). With two eccentric gantries with separate energy degraders and a beam splitter, around 6000 patients can be treated per year as illustrated in Figs. 8.36–8.38. The beam splitter will allow simultaneous treatment in one room each of the two eccentric gantries.

8.8.5. *Development of a Second-generation Ultra-compact Treatment System for Biologically Optimized Light Ion Therapy at the New Karolinska*

The clinical results from past 2 decades in Japan have shown that light ions are exceptionally useful in treating many tumors with almost 50% increase in cure for advanced tumors. We have, therefore, developed an efficient, compact, cost-effective treatment unit that can make such treatments available worldwide. To minimize the accelerator size, a compact superconducting cyclotron for carbon, boron, lithium,

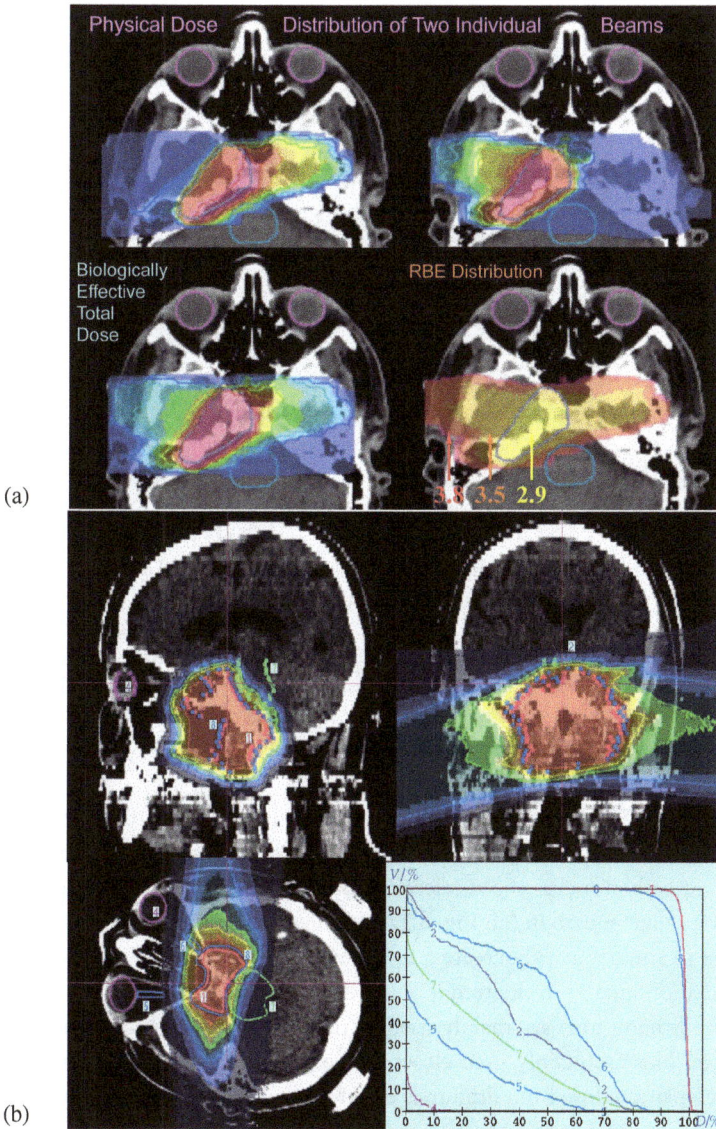

Figure 8.33(a). Comparison of the physical dose distribution of two individual beams on a brain tumor with the associated RBE distribution and biologically effective dose distribution. **(b)** Advanced carbon ion treatment of a brain tumor where the brain stem and optical nerves and chiasm are at risk. It is seen that the tumor gets a very high uniform dose and that most normal structures are well preserved (courtesy J Debus).

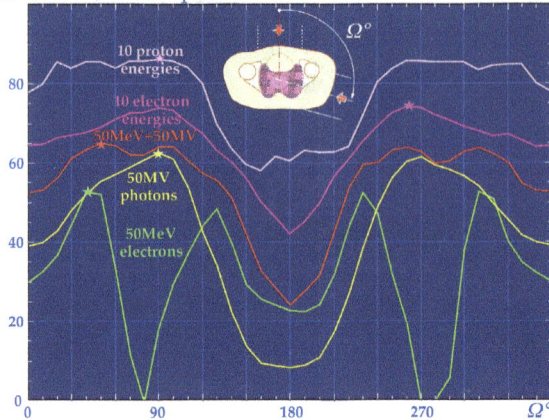

Figure 8.34(a). Illustration of how the complication-free tumor control, P_+, varies with the angle of incidence for a single portal of different radiation therapy modalities applied on an advanced cervix cancer. The maximum P_+ increases going from high-energy electrons to photons to simultaneously combined electrons and photons to 10 different electrons and proton energies. Interestingly, the variation with the angle of incidence is comparatively small for multiple protons and electron energies, due to the use of optimal range modulation from each beam direction. In this figure, all beam combinations are delivered through one and the same multileaf collimated beam portal and beam angle as opposed to Fig. 8.35 where two beam portals are used. With one single proton beam energy, the result is very poor and not even useful since good target coverage will require very high dose delivery to normal tissues.

helium, and hydrogen ions have been designed with peak energy of about 360 MeV/u and maximum boron range of about 30 cm ($\approx$26 cm carbon). The use of boron instead of carbon makes the accelerator about 5 m in diameter, instead of 6 m for the same range in carbon (cf. Fig. 8.37). In mixed-modality radiation therapy, it is possible to simultaneously create a uniform distribution of biological effect, dose and radiation quality in the tumor volume. For instance, with lithium and carbon ions, the lithium ions are mainly used in the distal tumor region, whereas boron or carbon ions are mainly used closer to the patient surface. By this method, the low LET plateau dose from lithium is elevated significantly by Bragg peak boron or carbon ion dose delivery in the proximal tumor region (Figs. 8.13(c) and (d); Vreede and Brahme 2010).

To make the facility compact and efficient one eccentric gantry serves four treatment rooms surrounding the gantry. This reduces costs for beam optical components, scanning systems, treatment heads, dosimetry systems, and patient positioning. In this way, a 35 by 15 m footprint is sufficient for four treatment rooms with a capability to treat 3000 patients per year at a cost for the equipment and building well below today's costs for a conventional proton-only installation. Interestingly, inside the gantry, there is also a very flexible beam-energy and particle-species selection system, allowing not only ^{12}C but also ^{11}C tumor irradiation. The use of ^{11}C permits a 50-fold increased specificity in PET-CT imaging of the Bragg peak delivery, thus, clearly visualizing in 3D the location of the really delivered dose distribution *in vivo* considering organ motions and setup uncertainties etc. allowing adaptive corrections in the following treatments. The patient is auto setup by a 4D imaging system based on a newly developed projection camera, a system assuring sub-millimeter accuracy in synchronizing the treatment with breathing movements, and even heart beats, to ensure accurate dose delivery in dynamic regions such as the thorax and pelvis.

The new technology provides a compact and very efficient system for curative treatment of several common malignant tumors of: head and neck, lung, liver, prostate, bone/soft-tissue sarcoma, cervix, and pelvis (Tsujii 2010). The system was initially designed with the New Karolinska University Hospital and the Karolinska Institutet in mind but many other centers have shown interest in it.

8.8.6. *Beam Delivery System, Treatment Rooms, and Gantries*

Since both light ions (cf. Figs. 8.17 and 8.38(c) and (d)) and high-energy photons (cf. Figs. 8.16 and 8.27) activate the tissues in the patient they all benefit from the possibility to use PET-CT-based dose delivery verification. A light ion center may, therefore, need to be equipped with two PET-CT cameras close to the light ion and photon scanned pencil beam treatment rooms so that patients can be rapidly scanned for the dose delivery after treatment as shown in Fig. 8.36. Two eccentric light ion beam gantries each of which covers four treatment rooms (cf. Figs. 8.37 and 8.38) with a wide range of possible beam angular intervals and anterior or posterior beam directions covering angles of incidence of $\pm 60°$ or about $120°$ in each room (see Figs. 8.38(f)–(h)). The $\pm$ is due to fact that the treatment couch can be rotated more than $180°$. These figures clearly illustrate the great flexibility in treatment setup with excentric gantries on a number of common treatment techniques currently used at NIRS in Chiba, Japan. Interestingly, the study in Figs. 8.34 and 8.35 shows that very specific beam angles are generally not required with 4D Bragg peak scanning for shaping of biologically optimized dose distributions. Once the right angular interval is available, as shown in these figures (cf. Figs. 8.38(f)–(h)), the exact angle of incidence is not very critical provided the angle between two different beam portals is more than about $30°$ and full flexibility in intensity and location (4D) of the Bragg peaks is available. In fact, Figs. 8.13 and 8.35 shows how the optimal

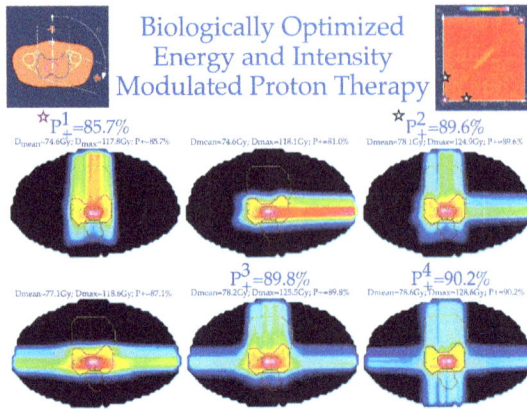

Figure 8.34(b). Illustration of some of the optimal dose plans in Figs. 8.34 and 8.35(b). It is seen here that when only 10 beam energies are used from each beam portal, the maximum achievable complication-free cure is almost reached by two beam portals since the biological effect to normal tissues is only marginally reduced going to three or four beam portals even if the entrance dose can be reduced with multiple beams. The extra dose reduction going to very many beam portals (cf. Brahme *et al.* 1992) does not reduce the severe damage to the normal tissue. It is interesting to note, that the optimal beam with a single portal almost remain the same when two, three, and four coplanar portals are used. This is normally not the case with photon and electron therapy. Furthermore, the new beam portals added has often a lower weight to avoid normal tissue damage along the new beam portals.

scanning pattern can be calculated using inverse therapy planning with physically or biologically optimized dose delivery. Interestingly, the biological optimization allows optimal partial irradiation with multiple beams to minimize normal tissue damage (cf. also Fig. 8.12). The stereotactic treatment couch (cf. Figs. 8.26(b), 8.36, and 8.38(a)) allows high-accuracy patient setup, not least in combination with the projection camera (cf. Figs. 8.16 and 8.28) both for patient auto setup in the treatment rooms and for PET-CT dose delivery imaging just outside the treatment rooms. An important research area is therefore to develop radiation biological and ion interaction physics to further improve the therapeutic possibilities in the future using radioactive or even chemo-toxic beams.

Interestingly, the clinically advantageous synergistic effect between the lower *LET* plateau regions and the high *LET* Bragg peaks shown in Figs. 8.8(e) and (f) should be considered when designing the 3D scanning pattern of the beam from the treatment units. It is highly desirable to deliver the majority of the local low and high *LET* dose fractions as closely in time as possible to maximize the therapeutic response (see Fig. 8.8(f)). The ridge filters used in Chiba as shown in Fig. 8.38(i) automatically do this. The ridge filter continuously and instantaneously shifts the

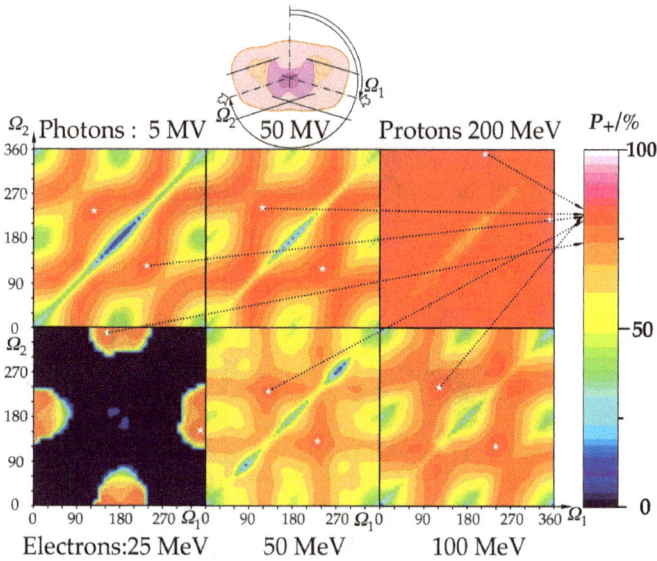

Figure 8.34(c). Comparison of the 2D phase spaces of complication-free cure with the angle of incidence of each beam portal as plotted on the horizontal, Ω_1, and vertical axis, Ω_2, respectively. When $\Omega_1 = \Omega_2$, that is, on the diagonal in these six diagrams, the one-dimensional case of Fig. 8.34(a) is obtained. It is seen that the range of variation between different modalities is even smaller when two beam portals are used than with just one. The small difference between the high-energy photons or electrons and the protons make the twofold to threefold increased cost per proton treatment much less attractive even though there is a few percentage increase in $\hat{P}_+$ with two beam portals. With three beam portals, a slight further increase may be possible, but protons and photons will now be almost equally good using modern biologically optimized IMRT dose delivery for both modalities. With light ions beyond protons such as lithium and carbon ions, a substantial further increase in treatment outcome is expected (cf. Fig. 8.24, for a similarly located prostate cancer), since these tumors are commonly hypoxic and rather radiation-resistant. It is interesting to note that it does not matter much how, in a proton plan, beam portals are selected since the P_+ value does not change much with the beam direction due to 3D Bragg peak scanning using biologically optimized IMRT. This indicates that a gantry is less important to use with protons than with photons and electrons where the peak values are reached only in rather small regions that may vary significantly for different patient anatomies, tumor locations, and radiation modalities. Owing to the increased biological specificity and degree of freedom in biological effect delivery not least using helium, lithium, and beryllium ions but also with boron, carbon ions the need of a gantry is even smaller, even though the extra flexibility, for example, with eccentric or isocentric gantries will always be an advantage, especially for certain target locations with multiple organs at risk.

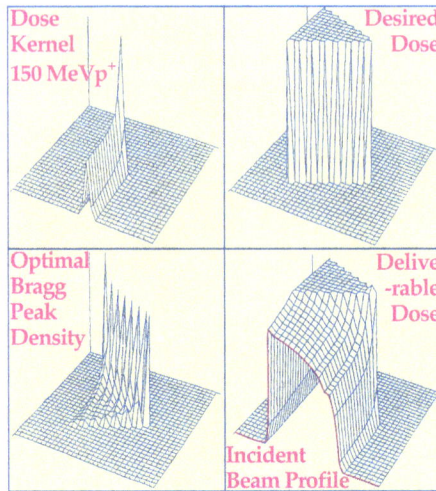

Figure 8.35(a). Illustration of optimization of the scanning pattern or Bragg peak density distribution of light ions (in this case 150 MeV protons) to produce a certain desired dose distribution with minimal overdosage if the exact distribution cannot be exactly produced somewhere, such as at the sharp edges of the triangular target. Interestingly, the dose profile of the incident beam really shows the surface dose problem with extended target volumes along the direction of the incident beam (lower right panel).

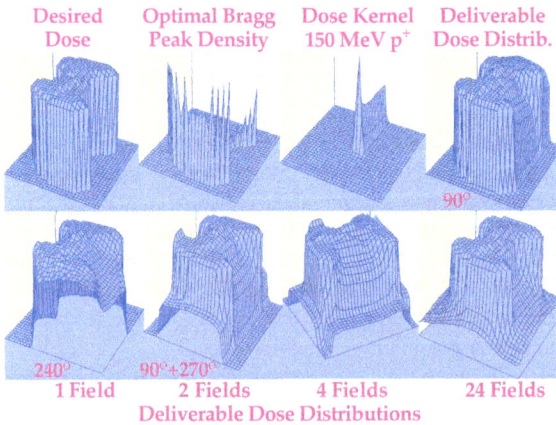

Figure 8.35(b). Illustration of optimization of the scanning pattern or Bragg peak density distribution of $\approx$200 MeV protons to cover the cervix target volume in Fig. 8.34 with a desired dose distribution with minimal overdosage. It is seen that more than three fields does not really lower the dose much in the normal tissue region around the target.

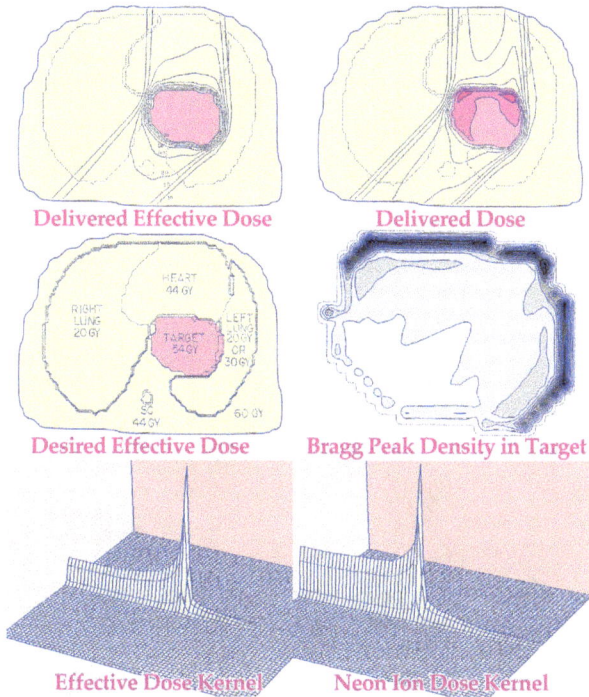

Figure 8.35(c). Illustration of optimization of the scanning pattern and Bragg peak density distribution of light ions (in this case neon ions) to produce the desired dose distribution in the esophagus with minimal overdosage if the exact distribution cannot be produced somewhere. Interestingly, both the absorbed dose and effect distribution can be calculated from the Bragg peak density distribution by convolution, similar to Fig. 8.30(a).

Bragg peak over the entire range of Bragg peak modulation, achieving maximum synergism. With 3D Bragg peak scanning (Fig. 8.38(j)), this is achieved only by fast longitudinal scanning, whereas lateral scanning starting at the most distal point may deliver the last anterior Bragg peaks 1–3 min later reducing the effective therapeutic high LET dose fraction by as much as 20–25% as seen in Fig. 8.8(f). This may explain the somewhat lower therapeutic effect seen in Darmstadt for cordomas treating tumors by fast lateral electromagnetic and slow longitudinal scanning compared to the result from Chiba which uses ridge filters. The difference of just over 20% seen in the clinical data summarized in Fig. 8.26(k) indicate a higher therapeutic effect as expected from Fig. 8.8(f) for treatment where the high- and low-LET dose fractions are delivered practically simultaneously as is the case with ridge filters. Unfortunately, it is slightly more difficult to rapidly vary the extraction energy and

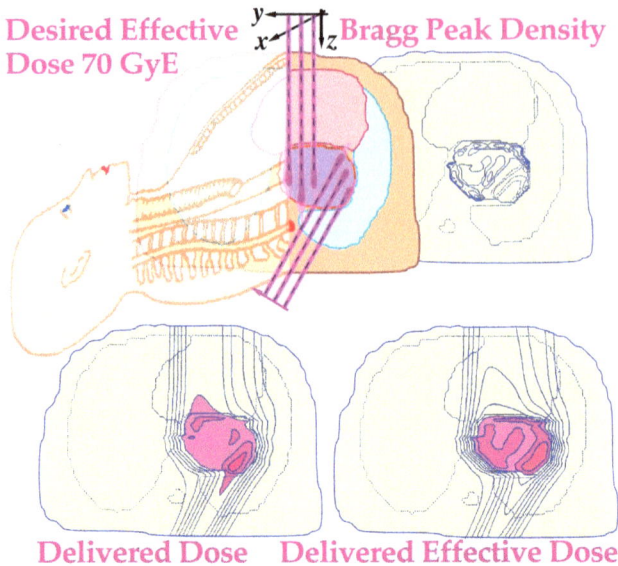

Figure 8.35(d). Illustration of optimization of the scanning pattern and Bragg peak density distribution derived using light ions (in this case neon ions) to produce a desired dose distribution in the esophagus. Minimal overdosage is obtained by partial irradiation from two different directions to minimize spinal cord and lung damage. Interestingly, both the absorbed dose and effect distribution can be calculated from the Bragg peak density distribution by convolution similar to Fig. 8.30(a) (cf. Figs. 8.35(c) and 8.12).

range using longitudinal scanning with a synchrotron of slow pulse repetition rate. A fast material range shifter at the end of the dose delivery system may be desirable or even needed to achieve fast longitudinal scanning of the Bragg peak. Interestingly, the considerable difference in slope of the DRR seen in Fig. 8.23(f) between carbon and proton or photon data indicates that these tumors are rather heterogeneous. As a consequence, the rather shallow slope seen with low LET photons and protons is significantly steepened for carbon ions where a smaller difference in radiation response is expected (cf. Figs. 8.5(m)–(o)). The slightly steeper response seen with the longitudinally scanning ridge filter is also consistent with a higher effective high LET dose fraction as discussed above and in Fig. 8.8(f). Again, this indicates the clinical desirability to treat the high and low LET regions locally as closely in time as possible preferably within 15–30 s to avoid undesirable low LET repair before the high LET component is also delivered. A slight improvement with slow longitudinal pencil beam scanning may be achieved by starting the treatment on the anterior side of the tumor and finishing it at the most distal part of the tumor. In this

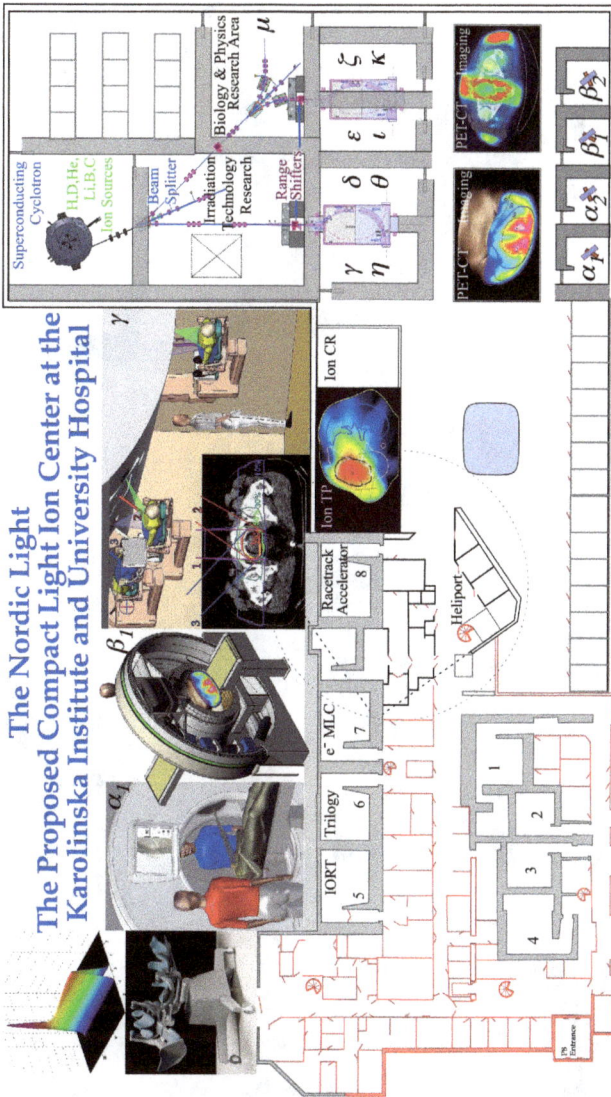

Figure 8.36 (below). Building and equipment locations in the proposed outline for an expansion of the existing radiotherapy department of the New Karolinska University Hospital is shown in the left part of the figure. The light ion treatment rooms at two levels are equipped with two eccentric gantries allowing treatments from practically any direction (cf. Fig. 8.38) via a 120° wide range of variability in each of the four rooms. PET-CT and MRSI-based tumor diagnostics as well as compact high-energy scanned photon therapy allowing PET-CT *in vivo* dose delivery verification (cf. Figs. 8.26–8.28) is also available. Furthermore, ^{11}C can be produced in the range shifter to produce positron emitting ion beams to allow more accurate *in vivo* Bragg peak density monitoring (cf. Brahme *et al.* 2001 and Figs. 8.37, 8.38).

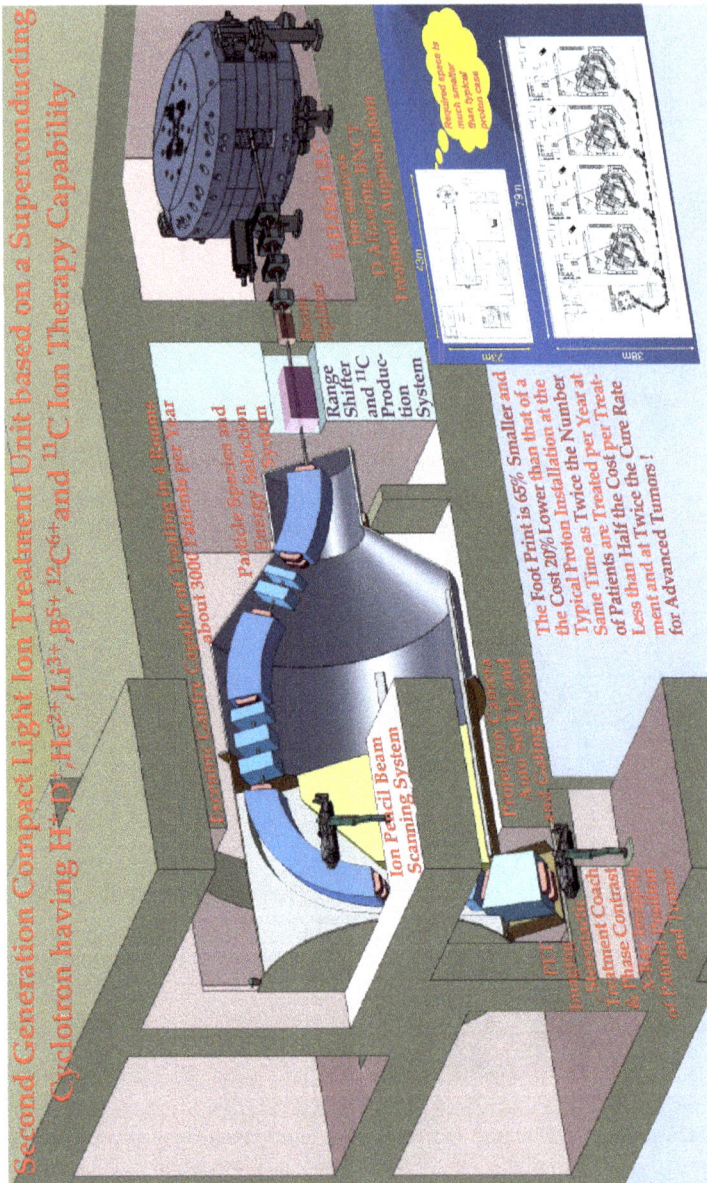

Figure 8.37 (a). CAD-drawing of the superconducting cyclotron and the beam transport to the excentric gantry and the four treatment rooms with the lower right one with a patient being setup for treatment. The performance of this first very flexible second-generation treatment unit is far beyond the traditional large synchrotron-based installations at a fraction of the cost and footprint. The ability to use ^{11}C ions for

(Continued)

Figure 8.37 (a). treatment increases the Bragg peak specificity to about 50-fold in dose delivery imaging. With two excentric gantries and a beam splitter (cf. Fig. 8.36), two beams can simultaneously be delivered on patients one with each gantry (the excentric gantry and superconducting cyclotron are available at LionessTherapeuticsAB, Sweden www.lightions.com and Sumitomo Heavy Industries Ltd, Japan; www.shi.co.jp/english/info/2011/6kgpsq0000001i60.html).

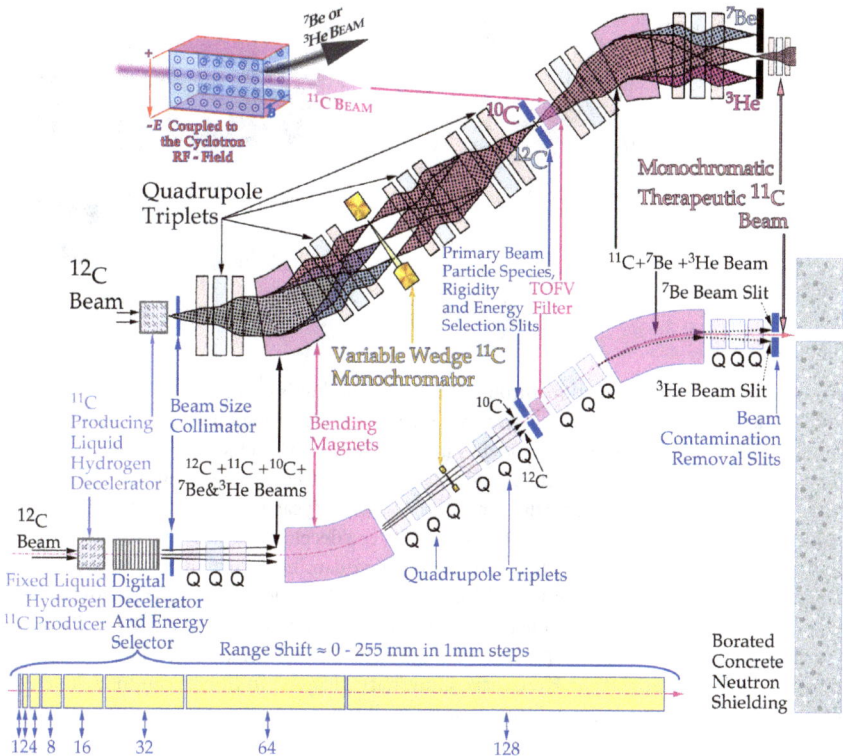

Figure 8.37 (b). Schematic illustration of the beam optical system at the entrance to the excentric gantry (Brahme 2010) used to produce therapeutic scanned ^{12}C and ^{11}C beams with less than 1% energy spread. The initial *liquid hydrogen and digital polyethylene decelerators* are optimal for making ^{11}C positron emitter beams of high brilliance (Lazzeroni and Brahme 2014). The following bending magnets, the quadrupole triplets, the *Time of Flight and Velocity filter (TOFV)* are used to both compress the ion beam phase space density and to maintain a high intensity and brilliance with minimum contamination. Beyond the last magnet shown here, there is a 90° bending and scanning magnet, which makes therapeutic beams of 30 × 40 cm^2 available in four surrounding treatment rooms each with about 120° freedom in selection of the incident beam direction.

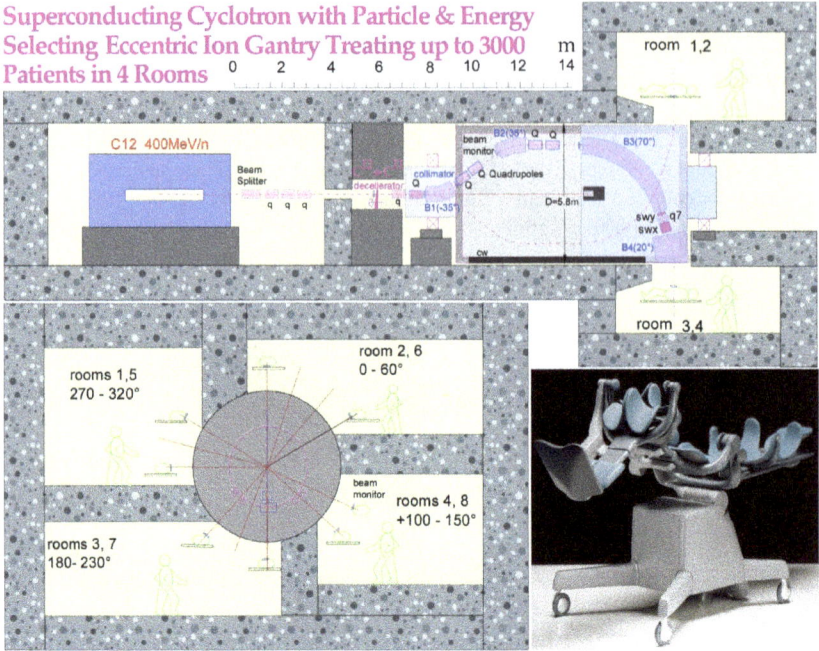

Figure 8.38(a). Cross-section through the superconducting cyclotron with the particle and energy selecting excentric ion gantry capable of treating patients in four surrounding rooms as shown in the lower left panel. The four treatment rooms with protons to carbon ions allow a 10–12 min setup time, and a 2–3 min treatment time/room. In this way, a total 12–16 pat/h, 100–120 pat/day, and 2500–3000 pat/year (at 10–12 fractions/pat) can conveniently be treated. In Figs. 8.36 and 8.38(a), the decelerating graphite range shifter is indicated where ^{11}C can be produced in therapeutic quantities allowing 50-fold accuracy in Bragg peak imaging as shown in Figs. 8.38(c) and (d). **(b)** The lower right panel shows a close up of the stereotactic treatment couch used in docking the patient both to the PET-CT camera (see Fig. 8.26(b)) and the therapy unit (cf. Figs. 8.36 and 8.38(e) and (f)). The structure is planned for installation at the Karolinska University Hospital in Stockholm to include two eccentric gantries and a beam splitter allowing at all times two treatment rooms out of eight to treat up to around 5000–6000 patients per year.

way, most of the low *LET* plateau part of the local dose delivery will arrive earlier in the part of the tumor where the low *LET* dose fraction is considerably higher than in the distal tumor with almost only high *LET* Bragg peaks. From this point of view, it would be desirable to scan the tumor at least twice and from different directions, for example, starting anteriorly treating toward the most distal tumor region and then back to minimize the time interval between low and high *LET* dose delivery.

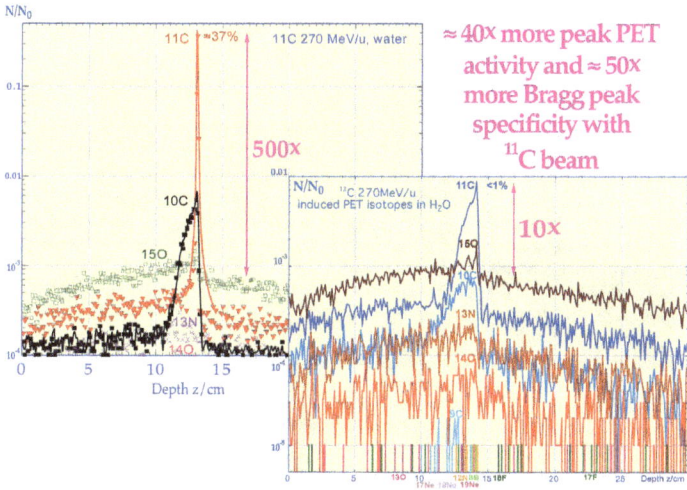

Figure 8.38(c). Comparison of the PET activity induced by ^{12}C and ^{11}C ion beams in water is shown. A considerably higher sensitivity to detect the primary carbon ion Bragg peak with the ^{11}C beam is clearly seen. This is of considerable importance for accurate *in vivo* dose delivery imaging far beyond what can be done with ^{12}C (cf. Figs. 8.16, 8.17, 8.26(b) and (d), 8.28, and 8.38(d)).

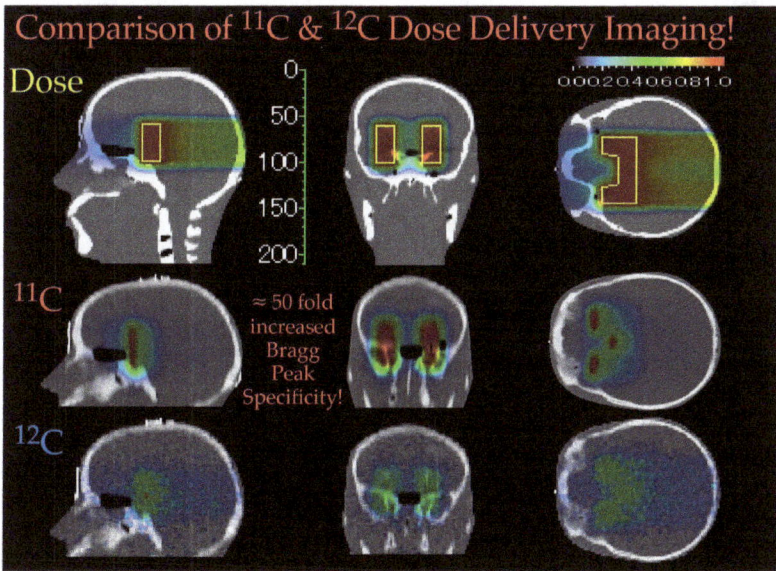

Figure 8.38(d). Comparison of the PET activity induced by ^{11}C and ^{12}C ion beams in a Rando phantom is shown. It is seen that the ^{11}C activity is much more clearly related to the planed dose delivery (modified from NIRS).

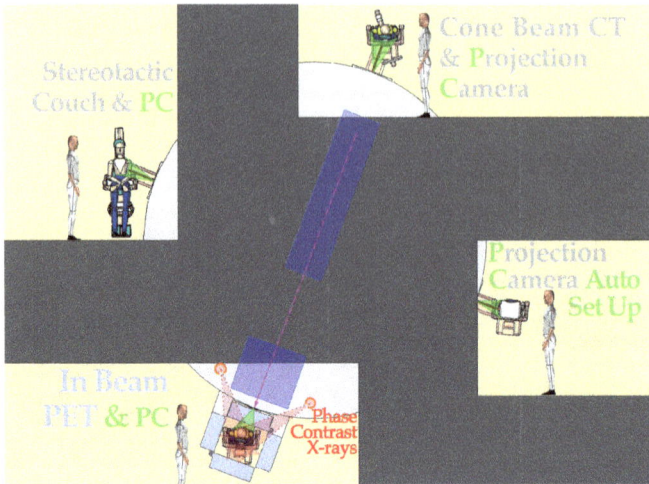

Figure 8.38(e). Cross-sectional view through the excentric gantry (≈6 m diameter) with the four treatment rooms surrounding it. Each room has about ±60° flexibility in selection of beam direction since the patient couch can be rotated more than 180°. Illustration in the lower right room depicts that the treatment will start in a few minutes; Illustration in the upper left room depicts that the patient is about to leave the treatment room; and illustration in the lower left room depicts that the patient is being studied by PET imaging during and after treatment, whereas, illustration in the upper right room depicts that the treatment setup is about to be finished. In this way, about 16 patients/h can be treated with one eccentric gantry.

8.8.7. *Accelerators*

The main acceleration of light ions will be produced by a 5–6 m in diameter superconducting cyclotron, delivering up to 330–400 MeV/u carbon ions, using any of five different ion sources. The lightest ions will be hydrogen or helium ions and an intermediate ion in the range of lithium, beryllium, or boron will also be available and finally carbon ions for the higher *LET* delivery. For effective low *LET* treatments, for example, with microscopically invasive tumors, four treatment units dedicated for IMRT are included as seen in Fig. 8.39: two use high-energy scanned photon beams that allow *in vivo* dose delivery imaging (Figs. 8.16 and 8.39(b)) and two are fast total beam area-modulated IMRT units to be quick and cost efficient (Fig. 8.39(a)). In this way, efficient low *LET* treatments can be performed using *in vivo* PET imaging of high-energy photon dose delivery verification with a similar size photon pencil beam size (pencil beam half width ≈10 mm) as used for light ion therapy.

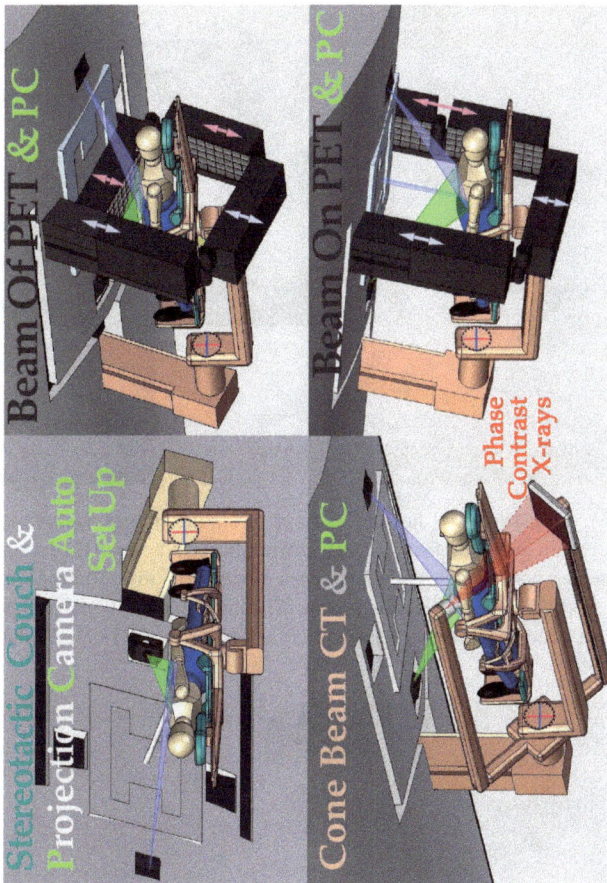

Figure 8.38(f) (below). Patient setup using both PCs with green beams for auto setup of the treatment based on the 3D shape of the patient's surface is illustrated. A cone beam CT facility for setup on internal reference points is included in the gantry for fast check that the patient's internal structures are accurately positioned. The cone beam CT device as well as the four PET detector arrays can be retracted into the excentric gantry during gantry rotation.

The cost-effectiveness of different radiation modalities is compared in Table 8.2. Interestingly, light ions from lithium to carbon are almost as cost effective as IMRT photons and they may even be more cost effective in a big clinic with excentric gantries since the cost per treatment room is substantially reduced and many more patients can be treated per year–up to about 3000 per excentric gantry and thus about 6000 with the installation proposed in Fig. 8.36.

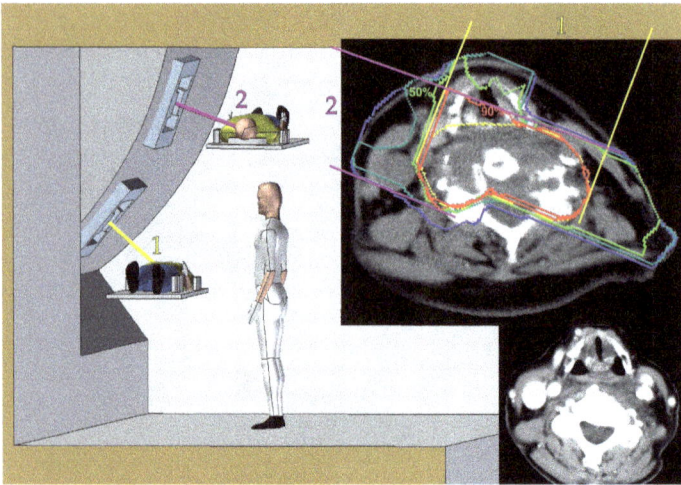

Figure 8.38(g). Illustration of how a treatment of a bone and soft tissue sarcoma in the head and neck region can be performed starting with the right beam in the lower left position (1). In the upper left position, the patient has been rotated 180° to deliver the left beam portal as also shown in the upper left treatment plan. The lower right CT image shows complete recovery of bone structure 5 years after the treatment (courtesy H Tsujii, Chiba).

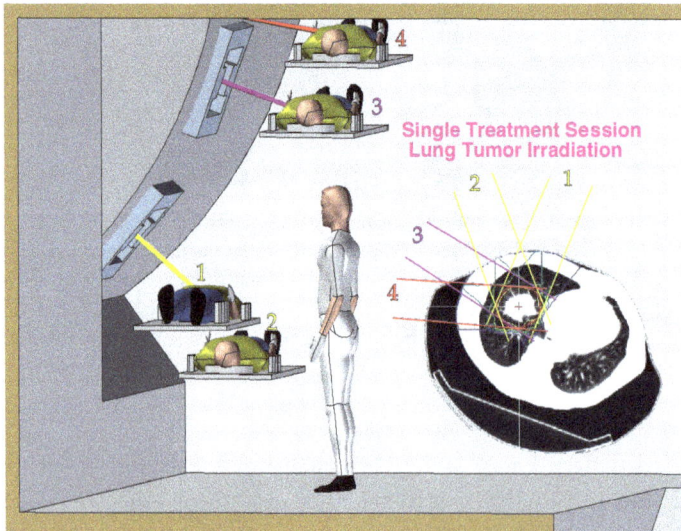

Figure 8.38(h). Treatment of a lung tumor according to Fig. 8.23 in one single session with four fields and 46 GyE carbon ions with the eccentric gantry according to Fig. 8.36 are shown. It is seen that a rather wide range of beam portals is possible by using the 180° rotation of the treatment couch even without patient rotation device used in Chiba (lower right insert).

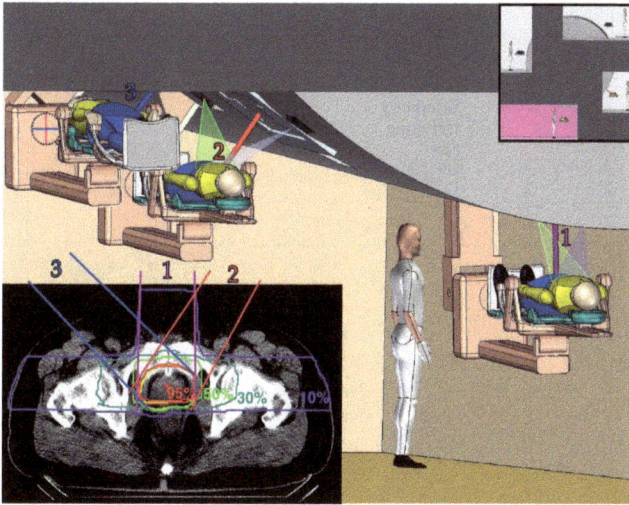

Figure 8.38(i). Treatment of a prostate cancer using the excentric gantry with the stereotactic treatment couch (cf. Fig. 8.38(a) and (f)) using a three-field technique is illustrated. The treatment takes place in the lower left treatment room (cf. Figs. 8.38(b) and (e)) to allow the first vertical beam (1). The rotation of the treatment couch is done automatically by supervising from the treatment room using an ordinary video camera and by the 3D projection camera with 0.5 mm resolution (cf. Fig. 8.28).

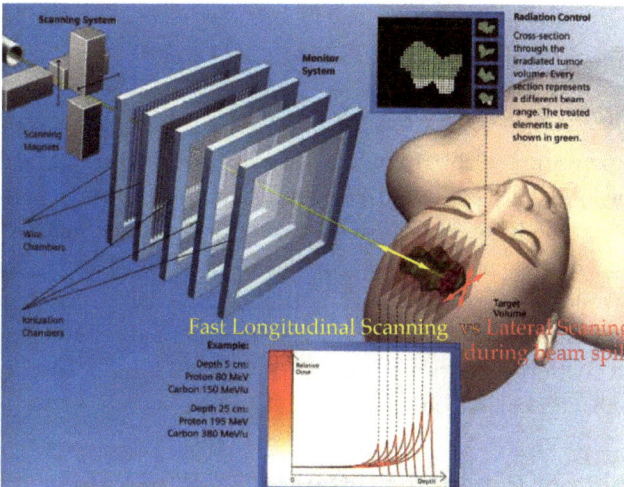

Figure 8.38(j). Illustration of the two possible fast scanning systems: longitudinal and lateral scanning. The latter is simpler for a synchrotron beam spill of 1–2 s as used in Darmstadt and Heidelberg. Longitudinal scanning allow real-time combinations of low and high *LET* dose contributions almost simultaneously to get highest possible synergistic biological effect of the low *LET*-sublethal plateau damage and the high *LET* Bragg peaks and about 20% higher effective *LET* (cf. Figs. 8.8(e) and (f)).

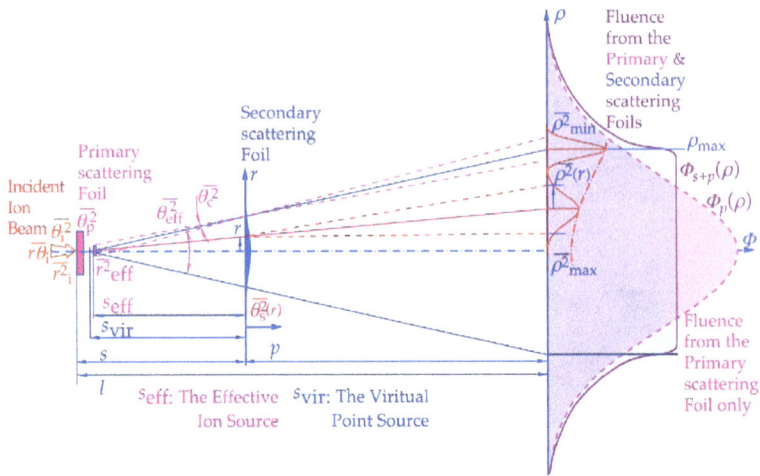

Figure 8.38(k). The design of a dual scattering foil system to effectively produce a broad flattened light ion beam (Brahme 1971, 1972, 1996) is shown. The ions that are only scattered a little in the primary foil are scattered even more in the secondary foil to achieve an almost uniform monoenergetic beam. By optimally designing the primary and secondary scatterers, the total foil thickness can be reduced almost by a factor of 10 (cf. Brahme 1977; Grusell *et al.* 1994). The effective ion source, size, and location can be calculated based on the scattering properties of the foils (Hollmark *et al.* 2004; Kempe and Brahme 2009).

8.9. Financial Aspects

8.9.1. *Capital Expenditure*

The total costs are available from the German, Italian, French, and Austrian facilities; the former, having three treatment rooms (one gantry), reportedly €88–105 million and the latter, having five treatment rooms (three gantries), reported €105 million. The planned facility at Karolinska University Hospital with a novel less-expensive design of the gantries is estimated to about 850 MSEK or €90 million with four treatment rooms. The costs of buildings and most of the infrastructure are included. With two gantries, the IMRT units, and PET-CT imaging, the total cost will be about €130 million.

8.9.2. *Cost per Patient for Different Types of Cancer Treatments*

The current cost for light ion therapy of the Highly Advanced Medical Treatment type (HAMT cf Fig. 8.18) at NIRS in Chiba Japan is about $30 000. In Austria, it

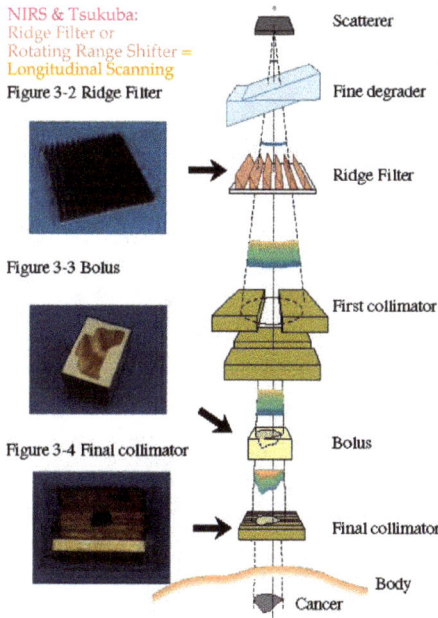

Figure 8.38(1). With a range-shifting ridge filter, all beam energies are present at the same time corresponding to ultra fast longitudinal scanning in Fig. 8.38(j) resulting in an 30% increased synergistic effect in the tumor but not in the normal tissue receiving almost only a low *LET* (cf. Fig. 8.8(f) and Chapter 1, Sec. 1.5.4).

is estimated that the cost for a complete ion therapy schedule will be €20,000 to be compared with an average €44,000 for chemotherapy and €18,000 for surgery. The French costs are lower but then a large part of the investment is from grants. In United States, each session using IMRT with conventional radiation are US$750 and in addition US$3000 for the planning, that is, for example, 35 sessions for a prostate costs about US$30,000. Proton treatments in United States are charged more than double that amount (US$75,000), due to the more expensive equipment and reimbursement regulation. The Swedish cost calculation for one ion session is about €850. As much fewer sessions are needed with light ions than with IMRT, a complete treatment would in average cost between €10 and €25,000 depending on the total number of patients treated. Owing to the fractionation advantage, the light ions are about three times more cost-effective than protons and comparable to IMRT photons and electrons at the same time as they are more effective and curative on resistant and hypoxic tumors.

In an European Union study, the situation of cancer treatment in 1991 Europe was presented, where it is shown that almost 50% of all patients were cured that

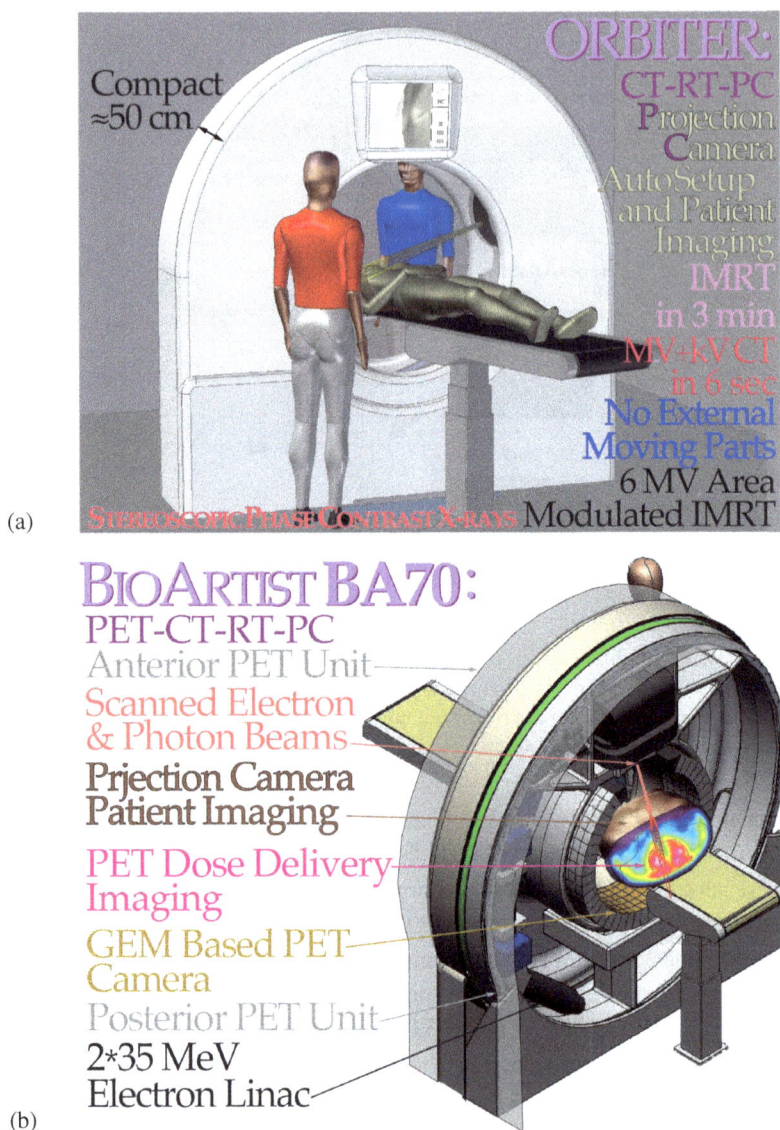

(a)

(b)

Figure 8.39. Illustration of external view of the 6 MV "ORBITER"-dedicated photon IMRT treatment unit in rooms $\beta1$ and $\beta2$ and in rooms $\alpha1$ and $\alpha2$ the 70 MV "BIOARTIST" unit in **(b)** with scanned photon and electron beams for BIOART type biologically optimized *in vivo* predictive assay-based radiation therapy including an open PET-CT camera (cf. Figs. 8.28 and 8.36). Both units include a 4D projection camera for auto setup of the patient with sub-millimeter accuracy and detectors capable of diagnostic and radiotherapeutic CT imaging (these treatment units are available through C-Rad Innovations AB, Sweden: www.c-rad.info/content.php?categoryID=12&parentID=0&pageID=270 and TopGrade Healthcare, China: www.topgradehc.com/html/fangshezhiliao%20-%20en.html).

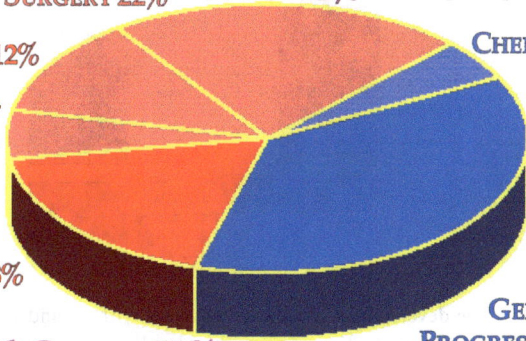

CANCER SITUATION PRESENTED BY EU 1991

LOCAL DISEASE 58% --> Total GENERALIZED
SURGERY 22% Cure 45% DISEASE 42%

CHEMO 5%

RADIATION 12%

RADIATION+
SURGERY 6%

LOCAL
PROGRESS 18%

GENERAL
PROGRESS 37%

(a) 2005 Total Cure ≈55%

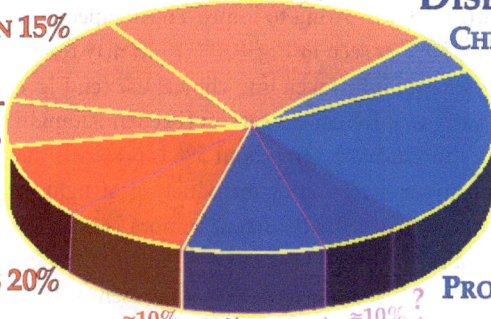

CANCER SITUATION ESTIMATED 2010

LOCAL DISEASE 65% GENERALIZED
SURGERY 20% DISEASE 35%

RADIATION 15% CHEMO 7%

RADIATION+
SURGERY 10%

LOCAL
PROGRESS 20% GENERAL
PROGRESS 28%

≈10% ≈10% ≈10% ?
Hypoxic & Oligo
Radiation Metastasis Saveable by
Resistant Molecular
Saveable by Tumors Therapies
IMRT & Light Ions

(b) -->Total Cure ≈70-75%

Figure 8.40. Overview of the cancer situation in 1991 Europe, when the cancer cure was 45%, and almost 20% of the patients were lost locally due to insufficient local treatment. **(b)** With early detection and improved local treatments using IMRT and light ions, it should be possible to reduce the locally progressing fraction to about 50% of its present value and increase the cure to almost 75%.

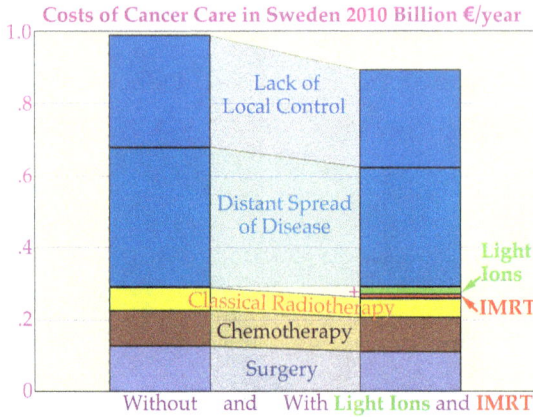

Figure 8.41. The development of cancer care costs without and with light ion therapy is illustrated. Through the reduction of the risk of locally recurrent tumors, the reduction of the expenditure for expensive palliative care may be reduced by around 1 billion SEK or €100 million each year by a large ion installation as in Figs. 8.37 and 8.38. This means that the total cost of the installation will be regained each year after the installation when it is in full use about 5 years after installation.

year as seen in Fig. 8.40(a). Owing to improved therapeutic methods, the situation today is well above 50% as seen in Fig. 8.40(a) and may reach as high as 75% when IMRT and light ions are fully taken into clinical use (cf. Fig. 8.40(b)).

The total yearly cost for cancer care in Sweden is rapidly reaching €1 billion (cf. Fig. 8.41). Of this total cost, only about 5% is presently due to radiation therapy and it may reach to about 7% after introduction of light ion therapy. The total cost for palliative care of locally recurrent tumors is as high as €200–300 million since about 20% of the treatments are today failing locally. However, with light ions, this fraction may be reduced to about 50% its present value or even lower leading to a yearly gain under full operation of the center of about the same size as the total cost of the equipment. Light ion therapy is, therefore, despite its large initial investment costs, one of the most cost-efficient advanced cancer therapy modalities available today as shown in Fig. 8.41 and the extremely promising clinical results in Figs. 8.19–8.24 above indicate that we urgently need to bring them into clinical practice.

8.10. Conclusion

Even if biologically optimized radiation therapy has proven difficult to introduce in most conventional radiation therapy departments, it is urgently needed with light ions in order to make full clinical use of the significant biological advantages of light ions. There is no doubt that QMRT and BIOART are very important developments for light ion therapy and they are procedures that are very important to introduce in the presently proposed compact second generation of clinically dedicated light ion installation. The largest benefits are expected for hypoxic tumors and generally radiation-resistant tumors for conventional low-*LET* beams, where we can expect the most important treatment improvements as clearly shown by the clinical results of Professor Tsujii and his team at NIRS in Chiba, Japan.

Bibliography

Aaltonen P, Brahme A, Lax I, Levernes S, Näslund I, Reitan JV, Turesson I (1997) Specification of dose delivery in radiation therapy. Recommendations by the NACP. *Acta Oncol* (Suppl. 10):1–32.

Andreassen CN, Alsner J, Overgaard M, Overgaard J (2003) Prediction of normal tissue radiosensitivity from polymorphisms in candidate genes. *Radiother Oncol* 69:127–135.

Antonovic L, Brahme A, Furusawa Y, Toma-Dasu I (2013) Multiparameter description of the *LET* dependence of cell survival for oxic and hypoxic cells irradiated with carbon ions. *J Radiat Res* 54:18–26.

Aoki M, Furusawa Y, Yamada T (2000) *LET* dependency of heavy-ion induced apoptosis in V79 cells. *J Radiat Res* 41:163–175.

Berger MJ, Seltzer SM (1969) Quality of radiation in a water medium irradiated with high-energy electron beams. In: *Book of Abstracts of XII International Congress of Radiology*, 127.

Bird RP (1979) Biophysical studies with spatially correlated ions. 3. Cell survival studies using diatomic deuterium. *Radiat Res* 78:3303–3309.

Blakely EA, Chang PY (2009) Biology of charged particles. *Cancer J* 15:271–284.

Brahme A (1977) Electron transport phenomena and absorbed dose distributions in therapeutic electron beams. *14th Int Congr Radiol*, Rio de Janeiro, Brazil.

Brahme A (1979) Scanning system for charged and neutral particles. *Swe Pat* 7904360-0.

Brahme A (1982) Physical and biologic aspects on the optimum choice of radiation modality. *Acta Radiol Oncol* 21:469–479.

Brahme A (1984) Dosimetric precision requirements in radiation therapy. *Acta Radiol Oncol* 23:379–391.

Brahme A (1988a) Optimal setting of Multileaf Collimators in stationary beam radiation therapy. *Strahlentherapie* 164:343–350.

Brahme A (1988b) Optimization of stationary and moving beam radiation therapy techniques. *Radiother Oncol* 12:129–140.

Brahme A (1992) Biological and physical dose optimization in radiation therapy. In: Fortner JG, and Rhoads JE (eds) *Accomplishments in Cancer Research*, 265–300. New York: Lippincott.

Brahme A (1999) Optimized radiation therapy based on radiobiological objectives. *Semin Oncol* 9:35–47.

Brahme A (2001) Radiobiological optimization of multimodality radiation therapy: from electrons, photons and neutrons to light ions. In: A Zanini A, Ongaro C (eds) *Neutron Spectrometry and Dosimetry: Experimental Techniques and MC Calculations* 15–34.

Brahme A (2003a) Biologically optimized 3-dimensional *in vivo* predictive assay based radiation therapy using positron emission tomography-computerized tomography imaging. *Acta Oncol* 42:123–136.

Brahme A (2003b) Biologically optimized light ion therapy. *Report to the Swedish Cancer Society*.

Brahme A (2004) Recent advances in light ion radiation therapy. *Int J Rad Onc Biol Phys* 58:603–616.

Brahme A (2005) Development of biologically Optimized light ion therapy. *Proc 7th Int Conf Time Dose Fractionation*, Madison, pp. 50–67.

Brahme A (2009) Potential developments of light ion therapy: the ultimate conformal treatment modality. First NIRS International Open Laboratory Workshop, November 17, 2008, Chiba, Japan, Tsujii H (ed.), *Radiol Sci* 52:8–31 (http://www.nirs.go.jp/info/report/rs-sci/pdf/200902.pdf).

Brahme A (2010) Optimal use of light ions for radiation therapy. *Radiol Sci* 53(8.9):35–61 (http://www.nirs.go.jp/publication/rs- sci/pdf/201008.pdf).

Brahme A (2011) Accurate description of the cell survival and biological effect at low and high doses and LETs. *J Rad Res* 52:389–407.

Brahme A, Ågren AK (1987) On the optimal dose distribution for eradication of heterogeneous tumors. *Acta Oncol* 26:377–385.

Brahme A, Eenmaa J, Lindbäck S, Montelius A, Wootton P (1983) Neutron beam characteristics of 50 MeV protons on beryllium with a continuously variable multileaf collimator. *Rad Ther Oncol* 1:65–76.

Brahme A, Gudowska I, Larsson S, Andreassen B, Holmberg R, Svensson R (2006) Application of Geant4 in the Development of New Radiation Therapy Treatment Methods. *Proc 9th Conf on Astroparticle, Particle and Space Physics, Detectors and Medical Physics Applications*, Barone, Borchi Gaddi *et al.* (eds) Singapore: World Scientific Publication, pp. 451–461.

Brahme A, Källman P, Lind BK (1988) Absorbed dose and biological effect planning in heavy ion therapy. *VIIIth ICMP*, San Antonio TX, *Phys Med Biol* 33 (Suppl. 1):73 (MP19.1).

Brahme A, Källman P, Lind BK (1989) Optimization of proton and heavy ion therapy using an adaptive inversion algorithm. *Radiother Oncol* 15:189–197.

Brahme A, Källman P, Lind BK (1991) Optimization of the Probability of Achieving Complication Free Tumor Control Using a 3D Pencil Beam Scanning Technique for Protons and Heavy Ions. *Proc NIRS Int Workshop on Heavy Charged Particle Therapy and Related Subjects*, AItano and Kanai T (eds) Chiba, Japan, pp. 124–142.

Brahme A, Kraepelien T, Svensson H (1980a) Electron and photon beam characteristics from a 50 MeV Racetrack Microtron. *Acta Radiol* 19:305–319.

Lazzeroni M and Brahme A (2014) Effective source size, radial, angular and energy spread of therapeutic ^{11}C positron emitter beams produced by ^{12}C fragmentation. *Submitted to Nuclear Instruments and Methods in Physics Research B*, 320:26–36.

Brahme A, Lewensohn R, Ringborg U, Amaldi U, Gerardi F, Rossi S (2001) Design of a centre for biologically optimized light ion therapy in Stockholm. *Nucl Instr Meth Phys Res B* 184:569–588.

Brahme A, Lind BK, Källman P (1990a) Inverse radiation therapy planning as a tool for 3D dose optimization. *Phys Med* 6:53–68.

Brahme A, Lind BK, Källman P (1990b) Physical and Biological Dose Optimization Using Inverse Radiation Therapy Planning. *Inter-Society Council for radiation oncology treatment planning*, Houston TX, March.

Brahme A, Montelius A, Nordell B, Reuthal M, Svensson H (1980b) Investigation of the possibility of using photoneutron beams for radiation therapy. *Phys Med Biol* 25:1111–1120.

Brahme A, Nilsson J, Belkic D (2001) Biologically optimized radiation therapy. Nobel Conf. 2000. *Acta Oncol* 40:725–734.

Brahme A, Nyman P, Skatt B (2008) 4D laser camera for accurate patient positioning, collision avoidance, image fusion and adaptive approaches during diagnostic and therapeutic procedures. *Med Phys* 35:1670–1681.

Brahme A, Roos J-E, Lax I (1982) Solution of an integral equation encountered in rotation therapy. *Phys Med Biol* 27:1221–1229.

Brahme A, Rydberg B, Blomqvist P (1997) Dual spatially correlated nucleosomal double strand breaks in cell inactivation. In: Goodhead DT, O'Neill P, Menzel HG (eds) *Microdosimetry: An Interdisciplinary Approach*, 125–128. Cambridge: The Royal Society of Chemistry.

Cera F, Cherubini R, DallaVecchia M, Favaretto S, Moschini G, Tiveron P, Belli M, Ianzini F, Levati L, Sapora O, Tabocchini MA, Simone G (1997) Cell inactivation, mutation and DNA damage induced by light ions: dependence on radiation quality. In: Goodhead DT, O'Neill P, Menzel HG (eds) *Microdosimetry: An Interdisciplinary Approach*, 191–194. Cambridge: The Royal Society of Chemistry.

Curtis SB (1976) The OER of mixed high- and low-LET radiation. *Radiat Res* 65:566–572.

Curtis SB (1986) Lethal and potential lethal lesions induced by irradiation: a unified repair model. *Radiat Res* 106:252–270.

Curtis SB (1988) The lethal and potentially lethal model — a review and recent development. In: Kiefer J (ed) *Quantitative Mathematical Models in Radiation Biology*, 137–148. Berlin, Heidelberg: Springer–Verlag.

Curtis SB (1992) Application of the LPL model to mixed radiations. In: Chadwick KH, Moschini G, Varma MN (eds) *Biophysical Modelling of Radiation Effects*, 21–28. Adam Hilger, Bristol.

Denekamp J, Harris SR, Morris C, Field SB (1976) The response of a transplantable tumor to fractionated irradiation. *Radiat Res* 68:93–103.

Eklöf A, Brahme A (1999) Composit energy deposition kernels for focused point monodirectional photon beams. *Phys Med Biol* 44:1655–1668.

Furusawa Y, Fukutsu K, Aoki M, Itsukaichi H, Eguchi-Kasai K, Ohara H, Yatagai F, Kanai T, Ando K (2000) Inactivation of aerobic and hypoxic cells from three different cell lines by accelerated 3He-, 12C- and 20Ne-Ion beams. *Radiat Res* 154:485–496.

Grusell E, Montelius A, Brahme A, Rikner G, Russel K (1994) A general solution to charged particle beam flattening using an optimized dual scattering foil technique, with application to proton therapy beams. *Phys Med Biol* 39:2201–2216.

Guerrero M, Allen Li X (2004) Extending the linear-quadratic model for large fraction doses pertinent to stereotactic radiotherapy. *Phys Med Biol* 49:4825–4835.

Hall EJ, Giaccia AJ (2006) *Radiobiology for the Radiologist*. Lippincott Williams & Wilkins, sixth edition.

Higgins PD, DeLuca PM Jr, Pearson DW, Gould MN (1983) V79 survival following simultaneous or sequential irradiation by 15-MeV neutrons and 60Co photons. *Radiat Res* 95:45–56.

Hoeijmakers JH (2001) Genome maintenance mechanisms for preventing cancer. *Nature* 411:366–374.

ICRU (1984) Radiation Dosimetry: Electron beams with energies between 1 and 50 MeV. *ICRU Report 35*.

Jakob B, Scholz M, Taucher-Scholz G (2003) Biological imaging of heavy charged particle tracks. *Radiat Res* 676–684.

Källman P, Ågren A, Brahme A (1992a) Tumor and normal tissue responses to fractionated non uniform dose delivery. *Int J Rad Biol* 62:249–262.

Källman P, Lind BK, Brahme A (1992b) An algorithm for maximizing the probability of complication free tumour control in radiation therapy. *Phys Med Biol* 37:871–890.

Källman P, Lind BK, Eklöf A, Brahme A (1988) Shaping of arbitrary dose distributions by dynamic multileaf collimation. *Phys Med Biol* 33:1291–1300.

Kellerer AM, Lam YP, Rossi HH (1980) Biophysical studies with spatially correlated ions. 4. Cell survival studies using diatomic deuterium. *Radiat Res* 83:522–528.

Kellerer AM, Mand A, Rossi HH (1974) The theory of dual radiation action. *Curr Topics in Radiation Res* VIII:111–158.

Kempe J, Brahme A (2010) Solution of the Boltzmann equation for primary light ions and the transport of their fragments. *Phys Rev ST Accel Beams* 13:104702-1-13.

Kempe J, Gudowska I, Brahme A (2007) Depth absorbed dose and *LET* distributions of therapeutic 1H, 4He, 7Li and 12C beams. *Med Phys* 34:183–192.

Kraft G, Kraft-Weyrather W (1987) Inactivation, Chromosome Aberrations and Strand Break Induction as Function of Radiation Quality. *Proc Third Workshop on Heavy Charged Particles in Biology and Medicine*, Darmstadt, July 13–15.

Lam GKY (1987) The survival response of a biological system to mixed radiations. *Radiat Res* 110:232–243.

Lind BK, Brahme A (1985) Generation of Desired Dose Distributions with Scanned Elementary Beams by Deconvolution Methods. *Proc VII ICMP*, Espoo, Finland, pp. 953–956.

Lind BK, Brahme A (1987) Optimization of Radiation Therapy Dose Distributions Using Scanned Photon Beams. *Proc 9th Int Conf on Comp in Rad. Therapy*, Bruinvis IAD, Van derGiessen PH, Van Kleffens HJ (eds), Elsevier, Amsterdam, pp. 235–239.

Lind BK, Brahme A (2007) The radiation response of heterogeneous tumors. *Phys Med* 23:91–99.

Lind BK, Brahme A (2010) A systems biology approach to radiation therapy optimization. *Radiat Environ Biophys* 49:111–124.

Lind BK, Nilsson J, Löf J, Brahme A (2001) Generalization of the normalized dose-response gradient to non-uniform dose delivery. Nobel Conf. 2000. *Acta Oncol* 40:718–724.

Lind BK, Persson LM, Edgren MR, Hedlöf I, Brahme A (2003) Repairable-Conditionally Repairable damage model based on dual Poisson processes. *Radiat Res* 160:366–375.

Lindborg L, Brahme A (1990) Influence of microdosimetric quantities on observed dose-response relationships in radiation therapy. *Rad Res* 124(Suppl 1):S23–S28.

Marples B, Wouters BG, Collis SJ, Chalmers AJ, Joiner MC (2004) Low-dose hyperradiosensitivity: a consequence of ineffective cell cycle arrest of radiation damaged g2-phas cells. *Radiat Res* 161:247–255.

Matsufuji N, Wada M, Kase Y, Uzawa A, Ando K (2008) Assessment of Biological Models for Hypofractionated Irradiation with Therapeutic Carbon Ion Beams.

Meijer AE, Jernberg AR, Heiden T, Stenerlöw B, Persson LM, Tilly N, Lind BK, Edgren MR (2005) Dose and time dependent apoptotic response in a human melanoma cell line exposed to accelerated boron ions at four different *LET*. *Int J Radiat Biol* 81(4):261–272.

Näfstadius P, Brahme A, Nordell B (1984) Computer assisted dosimetry of scanned electron and photon beams. *Radiother Oncol* 2:261–269.

Nakamura Y (2004) Isolation of p53-target genes and their functional analysis. *Cancer Sci* 95:7–11.

Ngo FQH, Blakely E, Tobias CA (1981) Sequential exposures of mammalian cells to low- and high-*LET* radiations. *Radiat Res* 87:59–78.

Nilsson J, Lind BK, Brahme A (2002) Radiation response of hypoxic and generally heterogeneous tissues. *Int J Radiat Biol* 78:389–405.

Paretzke H (1987) Radiation track structure theory. In: Freeman GR (ed.) *Kinetics of Nonhomogeneous Processes*, 89–170. New York: John Wiley & Sons.

Park C, Papiem L, Zhang S, Story M, Timmerman R (2008) Computation of cell survival in heavy ion beams for therapy. *Int J Rad Onc Biol Phys* 70:847–852.

Raju MR (1980) *Heavy Particle Radiotherapy*. Academic Press, London.

Raju MR, Eisen Y, Carpenter S, Jarrett K, Harvey WF (1993) Radiobiology of alpha particles. IV. Cell inactivation by alpha particles of energies 0.4–3.5 MeV. *Radiat Res* 133:289–296.

Ringborg U, Bergqvist D, Brorson B, Cavallin-Sthl E, Ceberg J, Einhorn N, Frödin JE, Järhult J, Lamnevik G, Lindholm C, Littbrand B, Norlund A, Nylén U, Rosén M, Svensson H, Möller TR (2003) The Swedish Council on Technology Assessment in Health Care (SBU) systematic overview of radiotherapy for cancer including a prospective survey of radiotherapy practice in Sweden 2001 — summary and conclusions. *Acta Oncol* 42:357–365.

Rydberg B, Heilbronn L, Holley WR, Löbrich M, Zeitlin C, Chatterjee A, Cooper PK (2002) Spatial distribution and yield of DNA double strand breaks induced by 3–7 MeV helium ions in human fibroblasts. *Radiat Res* 158:32–42.

Scholz M, Kellerer AM, Kraft-Weyrather W, Kraft G (1997) Computation of cell survival in heavy ion beams for therapy. The model and its approximation. *Radiat Environ Biophys* 36:59–66.

Singh B, Arrand JE, Joiner MC (1994) Hypersensitive response of normal human lung epithelial cells at low radiation doses. *Int J Radiat Biol* 65:457–464.

Svensson H, Ringborg U, Näslund I, Brahme A (2004) Development of light ion therapy at the Karolinska Hospital and Institute. *Radiother Oncol* 73(Suppl 2):206–210.

Tilikidis A, Brahme A (1994) Microdosimetric description of beam quality and biological effectiveness in radiation therapy. *Acta Oncol* 33:457–469.

Tilikidis A, Lind B, Näfstadius P, Brahme A (1996) An estimation of the relative biological effectiveness of 50 MV bremsstrahlung beams by microdosimetric techniques. *Phys Med Biol* 41:55–69.

Tilly N, Brahme A, Carlsson J, Glimelius B (1999) Comparison of cell survival models for mixed *LET* radiation. *Int J Radiat Biol* 75:233–243.

Tilly N, Fernandez-Varea JM, Grusell E, Brahme A (2002) Comparison of Monte Carlo calculated electron slowing-down spectra generated by 60Co gamma-rays, electrons, protons and light ions. *Phys Med Biol* 47:1303–1319.

Tobias CA (1985) The repair-misrepair model in radiobiology. *Radiat Res* 104:77–95.

Todd PW (1964) Reversible and Irreversible effects of ionizing radiations on the reproductive integrity of mammalian cells cultured *in vitro*, PhD thesis, Lawrence Radiation Laboratory, Berkeley, California, report UCRL-11614.

Tsujii H (2010) Overview of Carbon Ion Therapy at NIRS: 15 Years Experience. *Proc Japanese-European Joint Symposium on Ion Cancer Therapy*, Karolinska, Stockholm, pp. 11–17.

Virsik RP, Blohm R, Hermann K-P, Harder D (1981) Fast, Short-ranged and Slow, Distant-ranged Interaction Processes Involved in Chromosome Aberration Formation. *7th Symp Microdosimetry 1980*. Ebert HG (ed.). Harwood Publication, pp. 943–957.

Vreede P, Brahme A (2009) Development of biologically optimized radiation therapy: maximizing the apoptotic cell kill. *Rad Sci* 52:31–52.

Vreede P, Brahme A (2010) Uniform Tumor Cell Kill and Absorbed Dose with Mixed Modality Light Ion Beams. *Proc NIRS–KI Joint Symp Ion-Radiation Sciences*, Karolinska, Stockholm, pp. 124–125.

Wedenberg M, Lind BK, Toma-Daşu I, Rehbinder H, Brahme A (2010) Analytical description of the *LET* dependence of cell survival using the repairable-conditionally repairable damage model. *Rad Res* 174:517–525.

Wiklund K, Olivera GH, Brahme A, Lind BK (2008) Radial Secondary electron dose profiles and biological effects in light-ion beams based on analytical and Monte Carlo calculations using distorted wave cross sections. *Radiat Res* 170:83–92.

Wingate CL, Baum JW (1976) Measured radial distributions of dose and *LET* for alpha and proton beams in hydrogen and tissue-equivalent gas. *Radiat Res* 65:1–19.

Zaider M, Bird RP, Rossi HH, Marino S, Rohrig N (1983) A study of cell survival in mammalian cells exposed to spatially correlated triads of protons. *Radiat Environ Biophys* 22:239–249.

Zaider M, Rossi HH (1980) The synergistic effect of different radiations. *Radiat Res* 83:732–739.

Zhou SA, Brahme A (2008) Development of phase-contrast X-ray imaging techniques and potential medical applications. *Phys Med* 24:129–148.

Zhou SA, Brahme A (2010) Development of high-resolution molecular phase-contrast stereoscopic X-ray imaging for accurate cancer diagnostics. *Rad Pro Dos* 139:334–338.

Acknowledgments

Continued fruitful discussions on treatment plans, technical developments and clinical results with professor Hirohiko Tsujii in Chiba, Professor Jürgen Debus and Thomas Haberer in Heidelberg and Ugo Amaldi at TERA as well as with the many talented researchers at these centers and in Stockholm are gratefully acknowledged.

Index